ADENOSINE AND ADENINE NUCLEOTIDES: FROM MOLECULAR BIOLOGY TO INTEGRATIVE PHYSIOLOGY

ADENOSINE AND ADENINE NUCLEOTIDES: FROM MOLECULAR BIOLOGY TO INTEGRATIVE PHYSIOLOGY

Edited by

Luiz Belardinelli
UNIVERSITY OF FLORIDA

and

Amir Pelleg
HAHNEMANN UNIVERSITY

Kluwer Academic Publishers

BOSTON DORDRECHT LONDON

DISTRIBUTORS:

Distributors for North America:

Kluwer Academic Publishers
101 Philip Drive
Assinippi Park
Norwell, Massachusetts 02061 USA

Distributors for all other countries:

Kluwer Academic Publishers Group
Distribution Centre
Post Office Box 322
3300 AH Dordrecht, THE NETHERLANDS

Library of Congress Cataloging-in-Publication Data

Adenosine and adenine nucleotides: from molecular biology to integrative physiology / editors, Luiz Belardinelli, Amir Pelleg.

p. cm.

®"5th International Symposium on Adenosine and Adenine Nucleotides held in Philadelphia, Pennsylvania, U.S.A., May 9–13, 1994: – Pref.

Includes index.

ISBN 0-7923-3190-7 (alk. paper)

1. Adenosine – Physiological effect – Congresses. 2. Adenine nucleotides – Physiological effect – Congresses. I. Belardinelli, Luiz. II. Pelleg, Amir. III. International Symposium on Adenosine and Adenine Nucleotides (5th: 1994: Philadelphia, Pa.)

[DNLM: 1. Adenosine – congresses. 2. Adenine Nucleotides – congresses. QU 58 A2278 1994]

QP625.A27A324 1994

615.7 – dc20

DNLM/DLC

for Library of Congress

94-40345

CIP

Printed on acid-free paper.

Printed in the United States of America

CONTENTS

CONTRIBUTING AUTHORS

ABBRACCHIO, Maria P., Institute of Pharmacological Science, School of Pharmacy, University of Milan, Italy.

ABEBE, Worku, Department of Pharmacology, School of Medicine, East Carolina University, Greenville, NC 27858.

ABRAHAM, Edward, Radiation Oncology Branch, National Cancer Institute, NIH, Bethesda, MD 20892.

ADAMS, D.J., Department of Molecular and Cellular Pharmacology, University of Miami School of Medicine, Miami, FL 33101.

ARENAS, Ernest, Department of Neuroscience, Department of Physiology and Pharmacology, Department of Medicinal Biochemistry and Biophysics, Karolinska Institutet, 171 77, Stockholm, Sweden.

ASSENDER, Jean W., Department of Physiology, and Pharmacology, Karolinska Institutet, S-171 77 Stockholm, Sweden.

BALDWIN, Stephen A.L., Department of Biochemistry and Molecular Biology, University of Leeds, Leeds LS2 9JT, United Kingdom.

BEAUMONT, Nick, Department of Biochemistry and Molecular Biology, University of Leeds, Leeds LS2 9JT, United Kingdom.

BECKER, Bernhard F., Department of Physiology, University of Munich, Pettenkoferstr. 12, 80336 Munich, Germany.

BELARDINELLI, Luiz, Department of Medicine, University of Florida, Gainesville, FL 32610-0277.

BENNETT, W., Department of Pulmonary Medicine, Division of Pulmonary Diseases, The University of North Carolina at Chapel Hill, Chapel Hill, NC.

BERTRAND, Gyslaine, Laboratoire de Pharmacologie, Faculte de Medecine, 34060 Montpellier Cedex, France.

BOUCHER, R.C., Department of Pulmonary Medicine, Division of Pulmonary Diseases, The University of North Carolina at Chapel Hill, Chapel Hill, NC.

BOYER, Jose L., Department of Pharmacology, University of North Carolina, Chapel Hill, NC.

BRAUN, Norbert, Biozentrum der J.W. Goethe-Universitat, AK Neurochemie, Zoologisches Institut, Marie-Curie-Str. 9, 60439 Frankfurt am Main, Germany.

BRONTE, Vincenzo, Institute of Oncology, University of Padova, Italy.

BÜNGER, Rolf, Department of Physiology, F.E. Hebert School of Medicine, Uniformed Services University of the Health Sciences, Bethesda, MD.

BURNSTOCK, G., Department of Anatomy and Developmental Biology, University College London, Gower Street, London WC1E 6BT, United Kingdom.

BUXTON, Iain L.O., Department of Pharmacology, University of Nevada, School of Medicine, Reno, NV.

CAMAIONI, Emidio, Dipartimento di Scienze Chimiche, Via S. Agostino, 1, Universita di Camerino, 62032 Camerino (MC), Italy.

CASILLAS, Teresa, Departamento de Bioquimica, Facultad de Veterinaria, Universidad Complutense, 28040 Madrid, Spain.

CASS, Carol E., Department of Biochemistry 474 Medical Sciences Building, University of Alberta, Edmonton, Alberta, Canada T6G 2H7.

CATTABENI, Flaminio, Institute of Pharmacological Science, School of Pharmacy, University of Milan, Italy.

CERUTI, Stefania, Institute of Pharmacological Science, School of Pharmacy, University of Milan, Italy.

CHAPAL, Jeannie, Laboratoire de Pharmacologie, Faculte de Medecine, 34060 Montpellier Cedex, France.

CHEEK, Dennis, Department of Pharmacology, University of Nevada, School of Medicine, Reno, NV.

COHEN, Michael V., Departments of Physiology and Medicine, University of South Alabama, MSB 3024, Mobile, AL 36688.

CONTI, Anna, Research Laboratories, Schering-Plough S.p.A., I-20060 Comazzo, Milan, Italy.

CORREIA-DE-SÁ, Paulo, Laboratory of Pharmacology, Gulbenkian Institute of Science, 2781 Oeiras Codex, Portugal.

CREA, Filippo, Institute of Cardiology, Catholic University, Rome, Italy.

CRISTALLI, Gloria, Dipartimento di Scienze Chimiche, Via S. Agostino, 1, Universita di Camerino, 62032 Camerino (MC), Italy.

CUNHA, Rodrigo A., Laboratory of Pharmacology, Gulbenkian Institute of Science, 2781 Oeiras Codex, Portugal.

CURTIS, Anne B., Division of Cardiology, Box 100277, University of Florida, Gainesville, FL 32610.

DALY, John W., Laboratory of Bioorganic Chemistry, National Institute of Diabetes and Digestive and Kidney Diseases, National Institutes of Health, Bethesda, MD 20892.

DASGUPTA, Sarmila, Department of Neuroscience, Department of Physiology and Pharmacology, Department of Medicinal Biochemistry and Biophysics, Karolinska Institutet, 171 77, Stockholm, Sweden.

DECKING, U., Department of Physiology, Heinrich-Heine-Universitat, Universitatsstr. 1, 40225 Dusseldorf, Germany.

DELICADO, Esmerilda G., Departamento de Bioquimica, Facultad de Veterinaria, Universidad Complutense, 28040 Madrid, Spain.

DE MENDONÇA, Alexandre, Laboratory of Pharmacology, Gulbenkian Institute of Sciences, 2781 Oeiras Codex, Portugal.

DIONISOTTI, Silvio, Research Laboratories, Schering-Plough S.p.A., I-20060 Comazzo, Milan, Italy.

DI VIRGILIO, Francesco, Institute of General Pathology, University of Ferrara, Italy.

DOWNEY, James M., Departments of Physiology and Medicine, University of South Alabama, MSB 3024, Mobile, AL 36688.

ESKESEN, Karen, Novo Nordisk Pharmaceuticals Division, Novo Nordisk Park, DK 2760 Malov, Denmark.

FALZONI, Simonetta, Institute of General Pathology, University of Ferrara, Italy.

FERRARI, Davide, Institute of General Pathology, University of Ferrara, Italy.

FERRÉ, Sergi, Department of Neuroscience, Department of Physiology and Pharmacology, Department of Medicinal Biochemistry and Biophysics, Karolinska Institutet, 171 77, Stockholm, Sweden.

FISCHER, Bilha, Molecular Recognition Section, Laboratory of Bioorganic Chemistry, NIDDK, National Institutes of Health, Bethesda, MD.

FORREST, John N., Jr., Department of Internal Medicine, Yale University School of Medicine, New Haven, CT 06515.

FREDHOLM, Bertil, Departments of Neuroscience, Physiology and Pharmacology and Medicinal Biochemistry and Biophysics, Karolinska Institutet, S-171 77, Stockholm, Sweden.

FUKUNAGA, Atsuo F., Department of Anesthesiology, Harbor-UCLA Medical Center, Torrance, CA 90509.

FUXE, Kjell, Department of Neuroscience, Department of Physiology and Pharmacology, Department of Medicinal Biochemistry and Biophysics, Karolinska Institutet, 171 77, Stockholm, Sweden.

GARRATT, Kirk N., Mayo Clinic/Foundation, Rochester, MN 55905.

GERWINS, Pär, Department of Physiology and Pharmacology, Karolinska Institutet, S-171 77 Stockholm, Sweden.

GIBBONS, Raymond J., Mayo Clinic/Foundation, Rochester, MN 55905.

GNECCHI-RUSCONE, T., Servizio di Cardiologia, Ospedale di Merate (Co), Italy.

GORDON, Ellen, Department of Neurology Surgery, University of Washington School of Medicine, Seattle, WA 98104.

HARDEN, T. Kendall, Department of Pharmacology, University of North Carolina, Chapel Hill, NC.

HERZER, William A., Center for Clinical Pharmacology, 623 Scaife Hall, University of Pittsburgh Medical Center, 200 Lothrop Street, Pittsburgh, PA 15213–2582.

HILLAIRE-BUYS, Dominique, Laboratoire de Pharmacologie, Faculte de Medecine, 34060 Montpellier Cedex, France.

HOLMES, David R., Mayo Clinic/Foundation, Rochester, MN 55905.

HOPPE, Edmund, Julius-Maximillians-Universitat, Institut fur Pharmakologie und Toxikologie, Versbacher Str. 9, 97078 Wurzburg, Germany.

HORI, Masatsugu, The First Department of Medicine, Osaka University School of Medicine, 2-2 Yamadaoka, Suite 565, Japan.

HOYLE, Charles H.V., Department of Anatomy and Developmental Biology, University College, London.

HSU, Sean S-F, Department of Neurology Surgery, University of Washington School of Medicine, Seattle, WA 98104.

HURT, Carl M., Hahnemann University, Department of Medicine, Broad and Vine, MS 110, Philadelphia, PA 19102-1192.

HUSSAIN, Tahir, Department of Pharmacology, School of Medicine, East Carolina University, Greenville, NC 27858.

ICHINOHE, Tatsuya, Department of Anesthesiology, Tokyo Dental College, Chiba, 261, Japan.

IJZERMAN, Ad P., LACDR, Division of Medicinal Chemistry, P.O. Box 9502 2300RA Leiden, The Netherlands.

ILLES, P., Pharmakologisches Institut der Universitat, Hermann-Herder-Strasse 5, D-79104, Freiburg, Germany.

INOUE, Michitoshi, Department of the Information Science, Osaka University Hospital, Osaka, Japan.

IRENIUS, Eva, Department of Physiology and Pharmacology, Karolinska Institutet, S-171 77 Stockholm, Sweden.

JACKSON, Edwin K., Center for Clinical Pharmacology, 623 Scaife Hall, University of Pittsburgh Medical Center, 200 Lothrop Street, Pittsburgh, PA 15213-2582.

JACOBSON, Kenneth A., Laboratory of Bioorganic Chemistry, National Institute of Diabetes and Digestive and Kidney Diseases, National Institutes of Health, Bethesda, MD 20892.

JACOBSON, Marlene A., Merck Research Labs, Department of Pharmacology, WP44B-122, West Point, PA 19486.

JANIGRO, Damir, Department of Neurological Surgery, Department of Environmental Health, University of Washington Seattle, WA 98104.

JEANSONNE, Neil, Department of Pharmacology, School of Medicie, East Carolina University, Greenville, NC 27858.

JUDGE, Martin E., Novo Nordisk Pharmaceuticals Division, Novo Nordisk Park, DK 2760 Malov, Denmark.

KAMADA, Takenobu, The First Department of Medicine, Osaka University School of Medicine, 2-2 Yamadaoka, Suite 565, Japan.

KANEKO, Yuzuru, Department of Anesthesiology, Tokyo Dental College, Chiba, 261, Japan.

KATCHANOV, Guennadi, Hahnemann University, Department of Medicine, Broad and Vine, MS 110, Philadelphia, PA 19102-1192.

KELLEY, Grant G., The Mount Desert Island Biological Laboratory, Salsbury Cove, ME 04672.

KIKUTA, Yoshinori, Department of Anesthesiology, Tokyo University School of Medicine, Tokyo, 173, Japan.

KITAKAZE, Masafumi, The First Department of Medicine, Osaka University School of Medicine, 2-2 Yamadaoka, Suite 565, Japan.

KLITGAARD, Henrik, Novo Nordisk Pharmaceuticals Division, Novo Nordisk Park, DK 2760 Malov, Denmark.

KNOWLES, M.R., Department of Pulmonary Medicine, Division of Pulmonary Diseases, The University of North Carolina at Chapel Hill, Chapel Hill, NC.

KNUTSEN, Lars J.S., Novo Nordisk Pharmaceuticals Division, Novo Nordisk Park, DK 2760 Malov, Denmark.

KOLLIAS-BAKER, Cynthia, Department of Medicine, University of Florida, Gainesville, FL 32610-0277.

KOMAMURA, Kazuo, The First Department of Medicine, Osaka University School of Medicine, 2-2 Yamadaoka, Suite 565, Japan.

KURIHARA, Toshinao, The First Department of Medicine, Osaka University School of Medicine, 2-2 Yamadaoka, Suite 565, Japan.

LASLEY, Robert D., Department of Surgery, University of Wisconsin School of Medicine, Madison, WI.

LAU, Jesper, Novo Nordisk Pharmaceuticals Division, Novo Nordisk Park, DK 2760 Malov, Denmark.

LAUDON, Dennis A., Mayo Clinic/Foundation, Rochester, MN 55905.

LINDEN, Joel, Box 158 Health Sciences Center, University of Virginia, Charlottesville, VA 22908.

LOBL, Joseph K., Mayo Clinic/Foundation, Rochester, MN 55905.

LOHSE, Martin J., Laboratorium fur Moleckulare Biologie der Universitat Munchen, Max-Planck-Institut fur Biochemie, Am Klopferspitz 18a, 82152 Martinsried, Germany.

LOMBARDI, F., Centro Ricerche Cardiovascolari, CNR, Medicina Interna II, Ospedale "L. Sacco", Universita di Milano, Italy.

LOUBATIÈRES-MARIANI, Marie-Madeleine, Laboratoire de Pharmacologie, Faculte de Medecine, 34060 Montpellier Cedex, France.

MAHMARIAN, John J., Section of Cardiology, Department of Internal Medicine, Baylor College of Medicine, The Methodist Hospital, Houston, TX 77030.

MAILLARD, Michel, Molecular Recognition Section, Laboratory of Bioorganic Chemistry, NIDDK, National Institutes of Health, Bethesda, MD.

MALLIANI, A., Centro Ricerche Cardiovascolari, CNR, Medicina Interna II, Ospedale "L. Sacco", Universita di Milano, Italy.

MARALA, Ravi, Department of Pharmacology, School of Medicine, East Carolina University, Greenville, NC 27858.

MASON, S.J., Department of Pulmonary Medicine, Division of Pulmonary Diseases, The University of North Carolina at Chapel, Hill, Chapel Hill, NC.

MENO, Joseph, Department of Neurology Surgery, University of Washington School of Medicine, Seattle, WA 98104.

MENTZER, Robert M., Department of Surgery, University of Wisconsin School of Medicine, Madison, WI.

MI, Zaichuan, Center for Clinical Pharmacology, 623 Scaife Hall, University of Pittsburgh Medical Center, 200 Lothrop Street, Pittsburgh, PA 15213–2582.

MINAMINO, Tetsuo, The First Department of Medicine, Osaka University School of Medicine, 2-2 Yamadaoka, Suite 565, Japan.

MIRAS-PORTUGAL, M. Teresa, Departamento de Bioquimica, Facultad de Veterinaria, Universidad Complutense, 28040 Madrid, Spain.

MIYAMOTO, Tadsomi, Cardiovascular Research Institute, Kokura Memorial Hospital, Kiakyushu-shi, 802, Japan.

MONOPOLI, Angela, Research Laboratories, Schering-Plough S.p.A., I-20060 Comazzo, Milan, Italy.

MONTANO, N., Centro Ricerche Cardiovascolari, CNR, Medicina Interna II, Ospedale "L. Sacco", Universita di Milano, Italy.

MUNERATI, Maddalena, Institute of General Pathology, University of Ferrara, Italy.

MUSTAFA, S. Jamal, Department of Pharmacology, School of Medicine, East Carolina University, Greenville, NC 27858.

NGUYEN, Thien-Son, Department of Neurological Surgery, University of Washington, Seattle, WA 98104.

NIEBER, K., Pharmakologisches Institut der Universitat, Hermann-Herder-Strasse 5, D-79104, Freiburg, Germany.

NOCKEN, Frank, Biozentrum der J.W. Goethe-Universitat, AK Neurochemie, Zoologisches Institut, Marie-Curie-Str. 9, 60439 Frankfurt am Main, Germany.

NODE, Koichi, The First Department of Medicine, Osaka University School of Medicine, 2-2 Yamadaoka, Suite 565, Japan.

NORENBERG, W., Pharmakologisches Institut der Universitat, Hermann-Herder-Strasse 5, D-79104, Freiburg, Germany.

NUTTER, T.J., Department of Molecular and Cellular Pharmacology, University of Miami School of Medicine, Miami, FL 33101.

O'CONNOR, William T., Departments of Neuroscience, Physiology and Pharmacology, and Medicinal Biochemistry and Biophysics, Karolinska Institutet, 171 77, Stockholm, Sweden.

OLANREWAJU, Hammed, Department of Pharmacology, School of Medicine, East Carolina University, Greenville, NC 27858,

OLIVIER, K.N., Department of Pulmonary Medicine, Division of Pulmonary Diseases, The University of North Carolina at Chapel Hill, Chapel Hill, NC.

ONGINI, Ennio, Research Laboratories, Schering-Plough S.p.A., I-20060 Comazzo Milan, Italy.

O'REGAN, Michael, Department of Physiology, School of Medicine, Wayne State University, 540 E. Canfield Ave., Detroit, MI 48201.

PAJOR, Ana M., University of Arizona, Department of Physiology, College of Medicine, Tucson, AZ 85724.

PELLEG, Amir, Hahnemann University, Department of Medicine, Broad and Vine, MS 110, Philadelphia, PA 19102-1192.

PERSSON, Håkan, Departments of Neuroscience, Physiology and Pharmacology, and Medicinal Biochemistry and Biophysics, Karolinska Institutet, 171 77, Stockholm, Sweden.

PETIT, Pierre, Laboratoire de Pharmacologie, Faculte de Medecine, 34060 Montpellier, Cedex France.

PHILLIS, John W., Department of Physiology, School of Medicine, Wayne State University, 540 E. Canfield Ave., Detroit, MI 48201.

PINTOR, Jesus, Dept. Bioquimica, Fac. Veterinaria, U.C.M. Madrid, Spain.

REDMAN, R.S., Department of Molecular Pharmacology and Biological Chemistry, Northwestern University Medical School, 303 E. Chicago Ave., Chicago, IL 60611.

REEDER, Guy S., Mayo Clinic/Foundation, Rochester, MN 55905.

REN, Hongzu, Departments of Medicine and Pharmacology, Duke University Medical Center, Box 3444 Durham, NC 27710.

RIBEIRO, J. Alexandre, Laboratory of Pharmacology, Gulbenkian Institute of Science, 2781 Oeiras Codex, Portugal.

RIBES, Gérard, Laboratoire de Pharmacologie, Faculte de Medecine, 34060 Montpellier Cedex, France.

RICHARDT, Gert, Medical Clinical, Technical University of Munich, Ismaningerstr. 22, 81675 Munich, Germany.

RUDOLPHI, Karl A. Höechst AG, Frankfurt, Germany.

SCHIFFMANN, Serge N., Brain Research Unit and Laboratory of Neuropathology and Neuropeptide Research, School of Medicine, Universite Libre de Bruxelles Brussels, Belgium.

SCHRADER, J., Department of Physiology, Heinrich-Heine-Universitat, Universitatsstr. 1, 40225 Dusseldorf, Germany.

SCHUBERT, Peter, Max-Planck-Institute for Psychiatry, Department of Neuromorphology, Planegg-Martinsried, Germany.

SEBASTIÃO, Ana M., Laboratory of Pharmacology, Gulbenkian Institute of Science, 2781 Oeiras Codex, Portugal.

SEN, Raquel P., Departamento de Bioquimica, Facultad de Veterinaria, Universidad Complutense, 28040 Madrid, Spain.

SHEARDOWN, Malcolm J., Novo Nordisk Pharmaceuticals Division, Novo Nordisk Park, DK 2760 Malov, Denmark.

SHRYOCK, John C., Department of Medicine, University of Florida, Gainesville, FL 32610-0277.

SILINSKY, E.M., Department of Molecular Pharmacology and Biological Chemistry, Northwestern University Medical School, 303 E. Chicago Ave., Chicago, IL 60611.

SNAPRUD, Per, Departments of Neuroscience, Physiology and Pharmacology, and Medicinal Biochemistry and Biophysics, Karolinska Institutet, 171 77, Stockholm, Sweden.

SONG, Yejia, Department of Medicine, University of Florida, Gainesville, FL 32610-0277.

STANNESS, Kathe Ann, Department of Neurological Surgery, University of Washington, Seattle, WA 98104.

STEINBERG, Thomas H., Infectious Diseases Division, Department of Medicine, Washington University, St. Louis, MO.

STILES, Gary L., Departments of Medicine and Pharmacology, Duke University Medical Center, Box 3444, Durham, NC 27710.

STUMPE, T., Department of Physiology, Heinrich-Heine-Universitat, Universitatsstr. 1, 40225 Dusseldorf, Germany.

STUTTS, M.J., Department of Pulmonary Medicine, Division of Pulmonary Diseases, The University of North Carolina at Chapel Hill, Chapel Hill, NC.

SYLVÉN, Christer, Karolinska Institute at Department of Medicine, Huddinge University Hospital, Huddinge, Sweden.

THOMSEN, Christian, Novo Nordisk Pharmaceuticals Division, Novo Nordisk Park DK 2760 Malov, Denmark.

TINKLEPAUGH, David L., Department of Neurological Surgery, University of Washington, Seattle, WA 98104.

UNGERSTEDT, Urban, Departments of Neuroscience, Physiology and Pharmacology, and Medicinal Biochemistry and Biophysics, Karolinska Institutet, 171 77, Stockholm, Sweden.

VAN BELLE, Herman, Janssen Research Foundation, B-2340 Beerse, Belgium.

VANDERHAEGHEN, Jean-Jacques, Brain Research Unit and Laboratory of Neuropathology and Neuropeptide Research, School of Medicine, Universite Libre de Bruxelles, Brussels, Belgium.

VAN DER WENDEN, Nora M., LACDR, Division of Medicinal Chemistry, P.O. Box 9502, 2300RA Leiden, The Netherlands.

VAN GALEN, Philip J.M., LACDR, Division of Medicinal Chemistry, P.O. Box 9502, 2300RA Leiden, The Netherlands.

VERANI, Mario S., Section of Cardiology, Department of Internal Medicine, Baylor College of Medicine, The Methodist Hospital, Houston, TX 77030.

VILLALBA, Martin, Institute of General Pathology, University of Ferrara, Italy.

VITTORI, Sauro, Dipartimento di Scienze Chimiche, Via S. Agostino, 1, Universita di Camerino, 62032 Camerino (MC), Italy.

VIZIANO, Monica, Research Laboratories, Schering-Plough S.p.A., I-20060 Comazzo, Milan, Italy.

VOLPINI, Rosaria, Dipartimento di Scienze Chimiche, Via S. Agostino, 1, Universita di Camerino, 62032 Camerino (MC), Italy.

VON LUBITZ, Dag K.J.E., Molecular Recognition Section, Laboratory of Bioorganic Chemistry, NIH/NIDDK, Bethesda, MD.

WEIS, Jan U., Novo Nordisk Pharmaceuticals Division, Novo Nordisk Park, DK 2760 Malov, Denmark.

WHITE, Thomas D., Department of Pharmacology, Dalhousie University, Halifax, Nova Scotia, Canada B3H 4H7.

WILLIAMS, Michael, Neuroscience Discovery, Pharmaceutical Products Division, Abbott Laboratories, Abbott Park, IL 60064.

WINN, H. Richard, Department of Neurological Surgery, University of Washington, Seattle, WA 98104.

YOUNG, James D., Department of Physiology, 7-55 Medical Sciences Building, University of Alberta, Edmonton, Alberta, Canada T6G2H7.

ZAHLER, Stefan, Department of Physiology, University of Munich, Pettenkoferstr. 12, 80336 Munich, Germany.

ZAMBON, Annalisa, Institute of Oncology, University of Padova, Italy.

ZANOVELLO, Paola, Institute of Oncology, University of Padova, Italy.

ZHOU, Jie-Gang, Department of Neurology Surgery, University of Washington School of Medicine, Seattle, WA 98104.

ZIMMERMAN, Herbert, Biozentrum der J.W. Goethe-Universitat, AK Neurochemie, Zoologisches Institut, Marie-Curie-Str. 9, 60439 Frankfurt am Main, Germany.

ZOCCHI, Cristina, Research Laboratories, Schering-Plough S.p.A., I-20060 Comazzo, Milan, Italy.

PREFACE

This book contains 55 chapters that summarize lectures given during the Fifth International Symposium on Adenosine and Adenine Nucleotides held in Philadelphia, Pennsylvania, U.S.A., May 9–13, 1994. The symposium and its proceedings follow four previous symposia and their published proceedings. Previous symposia were held in Banfff, Canada (1978), Charlottesville, Virginia (1982), Munich, Germany (1986), and Lake Yamanaka, Japan (1990).

The program included a wide range of subjects from molecular and cellular biology to clinical applications. Great emphasis was placed on cutting edge information derived from studies using molecular and cellular biology techniques applied to the field of adenosine and adenine nucleotide research. Likewise, the program included a number of presentations on potentially promising developments in the therapeutic applications of adenosine-related drugs and ATP. The tradition of past meetings was maintained by including in the program presentations of new research in the areas of purine metabolism and physiology, and the pharmacology of adenosine and adenine nucleotides.

The symposium was truly an international meeting, attended by more than 500 people from 25 different countries. The final program consisted of 112 oral and 352 poster presentations given by basic and clinical investigators. There were too many excellent presentations, provocative statements, new ideas, and discoveries reported to recount them all here. For abstracts of all presentations, the reader is referred to the special issue of *Drug Development Research* (Volume 31, Number 4, April 1994). Measured by the number and quality of presentations, and by the record number of participants, the Fifth International Symposium on Adenosine and Adenine Nucleotides was clearly an intellectually stimulating meeting. It is safe to conclude that the purine research field is vibrant and still growing. Adenosine has rapidly become the drug of choice for the acute management of supraventricular tachyarrhythmias and was recently approved for use in combination with cardiac imaging for detection of ischemic heart disease. Additional clinical applications of adenosine and adenosine-related drugs and ATP seem closer than ever before.

The editors would like to express their gratitude to the distinguished contributors for their excellent chapters and to all participants. Our thanks also go to the sponsors and supporters, who made the symposium and the publication of this book possible.

We hope that the Proceedings of the Fifth International Symposium on Adenosine and Adenine Nucleotides will prove to be an excellent up-to-date reference book for basic and clinical scientists in the field of adenosine and adenine nucleotides. Equally important, it is our goal that the contents of this book will stimulate new research and attract an ever increasing number of investigators into this exciting field.

Luiz Belardinelli, M.D.
Amir Pelleg, Ph.D.

ACKNOWLEDGMENTS

The Fifth International Symposium on Adenosine and Adenine Nucleotides was very generously supported by the following contributors:

Sponsors: Fujisawa Pharmaceutical Co.
Medco Research Inc.

Further support was kindly provided by

Abbott Laboratories	American Cyanamid Co.
Burroughs Wellcome Co.	Crinos Industria Farmacobiologica S.p.A.
Discovery Therapeutics	Fisons Pharmaceuticals
Gensia Pharmaceuticals, Inc.	Glaxo Group Research Limited
Hahnemann University	Item Development
Janssen Research Foundation	Marion Merrell Dow Inc.
Merck & Co., Inc.	Novo Nordisk A/S
Pfizer Inc.	Sandoz Pharmaceuticals Co.
Schering Plough S.p.A.	Technical Resources Inc.
University of Florida, College of Medicine	The University of Virginia Patents Foundation
The University of Virginia, College of Medicine	

I. MOLECULAR AND CELLULAR BIOLOGY OF ADENOSINE AND ATP

A. MOLECULAR BIOLOGY

1. MOLECULAR BIOLOGY OF ADENOSINE RECEPTORS

Marlene A. Jacobson

Prior to the application of recombinant DNA technology, the characterization of adenosine receptor subtypes was limited to information obtained from classic pharmacological analysis. Through the use of gene cloning and molecular biologic techniques, we have significantly advanced our understanding of the number of adenosine receptor subtypes and their molecular characteristics, such as primary sequence, structural homology between species, pharmacologic profiles of agonists and antagonists, and distribution of tissue and cellular expression. Molecular cloning has confirmed the existence of three adenosine receptor subtypes predicted from affinity order profiles of agonists and antagonists, and coupling to adenylyl cyclase, and, in addition, has identified a fourth subtype that had previously remained pharmacologically undetected. In this overview progress in the cloning and molecular characterization of adenosine receptors will be summarized.

The initial classification of adenosine receptors into A_1 and A_2 subtypes was based on functional coupling to adenylyl cyclase in which inhibition or stimulation of intracellular cAMP levels was determined, respectively [1]. The A_2 class was subdivided into A_{2a} and A_{2b} on the basis of high and low affinity for adenosine and adenosine analogs [2,3]. The existence of distinct A_1, A_{2a}, and A_{2b} adenosine receptor subtypes in various tissues was supported by pharmacologic evidence resulting from rank-order potencies of agonists and antagonists and coupling to adenylyl cyclase [4].

Chronology of Adenosine Receptor Cloning

A pivotal point in the purinergic receptor field was the cloning of two orphan receptors, designated RDC7 and RDC8, using a screening strategy based on the sequence homology of cloned G-protein–coupled receptors. Degenerate oligonucleotide primers were designed corresponding to conserved sequences within the third and sixth transmembrane domains of cloned G-protein–coupled receptors and utilized in the polymerase chain reaction to amplify homologous sequences [5]. The partial cDNAs were used to screen a canine thyroid cDNA library and identified two cDNAs that exhibited 51% overall identity to each other. RDC7 and RDC8 were determined to encode the A_1 and A_{2a} adenosine receptor subtypes, respectively, through expression in COS and CHO cells in which functional coupling to adenylyl cyclase and binding of agonists and antagonists was demonstrated [6,7]. The canine A_1 and A_{2a} receptors were the first of the adenosine receptor subtypes to be cloned and have been followed by the cloning of receptor subtypes from additional species, including human. The A_{2b} subtype was cloned from rat brain using a probe generated by a polymerase chain reaction methodology using degenerate oligonucleotide primers based on conserved regions of G-protein–coupled receptor transmembrane domains. The cloned sequence encoded a protein that exhibited 46% and 45% homology to the rat A_1 and A_{2a} receptors and 62% and 73% homology within the transmembrane regions of the rat A_1 and A_{2a} receptors, respectively [8]. The identity was confirmed to be the A_{2b} receptor on the basis of the positive coupling to adenylyl cyclase, lower affinity for adenosine agonists, and higher affinity for adenosine antagonists than the A_{2a} receptor, and the similar cAMP dose-response curves for adenosine agonists and antagonists measured for the expressed receptor in CHO cells in comparison with the endogenous A_{2b} receptor in VA 13 fibroblasts [9]. A fourth subtype, designated A_3, had remained pharmacologically obscure until its identification by molecular cloning in the rat [10,11]. The rat sequence exhibited 47% and 42% overall identity with the canine A_1 and A_{2a} receptors, respectively, and 58% homology within the transmembrane domains. Through functional expression in CHO cells, the A_3 subtype was determined to have an affinity order profile for agonists and antagonists distinct from A_1, A_{2a}, and A_{2b} receptors and was coupled to inhibition of adenylyl cyclase. The cloned A_3 receptor is distinct from an A_3 receptor postulated by Ribeiro and Sebastiao to exist on prejunctional sites on neurons and in atrial cells that was not found to be coupled to adenylyl cyclase but linked to calcium [12].

Characteristics of Cloned Adenosine Receptor Subtypes

The molecular characteristics of the four cloned adenosine receptor subtypes are summarized in Table 1-1. Accession numbers for sequences deposited into GenBank are indicated. The majority of the sequences encoding adenosine receptor subtypes were obtained from cDNA libraries prepared from brain tissue and may reflect the abundance of adenosine receptor subtype expression in the brain suggested from radioligand binding [13,14]. In addition, adenosine receptor cDNAs have been isolated from other tissues such as the thyroid (A_1 and A_{2a}), heart (A_1, A_{2a} and A_3), testis (A_3), and bone marrow–derived mast cells (A_{2a} and A_{2b}).

The cloning of adenosine receptors from multiple tissues provided the opportunity to evaluate whether specific sequence differences exist between tissues. A comparison of human A_1 adenosine receptor sequences cloned from heart ventricle, brain, and kidney showed no differences in sequence and suggested the corresponding cardiac, brain, and renal receptors were identical [15]. In addition, no tissue-specific differences were determined in A_{2a} receptors cloned from heart ventricle and striata.

Sequences for the A_1 adenosine receptor have been reported from six species, the A_{2a} from five and the A_{2b} and the A_3 from three. The percent overall identity between species relative to the human sequence is reported in Table 1-1 for the four adenosine receptor subtypes. The A_1 sequence is the most highly conserved and exhibits greater than 90% overall identity within the six species cloned. The human and canine A_{2a} sequences are more similar to each other than the guinea pig, rat, and mouse A_{2a} sequences. The mouse and rat A_{2b} sequences are 87% and 86% identical to the human, respectively. The sheep and human A_3 sequences have a higher percentage of identity (85%) in comparison to the rat A_3 sequence (72%). Although the amino acid sequences of the adenosine receptor subtypes are considerably conserved between species, they contain regions where a number of sequence differences are observed. In general, the most variation in sequence among species occurs in the region between the fourth and fifth transmembrane domains comprising the second extracellular loop and in the carboxy-terminal tail. Distinct species differences in adenosine receptor pharmacology have been well documented [16,17]. The sequences determined as

TABLE 1-1. Characteristics of cloned adenosine receptor subtypes

Receptor subype	Species	Origin	No. AA	Percent identity	Accession no.	Ref.
A_1	Human	Brain, heart	326	100	S45235, X68485,	35, 15
		Genomic			L22214	26
	Canine	Thyroid	326	94	X14051	5
	Rat	Brain	326	94	M64299, M69045	36, 37
	Bovine	Brain	326	94	X63592, M86261	38, 39
	Rabbit	Genomic	328	94	L01700	25
	Mouse	Brain	326	94	U05671	39
A_{2a}	Human	Brain, heart	412	100	M97370, X68486,	unpublished,
					S46950	15, 40
	Canine	Thyroid	412	93	X14052	5
	Guinea pig	Brain	409	87	U04201	41
	Rat	Brain	410	84	L08102, M91214	42, 43
	Mouse	Bone-marrow mast cells	410	84	U05672	39
A_{2b}	Human	Brain	332	100	M97759, X68487	44, 15
	Mouse	Bone-marrow mast cells	332	87	U05673	39
	Rat	Brain	332	86	M91466	8
A_3	Human	Brain, heart	318	100	L20463, L22607	45, 23
	Sheep	Brain	317	85	S65334	24
	Rat	Testis, brain	320	72	X59249, M94152	10, 11

AA = amino acids.

the result of cloning adenosine receptors from multiple species may form the basis for correlating specific differences in amino acid sequence to species variation in pharmacologic profiles determined in whole tissues.

The four human adenosine receptor subtype sequences are compared in Figure 1-1. Amino acid residues that are identical are indicated in the boxed regions. The highest degree of identity between the subtypes is found within the transmembrane (TM) domains. Specific regions of near identity are found in TM2 (FXVSLAXADXAVG), TM3 (LLAXADRY), TM4 (GLTPXXGWN), TM5 (YMVYFXF), and TM6 (FALXWLP). The percent overall identity between subtypes within a single species is lower than the percent values for a given subtype between species. In the human, a comparison of the subtype amino acid sequences yielded percent overall identities of $A_1 - A_{2a} = 50\%$, $A_1 - A_{2b} = 46\%$, $A_1 - A_3 = 49\%$, $A_{2a} - A_{2b} = 58\%$, $A_{2a} - A_3 = 43\%$, and $A_{2b} - A_3 = 40\%$. The amino-terminal sequence of adenosine receptors is short relative to the other members of the G-protein–coupled receptor family [18]. The A_{2a} receptor is the largest and the A_3 receptor is the smallest of the four subtypes. Adenosine receptors lack an aspartic acid residue in TM III that has been shown to be important in receptors that bind cationic amines [18]. Potential N-linked glycosylation sites exists in all four subtypes within the second extracellular loop. In addition, the A_3 subtype contains potential glycosylation sites within the amino-terminal sequence. A conserved cysteine residue is found in the carboxy-terminal region corresponding to positions 309, 311, and 303 within the A_1, A_{2b}, and A_3 receptors, respectively, but not in the A_{2a} receptor. This residue has been proposed to

```
                                           I
A3   MPNNSTALSL  ANVTYITMEI  FIGLCAIVGN  VLVICVVKLN  PSLQTTTFYF   50
A1   MPPSISAFQA  A...YIGIEV  LIALVSVPGN  VLVIWAVKVN  QALRDATFCF   47
A2a  MP......IM  GSSVYITVEL  AIAVLAILGN  VLVCWAVWLN  SNLQNVTNYF   44
A2b  .....MLLET  QDALYVALEL  VIAALSVAGN  VLVCAAVGTA  NTLQTPTNYF   45

                 II                                       III
A3   IVSLALADIA  VGVLVMPLAI  VVSLGITIHF  YSCLFMTCLL  LIFTHASIMS  100
A1   IVSLAVADVA  VGALVIPLAI  LINIGPQTYF  HTCLMVACPV  LILTQSSILA   97
A2a  VVSLAAADIA  VGVLAIPFAI  TISTGFCAAC  HGCLFIACFV  LVLTQSSIFS   94
A2b  LVSLAAADVA  VGLFAIPFAI  TISLGFCTDF  YGCLFLACFV  LVLTQSSIFS   95

                                                IV
A3   LLAIAVDRYL  RVKLTVRYKR  VTTHRRIWLA  LGLCWLVSFL  VGLTPMFGWN  150
A1   LLAIAVDRYL  RVKIPLRYKM  VVTPRRAAVA  IAGCWILSFV  VGLTPMFGWN  147
A2a  LLAIAIDRYI  AIRIPLRYNG  LVTGTRAKGI  IAICWVLSFA  IGLTPMLGWN  144
A2b  LLAVAVDRYL  AICVPLRYKS  LVTGTRARGV  IAVLWVLAFG  IGLTPFLGWN  145

                                                        V
A3   MK....LTSE  YHRNVTFLSC  QFVS......  VMRMDYMVYF  SFLTWIFIPL  190
A1   NL....SAVE  RAWAANGSMG  EPVIKCEFEK  VISMEYMVYF  NFFVWVLPPL  193
A2a  NC....GQPK  EGKNHSQGCG  EGQVACLFED  VVPMNYMVYF  NFFACVLVPL  190
A2b  SKDSATNNCT  EPWDGTTNES  CCLVKCLFEN  VVPMSYMVYF  NFFGCVLPPL  195

A3   VVMCAIYLDI  FYIIRNKLSL  NLSNSKETGA  FY...GREFK  TAKSLFLVLF  237
A1   LLMVLIYLEV  FYLIRKQLNK  KVSASSGDPQ  KYY..GKELK  IAKSLALILF  241
A2a  LLMLGVYLRI  FLAARRQLKQ  MESQPLPGER  ARSTLQKEVH  AAKSLAIIVG  240
A2b  LIMLVIYIKI  FLVACRQL.Q  ...RTELMDH  SRTTLQREIH  AAKSLAMIVG  241

                 VI                                 VII
A3   LFALSWLPLS  IINCIIYFNG  ....EVPQLV  LYMGILLSHA  NSMMNPIVYA  283
A1   LFALSWLPLH  IINCITLFCP  ..SCHKPSIL  TYIAIFLTHG  NSAMNPIVYA  289
A2a  LFALCWLPLH  IINCFTFFCP  .DCSHAPLWL  MYLAIVLSHT  NSVVNPFIYA  289
A2b  IFALCWLPVH  AVNCVTLFQP  AQGKNKPKWA  MNMAILLSHA  NSVVNPIVYA  291

A3   YKIKKFKETY  LLILKACVVC  HPSDSLDTSI  EKNSE                    318
A1   FRIQKFRVTF  LKIWNDHFRC  QPAPPIDEDL  PEERPDD                  326
A2a  YRIREFRQTF  RKIIRSHVLR  QQEPFKAAGT  SARVLAAHGS  DGEQVSLRLN  339
A2b  YRNRDFRYTF  HKIISRYLLC  QADVKSGNGQ  AGVQPALGVG  L            332

A2a  GHPPGVWANG  SAPHPERRPN  GYALGLVSGG  SAQESQGNTG  LPDVELLSHE  389

A2a  LKGVCPEPPG  LDDPLAQDGA  GVS                                  412
```

FIGURE 1-1. Primary sequence homology between the four human adenosine receptor subtypes. The sequence alignment is presented on the basis of the greater homology between A_1 and A_3, or A_{2a} and A_{2b}, subtypes. Boxed amino acids indicate identical amino acids. Putative transmembrane domains I–VII are designated by solid overlines. Potential N-linked glycosylation sites are in bold and underlined.

serve as a palmitoylation site involved in the formation of a fourth intracellular loop in other G-protein–coupled receptors [19].

The availability of cloned adenosine receptor sequences provides the opportunity to examine structure-function relationships of amino acid residues in ligand binding using site-directed mutagenesis. This methodology has been applied to the bovine and human A_1 adenosine receptors. Two histidine residues, His 251 and His 278, present in TM VI and TM VII, respectively, were individually replaced by leucine residues into the bovine A_1 receptor [20]. This TM VII histidine is conserved in all four adenosine receptor subtypes, and the TM VI histidine is absent in the A_3 sequence. Functional expression of mutant receptors in COS cells showed that modification at position 278 resulted in a decrease in both agonist and antagonist affinity of >90%. Replacement at position 251 decreased antagonist affinity by fourfold with no effect on agonist affinity.

Mutation of His^{251} resulted in a decreased number of agonist and antagonist binding sites and suggests that an alteration in protein folding occurred that could have interfered with proper insertion of the receptor into the membrane. The mutagenesis studies confirmed the importance of the histidine residues in ligand binding initially proposed from chemical modification [21]. Mutation of threonine 277, proximal to the histidine in TM VII, of the human A_1 receptor to either a serine or alanine had no effect on the affinity for the antagonist 8-cyclopropyl-1,3,-dipropylxanthine (DPCPX) [22]. However, removal of the hydroxyl group in the $Thr^{277}Ala$ mutant resulted in a significant decrease in the affinity for 5′-N-ethylcarboxamidoadenosine (NECA) but did not affect the affinity for N^6-modified analogues. These results suggested that the threonine hydroxyl in TM VII of the native A_1 receptor forms a contact with the 5′ substitution of NECA.

The expression of cloned adenosine receptors in transfected cell lines has greatly facilitated the pharmacologic characterization of the individual subtypes. Previously, analysis of the A_{2b} receptor has been limited by the lack of selective and high-affinity ligands, and, therefore, evaluation has been difficult in whole tissue in which multiple adenosine receptor subtypes are present. The ability to individually express the receptors provides a method to overcome difficulties in the pharmacologic evaluation of subtypes that have low representation or undefined distribution of expression in whole tissue. The expression of recombinant human A_3 adenosine receptors was crucial in defining the pharmacologic profile of this subtype, which had previously remained pharmacologically undetected. The human A_3 affinity order profiles of agonist and antagonist binding differ substantially from A_1, A_{2a}, and A_{2b} receptors and can be distinguished from the other subtypes by a preference for N^6-(4-amino-3-iodobenzyl)adenosine (I-ABA), relative to the agonists, 5′-N-ethylcarboxamidoadenosine (NECA) and (R)-N^6-phenyl-2-propyladenosine (R-PIA). We have shown that the human A_3 receptor has high affinity for 8-phenylxanthine antagonists with para-acidic groups, such as 3-(3-iodo-4-aminobenzyl)-8-(4-oxyacetate)phenyl-1-propylxanthine, I-ABOPX [23]. In comparison, the sheep A_3 receptor also has a similar potency for 8-phenylxanthines [24]. The antagonist profiles of the human and sheep A_3 receptors are different from the rat A_3 receptor, which has been reported to be insensitive to xanthine antagonists [11].

Organization of Adenosine Receptor Genes

The genomic organization of adenosine receptors has recently been determined through the cloning and characterization of A_1 and A_{2a} receptor genes. Partial characterization of a rabbit A_1 genomic clone revealed the presence of an intron that interrupted the coding sequence [25]. The intron occurred within the region corresponding to the second intracellular loop. The human A_1 gene was also found to contain an intron within the coding region, and, in addition, multiple introns were found within the 5′ untranslated region [26].

The human A_1 gene was determined to be comprised of a total of five introns and six exons. Two types of spliced transcripts were detected in mature transcripts containing either exons 4, 5, and 6 or exons 3, 5, and 6, and no evidence for the expression of the first two exons was found. The presence of multiple intron/exon structure in the 5′-untranslated region suggests that alternative splicing may occur in mRNA expression and has potential significance in the regulation of A_1 adenosine receptor gene expression. The human A_{2a} gene also contains an intron that interrupts the coding region. Similar to the A_1 gene, the intron occurs in a region corresponding to the second intracellular loop. The size of the intron within the A_{2a} gene has been estimated at greater than 4–6 kb [27]. Most of the genes cloned for other members of the G-protein–coupled family have been shown to be intronless, with the exception of substance P, substance K, opsins, and D_2 dopamine receptors [18]. The adenosine receptors, therefore, belong to a subfamily of intron-containing G-protein–coupled receptors.

Chromosomal Localization

The A_1 and A_{2a} receptors were initially mapped to chromosomes 22q11.2–q13.1 and 11q11–q13 by in situ hybridization employing [^{3}H]-labeled partial cDNAs of 463 and 469 bp in length [28]. Recently, the A_{2a} receptor was localized to chromosome 22 through Southern blot analysis of cDNAs isolated from a somatic cell hybrid panel and from cosmid cloning [29]. The A_1 receptor has been subsequently mapped to chromosome 1 [30]. The source of the discrepancy in the chromosomal localization of A_1 and A_{2a} receptors remains unclear. It may be possible that the partial cDNAs utilized in the in-situ hybridization analysis crosshybridized to a homologous region of another subtype or, alternatively, another member of the G-protein–coupled family of receptors.

Tissue Distribution of Transcript Expression

Comparison of the tissue distribution of human adenosine receptor transcripts revealed that the A_1, A_{2a}, A_{2b}, and A_3 subtypes are expressed in multiple tissues and the abundance of each subtype differs between tissues [23]. The tissues in which the most abundant expression has been reported for the individual adenosine receptor subtypes and results from in situ hybridization analysis are summarized in Table 1-2. A_1 transcripts (2.9 kb) were detected in human brain, heart, kidney, and lung, with the most abundant expression observed in the brain. A second hybridizing band of 4.3 kb was also observed in lower amounts in the brain. High levels of A_1 expression have also been reported in human testis [30]. A_{2a} transcripts (2.8 kb) were found in brain, heart, kidney, and lung. In situ hybridization of human brain has shown that A_{2a} transcripts are detected in the caudate, putamen, accumbens nuclei, and olfactory tubercle and are exclusively expressed in medium-sized neurons, which also express dopamine D_2 receptors and enkephalin [31,32]. In contrast, A_1 transcripts are highly expressed throughout the cortex, hippocampus, and cerebellum [32]. The Northern blot analysis results agree with the distribution of A_1 and A_{2a} receptors determined in the brain by radioligand binding studies [33,34]. For the A_{2b} receptor, two

TABLE 1-2. Distribution of abundant adenosine receptor transcript expression

Subtype	Tissue	In situ localization
A_1	Cortex, hippocampus, cerebellum testis (human, rat)	Widespread in CNS
A_{2a}	Striatum (human, rat, canine)	Caudate, putamen, accumbens nuclei, olfactory tubercle
A_{2b}	Colon (human)	
	Cecum, large intestine, urinary bladder (rat)	Pars tuberalis
A_3	Lung, liver (human)	
	Lung, spleen, pineal gland (sheep)	
	Testis (rat)	Spematocytes, spematids

transcripts (1.7 and 2.1 kb) are expressed in brain, heart, and lung, and no expression was observed in the kidney. High levels of A_{2b} transcript expression have been determined in human colon. In the rat, the highest levels of A_{2b} expression have been reported in the cecum, large intestine, and urinary bladder [8]. In situ hybridization analysis of rat brain A_{2b} receptors localized expression to the hypophyseal pars tuberalis. The physiologic action of A_{2b} receptors in these tissues remains to be defined. In contrast to the human A_1, A_{2a}, and A_{2b} subtypes, the A_3 receptor exhibits a different distribution of transcript expression. The most abundant expression of A_3 transcripts is found in the lung and liver. Moderate expression is found in the brain and aorta, and low levels are detected in the heart and testis. The pattern of human A_3 transcript expression is similar to sheep A_3 receptor transcripts, which are widely distributed and are highly expressed in the lung [24]. In contrast, the expression of rat A_3 transcripts occurs primarily in testis. On the basis of in situ localization in spermatocytes and spermatids, the involvement of A_3 receptors in spermatogenesis has been postulated [10].

Summary

The cloning of adenosine receptor subtypes is a significant contribution to purinergic receptor research and provides the foundation for the further analysis of structure-function relationships of ligand binding through site-directed mutagenesis, definition of physiologic roles for individual subtypes through localization of tissue and cellular expression, and evaluation of mechanisms for regulation of gene expression. The human adenosine receptor cDNAs expressed in mammalian cell lines are invaluable tools for use in the screening and development of subtype-selective agents that may have potential therapeutic utility. Because of the species variability in adenosine receptor pharmacology, extrapolation of affinity order profiles of agonists and antagonists established in one species may not always be relevant to human pharmacology. With the availability of cloned adenosine receptors from human and multiple species, the selection of appropriate animal models for the evaluation of agents for human therapy may now be achieved.

Note Added in Proof

1. The guinea pig A_1 receptor has been cloned (GenBank accession no. U04279)
2. An error in the interpretation of the chromosomal mapping of A_1 and A_{2a} receptors has been reported (Genomics (1994) 23:305). The probe hybridizing to chromosome 22q was verified to correspond to the A_{2a} receptor (in agreement with ref [29]) and that hybridizing to chromosome 11q did not correspond to the A_1 receptor.

References

1. Van Calker D, Muller M, Hamprecht B (1979) Adenosine regulates via two different types of receptors, the accumulation of cyclic AMP in cultured brain cells. J Neurochem 33:999–1005.
2. Daly JW, Butts-Lamb P, Padgett W (1983) Subclasses of adenosine receptors in the central nervous system: Interaction with caffeine and related methylxanthines. Cell Mol Neurobiol 3:69–80.
3. Bruns RF, Lu GH, Pugsley TA (1986) Characterization of the A_2 adenosine receptor labeled by [^{3}H]NECA in rat striatal membranes. Mol Pharmacol 29:331–346.
4. Jacobson KA, van Galen PJM, Williams M (1992) Adenosine receptors: Pharmacology, structure-activity relationship and therapeutic potential. J Med Chem 35:407–422.
5. Libert F, Parmentier M, Lefort A, Dinsart C, Van Sande J, Maenhaut C, Simons M, Dumont JE, Vassart G (1989) Selective amplification and cloning of four new members of the G protein-coupled receptor family. Science 244:569–572.
6. Libert F, Schiffmann SN, Lefort A, Parmentier M, Vanderhaeghen J-J, Vassart G (1991) The orphan receptor cDNA RDC7 encodes an A_1 adenosine receptor. EMBO J 10:1677–1682.
7. Maenhaut C, Van Sande J, Libert F, Abramowicz M, Parmentier M, Vanderhaeghen J-J, Dumont JE, Vassart G, Schiffmann S (1990) RDC8 codes for an adenosine A_2 receptor with

physiological constitutive activity. Biochem Biophys Res Comm 173:1169–1178.

8. Stehle JH, Rivkees SA, Lee JJ, Weaver DR, Deeds JD, Reppert SM (1992) Molecular cloning and expression of the cDNA for a novel A_2-adenosine receptor subtype. Mol Endocrinol 6:384–393.
9. Rivkees SA, Reppert SM (1992) RFL9 encodes an A_{2b}-adenosine receptor. Mol Endocrinol 6: 1598–1604.
10. Meyerhof W, Muller-Brechlin R, Richter D (1991) Molecular cloning of a novel putative G-protein coupled receptor expressed during rat spermiogenesis. FEBS Lett 284:155–160.
11. Zhou Q-Y, Chuanyi L, Olah ME, Johnson RA, Stiles GL, Civelli O (1992) Molecular cloning and characterization of an adenosine receptor: The A_3 adenosine receptor. Proc Natl Acad Sci USA 89:7432–7436.
12. Ribeiro JA, Sebastiao AM (1986) Adenosine receptors and calcium: Basis for proposing a third (A_3) adenosine receptor. Prog Neurobiol 26:179–209.
13. Bruns RF, Daly JW, Snyder SH (1980) Adenosine receptors in brain membranes: Binding of N^6-cyclohexyl[^{3}H]adenosine and 1,3-diethyl-8-[^{3}H]phenylxanthine. Proc Natl Acad Sci USA 77:5547–5551.
14. Wan W, Sutherland GR, Geiger JD (1990) Binding of the adenosine A_2 receptor ligand [^{3}H]CGS21680 to human and rat brain: Evidence for multiple affinity sites. J Neurochem 55:1763–1771.
15. Salvatore CA, Luneau CJ, Johnson RG, Jacobson MA (1992) Molecular cloning and characterization of human A_1 and A_2 adenosine receptors. Int J Pur Pyrid Res 3:82.
16. Ukena D, Jacobson KA, Padgett WL, Ayala C, Shamim MT, Kirk KL, Olsson RO, Daly JW (1986) Species differences in the structure-activity relationships of adenosine agonists and xanthine antagonists at brain A_1 adenosine receptors. FEBS Lett 209:122–128.
17. Stone GA, Jarvis MF, Sills MA, Weeks B, Snowhill EW, Williams M (1988) Species differences in high-affinity adenosine A_2 binding sites in striatal membranes from mammalian brain. Drug Dev Res 15:31–46.
18. Probst WC, Snyder LA, Schuster DI, Brosius J, Sealfon SC (1992) Sequence alignment of the G-protein coupled receptor superfamily. DNA Cell Biol 11:1–20.
19. O'Dowd BF, Hnatowich M, Caron MG, Lefkowitz RJ, Bouvier M (1989) Palmitoylation of the human β_2-adrenergic receptor. J Biol Chem 264:7564–7569.
20. Olah ME, Ren H, Ostrowski J, Jacobson KA, Stiles GL (1992) Cloning, expression and characterization of the unique bovine A_1 adenosine receptor. J Biol Chem 267:10764–10770.
21. Klotz K-N, Lohse MJ, Schwabe U (1988) Chemical modification of A_1 adenosine receptors in rat brain membranes. J Biol Chem 263: 17522–17526
22. Townsend-Nicholson A, Schofield PR (1994) A threonine residue in the seventh transmembrane domain of the human A_1 adenosine receptor mediates specific agonist binding. J Biol Chem 269:2373–2376.
23. Salvatore CA, Jacobson MA, Taylor HE, Linden J, Johnson RG (1993) Molecular cloning and characterization of the human A_3 adenosine receptor. Proc Natl Acad Sci USA 90:10365–10369.
24. Linden J, Taylor HE, Robeva AS, Tucker AL, Stehle JH, Rivkees SA, Fink SJ, Reppert SM (1993) Molecular cloning and functional expression of a sheep A_3 adenosine receptor with widespread tissue distribution. Mol Pharm 44: 524–532.
25. Bhattacharya S, Dewitt DL, Burnatowska-Hledin M, Smith WL, Spielman WS (1993) Cloning of an adenosine A_1 receptor-encoding gene from rabbit. Gene 128:285–288.
26. Ren H, Stiles GL (1994) Characterization of the human A_1 adenosine receptor gene. J Biol Chem 269:3104–3110.
27. Peterfreund RA, MacCollin M, Gusella J, Fink JS (1994) Structure of the human A_{2a} receptor gene. Drug Dev Res 31:306.
28. Libert F, Passage E, Parmentier M, Simons M, Vassart G, Mattei M (1991) Chromosomal mapping of A_1 and A_2 adenosine receptors, VIP receptor and a new subtype of serotonin receptor. Genomics 11:225–227.
29. MacCollin M, Peterfreund R, MacDonald M, Fink JS, Guesella J (1994) Mapping of a human A_{2a} adenosine receptor (ADORA2) to chromosome 22. Genomics 20:332–333.
30. Rivkees SA, Reppert SM, Wells HB (1994) Molecular cloning and characterization of a human A_1-adenosine receptor. Drug Dev Res 31:314.
31. Schiffman SN, Libert F, Vassart G, Vanderhaeghen J-J (1991) Distribution of adenosine A_2 receptor mRNA in the human brain. Neurosci Lett 130:177–181.

32. Schiffman SN, Libert F, Vassart G, Dumont JE, Vanderhaeghen JJ (1990) A cloned G protein-coupled protein with a distribution restricted to striatal medium-sized neurons. Possible relationship with D_1 dopamine receptor. Brain Res 519:333–337.
33. Fastbom J, Pazos A, Probst A, Palacious JM (1986) Adenosine A_1-receptors in human brain: Characterization and autoradiographic visualization. Neurosci Lett 65:127–132.
34. Martinez-Mir MI, Probst A, Palacious JM (1991) Adenosine A_2 receptors: Selective localization in the human basal ganglia and alterations with disease. Neuroscience 42:697–706.
35. Libert F, Van Sande J, Lefort A, Czernilofsky A, Dumont JE, Vassart G, Ensinger HA, Mendla KD (1992) Cloning and functional characterization of a human A_1 adenosine receptor. Biochem Biophys Res Comm 187:919–926.
36. Mahan LC, McVittie LD, Smyk-Randall EM, Nakata H, Monsma FJ, Gerfen CR, Sibley DR (1991) Cloning and expression of an A_1 adenosine receptor from rat brain. Mol Pharmacol 40:1–7.
37. Reppert SM, Weaver DR, Stehle JH, Rivkees SA (1991) Molecular cloning and characterization of a rat A_1 adenosine receptor that is widely expressed in brain and spinal cord. Mol Endocrinol 5:1037–1048.
38. Tucker AL, Linden J, Robeva AS, D'Angelo DD, Lynch KR (1992) Cloning and expression of a bovine adenosine A_1 receptor cDNA. FEBS Lett 297:107–111.
39. Marquardt DL, Walker LL, Heinemann S (1994) Cloning of two adenosine receptor subtypes from mouse bone marrow-derived mast cells. J Immunol 152:4508–4515.
40. Furlong TJ, Pierce KD, Selbie LA, Shine J (1992) Molecular characterization of a human brain adenosine A_2 receptor. Mol Brain Res 15:62–66.
41. Meng F, Xie G, Chalmers D, Morgan C, Watson SJ, Akil H (1994) Cloning and expression of the A_{2a} adenosine receptor from guinea pig brain. Neurochem Res 19:613–621.
42. Fink JS, Weaver DR, Rivkees SA, Peterfreund RA, Pollack AE, Adler EM, Reppert SM (1992) Molecular cloning of the rat A_2 receptor: Selective co-expression with D_2 dopamine receptors in rat striatum. Mol Brain Res 14:186–195.
43. Chern Y, King K, Lai H (1992) Molecular cloning of a novel adenosine receptor gene from rat brain. Biochem Biophys Res Comm 185: 304–309.
44. Pierce KD, Furlong TJ, Selbie LA, Shine J (1992) Molecular cloning and expression of an adenosine A_{2b} receptor from human brain. Biochem Biophys Res Comm 187:86–93.
45. Sajjadi F, Firestein GS (1993) cDNA cloning and sequence analysis of the human A_3 adenosine receptor. Biochem Biophys Acta 1179:105–107.

2. RECOMBINANT TECHNIQUES AS APPLIED TO THE STUDY OF A_1 ADENOSINE RECEPTORS

Joel Linden

One of the great utilities of the use of recombinant techniques as applied to the study of receptors is the ability to readily change receptor structure. This has been done in a number of different ways. Receptor chimera, that is, receptors composed partly of one type of receptor and partly of another, can readily be constructed. For example, chimeric alpha$_2$-beta$_2$–adrenergic receptors were constructed to delineate domains involved in ligand binding and effector coupling [1]. The impact of individual amino acids on receptor function can be examined by site-directed mutagenesis. For example, site-directed mutagenesis of asp^{113} of the beta$_2$-adrenergic receptor was used to identify a key amino acid involved in catecholamine recognition [2]. As examples of these types of approaches applied to adenosine receptors, the identification of regions of the A_1 adenosine receptors responsible for conferring species specificity in ligand binding to A_1 adenosine receptors by our laboratory is summarized.

The A_1 adenosine receptor has been cloned from six species—dog, cow, rat, rabbit, mouse, and human [3–6]. Despite >90% receptor identity at the amino acid level, the ligand binding characteristics of adenosine A_1 receptors differ between species [7,8]. Bovine and canine receptors differ the most, with bovine A_1 adenosine receptors having relative high affinity for N^6-substituted adenine analogs such as the agonist R-N^6-phenylisopropyladenosine (R-PIA) and the antagonist (S)-N^6-endonorbornan-2-yl-9-methyladenine (N-0861). Bovine A_1 receptors also have relatively high affinity for C8-substituted xanthine antagonists such as 8-cyclopentyl-1,3-dipropylxanthine (CPX or DPCPX) [4,7,9]. Our initial approach to identifying the region(s) of receptors responsible for these species differences was to construct chimeric canine/bovine receptors. This is illustrated schematically in Figure 2-1A. As a follow-up to these studies, we have used site-directed mutagenesis to target individual amino acids, as illustrated schematically in Figure 2-1B.

Another use of recombinant technology has been the introduction of an extra sequence into proteins to facilitate their purification. The introduction of six adjacent histidines (hexahistidine) has been found to result in proteins that will adhere tightly to nickel adsorbed to chelator affinity columns [10–12]. Similarly, antigenic peptides such as the FLAG™ octapeptide (DYKDDDDK) can be engineered into recombinant receptors to facilitate purification by application to antibody columns or by immunoprecipitation [13]. We describe here the con-

FIGURE 2-1. Schematic representation of recombinant A_1 adenosine receptors. Receptors are depicted in a serpentine seven structure thought to be characteristic of G-protein–coupled receptors. **A**: Schematic structure of a chimeric receptor composed partly of amino acids of one species (solid) and partly of amino acids of another species (striped). **B**: Schematic structure of a receptors with two amino acids in the seventh transmembrane helix (circles) modified by site-directed mutagenesis. **C**: Schematic structure of a receptor modified by the addition of extra amino acids (boxed), such as hexahistidine or the FLAG epitope, added to facilitate purification.

struction of A_1 adenosine receptors modified to contain hexahistidine and/or the FLAG epitope, as illustrated schematically in Figure 2-1C.

Results and Discussion

CHIMERIC A_1 ADENOSINE RECEPTORS

The results of experiments on the binding of ligands to chimeric bovine/canine A_1 adenosine receptors, such as the receptor pictured schematically in Figure 2-1A, indicates that species differences can be localized to the carboxy-terminal third of the receptor [14]. For several G-protein–coupled receptors, including the alpha$_2$-adrenergic, the 5-HT$_2$, and the 5-HT$_{1B}$ receptors, marked pharmacologic differences between homologues from different species can be attributed to single amino acid differences [15–17]. For the alpha$_2$-adrenergic and the 5-HT$_2$ receptors, the critical amino acid is in TM5, but for 5-HT$_{1B}$ it is in TM7. The bovine and canine A_1 adenosine receptors are identical in TM5 and TM6 but have two animo acid differences in TM7. These two amino acids were targeted for site-directed mutagenesis.

SITE-DIRECTED MUTATIONS OF A_1 ADENOSINE RECEPTORS

The results of site-directed mutagenesis of TM7 amino acids to change them from bovine to canine or vice versa are described in detail elsewhere [14] and are summarized qualitatively in Table 2-1. Adenine and xanthine compounds substituted with aryl or cycloalkyl substituents at the N^6 or the C8 positions, respectively, bind preferentially to bovine as opposed to canine receptors, primarily due to amino acid 270, isoleucine in the bovine receptor, and methionine in the canine receptor. This is entirely consistent with the idea that the N^6 position of adenines and the C8 position of xanthines occupy the same position in the binding pocket of the A_1 adenosine receptor. This N^6/C8 model also is supported by an analysis of steric and electrostatic properties of ligands [18].

5′-N-ethylcarboxamidoadenosine (NECA), which like adenosine is unsubstituted in the N^6 position, binds preferentially to receptors that contain Thr^{277} (canine) as opposed to Ser^{277} (bovine). Changing amino acid 277 from threonine to serine markedly reduces the affinity of receptors for NECA only if amino acid 270 is isoleucine rather than methionine. These data are consistent with the recent finding that mutation of Thr^{277} in the human A_1 adenosine receptor to Ser or Ala causes a relatively selective

TABLE 2-1. Influence on ligand binding of mutations in amino acids 270 and 277 of recombinant A_1 adenosine receptors

AA-270	AA-277	CPX	N-0861	R-PIA	NECA
I	S	High	High	High	Low
I	T	High	High	High	High
M	S	Low	Low	Low	High
M	T	Low	Low	Low	High

Species differences (bovine/canine) in the affinity of the C8-substituted xanthine, CPX, and two N^6-substituted adenine compounds, N-0861 and R-PIA, depend almost exclusively on the composition of AA-270. NECA binding affinity is high if AA-270 is Met but depends on the composition of AA-277 if AA-270 is Ile. In the wild-type bovine receptor AA-270 and AA-277 are Ile and Ser, respectively.

decrease in NECA binding affinity [19]. These data indicate some interplay between amino acids in positions 270 and 277 in determining affinities for ligands without bulky N^6 or C8 substitutions. In the absence of the Met^{270} (such as in the bovine receptor, which has an Ile, or the human receptor, containing a Thr), the amino acid at position 277, via interactions with a different region of the ligand, apparently has a larger influence on ligand affinity for 5′ or C2-substituted ligands without N^6 substitutions. A threonine at position 277 favors high-affinity binding to such ligands. The rat A_1 adenosine receptor has a ligand binding profile that is consistent with this theory; this receptor displays high affinity for N^6- and C8-substituted ligands and for ligands with 5′- or C2 substitutions. It has an isoleucine at position 270, as does the bovine A_1 receptor, but a threonine at position 277, like the canine A_1 receptor. Our data support a model for ligand binding in which the N^6 or C8 substituents of the ligand interact with a region of the receptor containing amino acid 270, while a different region of the ligand, perhaps on the ribose, interacts with a hydroxyl on amino acid 277. Because the antagonists, CPX and N-0861, do not have ribose moieties and their binding is not significantly affected by the T/S mutations at position 277, the interaction of the ribose at this position is an attractive hypothesis. Either the 5′ N-ethylcarboxyl substituent or some other portion of the ribose moiety of NECA may interact at position 277.

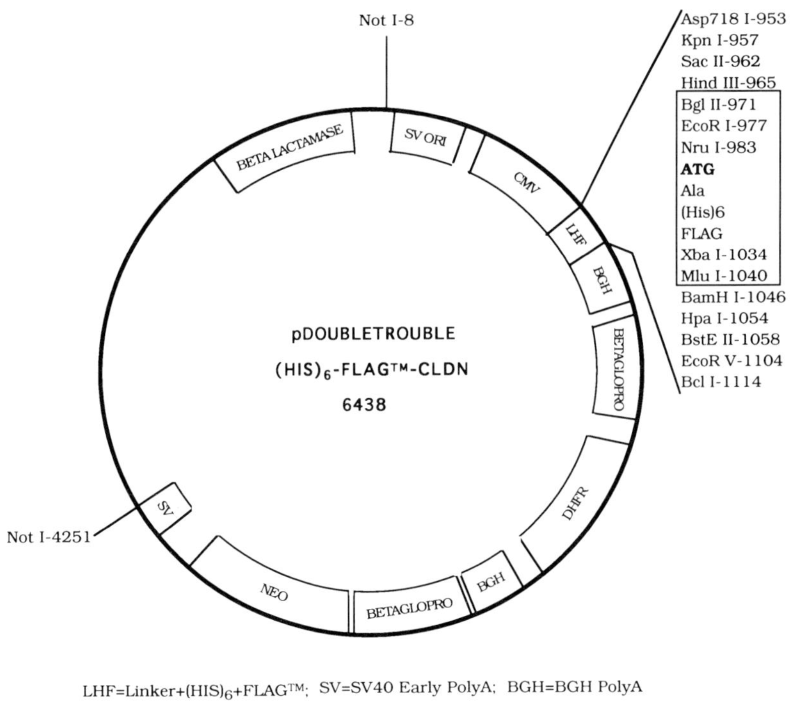

FIGURE 2-2. Map of pDOUBLETROUBLE. The plasmid CLDN10B that contains an SV40 ori suitable for transient COS cell transfections and neomycin (neo) resistance suitable for selecting stable transfectants with G418 has been modified to contain a new start codon, hexahistidine, the FLAG epitope, and additional restriction sites as indicated by the box.

RECOMBINANT RECEPTORS ENGINEERED TO CONTAIN HEXAHISTIDINE AND/OR THE FLAG EPITOPE

We have produced recombinant human A_1 adenosine receptors containing extra amino acid sequence to facilitate purification. Addition of extra amino acids at either the amino- or carboxy-terminal regions has little or no effect on ligand binding or G-protein coupling. Although recombinant receptors modified to contain hexahistidine and solubilized in detergents (CHAPS or digitonin) can be purified in high yield on Ni-chelate affinity columns, the degree of purification has been disappointing (300- to 500-fold) due to nonspecific binding of other proteins to the metal chelate columns. Much greater purification can be achieved by using recombinant receptors containing the FLAG epitope and purification with immobilized anti-FLAG antibodies that are commercially available. We have constructed an expression vector called pDOUBLETROUBLE because it introduces both hexahistidine and the FLAG epitope onto the amino terminus of receptors subcloned into it (Fig. 2-2). The polymerase chain reaction is used to extend receptor cDNAs with appropriate restriction sites to facilitate directional subcloning into this expression plasmid.

Figure 2-3 shows the elution of a recombinant human A_1 adenosine receptor extended to contain the FLAG epitope from an anti-FLAG antibody column. In a single step the receptor is purified over 1000-fold in high yield (>50%). Moreover, G proteins associated with recombinant receptors can be copurified with the receptor provided that adenosine is included during the purification to stabilized receptor–G-protein complexes. G proteins can be sensitively detected by [^{35}S]GTPγS binding (Fig. 2-3). G proteins have previously been copurified with A_1 adenosine receptors purified by agonist affinity chromatography [20]. Since these extended recombinant receptors can be readily purified, they hold great promise for permitting an analysis of putative post-translational modifications, such as glycosylation, acylation, and phosphorylation.

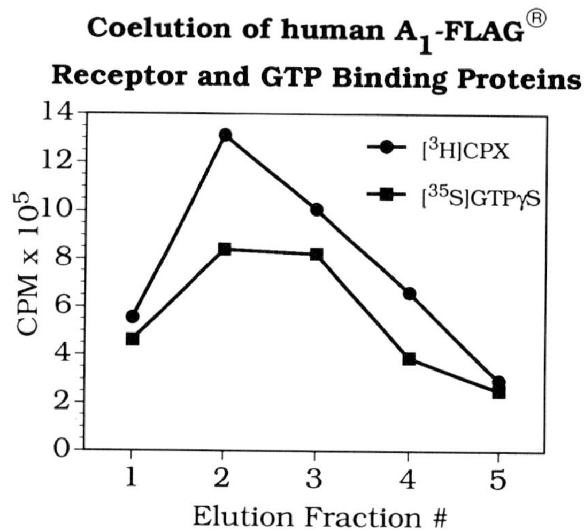

FIGURE 2-3. Elution of modified human A_1 adenosine receptors and associated G proteins from an anti-FLAG affinity column. The human A_1 adenosine receptor was subcloned into pDOUBLETROUBLE and stably expressed in CHO-K1 cells. Receptors were solubilized in digitonin and applied to an anti-FLAG affinity column with 10 μm adenosine. Following extensive washing, receptors and associated G proteins were eluted with five 1-ml fractions of the FLAG octapeptide (200 μg/ml). Eluted receptors were reconstituted in membranes prepared from untransfected CHO cells and quantified by [^{3}H]CPX binding. G proteins were quantified by [^{35}S]GTPγS binding.

Conclusions

The use of recombinant techniques has begun to have an impact on our understanding of the structure and function of adenosine receptors. It is likely that in the next few years recombinant receptors will be of great value in assessing receptor fine structure, post-translational modifications, distribution, and interactions with G proteins and possibly other proteins.

References

1. Kobilka BK, Kobilka TS, Daniel K, Regan JW, Caron MG, Lefkowitz RJ (1988) Chimeric α_2-, β_2-adrenergic receptors: Delineation of domains involved in effector coupling and ligand binding specificity. Science 240:1310–1316.
2. Strader CD, Sigal IS, Candelore MR, Rands E, Hill WS, Dixon RAF (1988) Conserved aspartic acid residues 79 and 113 of the β-adrenergic receptor have different roles in receptor function. J Biol Chem 263:10267–10271.
3. Linden J, Tucker AL, Robeva AS, Graber SG, Munshi R (1993) Properties of recombinant adenosine receptors. Drug Dev Res 28: 232–236.
4. Tucker AL, Linden J (1993) Cloned receptors and cardiovascular responses to adenosine. Cardiovas Res 27:62–67.
5. Bhattacharya S, Dewitt DL, Burnatowska-Hledin M, Smith WL, Spielman WS (1993) Cloning of an adenosine A1 receptor-encoding gene from rabbit. Gene 128:285–288.
6. Marquardt DL, Walker LL, Heinemann S (1994) Cloning of two adenosine receptor subtypes from mouse bone marrow-derived mast cells. J Immunol 152:4508–4515.
7. Klotz K-N, Vogt H, Tawfik-Schlieper H (1991) Comparison of A_1 adenosine receptors in brain from different species by radioligand binding and photoaffinity labelling. Naunyn Schmiedebergs Arch Pharmacol 343:196–201.
8. Ukena D, Jacobson KA, Padgett WL, Ayala C, Shamin MT, Kirk KL, Olsson RO, Daly JW (1986) Species differences in structure-activity relationships of adenosine agonists and xanthine antagonists at brain A_1 adenosine receptors. FEBS Lett. 209:122–128.
9. Tucker AL, Linden J, Robeva AS, D'Angelo DD, Lynch KR (1992) Cloning and expression of a bovine adenosine A_1 receptor cDNA. FEBS Lett 297:107–111.
10. Hochuli E, Dobeli H, Schacher A (1977) New metal chelate adsorbent selective for proteins and peptides containing neighboring histidine residues. J Chromatog 411:177–184.
11. Ljungquist C, Breitholtz A, Brink-Nilsson H, Moks T, Uhlen M, Nilsson B (1993) Immobilization and affinity purification of recombinant proteins using histidine peptide fusions. Eur J Biochem 186:563–569.
12. Van Dyke MW, Sirito M, Sawadogo M (1993) Single-step purification of bacterially expressed polypeptides containing an oligo-histidine domain. Gene 111:99–104.
13. Hopp TP, Prickett KS, Price VL, Libby RT, March CJ, Cerretti DP, Urdal DL, Conlon PJ (1988) A short polypeptide marker sequence useful for recombinant protein identification and purification. Biotechnology 6:1204–1210.
14. Tucker AL, Robeva AS, Taylor HE, Holeton D, Bockner M, Lynch KR, Linden J (1994) A_1 adenosine receptors: Two amino acids are responsible for species differences in ligand recognition. J Biol Chem, in press.
15. Kao HT, Adham N, Olsen MA, Weinshank RL, Branchek TA, Hartig PR (1992) Site-directed mutagenesis of a single residue changes the binding properties of the serotonin 5-HT2 receptor from a human to a rat pharmacology. FEBS Lett 307(3):324–328.
16. Oksenberg D, Marsters SA, O'Dowd BF, Jin H, Havlik S, Peroutka SJ, Ashkenazi A (1992) A single amino-acid difference confers major pharmacological variation between human and rodent 5-HT1B receptors. Nature 360: 161–163.
17. Link R, Daunt D, Barsh G, Chruscinski A, Kobilka BK (1992) Cloning of two mouse genes encoding alpha$_2$-adrenergic receptor subtypes and identification of a single amino acid in the mouse alpha$_2$-C10 homolog responsible for an interspecies variation in antagonist binding. Mol Pharmacol 42:16–27.
18. Van der Wenden EW, Ijzerman AP, Soudijn W (1992) A steric and electrostatic comparison of three models for the agonist/antagonist binding site on the adenosine A_1 receptor. J Med Chem 35:629–635.
19. Townsend-Nicholson A, Schofield PR (1994) A threonine residue in the seventh transmembrane domain of the human A_1 adenosine receptor mediates specific agonist binding. J Biol Chem 269:2373–2376.
20. Munshi R, Linden J (1989) Co-purification of A_1 adenosine receptors and guanine nucleotide-binding proteins from bovine brain. J Biol Chem 264:14853–14859.

3. IN SITU HYBRIDIZATION OF ADENOSINE RECEPTORS IN BRAIN

Jean-Jacques Vanderhaeghen
Serge N. Schiffmann

Recent progress in molecular biology of receptors has led to the cloning of the different adenosine receptors subtypes. Indeed, the A_1 and A_{2a} receptors, and more recently the A_{2b} and A_3 receptors, have been cloned in several species [1]. In the late 1980s, the technique of receptors cloning using the polymerase chain reaction (PCR) and degenerated oligonucleotides led to the identification of some already known receptors and to the discovery of new but orphan receptors, in quest for their ligands [2]. Since study of their distribution may help in their identification, it was taken advantage of by using the in situ hybridization technique to localize in the brain [3–5] the mRNAs encoding two orphan receptors, RDC7 and RDC8, which belong to the G-protein–coupled receptor superfamily. Based on these established distributions [3–5], RDC7 and RDC8 were definitively identified and pharmacologically characterized as the canine A_1 [6] and A_{2a} [6,7] adenosine receptors, respectively. The cloning of these canine A_1 and A_{2a} receptors, identified using in situ hybridization as a key technique, further allowed their cloning in different species and the cloning of the other adenosine receptors subtypes [1].

The A_1 and A_{2a} receptor distributions in the central nervous system (CNS) have been fully documented using either binding techniques before cloning or in situ hybridization after cloning. Conversely, the distributions of the A_{2b}, and certainly that of the A_3, receptor are poorly characterized at the present time. Indeed, although the existence of an A_{2b} subtype has been proposed for a longer time based on pharmacologic or physiologic data [8], the lack of specific ligands, its recent cloning, and its relatively low expression in the brain [9] explain the small amount of available neuroanatomic data on this receptor. Finally, the recent cloning of an unexpected A_3 receptor subtype expressed at a relatively low level in the brain led to the same conclusion [10,11].

Distribution of Adenosine A_1 Receptor

The A_1 receptor is widely expressed throughout the CNS of different species, such as the rat and dog, as detected by in situ hybridization [3,12, 13], therefore confirming the widespread detection of physiologic responses to A_1 agonists. Indeed, in most areas at least a faint specific labeling was apparent, with very few regions remaining completely devoid of labeling [see ref. 12 for an exhaustive description]. At the cellular level, it appears that most neurons exhibit a weak labeling, indicating a low level of expression per cell, although some neuronal populations are more intensely labeled. Among

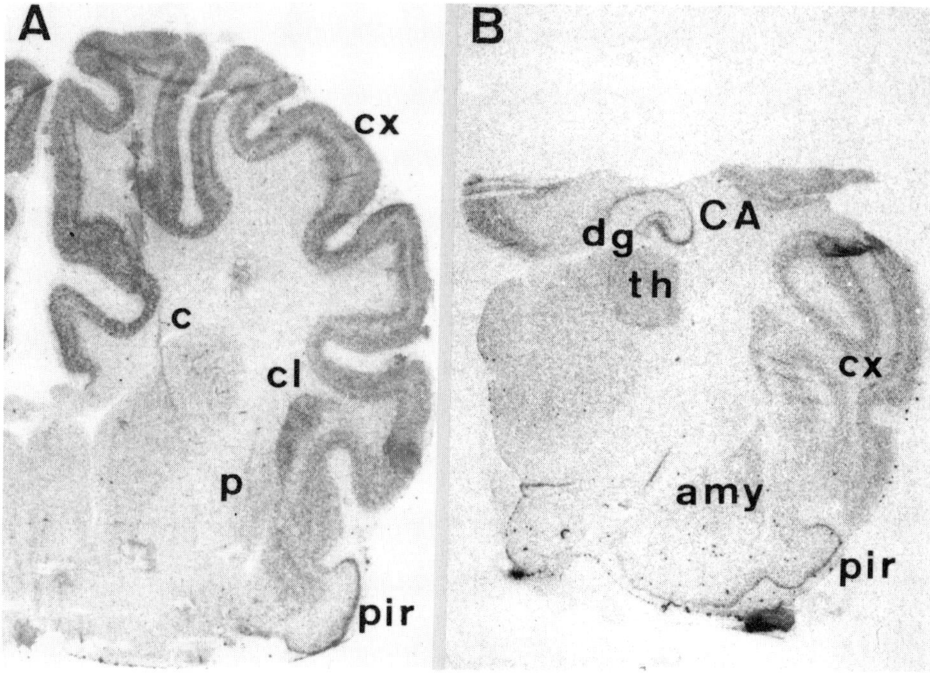

FIGURE 3-1. Localization of the A_1 receptor mRNA in the canine brain using in situ hybridization. Amy = amygdala; c = caudate; CA = Ammon's horn of the hippocampus; cl = claustrum; cx = cerebral cortex, dg = dentate gyrus of the hippocampus; pir = piriform cortex; p = putamen; th = thalamus.

them, one should note high expression of this receptor in the pyramidal layer of Ammon's horn and the granular layer of the dentate gyrus in the hippocampal formation (Fig. 3-1B); in layers 3 and 6 of the cerebral cortex (Fig. 3-1); in the granular layer of the cerebellum; in most of the thalamic nuclei (Fig. 3-1B) [12]; in nuclei of cranial nerves III, V, VII and XII; and in motoneurons of the ventral horn of spinal cord. In some areas the labeling is so faint that divergent conclusions have been inferred from the data since, for instance, the labeling in the striatum has been reported either as very faint but significant [3,13] or insignificant [12]. Therefore, near or below the threshold of detection for in situ hybridization, other techniques such as Northern blot and reverse-transcribed PCR appear to be very useful to explore this anatomic analysis further. Indeed, these techniques reveal an expression of the A_1 receptor in the striatum [13; Schiffmann, unpublished].

Brain Distribution of Adenosine A_{2a} Receptor

Converse to widespread distribution of the A_1 receptor, in the rat, canine, and human brain, expression of the A_{2a} receptor is abundantly and exclusively detected in different components of the striatum; the caudate, putamen, and accumbens nuclei; and the olfactory tubercle (Fig. 3-2) [3–5,9,12,14]. The expression of this receptor is present in the whole rostrocaudal extent of the putamen and the caudate nucleus (Fig. 3-2). In these studies [3–5,9,12,14], using standard protocols for in situ hybridization, regions such as the cerebral cortex, hippocampus, amygdala, thalamus, hypothalamus, globus pallidus, mesencephalon including the

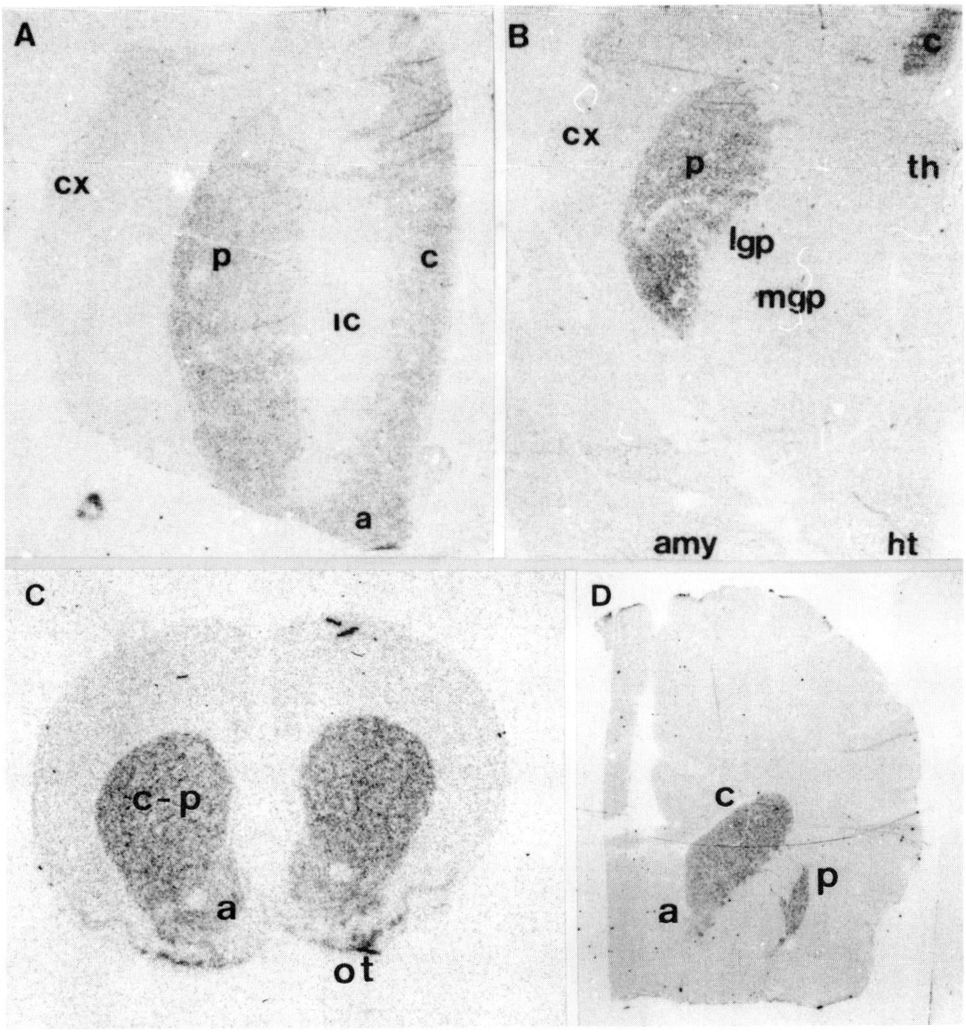

FIGURE 3-2. Localization of A_{2a} receptor mRNA in human (**A**, **B**), rat (**C**), and canine (**D**) brains using in situ hybridization. A_{2a} receptor is exclusively expressed in the caudate, putamen, accumbens nucleus, and olfactory tubercle. Abbreviations as in Figure 3-1. a = accumbens nucleus; c-p = caudate-putamen; ht = hypothalamus; ic = internal capsule; lgp = lateral globus pallidus; mgp = medial globus pallidus; ot = olfactory tubercle.

substantia nigra, pons, medulla oblongata, and cerebellum appear unlabeled (Fig. 3-2). However, Northern blot and reverse-transcribed PCR give new insights by revealing that this receptor is also expressed, at a much lower level, in other regions such as the cerebral cortex [14; Schiffmann, unpublished], therefore confirming some physiological data. At the cellular level, labeling is detected on 50–60% of medium-sized neurons in the putamen, caudate, and accumbens and in layers II–III of the olfactory tubercle (Fig. 3-3B). Glial cells and large striatal neurons remain unlabeled [3,5].

In the three studied species, a differential compartmentalization of A_{2a} receptor expression in striosomes or matrix could not be detected during adulthood [3–5,14,15]. However, during brain development the striosomes were transiently enriched in A_{2a} receptors [15,16], and this corresponds at the cellular level to a

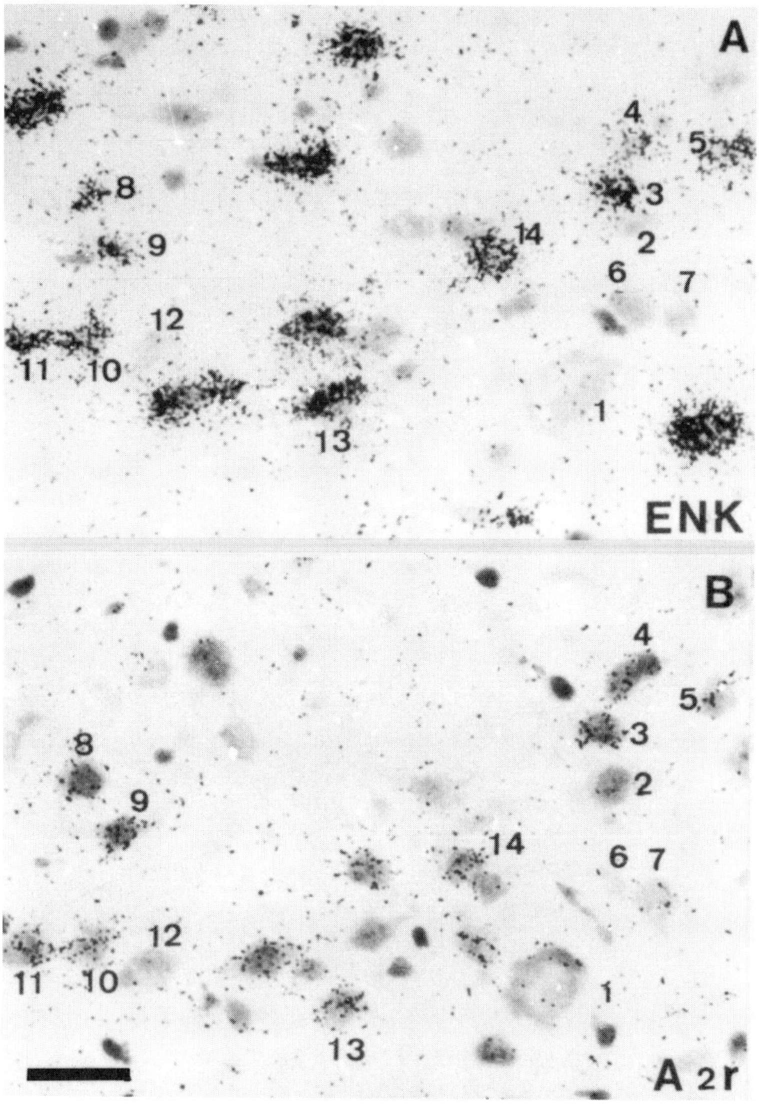

FIGURE 3-3. Demonstration of a substantial coexpression of enkephalin (ENK) (**A**) and adenosine A_{2a} receptor (A_{2r}) mRNAs (**B**) using in situ hybridization on two consecutive sections of the rat caudate-putamen (neurons identified on both adjacent sections have been numbered). Bar = 25 μm.

higher density of labeled neurons in the patch areas together with a higher level of expression per labeled patch neuron as compared with the matrix neurons. Like other striatal components, the A_{2a} receptor could thus have a role in the generation, maturation, and/or maintenance of striatal compartmentalization. In addition to this developmentally regulated compartmentalization of A_{2a} striatal expression, this receptor is transiently expressed in several regions devoid of any detectable labeling at adulthood, such as the cerebral cortex, the subiculum, the locus coeruleus, the area postrema, and several cranial nerve ganglia [16], also suggesting putative influences of adenosine acting at the A_{2a} receptor on the developmental processes of the brain.

Selective Expression of Adenosine A_{2a} Receptors in a Striatal Neuronal Subpopulation

Striatal neurons are heterogeneous in terms of their morphology, phenotype, and connections. The efferent neurons of the striatum are medium-sized spiny neurons and account for more than 90% of the striatal neurons. Although all utilize gamma-aminobutyric acid (GABA) as neurotransmitter, two subpopulations of these efferent neurons have been identified. The neurons projecting to the globus pallidus, striatopallidal, express enkephalin and mostly the dopamine D_2 receptor, whereas the neurons projecting to the susbtantia nigra pars reticulata, striatonigral, express substance P, dynorphin, and the dopamine D_1 receptor [17]. These two subpopulations give rise to different loops and act, respectively, as inhibitory and an excitatory pathways to the thalamus [18]. Two classes of interneurons represent most of the remaining striatal neurons: the cholinergic and somatostatin-expressing neurons.

The projection site of the A_{2a} receptor expressing striatal neurons has been assessed by using in situ hybridization combined with retrograde fluorogold labeling of striatonigral neurons [19]. The A_{2a} receptor is expressed by less than 5% of the flurogold-positive striatonigral neurons, and more than 95% of the neurons expressing this receptor are flurogold-negative striatopallidal neurons [19].

Analysis of the putative coexpression of the A_{2a} receptor with either enkephalin, substance P, choline acetyltransferase, somatostatin, dopamine D_1, or dopamine D_2 receptor has been performed using in situ hybridization performed on thin adjacent sections or with a combination of radioactive and nonradioactive probes on the same section [5,14,19,20]. The A_{2a} receptor is expressed in more than 90% of the neurons expressing enkephalin (Fig. 3-3) or the dopamine D_2 receptor, whereas its expression is nearly never detected in neurons expressing substance P, choline acetyltransferase, somatostatin, or the dopamine D_1 receptor [5,14,19,20]. The A_{2a} receptor is therefore mainly expressed by the striatopallidal neurons that also express enkephalin and the dopamine D_2 receptor, and this should represent the anatomic substratum for interactions between adenosine and dopamine receptors, especially the A_{2a} and D_2 subtypes described in the basal ganglia system [21,22].

Brain Distribution of Adenosine A_{2b} and A_3 Receptors

As stated in the introduction, using in situ hybridization few if any data are available concerning the brain distribution of A_{2b} and A_3 receptors. Using this technique expression of the A_{2b} receptor is only clearly detected in the hypophyseal pars tuberalis at the ventral surface of the median eminence. Northern blot or reverse-transcribed PCR experiments reveal that both receptors are expressed at a relatively low level in the brain as compared with several peripheral tissues [9–11]. Using these techniques, expression of the A_{2b} receptor could be detected in the cerebral cortex, striatum, and ventral mesencephalon [Schiffmann, unpublished] and that of the A_3 receptor in the cerebral cortex, cerebellum, striatum, hippocampus, and hypothalamus [11; Schiffmann, unpublished].

Acknowledgments

We are thankful to J.-L. Conreur, P. Halleux, and R. Menu for photography and excellent technical support. This work was supported by Belgian grants from Fonds de la Recherche Scientifique Medicale, Fondation Médicale Reine Elisabeth, Loterie Nationale, and Ministère de la Politique Scientifique (PAI 1990–95). S.N. Schiffmann is a Research Associate of the FNRS (Belgium).

References

1. Jacobson M, et al., present book.
2. Libert F, Parmentier M, Lefort A, Dinsart C, Van Sande J, Maenhaut C, Simons M-J, Dumont JE, Vassart G (1989) Selective amplification and

cloning of four new members of the G protein-coupled receptor family. Science 244:569–572.
3. Schiffmann SN, Libert F, Vassart G, Dumont JE, Vanderhaeghen JJ (1990) A cloned G protein-coupled protein with a distribution restricted to striatal medium sized neurons. Possible relationship with D1 dopamine receptor. Brain Res 519:333–337.
4. Schiffmann SN, Jacobs O, Vanderhaeghen JJ (1991a) The striatal restricted adenosine A_2 receptor (RDC8) is expressed by enkephalin but not by substance P neurons. An in situ hybridization histochemistry study. J Neurochem 57: 1062–1067.
5. Schiffmann SN, Libert F, Vassart G, Vanderhaeghen JJ (1991b) Distribution of adenosine A_2 receptor mRNA in the human brain. Neurosci Lett 130:177–181.
6. Libert F, Schiffmann SN, Lefort A, Parmentier M, Gerard C, Dumont JE, Vanderhaeghen JJ, Vassart G (1991) The orphan receptor cDNA encodes and A adenosine receptor. EMBO J 10:1677–1682.
7. Maenhaut C, Van Sande J, Libert F, Abramowicz, Parmentier M, Vanderhaeghen JJ, Dumont JE, Vassart G, Schiffmann SN (1990) RDC8 codes for an adenosine A_2 receptor with physiological constitutive activity. Biochem Biophys Res Comm 173:1169–1178.
8. Bruns RF, Lu GH, Pugsley TA (1986) Characterization of the A_2 adenosine receptor labeled by $[^3H]$NECA in rat striatal neurons. Mol Pharmacol 29:331–346.
9. Stehle JH, Rivkees SC, Lee JJ, Weaver DR, Deeds JD, Reppert SM (1992) Molecular cloning and expressin of the cDNA for a novel A2-adenosine receptor subtype. Mol Endocrinol 6:384–393.
10. Zhou Q-Y, Li C, Olah ME, Johnson RA, Stiles GL, Civelli O (1992) Molecular cloning and characterization of an adenosine receptor: The A3 adenosine receptor. Proc Natl Acad Sci USA 89:7432–7436.
11. Linden J, Taylor HE, Robeva AS, Tucker AL, Stehle JH, Rivkees SA, Fink JS, Reppert SM (1993) Molecular cloning and functional expression of a sheep A3 adenosine receptor with widespread tissue distribution. Mol Pharmacol 44:524–532.
12. Reppert SM, Weaver DR, Stehle JH, Rivkees SA (1991) Molecular cloning and characterization of a rat A1-adenosine receptor that is widely expressed in brain and spinal cord. Mol Endocrinol 5:1037–1048.
13. Mahan LC, McVittie LD, Smyk-Randall EM, Nakata H, Monsma FJ Jr, Gerfen CR, Sibley DR (1991) Cloning and expression of an A1 adenosine receptor from rat brain. Mol Pharmacol 40:1–7.
14. Fink JS, Weaver DR, Rivkees SA, Peterfreund RA, Pollack AE, Adler EM, Reppert SM (1992) Molecular cloning of the rat A2 adenosine receptor: Selective co-expression with D2 dopamine receptors in rat striatum. Mol Brain Res 14:186–195.
15. Schiffmann SN, Vanderhaeghen JJ (1992) Ontogeny of gene expression of adenosine A2 receptor in the striatum: Early localization in the patch compartment. J Comp Neurol 317: 117–128.
16. Weaver DR (1993) A2a adenosine receptor gene expression in developing rat brain. Mol Brain Res 20:313–327.
17. Gerfen CR, Engber TM, Mahan LC, Susel Z, Chase TN, Monsma FJ Jr, Sibley DR (1990) D1 and D2 dopamine receptor-regulated gene expression of striatonigral and striatopallidal neurons. Science 250:1429–1432.
18. Albin RL, Young AB, Penney B (1989) The functional anatomy of basal ganglia disorders. Trends Neurosci 12:366–375.
19. Schiffmann SN, Vanderhaeghen J-J (1993) Adenosine A2 receptors regulate the gene expression of striatopallidal and striatonigral neurons. J Neurosci. 13:1080–1087.
20. Augood SJ, Emson PC (1994) Adenosine A2a receptor mRNA is expressed by enkephalin cells but not by somatostatin cells in rat striatum: A co-expression study. Mol Brain Res 22:204–210.
21. Schiffmann et al., present book.
22. Fuxe et al., present book.

4. MOLECULAR MODELING OF ADENOSINE A_1 AND A_{2a} RECEPTORS

Ad P. IJzerman
Nora M. van der Wenden
Philip J.M. van Galen
Ken A. Jacobson

The superfamily of G-protein–coupled receptors encompasses over 300 cloned and sequenced members, and this number is still rapidly growing. Among them are the adenosine receptor subtypes—A_1, A_{2a}, A_{2b}, and A_3—some of which originate from various species. Adenosine receptors also show the characteristic secondary structure of all G-protein–coupled receptors, that is, seven predominantly hydrophobic stretches of 20–25 amino acids that could span the cell membrane as α-helices (Fig. 4-1) [1–3]. There is much uncertainty about the atomic coordinates of the receptor proteins. At present we have to rely on their suggested three-dimensional structural analogy with the related membrane proteins, rhodopsin and bacteriorhodopsin. Only the latter protein has been analyzed at sufficient resolution [4] to allow its use as a template for receptor models [5].

In the present study we propose a model for ligand binding and activation of the human A_1 and A_{2a} receptor, taking into account conformational characteristics of the ligands, the involvement of histidine residues, as has become evident from chemical modification and mutagenesis studies [6–9], as well as the putative bacteriorhodopsinlike architecture of the receptors. It should be emphasized that these models are not meant to represent the definitive structural characterization of the receptors. Hopefully, however, they will serve as a conceptual framework for the design of new experiments in the triad of organic chemistry, molecular modeling, and molecular biology.

Computational Methods

All methods used have been described extensively previously [10,11]. All model building, docking, energy minimizations, and molecular dynamics calculations were carried out using the software package BIOGRAF, version 3.1 (Molecular Simulations Inc., Waltham, MA). All manipulations were performed on a Silicon Graphics 4D/25GT or 4000XZ workstation. The atomic coordinates of bacteriorhodopsin were retrieved from the Brookhaven Protein Databank (reference code 1BRD). Transmem-

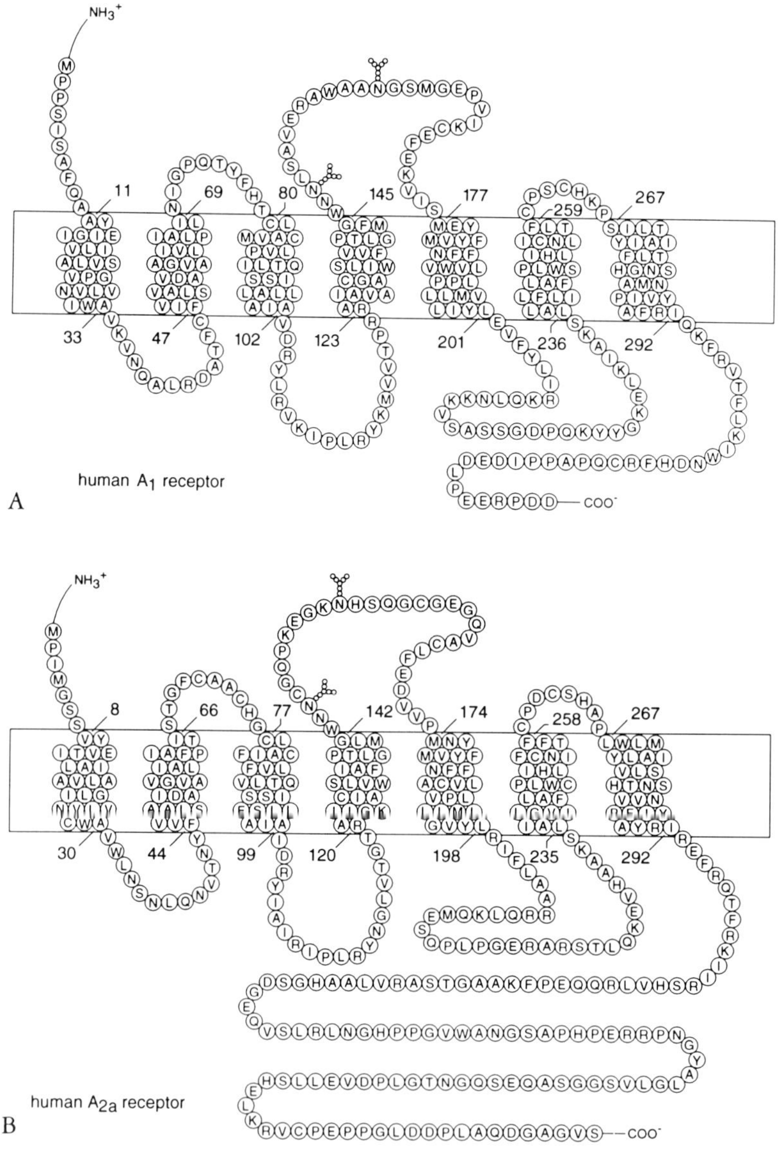

FIGURE 4-1. Two-dimensional representation of the human adenosine A_1 (**A**) and A_{2a} (**B**) receptor. The N terminus is located on the extracellular side, and the C terminus on the cytosolic side.

brane domains in the adenosine receptors were predicted from standard routines available in the Sequence Analysis Software Package [12].

Results

ADENOSINE RECEPTOR TOPOLOGY

For the construction of the receptor models we used the atomic coordinates of bacteriorhodopsin. Since no data for the nonmembrane parts of this protein (i.e., extracellular and intracellular loops and tails) are available, we refrained from modeling equivalent parts of the adenosine receptors. The proline residues in the transmembrane domains of the three proteins are partly in different positions, which renders the topology of the receptor models slightly different from bacteriorhodopsin because prolines may act as "helix benders". Due to the procedure followed ("mutation" with subsequent minimization), however, the adenosine receptor models retain the overall appearance of bacteriorhodopsin, with only slight changes in helical architecture.

DOCKING OF AGONISTS

The selective agonists N^6-cyclopentyladenosine (CPA; A_1) and 2-(cyclohexylmethylidinehydrazino)adenosine (SHA) 174 (A_{2a}; Fig. 4-2) were used as starting points for docking purposes. CPA was docked in several orientations into the A_1 receptor pore, in all instances close to the two histidine residues in helices VI and VII. The orientations of CPA with the cyclopentyl substituents between the helices led to an overall distortion of receptor architecture and were not considered further. Of the possibilities available when CPA is entirely within the pore, it appeared that the gain in interaction energy between ligand and receptor was not identical. The energetically most favorable orientation was with the cyclopentyl substituent pointing to the extracellular side of the protein, close to helices IV, V, and VI. Figure 4-3a is a representation of CPA surrounded by the seven alpha-helices (the main chains shown only).

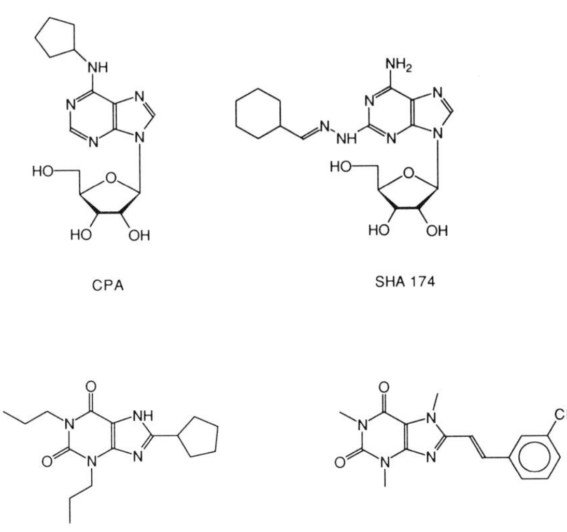

FIGURE 4-2. Structures of various adenosine receptor ligands. CPA = N^6-cyclopentyladenosine (A_1 agonist); SHA 174 = 2-(cyclohexylmethylidinehydrazino)adenosine (A_{2a} agonist); DPCPX = 1,3-dipropyl-8-cyclopentylxanthine (A_1 antagonist); CSC = 8-(3-chlorostyryl)caffeine (A_{2a} antagonist).

The ligand binding site on the A_1 receptor (arbitrarily defined as the amino acid residues within 4.5 Å from N^6-cyclopentyladenosine) is represented in Figure 4-4, showing more hydrophilic residues in close proximity to the hydrophilic purine and ribose moieties of CPA. The two histidine residues (251 and 278) may form hydrogen bonds with CPA, via N^6-H, and 2′- and 3′-OH, respectively, as does Ser^{281} with 5′-OH. It should be stressed that a slightly different orientation of CPA would probably lead to other hydrogen bonds formed, for example, with other alcoholic amino acid residues (Thr^{91}, Ser^{94}, and Ser^{246}). The presence of Asn^{254} at the interface between the purine ring and the cyclopentyl substituent could also be important for interaction. The hydrophobic cyclopentyl substituent, in contrast, is surrounded by three hydrophobic amino acid side chains, that is, Val^{138}, Phe^{185}, and Cys^{255}.

Similarly, SHA 174 was docked in the adenosine A_{2a} receptor topology. It turned out that SHA 174, with its C2 side chain pointing downwards (i.e., to the intracellular side of the protein), could not be accommodated by the receptor. The cavity within the receptor has funnel-like characteristics, just as in bacteriorhodopsin, the diameter of the pore at the extracellular side being larger than at the intracellular side. An energetically favorable orientation could be obtained with the C2 side chain pointing to the extracellular side of the protein, close to helices III, IV, and V. Figure 4-3B is a representation of SHA 174 surrounded

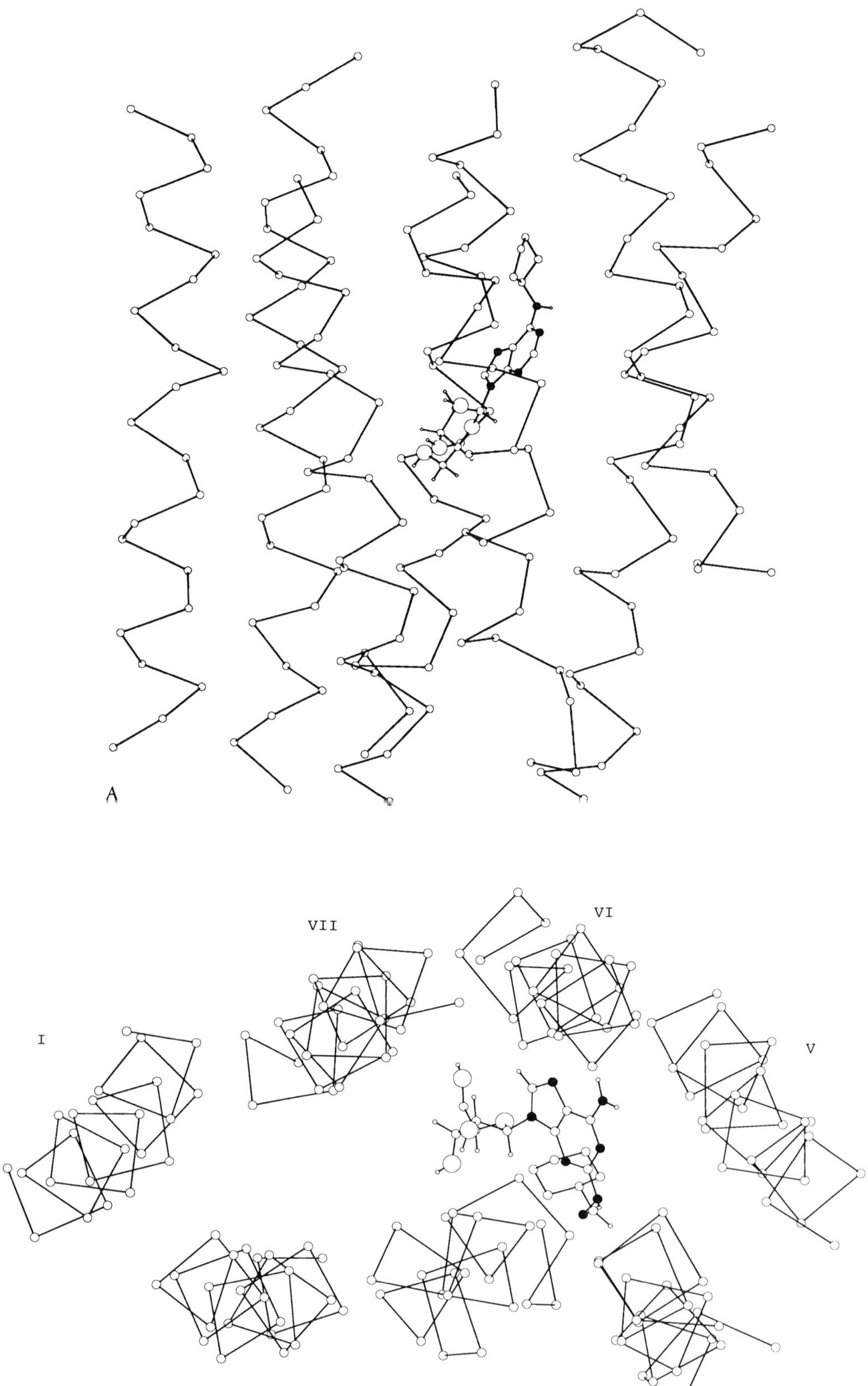
A
VII
VI
I
V
B
II
III
IV

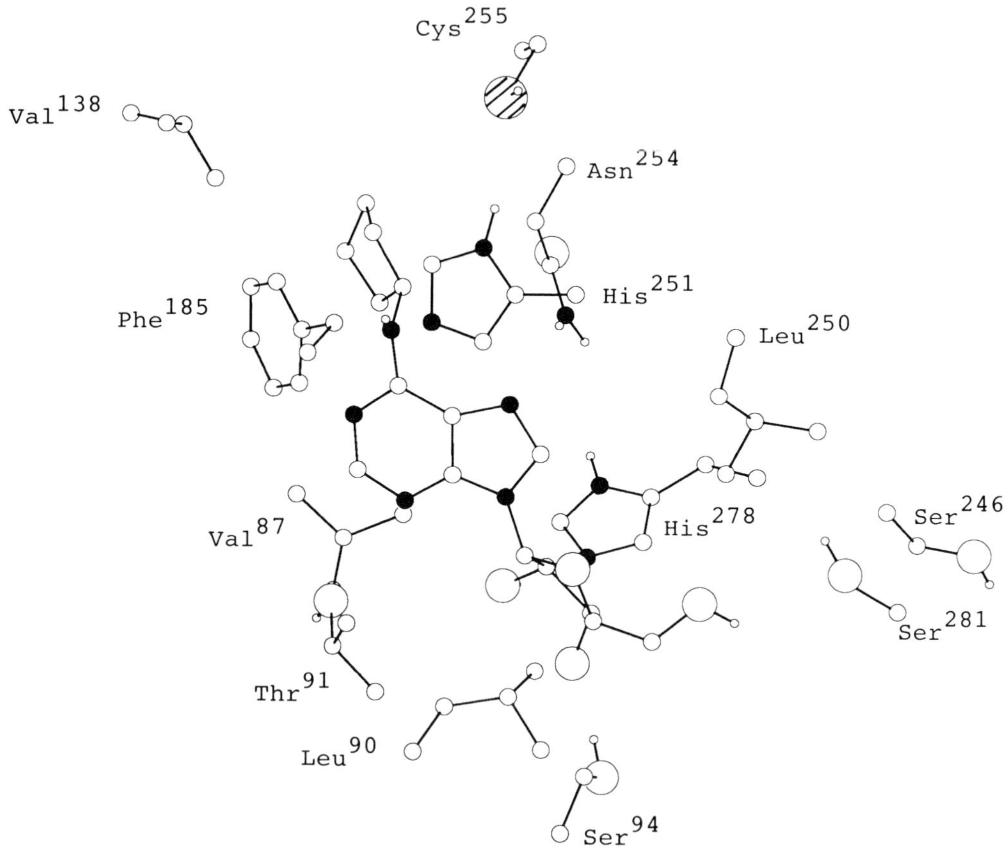

FIGURE 4-4. Representation of the proposed CPA binding site on the human adenosine A_1 receptor. The amino acids, the side chains of which are shown only, have been numbered consecutively. Residues involved are valine (87, 138), leucine (90, 250), threonine (91), serine (94, 246, 281), phenylalanine (185), histidine (251, 278), asparagine (254), and cysteine (255). Tryptophan (247), being in front of the ribose moiety, was omitted for reasons of clarity. All amino acids are within 4.5 Å from CPA. Atoms as in Figure 4-3A; dashed circle indicates sulfur.

FIGURE 4-3. **A:** Binding of CPA to the human A_1 receptor. The receptor backbone is shown only, without side chains. Top: Extracellular side, large open circles indicate oxygen atoms; medium open circles indicate carbon atoms; small open circles indicate hydrogen atoms; medium filled circles indicate nitrogen atoms. **B:** Binding of SHA 174 to the human adenosine A_{2a} receptor (direction of view is from the outside to the inside of the cell). Only the receptor backbone is shown, without side chains. Atoms as in Figure 4-3A.

by the seven α-helices, as seen from the extracellular side. The nonprotonated nitrogen atom of His^{250} forms a hydrogen bond with the exocyclic N^6-H in SHA 174. His^{278} may form a double hydrogen bond with ribose, as in CPA.

The ligand binding site on the adenosine A_{2a} receptor (again arbitrarily defined as the amino acid residues with atoms within 4.5 Å from SHA 174) consists, apart from the histidines, of other hydrophilic residues close to the hydrophilic purine and ribose moieties of SHA 174. The amino group of the carboxamide function in

Asn[253] interacts with N1 of the purine ring system. The nonprotonated nitrogen atom in the (hydrophilic) hydrazino part of the C2 side chain of SHA 174 interacts with the OH group in Ser[132]. Finally, Ser[281] and the 5′-OH group of the ribose moiety appear to interact, again via hydrogen bonding. In contrast, the hydrophobic cyclohexyl group of SHA 174 is indeed in a lipophilic environment. Ile[135], Phe[182], and Cys[254] surround the cyclohexyl group within a distance of 4.5 Å. A comparison of the two binding sites reveals that corresponding amino acids in the receptor models are involved. This is evident from Figure 4-5A, in which the receptor bound conformations of CPA and SHA 174 show a great deal of overlap.

DOCKING OF ANTAGONISTS

How might the potent and A_1-selective antagonist 1,3-dipropyl-8-cyclopentylxanthine (DPCPX; Fig. 4-2) fit in this A_1 binding pocket? From our present modeling studies it appears that the mode in which both cyclopentyl substituents of DPCPX and CPA are more or less overlapping (the so-called N^6/C8 model) is the most obvious possibility (see also Fig. 4-5B). This particular orientation appears to have more general validity, since longer C8 substituents,

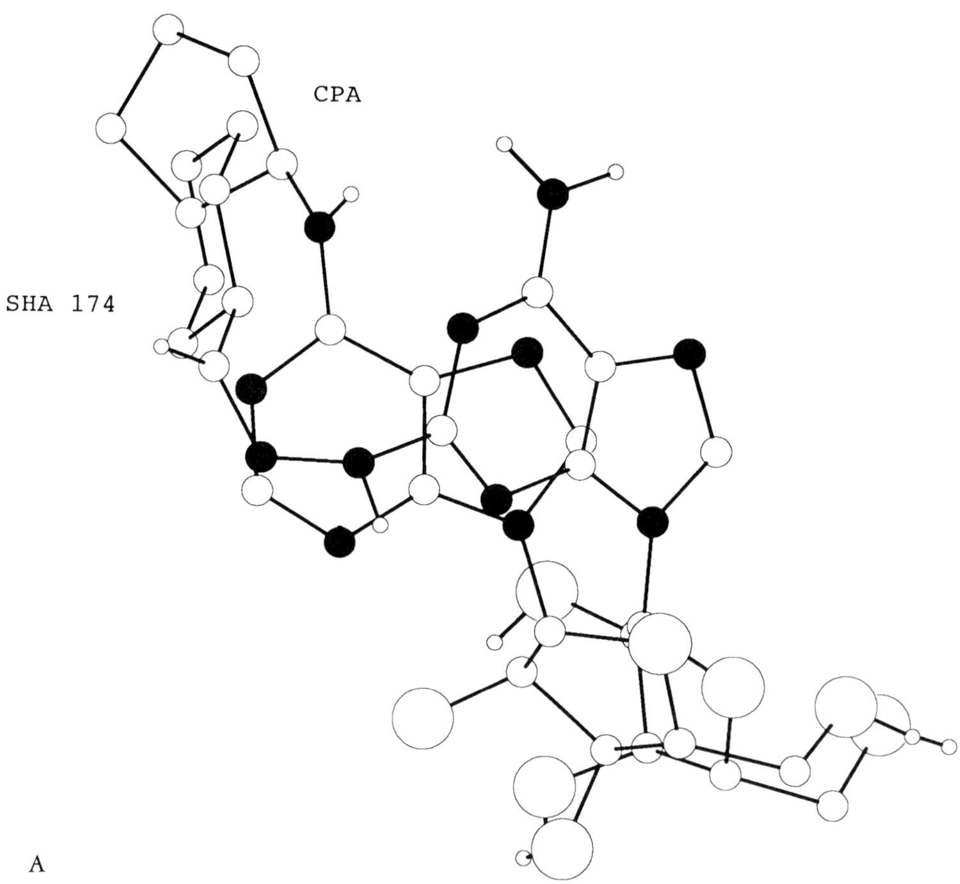

FIGURE 4-5. **A:** Receptor-bound conformations of CPA (A_1 receptor) and SHA 174 (A_2 receptor). Atoms as in Figure 4-3A. **B:** Receptor-bound conformations of N^6-cyclopentyladenosine (CPA) and 1,3-dipropyl-8-cyclopentylxanthine (DPCPX). Atoms as in Figure 4-3A.

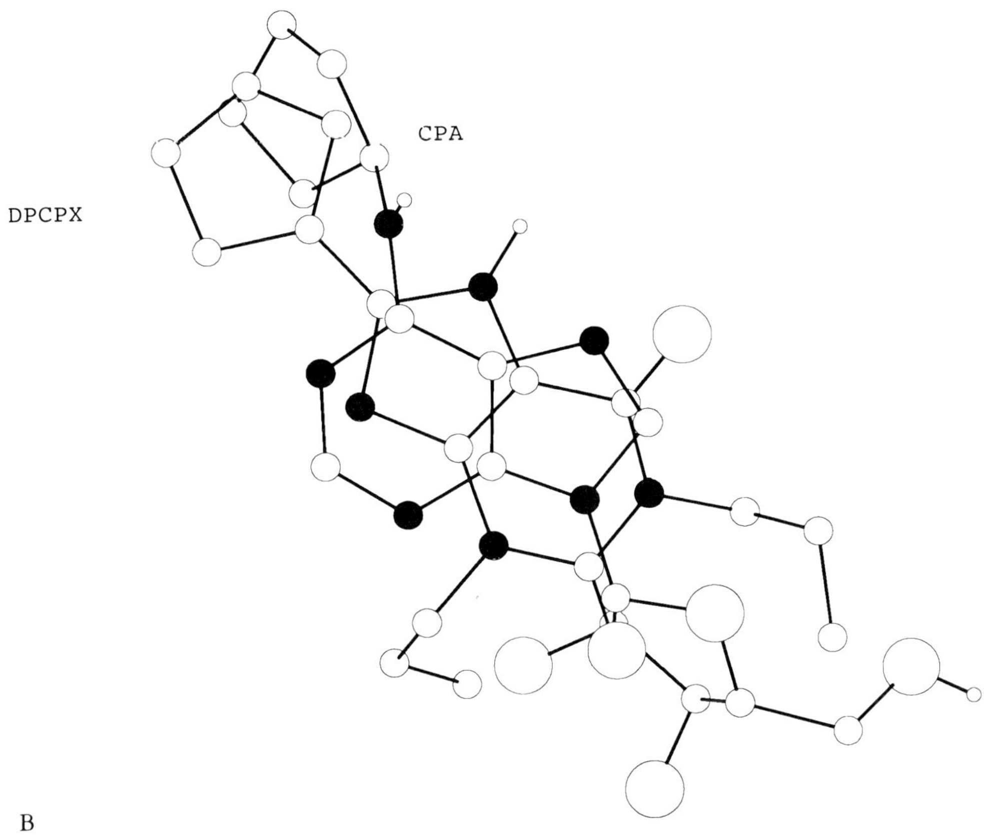

FIGURE 4-5. *Continued*

as in the xanthine amine congener (XAC), can also be accommodated.

Very recently a new series of A_2 receptor-selective antagonists has been described [13,14]. The most selective member of this class, all 8-styrylxanthines, is 8-(3-chlorostyryl) caffeine (CSC) (Fig. 4-2). Again, the N^6/C8 model seems applicable here, since N^6-substituted agonists that are A_{2a} selective are known. The styryl substituent in the xanthines may partially overlap in that case with these N^6 substituents. The conformational characteristics of CSC allow ample interactions within the agonist receptor binding domain. The rather hydrophilic xanthine moiety is surrounded by largely hydrophilic residues, whereas the hydrophobic 3-chlorostyryl substituent is buried in a hydrophobic domain. It should be emphasized that due to the larger size of the agonists used, the defined binding domain has ample space to accommodate the smaller antagonists, leaving some uncertainty with respect to localization of these ligands.

RECEPTOR ACTIVATION

The basic difference between adenosine receptor agonists and antagonists resides in the presence of a ribose moiety. Thus it is conceivable that the histidine-ribose interaction described above could play a role in adenosine receptor activation. Since proton transfer seems to be a trigger for activation in several proteins, such a transfer from the ribose group to the histidine residue in helix VII was mimicked by analyzing the methanol–4-methylimidazole interaction as a model. The BIOGRAF force field was applied to the two extremes in this interaction, that is, the uncharged methanol–4-methylimidazole complex versus the charged (but neutral) methanolate–4-methylimidazolium complex. For example (dielectric constant $\varepsilon = 5$, a value generally considered relevant to a protein environment), the gain in interaction energy for the uncharged complex is 6 kcal/mol and 11 kcal/mol for the charged complex. Also, the charged complex appears to be more stable (8 kcal/mol) than the uncharged one (10 kcal/mol). Although these calculations may very well be an oversimplification, the transfer of a proton in this typical interaction appears possible, thereby drastically changing the electrostatic characteristics of the groups involved.

Discussion

ADENOSINE RECEPTOR TOPOLOGY

Experimental data on structural aspects of membrane-bound proteins is scarce. Recently, Schertler et al. [15] determined a projection map of rhodopsin, the mammalian G-protein–coupled visual pigment, at 9 Å resolution. In this study, the authors confirmed the existence of seven transmembrane domains, as had been evident from the atomic coordinates of bacteriorhodopsin, a similar bacterial protein, although not coupled to a G protein. The three-dimensional structure of the transmembrane segments of bacteriorhodopsin had been determined earlier at a better resolution and served as the first indication for the general architecture of G-protein–coupled receptors [4]. There is considerable debate as to the degree of similarity of both protein structures [16,17]. In this study we employed bacteriorhodopsin as a template for a three-dimensional model of the membrane-spanning parts of the human adenosine receptors. These models can only be approximate at best. Hopefully, though, they can serve as a starting point for site-directed mutagenesis and other experiments.

LIGAND BINDING SITES

The binding site for the highly hydrophobic ligand retinal in bacteriorhodopsin has been shown to be buried in the membrane, just above the center of the cavity formed by the seven transmembrane domains [4]. From chemical modification and site-directed mutagenesis studies, two histidines (in helices VI and VII, respectively) appear particularly important for ligand binding to adenosine receptors [6–9], suggesting that the binding site for adenosine receptor ligands is also somewhere within the receptor cavity. Thus, these histidines were used as interaction sites, leading to the models proposed above. Upon relaxation of the receptor-bound conformation of N^6-cyclopentyladenosine, the reference A_1 agonist, the purine-cyclopentyl torsion angle was $-89°$, not very different from the value of $-75°$ determined in a ligand modeling study [18]. The hydrogen bond formed between the histidine residue in helix VI and N^6-H may well explain the fact that N^6-disubstitution of adenosine derivatives is detrimental for affinity. The ribose moiety, in an *anti* conformation, may interact with the histidine residue in helix VII via the *cis*-diol moiety. This is compatible with the SAR known for the ribose group. For example, removal of both hydroxyl groups, as in 2′,3′-dideoxy-N^6-cyclohexyladenosine, yielded an antagonist with only moderate affinity [19]. The docking of DPCPX according to the N^6/C8 overlap model would explain the similar increases in affinity for various, even bulky, substituents in both N^6-substituted agonists and C8-substituted antagonists [20].

The biochemical evidence that the two histidine residues mentioned earlier are also critical for ligand interaction on the adenosine A_{2a} receptor led to a model that is largely compatible with the A_1 receptor model. The finding that SHA 174, a conformationally relatively restricted compound, is a potent and A_{2a}-selective agonist proved of great value, because its C2 side chain could only be directed into a hydrophobic pocket of similar (but not identical) nature, as had been found for the so-called N^6 region in the adenosine A_1 receptor. This finding could also explain why ligands in which both N^6 and C2 substituents have been introduced do not have improved affinities over the mono-substituted analogs [21].

Some N^6-substituted agonists, such as N^6-2-(3,5-dimethoxyphenyl)-2-(2-methylphenyl) ethyl]adenosine (DPMA), are A_{2a} selective as well [22]. The bulky N^6 substituent of this compound may very well overlap with the styryl moieties in C8-substituted xanthines, which are also A_2 selective. This provides evidence for the general validity of the so-called N^6/C8 model [20] to explain the competitive interaction between agonists and antagonists on both adenosine A_1 and A_{2a} receptors.

RECEPTOR ACTIVATION

Since the histidine residue in helix VII in our models coordinates the ribose group present in adenosine receptor agonists, it seems that adenosine receptor agonism has at least part of its molecular origin here. In both bacteriorhodopsin and mammalian rhodopsin, a lysine residue is at the corresponding position on helix VII. This lysine residue forms a protonated Schiff's base linkage with retinal. Subsequently, proton transfer may take place, as the trigger of the opsins' activation process [23]. Additional histidine residues appear to be involved too, which are thought to be effective upon their protonation [24].

Could proton transfer play a role as a trigger in the activation of the adenosine receptor, as it does in the opsins? The energy data mentioned above suggest that a proton transfer from ribose to histidine is feasible. Do such processes occur in "biochemical reality"? An intriguing parallel is seen with the functioning of the catalytic triad (Ser-His-Asp) in serine proteases, a large and diverse enzyme family that catalyze the hydrolysis of peptide bonds. The serine's hydroxy group is always within the hydrogen-bonding distance of the (nonprotonated) nitrogen of the imidazole ring system of the histidine residue. The other nitrogen atom is hydrogen bonded to the carboxylate moiety of the aspartic acid residue [25]. Speculatively, the ribose group could have a similar role as the serine, since 2′- or 3′-deoxyribose (even more resembling a serine), replacing an intact ribose in adenosine derivatives, leads to receptor activation. As an example, 3′-deoxy-R-PIA is a fairly potent adenosine receptor agonist [26].

Which carboxylate group in the receptor may be available for proton transfer? There is an aspartate residue in helix II that is present in (virtually) all G-protein–coupled receptors and indeed implicated in regulatory aspects of receptor activation. However, due to its presence in all receptors, this residue is probably not directly linked to the histidine that occurs in adenosine receptors only. The other acidic residue is a glutamate in helix I that is specific for the adenosine receptors. This glutamate residue is in the trans-sectional plane of the ribose-OH groups and the histidine residue in helix VII, and can be brought in the vicinity (within hydrogen-bonding distance) of the histidine residue. Thus, a functional element that resembles the catalytic triad in serine proteases could be present in the adenosine receptors.

Conclusions

In this report we describe a model for the ligand binding site on the human adenosine A_1 and A_{2a} receptor. Two histidine and several serine and threonine residues appear to play a role in the coordination of the hydrophilic purine and ribose moieties in the agonists CPA and SHA 174. The region around the cyclopentyl and cyclohexyl rings in both ligands is much more hydrophobic in nature, allowing van der Waals interactions with these two substituents. The models could serve as a starting point for site-directed mutagenesis studies. Combined efforts in computational chemistry and molecular biology may lead to validation and optimization of the proposed models, eventually enabling the rational design of new chemical entities.

References

1. Libert F, Parmentier M, Lefort A, Dinsart C, Van Sande J, Maenhaut C, Simons MJ, Dumont JE, Vassart G (1989) Selective amplification and cloning of four new members of the G protein-coupled receptor family. Science 244:569–572.
2. Furlong TJ, Pierce KD, Selbie LA, Shine J (1992) Molecular characterization of a human brain adenosine A_2 receptor. Mol Brain Res 15:62–66.
3. Jacobson MA, Johnson RG, Luneau CJ, Salvatore CA (1993) Human adenosine receptors. UK patent application GB 2264948A.
4. Henderson R, Baldwin JM, Ceska TA, Zemlin F, Beckmann E, Downing KH (1990) Model for the structure of bacteriorhodopsin based on high-resolution electron cryo-microscopy. J Mol Biol 213:899–929.
5. Hibert MF, Trumpp-Kallmeyer S, Bruinvels A, Hoflack J (1991) Three-dimensional models of neurotransmitter G-binding protein-coupled receptors. Mol Pharmacol 40:8–15.
6. Jacobson KA, Stiles GL, Ji XD (1992) Chemical modification and irreversible inhibition of striatal A_{2a} adenosine receptors. Mol Pharmacol 42:123–133.
7. Klotz K-N, Lohse MJ, Schwabe U (1988) Chemical modification of A_1 adenosine receptors in rat brain membranes. J Biol Chem 263: 17522–17526.
8. Garritsen A, IJzerman AP, Beukers MW, Soudijn W (1990) Chemical modification of adenosine A_1 receptors. Implications for the interaction with R0PIA, DPCPX and amiloride. Biochem Pharmacol 40:835–842.
9. Olah ME, Ren H, Ostrowski J, Jacobson KA, Stiles GL (1992) Cloning, expression, and characterization of the unique bovine A_1 adenosine receptor. J Biol Chem 267:10764–10770.
10. IJzerman AP, van Galen PJM, Jacobson KA (1992) Molecular modelling of adenosine receptors. I. The ligand binding site on the A_1 receptor. Drug Design Discov 9:49–67.
11. IJzerman AP, van der Wenden EM, van Galen PJM, Jacobson KA (1994) Molecular modeling of adenosine receptors. 2. The ligand binding site on the rat adenosine A_{2a} receptor. Eur J Pharmacol Mol Pharm Sect, in press.
12. Devereux J, Haeberli P, Smithies O (1984) A comprehensive set of sequence analysis programs for the VAX. Nucleic Acids Res 12:387–395.
13. Shimada J, Suzuki F, Nonaka H, Ishii A, Ichikawa S (1992) (E)-1,3-dialkyl-7-methyl-8-(3,4,5-trimethoxystyryl)xanthines: Potent and selective adenosine A_2 antagonists. J Med Chem 35:2342–2345.
14. Jacobson KA, Gallo-Rodriguez C, Melman N, Fischer B, Maillard M, van Bergen A, van Galen PJM, Karton Y (1993) Structure-activity relationships of 8-styrylxanthines as A_2-selective adenosine antagonists. J Med Chem 36: 1333–1342.
15. Schertler GFX, Villa C, Henderson R (1993) Projection structure of rhodopsin. Nature 362: 770–772.
16. Baldwin JM (1993) The probable arrangement of the helices in G protein-coupled receptors. EMBO J 12:1693–1703.
17. Hoflack J, Trumpp-Kallmeyer S, Hibert M (1994) Re-evaluation of bacteriorhodopsin as a model for G protein-coupled receptors. Trends Pharmacol Sci 15:7–9.
18. Van Galen PJM, Leusen FJJ, IJzerman AP, Soudijn W (1989) Mapping the N^6-region of the adenosine A_1 receptor with computer graphics. Eur J Pharmacol-Mol Pharm Sect 172:19–27.
19. Lohse MJ, Klotz K-N, Diekmann E, Friedrich K, Schwabe U (1988) 2′,3′-dideoxy-N^6-cyclohexyladenosine: An adenosine derivative with antagonist properties at adenosine receptors. Eur J Pharmacol 156:157–160.
20. Van der Wenden EM, IJzerman AP, Soudijn W (1992) A steric and electrostatic comparison of three models for the agonist/antagonist binding site on the adenosine A_1 receptor. J Med Chem 35:629–635.
21. Thompson RD, Secunda S, Daly JW, Olsson RA (1991) Activity of N^6-substituted 2-chloroadenosines at A_1 and A_2 adenosine receptors. J Med Chem 34:3388–3391.
22. Bridges AJ, Bruns RF, Ortwine DF, Priebe SR, Szotek DL, Trivedi BK (1988) N^6-[2-(3,5-dimethoxyphenyl)-2-(2-methylphenyl)ethyl] adenosine and its uronamide derivatives. Novel adenosine agonists with both high affinity and high selectivity for the adenosine A_2 receptor. J Med Chem 31:1282–1285.
23. Khorana HG (1992) Rhodopsin, photoreceptor of the rod cell. J Biol Chem 267:1–4.
24. Weitz CJ, Nathans J (1992) Histidine residues regulate the transition of photoexcited rhodopsin to its active conformation, metarhodopsin II. Neuron 8:465–472.
25. Branden C, Tooze J (1991) Introduction

to Protein Structure. New York: Garland Publishing, pp. 231–246.
26. Taylor MD, Moos WH, Hamilton HW, Szotek DS, Patt WC, Badger EW, Bristol JA, Bruns RF, Heffner TG, Mertz TE (1986) Ribose-modified adenosine analogues as adenosine receptor agonists. J Med Chem 29:346–353.

5. PURINOCEPTOR NOMENCLATURE: CHALLENGES FOR THE FUTURE

Michael Williams

The development of a systematic nomenclature for purinoceptors has for a considerable period been limited by a paucity of selective, potent, and bioavailable agonist and antagonist ligands, especially in the area of P_2 receptors. This situation has largely occurred because of the widely held belief, now historical, that neither adenosine nor its nucleotide, ATP, could function as neurohumoral agents and were in fact more likely to act as cytotoxic agents. This significantly restricted efforts in the area of medicinal chemistry targeted towards the development of needed ligands and those that have been available have, to an overwhelming extent, been agonists. These compounds have also been limited in terms of potency, selectivity, physical availability, and structural diversity.

This did not, however, preclude the use of a limited number of available receptor ligands in classic physiologic and pharmacologic paradigms [1–6] in developing an evolving system for purinoceptor nomenclature. The latter has been subsequently complimented by the cloning of several of these receptors [7,8]. As pharmacology has synergized with advances in the molecular biology of purinoceptors, the derived molecular targets are being used as templates for renewed medicinal chemistry efforts [9,10], which, in turn, are providing important new ligands to define and redefine the purinoceptors. These tools, together with receptor characterization in diseased tissues, may then be used to determine the selective function of the receptors and their involvement in tissue pathophysiology.

Historical Background

In 1978, Burnstock proposed [2], on the basis of available information, the existence of two classes of purinoceptor: P_1 receptors that were responsive to adenosine and P_2 receptors responsive to ATP and related nucleotides. As with all receptor nomenclature schemes [11], the P_1/P_2 system was subject to further modification as 1) new responses and/or receptors responsive to either adenosine or ATP were identified; 2) new ligands, especially antagonists, were synthesized and characterized for their functional effects; and 3) the receptors were cloned and their structure, transduction systems, and distribution elucidated.

Concomitant with the Burnstock P_1/P_2 schema, two groups, Londos et al. [12] and Van Calker and coworkers [13], identified subtypes of the P_1 receptor using receptor-mediated second messenger assays. These were initially termed R_i and R_a [12] to reflect the need for intact ribose groups on the adenosine pharmacophore to inhibit (i) and activate (a) adenylate cyclase activity, respectively. These same receptors were termed A_1 and A_2 by Van Calker et al. [13], a nomenclature that has been widely accepted and forms the basis of the IUPHAR Committee on Purinoceptor Nomenclature's recommendations for P_1 purinoceptor nomenclature [14]. Both classes of P_1 receptor could be effectively antagonized by theophylline and related 8-phenylxanthines. Nonetheless, purinoceptor nomenclature has relied heavily on the rank-order potency of agonists in eliciting biochemical or physiologic responses, a sytem that is far more prone to artifact than antagonist-derived classification [11]. The nomenclature system for P_2 purinoceptors has evolved in a somewhat less logical manner, as discussed below [15,16].

Adenosine Purinoceptors

The original A_1 and A_2 receptor classification has been extended to reflect the existence of four distinct (A_1, A_{2a}, A_{2b}, A_3) adenosine receptors that have been cloned from a variety of mammalian species and the putative A_4 adenosine receptor, all of which are members of the P_1 purinoceptor family (Table 5-1). Transduction systems for the A_1 receptor include inhibition of adenylate cyclase, stimulation of K^+ flux, inhibition of calcium flux, and increased IP_3 turnover [14]. Additional data from rat and guinea pig brain and peripheral tissues led to the proposal [17] of subclasses of the A_1 receptor termed A_{1a} and A_{1b}. While the A_1 receptor has been cloned [7] (Table 5-1), there is currently no evidence at the molecular level to support the existence of A_{1a} and A_{1b} receptors. Thus this classification has not been widely accepted.

The A_2 receptor exists as two subtypes [5,6], the A_{2a}, a high-affinity receptor highly localized to the striatum, activation of which results in an increase in cAMP formation and that is potentially involved in modulating dopaminergic function [18], and the A_{2b}, a lower affinity form of the A_2 receptor, also linked to adenylate cyclase activation [5,19]. Adenosine A_{2a} and A_{2b} receptors have been cloned from a number of species [7] (Table 5-1).

The existence of a prejunctional adenosine receptor, termed A_3, was originally proposed on the basis of antagonist effects in the frog neuromuscular junction by Riberio and Sebastiao [20], since the rank-order activity of antagonists in this preparation did not correlate with either A_1 or A_2 receptor pharmacology. At the Purines '92 meeting in Milan [15], the evidence for the existence of this receptor was discussed and,

TABLE 5-1. Adenosine purinoceptors

Adenosine receptor	Human clone aa[a] (SwissProt Number)	Agonists	Antagonists
A_1	326 aa (545235)	CHA, CPA, CCPA, NECA, NNC 21–0041	CPX, CPT, KFM 19, MDL 102,234, CGS 15943
A_{2a}	412 aa (P29274)	CGS 21680, MPEA	KF 17837, CSC, CGS 15943, CP 66713, AMBP
A_{2b}	328 aa (P29275)	NECA	Xanthines[b]
A_3	318 aa (L22607)	N^6-benzyl NECA, N^6-iodobenzyl MECA	I-ABPOX
A_4	Not yet cloned	CV 1808	CGS 15943A[c]

[a] Amino acids.
[b] A_3 receptors isolated from sheep and human cDNA libraries are sensitive to blockade by xanthines. Those from rat are xanthine insensitive.
[c] The A_4 receptor is insensitive to blockade by xanthines.

given the molecular cloning of a rat adenosine receptor, also termed A_3 [21], was considered [15] to be subject to re-evaluation following the cloning of a receptor with the characteristics of the "prejunctional A_3." Subsequent work [22] has indicated that the prejunctional A_3 receptor may represent an A_1 receptor. The nature of the prejunctional A_3 receptor, however, remains the subject of debate related to receptor signaling processes [23] and has consequently been described as a "non-A_1, non-A_2, non-A_3" receptor [24].

The rat brain A_3 receptor cloned by Stiles and coworkers [21] had 58% sequence homology with the dog A_1 and A_2 receptors [7] and was identical in structure to a previously isolated G-protein–coupled receptor (GPCR), *tgpcrl*, found in the testis. The rat A_3 receptor was unique for a P_1 purinoceptor in that radioligand binding was insensitive to xanthine blockade. The subsequent cloning of the A_3 receptor from sheep [25] and human [26] brain showed that in these species this receptor was sensitive to acidic xanthines. Given these discrepancies in xanthine sensitivity and the approximate 75% sequence homology between the human and rat A_3 receptors, it is conceivable that the observed species differences in pharmacology could be considered a reflection of distinct receptor subtypes. An important point in regard to receptor structure as determined by cloning and expression, and function, as determined initially by pharmacologic profile, is whether sequence homology, or lack thereof, can be used to determine probable function. Receptor chimeras or mutated receptors in which a single amino acid alteration in 400 or so amino acids can markedly alter receptor pharmacology and/or transductional coupling underline this question. In the latter instance, the wild-type receptor and the point mutation have a sequence homology of 99.75%, yet the pharmacology of the receptor can differ to the extent that binding studies would argue against the wild-type and the point mutation belonging to the same family.

An adenosine receptor termed A_4 has been characterized in rat brain tissue using the radioligand [^{3}H]CV 1808 (2-phenylaminoadenosine) [27]. The A_4 receptor, like the rat A_3, was insensitive to xanthines but did bind triazoloquinazoline adenosine antagonists, including CGS 15943. The proposed A_4 receptor is linked to K^+ channels. Like the prejunctional A_3 receptor, the discussion at the Purines '92 meeting in Milan [15] concluded that the existence of the proposed A_4 receptor would be dependent on its cloning, which has not occurred to date. Subsequent binding studies [28] using A_{2a}-receptor–transfected COS cells has indicated that the A_4 receptor may be a temperature-sensitive form of the A_{2a} receptor, although the xanthine insensitivity of the A_4 receptor has yet to be resolved. Finally, another putative adenosine receptor has been identified in cerebral cortex using [^{3}H]CGS 21680 [29] that is different pharmacologically from the striatal A_{2a} receptor. This receptor, fortunately, has not been designated the A_5 receptor.

Species differences in both the A_1 and A_{2a} receptors [30,31] have been observed in radioligand binding studies; and it appears likely, based on the differences between expressed rat and sheep and human clones, and A_3 binding in rat, gerbil, and rabbit brain membranes [32], that the A_3 receptor is highly species dependent in its pharmacology. Work with antibodies to the adenosine A_{2a} receptor suggests that there may also be tissue differences in this receptor class [33]. The minimal cardiovascular liabilities seen in animal models with the A_1-selective antagonist MDL 102,234, targeted as a cognitive enhancer [34]; the A_1-selective agonist NNC 21-0041, targeted as a neuroprotective agent/antiepileptic [35]; and the A_{2a}-selective agonist MDL 101,483, which mimics the actions of neuroleptics in animal models [36], indicate that anticipated nonspecific actions on cardiovascular as well as central nervous system (CNS) adenosine receptors are not seen in whole animal models. The possibility that pharmacokinetic and pharmacodynamic differences contribute to tissue selectivity cannot be overlooked, however.

Adenosine Purinoceptor Ligands

Ligands selective for the characterization of the various adenosine receptors are discussed elsewhere in this volume [37]. Some comment is necessary, however, with regard to selective agonists and antagonists. Despite the finding that 5′-N-ethylcarboxamidoadenosine (NECA) is an agonist active at A_1, A_{2a}, A_{2b}, and A_3 receptors and thus nonselective, it is frequently described in the literature as an A_2-"selective" agonist. Similarly, especially in the cardiovascular field, theophylline and caffeine are still routinely used as adenosine antagonists despite their very limited efficacy, their activity as phosphosphodiesterase inhibitors, and the availability of an abundance of potent 8-substituted xanthines [9].

Selective agonist ligands for the A_1 receptor that are commercially available include CHA (N^6-cyclohexyl-adenosine), CPA (N^6-cyclopentyladenosine), and CCPA (2-chloro-N^6-cyclopentyl-adenosine). Agonists for the A_{2a} receptor include a large number of 2-substituted adenosine analogs [9], including CGS 21680 and 2-CHEA, although these agents are less selective when comparing A_{2a} and A_3 receptor activity [38]. For the A_{2b} receptor, NECA and 2-CADO (2-chloroadenosine) are potent ligands [19]. At the rat A_3 receptor, N^6-benzyl NECA and 3-iodo-N^6-benzyl MECA are selective agonists [38].

A large number of 8-substituted xanthines, including CPT (cyclopentyltheophylline), CPX (cyclopentylxanthine), XAC (xanthine amine congener), KFM 19, and MDL 102,234, are selective antagonists of the A_1 purinoceptor [9]. Among the nonxanthine heterocycles active as A_1 antagonists are the triazoloquinazolines, which include CGS 15943 and the 7-deaza-2-phenyladenines. For the A_{2a} receptor, the styryl xanthines, KF 17837 and CSC, the triazoloquinoxalinamine, CP 66713, and the benzopyranopyrazolone, AMBP, are potent and selective antagonists [9,39]. At the sheep A_3 receptor, I-ABPOX(BWA 522) has been reported to be a moderately selective antagonist [25].

P_2 Purinoceptors

The nomenclature for the P_2 purinoceptor family has evolved in a far less systematic manner than that for the P_1 receptor class. This, as already noted, can be ascribed to a very limited number of selective ligands, the majority of these being agonists, and a limited database involving primarily intact tissue rather than biochemical studies [15,40]. Potent and selective radioligands for the various members of the P_2 receptor family have yet to be developed, which has limited the use of high through-put screening assays to identify novel P_2 purinoceptor pharmacophores. Thus data from different tissues using different agonists structurally related to ATP and different assay conditions have been used as the basis for the development of P_2 purinoceptor structure-activity relationships and subsequent nomenclature.

Clouding the classification of P_2 purinoceptors is the fact that many of the ATP analogs used as agonists undergo metabolism via ectonucleotidase activity to differing degrees. Thus the different rank-order potencies of agonists used to define a particular receptor subtype may also reflect the structure-activity relationship and activity of the ectonucleotidases present in a given tissue [41]. A recently described selective ectonucleotidase inhibitor, FPL 67156, has been found to markedly affect rank-order agonist potencies at P_2 receptors [42]. The situation is further compounded by a paucity of selective and potent antagonists. Suramin, $ANAAP_3$, reactive blue 2, and PPADS have varying activity as P_2 receptor antagonists [14]. It should be noted, however, that suramin is nonselective in its actions, also being an antagonist at bFGF receptors. FPL 66096 and FPL 67085 [42] are stable ATP analogs that are potent and selective at P_{2t} receptors.

Following the description of the original P_{2x}/P_{2y} receptor classes [3] (Table 5-2), P_{2t} and P_{2z} receptors were also described [4], the former being an ADP-sensitive receptor present in platelets and the latter a nonselective ion pore sensitive to ATP^{4-}. Subsequently, P_{2u}/P_{2n} receptors responsive to UTP and ATP have been described [43], as has the P_{2D} receptor, which is sensitive to the effects of adenine dinucleotides (Ap_xA), specifically Ap_{4A}, Ap_{5A}, and Ap_{6A} [44,45] (Table 5-2).

The somewhat random nature of the alphabetic designation for P_2 receptors following the initial P_{2x}/P_{2y} classification gave cause for concern to the IUPHAR Purinoceptor Committee, whose mandate from the IUPHAR Committee on Receptor Nomenclature was to recommend a system consistent with that in use for other receptor classes [14]. The possibility of replacing the P_2 nomenclature with an N (for nucleotide receptor) system (N_1, N_2, N_3, etc.) was discussed and eventually discounted, given historical precedent and the many thousands of papers in the literature using the P_{2x} and P_{2y} nomenclature system. Recognizing the need to provide some consistent yet flexible framework, the committee provisionally recognized a proposal by Abbracchio and Burnstock [16] that classified the actions of adenine nucleotides into two broad receptor superfamilies. The P2X family, a receptor class mediating the "fast" actions of ATP and related nucleotides through ligand-gated ion channels (LGICs), and the P2Y family, acting via metabotropic GPCRs. This proposal, based on an in-depth evaluation of existing pharmacologic and physiologic literature, has parallels with the cholinergic receptor superfamily, in which nicotinic receptors (nAChRs) mediate fast responses to acetylcholine (ACh) and muscarinic receptors constitute the metabotropic family [46], and also with the LGIC and metabotropic classes of the glutamate receptor superfamily [47].

TABLE 5-2. P_2 purinoceptor classification—1985–1992

P_2 purinoceptor	Agonists	Antagonists
P_{2X}	α,β-meATP > β,γ-meATP > ATP	Suramin, PPADS
P_{2Y}	2MeSATP > ATP	Suramin
P_{2t}	2MeSADP > ADP	ATP, FPL 66096, FPL 67085
P_{2Z}	ATP^{4-}	None known
$P_{2u/n}$	UTP ≥ ATP	Suramin
P_{2D}	Ap_4A > ADPβS	None known

The use of the superfamily nomenclature requires that the P2 descriptor for the pharmacologically defined receptor be on the same line, with the X or Y designation in capitals followed by the arabic numeral, again all on the same line for those receptors defined on a pharmacologic basis, for example, P2X1 rather than the previously used P_{2x} nomenclature. When the receptor is cloned, however, the IUPHAR Purinoceptor Committee [14] has recommended that the designator become a subscript, for example, P_{2X1}. Within the superfamily schema (Table 5-3), the numerical sequence for receptor identification has been used to reflect the order in which the receptors have been cloned. Thus the P2X receptor recently reported to have been cloned by the Glaxo group in Geneva from rat vas deferens [48] has been defined as the P_{2X1} receptor and may exist in different forms in different tissues. The definition of the proposed P2X2 and P2X3 receptors remains at the pharmacologic level because these had not been cloned at the time of writing.

Members of the P2Y superfamily have also been cloned. The first of these, P_{2Y1}, was cloned

TABLE 5-3. Proposed P_2 purinoceptor classification—1994 (Abbracchio and Burnstock)

P_2 Purinoceptor (previous designation)	Human clone SwissProt no.	Agonists	Antagonists
P_{2X1}	399 aa	α,β-meATP; 5 fluro UTP > 2-hexnylthioATP	Suramin, PPADS
P_{2X2}		α,β-meATP; 2-(4-nitrophenylethylthio)ATP	Suramin, PPADS
P_{2X3}		α,β-meATP	Suramin, PPADS
P_{2Y1}	362 aa	2MeSATP	Suramin
P_{2Y2} ($P_{2u/n}$)	378 aa	UTP, ATP, $UTP_{\gamma}S$	ATP, suramin
P_{2Y3} (P_{2t}-like)		2MeSADP > ADP	FPL 66096, FPL 67085
P_{2Y4}		2-(6-cyanohexylthio) ATP > 2MeSATP ≫ ATP	None known
P_{2Y5}		2MeSATP ≥ ATP 8-(6-aminohexyl)amino ATP	None known
P_{2Y6}		2MeSATP Inactive: 2-(6-cyanohexylthio) ATP; 8-(6-amino hexyl)amino ATP	None known
P_{2Y7} (P_{2D})		$Ap_4A \geq Ap_5A \geq Ap_6A$	None known
P_{2Z}		$ATP4^-$	

from chick brain [49], while an identical P_2 receptor has been cloned from turkey brain and has been shown to utilize phospholipase C as a second messenger [50]. ATP/UTP-sensitive P_2 receptors have been cloned from the NG108-15 neuroblastoma cell line [51], human epithelium [52], and rat heart [53]. This receptor has been classified as the P_{2Y2} receptor [16]. A P_{2Y3} receptor with characteristics of the platelet P_{2T} receptor has also been reported to have been isolated from chick brain [8]. Of the remaining members of the proposed P2Y superfamily, the P2Y4-P2Y6 receptors (Table 5-3) are characterized exclusively on the basis of agonist-evoked functional responses in various tissues, while the P_{2D} receptor has been designated the P2Y7 receptor. The original P_{2Z} receptor classification remains unchanged in the new scheme.

Like the original P_1/P_2 and P_{2x}/P_{2y} proposals, the P2X/P2Y superfamily concept is a dynamic framework and inevitably requires continual re-evaluation as new data using as yet undiscovered potent and selective agonist and antagonist ligands become available and as various receptors are cloned. Thus the P2X/P2Y superfamily concept represents a very logical alternative to the continued arbitary appearance of, for instance, P_{2e}, P_{2g}, P_{2j}, P_{2o}, P_{2q} receptors and has precedent from other receptor superfamilies, as has already been discussed.

Nonetheless, there may be a need to recognize new classes of receptor associated with the P2 superfamilies. At the P_{2u}/P_{2Y2} receptor, UTP, in the majority of instances, is equiactive with ATP as an agonist [16], although this receptor has a low sequence homology with that of the P_{2Y1} receptor. More recent data in the C6-2B rat glioma cell line have identified a receptor linked to PI turnover that is selectively activated by UTP and related uridine nucleotides but is insensitive to ATP, thus differentiating it from the P_{2U} receptor [54]. This receptor has been termed the *uridine nucleotide receptor* and may represent a new class of P2-like receptors.

Another point of confusion related once again to the use of agonists to define receptor nomenclature reflects the multiple actions of ligands at different P_1 and P_2 receptors. For instance, while Ap5A is a ligand for the P_{2D}/P2Y7 receptor, it is also active at P_1 receptors [55], as are a number of other polyphosphate analogs [56]. Similarly, the production of ADP, AMP, and adenosine from ATP in the extracellular milieu via the purinergic cascade [57] can lead to confusing results in intact tissues, which only the use of antagonist ligands can satisfactorily resolve.

Of additional note in the context of receptor evolution are the proposal of Westfall et al. [58] for the existence of a P_3 receptor based on responses evoked by both P_1 and P_2 receptor ligands in sympathetic nerve terminals and the findings of Rathbone et al. [59] using PC12 cells, where a unique guanosine-dependent effect on neurite outgrowth has been demonstrated that appears to involve a receptor linked to the induction of the heat-shock protein, heme oxygenase, that is insensitive to either adenosine or ATP.

Future Directions

The evolution of receptor nomenclature is ultimately dependent on the generation and integration of data from numerous laboratories throughout the world, a process that is ever more complex as data generation continues in an exponential mode. Nonetheless, to even begin the evolution of a workable yet systematic nomenclature, it is imperative that a framework be in place to both integrate new results and focus discussion. In the purinoceptor area, this has worked admirably as the P_1/P_2 proposal of Burnstock has evolved to include the various P_1 receptors, the P_{2x}/P_{2y} nomenclature, and now the P2X and P2Y superfamilies. To operate without such a framework can only lead to greater confusion and controversy, especially to those who observe purinoceptors from a distance. One can but marvel at the way in which enzyme nomenclature evolved, although the era was far less hectic, the data less voluminous, and the fax machine absent.

Acknowledgments

The author would like to thank Maria Abbracchio, Geoff Burnstock, and Ken Harden for preprints of articles in press and John Daly, Bertil Fredholm, Ken Jacobson, and Joel Linden for helpful discussions. The viewpoints presented represent the author's distillation of discussions related to his presentation on purinoceptor nomenclature at the Fifth International Symposium on Adenosine and Adenine Nucleotides in Philadelphia and do not reflect a consensus opinion of the IUPHAR Committee.

References

1. Burnstock G (1972) Purinergic nerves. Pharmacol Rev 24:509–581.
2. Burnstock G (1978) A basis for distinguishing two types of purinergic receptor. In Bolis L, Stranb R (eds.), Cell Membrane Receptors for Drugs and Hormones. New York: Raven Press, pp. 107–118.
3. Burnstock G, Kennedy C (1985) Is there a basis for distinguishing two types of P_2-purinoceptor? Gen Pharmacol 16:433–440.
4. Gordon JL (1986) Extracellular ATP: Effects, sources and fate. Biochem J 233:309–319.
5. Daly JW, Butts-Lamb P, Padgett W (1983) Subclasses of adenosine receptors in the central nervous system; interaction with caffeine and related methylxanthines. Cell Mol Neurobiol 3:69–80.
6. Bruns RF, Lu G, Pugsley TA (1986) Characterization of the A_2 adenosine receptor labeled by [^{3}H] NECA in rat striatal membranes. Mol Pharmacol 29:331–346.
7. Linden J, Jacobson MA, Hutchins C, Williams M (1994) Adenosine receptors. In Peroutka SJ (ed.), Handbook of Receptors and Channels, Vol. 1. G-Protein-Coupled Receptors. Boca Raton, FL: CRC Press, pp. 29–44.
8. Barnard EA, Burnstock G, Webb TE (1994) G-protein-coupled receptors for ATP and other nucleotides: A new receptor family. Trends Pharmacol Sci 15:67–70.
9. Jacobson KA, van Galen PJM, Williams M (1992) Adenosine receptors: Pharmacology, structure–activity relationships and therapeutic potential. J Med Chem 35:407–415.
10. Burnstock G, Fischer B, Hoyle CHV, Maillard

M, Ziganshin AU, Brizzolara AL, von Isakoviks A, Boyer JL, Harden TK, Jacobson KA (1994) Structure activity relationships for derivatives of adenosine-5′-triphosphate as agonists at P_2 purinoceptors: Heterogeneity within P_{2x} and P_{2y} subtypes. Drug Dev Res 31:206–219.
11. Kenakin TP, Bond RA, Bonner TI (1992) Definition of pharmacological receptors. Pharmacol Rev 44:351–361.
12. Londos C, Cooper DMF, Wolff J (1980) Subclasses of external adenosine receptors. Proc Natl Acad Sci USA 77:2251–2554.
13. Van Calker D, Muller M, Hamprecht B (1979) Adenosine regulates via two different types of receptors, the accumulation of cyclic AMP in cultured brain cells. J Neurochem 3:999–1005.
14. Fredholm BB, Abbracchio MP, Burnstock G, Daly JW, Harden TK, Jacobson KA, Leff P, Williams M (1994) Nomenclature and classification of purinoceptors: A report from the IUPHAR subcommittee. Pharmacol Rev 46: 143–156.
15. Abbracchio MP, Cattabni F, Fredholm BB, Williams M (1993) Purinoceptor nomenclature: A status report. Drug Dev Res 28:207–213.
16. Abbracchio MP, Burnstock G (1994) Purinoceptors: Are there families of P2x and P2y purinoceptors? Pharm Therap, in press.
17. Gustaffson LE, Wiklund CU, Wiklund NP, Stelius L (1990) Subclassification of neuronal adenosine receptors. In Jacobson KA, Daly JW, Manganiello V (eds.), Purines in Cellular Signalling. Targets for New Drugs. New York: Springer-Verlag, pp. 200–205.
18. Ferre S, Fuxe K, Von Euler G, Johansson B, Fredholm BB (1992) Adenosine-dopamine interactions in the brain. Neuroscience 51: 501–512.
19. Brackett LE, Daly JW (1994) Functional characterization of the A_{2b} adenosine receptor in NIH 3T3 fibroblasts. Biochem Pharmacol 47:801–814.
20. Riberio JA, Sebastiao AM (1986) Adenosine receptors and calcium: Basis for proposing a third (A_3) adenosine receptor. Prog Neurobiol 26:179–209.
21. Zhou FQ-Y, Olah ME, Li C, Johnson RA, Stiles GA, Civelli O (1992) Molecular cloning and characterization of a novel adenosine receptor: The A_3 adenosine receptor. Proc Natl Acad Sci USA 89:7432–7436.
22. Redman RS, Silinsky E (1993) A selective adenosine antagonist (8-cyclopentyl-1,3-dipropylxanthine) eliminates both neuromuscular depression and the action of exogenous adenosine by an effect on A_1 receptors. Mol Pharmacol 44:835–840.
23. Riberio JA, Sebastiao AM (1994) Further evidence for adenosine A_3 receptors. Trends Pharmacol Sci 15:13.
24. Carruthers AM, Fozard J (1994) The cloned adenosine A_3 receptor should take priority. Trends Pharmacol Sci 15:14.
25. Linden J, Taylor HE, Robeva AS, Tucker AL, Stehle JH, Rivkees SA, Fink SJ, Reppert SM (1993) Molecular cloning and functional expression of a sheep A_3 adenosine receptor with widespread tissue distribution. Mol Pharmacol 44:524–532.
26. Salvatore CA, Jacobson MA, Taylor HE, Linden J, Johnson RG (1993) Molecular cloning and characterization of the human A_3 adenosine receptor. Proc Natl Acad Sci USA 90:10365–10369.
27. Cornfield L, Hu S, Hurt SD, Sills MA (1992) [^{3}H]2-Phenylaminoadenosine ([^{3}H] CV 1808) labels a novel adenosine receptor in rat brain. J Pharmacol Exp Ther 363:552–561.
28. Luthin D, Linden J (1994) Temperature dependent binding of [^{3}H]CV 1808: Mimicry of "A_4" binding in COS cells expressing only recombinant A_{2a} adenosine receptors. Drug Dev Res 31:292.
29. Johansson B, Georgiev V, Parkinson FE, Fredholm BB (1993) The binding of the adenosine A_2 receptor selective agonist [^{3}H]CGS 21680 to rat cortex differs from its binding in rat striatum. Eur J Pharmacol Mol Pharmacol Sec 247:103–110.
30. Ferkany JW, Valentine HL, Stone GA, Williams M (1986) Adenosine A_1 receptors in mammalian brain: Species differences in their interactions with agonists and antagonists. Drug Dev Res 9:85–93.
31. Stone GA, Jarvis MF, Sills MA, Weeks B, Snowwhill EW, Williams M (1988) Species differences in high affinity adenosine A_2 receptor binding in striatal membranes from mammalian brain. Drug Dev Res 15:31–46.
32. Ji X-D, von Lubitz D, Olah ME, Siles GL, Jacobson KA (1994) Species differences in ligand affinity at central A_3-adenosine receptors. Drug Dev Res 33:51–59.
33. Palmer TM, Jacobson KA, Stiles GL (1992) Immunological identification of A_2 adenosine receptors by two antipeptide antibody preparations. Mol Pharmacol 42:391–397.
34. Dudley M, Racker M, Ogden AM, Peet N,

Secreset R, McDermott R (1992) MDL 102, 234: A selective adenosine A1 receptor antagonists reflecting a new binding mode to the receptor. Abstr Soc Neurosci 18:998.
35. Knutsen LJ, Lau J (1993) WO 9323418, 11/25/1993.
36. Dudley MW (1993) MDL 101483, a selective adenosine A_{2a} receptor agonist with potent activity in rodent models of psychosis. Abst Am Coll Neuropsychopharmacol, Honolulu, Hawaii, December, 1993.
37. Daly JW, Jacobson KA (1994) Adenosine receptor selective agonists and antagonists. In Belardinelli L, Pelleg A (eds.), Adenosine and Adenine Nucleotides: From Molecular Biology to Integrative Physiology. Norwell, MA: Kluwer, pp. 157–166.
38. Gallo-Rodriguez C, Ji X-D, Melman N, Siegman BD, Sanders LH, Orlina J, Fischer B, Pu Q, Olah ME, van Galen PJM, Stiles GL, Jacobson KA (1994) Structure-activity relationships of N^6-benzyladenosine-5′-uronamides as A_3-selective adenosine agonists. J Med Chem 37:636–646.
39. Colotta V, Cecchi L, Catarzi D, Melani FF, Filacchioni G, Martini G, Tacchi P, Lucacchini A (1993) Novel adenosine receptor ligands: 1,3-disubstituted [1] benzopyrano [2,3-c] pyrazolo-4-ones. Synthesis and structure activity relationships. Recept Channels 1:111–119.
40. Cusack N (1993) P_2 receptors; subclassification and structure-activity relationships. Drug Dev Res 28:244–252.
41. Ziganshin AU, Hoyle CHV, Burnstock G (1994) Ecto-enzymes and metabolism of extracellular ATP. Drug Dev Res 32:134–146.
42. Leff P, Humphries RG, Crack BE, Pollard CE, Mckechnie K (1994) New ATP analogues: Pharmacological tools and potential drugs. In Belardinelli L, Pelleg A (eds.), Adenosine and Adenine Nucleotides: From Molecular Biology to Integrative Physiology. Norwell, MA: Kluwer, Drug Dev Res 31:290.
43. O'Connor SE, Dainty IA, Leff P (1991) Further subclassification of ATP receptors based on agonist studies. Trends Pharmacol Sci 12: 137–141.
44. Hoyle CHV (1990) Pharmacological actions of adenine dincleotides in the periphery; possible receptor classes and transmitter function. Gen Pharmacol 21:827–831.
45. Pintor J, Miras-Portugal MT (1993) Diadenosine polyphosphates (ApxA) as new transmitters. Drug Dev Res 28:259–262.
46. Arneric SP, Sullivan J, Williams M (1995) Neuronal nicotinic acetylcholine receptors. Targets for new therapeutic agents. In Bloom FE, Kupfer DJ (eds.), Psychopharmacology. The Fourth Generation of Progress. New York: Raven, in press.
47. Hollman M, Heinemann S (1994) Cloned glutamate receptors. Ann Rev Neurosci 17:31–108.
48. Valera S, Hussy N, Evans RJ, Adami N, North RA, Surprenant A, Buell G (1994) A new class of ligand-gated channel defined by P_{2x} receptor for extracellular ATP. Nature 371:516–519.
49. Webb T, Simon J, Krishek BJ, Bateson AN, Smart TG, King BF, Burnstock G, Barnard EA (1993) Cloning and functional expression of a brain G-protein coupled ATP receptor. FEBS Lett 324:219–225.
50. Filtz TM, Li Q, Boyer JL, Nicholas RA, Harden TK (1994) Expression of a cloned P2Y-purinergic receptor that couples to phospholipase C. Mol Pharmacol 46:8–14.
51. Lustig KD, Shaiu AK, Brake AJ, Julius D (1993) Expression cloning of an ATP receptor from mouse neuroblastoma cells. Proc Natl Acad Sci USA 90:5113–5117.
52. Parr CE, Sullivan DM, Paradiso AM, Lazarowksi ER, Burch LH, Olsen JC, Erb L, Weisman GA, Boucher RC, Turner JT (1994) Cloning and expression of a human P_{2U} nucleotude receptor, a target for cystic fibrosis pharmacotherapy. Proc Natl Acad Sci USA 91: 3275–3279.
53. Goedecke S, Schrader J (1994) Cloning and expression analysis of the rat ATP receptor. Drug Dev Res 31:273.
54. Lazarowksi ER, Harden TK (1994) Identification of a uridine nucleotide-selective G-protein-linked receptor that activates phospholipase C. J Biol Chem 269:11830–11836.
55. Klishin A, Lozovaya N, Pintor J, Miras-Portugal MT, Krishtal O (1994) Possible functional role of diadenosine polyphosphates: Negative feedback for excitation in hippocampus. Neuroscience 58:235–236.
56. Wollack JB, Collins RM Jr (1994) Dimeric analogs of adenosine: Effects of longer bridging groups on adenosine receptor binding. Drug Dev Res 31:334.
57. Williams M (1995) Purinoceptors in central nervous system function. Targets for new therapeutic agents. In Bloom FE, Kupfer DJ (eds.), Psychopharmacology. The Fourth Generation of Progress. New York: Raven.
58. Westfall DP, Shinozuka K, Forsyth KN, Bjur

RA (1990) Presynaptic purine receptors. Ann NY Acad Sci 604:130–135.
59. Rathbone MP, Gysbers JW, Hindley S, Middlemiss PJ, Juurlink BHJ (1994) Guanosine and GTP stimualte neurite outgrowth from PC12 pheochromocytoma cells and from hippocampal neurons through different mechanisms. Drug Dev Res 31:312.

6. MOLECULAR CLONING AND EXPRESSION OF SNST1, A RENAL SODIUM/NUCLEOSIDE COTRANSPORTER

Ana M. Pajor

The general mechanisms for transport of nucleosides across cell membranes include both sodium-dependent, or concentrative, and sodium-independent, or equilibrative, pathways. The sodium-coupled concentrative transporters couple the movement of a substrate against its electrochemical gradient to the downhill movement of sodium. Mammalian cells typically have an inwardly directed electrochemical gradient for sodium across the cell membrane (the sodium activity is greater outside than inside cells, and the membrane potential is negative inside), and this gradient is harnessed by sodium-coupled transporters to drive concentrative transport of the substrate. Thus these transporters can scavenge substrates from low concentrations outside the cells and accumulate them inside cells.

At least four general classes of sodium-coupled nucleoside transporters have been characterized in mammalian cells. This classification is based on substrate specificity, and there is still very little information about the transport proteins themselves. The N1 class of Na^+/nucleoside transporters prefers uridine and purine nucleosides as substrates [1]. The test substrate for this class of transporters is an analog of inosine called Formycin B. The N1 class is also refered to as *cif* (concentrative, insensitive to the inhibitor nitrobenzylthioinosine [NBMPR], and transports Formycin B) [2]. The coupling coefficient or number of sodium ions transported per substrate appears to be 1 [1,3]. The N1 class of nucleoside transporters has been observed in a number of epithelia, such as intestine [2,4], kidney [1,5,6], and liver [3]. Examples of N1-type nucleoside transporters have also been reported for nonepithelial cells, including splenocytes [7], macrophages [8], and several cell lines such as leukemia L1210 cells [9] and Walker 256 carcinosarcoma cells [10]. An N1-type transporter has been expressed in *Xenopus* oocytes injected with rat intestinal mRNA [11].

The N2 class of transporters prefers adenosine and pyrimidine nucleosides as substrates. The test substrate for this class of transporters is thymidine. For this reason, the N2 class is also known as *cit* (concentrative, insensitive, and transports thymidine). These transporters also couple one sodium ion to the transport of each nucleoside. N2 transporters have been observed in intestine [2] and kidney [1]. N2-type transport has been expressed in *Xenopus* oocytes injected with rabbit intestinal mRNA [12].

Recently, a rat intestinal N2 transporter cDNA was cloned by functional expression in *Xenopus* oocytes [19]. This protein has been named cNT1 and bears no homology to any of the other mammalian sodium-coupled transporters.

The N3 class of transporters has a very broad substrate specificity and will accept both purine and pyrimidine nucleosides. It is also called *cib* for its broad substrate specificity. This class of transporters appears to couple two sodium ions to the transport of each nucleoside [13]. N3 transporters have been reported in choroid plexus [13], small intestine [11], kidney [14], HL-60 cells [15], and Caco-2 cells [16]. The cloned sodium-nucleoside transporter (SNST1) appears to be an N3 transporter [14]. Finally, there has been one report of a human renal N4 transporter [17], which has a substrate specificity similar to the N2 class but also accepts guanosine.

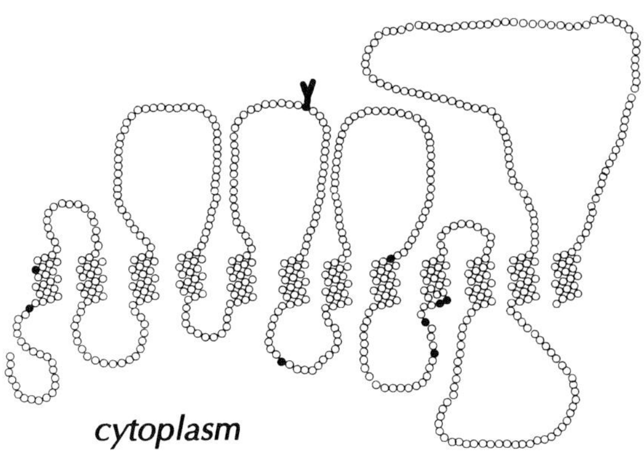

FIGURE 6-1. Secondary structure model of SNST1 [14]. There are 12 putative transmembrane domains and a single N-linked glycosylation site (shown by the Y symbol). Also highlighed are amino acids conserved with other members of the SGLT family, including Asp^{28}, Gly^{43}, Arg^{300}, and the motif Gly-Ala-X-X-X-X-Leu-X-X-X-Gly-Arg.

SNST1 Sequence

The rabbit renal Na^+/nucleoside cotransporter, SNST1, was cloned by hybridization with the rabbit renal high-affinity Na^+/glucose cotransporter, SGLT1 [14]. SNST1 codes for a 672 amino acid protein that is 61% identical and 80% similar in sequence to SGLT1 [14]. The sequence similarity to the Na^+/glucose cotransporter, SGLT1, places SNST1 in the SGLT family. This family includes a number of sodium-coupled solute transporters for substrates including glucose, myoinositol, and neutral amino acids [18]. Interestingly, SNST1 is not related in sequence to the other cloned Na^+/nucleoside cotransporter, cNT1 [19].

The predicted secondary structure of SNST1 protein is shown in Figure 6-1. SNST1 contains 12 putative transmembrane domains, like the other members of the SGLT family. SNST1 contains three consensus sites for N-linked glycosylation, two of which are shared with SGLT1 and a third that is in a putative transmembrane domain. A single residue, Asn^{248}, is glycosylated in SGLT1 [20], and it is likely that the corresponding residue is glycosylated in SNST1, since the extent of glycosylation seen in vitro translation is the same in both proteins [14]. Finally, SNST1 contains a number of amino acids conserved in all members of the SGLT family, including Asp^{28} [21], which is thought to be essential for function; Gly^{43} and Arg^{300}, which have been implicated in sodium binding or translocation [22]; as well as a sequence of five amino acids (see Fig. 6-1) [23] present in many sodium-coupled transporters, including those not related to SGLT1.

SNST1 Tissue Distribution

The tissue distribution of the message for SNST1 was identified by Northern blotting (Fig. 6-2) [14]. The predominant hybridization signal was a single band at around 2.3 kb, and the signal was seen only in two organs, kidney and heart. The hybridization signal in heart was unexpected because we have seen no reports of sodium-coupled nucleoside transport in heart. In preliminary experiments using canine cardiac

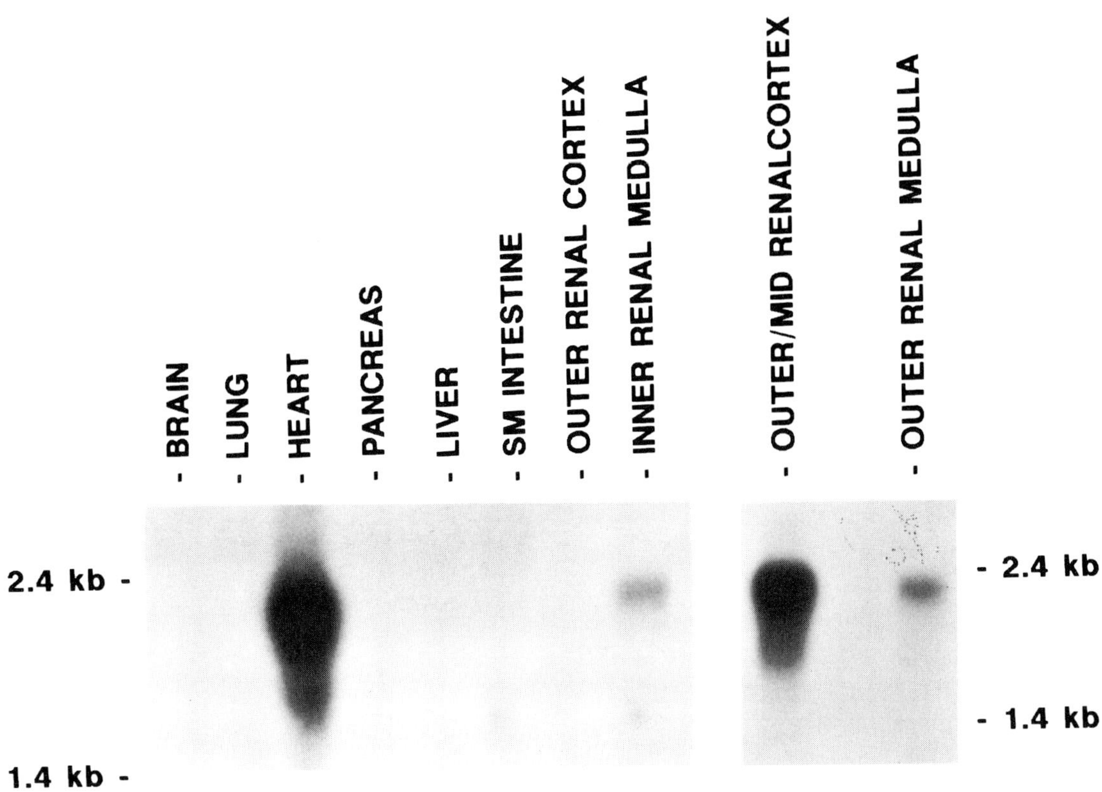

FIGURE 6-2. Tissue distribution of SNST1 mRNA [14]. Rabbit mRNA samples (5 μg each) were separated on a formaldehyde-containing agarose gel and transferred to nylon-reinforced nitrocellulose. The blots were probed and washed at high stringency with ^{32}P-labeled SNST1 cDNA as described in Pajor and Wright [14].

sarcolemmal vesicles, the transport of both uridine and Formycin B was sodium independent and largely resistant to NBMPR. Any sodium-dependent transport could be masked by the large sodium-independent fluxes. Confirmation of the signal in heart may have to be done using antibodies to detect the presence of protein.

SNST1 Function

The initial transport studies to determine the function of SNST1 were done using a chimera, called SNST1c, of SNST1 and the sodium-glucose transporter, SGLT1. This chimera was made because the first SNST1 cDNA isolated was missing the start codon. SNST1c differs from SNST1 by 28 amino acids [14] in the N terminus and first transmembrane domain. However, it does not appear that these residues determine substrate specificity since SNST1c does not transport α-methylglucose, a substrate of the sodium-glucose transporter. Furthermore, later studies have indicated that there are no functional differences between SNST1 and SNST1c.

The function of SNST1 was determined by expression of RNA in *Xenopus* oocytes. Because of the sequence similarity to SGLT1, it was very likely that SNST1 also coded for a sodium cotransporter. Therefore, oocytes injected with SNST1c RNA were exposed to pools of substrates and ^{22}Na. SNST1c-injected oocytes had increased transport of ^{22}Na in the presence of nucleosides (adenosine, uridine, guanosine, and cytidine) [14]. These oocytes also showed sodium-dependent transport of radiolabelled uridine [14]. The transport of uridine in oocytes injected with SNST1c RNA was inhibited by a broad range of nucleosides, both purine and pyrimidine nucleosides [14], suggesting that SNST1 belongs to the N3 class of nucleoside transporters. Further studies in oocytes showed that SNST1 transports both Formycin B and thymidine, although the rate of transport is highest with uridine as a substrate.

SNST1c has also been expressed in the insect cell line, *Sf9*. This expression system is under the control of the baculovirus polyhedrin promoter, which allows the production of extremely high recombinant protein levels [24]. The drawback to this system for transport studies is the high (Fig. 6-3), and also variable (not shown), background transport of nucleosides in control *Sf9* cells infected with wild-type baculovirus. However, the extremely high protein expression makes this system very useful for protein purification. Figure 6-3 shows the transport of uridine in Sf9 cells infected with wild-type baculovirus (controls) or with recombinant baculovirus containing SNST1c cDNA. There was more than a twofold increase in uridine uptake in *Sf9* cells infected with recombinant baculovirus compared with the controls. *Sf9* cells expressing SNST1c also had an increased transport of thymidine and Formycin B (not shown), again suggesting that SNST1 is a broad-specificity nucleoside transporter of the N3 class.

The sequence of SNST1 contains only a single consensus site for phosphorylation, for protein kinase C, located in a putative cytoplasmic location. A number of studies have shown that the function of both transporters [25–27] and ion channels [28] can be regulated by phosphorylation by protein kinases. The effects of phorbol esters, which stimulate the activity of protein kinase C, were tested. As

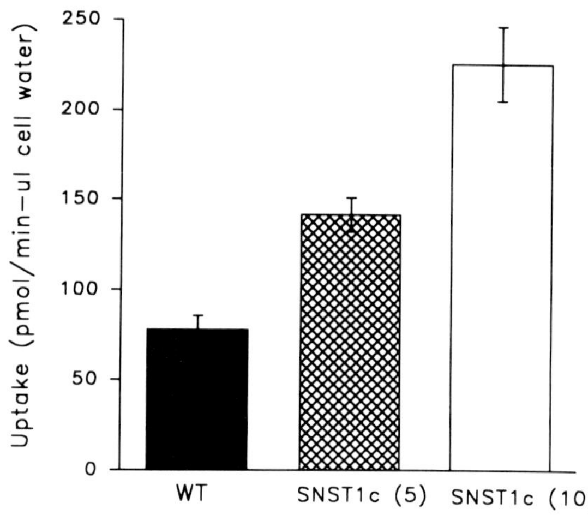

FIGURE 6-3. Uridine transport in *Sf9* cells infected with wild-type baculovirus (WT) or recombinant baculovirus containing SNST1c cDNA. The multiplicity of infection is shown in parentheses. Cells were assayed for transport activity 2 days after infection. The transport of 50 μM ^{14}C-uridine was measured over a 5-minute time period using an inhibitor-oil stop mehod [29].

shown in Figure 6-4, there was a dramatic stimulation of Formycin B transport in oocytes expressing either the complete SNST1 or the chimera SNST1c, but not in control oocytes, after incubation with phorbol 12-myristate, 13-acetate (PMA). In contrast, oocytes injected with RNA coding for the Na^+/glucose cotransporter, SGLT1, (not shown) showed a decrease in sugar transport by about 50% in the presence of PMA. The effect of protein kinase C activation on SNST1 activity could be due to insertion of transporter protein into the oocyte plasma membrane or to direct phosphorylation of SNST1. This remains to be determined. The different effects of phorbol esters on SNST1 and SGLT1 indicate that the effects are specific to the transporters. In addition, the stimulation of SNST1 transport activity by phorbol esters may explain the rather low basal expression in

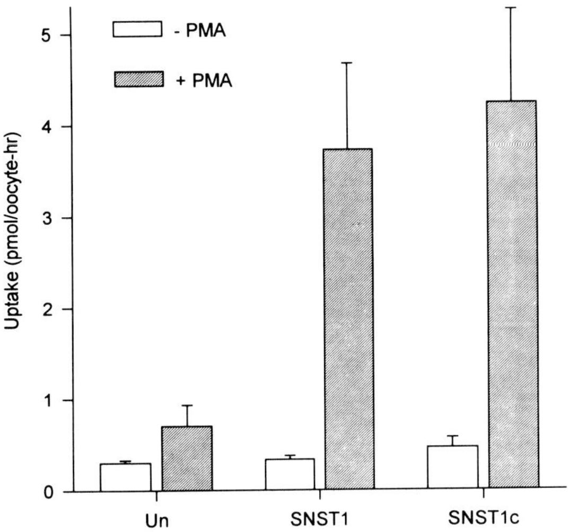

FIGURE 6-4. Effect of phorbol esters on the transport of Formycin B in oocytes injected with SNST1 or SNST1c mRNA. *Xenopus* oocytes were injected with 50 ng RNA for SNST1 or SNST1c (as in Pajor and Wright [14]). Control oocytes were uninjected. Uptakes were measured 5 days after injection. Oocytes were preincubated in choline transport solution with or without phorbol 12-myristate 13-acetate (PMA) for 30 minutes, then the uptake of 50 μM ^{3}H-Formycin B was measured over 30 minutes in sodium-containing transport buffer with or without PMA.

oocytes. It is possible that this transporter is targeted to an intracellular location in the cell and requires activation for the transporters to be translocated to the plasma membrane, or the transporter may normally be found in a relatively inactive state and requires phosphorylation for activity.

In summary, SNST1 is one of two cloned sodium-nucleoside transporters. SNST1 is related to the sodium-glucose transporter related gene family but not related to the other cloned sodium-nucleoside transporter, cNT1 [20]. SNST1 mRNA is found in kidney as well as heart. SNST1 appears to be an N3 type of nucleoside transporter, as it shows a relatively broad substrate specificity. Finally, the function of this transporter appears to be regulated by protein kinase C.

References

1. Williams TC, Jarvis SM (1991) Multiple sodium-dependent nucleoside transport systems in bovine renal brush-border membrane vesicles. Biochem J 274:27–33.
2. Vijayalakshmi D, Belt JA (1988) Sodium-dependent nucleoside transport in mouse intestinal epithelial cells. J Biol Chem 263: 19419–19423.
3. Ruiz-Montasell B, Casado FJ, Felipe A, Pastor-Anglada M (1992) Uridine transport in basolateral plasma membrane vesicles from rat liver. J Membr Biol 128:227–233.
4. Roden M, Paterson ARP, Turnheim K (1991) Sodium-dependent nucleoside transport in rabbit intestinal epithelium. Gastroenterology 100:1553–1562.
5. Le Hir M (1990) Evidence for separate carriers for purine nucleosides and for pyrimidine nucleosides in the renal brush border membrane. Renal Physiol Biochem 13:154–161.
6. Le Hir M, Dubach UC (1985) Concentrative transport of purine nucleosides in brush border vesicles of the rat kidney. Eur J Clin Invest 15:121–127.
7. Darnowski JW, Holdridge C, Handschumacher RE (1987) Concentrative uridine transport by murine splenocytes: Kinetics, substrate specificity, and sodium dependency. Cancer Res 47:2614–2619.
8. Plagemann PGW (1991) Na^+-dependent, concentrative nucleoside transport in rat macrophages. Biochem Pharmacol 42:247–252.
9. Crawford CR, Ng CYC, Noel D, Belt J (1990) Nucleoside transport in L1210 murine leukemia cells. J Biol Chem 265:9732–9736.
10. Crawford CR, Belt JA (1991) Sodium-dependent, concentrative nucleoside transport in Walker 256 rat carcinosarcoma cells. BBRC 175:846–851.
11. Huang QQ, Harvey CM, Paterson ARP, Cass CE, Young JD (1993) Functional expression of Na^+-dependent nucleoside transport systems of rat intestine in isolated oocytes of *Xenopus laevis*. J Biol Chem 268:20613–20619.
12. Jarvis SM, Griffith DA (1991) Expression of the

rabbit intestinal N2 Na^+/nucleoside transporter in *Xenopus laevis* oocytes. Biochem J 278:605–607.
13. Wu X, Yuan G, Brett CM, Hui AC, Giacomini KM (1992) Sodium-dependent nucleoside transport in choroid plexus from rabbit. J Biol Chem 267:8813–8818.
14. Pajor AM, Wright EM (1992) Cloning and functional expression of a mammalian Na^+/nucleoside cotransporter. J Biol Chem 267:3557–3560.
15. Lee CW, Sokoloski JA, Sartorelli AC, Handschumacher RE (1991) Induction of the differentiation of HL-60 cells by phorbol 12-myristate 13-acetate activates a Na^+-dependent uridine transport system. Biochem J 274:85–90.
16. Belt JA, Phelps DA, Crawford CR (1991) Na^+-dependent, concentrative nucleoside transport in human colon carcinoma cells (abstr). Proc Am Assoc Cancer Res 32:96.
17. Gutierrez MM, Brett CM, Ott RJ, Hui AC, Giacomini KM (1992) Nucleoside transport in brush border membrane vesicles from human kidney. Biochim Biophys Acta 1105:1–9.
18. Wright EM, Hager KM, Turk E (1992) Sodium cotransport proteins. Curr Opin Cell Biol 4:696–702.
19. Huang QQ, Yao SYM, Ritzel MWL, Paterson ARP, Cass CE, Young JD (1994) Cloning and functional expression of a complementary DNA encoding a mammalian nucleoside transport protein. J Biol Chem 269:17757–17760.
20. Hediger MA, Mendlein J, Lee HS, Wright EM (1991) Biosynthesis of the cloned intestinal Na^+/glucose cotransporter. Biochim Biophys Acta 1064:360–364.
21. Turk E, Zabel S, Mundlos S, Dyer J, Wright EM (1991) Glucose/galactose malabsorption caused by a defect in the Na^+/glucose cotransporter. Nature 350:354–356.
22. Yamato I, Ohsawa M, Anraku Y (1990) Defective cation-coupling mutants of Escherichia coli Na^+/proline symport carrier. J Biol Chem 265:2450–2455.
23. Deguchi Y, Yamato I, Anraku Y (1990) Nucleotide sequence of gltS, the Na^+/glutamate
symport carrier gene of *Escherichia coli*. J Biol Chem 265:21704–21708.
24. O'Reilly DR, Miller LK, Luckow VA (1992) Baculovirus Expression Vectors. New York: W.H. Freeman and Co., 347 pp.
25. Ruiz OS, Arruda JAL (1992) Regulation of the renal Na-HCO_3 cotransporter by cAMP and Ca-dependent protein kinases. Am J Physiol 262:F560–F565.
26. Vasilets LA, Schwarz W (1992) Regulation of endogenous and expressed Na^+/K^+ pumps in Xenopus oocytes by membrane potential and stimulation of protein kinases. J Membr Biol 125:119–132.
27. Lytle C, Forbush B III (1992) The Na-K-Cl cotransport protein of the shark rectal gland. II. Regulation by direct phosphorylation. J Biol Chem 267:25438–25443.
28. Shearman MS, Sekiguchi K, Nishizuka Y (1989) Modulation of ion channel activity: A key function of the protein kinase C enzyme family. Pharmacol Rev 41:212–237.
29. Domin BA, Mahony WB, Zimmerman TP (1988) Purine nucleobase transport in human erythrocytes. J Biol Chem 263:9276–9284.

7. ANTIBODIES AS PROBES OF NITROBENZYLTHIOINOSINE-SENSITIVE NUCLEOSIDE TRANSPORTERS

Nick Beaumont
Stephen A. Baldwin
Carol E. Cass
James D. Young

The passage of nucleosides across the membranes of mammalian cells is mediated by a variety of different transport systems [1]. In the kidney and small intestine active, sodium-linked uptake systems are present, at least four distinct transporters having been identified by virtue of their differing substrate specificities [2]. However, in most other mammalian cells nucleoside uptake and efflux occurs via the passive process of facilitated diffusion [1]. The passive transporters, like the active, can also be subdivided into classes, in this case by virtue of their sensitivity to inhibition by the 6-thiopurine analogue nitrobenzylthioinosine (NBMPR), to which the sodium-dependent transporters are insensitive. Passive transporters designated *es* (equilibrative-sensitive) bind and are inhibited by NBMPR with affinities in the 0.1–1 nM range, whereas transporters of the *ei* class (equilibrative-insensitive) are unaffected by micromolar concentrations of NBMPR [1,3,4].

Nitrobenzylthioinosine-sensitive *es* transporters are the most widely distributed of the nucleoside transporters and exhibit a broad substrate specificity toward both purines and pyrimidines [1,3]. However, the kinetic properties of the *es* transporters from different tissues and species do vary, as does the apparent M_r of the proteins identified as *es* transporters by photolabeling with [^{3}H]NBMPR. For example, in most mammals, including pigs and humans, NBMPR-sensitive nucleoside transport is also potently inhibited by the vasodilators dilazep and dipyridamole [3]. However, transport in rodent tissues is much less sensitive to inhibition by these two agents [5,6]. Despite this kinetic difference, the enzymically deglycosylated *es* transporters of rat tissues exhibit an apparent M_r that is identical to that of the similarly treated human erythrocyte transporter, as revealed by photoaffinity labeling with [^{3}H]NBMPR [5–7].

In contrast, the kinetically similar *es* transporters of human and pig erythrocytes differ substantially in apparent size. In human erythrocytes the NBMPR-photolabeled protein migrates on SDS/polyacrylamide gels as a broad band of average apparent M_r 55,000, which can be converted by deglycosylation with endoglycosidase F to a species that migrates as a sharp band of apparent M_r 45,000 [8]. By comparison, the photolabeled pig erythrocyte transporter migrates with apparent M_r values of 64,000 and 57,000 before and after deglycosylation, respectively [8].

The kinetic and structural differences described above suggest the possible existence of a family of *es* transporter isoforms. In order to investigate this possibility, in the present work we describe the use of antibodies as a means of both studying the tissue and species distribution of different transporter isoforms and of purifying them for detailed structural studies.

Results and Discussion

MONOCLONAL ANTIBODIES AS PROBES OF *es* TRANSPORTER DISTRIBUTION IN PIG TISSUES

In order to generate antibody probes for studying the structure, function, and tissue distribution of *es* nucleoside transporters, it was first necessary to obtain a suitable immunogen. Pig erythrocyte membranes were chosen as the source because they are particularly rich in transporters. After stripping off the peripheral membrane proteins by treatment at alkaline pH, the integral membrane proteins were solubilized in the nonionic detergent octyl glucoside and then fractionated by ion-exchange chromatography on DEAE-cellulose to yield a preparation of band 4.5 membrane proteins [8]. Although by no means pure, this preparation proved sufficiently enriched in nucleoside transporters to enable the successful preparation of transporter-specific monoclonal antibodies, as previously described [9]. The antibody used in the present study (11C4) recognizes a single species of apparent M_r 64,000, identical to that of the [^{3}H]NBMPR-photolabeled transporter on Western blots of pig erythrocyte membranes (Fig. 7-1). The specificity of the antibodies for the transporter has been confirmed by their ability to immunoprecipitate the [^{3}H]NBMPR-photolabeled protein [9].

The monoclonal antibodies were used to investigate the distribution of immunologically crossreactive nucleoside transporters in a variety of pig tissues. It was not possible to use Western blotting for this investigation, given the high concentration of nucleoside transporters in erythrocytes and the likely contamination of tissue samples with the latter. However, we had previously shown that the antibodies recognized not only the denatured transporter on blots, but also the native, membrane-embedded protein [9]. An immunocytochemical approach was therefore possible. To perform the latter, unfixed frozen sections of pig tissues were employed to minimize damage to the antigen. For staining, the cryostat sections (10 μm thick) were incubated with a 1:50 dilution of ascitic fluid containing the monoclonal antibody overnight at 20°C, followed by incubation for 1 hour at 20°C with a 1:50 dilution of swine anti-mouse second antibody, conjugated to rhodamine (Dako). As a control, identical sections were incubated in parallel with a monoclonal antibody specific for the GLUT1 passive glucose transporter isoform [10]. After washing, the sections were mounted in antifade mountant (Citifluor, City University, London) and examined using a Bio-Rad MRC-600 confocal microscope.

Sections of pig spleen that exhibited no fluorescence when incubated with anti-GLUT1 monoclonal antibodies showed a very bright fluorescence when stained with the anti-pig nucleoside transporter monoclonal antibodies (data not shown). Staining was particularly intense in the marginal zone and is likely to reflect the presence of endogenous transporters. Although the possibility that staining stems from erythrocyte antigens being degraded by the spleen cannot be discounted, these results do demonstrate the utility of the immunocytochemical approach and preservation of transporter immunogenicity. It was therefore surprising

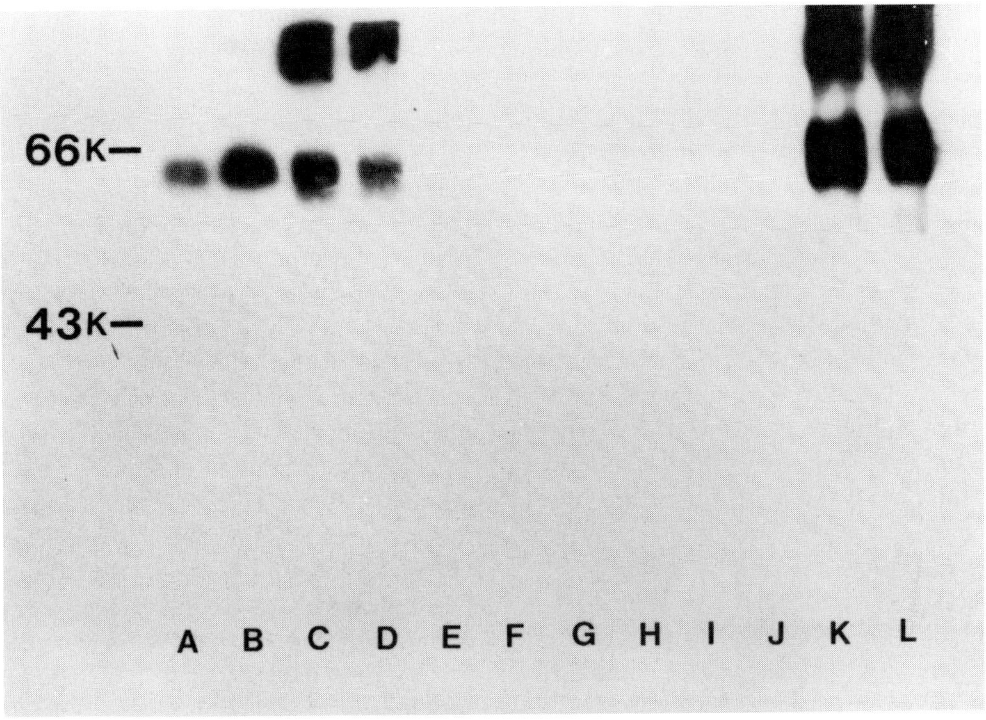

FIGURE 7-1. Western blot of fractions obtained during the purification of the pig erythrocyte nucleoside transporter, stained with the monoclonal antibody 11C4. Lane A, pig erythrocyte ghosts; lane B, alkali-stripped pig erythrocyte membranes; lanes C, D, band 4.5 membrane proteins eluted from the DEAE-cellulose column with 70 mM NaCl; lanes E, F, proteins remaining in solution after adsorption of the band 4.5 proteins with immobilized monoclonal antibodies; lanes K, L, purified nucleoside transporter eluted from the immobilized monoclonal antibody adsorbant with SDS and urea. The positions of marker proteins of M_r 66,000 and 43,000 are indicated.

that the monoclonal antibodies failed to recognize nucleoside transporters in any other pig tissue tested, including liver, lung, kidney, adrenal gland, skeletal muscle, and heart. In contrast, in those tissues where the GLUT1 transporter is present (e.g., lung), strong staining by the anti-GLUT1 monoclonal antibodies used as a control was obtained (data not shown). Several of these tissues, in particular, liver and lung, are known to be rich in *es*-type transporters [5,11]. It follows that pigs must possess at least two antigenically distinct *es*-type nucleoside transporter isoforms.

POLYCLONAL ANTIBODIES AS PROBES OF *es* TRANSPORTER DISTRIBUTION IN HUMAN TISSUES

It was not possible to use the monoclonal antibodies to examine the distribution of *es* transporters in other species because they proved to be species specific, failing to recognize even the erythrocyte *es* transporters of mouse, rabbit, or human [9]. In order to discover whether humans, like pigs, also contained more than one *es* nucleoside transporter isoform, it was therefore necessary to raise antibodies against the human protein. It was not possible to raise transporter-specific monoclonal antibodies using human erythrocyte membrane band 4.5 membrane proteins as immunogens, because the transporter represents only a very minor component of this fraction. However, it proved possible to separate the nucleoside transporter from the major component, the glucose transporter GLUT1, by chromatography of the solubilized proteins on a column of immobilized antibodies against the latter. The resultant preparation was demonstrated by its NBMPR-binding site content to be at least 60% pure [12].

Polyclonal antibodies were raised in rabbits against this material and proved to be transporter specific, labeling solely a band of apparent M_r 55,000, identical to that of the [^{3}H] NBMPR-photolabelled transporter, on Western blots of human erythrocyte membranes [13]. The antibodies also recognized the native transporter, and so were suitable for immunocytochemical investigations. When such studies were performed on sections of human placenta, crossreactivity was detected only at the brush-border surface of the syncytiotrophoblast and not at the basal surface, despite the presence of approximately equal concentrations of NBMPR binding sites at these two locations [14,15]. It follows that humans, like pigs, probably contain at least two antigenically distinct *es*-type nucleoside transporter isoforms.

IMMUNOLOGIC AND STRUCTURAL EVIDENCE OF HOMOLOGY BETWEEN THE *es* TRANSPORTERS OF DIFFERENT SPECIES

Despite the specificity of the polyclonal anti-nucleoside transporter antibodies, as shown by their failure to recognize the transporters of the syncytiotrophoblast basal membranes, the antibodies were found to be species crossreactive. Indeed, they even recognized the kinetically distinct *es* transporters of rodent tissues, as evidenced by their staining of a band of M_r 60,000, identical to that of the [^{3}H]NBMPR-photolabeled transporter, on Western blots of rat liver membranes [13]. In addition, the polyclonal anti-human erythrocyte nucleoside transporter antibodies were also found to recognize the pig erythrocyte transporter, despite the failure of monoclonal antibodies against the latter to recognize the human protein [13]. This finding suggested that, despite their difference in size, the human and pig nucleoside transporters might be structurally related.

In order to confirm this relationship it was necessary to purify sufficient amounts of the two proteins for sequence analysis. Purification of the human transporter was achieved as previously described [12]. For purification of the pig protein, a band 4.5 membrane protein preparation was first prepared by chromatography of pig erythrocyte integral membrane proteins (2 mg/ml) on a DEAE-cellulose column at 4°C after solubilization in 10 mM Tris-HCl, pH 7.4, containing 1% of the detergent octaethyleneglycol n-dodecyl ether ($C_{12}E_8$). It was necessary to use this detergent, as opposed to octyl glucoside, in order to minimize non-specific aggregation of proteins in the following immunoaffinity adsorption step. After loading, the column was washed with 10 mM Tris-HCl, pH 7.4, containing 50 mM NaCl and 1% $C_{12}E_8$, and then a band 4.5 protein fraction, containing the bulk of the nucleoside transporters detectable by slot blotting with the antibodies, was eluted in 10 mM Tris-HCl, pH 7.4 containing 70 mM NaCl and 1% $C_{12}E_8$.

Further purification was achieved by adsorption of the nucleoside transporter onto monoclonal antibodies covalently immobilized on protein A sepharose CL-4B. This procedure removed all of the immunologically detectable nucleoside transporter from solution (Fig. 7-1), but several components of the band 4.5 proteins remained unbound, as revealed by SDS/polyacrylamide gel electrophoresis (Fig. 7-2). After washing the immunoadsorbant with 10 mM Tris-HCl, pH 6.8, containing 1% $C_{12}E_8$, the bound nucleoside transporter was eluted in 40 mM Tris-HCl, pH, 6.8, containing 0.8 mM EDTA, 6 M urea, and 1% SDS. The purified protein migrated as a broad band of apparent M_r 64,000 on SDS/polyacrylamide gels (Fig. 7-2), with the pattern of silver staining exactly matching the pattern of staining by the monoclonal antibodies on Western blots (Fig. 7-1).

Amino acid sequencing of both the human and pig proteins was performed using an Applied Biosystems 477ALP sequencer, and the sequence of the N-terminal 21 residues was determined in each case. For both proteins the first 13 residues

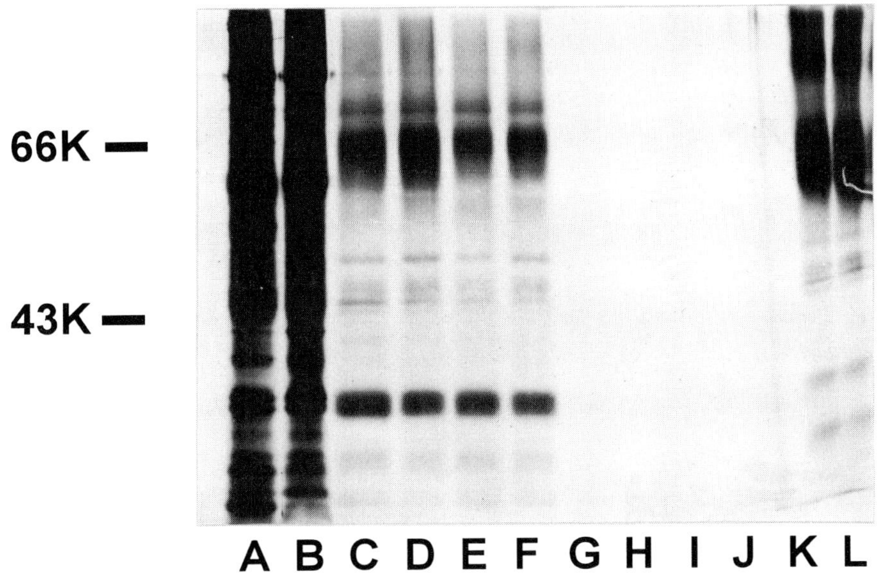

FIGURE 7-2. SDS/polyacrylamide gel electrophoresis of fractions obtained during the purification of the pig erythrocyte nucleoside transporter. The gel was stained with silver by the method of Wray et al. [16]. Lane A, pig erythrocyte ghosts; lane B, alkali-stripped pig erythrocyte membranes; lanes C, D, band 4.5 membrane proteins eluted from the DEAE-cellulose column with 70 mM NaCl; lanes E, F, proteins remaining in solution after adsorption of the band 4.5 proteins with immobilized monoclonal antibodies; lanes K, L, purified nucleoside transporter eluted from the immobilized monoclonal antibody adsorbant with SDS and urea. The positions of marker proteins of M_r 66,000 and 43,000 are indicated.

of the sequence were hydrophilic in character and were followed by a segment of hydrophobic residues typical of a membrane-spanning segment in an integral membrane protein. Interestingly, the sequences of the two proteins were identical, apart from a single conservative substitution in the hydrophilic region. It follows that the pig and human transporters, despite their difference in size, must be evolutionarily related. The somewhat larger pig protein (M_r 57,000 compared with M_r 45,000 for the deglycosylated proteins [8]) presumably contains an additional domain(s) either internally or at the C-terminus. A search of the protein sequence databases revealed the existence of no sequences closely similar to those of the N-terminal regions of the pig and human nucleoside transporters, suggesting that these are members of a novel family of transport proteins.

Conclusion

The use of both monoclonal and polyclonal antibodies specific for the pig and human erythrocyte nucleoside transporters has shown that both human and pig tissues contain at least two antigenically distinct *es*-type nucleoside transporter isoforms. Given the apparent kinetic similarities between the *es* transporters of different tissues, the physiologic significance of this diversity of isoforms remains unclear. Immunologic studies have similarly revealed that the kinetically distinct *es*-type transporters of rodent tissues are likely to be homologous to at least one of the *es*-type transporter isoforms of pigs, humans, and other mammals. Direct evidence for homology between the structurally distinct but kinetically similar *es* transporters of pig and human erythrocytes has been obtained by protein sequencing, which has revealed that these proteins are apparently unrelated to any other protein in the sequence databases. The availability of this sequence information should now enable the production of additional immunologic probes for the transporters in the form of antipeptide antibodies and will facilitate cloning of the proteins. Work toward these objectives is currently in progress.

Acknowledgments

This work was supported by the Yorkshire Cancer Research Campaign, the Cancer Reasearch Campaign, and the Medical Research Council of the United Kingdom.

References

1. Plagemann PG, Wohlhueter RM, Woffendin C (1988) Nucleoside and nucleobase transport in animal cells. Biochim Biophys Acta 947: 405–443.
2. Huang Q-Q, Yao SYM, Ritzel MWL, Paterson ARP, Cass CE, Young JD (1994) Cloning and functional expression of a complementary DNA encoding a mammalian nucleoside transport protein. J Biol Chem 269:17757–17760.
3. Young JD, Jarvis SM (1983) Nucleoside transport in animal cells. Biosci Rep 3:309–322.
4. Paterson ARP, Clanachan AS, Craik JD, Gati WP, Jakobs ES, Wiley JS, Cass CE (1991) Plasma membrane transport of nucleosides, nucleobases and nucleotides: An overview. In Imai S, Nakazawa M (eds.) Role of Adenosine and Adenine Nucleotides in the Biological Systems. Oxford: Elsevier, pp. 133–150.
5. Shi MM, Wu J-SR, Lee C-M, Young JD (1984) Nucleoside transport. Photoaffinity labelling of high-affinity nitrobenzylthioinosine binding sites in rat and guinea-pig lung. Biochem Biophys Res Commun 118:594–600.
6. Jarvis SM, Young JD (1986) Nucleoside transport in rat erythrocytes: Two components with differences in sensitivity to inhibition by nitrobenzylthioinosine and p-chloromercuriphenyl sulfonate. J Membr Biol 93:1–10.
7. Jarvis SM, Young JD (1987) Photoaffinity labelling of nucleoside transporter polypeptides. Pharmacol Ther 32:339–359.
8. Kwong FYP, Baldwin SA, Scudder PR, Jarvis SM, Choy MYM, Young JD (1986) Erythrocyte nucleoside and sugar transport. Endo-β-galactosidase and endoglycosidase-F digestion of partially purified human and pig transporter proteins. Biochem J 240:349–356.
9. Good AH, Craik JD, Jarvis SM, Kwong FYP, Young JD, Paterson ARP, Cass, CE (1987) Characterization of monoclonal antibodies that recognize band 4.5 polypeptides associated with nucleoside transport in pig erythrocytes. Biochem J 244:749–755.
10. Boyle JM, Whetton AD, Dexter TM, Meeran K, Baldwin SA (1985) Characterisation of monoclonal antibodies which specifically recognise the human erythrocyte glucose transport protein. EMBO J 4:3093–3098.
11. Wu J-SR, Young JD (1984) Photoaffinity labelling of nucleoside-transport proteins in plasma membranes isolated from rat and guinea-pig liver. Biochem J 220:499–506.
12. Kwong FYP, Davies A, Tse CM, Young JD, Henderson PJF, Baldwin SA (1988) Purification of the human erythrocyte nucleoside transporter by immunoaffinity chromatography. Biochem J 255:243–249.
13. Kwong FYP, Fincham HE, Davies A, Beaumont N, Henderson PJF, Young JD, Baldwin SA (1992) Mammalian nitrobenzylthioinosine-sensitive nucleoside transport proteins: Immunological evidence that transporters differing in size and inhibitor-sensitivity share sequence homology. J Biol Chem 267:21954–21960.
14. Barros LF, Bustamante JC, Yudilevich DL, Jarvis SM (1991) Adenosine transport and nitrobenzylthioinosine binding in human placental membrane vesicles from brush-border and basal sides of the trophoblast. J Membr Biol 119: 151–161.
15. Barros LF, Beaumont N, Jarvis SM, Young JD, Henderson PJF, Yudilevich DL, Thrasivoulou C, Baldwin SA (1992) Nucleoside transporters in human placenta. Biochem Soc Trans 20: 244S.
16. Wray W, Boulikas T, Wray VP, Hancock R (1981) Silver staining of proteins in polyacrylamide gels. Anal Biochem 118:197–203.

B. CELLULAR BIOLOGY

8. REGULATION OF THE HUMAN A_1 ADENOSINE RECEPTOR GENE

Hongzu Ren
Gary L. Stiles

Adenosine is a physiologically important metabolite and a clinically useful drug. Endogenous release of adenosine mediates a variety of physiologic effects in a wide range of organ systems [1]. These effects are largely transduced via specific adenosine receptors (AR). At present, four subtypes of adenosine receptors have been reported in the literature—A_1, A_{2a}, A_{2b}, and A_3 receptors—and all have been cloned. The A_1 receptor mediates a wide range of physiologic effects, including inhibition of neurotransmitter release, suppression of heart rate and contractility, inhibition of lipolysis, and regulation of smooth muscle tone [2]. Activation of the A_1 receptor may produce a dramatic cardioprotective effect against ischemic heart damage [3]. Factors that regulate A_1 receptor expression in tissues thus become critically important in understanding physiologic and therapeutic processes.

We have recently cloned the human A_1 receptor gene [4] and found it to contain six exons in which exons 1–4 and part of exon 5 represent 5′-untranslated regions, while the 3′ end of exons 5 and 6 represents coding and 3′-untranslated sequences. The intron separating the coding sequence is located in the second intracellular loop between the third and the fourth transmembrane regions. Exons 1 and 2, which are present in a human brain cDNA clone, are not detected in mature transcripts in any human tissue studied thus far. Exons 3 and 4 are mutually exclusive, giving rise to two classes of mRNA containing either exons 3, 5, and 6 or exons 4, 5, and 6. All human tissues known to express even low levels of A_1 receptor contain mRNA with exons 4, 5, and 6, whereas tissues such as brain, kidney, or testis, which express high levels of A_1 receptor, also contain mRNAs with exons 3, 5, and 6. Since there are two upstream ATG codons in exon 4 and upstream ATG codons have been reported to have an inhibitory effect on translation, we compared the levels of human A_1 receptor expression in COS-7 cells transfected with plasmids containing exon 4 or exon 3, or mutated exon 4, in which ATG codons were mutated to GGG. Our data show that upstream ATG codons in exon 4 inhibit A_1 receptor expression and that mutation of those ATG codons leads to the relief of inhibition such that A_1 receptor expression equals that seen with exons 3, 5, and 6.

Methods

REVERSE TRANSCRIPTION-POLYMERASE CHAIN REACTION (RT-PCR)

Total RNA was prepared from frozen human brain tissue that was collected a few hours after the patient's death and stored in a −70°C freezer. The samples were obtained through the Human Brain Bank of the Bryan Neurosciences program at Duke University Medical Center.

The frozen tissue was suspended in buffer (4 M guanidine thiocyanate, 0.1 M Tris-HCI, pH 7.5, and 1% beta-mercaptoethanol) and immediately homogenized with Polytron at full speed for two to three 15-second bursts. Sodium lauryl sarcosinate (10%) was added to the homogenate to give a final concentration of 0.5%, and the suspension was layered onto a CsCl cushion (5.7 M CsCl, 0.01 M EDTA, pH 7.5) in an ultracentrifuge tube. Samples were centrifuged at 175,000 × g for 20 hours. The RNA pellet was resuspended in water and stored in 75% ethanol at −20°C for future use. Human testis total RNA and human fetal brain total RNA were obtained from Clontech. Total RNAs from fat, skeletal muscle, and kidney were a gift from Dr. Stephen Liggett of the University of Cincinnati. Total RNAs from heart, liver, and spleen were a gift from Dr. Debra Schwinn of Duke University Medical Center.

Approximately 5 µg of total RNA was resuspended in 8 µl of water, heated at 65°C for 10 minutes, and cooled on ice. Then oligonucleotide, dithiothreitol (DTT), and the bulk reaction mixture of Pharmacia's first-strand cDNA synthesis kit were added according to the manufacturer's instruction. The reverse transcription was carried out in 15 µl at 37°C for 60 minutes. Three microliters of this reaction mixture was then mixed with water, 10× PCR reaction buffer, dNTPs, Mg^{2+}, the appropriate oligonucleotide, and 2.5 units of Taq polymerase (Perkin-Elmer Cetus) in a 50 µl final volume. The polymerase chain reaction was carried out in a Perkin-Elmer DNA Thermal Cycler 480 programmed as follows: the first two cycles at 95°C for 5 minutes first and 1 minute second, and 70°C for 5 minutes; the next 42 cycles at 95°C for 1 minute, and at 70°C for 2 minutes; the last cycle at 95°C, at 1 minute, and at 70°C for 10 minutes.

CONSTRUCTION OF PLASMIDS USED FOR TRANSFECTION

PCR cloning was used to construct a series of plasmids for transient transfection of COS-7 cells. The vector used was pCMV5 digested with *Eco*RI and *Xba*I [4]. Plasmids were named according to the exons of the A_1 receptor gene that were included in the inserts. For example, the Ex356 insert includes exons 3, 5, and 6.

The template for inserts of Ex1–6, Ex3–6, Ex4–6, and Ex56 was human A_1-receptor cDNA clone 7A [4]. The template for insert of Ex356 was the first-strand cDNA reverse transcribed from human testis total RNA (Clontech). The downstream primer used in PCR for all five plasmid inserts was 3UTXb (5′ACCCCTCTAGATGTGGGCTGGTGG-GA3′), which had an artificial *Xba*I site (underlined) for cloning and the remaining sequence was complementary to the human A_1 receptor cDNA sequences 1406–1431. The upstream primers used in PCR are as follows: For Ex1–6: primer KS (5′-CGAG-GTCGACGGTATCG3′) (Stratagene). For Ex3–6 and Ex356: Ex3Eco (5′-TGGA-AGGAATTCCTGGAGCTAGCGGCTGCT GAA3′). For Ex4–6: Ex4Eco (5′TTGGTGA GAATTCGCCGGGCTGGGAGCGCTGCG-3′). For Ex56: Ex5Eco (5′GCCTGTG-GAATTCATGCCGCCCTCCATC TCAGCTT-3′).

All of the above-mentioned primers except KS have artificial *Eco*RI site (underlined) for cloning after PCR. The PCR product of Ex1–6 has an original *Eco*RI cloning site. The PCR was performed in a volume of 100 µl with 2 units of Vent DNA polymerase (New England Biolabs). The cycle program used was 95°C for 1 minutes, and 70°C for 2 minutes, for 45 cycles.

The PCR fragments were purified with the Qiaex DNA purification kit (Qiagen, Chatsworth, CA) after being separated on 1% agarose gel. Then they were digested with *Eco*RI and *Xba*I, and again purified with Qiaex. The digested inserts and vector were ligated and used to transform to *Escherichia coli* XL1-Blue cells (Stratagene). The isolated colonies contained the expected recombinant plasmids, which were confirmed by DNA sequencing.

CONSTRUCTION OF MUTANT PLASMIDS
To study the effect of upstream AUG codons in exon 4 on the expression of human A$_1$ receptors, three mutant plasmids were constructed with PCR cloning. The mutant ATG1 has the first AUG codon in exon 4 (base 237–239) mutated to GGG. The mutant ATG2 has the second AUG codon in exon 4 (base 351–353) mutated to GGG. The mutant ATG12 has both of those AUG codons mutated to GGG. The primers used in PCR were as follows: For ATG1, the upstream primer was Ex4MUT (5′CTGGGAGC-GCTGCGGCGGGAGCCGGA GGACTGGG-AGCTGC3′) and the downstream primer was 3UTXb; for ATG2, the upstream primer was Ex4Eco and the downstream primer was Ex54 (5′-GGGCACCAAGGGGCACGAC GCAAGCAC CCCCCTGCCTGCT3′), which is complementary to the cDNA sequence 341–380; for ATG12, the upstream primer was Ex4MUT and the downstream primer was Ex54. The template used in PCR was the plasmid Ex4–6 described earlier, and PCR conditions were as described except that 35 cycles were programmed.

The PCR fragment ATG1 was digested with *Eco*47III and *Xba*I, purified with Qiaex and ligated to plasmid Ex4–6 that hads been digested with the same enzymes. The PCR fragment ATG2 was digested with *Eco*RI and *Sty*I, purified with Qiaex, and ligated to pCMV5 digested with *Eco*RI and *Xba*I, and a fragment (~1 kb) isolated from the *Sty*I/*Xba*I digestion of Ex4–6. The PCR fragment ATG12 was digested with *Eco*47III and *Sty*I, purified with Qiaex, and ligated with a fragment isolated from the *Sty*I/*Xba*I digestion of plasmid Ex4–6 (same as for ATG2), and the same plasmid was used for ATG1 ligation. The sequences of PCR-generated fragments in all three mutant plasmids were confirmed with DNA sequencing.

TRANSFECTION OF COS-7 CELLS
In all experiments, 15 μg of plasmid DNA was transfected into COS-7 cells, as described earlier [5].

RADIOLIGAND BINDING OF TRANSFECTED COS-7 CELL MEMBRANES
About 72 hours after transfection, the cell membranes were prepared for radioligand binding [5]. The procedure used was essentially the same as in Olah et al. [5], except that the radioligand used was the A$_1$ receptor antagonist [^{3}H]DPCPX (DPCPX = 8-cyclopentgl-1,3-dipropylxanthine instead of the agonist [^{125}I] PNEA (N^6-2-(4-aminophengl)ethyladenosine).

NORTHERN BLOT ANALYSIS
Total RNA was isolated from the transfected COS-7 cells 48 hours after transfection with Trizol reagent (Gibco/BRL) according to the manufacturer's instruction. Poly(A)$^+$RNA was isolated according to Celano et al. [6] and was loaded onto a 1% agarose/formaldehyde gel [7]. After electrophoresis, RNA was transferred to a Zeta-Probe membrane (Bio-Rad) [8]. The membrane was then prehybridized and hybridized as described in Olah et al. [5]. The hybridization probe labeled by random priming was the entire coding sequence of human A$_1$ receptor. After overnight hybridization, the membrane was washed in 15 mM NaCl/1.5 mM trisodium citrate, pH 7/0, 1% SDS at 60°C twice for 10 miutes each and once for 30 miutes. The blot was later reprobed with random-priming–labeled human S-adenosyl-homocysteine hydrolase cDNA [9] and washed twice in 150 mM NaCl/15 mM trisodium citrate, PH 7, 0.1% SDS at 55°C twice for 10 miutes.

Results and Discussion

The genomic structure of human A$_1$ adenosine receptor consists of six exons with exons 1, 2, 3, and 4 in the 5′-untranslated region [4]. Although exons 1 and 2 are present in a cDNA clone isolated from the human brain cDNA

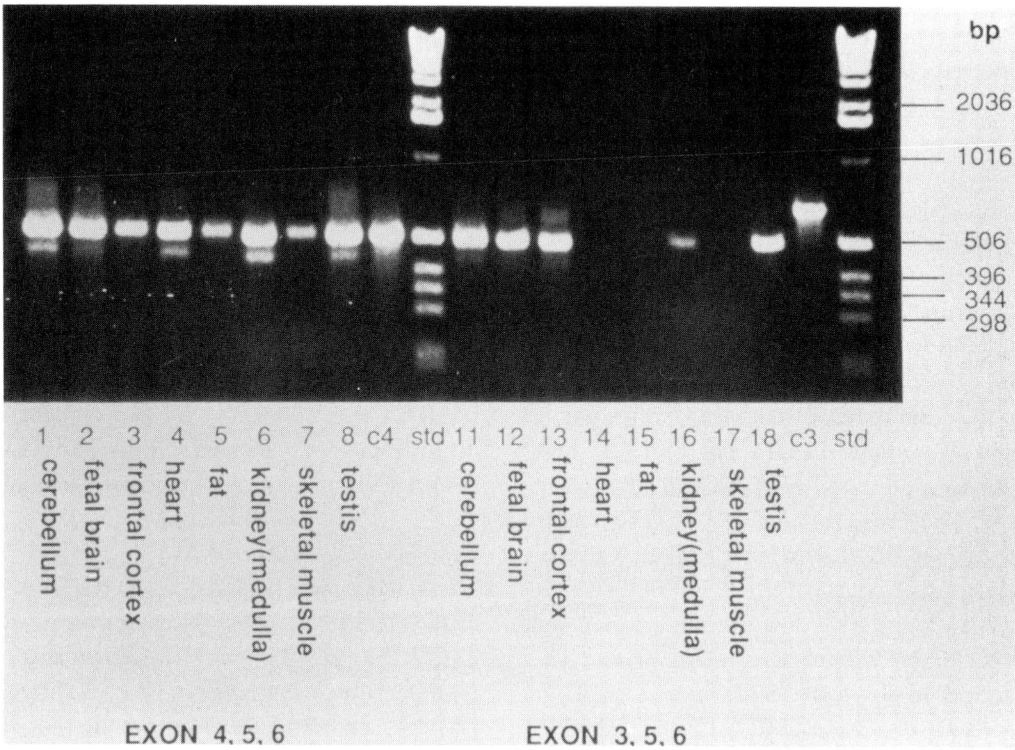

FIGURE 8-1. RT-PCR products obtained from a number of human tissues and separated on a 2.0% agarose gel. RNA was prepared from a variety of human tissues as shown below, subjected to reverse transcriptase, and then amplified with upstream primers specific for exons 3 or 4. The downstream primer used in all experiments are in the coding region, representing antisense sequence of bases 755–804. Lanes 1–c4 represent DNA amplified with upstream primer in exon 4, with lanes 11–c3 representing DNA amplified with upstream primer in exon 3. The lane labeled std (standard) is a 1-kb DNA ladder from Life Technologies, Inc. Lanes c4 and c3 are plasmid 7A cDNA used as positive controls. The product in lane c4 has 517 bp (bases 288–804), and the product in lane c3 has 658 bp (bases 147–804, including exon 4). The products in lanes 11–18 have 504 bp (bases 147–804, subtracting the 154 bases out of exon 4).

library, they have not been detected in any mRNA isolated from any human tissues. Moreover, we found that exons 3 and 4 are mutually exclusive in mature transcripts, thus producing two types of transcripts containing either exons 3, 5, and 6 or exons 4, 5, and 6. In Figure 8-1 RT-PCR with specific primers corresponding to sequences in exon 3 or exon 4 showed that most tissues express mRNA with exons 4, 5, and 6 but only a limited number of tissues, such as brain, testis, and kidney, express mRNA with exons 3, 5, and 6. This indicates that a tissue-specific regulatory mechanism may be present for human A_1 receptor gene expression.

In order to study the effect of exons 3 and 4 on the expression of human A_1 receptor, we tested the pCMV5 constructs containing various combinations of 5′ untranslated regions along with the coding sequence for their translatability in COS-7 cells [10]. The "natural" constructs containing exons 3, 5, and 6 or exons 4–6 were compared with constructs containing only the coding sequence—exons 5 and 6—or exons 1–6 or exons 3–6. These latter motifs are not found in human tissues but allow analysis of the effect exons 3 and 4 in tandem. As seen in

Figure 8-2, constructs without exon 4 (i.e., Ex56 and Ex356) expressed higher levels of A_1 adenosine receptors than any construct with exon 4. Ex356 expressed A_1 receptors at three to seven times the level seen with other constructs. Exons 3 and 4 contain 100 and 154 bases, respectively, so that overall length is not likely to be a critical factor.

The levels of mRNA observed with the five constructs were essentially the same (Fig. 8-3), indicating that the difference in receptor expression was not a consequence of differences in mRNA quantity or half-life. S-adenosylhomocysteine hydrolase mRNA served as a control.

By examining the sequence of exons 3 and 4, we noticed that there are two upstream ATG codons in exon 4 but none in exon 3. The location of upstream ATG codons in exon 4 are shown in Figure 8-4. The upstream ATG

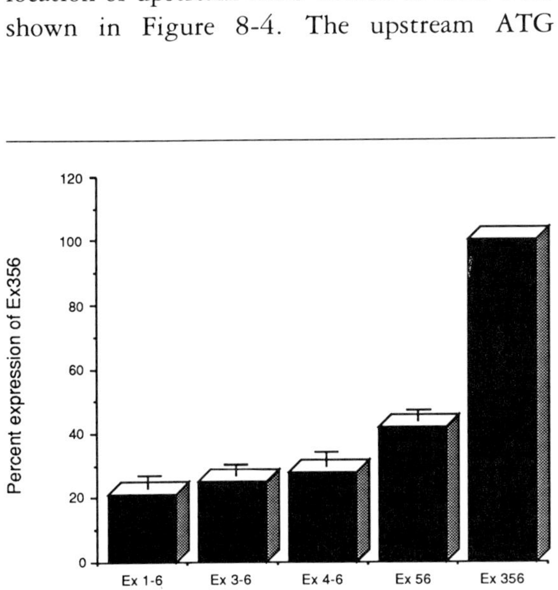

FIGURE 8-2. A_1 receptor expression in transfected COS-7 cells. The expression was measured as maximal [^{3}H]-DPCPX binding capacity in membrane preparations from cells transfected with pCMV5 vectors containing the human A_1 receptor coding sequence plus various lengths of the 5′-untranslated sequence. The exons (Ex) represented in the pCMV5 expression vectors are shown under each bar representing the calculated receptor expression level relative to that of Ex356 (100%). The standard error (n = 4) is indicated above each bar, except for Ex356, which is expressed as 100%.

codons in exon 4 code for open reading frames of 50 and 12 amino acid residues, respectively, which are in frame with the receptor open reading frame, and the stop codon shared by the two upstream open reading frames is located 21 bases upstream of the A_1 receptor start codon. Therefore, there is no possibility of creating A_1 receptors with extended amino-terminal tails. Also shown is an ATG codon upstream of the ATG initiation codon in exon 5, which has none of the typical Kozak consensus sequence [11] and appears to have no dramatic effect on expression, as shown in Figure 8-2, because its removal in construct Ex56 did not enhance A_1 receptor expression compared with that from construct Ex356. When the ATG codons were mutated to GGG individually and concurrently, the expression of the A_1 adenosine receptor increased dramatically (Fig. 8-5). ATG1 is the most 5′ ATG and ATG2 is at the junction of exons 4 and 5. The effects of the mutation are additive, because the construct with both ATG codons mutated demonstrates the highest expression and closely approaches the A_1 receptor level observed with the construct Ex356. Five-prime untranslated regions that contain G+C-rich leader sequences have been shown to contribute to suppression of translation in a number of systems [12]. Although the G+C content of exon 4 is very high at 73%, our data show that G+C richness is not a likely source of receptor suppression, because the ATG12 mutant was expressed equally as well as the Ex356 construct, even though it contained the G+C-rich region. Moreover, Northern analysis of poly(A)$^+$RNA isolated from transfected COS-7 cells showed nearly equal amounts of transcripts for all constructs in Figure 8-5 (data not shown).

Receptor regulation is known to occur in response to hormonal, environmental, or developmental changes and to provide tissue specificity [2]. Regulation has been demonstrated at the level of protein modification, transcription, and message stability [2]. These data represent an example in which alternative forms of receptor mRNA are utilized to regulate the

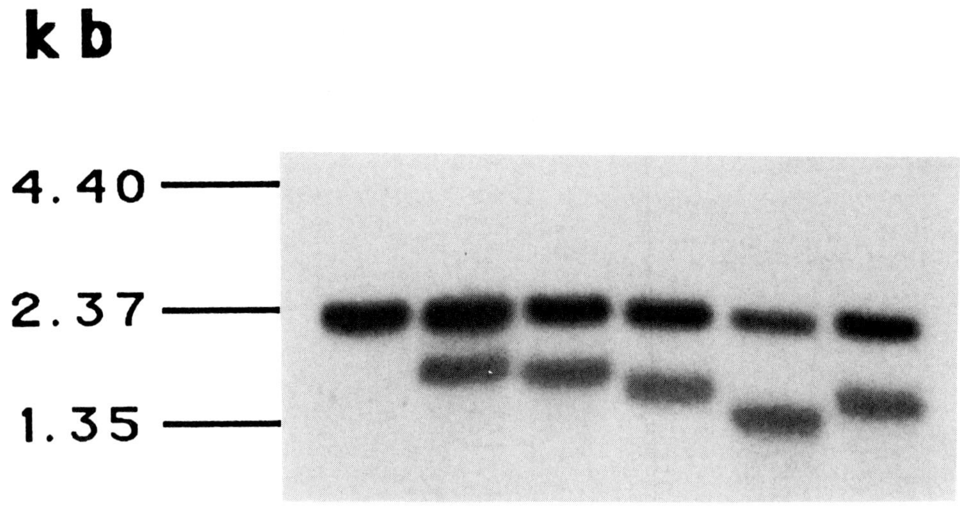

FIGURE 8-3. Northern blot analysis of the poly(A)$^+$RNAs isolated from COS-7 cells transfected with expression vectors with or without exon 4. The upper transcripts were visualized with ^{32}P-labeled human S-adenosylhomocysteine hydrolase cDNA, and the lower transcripts were visualized with the ^{32}P-labeled human A_1 receptor coding sequence cDNA. The order of samples loaded is, from left to right, untransfected COS-7 cells as a negative control and cells transfected with Ex1–6, Ex3–6, Ex4–6, Ex56, or Ex356.

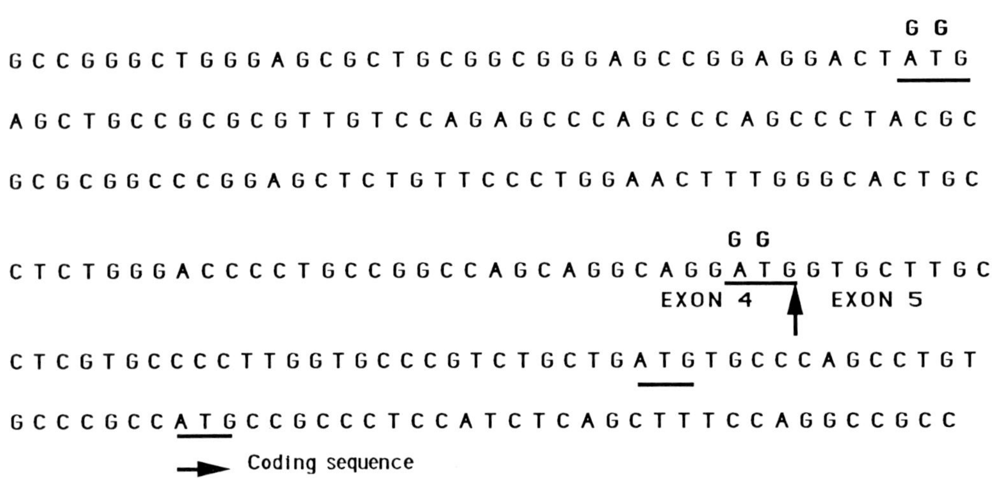

FIGURE 8-4. Complete sequence of exon 4 and partial sequence of exon 5 of human A_1 adenosine receptor cDNA. All four ATG codons, including the A_1 receptor initiation codon, are underlined. The first and the second upstream ATG codons in exon 4 were mutated to GGG as indicated. The vertical arrow indicates the border between exons 4 and 5. The horizontal arrow indicates the beginning of the coding sequence.

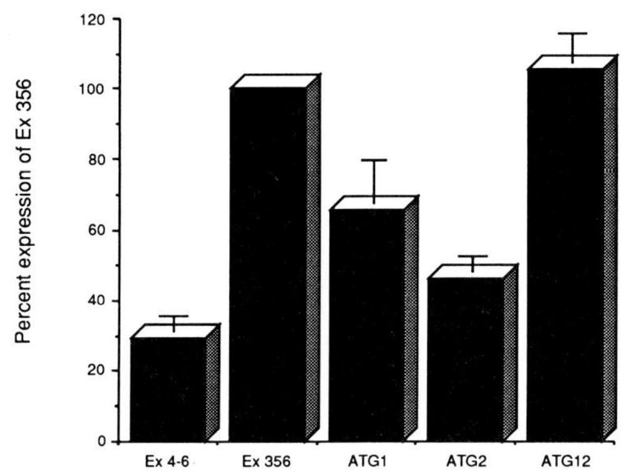

FIGURE 8-5. A_1 receptor expression in COS-7 cells transfected with vectors containing mutations in exon 4. The level of expression was measured by the maximal [^{3}H]-DPCPX binding capacity of membrane from COS-7 cells transfected with vectors. The exons or mutations in exon 4 represented in the expression vector (pCMV5) transfected into COS-7 cells are listed under each individual bar. Expression of A_1 receptor is shown as the percentage relative to the expression level of Ex356 (100%). The standard error (n = 3) is indicated above each bar except for Ex356.

level of receptor in a tissue-specific manner, presumably by translational suppression. The presence of two alternative forms of transcripts could be the result of either alternative splicing or the use of multiple promoters. Upstream ATG codons are well known to be involved in the translational regulation of gene expression [13–15]. These ATG codons tend to be inhibitory and suppress translation of highly active proteins such as proto-oncoproteins, transcription factors, and inflammatory mediators [12,16–22]. For example, in human placenta growth factor mRNA, a small upstream open reading frame has been found to be a translational inhibitory element [22] and mutation in its AUG codons resulted in a 2- to 2.5-fold increase in expression in transfected COS-1 cells. However, this repression is constant and is not modulated by alternative transcripts or in a tissue-specific manner. Thus, although inhibition of translation by upstream ATG codons, G+C-rich secondary structure, or leader length is well documented in yeast and some higher organisms, this specific regulation of the human A_1 adenosine receptor by use of a selective exon-induced repression of expression (most likely translation) appears to provide another mechanism for receptor control.

References

1. Olah ME, Stiles GL (1992) Adenosine receptors. Annu Rev Physiol 54:211–225.
2. Stiles GL (1992) Adenosine receptors. J Biol Chem 267:6451–6454.
3. Ely SW, Berne RM (1992) Protective effects of adenosine in myocardial ischemia. Circulation 85:893–904.
4. Ren H, Stiles GL (1994) Characterization of the human A_1 adenosine receptor gene: Evidence for alternative splicing. J Biol Chem 269: 3104–3110.
5. Olah ME, Ren H, Ostrowski J, Jacobson KA, Stiles GL (1992) Cloning, expression, and characterization of the unique bovine A_1 adenosine receptor. J Biol Chem 267:10764–10770.
6. Celano P, Vertino PM, Casero RA Jr (1993) Isolation of polyadenylated RNA from cultured cells and intact tissues Bio Techniques 15: 26–28.
7. Tsang SS, Yin X, Guzzo-Arkuran C, Jones VS, Davison AJ (1993) Loss of resolution in gel electrophoresis of RNA: A problem associated with the presence of formaldehyde gradients. Bio Techniques 14:380–381.
8. Chomczynski P (1992) One-hour downward alkaline capillary transfer for blotting of DNA and RNA. Analyt Biochem 201:134–139.
9. Coulter-Karis DE, Hershfield MS (1989) Sequence of full length cDNA for human S-adenosylhomocysteine hydrolase. Ann Hum Genet 53:169–175.
10. Ren H, Stiles GL (1994) Posttranscriptional mRNA processing as a mechanism for regulation of human A_1 adenosine receptor expression. Proc Natl Acad Sci USA 91:4864–4866.
11. Kozak M (1986) Point mutations define a sequence flanking the AUG initiator codon that modulates translation by eukaryotic ribosomes. Cell 44:283–292.
12. Kozak M (1992) Regulation of translation in eukaryotic systems. Annu Rev Cell Biol 8: 197–225.

13. Mueller PP, Hinnebusch AG (1986) Multiple upstream AUG codons mediate translational control of GCN4. Cell 45:201–207.
14. Reynolds GA, Basu SK, Osborne TF, Chin DJ, Gil G, Brown MS, Goldstein JL, Luskey KL (1984) HMG CoA reductase: A negatively regulated gene with unusual promoter and 5′ untranslated regions. Cell 38:275–285.
15. Reynolds GA, Goldstein JL, Brown MS (1985) Multiple mRNAs for 3-hydroxy-3-methylglutaryl coenzyme A reductase determined by multiple transcription initiation sites and intron splicing sites in the 5′-untranslated region. J Biol Chem 260:10369–10377.
16. Horiuchi T, Macon KJ, Kidd VJ, Volanakis JE (1990) Translational regulation of complement protein C2 expression by differential utilization of the 5′-untranslated region of mRNA. J Biol Chem 265:6521–6524.
17. Kozak M (1991) An analysis of vertebrate mRNA sequences: Intimations of translational control. J Cell Biol 115:887–903.
18. Kozak M (1987) An analysis of 5′-noncoding sequences from 699 vertebrate messenger RNAs. Nucleic Acids Res 15:8125–8148.
19. Marth JD, Overell RW, Meier KE, Krebs EG, Perlmutter RM (1988) Translational activation of the lck proto-oncogene. Nature 332: 171–173.
20. Williams NP, Hinnebusch AG, Donahue TF (1989) Mutations in the structural genes for eukaryotic initiation factors 2a and 2b of Saccharomyces cerevisiae disrupt translational control of GCN4 mRNA Proc Natl Acad Sci USA 86:7515–7519.
21. Pekhletsky RI, Chernov BK, Rubtsov PM (1992) Variants of the 5′-untranslated sequence of human growth hormone receptor mRNA. Mol Cell Endocrinol 90:103–109.
22. Maglione D, Guerriero V, Rambaldi M, Russo G, Persico MG (1993) Translation of the placenta growth factor mRNA is severely affected by a small open reading frame localized in the 5′ untranslated region. Growth Factors 8: 141–152.

9. ADENOSINE A_2 RECEPTOR REGULATION OF STRIATAL GENE EXPRESSION

Serge N. Schiffmann
Jean-Jacques Vanderhaeghen

Any perturbations within the basal ganglia system lead to movement disorders [1]. The striatum, receiving its main input from the cerebral cortex, is the first relay of this system, and its medium-sized spiny neurons give rise to the output pathways. Two subpopulations of these efferent GABAergic neurons have been identified. The neurons projecting to the globus pallidus, striatopallidal, express enkephalin and mostly the dopamine D_2 receptor, whereas the neurons projecting to the susbtantia nigra pars reticulata, striatonigral, express substance P, dynorphin, and the dopamine D_1 receptor [2]. These two subpopulations give rise to different loops, acting respectively as an inhibitory and an excitatory pathway to the thalamus [1]. In addition, two classes of interneurons mostly represent the remaining striatal neurons: the cholinergic and the somatostatin-expressing neurons.

The dopaminergic nigrostriatal pathway influences these two striatal subpopulations differentially. Reduction of dopamine input into the striatum decreases substance P expression in striatonigral neurons, whereas it increases enkephalin expression in the striatopallidal subpopulation [2]. Moreover, these actions of dopamine are differentially mediated by dopamine D_1 and D_2 receptors [2], the former being mainly expressed in striatonigral neurons and the latter mainly in striatopallidal neurons [2]. It has been proposed that hyperkinetic states such as Huntington's disease or hemiballism, and hypokinetic states such as parkinsonism, all result from disequilibria between these two striatal subpopulations and/or the specific pathways originating from them [1].

The expression of adenosine A_{2a} receptor is selectively detected in the striatum [3–8]. Moreover, this receptor is mainly expressed by striatopallidal neurons that also express enkephalin and the dopamine D_2 receptor [4,6,8]. Conversely, it is not expressed by the striatonigral neurons expressing substance P and the dopamine D_1 receptor, by the cholinergic interneurons, nor by the somatostatin-expressing interneurons [4,6,8,9]. These neuroanatomical data constitute the substratum for involvement(s) of this receptor in the basal ganglia physiology because the adenosine analogues have been recognized as strong depressors of the locomotor activity, most probably by acting at this striatal A_2 receptor [10–12]. Moreover, interactions with the dopamine D_2 receptor occur in controlling this behavior [11–13] and A_{2a}/D_2 receptor interactions have been also observed in the rat striatum at either the membrane level [14,15] or in the regulation of the GABA release [15–17].

Adenosine receptors are also involved in the regulation of gene expression in the rat striatum, and this effect is also subject to interactions between adenosinergic and dopaminergic systems. The present chapter will summarize these recent data.

Regulation of Gene Expression in the Dopamine-Depleted Striatum

In the case of dopamine depletion in the striatum, such as encountered in Parkinson's disease, enkephalin expression is increased while substance P expression is decreased [2]. In that condition D_2 agonist selectively normalizes expression of enkephalin and D_1 agonist selectively normalizes expression of substance P [2]. In the control of adenylyl cyclase activity and locomotor activity, A_{2a} and D_2 receptors have opposite effects. It was therefore hypothesized that blockade of this A_{2a} receptor, selectively expressed by the striatopallidal neurons, would have the same effect as activation of the D_2 receptor.

To test this hypothesis, rats unilaterally depleted of dopamine in the right side subsequently received chronic saline or caffeine treatments [6]. As expected, in saline-treated animals expression of enkephalin and substance P were increased (Figs. 9-1A, 9-4B) and decreased (Figs. 9-1C, 9-4B), respectively, in the dopamine-depleted striatum. In this dopamine-depleted striatum, caffeine treatment decreased and therefore tended to normalize the level of enkephalin expression (Figs. 9-1B, 9-4C), without any effect on the decreased level of substance P (Figs, 9-1D, 9-4C) [6]. This validates the hypothesis because, indeed, this action of caffeine is totally similar to that observed for a D_2 agonist [2]. In hypokinetic disorders such as Parkinson's disease, all treatments that reduced the hyperactivity of the loop arising from the striatopallidal neurons completely resolve the parkinsonian syndrome. This includes D_2 agonist administration and lesion of the subthalamic nucleus [18]. Therefore, selective adenosine A_{2a} antagonists, which also decrease the hyperactivity of this loop in dopamine-depleted rats, should widen the therapeutic arsenal of Parkinson's disease and other syndromes characterized by parkinsonism. Moreover, the opposite effects of A_{2a} and D_2 agonists on the same neuronal target, suggest that, like D_2 antagonists, selective A_{2a} agonists could be useful in the treatment of schizophrenia.

Regulation of Gene Expression in the Intact Striatum

From the previous series of experiments, it also appeared that, unexpectedly, the same caffeine treatment regulates differentially the expression of substance P and enkephalin in the normal striatum (left side) (Fig. 9-1B, 9-1D) [6]. Indeed, in the normal striatum caffeine increased and decreased the expression of enkephalin and that of substance P, respectively (Figs. 9-1B, 9-1D, 9-4D) [6]. These effects were significantly reversed by concomitant treatment with 5′-N-carboxyamidoadenosine (NECA) but not by concomitant treatment with N^6-cyclohexyladenosine (CHA) [6], suggesting that caffeine at least partially acts at an A_2 receptor. This caffeine-induced modifications of gene expression are identical to that observed in the case of dopamine depletion (see Fig. 9-1 and compare Fig. 9-4B and 9-4D).

Moreover, the same chronic caffeine treatment also induces the expression of neurotensin and cholecystokinin mRNAs (Figs. 9-2, 9-3) in different subsectors of the striatum, the subcallosal region (Fig. 9-2), and the dorso- and ventro-lateral quadrants (Fig. 9-3), respectively [19]. Once again, these peculiar patterns of expression are very similar to those observed in the case of dopamine depletion [20,21].

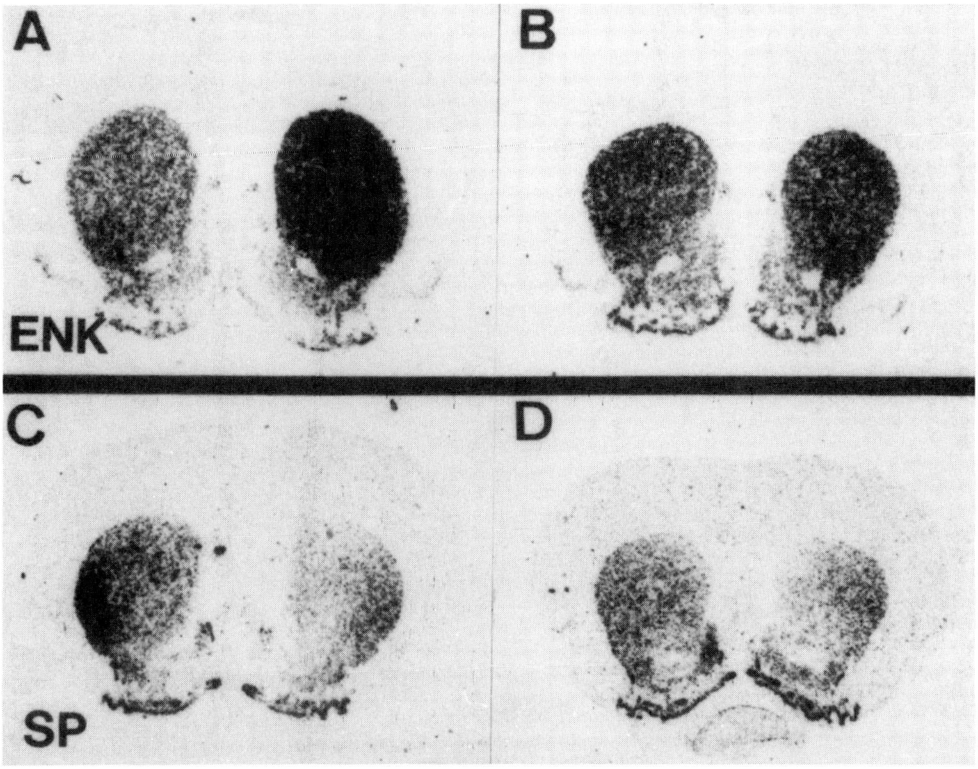

FIGURE 9-1. Autoradiograms generated by hybridization in the striatum with an enkephalin probe (A,B) or a substance P probe (C,D) in right-side dopamine-depleted animals subsequently treated with saline (A,C) or caffeine (B,D). The increase in enkephalin mRNA induced by the dopamine depletion (**A**) was partly reversed by caffeine (**B**), which did not affect the decrease in substance P mRNA (**C**, **D**). Caffeine increased enkephalin mRNA (B) and decreased substance P mRNA (D) in the intact contralateral striatum.

A chronic caffeine treatment could result in relative dopamine depletion by the blockade of presynaptic A_2 receptors located on nigrostriatal terminals (Fig. 9-4D). Indeed, it has been reported that these receptors stimulate synaptic tyrosine hydroxylase activity [22] and striatal dopamine release [23]. It remains to be evaluated whether or not these presynaptic receptors are of the A_{2b} subtype. Although more doubtful for several reasons, other mechanisms leading to the effects of caffeine in the intact striatum could not be definitively ruled out. Indeed, gene expression in striatal neurons could also be regulated by glutamatergic and cholinergic fibers, which, such as the dopaminergic terminals, exhibit presynaptic adenosine receptors regulating transmitter release.

Finally, acute administration of caffeine induces the expression of the immediate early gene c-*fos* [24]. Although the mechanism leading to this induction is not known at the present time, it appears that it could involved both dopamine-dependent and -independent pathways.

Therefore, in conclusion, adenosine through A_2 receptors is involved in at least two different regulatory mechanisms in basal ganglia physiology. First, by acting on the postsynaptic A_{2a} receptors adenosine specifically counteracts the inhibitory action exerted by dopamine on striatopallidal neurons through the D_2 receptors. This effect is particularly evident in the case of striatal dopamine depletion. On the other hand, by acting presynaptically adenosine could

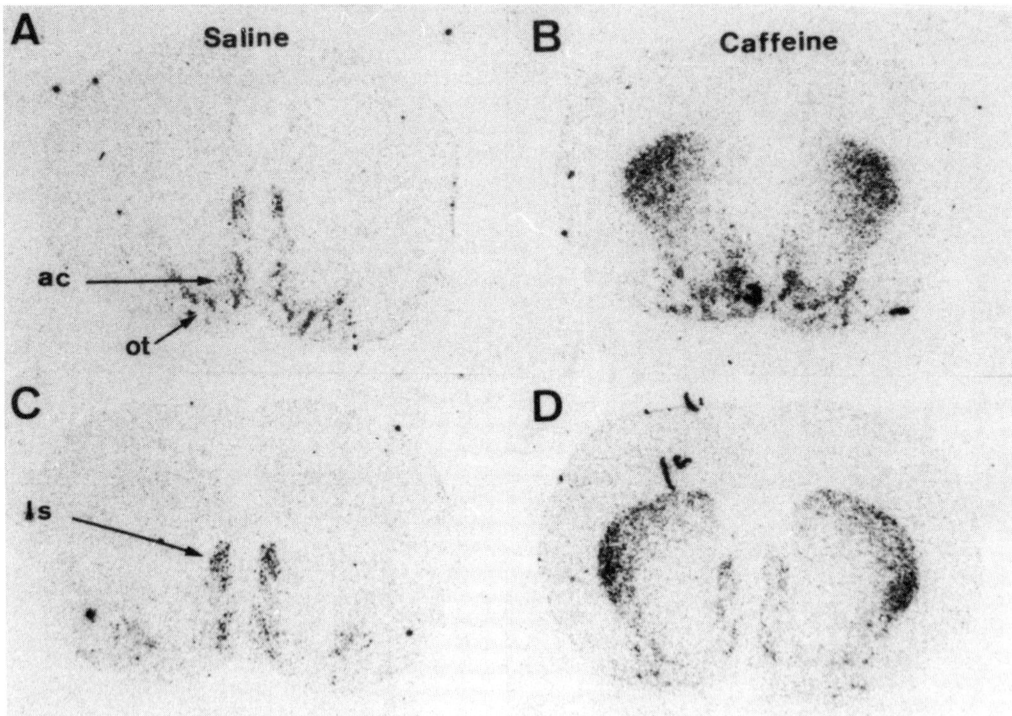

FIGURE 9-2. Autoradiograms generated by hybridization in the rostral (**A, B**) or midddle (**C, D**) striatum with a neurotensin probe in animals chronically treated with saline (A,C) or caffeine (B,D). Caffeine bilaterally induced the expression of neurotensin mRNA in the subcallosal region of the caudateputamen (B), although in the more rostral aspect of the striatum, it was expressed in the whole dorsolateral quadrant (B) and was increased in the accumbal shell (B). A: ac = accumbens; ls = lateral septum; ot = olfactory tubercle.

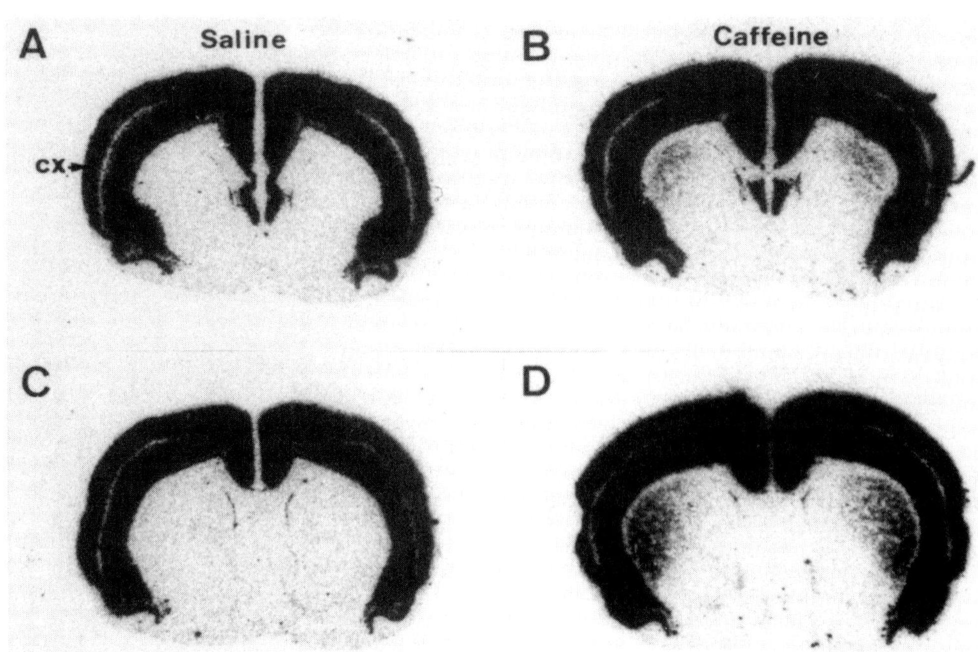

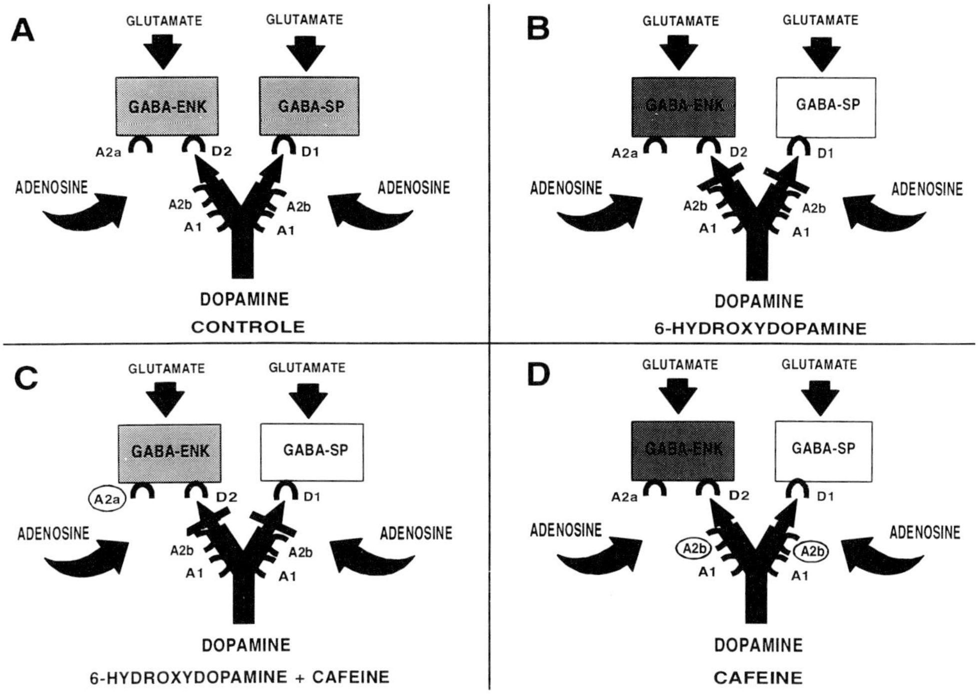

FIGURE 9-4. Schematic representation of the two striatal subpopulations and levels of their gene expression in control conditions (**A**), in dopamine depletion (**B**), after chronic caffeine treatment (**D**), and after chronic caffeine treatment with dopamine depletion (**C**). Note that caffeine trends to normalize the levels of enkephalin expression in the dopamine-depleted striatum (C). Note also that changes in substance P and enkephalin expression in the intact striatum following caffeine (D) are identical to those observed with dopamine depletion (B). (For clarity, only the dopaminergic nigrostriatal pathway is represented; other afferent fibers have been deliberately omitted.)

FIGURE 9-3. Autoradiograms generated by hybridization in the rostral (**A,B**) and middle (**C,D**) striatum with cholecystokinin probes in animals chronically treated with saline (A,C) or caffeine (B,D). Caffeine bilaterally induced the expression of cholecystokinin mRNA in the lateral caudateputamen (B,D). A: cx = cerebral cortex.

stimulate dopamine release and thus reinforce the opposite effect of dopamine on striatonigral and striatopallidal neurons mediated by D_1 and D_2 receptors, respectively.

Acknowledgments

We are thankful to J.-L. Conreur, P. Halleux, and R. Menu for photography and excellent technical support. This work was supported by Belgian grants from the Fonds de la Recherche Scientifique Médicale, Fondation Médicale Reine Elisabeth, Loterie Nationale, and Ministère de la Politique Scientifique (PAI 1990–95). S.N. Schiffmann is a Research Associate of the FNRS (Belgium).

References

1. Albin RL, Young AB, Penney B (1989) The functional anatomy of basal ganglia disorders. Trends Neurosci 12:366–375.
2. Gerfen CR, Engber TM, Mahan LC, Susel Z, Chase, TN, Monsma FJ Jr, Sibley DR (1990) D1 and D2 dopamine receptor-regulated gene

expression of striatonigral and striatopallidal neurons. Science 250:1429–1432.
3. Schiffmann SN, Libert F, Vassart G, Dumont JE, Vanderhaeghen JJ (1990) A cloned G protein-coupled protein with a distribution restricted to striatal medium sized neurons. Possible relationship with D1 dopamine receptor. Brain Res 519:333–337.
4. Schiffmann SN, Jacobs O, Vanderhaeghen JJ (1991a) The striatal restricted adenosine A2 receptor (RDC8) is expressed by enkephalin but not by substance P neurons. An in situ hybridization histochemistry study. J Neurochem 57:1062–1067.
5. Schiffmann SN, Libert F, Vassart G, Vanderhaeghen JJ (1991) Distribution of adenosine A2 receptor mRNA in the human brain. Neurosci Lett 130:177–181.
6. Schiffmann SN, Vanderhaeghen J-J (1993) Adenosine A2 receptors regulate the gene expression of striatopallidal and striatonigral neurons. J Neurosci 13:1080–1087.
7. Vanderhaeghen et al., present book.
8. Fink JS, Weaver DR, Rivkees SA, Peterfreund RA, Pollack AE, Adler EM, Reppert SM (1992) Molecular cloning of the rat A2 adenosine receptor: Selective co-expression with D2 dopamine receptors in rat striatum. Mol Brain Res 14:186–195.
9. Augood SJ, Emson PC (1994) Adenosine A2a receptor mRNA is expressed by enkephalin cells but not by somatostatin cells in rat striatum: A co-expressin study. Mol Brain Res 22:204–210.
10. Durcan MJ, Morgan PF (1989) Evidence for adenosine A2 receptor involvement in the hypomobility effects of adenosine analogues in mice. Eur J Pharmacol 168:285–290.
11. Ferre S, Herrera-Marschitz M, Grabowska-Anden M, Ungerstedt U, Casas M, Anden N-E (1991) Postsynaptic dopamine/adenosine interaction: I. Adenosine analogues inhibit dopamine D2-mediated behaviour in short-term reserpinized mice. Eur J Pharmacol 192:25–30.
12. Ferre S, Herrera-Marschitz M, Grabowska-Anden M, Casas M, Ungerstedt U, Anden N-E (1991) Postsynaptic dopamine/adenosine interaction: II. Post-synaptic dopamine agonism and adenosine antagonism of methylxanthines in short-term reserpinized mice. Eur J Pharmacol 192:31–37.
13. Jarvis MF, Williams M (1987) Adenosine and dopamine function in the CNS. Trends Pharmacol Sci 8:330–331.
14. Ferre S, Von Euler G, Johanson B, Fredholm BB, Fuxe K (1991) Stimulation of adenosine A2 receptors decreases the affinity of dopamine D2 receptors in rat striatal membranes. Proc Natl Acad Sci USA 88:7238–7241.
15. Fuxe et al., present book.
16. Ferre S, O'Connor WT, Fuxe K, Ungerstedt U (1993) The striatopallidal neuron: A main locus for adenosine-dopamine interactions in the brain. J Neurosci 13:5402–5406.
17. Mayfield RD, Suzuki F, Zahniser N (1993) Adenosine A2a receptor modulation of electrically evoked endogenous GABA release from slices of rat globus pallidus. J Neurochem 60: 2334–2337.
18. Bergman H, Wichmann T, DeLong MR (1990) Reversal of experimental parkinsonism by lesions of the subthalamic nucleus. Science 249:1436–1438.
19. Schiffmann SN, Vanderhaeghen J-J (1993b) Caffeine regulates neurotensin and cholecystokinin messenger RNA expression in the rat striatum. Neuroscience 54:681–689.
20. Schiffmann SN, Vanderhaeghen J-J (1992) Lesion of the nigrostriatal pathway induces cholecystokinin mRNA expression in the rat striatum. An in situ hybridization histochemistry study. Neuroscience 50:551–558.
21. Merchant KM, Dobner PR, Dorsa DM (1992) Differential effects of haloperidol and clozapine on neurotensin gene transcription in rat neostriatum. J Neurosci 12:652–663.
22. Onali P, Olianas MC, Bunse B (1988) Evidence that adenosine A_2 and dopamine autoreceptors antagonistically regulate tyrosine hydroxylase activity in rat striatal synaptosomes. Brain Res 456:302–309.
23. Zetterström T, Fillenz M (1990) Adenosine agonists can both inhibit and enhance in vivo striatal dopamine release. Eur J Pharmacol 180:137–143.
24. Johanson B, Lindström K, Fredholm BB (1994) Differences in the regional and cellular localization of c-fos messenger RNA induced by amphetamine, cocaine and caffeine in the rat. Neuroscience 59:837–849.

10. NEURONAL ATP RECEPTORS

P. Illes
K. Nieber
W. Nörenberg

ATP stores and supplies energy in neurons but also acts as an extracellular signal molecule. It may be released from nerve terminals either alone or in combination with classic transmitters by a Ca^{2+}-dependent exocytotic mechanism. Alternatively, it may reach the extracellular space by passing the intact cellular membrane via a transporter or leaking out through the damaged membrane. Irrespective of its origin, ATP activates P_2 purinoceptors belonging either to the P_{2X} or P_{2Y} subtype. The P_{2X} purinoceptor is a typical ligand-gated cation channel, and the P_{2Y} purinoceptor is coupled via a G protein to ion channels or diverse intracellular second messengers [1]. In non-neuronal, multicellular preparations, such as smooth muscle, receptor classification was based originally on the rank order of agonist potencies of ATP analogues [2]. These were α,β-methylene ATP (α,β-meATP) > ATP = 2-methylthio ATP (2-MeSATP) at the P_{2X} purinoceptor and 2-MeSATP > ATP > α,β-meATP at the P_{2Y} purinoceptor. This pharmacologic approach is only of limited value in single non-neuronal [3] and neuronal cells [4]. It is interesting to note that ATP is degraded by ectonucleotidases to ADP and AMP, and finally by ecto-5′-nucleotidases to adenosine. Adenosine activates its own purinoceptor called P_1 [2].

Hence, receptor characterization based on rank order of agonist potencies has serious limitations because ATP and some of its analogues are enzymatically converted to products having activities lower than or even opposite to that of their mother compounds. Nonetheless, neuronal P_{2X} purinoceptors, in contrast to muscular P_{2X} purinoceptors, had a higher sensitivity to 2-MeSATP than to α,β-meATP. Because α,β-meATP is enzymatically stable while 2-MeSATP is a substrate of ectonucleotidases, the difference in potency would increase rather than decrease after blockade of enzymatic degradation. Hence, it has recently been proposed that neuronal and muscular P_{2X} purinoceptors represent different functional and probably structural entities [4,5]. Apparently similar differences exist between neuronal and muscular nicotinic acetylcholine receptors, both with respect to ligand selectivities and amino acid sequences [6].

Finally, a separate receptor that recognizes pyrimidine nucleotides, for example, uridine 5′-triphosphate (UTP), has been described in some tissues [7]. In contrast, responses to ATP and UTP are mediated in other tissues by a common nucleotide receptor.

Localization of P_{2X} and P_{2Y} Purinoceptors and/or Pyrimidinoceptors on the Same Type of Neuron

Both P_{2X} and P_{2Y} purinoceptors, and probably also pyrimidinoceptors, are widely distributed over the peripheral and central nervous systems. However, their characteristics can be compared particularly well in a few types of neurons in which they are colocalized.

SENSORY NEURONS

ATP and its analogues produced in dorsal root ganglion neurons (DRG) of bullfrogs ($<35\,\mu m$ diameter) an inward current that reversed polarity near 0 mV and showed strong inward rectification [8]. This conductance was highly selective for cations over anions; Na^+ and K^+ were equally permeable, while Ca^{2+} had a somewhat lower permeability [9]. In single-channel measurements the mean conductance was 17 pS at a holding potential of -75 mV [9]. The macroscopic whole-cell and microscopic single-channel currents exhibited the same voltage dependence. When inward current responses evoked by ATP and acetylcholine were compared on the same cell, only the acetylcholine-induced response desensitized [8]. In fact, the purinoceptor involved was a typical ligand-activated cationic channel, which belonged to the P_{2X} type. The rank order of agonist potencies was not determined in bullfrog DRG neurons. However, in rat DRG neurons that in all respects behaved similar to their bullfrog counterparts, β,γ-meATP was a weaker agonist than ATP itself [10].

In bullfrog DRG neurons with a larger diameter ($>50\,\mu m$) than those chosen for the previous study, P_{2Y} purinoceptors inhibited the outward M current via a guanine-nucleotide binding (G) protein [11]. This current is a particular type of voltage-sensitive, non-inactivating K^+ conductance, which contributes to the resting membrane potential and is blocked by acetylcholine via muscarinic acetylcholine receptor activation. In contrast to the nonselective cationic current described earlier, the M current does not show inward rectification. The P_{2X}-preferential agonist α,β-meATP was in these experiments about one tenth as potent as ATP.

SUBMUCOUS NEURONS

Neurons of the guinea-pig submucous plexus belong to two groups: S cells have extensive synaptic inputs, while AH cells show a prominent afterhyperpolarization after each action potential. ATP evoked a biphasic depolarization in S cells; the first component occurred with a rapid onset and lasted only for a few seconds, while the second component was much slower in onset and was longer lasting [12]. Patch-clamp recordings revealed that the first component is due to a rapidly desensitizing inward current, which reverses polarity at about 0 mV and is much reduced in low extracellular sodium concentration. The second component was associated with an increase in input resistance and reversed polarity near the potassium equilibrium potential. Hence, the receptors involved belonged to the P_{2X} and P_{2Y} subtypes, respectively. At both receptors ATP, 2-MeSATP, and ATP-γ-S had similar potencies, and α,β-meATP was nearly or completely inactive.

LOCUS COERULEUS NEURONS

Central noradrenergic neurons of the rat nucleus locus coeruleus (LC) possess inhibitory P_1 purinoceptors [13]. The neurons were spontaneously active at the resting membrane potential, and adenosine decreased the firing rate by slightly hyperpolarizing the membrane. 8-Cyclopentyl-1,3-dipropylxanthine (DPCPX) antagonized the effect of adenosine.

ATP caused no consistent change when given alone, but increased the firing rate when given in the presence of DPCPX (Fig. 10-1), indicating a balance between a direct excitatory P_2 effect and an indirect (mediated by the degradation product adenosine) inhibitory P_1 effect [14]. Noradrenaline abolished the discharge of

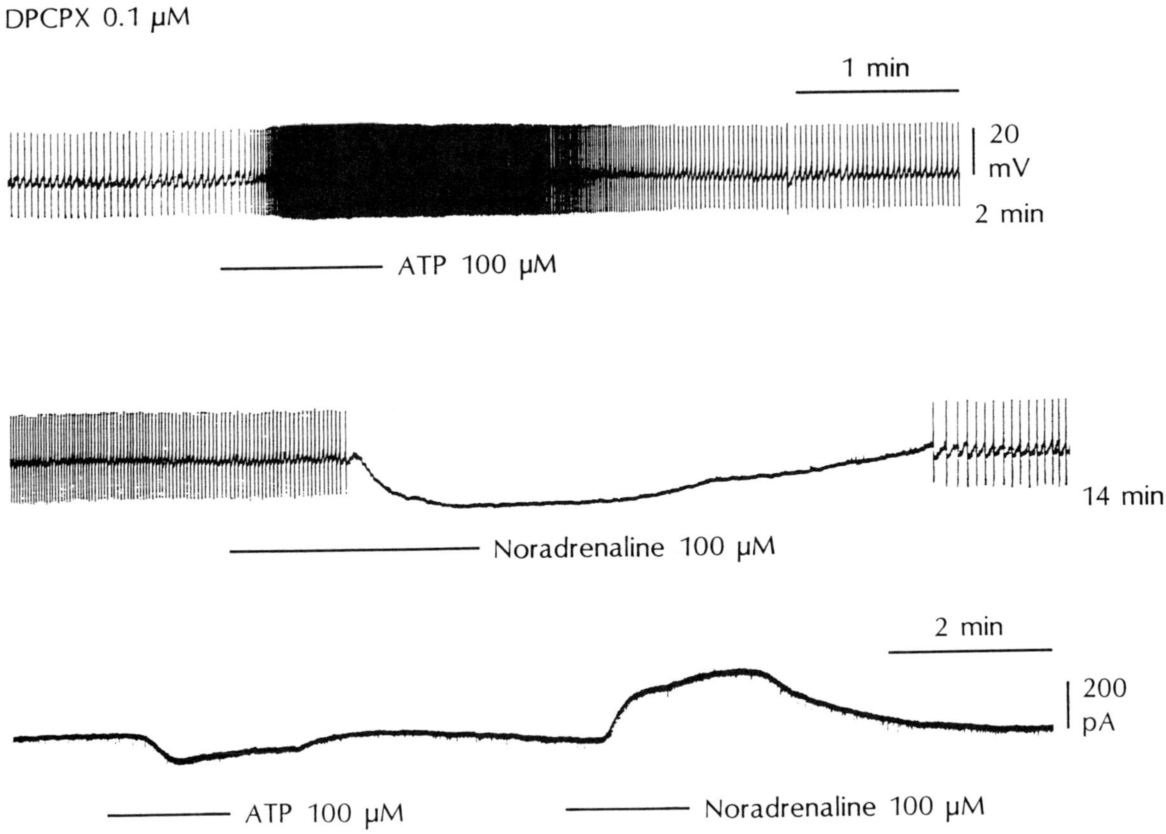

FIGURE 10-1. Effects of ATP and noradrenaline in a rat locus coeruleus neuron. **Upper two traces**: Recording of the membrane potential and firing rate. The full height of action potentials was not reproduced by the pen recorder. ATP (100 μM) increased the firing rate, while noradrenaline (100 μM) abolished it and, in addition, hyperpolarized the cell. **Lower trace**: Current recording from the same neuron voltage clamped at −60 mV. 8-Cyclopentyl-1,3-dipropylxanthine (DPCPX; 0.1 μM) was added 15 minutes before the first application of ATP (100 μM) and was present throughout the experiment. The superfusion times to ATP and noradrenaline are indicated by the horizontal bars. The intervals between the traces are shown. Modified from Harms et al. [15], with permission.

spontaneous action potentials and produced hyperpolarization (Fig. 10-1). The underlying currents were inward for ATP and outward for noradrenaline.

When LC neurons were hyperpolarized by current injection in order to prevent spontaneous firing, α,β-meATP, applied by pressure from a micropipette, depolarized the membrane (Fig. 10-2). This effect was concentration-dependently and reversibly counteracted by the P_2-purinoceptor antagonist suramin. α,β-meATP superfused onto LC neurons also caused depolarization and, in addition, slightly increased the input resistance [15]. A gradual shift of the membrane potential by hyperpolarizing current injection led to a corresponding decrease, but no reversal, of the α,β-meATP effect. In the hyperpolarized region, the current-voltage curve of α,β-meATP came into close apposition with, but did not cross, the control curve (Fig. 10-3). Both a high K^+ and a low Na^+ medium markedly reduced the

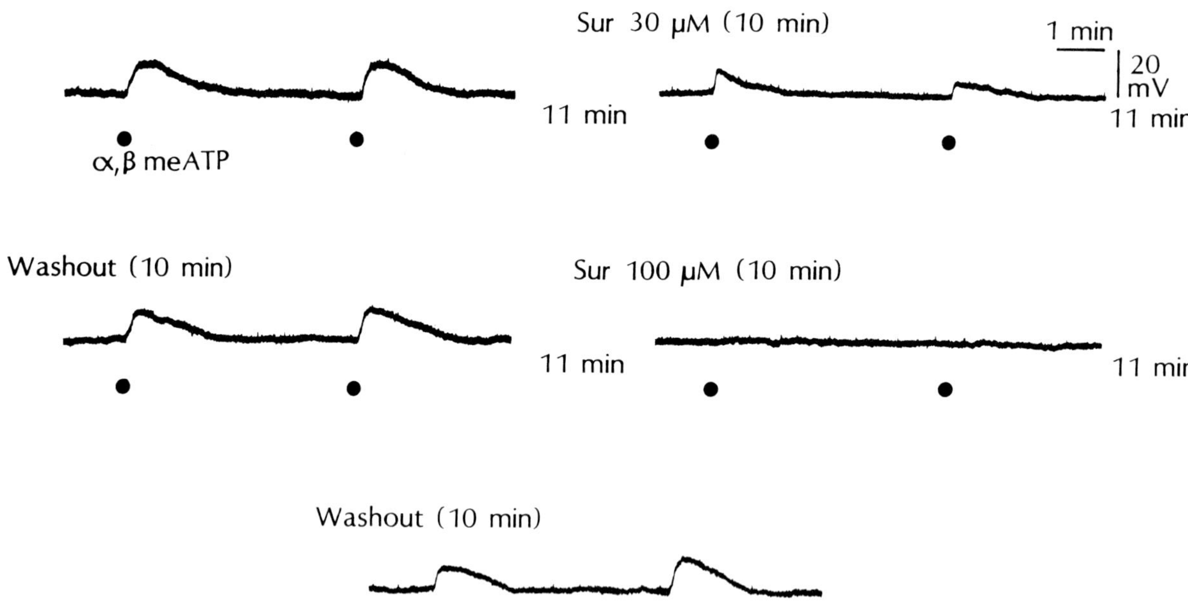

FIGURE 10-2. Effect of α,β-meATP on the membrane potential of a locus coeruleus neuron and interaction with suramin. The neuron was hyperpolarized (about 20 mV) by continuously passing current through the microelectrode. α,β-meATP (1 mM) was pressure ejected (33 kPa, 640 ms) every 5 minutes from a micropipette positioned near the recording electrode. Application was either in the absence or presence of increasing concentrations of suramin (Sur; 30, 100 μM). The superfusion times and the duration of washout are indicated in brackets. The inhibitory effect of suramin was reversible on washout. The intervals between traces are shown.

depolarization by α,β-meATP. On the basis of all these results, it was suggested that α,β-meATP inhibits a resting K^+ conductance (probably via G protein activation) and, at the same time, opens nonselective cationic channels. Direct measurements of ion currents confirmed these assumptions [16].

Hence, LC neurons may be endowed both with P_{2X} and P_{2Y} purinoceptors. At the resting membrane potential and in the presence of DPCPX, α,β-meATP had the highest potency, followed by 2-MeSATP and ATP [14]. However, these results are not really conclusive because both subtypes of P_2 purinoceptors appear to be involved in facilitation of the firing rate.

SUPERIOR CERVICAL GANGLION

When DC potentials were measured from the whole-rat superior cervical ganglion, purines and pyrimidines were shown to cause depolarization, with the rank-order UTP > α,β-meATP ≫ 2-MeSATP [17]. Adenosine and ATP hyperpolarized the neurons. The P_1 antagonist 8-phenyltheophylline abolished the adenosine-induced hyperpolarization and converted the ATP-induced hyperpolarization into depolarization. Hence, it was concluded that ATP itself produced depolarization via activation of P_2 purinoceptors, while its degradation product adenosine caused an opposite effect via activation of P_1 purinoceptors. The high potency of α,β-meATP suggests the presence of a P_{2X} purinoceptor, which probably initiates a cationic conductance increase. Separate pyrimidinoceptors may also exist, because prolonged contacts with either α,β-meATP or UTP strongly

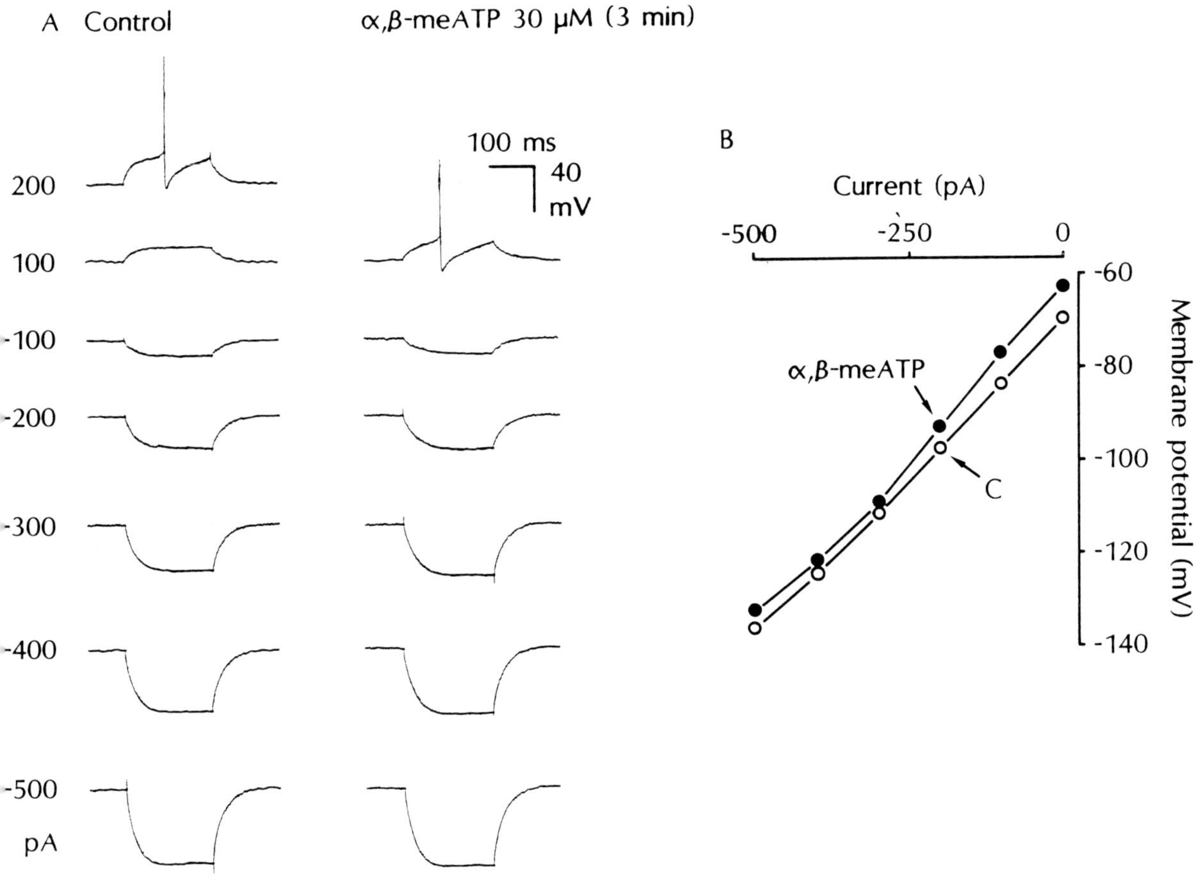

FIGURE 10-3. The effect of α,β-methylene ATP on the current-voltage (I–V) relation of a locus coeruleus neuron. In this neuron, which did not fire spontaneously, electrotonic potentials were evoked by injection of current pulses of 200 ms duration and the indicated amplitudes. **A: Left panel:** At the resting membrane potential of −70 mV, 200 pA current evoked an action potential. **Right panel:** Three minutes after the application of α,β-methylene ATP (α,β-meATP; 30 μM), the membrane potential was depolarized by 7 mV to −63 mV, and already 100 pA current evoked and action potential. **B:** In the hyperpolarized region, the I–V curve of α,β-meATP (30 μM) comes into close contact with, but does not cross, the control curve (c). Data were replotted from A. Modified from Harms et al. [15], with permission.

depressed responses to a second application of the respective agonist without any sign of cross-desensitization between α,β-meATP and UTP [18].

These results are interesting because they show for the first time the existence of a neuronal P_2 purinoceptor that is much more sensitive to α,β-meATP than to 2-MeSATP. Another novelty is the demonstration of neuronal pyrimidinoceptors in the superior cervical ganglion. However, forthcoming patch-clamp experiments are needed to clarify the transduction mechanisms and to determine whether the same neuron is endowed with more than one type of receptor.

P_2 Purinoceptor Function

In cultured celiac ganglion neurons of guinea-pigs, electrical stimulation of the fiber network evoked fast excitatory postsynaptic potentials (EPSPs) and currents (EPSCs) [19]. These potentials were mimicked by ATP and blocked by the P_2-purinoceptor antagonist suramin. The reversal potential for both the EPSC and the ATP current was about 5 mV. Reducing the external Na^+ concentration shifted the reversal potential for the ATP current towards negative potentials. Both these results and the fast onset of the ATP current indicate the involvement of P_{2X} purinoceptors. An increase in the frequency of spontaneous EPSPs induced by high extracellular potassium was also antagonized by suramin [5].

Electric stimulation of inputs to medial habenular neurons evoked suramin-sensitive EPSCs [20]. ATP itself was ineffective (probably because of its rapid enzymatic degradation), but α,β-meATP caused a small inward current, which was also inhibited by suramin. Hence, ATP by acting at P_{2X} purinoceptors is a fast transmitter in some peripheral and central neurons. A slow transmitter function of ATP via P_{2Y} purinoceptor signaling may also be assumed, although the experimental evidence is still missing.

Ionotropic and Metabotropic P_2 Purinoceptors

Chromaffin cells of the adrenal medulla and sympathetic neurons derive from the same part of the neural crest, but their phenotype becomes neuroendocrine during development. In rat pheochromocytoma (PC12) cells, ATP and its structural analogues activated an inward current, which probably passes through similar nonselective cationic channels as found in sensory neurons [21]. As in the latter, the ATP-induced current reversed near 0 mV and exhibited inward rectification, indicating the involvement of P_{2X} purinoceptors. Suramin and another P_2-purinoceptor antagonist, reactive blue 2, both blocked the ATP-activated current [22]. Since the cationic channels were highly permeable to Ca^{2+} [21], the rise in intracellular Ca^{2+} was exclusively due to entry from the extracellular space [23].

In bovine chromaffin cells, ATP had a dual effect mediated by P_{2Y} purinoceptors. On the one hand, ATP led to the formation of inositol trisphosphate (IP_3) with the subsequent mobilization of Ca^{2+} from intracellular storage sites [24]. On the other hand, ATP inhibited the voltage-dependent Ca^{2+} current [25]. This effect was mediated by a pertussis toxin-sensitive G protein.

The application of ATP leads, via an increase in intracellular Ca^{2+}, to catecholamine release in both PC12 cells and bovine chromaffin cells. In the first case an ionotropic P_{2X} purinoceptor is involved [23], while in the second case a metabotropic P_{2Y} purinoceptor is involved [24]. In bovine chromaffin cells, ATP has an additional effect; it may decrease the opening of voltage-dependent Ca^{2+} channels [25] evoked by other types of depolarizing stimuli, that is, acetylcholine. It is interesting to note that metabotropic P_{2Y} purinoceptors were shown to exist also in hippocampal and thalamic neurons [26].

Binding Studies and Expression Cloning

Radioligand binding studies were carried out with the P_{2X}-agonist [^{3}H]α,β-meATP in membranes prepared from several brain regions of the rat [27]. A high-affinity binding site was identified in the striatum, cerebral cortex, and hippocampus; α,β-meATP had a consistently higher activity than 2-MeSATP. In contrast, [^{3}H]diadenosine tetraphosphate labeled a site in bovine chromaffin cells, that exhibited a typical P_{2Y} profile in displacement studies; adenosine 5′-O-(2 thiodiphosphate) (ADP-β-S) was much more potent than α,β-meATP [28].

Isolation of cDNAs encoding two different G-protein–coupled P_{2Y} purinoceptors was recently reported [29,30]. One of these receptors was cloned from embryonic chick brain and after expression in oocytes responded to ATP with a slowly developing inward current [29]. The other receptor was cloned from a mouse neuroblastoma cell line and after expression in oocytes responded equally well to both ATP and UTP [30].

Conclusions

It was not intended to give a full report on neuronal P_2 purinoceptors. Instead, a few relevant points were discussed in depth, especially those that are not included in recent reviews [3,4]. There is no doubt that neuronal P_2 purinoceptors may utilize either ligand-gated cation channels or G proteins, respectively [4]. However, the rank order of agonist potencies at both receptor types is usually 2-MeSATP > ATP > α,β-meATP and, thereby, conforms with that at non-neuronal P_{2Y} purinoceptors. Therefore, we suggested some time ago to name P_2 purinoceptors utilizing ligand-gated cationic channels $P_{2Y\alpha}$, and P_2 purinoceptors coupled to G proteins $P_{2Y\beta}$, in neuronal tissue [4]. An alternative terminology proposed by the IUPHAR Purinoceptor Subcommittee [1] was adapted in the present review. This terminology neglects the sensitivities of P_2-purinoceptor subtypes to the presently known agonists and classifies them on the grounds of functional and, if possible, structural characteristics. Hence, P_{2X} purinoceptors are ligand-gated cation channels, while P_{2Y} purinoceptors are coupled via a G protein to ion channels or to intracellular second messengers.

In fact, recently α,β-meATP has been found to be much more potent than 2-MeSATP in the rat superior cervical ganglion [17], in accordance with the reported agonist potency order at typical non-neuronal P_{2X} purinoceptors. Unfortunately, in this study only extracellular DC potentials were measured, and, therefore, there is no information available on the post-receptor transduction mechanisms.

It appears that in neurons, P_{2X} purinoceptors mediate fast and P_{2Y} purinoceptors mediate slow synaptic responses, although at the present time experimental evidence supports only the operation of the first type of interneuronal communication. P_{2X} and P_{2Y} purinoceptors can be studied particularly well in some neurons, where they coexist. P_{2X} purinoceptors are cationic channels with a considerable permeability to Ca^{2+}; they represent a possible pathway for Ca^{2+} entry into the cell. P_{2Y} purinoceptors are coupled to G proteins, which in turn may close voltage-dependent K^+ or Ca^{2+} channels. Alternatively, these G proteins may be linked to the enzyme phospholipase C and finally lead, via IP_3 production, to the release of Ca^{2+} from intracellular storage sites. In this latter case, the transmembrane fluxes of ions are unaltered and, therefore, the P_{2Y} purinoceptor involved may be termed *metabotropic* in contrast to the *ionotropic* P_{2X} purinoceptor.

Radioligand binding studies prove the existence of both P_{2X} and P_{2Y} sites at neurons. Up until now only the P_{2Y} purinoceptor has been cloned from neuronal tissue and was reported to be a member of the G-protein–coupled receptor superfamily.

References

1. Abbracchio MP, Cattabeni F, Fredholm BB, Williams M (1993) Purinoceptor nomenclature: A status report. Drug Dev Res 28:207–213.
2. Burnstock G, Buckley NJ (1985). The classification of receptors for adenosine and adenine nucleotides. In Paton DM (ed.), Methods in Pharmacology, Vol. 6. New York: Plenum Publishing, pp. 193–212.
3. Bean BP (1992) Pharmacology and electrophysiology of ATP-activated ion channels. Trends Pharmacol Sci 13:87–90.
4. Illes P, Nörenberg W (1993) Neuronal ATP receptors and their mechanism of action. Trends Pharmacol Sci 14:50–54.
5. Silinsky EM, Gerzanich V (1993) On the excitatory effects of ATP and its role as a neurotransmitter in coeliac neurons of the guinea-pig. J Physiol 464:197–212.
6. Sargent PB (1993) The diversity of neuronal

nicotinic acetylcholine receptors. Ann Rev Neurosci 16:403–443.
7. Kügelgen Iv, Häussinger D, Starke K, (1987) Evidence for a vasoconstriction-mediating receptor for UTP, distinct from the P_2-purinoceptor, in rabbit ear artery. Naunyn Schmiedebergs Arch Pharmacol 336:556–560.
8. Bean, BP (1990) ATP-activated channels in rat and bullfrog sensory neurons: Concentration dependence and kinetics. J Neurosci 10:1–10.
9. Bean BP, Williams CA, Ceelen PW (1990) ATP-activated channels in rat and bullfrog sensory neurons: Current-voltage relation and single-channel behavior. J Neurosci 10:11–19.
10. Krishtal OA, Marchenko SM, Pidoplichko VI (1983) Receptor for ATP in the membrane of mammalian sensory neurones. Neurosci Lett 35:41–45.
11. Tokimasa T, Akasu T (1990) ATP regulates muscarine-sensitive potassium current in dissociated bull-frog primary afferent neurones. J Physiol 426:241–264.
12. Barajas-Lopez C, Espinosa-Luna R, Gerzanich V (1994) ATP closes a potassium and opens a cationic conductance through different receptors in neurons of guinea pig submucous plexus. J Pharmacol Exp Ther 268:1396–1402.
13. Regenold JT, Illes P (1990) Inhibitory adenosine A_1-receptors on rat locus coeruleus neurones. An intracellular study. Naunyn Schmiedebergs Arch Pharmacol 341:225–231.
14. Tschöpl M, Harms L, Nörenberg W, Illes P (1992) Excitatory effects of adenosine 5′-triphosphate on rat locus coeruleus neurones. Eur J Pharmacol 213:71–77.
15. Harms L, Finta EP, Tschöpl M, Illes P (1992) Depolarization of rat locus coeruleus neurons by adenosine 5′-triphosphate. Neuroscience 48: 941–952.
16. Shen K-Z, North RA (1993) Excitation of rat locus coeruleus neurons by adenosine 5′-triphosphate: Ionic mechanism and receptor characterization. J Neurosci. 13:894–899.
17. Connolly GP, Harrison PJ, Stone TW (1993) Action of purine and pyrimidine nucleotides on the rat superior cervical ganglion. Br J Pharmacol 110:1297–1304.
18. Connolly G (1994) Evidence from desensitization studies for distinct receptors for ATP and UTP on the rat superior cervical ganglion. Br J Pharmacol 112:357–359.
19. Evans RJ, Derkach V, Surprenant A (1992) ATP mediates fast synaptic transmission in mammalian neurons. Nature 357:503–505.
20. Edwards FA, Gibb AJ, Colquhoun D (1992) ATP receptor-mediated synaptic currents in the nervous system. Nature 359:144–147.
21. Nakazawa K, Fujimori K, Takanaka A, Inoue K (1990). An ATP activated conductance in pheochromocytoma cells and its suppression by extracellular calcium. J Physiol 428:257–272.
22. Nakazawa K, Inoue K, Fujimori K, Takanaka A (1991) Effects of ATP antagonists on purinoceptor-operated inward currents in rat pheochromocytoma cells. Pflügers Arch 418: 214–219.
23. Reber BFX, Neuhaus R, Reuter, H (1992) Activation of different pathways for calcium elevation by bradykinin and ATP in rat pheochromocytoma (PC12) cells. Pflügers Arch 420:213–218.
24. Sasakawa N, Nakaki T, Yamamoto S, Kato R (1989) Stimulation by ATP of inositol trisphosphate accumulation and calcium mobilization in cultured adrenal chromaffin cells. J Neurochem 52:441–447.
25. Gandia L, Garcia AG, Morad M (1993) ATP modulation of calcium channels in chromaffin cells. J Physiol 470:55–72.
26. Mironov SL (1994) Metabotropic ATP receptor in hippocampal and thalamic neurones: Pharmacology and modulation of Ca^{2+} mobilizing mechanisms. Neuropharmacology 33:1–13.
27. Michel AD, Humphrey PPA (1993) Distribution and characterisation of [^{3}H] α,β-methylene ATP binding sites in the rat. Naunyn Schmiedebergs Arch Pharmacol 348:608–617.
28. Pintor J, Torres M, Castro E, Miras-Portugal MT (1991) Characterization of diadenosine tetraphosphate (Ap_4A) binding sites in cultured chromaffin cells: Evidence for a P_{2y} site. Br J Pharmacol 103:1980–1984.
29. Webb TE, Simon J, Krishek BJ, Bateson AN, Smart TG, King BF, Burnstock G, Barnard EA (1993) Cloning and functional expression of a brain G-protein-coupled ATP receptor. FEBS Lett 324:219–225.
30. Lustig KD, Shiau AK, Brake AJ, Julius D (1993) Expression cloning of an ATP receptor from mouse neuroblastoma cells. Proc Natl Acad Sci USA 90:5113–5117.

11. POSSIBLE ROLE OF GLIA IN THE INDUCTION OF CNS-LIKE PROPERTIES IN AORTIC ENDOTHELIAL CELLS: ATP-ACTIVATED CHANNELS

Damir Janigro
Kathe Ann Stanness
Thien-Son Nguyen
David L. Tinklepaugh
H. Richard Winn

Vascular endothelial cells provide a monolayer surface that lines the lumen of blood vessels. Due to their anatomic proximity to both smooth muscle and brain parenchyma cells, brain endothelial cells play a dual physiologic role: They modulate the caliber of brain vessels by releasing relaxing and constricting factors and constitute the blood-brain barrier (BBB). It is thus not surprising that nitric oxide, originally defined as endothelium-derived relaxing factor [1], has been implicated in both the regulation of cerebral blood flow and BBB permeability [2–5]. Endothelial cells (EC) mediate the actions of a variety of vasoactive substances, including adenosine 5′-trisphosphate (ATP). ATP vasodilatatory actions are believed to depend on a calcium-dependent release of endothelium-derived relaxing factor (EDRF). These effects appear to involve nitric oxide (NO) released by endothelial cells [6]. Nitric oxide production by endothelial cells critically depends on the intracellular concentration of calcium because constitutive nitric oxide synthase, the enzyme responsible for NO synthesis, requires Ca^{2+} [7]. However, the regulation of endothelial $[Ca^{2+}]_i$ does not depend on influx of calcium through voltage-operated channels (VOCs) [8], and it is generally accepted that EC do not express voltage-dependent calcium channels [9]. It therefore appears that the regulation of $[Ca^{2+}]_i$ in EC depends upon 1) calcium release from intracellular stores, 2) calcium transport mechanisms, and/or 3) calcium influx through receptor-operated calcium channels (ROCCs).

Among the nearest neighbors of brain EC are the processes of astrocytes that surround the capillaries. There is a growing body of evidence suggesting that EC from non-central nervous system (CNS) blood vessels may undergo significant transformations when exposed to glial cells [10,11]. In particular, BBB properties have been found in bovine aortic EC (BAEC) after exposure to either rat glioma or rat astrocytes [12]. The possibility that coculturing glial cells with peripheral EC may induce in the latter central nervous system properties other than specific markers for BBB has not been investigated in detail, but we have reported preliminary results suggesting that BAEC may be induced to express a CNS response to ATP [13].

Methods

The isolated vessel was cannulated using a system of concentric glass pipettes as previously described [14]. After an equilibration period of approximately 40 minutes, during which time the bath solution was changed once every 5–10 minutes, viable arterioles developed vasomotion and spontaneously contracted; the control diameter was measured after the development of steady tone. We cultured rat brain microvascular endothelial cells (RBMEC) and bovine aortic cells (BAEC) as previously described [8,15]. These cells were identified as endothelial cells based on positive immunoreactivity for Factor VIII antigen and specific uptake of DiI-Ac-LDL. Primary cultures of rat astrocytes were obtained as previously described by others [16]. Prior to insertion in the cell culture cartridge (see below), cells were shaken for 4 hours and the cells remaining attached were trypsinized. C6 rat glioma cells were obtained from ATCC and cultured in the same media used for astrocytes. Approximately 85% of the cells in primary glial cultures were positive for glial fibrillary acidic protein (GFAP) stain, while the remaining cells were microglial and positive for OX-42.

Cells were cocultured using hollow fiber tubes sealed in a chamber accessible by communication ports. Media were pumped through the "intraluminal" portion of the apparatus (i.e., through polypropylene capillaries coated with ProNectin○™ F, inner diameter 330 μM, total surface 0.023 m^2) and nutrients freely diffused to the extraluminal space (surface 0.04 m^2) through 0.5 μM pores. Flow rate was controlled by a rotor pump (set at 1–10 dyne/cm^2) and was maintained throughout the culturing period. Cartridges were kept in a cell culture incubator and were monitored for lactic acid production and glucose consumption at daily intervals. BAEC (approximately 6×10^6) were inserted in the intraluminal portion and allowed to grow for 10 days prior to insertion of either C6 or astrocytes to the extraluminal space (7 ml volume). After variable periods of time (10–27 days) cells were removed from the cartridge by gentle trypsinization and replated either in coculture or alone in 35-mm tissue culture dishes. For patch-clamp experiments, cells were incubated for 4 hours in acetylated low density lipoprotein (Ac-LDL) to test for their endothelial nature.

Patch-clamp experiments were performed as previously described [13]. Cells were continuously perfused with medium containing (mM/l): NaCl 140, KCl 5, $MgCl_2$ 2, $CaCl_2$ 2, Hepes-OH 10, and dextrose 10; the intracellular solution used in whole-cell recordings consisted of (mM/l): KCl 135, $MgCl_2$ 1, $CaCl_2$ 1, EGTA-KOH 10, Hepes-KOH 5, Na_2ATP 5, Na_2GTP 0.5, pH 7.30 with KOH. Voltage steps were delivered via a Labmaster-TL1 board (Scientific Solutions) driven by pClamp software (Axon). The acquired data were stored on a PC (Delphi 486) hard disk for off-line analysis [19].

For nitric oxide polarographic measurements, cells were isolated and cultured as described earlier. BAEC and RBMEC were then plated on 35 mm polystyrene culture dishes and grown to confluence (approximately 2×10^6 cells/dish, as described earlier). NO measurements were performed with an isolated NO probe (World Precision Instruments, Iso-NO probe) [17]. For each experiment plated cells were gently washed twice with buffered solution in order to remove cellular debris. The composition of the solution bathing the cells and the apparatus used for solution exchange were identical to that described for patch-clamp experiments (as described earlier), except the extracellular solution additionally contained 2 mM L-arginine and 8–150 U/ml superoxide dismutase.

Results

We tested the hypothesis that ATP may act on brain EC ion channels, resulting in calcium-dependent synthesis of NO and thus causing vasodilation. First, we applied ATP to cannulated brain arterioles: ATP (10 μM) caused vasodilatation when applied intraluminally (Fig. 11-1A, ATP = 10 μM). The nitric oxide synthase inhibitor N^G-methyl-L-arginine (L-NMA) caused a significant (20%) reduction in the resting diameter without abolishing the response to ATP, which was reduced by approximately 50%. Removal of nitric oxide synthase (NOS) blockade by perfusion of the intraluminal portion of the vessel with L-Arg (2 mM for 15 minutes) restored both basal vessel diameter and the response to ATP to pre-L-NMA levels (Fig. 11-1). ATP (1–10 μM had little effect on vessel diameter when applied extraluminally but never caused vasoconstriction. Thus ATP caused an endothelium-dependent and NO-mediated vasodilatation.

We further investigated the possible role of ATP in the generation of NO from EC by directly measuring nitric oxide by means of a polarographic NO electrode. Figure 11-1-B1 shows the results obtained from native cultured BAEC. Cells were perfused in control buffer for approximately 10 minutes prior to establishing a baseline. Endothelial cells were then challenged with 100 μM ATP (black arrow) in solutions containing 2 mM $CaCl_2$. Exposure of the cells to ATP resulted in a marked increase in the measured NO. The dependency of the BAEC responsiveness to ATP on extracellular calcium was investigated in a set of experiments in which Ca^{2+} was removed from the media and residual calcium was chelated by 5 mM EGTA (Fig. 11-1-B1). In 0 calcium/EGTA solutions, the response to ATP was still observed, albeit reduced in size (by 18%, n = 7). Similar experiments were performed on cultured RBMEC (Fig. 11-1-B2). As in BAEC, ATP (10–100 μM) caused an increase in measurable NO; however, removal of extracellular calcium effectively prevented NO release in response to ATP.

We also investigated the electrophysiologic actions of extracellular ATP on brain microvascular and bovine aortic endothelial cells (RBMEC and BAEC, respectively) by using the patch-clamp technique. While bovine aortic endothelial cells were hyperpolarized by ATP (10 μM at cell resting potential, mean = −34 mV), ATP caused the activation of a cation-permeant current in brain endothelial cells (i.e., cell depolarization; mean = −38 mV) [18]. The results of these experiments are summarized in Figure 11-2. In order to understand whether the physiologic differences between BAEC and RBMEC responses to extracellular ATP resulted from differences due to organ- or species-specific variability, we cocultured BAEC with glial cells in a hollow-fiber artificial capillary system (see Methods). BAEC were grown alone for 7 days in the intraluminal portion of the cartridge, and C6 rat glioma or rat astrocytes were then added to the extraluminal space. After several days in culture (10–27), cells were removed from the cartridge and replated on 35-mm tissue culture dishes. Upon light microscopic examination of the cells following hollow-fiber culturing, we observed a novel pattern of BAEC growth, because BAEC did not retain (after exposure to either C6 or rat astrocytes) their normal cobblestone morphology (Fig. 11-3). BAEC grew close to or associated with glial cells and frequently had a polygonal shape, characterized by the outgrowth of elongations. When cells were used for patch-clamp experiments, the endothelial nature of the individual cells was confirmed by Ac-LDL positivity. No obvious difference was found in

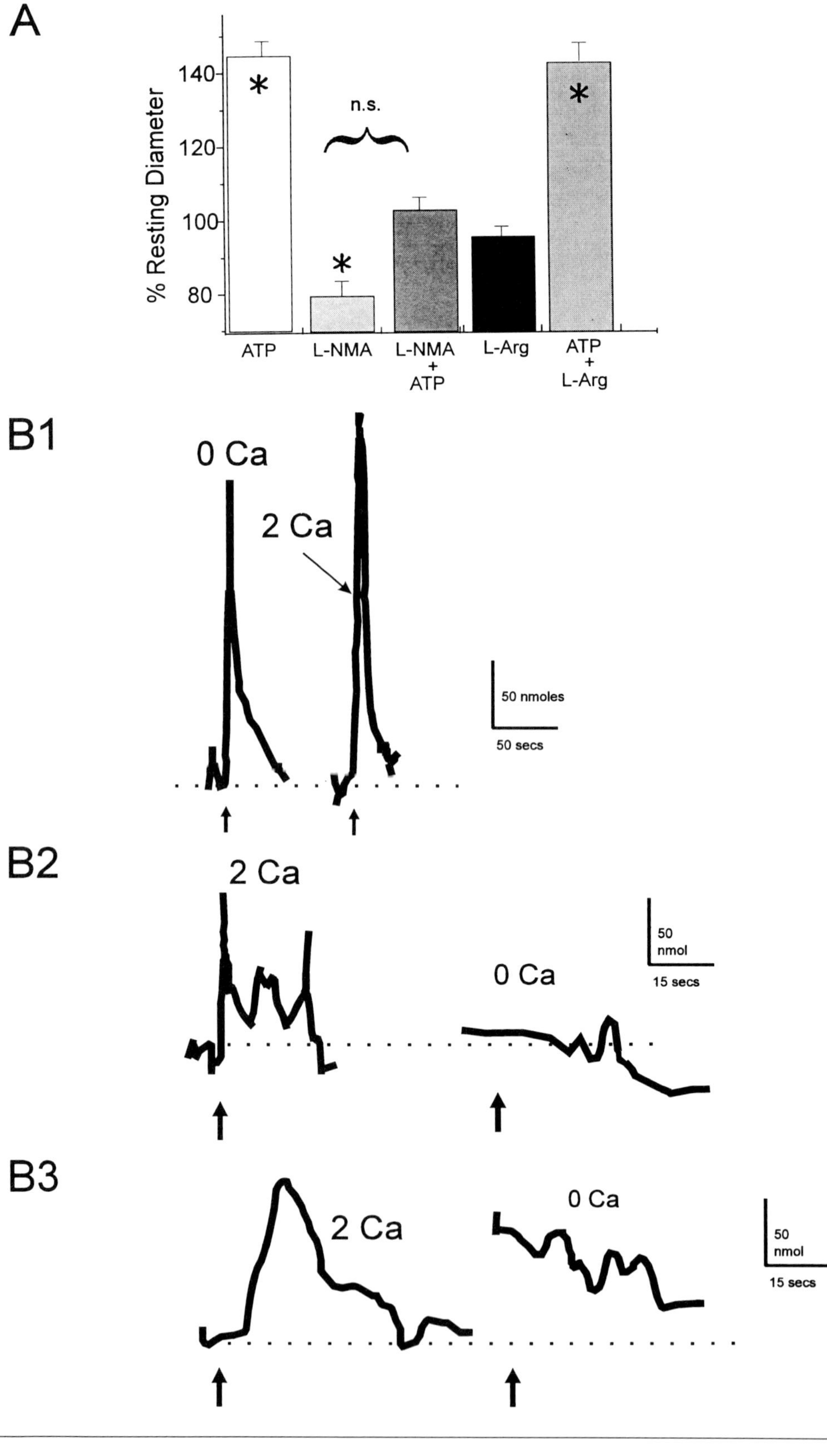

A
140
120
100
80
% Resting Diameter
n.s.
*
*
*
ATP
L-NMA
L-NMA + ATP
L-Arg
ATP + L-Arg
B1
0 Ca
2 Ca
50 nmoles
50 secs
B2
2 Ca
0 Ca
50 nmol
15 secs
B3
2 Ca
0 Ca
50 nmol
15 secs

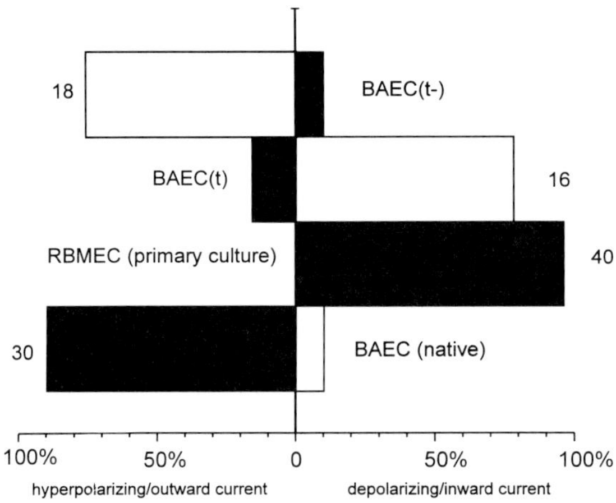

FIGURE 11-2. Comparison of the responses to 10 μM ATP in native bovine aortic EC (BAEC), rat brain microvascular EC (RBMEC), cocultured BAEC replated with C6 [BAEC(T+)], and cocultured BAEC replated in the absence of glial cells [BAEC(T−)]. Cells were either voltage (at −40/−90 mV) or current clamped (at resting potential), and the response to ATP was defined as outward/hyperpolarizing if either a >2 mV hyperpolarization or an outward current occurred. Conversely, a depolarizing response from resting potential or the development of an inward current in response to ATP was grouped as inward/depolarizing. Cell resting potentials were not significantly different in either group [BAEC = −34 mV, RBMEC = −38 mV, BAEC(T+) = −41 mV, and BAEC(T−) = −47]. Statistical analysis was performed by a Z comparison of proportions; * indicates data significant at $p < 0.05$.

endothelial cell morphology when BAEC were cocultured with C6 rat glioma or astrocytes.

Bovine aortic cells cocultured and transformed by either C6 or rat astrocytes ($BAEC_t$) displayed marked electrophysiologic differences when compared to their native BAEC counterparts and challenged with extracellular ATP applications (see Figs. 11-2 and 11-4). In fact, ramp voltage protocols applied to native BAEC before and during application of ATP revealed that an *outward* potassium current developed in response to 10 μM ATP (Fig. 11-4A,B). However, while ATP caused the activation of an outward current associated with cell hyperpolarization in native BAEC, $BAEC_t$ responded to ATP with cell depolarization (mean resting potential = −41 mV; see Figs. 11-2 and 11-4). This depolarization was due to the activation of an inward current and was also observed after removal of extracellular sodium (Fig. 11-4D) and after blockade of outward potassium currents with intracellular CsCl (140 mM). Thus, similar to what observed after application to RBMEC [13,18], 10 μM ATP caused the activation of a mixed cation current in $BAEC_t$. No significant difference was found when comparing VOCs or passive membrane properties expressed in BAEC versus $BAEC_t$, including the expression in both cells of the newly described inward I_{HA} current [8]. In another set of experiments, $BAEC_t$ were col-

FIGURE 11-1. ATP applied intraluminally causes dilatation of cerebral microvessels and triggers NO release from EC. **A:** ATP (10 μM) caused a large dilatation of the vessels ($p < 0.05$), while L-NMA alone (0.1 mM) caused a marked and significant increase in tone. The response to ATP was only partially blocked by L-NMA. L-Arg (2 mM) reversed the effects of L-NMA on basal tone and fully restored the response to ATP. The asterisk denotes differences from control (100% diameter), $p < 0.05$; n.s. refers to the nonsignificant difference between L-NMA and ATP+L-NMA diameter changes. **B1:** Nitric oxide measurements from BAEC. Application of 100 μM ATP to the cells (dark arrow) caused a transient increase in NO, and extracellular Ca^{2+} is not required to elicit this response. Cells were initially bathed in the absence of calcium (with EGTA present in the medium) and ATP (100 μM) was applied (filled arrows). Readmission of calcium to the medium did not significantly change the overall pattern of the measured NO response. **B2:** NO measurements from RBMEC. ATP (100 μM, arrow) causes a prolonged increase in NO concentrations, but calcium is required for the generation of the response to ATP in RBMEC. **B3:** NO responses to ATP (100 μM) in cocultured $BAEC_t$ ATP application (arrow) caused an increase in measurable NO; this effect was entirely prevented by removal of extracellular Ca^{2+}.

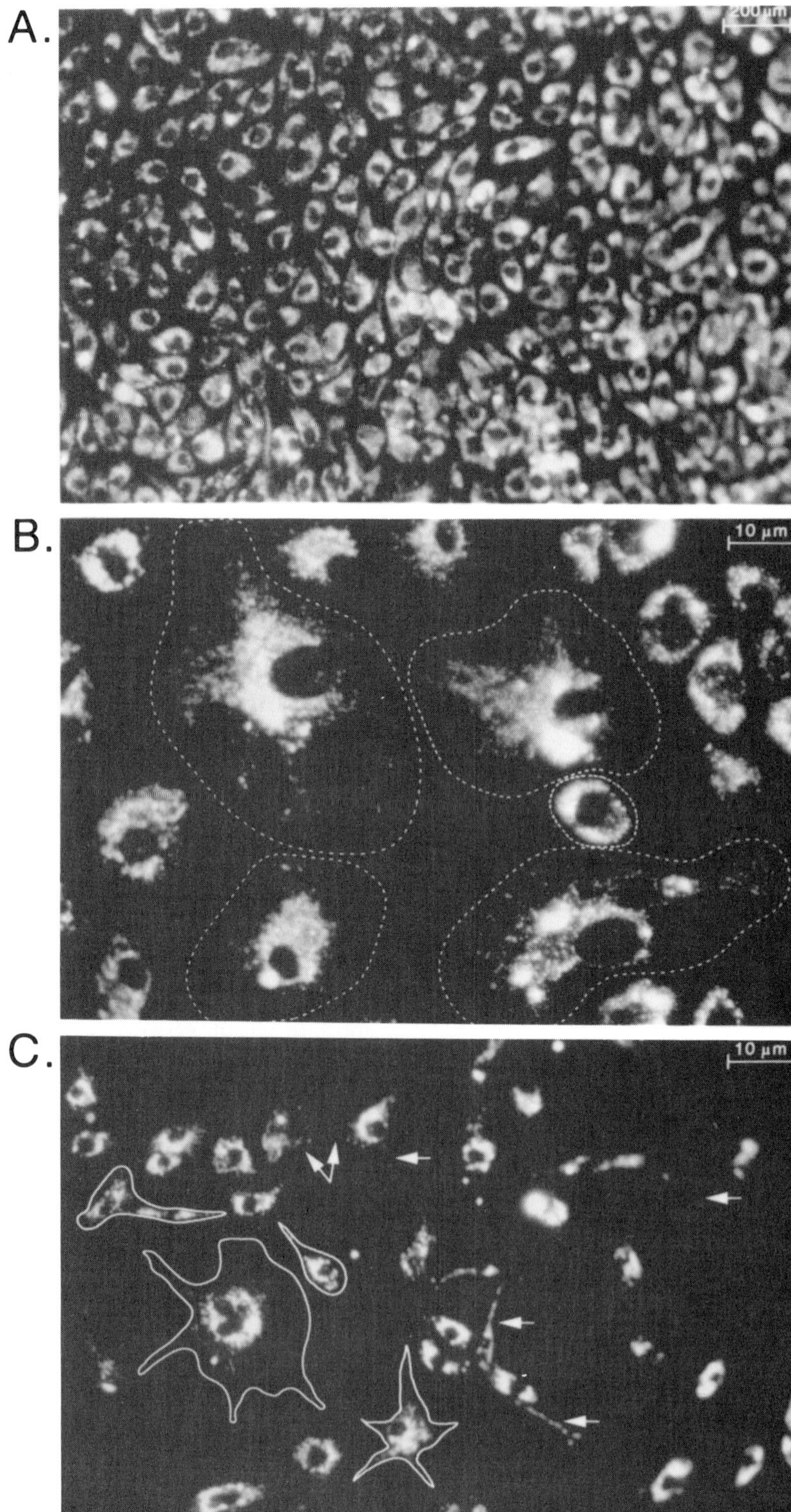
A.
200 µm
B.
10 µm
C.
10 µm

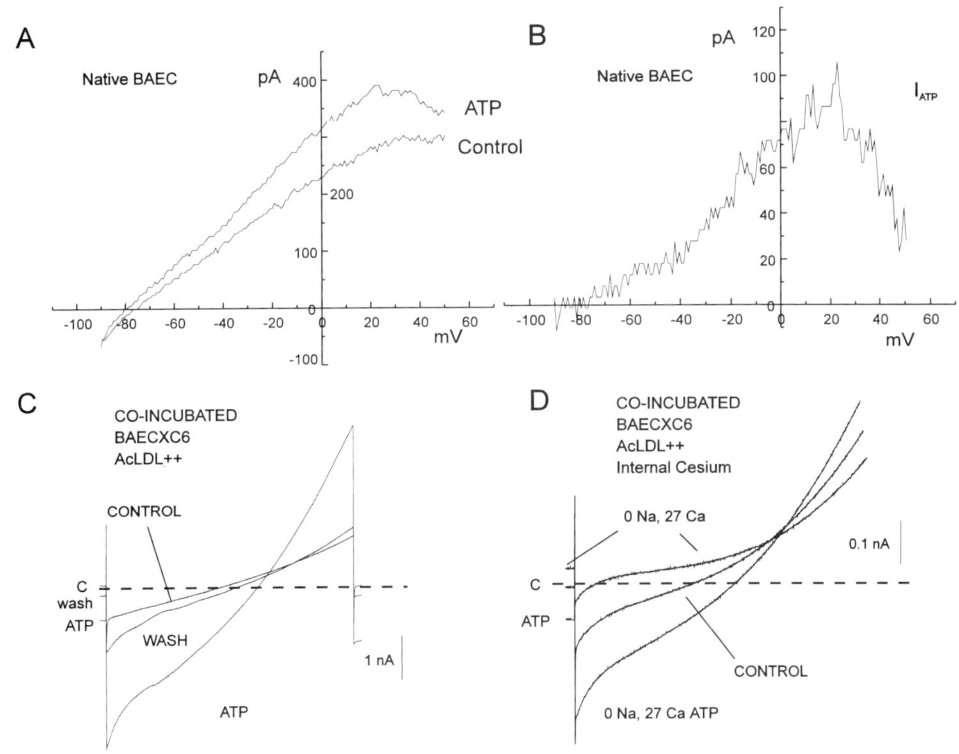

FIGURE 11-4. Voltage-clamp records from native BAEC and coincubated $BAEC_t$. The predominant response to 10 μM ATP in BAEC was the activation of an inwardly rectifying potassium current (**A**, **B**), while $BAEC_t$ were depolarized by ATP due to the activation of a cation conductance (**C**, **D**).

lected from the artificial capillary culture system and replated in the absence of glial cells on petri dishes for variable periods of time (1–5 days). These cells responded to ATP (10 μM) as native BAEC, that is, the prominent response to ATP was the activation of an *outward* current (Fig. 11-2). Taken together, these findings suggest that the electrophysiologic responses to ATP normally found in brain microvascular endothelial cells can be reversibly induced in aortic cells by coculturing with glia.

FIGURE 11-3. Morphologic properties of BAEC and $BAEC_t$. Cells were visualized in phase contrast, and cells shapes were outlined from photographs. The same cells were then observed and photographed again by visualizing direct Ac-LDL fluorescence. Note the typical growth pattern of native BAEC (**B**, **C**) and the fusiform, polygonal shape of $BAEC_t$.

Since coculturing of native BAEC with glial cells induced a significant change in the electrophysiologic response to extracellularly applied ATP, we have also measured the response of $BAEC_t$ to ATP in NO detection experiments. Bovine aortic endothelial cells cocultured with either C6 rat glioma cells or rat astrocytes responded to the application of 100 μM ATP with an increase in NO release (Fig. 11-1-B3). Removal of extracellular calcium and chelation of residual Ca^{2+} with EGTA caused abolishment of the NO increases following application of ATP. Thus, $BAEC_t$ also behaved as brain microvascular endothelial cells with regard to NO production.

Discussion

Induction of CNS properties in non-CNS cells has been investigated in a variety of cellular models aimed at describing the factors responsible for the unique properties of the cellular components of the mammalian brain. By the use of chimeras it has been demonstrated that blood cells of bone marrow origin can adopt the morphology of microglial cells [20]. Similarly, BBB phenotypes have been described in peripheral endothelial cells after exposure to astrocytes [21–23]. Our results show that other functional properties, such as the expression of an ATP-activated ion channel in BAEC, can be modulated by glial cells. In fact, while native BAEC responded to extracellular ATP with the activation of a hyperpolarizing potassium conductance, $BAEC_t$ behaved in a manner identical to native brain microvascular cells, that is, they were depolarized by ATP. As a consequence, ATP-induced release of NO from BAEC was virtually independent of extracellular calcium, while RBMEC and $BAEC_t$ required calcium influx from the extracellular space. Interestingly, the induction process could be reversed by removal of glial elements from the culture. Figure 11-5 illustrates our interpretation of the results obtained from coculture experiments.

There are several mechanisms that may underlie the observed effects on EC by glial cells: 1) BAEC may have a "dormant" potential for phenotypic expression that is revealed by coculturing with glia; 2) BAEC normally express both cation nonselective and potassium channels gated by ATP, but the latter are predominant; 3) astrocytes promote ex novo transcription of the gene responsible for brain I_{ATP}. Our results address these different hypotheses only in part, but the presence, albeit in a small percentage of cells (<10%), of a depolarizing response to ATP in BAEC makes the last hypothesis unlikely [24]. Since native RBMEC also expressed a small percentage of "aortic" responses to ATP, we believe that both channels (and their transcription material) coexist in either native RBMEC and BAEC. The influence of glial cells appears to be limited to a signal capable of favoring expression of a cation channel over a potassium conductance mechanism. As to the nature of this "glia-derived factor," our results suggest the close interactions of glial cells and EC is necessary to maintain the brainlike responsiveness to ATP found in $BAEC_t$. However, diffusible factors may be sufficient to establish the morphologic alterations induced by glia because these were also observed in hollow-fiber–grown $BAEC_t$ after two-dimensional culturing in the absence of either C6 or astrocytes.

We ruled out the possibility that different culturing conditions (exposure of the cells to continuous flow in the hollow fiber, different substrate for growth, etc.) were responsible for this transformation. In fact, BAEC cultured under hollow-fiber conditions, but in the absence of glia, did not develop brain endothelial characteristics (not shown). In addition, it appears unlikely that the observed differences were species specific because rat aortic cells responded to ATP in a fashion indistinguishable from BAEC.

Our results also demonstrate that culturing brain microvascular endothelial cells does not significantly alter their in vivo (or ex situ) properties related to BBB function and NO release [10,12,15]. Both isolated cannulated arterioles and cultured RBMEC responded to ATP by increased NO release. While we have directly demonstrated that Ca^{2+} influx was required for the in vitro response from cultured cells, our isolated vessel experiments did not directly determine whether the dilation following intraluminal application required calcium influx. We did, however, rule out involvement of the release of calcium from intracellular stores, because exposure of the vessel to thapsygargin did not affect vessel responses to ATP (data not shown).

Coculturing of BAEC with glia caused dramatic changes in EC morphology. Endothelial cells grown in hollow fibers and replated

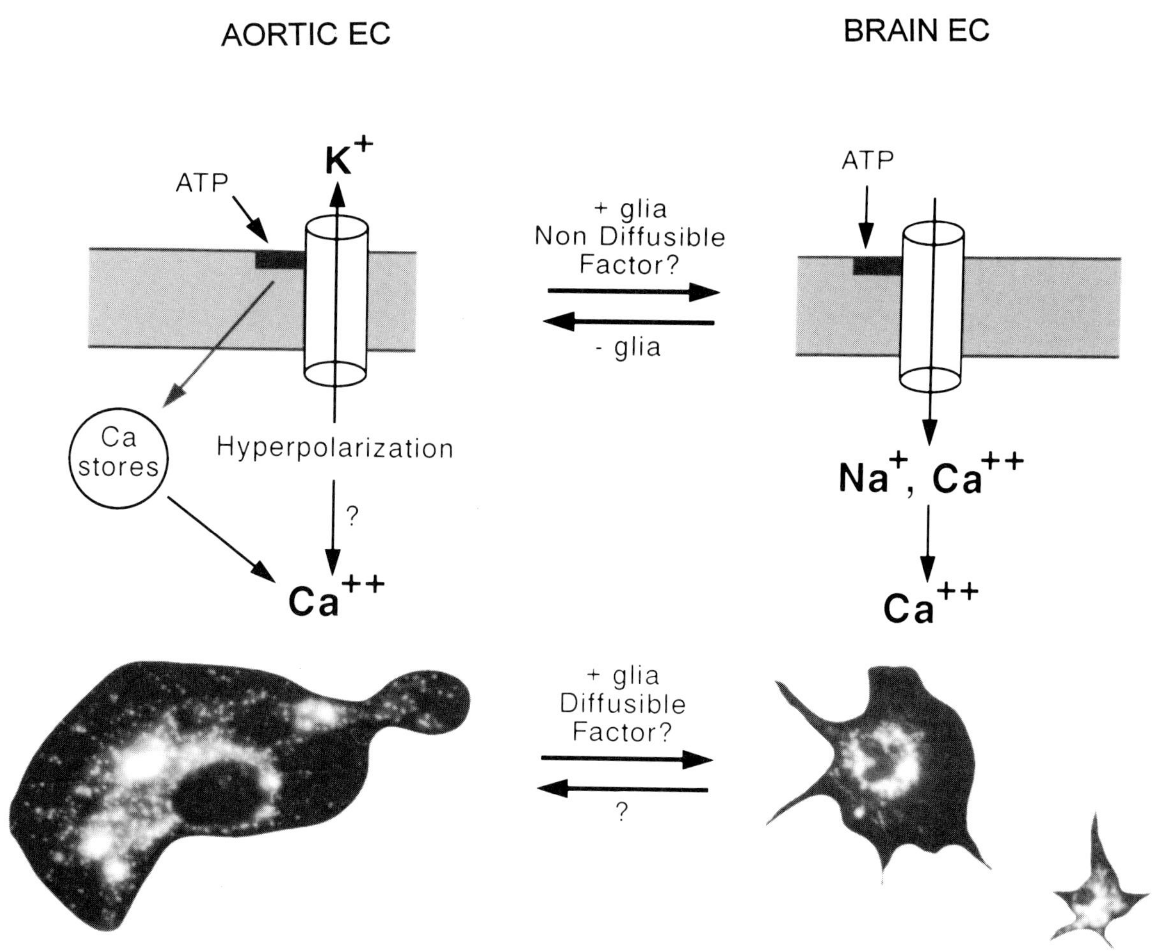

FIGURE 11-5. Summary and interpretation of the results. See text for details.

as two-dimensional cultures displayed abnormal morphologic properties, such as polymorphic nuclei, polygonal shape, and outgrowth of processes. While these modifications are not suggestive of a development of BAEC towards BMEC morphology per se, they suggest that the effects of glial cells were not limited to the induction of electrophysiologic properties in BAEC.

Induction of BBB properties in nonbrain EC has been shown by several investigators [21]. It has also been suggested that BBB properties are induced by cell-to-cell interactions and that diffusible substances released by astrocytes are not sufficient to cause BBB induction. Our experimental results confirm these findings. In fact, when BAEC grown in the presence of astrocytes (or C6 cell) were replated under glia-free conditions, the brainlike properties were not observed (data not shown). Since our results focused primarily on differences concerning NO release mechanisms, it may appear that the observed effects of glia on BAEC have little relationship to BBB properties. We have recently described an ionic current possibly involved in transendothelial transport of Na^+ and K^+ [25]. This current is modulated by NO, suggesting a possible autocrine effect of NO on

BBB function [25]. It is thus possible that the unique properties of RBMEC described here may relate to both regulation of brain vascular tone and BBB-regulated ionic homeostasis.

Acknowledgments

We would like to thank Marina Guizzetti for introducing us to the astrocyte culture technique and M.L. Joseph for continuous assistance in data analysis and preparation of the figures. This work was supported by National Institute of Health grants NS 51614, NS30305, NS21076, and NS07144.

References

1. Furchgott RF (1984) The role of endothelium in the responses of vascular smooth muscles to drugs. Rev Pharmacol Toxicol 24:175–197.
2. Moncada S, Palmer RMJ, Higgs EA (1991) Nitric oxide: Physiology, pathophysiology and pharmacology. Pharmacol Rev 43:109–142.
3. Prado R, Watson BD, Kuluz J, Dietrich WD (1992) Endothelium-derived nitric oxide synthase inhibition. Effects on cerebral blood flow, pial artery diameter, and vascular morphology in rats. Stroke 23:1118–1123.
4. Rubin LL, Hall DE, Porter S, Barbu K, Cannon C, Horner HC, Janatpour N, Liaw CW, Manning K, Morales J, Tanner LI, Tommaselli KJ, Bard F (1991) A cell culture model of the blood-brain barrier. J Cell Biol 115:1725–1735.
5. Smith QR, Nagura H, Takada Y, Duncan MW (1992) Facilitated transport of the neurotoxin, beta-N-methylamino-L-alanine, across the blood-brain barrier. J Neurochem 58:1330–1337.
6. Burnstock G, Kennedy C (1986) A dual function for adenosine 5′ trisphosphate in the regulation of vascular tone. Circ Res 58:319–330.
7. Ignarro LJ (1989) Biological actions and properties of endothelium-derived nitric oxide formed and released from artery and vein. Circ Res 65:1–21.
8. Janigro D, West GA, Gordon EL, Winn HR (1993) ATP-sensitive potassium channels in rat brain microvascular endothelial cells. Am J Physiol 265:C812–C821.
9. Adams DJ (1994) Ion channels in vascular endothelial cells. Trends Cardiovasc Med 4:18–26.
10. Tao-Cheng J-H, Nagy Z, Brightman MW (1987) Tight junctions of brain endothelium in vitro are enhanced by astroglia. J Neurosci 7:3293–3299.
11. Nakagawa T, Akaike N, Kimitsuki T, Komune S, Arima T (1990) ATP-induced current in isolated outer hair cells of guinea pig cochlea. J Neurophysiol 63:1068–1074.
12. Janzer RC, Raff MC (1987) Astrocytes induced blood-brain barrier properties in endothelial cells. Nature 325:253–257.
13. Janigro D, Nguyen T-S, Gordon EL, Winn HR (1994) ATP-activated cation channels in rat brain microvascular endothelial cells. Drug Dev Res 31:282.
14. Dacey RG, Duling BR (1982) A study of rat intracerebral arterioles: Methods, morphology and reactivity. Am J Physiol 242:H683–H687.
15. Gordon EL, Danielsson PE, Nguyen T, Winn HR (1991) A comparison of primary cultures of rat cerebral microvascular cells to aortic endothelial cells. In Vitro Cell Dev Biol 27A:312–326.
16. McCarthy KD, DeVellis J (1980) Preparation of separate astroglial and oligodendroglial cell cultures from rat cerebral tissues. J Cell Biol 85:890–899.
17. Tsukhahara H, Gordienko DV, Goligorsky MS (1993) Continuous monitoring of nitric oxide release from human umbilical vein endothelial cells. Biochem Biophys Res Comm 193:722–729.
18. Stanness KA, Winn HR, Janigro D (1994) Electrophysiological and functional properties of in vitro blood-brain barrier cells. Society for Neuroscience Annual Meeting, November 1994.
19. Janigro D, Joseph M (1994) Cost effective alternatives to high priced software for the small research laboratory, Scient Comput Automat 10:13–16.
20. Hickey WF, Kimura H (1988) Perivascular microglia are bone marrow derived and present antigen in vivo. Science 239:290–292.
21. Cancilla PA, Bready J. Berliner J (1993) Brain endothelial-astrocyte interactions. In Pardridge WM (ed.), The Blood-Brain Barrier, Cellular and Molecular Biology. New York: Raven Press pp. 25–46.
22. Janzer RC, Raff MC (1987) Astrocytes induced blood-brain barrier properties in endothelial cells. Nature 325:253–257.
23. Beck DW, Roberts RL, Olson JJ (1986) Glial cells influence membrane-associated enzyme

activity at the blood-brain barrier. Brain Res 381:131–137.
24. Sauve R, Parent L, Simoneau C, Roy G (1988) External ATP triggers a biphasic activation process of a Ca^{2+}-dependent K^{+} channel in cultured bovine aortic endothelial cells. Pflügers Arch 412:469–481.
25. Janigro D, West GA, Winn HR (1994) Regulation of blood-brain barrier endothelial cells by nitric oxide. Circ Res 75:528–538.

12. MODULATION OF CARDIOMYOCYTE MEMBRANE CURRENTS BY A_1 ADENOSINE RECEPTORS

Yejia Song
John Shryock
Luiz Belardinelli

This review summarizes the direct and indirect effects of adenosine on cardiomyocyte membrane ion currents and action potential duration. Modulation of specific membrane ion currents (Fig. 12-1) underlies the anti-beta-adrenergic and the negative chronotropic, dromotropic, and inotropic actions of adenosine on cardiac myocytes. These actions of adenosine are mediated by A_1 receptors and are blocked by selective A_1 receptor antagonists [1–3]. A_1 receptors in the heart are coupled to effectors by guanine nucleotide-binding proteins (G proteins) that are subject to ADP ribosylation catalyzed by pertussis toxin (PTX) [4,5].

The effects of adenosine on individual ion currents appear to be mediated by one of two mechanisms: (1) a direct adenosine receptor–G protein–ion channel interaction, or (2) an indirect adenosine receptor–G protein–adenylate cyclase inhibition-channel protein dephosphorylation pathway. By the direct pathway, adenosine can modulate "basal" ion currents. By the indirect (anti-adrenergic) pathway, adenosine can alter ion currents only when the cellular cAMP concentration has been elevated above basal, as by catecholamines or forskolin.

Direct Effects of Adenosine

A direct linkage by PTX-sensitive G proteins between A_1 adenosine receptors and K_{Ado} channels in atrial myocytes was demonstrated by Kurachi et al. [4]. In their single-channel recording experiments, I_{KAdo} was activated in cell-attached patch membranes by adenosine present in the pipette solution. The K_{Ado}-channel activity disappeared after formation of an inside-out patch, and it was quickly reactivated by application of GTP to the intracellular side of the membrane and was inhibited by addition of PTX in the presence of GTP.

Adenosine activates the inwardly rectifying K^+ current, I_{KAdo}, in atrial myocytes [6], sinoatrial (SA) nodal cells [7], and atrioventricular (AV) nodal cells [1]. I_{KAdo} is the major current responsible for adenosine-induced shortening of action potential duration [8], hyperpolarization of resting membrane potential

[6,9], and reduction of contractile force of atrial myocytes [10]. Likewise, activation of I_{KAdo} plays a major mechanistic role in the negative chronotropic [7] and dromotropic [1] effects of adenosine. Therefore, activation of I_{KAdo} is the most important action of adenosine on supraventricular tissues. Although adenosine-activated I_{KAdo} is pronounced in atrial myocytes and in SA and AV nodal cells, it is small or undetectable in ventricular myocytes of most species.

Both whole-cell current and single-channel properties of I_{KAdo} are similar to those of the potassium current activated by the muscarinic agonist acetylcholine (i.e., I_{KACh}). The maximum effects of adenosine and acetylcholine in activating these K^+ channels are not additive. That is, when the potassium current had been maximally activated by the muscarinic agonist carbachol, addition of adenosine and/or of the A_1 adenosine receptor agonist cyclopentyladenosine did not cause a further increase in the carbachol-activated current [Thedford and Belardinelli, unpublished]. The actions of both adenosine and acetylcholine to activate K^+ channels are blocked by pertussis toxin [4]. Single K^+ channels activated by adenosine and by acetylcholine have the same unitary slope conductance of about 45 pS (when $[K^+]_i$ = $[K^+]_o$) in human [11] and guinea-pig atrial myocytes [4]. These findings suggest that adenosine and acetylcholine activate the same population of K^+ channels.

Basal (i.e., unstimulated) L-type Ca^{2+} current ($I_{Ca,L}$) can be reduced by adenosine, apparently by a direct pathway. In guinea-pig atrial myocytes, adenosine reduced basal $I_{Ca,L}$ by about 12–35% [8,12]. However, compared with the activation of I_{KAdo}, this inhibitory effect of adenosine on $I_{Ca,L}$ is minor and appears to contribute little to adenosine-induced shortening of action potential duration and reduction of contractile shortening of atrial myocytes [8,12]. In ventricular myocytes of most mammalian species, adenosine has little or no effect on basal $I_{Ca,L}$. A known exception is the ferret ventricular myocyte, where adenosine has recently been reported to inhibit basal $I_{Ca,L}$ by as much as 35% [13].

	Supraventricular	Ventricular
I_{KAdo}	↑	↑ or –
$I_{Ca,L}$	↓	↓ (ISO)
I_{Ti}		↓ (ISO)
I_f	↓	
I_{Cl}		↓ (ISO)
I_K		↓ (ISO)

I_{KAdo}, adenosine-activated K^+ current. $I_{Ca,L}$, L-type Ca^{2+} current. I_{Ti}, transient inward current. I_f, pacemaker current. I_{Cl}, time-independent Cl^- current. I_K, delayed rectifier K^+ current. ISO = isoproterenol-stimulated.

FIGURE 12-1. Effects of adenosine on cardiomyocyte ion currents.

It was recently demonstrated that adenosine inhibited the pacemaker current, I_f, in rabbit isolated AV nodal cells [2]. Because adenosine inhibits this current in the absence of catecholamines, a direct coupling of A_1 receptor to inhibitory G protein to I_f is assumed. The inhibition of I_f and the activation of I_{KAdo} by adenosine in pacemaker cells both contribute to a depression of phase 4 depolarization and pacemaking activity of these cells.

A direct effect of adenosine in activating the ATP-sensitive K^+ current, I_{KATP}, has been postulated [14]. Kirsch et al. [14] reported that application of adenosine facilitated the activation of tolbutamide-sensitive K^+ channels in neonatal rat ventricular myocytes when intracellular ATP level was reduced [14]. However, we failed to observe activation by adenosine or by the A_1 agonist cyclopentyladenosine of I_{KATP} in guinea-pig atrial and ventricular myocytes [15]. Consistent with our observations, a recent report showed that endogenous adenosine released during hypoxia did not activate I_{KATP} in guinea-pig hearts [16]. The reason for this apparent discrepancy between our results and those of Kirsch et al. [14] remains to be determined. The effects of adenosine on the Na^+ current and T-type Ca^+ current were studied by Tytgat et al. [17], who found that neither of these currents was significantly affected by adenosine.

Indirect Effects of Adenosine

In cardiac myocytes, adenylate cyclase activity is stimulated by catecholamines, forskolin, histamine, and other agonists. The increased production of cAMP, in turn, activates a variety of membrane currents (e.g., I_{Ca}, I_K, I_{Cl}) via phosphorylation of channel proteins. Adenosine "indirectly" attenuates catecholamine-stimulated currents by inhibiting adenylate cyclase activity. This action of adenosine has been observed in both supraventricular and ventricular tissues [3,7,18,19]. The indirect action of adenosine to inhibit catecholamine-stimulated ion currents can explain most of the effects of this nucleoside on the ventricular myocardium and thus is the most important action of adenosine on ventricular myocytes.

Although in ventricular myocytes of most mammalian species adenosine has little or no effect on basal Ca^{2+} current, it effectively attenuates isoproterenol-stimulated L-type Ca^{2+} current without apparently altering the current kinetics [19]. The time course of the action of adenosine to attenuate isoproterenol-stimulated I_{Ca} in the whole cell is slower than the time course for activation of I_{KAdo}. Application of adenosine in the presence of isoproterenol caused a rapid activation of potassium current (I_{KAdo}) and a slower onset of inhibition of isoproterenol-stimulated I_{Ca} [Wu and Belardinelli, unpublished], consistent with the idea that inhibition by adenosine of isoproterenol-stimulated I_{Ca} depends on the relatively slow reduction of the level of a second messenger, cAMP. Consistent with the effect of adenosine on isoproterenol-stimulated whole-cell Ca^{2+} current are the results obtained from studies of single Ca^{2+} channels. Using the single-channel recording technique to study guinea-pig ventricular myocytes, Kato et al. [20] found that Ca^{2+}-channel availability and mean Ca^{2+} current were increased by isoproterenol and decreased by the addition of adenosine in the presence of isoproterenol. The Ca^{2+}-channel open probability was not significantly altered by either isoproterenol or adenosine. Thus, it was concluded that at the single-channel level adenosine inhibits isoproterenol-stimulated Ca^{2+} current in ventricular myocytes, mainly by reducing the availability of Ca^{2+} channels.

Stimulation of L-type Ca^{2+} channels by isoproterenol increases Ca^{2+} influx. Because intracellular Ca^{2+} overload causes activation of a transient inward current, I_{Ti}, inhibition by adenosine of isoproterenol-induced Ca^{2+} overload results in an attenuation of I_{Ti} induced by isoproterenol [3]. Attenuation by adenosine of isoproterenol-induced I_{Ti}, afterdepolarizations and increases in cellular cAMP and contraction were abolished in PTX-pretreated ventricular myocytes [3,21].

In addition to the L-type Ca^{2+} current, isoproterenol activates the delayed rectifier K^+ current (I_K) in guinea-pig ventricular myocytes and activates both a transient outward Cl^- current (I_{TO}) and a time-independent Cl^- current (I_{Cl}) in rabbit ventricular myocytes. The mechanism(s) underlying the effects of isoproterenol on these currents may not be the same. Activation of I_K [22] and I_{Cl} [23] have been shown to be cAMP dependent, whereas activation of I_{TO} is Ca^{2+} dependent [24]. Recently, the effects of adenosine on isoproterenol-stimulated I_K in guinea-pig ventricular myocytes were observed in our laboratory [25]. In these experiments, Na^+ and Ca^{2+} currents were minimized, respectively, by holding the cell membrane potential at -40 mV and by bath application of the Ca^{2+}-channel blocker nitrendipine. Application of 1-second long depolarizing pulses to the positive membrane potentials elicited a time-dependent outward current, I_K, and an outward tail current that were not affected by adenosine alone. The amplitudes of I_K and tail currents, however, were increased by isoproterenol (10 nM) and decreased by adenosine (100 μM) in the presence of isoproterenol.

The effects of adenosine on Cl^- currents of rabbit ventricular myocytes were tested in our laboratory. To minimize K^+ currents, experiments were performed at 22°C, with short voltage-clamp pulses, and by using Cs^+ to replace K^+. A transient outward current, I_{TO}, and a sustained outward current, I_{Cl}, were elicited by applying 100-ms depolarizing pulses from a holding potential of −40 mV. Under basal conditions, I_{Cl} was very small. In the presence of isoproterenol (100 nM), the amplitudes of both I_{Cl} and I_{TO} were markedly increased. Addition of adenosine (100 μM) in the presence of isoproterenol attenuated I_{Cl} but did not significantly affect I_{TO}.

The inhibition by adenosine of isoproterenol-stimulated cAMP-dependent I_K and I_{Cl}, and the lack of effect of adenosine on Ca^{2+}-dependent I_{TO}, support the idea that adenosine specifically inhibits those currents that are activated via a cAMP-mediated phosphorylation secondary to a stimulation of adenylate cyclase activity in ventricular myocytes. This specificity of the action of adenosine could be utilized to distinguish currents activated upon stimulation of adenylate cyclase activity from those activated by other mechanisms, such as calcium ions. Similar to the inhibition of isoproterenol-stimulated ion currents, adenosine also attenuates currents that are activated by agents known to stimulate adenylate cyclase activity, such as forskolin-stimulated Ca^{2+} current [26] and histamine-stimulated I_{Cl} [27].

Modulation of Action Potential Duration

Adenosine shortens the atrial action potential by directly activating I_{KAdo} [8]. On the other hand, adenosine has little or no effect on duration of the ventricular action potential under basal conditions [28]. The effects of adenosine on the duration of the ventricular action potential in the presence of isoproterenol, however, are complex. Isoproterenol alone may increase or decrease the action potential duration, depending on which current, I_{Ca}, I_K, or I_{Cl}, is predominantly affected. Because adenosine attenuates isoproterenol-stimulated I_{Ca}, I_K, and I_{Cl}, it may also either prolong or shorten the ventricular action potential in the presence of isoproterenol. This assumption was confirmed by results of experiments performed on guinea-pig ventricular myocytes [25]. The predominant effect of 10 nM isoproterenol in these experiments was to stimulate inward I_{Ca} and thereby to increase the action potential duration. The inhibition of the effects of isoproterenol by adenosine resulted in a decrease of the action potential duration. However, in the presence of the Ca^{2+}-channel blocker nitrendipine, which partially blocks I_{Ca}, 10 nM isoproterenol caused a shortening, instead of a prolongation, of the action potential, presumably because the major effect of isoproterenol under such conditions was to stimulate outward I_K. In this case, addition of adenosine (100 μM) in the presence of isoproterenol resulted in an increase of the action potential duration. Thus, in the presence of isoproterenol modulation by adenosine of the inward and outward ion currents may cause various effects on duration of the ventricular action potential. Because the cardiac refractory period and intracellular Ca^{2+} loading are largely determined by the duration of the action potential, adenosine may therefore have either an anti- or a pro-arrhythmic effect on cardiac electrophysiologic activity.

Conclusions

The effects of adenosine on membrane currents of cardiac myocytes, as shown in Figure 12-1, include 1) activation of I_{KAdo}, inhibition of I_f, and a small reduction of basal I_{Ca} in supraventricular tissues; and 2) attenuation of I_{Ca}, I_{Ti}, I_{Cl}, I_K, and I_f stimulated by isoproterenol, or by other agents known to stimulate adenylate cyclase activity in both ventricular and supraventricular tissues. The effects of adenosine on membrane currents are mediated by the A_1 adenosine receptor coupled to PTX-sensitive G proteins. Recent studies indicate that the effects of adenosine on ventricular action potentials are far more complicated than previously recognized. The net effect of adenosine on the duration of the ventricular action potential depends on the predominant current being affected by isoproterenol. Because the duration of the action potential is a major determinant of ventricular refractoriness and intracellular calcium loading, modulation by A_1 adenosine receptors of membrane currents, and therefore of action potential duration, provides potential mechanisms to explain pro-arrhythmic, antiarrhythmic, and cardioprotective actions of adenosine.

References

1. Martynyuk A, Kane KA, Cobbe SM, Rankin AC (1994) The effects of adenosine on potassium and calcium conductances in isolated rabbit atrioventricular nodal myocytes (abstr). Drug Dev Res 31:295.
2. Wang D, Belardinelli L (1994) Effects of adenosine on phase 4 depolarization and pacemaker current (I_f) in single rabbit atrioventricular nodal myocytes (abstr). FASEB J 8:A611.
3. Song Y, Thedford S, Lerman BB, Belardinelli L (1992) Adenosine-sensitive afterdepolarizations and triggered activity in guinea pig ventricular myocytes. Circ Res 70:743–753.
4. Kurachi Y, Nakajima T, Sugimoto T (1986) On the mechanisms of activation of muscarinic K^+ channels by adenosine in isolated atrial cells: Involvement of GTP-binding proteins. Pflügers Arch 407:264–274.
5. Bohm M, Schmitz W, Scholz H, Wilken A (1989) Pertussis toxin prevents adenosine receptor- and m-cholinoceptor-mediated sinus rate slowing and AV conduction block in the guinea-pig heart. Naunyn Schmiedebergs Arch Pharmacol 339:152–158.
6. Belardinelli L, Isenberg G (1983) Isolated atrial myocytes: Adenosine and acetylcholine increase potassium conductance. Am J Physiol 244: H734–H737.
7. Belardinelli L, Giles WR, West A (1988) Ionic mechanisms of adenosine actions in pacemaker cells from rabbit heart. J Physiol 405:615–633.
8. Visentin S, Wu SN, Belardinelli L (1990) Adenosine-induced changes in atrial action potential: Contribution of Ca and K current. Am J Physiol 258:H1070–H1078.
9. Jochem G, Nawrath H (1983) Adenosine activates a potassium conductance in guinea-pig atrial heart muscle. Experientia 39:1347–1349.
10. Wang D, Thedford S, Belardinelli L (1990) Mechanism of the negative inotropic effect of adenosine in guinea pig atrial myocytes (abstr). FASEB J 7:A134.
11. Heidbüchel H, Vereecke J, Carmeliet E (1990) Three different potassium channels in human atrium: Contribution to the basal potassium conductance. Circ Res 66:1277–1286.
12. Cerbai E, Klöckner U, Isenberg G (1988) Ca-antagonistic effects of adenosine in guinea pig atrial cells. Am J Physiol 255:H872–H878.
13. Qu Y, Campbell DL, Whorton AR, Strauss HC (1993) Modulation of basal L-type Ca^{2+} current by adenosine in ferret isolated right ventricular myocytes. J Physiol 471:269–293.
14. Kirsch GE, Codina J, Birnbaumer L, Brown AM (1990) Coupling of ATP-sensitive K^+ channels to A_1 receptors by G proteins in rat ventricular myocytes. Am J Physiol 259: H820–H826.
15. Song Y, Belardinelli L (1994) Evidence against the existence of adenosine-induced activation of ATP-sensitive K^+ channels in guinea pig cardiac myocytes (abstr). J Am Coll Cardiol February:467A.
16. Xu J, Wang L, Hurt CM, Pelleg A (1994) Endogenous adenosine does not activate ATP-sensitive potassium channels in the hypoxic guinea pig ventricle in vivo. Circulation 89: 1209–1216.
17. Tytgat J, Vereecke J, Carmeliet E (1990) A combined study of sodium current and T-type calcium current in isolated cardiac cells. Pflügers Arch 417:142–148.
18. Wu SN, Yu SH (1993) Characterization of functional antagonism of adenosine and isoproterenol in guinea pig atrial myocytes. Jpn J Physiol 43:177–189.
19. Isenberg G, Belardinelli L (1984) Ionic basis for the antagonism between adenosine and isoproterenol on isolated mammalian ventricular myocytes. Circ Res 55:309–325.
20. Kato M, Yamaguchi H, Ochi R (1990) Mechanism of adenosine-induced inhibition of calcium current in guinea pig ventricular cells. Circ Res 67:1134–1141.

21. Brown LA, Humphrey SM, Harding SE (1990) The anti-adrenergic effect of adenosine and its blockade by pertussis toxin: A comparative study in myocytes isolated from guinea-pig, rat and failing human hearts. Br J Pharmacol 101: 484–488.
22. Tsien RW, Giles W, Greengard P (1972) Cyclic AMP mediates the effects of adrenaline on cardiac Purkinje fibers. Nature (New Biol) 240: 181–183.
23. Harvey RD, Hume JR (1989) Autonomic regulation of a chloride current in heart. Science 244:983–985.
24. Zygmunt AC, Gibbons WR (1991) Calcium-activated chloride current in rabbit ventricular myocytes. Circ Res 68:424–437.
25. Song Y, Srinivas M, Belardinelli L (1994) Effects of adenosine on isoproterenol-induced outward currents in cardiac ventricular myocytes (abstr). Drug Dev Res 31:324.
26. West GA, Isenberg G, Belardinelli L (1986) Antagonism of forskolin effects by adenosine in isolated hearts and ventricular myocytes. Am J Physiol 250:H769–H777.
27. Harvey RD, Hume JR (1990) Histamine activates the chloride current in cardiac ventricular myocytes. J Cardiovasc Electrophysiol 1:309–317.
28. Belardinelli L, Isenberg G (1983) Actions of adenosine and isoproterenol on isolated mammalian ventricular myocytes. Circ Res 53: 287–297.

13. REGULATION OF PHOSPHOLIPASES C AND D, CALCIUM, AND PROTEIN KINASE C BY ADENOSINE A_1 RECEPTORS

Bertil B. Fredholm
Pär Gerwins
Jean W. Assender
Eva Irenius

The adenosine A_1 receptor was first recognized on the basis of its being negatively coupled to adenylyl cyclase [1,2]. Later it was conclusively shown that the A_1 receptor is coupled to potassium channels [3] and to several types of calcium channels [4]. In this chapter, evidence that A_1 receptors, at least in some cell types, can stimulate phospholipases C and D, raise intracellular Ca^{2+} levels, and activate protein kinase C is summarized.

Materials and Methods

Cultured human renal artery smooth muscle cells (SMC) were a gift from Dr. J. Nilsson at the Karolinska Hospital, Stockholm. Human saphenous vein was obtained from the Department of Cardiology, Cardiff. Rat vas deferens were freshly dissected out. DDT_1 MF-2 smooth muscle cells, originating from a steroid-induced leiomyosarcoma of Syrian hamster vas deferens, were grown in suspension and maintained in Dulbecco's modified Eagle's medium with 4.5 g/l of glucose, containing 5% fetal calf serum (FCS), 100 U/ml penicillin, 100 µg/ml streptomycin, and 2 mM L-glutamine at 37°C in 5% CO_2/95% air. The cells were subcultured three times weekly and used at a density of approximately 10^5 cells/ml. SMC were cultured under similar conditions except that FCS was 10% cells were passaged using trypsin (0.05%)-EDTA (0.02% w/v in normal saline) as required, and batches were frozen in liquid nitrogen. Two timepoints for analysis were selected, namely, after passage 6–8 or after passage 27–30, to differentiate between cell populations containing some "reversibly" synthetic cells and SMC selected by multiple passage for their synthetic ability. All cell populations were confirmed to be pure cultures of SMC by positive staining with a monoclonal antibody to α-SMC actin (Clone 1A4, Sigma).

RECEPTOR BINDING EXPERIMENTS

Binding experiments were performed both on membranes, prepared by sonication, and on intact cells, using tritiated 1,3-dipropyl-8-cyclopentyl xanthine (DPCPX) as the labeled ligand [5].

DETERMINATION OF $INS(1,4,5)P_3$

We used a modification of a sensitive and specific protein-binding method in which $[^3H]$-$Ins(1,4,5)P_3$ competes with $Ins(1,4,5)P_3$ in the samples, or in standards, for binding to a binding protein prepared from bovine adrenals [6]. The assay is carried out in 96-well microtiter plates and bound $[^3H]$-$Ins(1,4,5)P_3$ is separated from free using a cell harvester for filtration.

DETERMINATION OF INTRACELLULAR LEVELS OF FREE Ca^{2+}

Cells were labeled with 5 μM Fura 2-AM for 40 minutes in Hank's balanced salt solution at 37°C, washed, and intracellular levels of Ca^{2+} were determined at 30°C in a dual-wavelength Sigma ZFP22 fluorometer by using the ratio of the fluorescence intensity obtained with an excitation wavelength of 334 nm to that obtained with an excitation wavelength of 366 nm with an emission cutoff at 500 nm. Free Ca^{2+} concentration was calculated as described in Grynkiewicz et al. [7]. In addition, measurements at the single cell level were carried out using a Zeiss system consisting of an inverted microscope equipped with a sensitive photomultiplier.

PHOSPHOLIPASE D ASSAY

Determination of phospholipase D (PLD) activity was done by measuring the formation of phosphatidylethanol, a specific reaction catalyzed by PLD in the presence of ethanol. DDT_1 MF-2 cells were labeled with $[^3H]$-palmitic acid (5 μCi/ml) for 24 hours, activated in the presence of 0.5% ethanol, and the amounts of phosphatidylethanol formed were analyzed by thin layer chromatography and liquid scintillation spectrometry.

DETERMINATION OF PROTEIN KINASE C ACTIVITY

Protein kinase C (PKC) activity was measured using a permeabilized cell assay [8] and a synthetic peptide substrate based on the amino acid sequence 4–14 of myelin basic protein (MBP_{4-14}) [9]. After activation, cells were pelleted and the supernatant was replaced by a phosphorylation solution that among other things contained MBP_{4-14} peptide, $[\gamma\text{-}^{32}P]$-ATP, and digitonin (to let the peptide enter the cells). After centrifugation, supernatants were spotted on Whatman P-81 phosphocellulose filters that were washed in phosphoric acid and counted in a liquid scintillation spectrometer. The addition of the lysing phosphorylation solution completely prevented subsequent receptor activation.

DIACYLGLYCEROL DETERMINATION

Cellular content of diacylglycerol was determined by measuring the formation of $[^{32}P]$-phosphatidic acid from diacylglycerol, catalyzed by *sn*-1,2-diacylglycerol kinase from *Escherichia coli* in the presence of $[\gamma\text{-}^{32}P]$-ATP. Lipids were separated by thin layer chromatography, and areas containing $[^{32}P]$-phosphatidic acid were identified by autoradiography. These areas were quantified by liquid scintillation spectrometry.

PROTEIN ANALYSIS BY WESTERN BLOTTING

Cellular protein was extracted by washing the cells with phosphate buffered saline (PBS) and lysing them in Triton buffer. The supernatant after centifugation was acetone precipitated, and the resulting protein pellet was resuspended in Laemmli sample buffer, boiled for 5 minutes, and then run on an 8% polyacrylamide denaturing gel. Proteins were blotted onto nitrocellulose membranes, and nonspecific binding was blocked by an overnight incubation in Tris-buffered saline (TBS) containing Tween-20 (0.5%) and 5% FCS. Membranes were incubated with PKC specific antibodies for 3 hours, before washing 3 × 10 minutes with TBS-Tween. Antibody binding was detected with a second (horseradish peroxidase conjugated anti-rabbit) antibody and was detected with an enhanced chemiluminescence system (Amersham, GB). Antibodies to PKC α, β_c (common to β_c and β_{II}), β_I and β_{II} (the two splice variants), γ and ε were kindly provided by Prof. Nishizuka, Kobe University, Japan; antibodies to PKC δ, ε, and ζ were purchased from Gibco (BRL, Grand Island, NY). All antibodies were raised against the CK_pV_5 region, except those to γ (CK_pV_3) and to the common region of the β isoforms (β_c, CK_pV_1).

MOLECULAR ANALYSIS

Cytoplasmic RNA was prepared using a "mini-prep" method and mRNA was selected using an oligo d(T) column. cDNA was produced from approximately 0.06 µg mRNA, using Moloney murine leukemia virus reverse transcriptase (Gibco-BRL) 10 units, 40°C for 50 minutes, in a buffer containing 6.9 mM $MgCl_2$, 50 mM KCl, 10 mM Tris-HCl pH 9.0, 0.1% Triton X-100, 4 mM dNTP, 1 unit rRNasin ribonuclease inhibitor, and 19 µM $pd(N)_6$ random hexamers, in a final volume of 20 µl. This cDNA was then analyzed by thermocycle amplification of sequences specific for the various isoforms of PKC. The sequences amplified were between base pairs 875–1199 for α, 925–1188 for β, 713–1257 for γ, 1013–1364 for δ, 577–1308 for ε, and 404–1085 for ζ. Products were run though a 2% agarose gel containing 3% (v/v) ethidium bromide and detected by trans-illumination with UV light.

Results and Discussion

DDT_1 MF-2 cells possess adenosine A_1 receptors with typical pharmacology. The cells possess high-affinity (K_D −0.5 ± 0.16 nM) binding sites for DPCPX, which are quite abundant (approximately 110,000 binding sites/cell). The potency of adenosine agonists and antagonists to displace the bound [^{3}H]-DPCPX also corresponds closely to that expected for classical A_1 receptors [5]. Using Western blot and pertussis toxin two putative G-protein α-subunits, $G_{i2\alpha}$, and $G_{i3\alpha}$, were identified as possible mediators [10]. There was a marked inhibition of forskolin- or isoproterenol-stimulated cAMP accumulation by adenosine analogs, which was blocked by pertussis toxin.

EFFECT OF ADENOSINE ANALOGS ALONE AND IN COMBINATION WITH BRADYKININ OR ATP

As expected DDT_1 MF-2 cells responded with a transient increase of Ins(1,4,5)P_3 to ATP and several analogs. ATP and UTP were approximately equipotent (EC_{50} approx. 20 µM). ADP was about three times less potent but possibly more efficacious. AMP, 2-methylthio ATP, and AMP-CPP were virtually inactive. Thus, these cells appear to possess receptors that could be classified as P_{2u} receptors [2,11]. The peak increase in Ins(1,4,5)P_3 due to ATP or UTP was approximately threefold over basal.

As seen in Figure 13-1, adenosine, N^6-cyclopentyladenosine (CPA) and bradykinin also increased Ins(1,4,5)P_3 accumulation. The magnitude of the response was less than that induced by ATP or UTP. The response to bradykinin had an EC_{50} value of 150 ± 60 nM, but no attempt was made to characterize the receptor involved. CPA was more potent than adenosine (EC_{50} 9.4 ± 0.4 versus 246 ± 8 nM), and the responses to both drugs were blocked by DPCPX with a K_i of 4.3 nM. All these results are indicative of typical A_1 receptors.

Pertussis toxin in a dose that completely ADP ribosylated the G_i proteins had no effect at all on the response to bradykinin; it reduced, but did not eliminate, the response to ATP and UTP, but completely eliminated the effect of the adenosine receptor agonists. This suggests that adenosine acts via one of the G_i proteins. It has

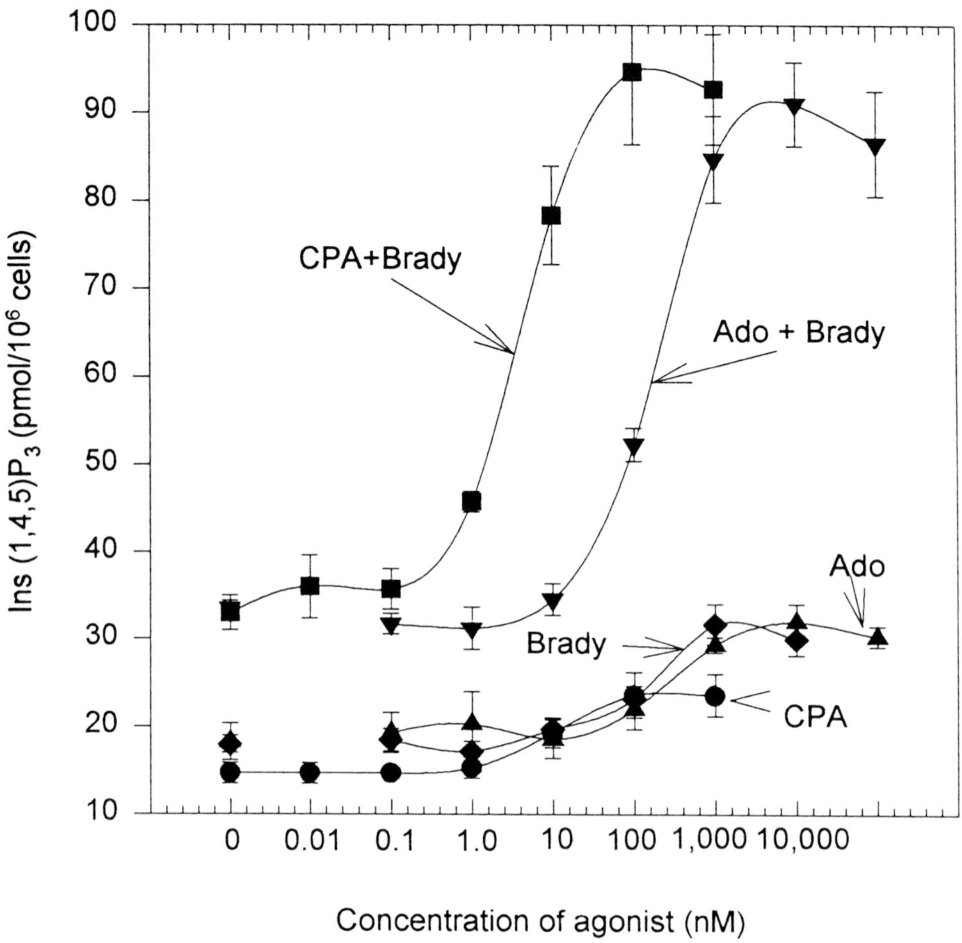

FIGURE 13-1. The effect of adenosine and CPA alone and in combination on the accumulation of Ins(1,4,5)P_3 in DDT_1 MF-2 cells.

been shown that the α-subunits are unable to stimulate phopspholipase C, but that the β,γ subunits can [12]. Bradykinin and ATP exclusively, or mainly, act via one of the members of the G_q/G_{11} family.

As seen in Figure 13-1 adenosine receptor agonists acted synergistically with bradykinin. The combined treatment was much more active than either agent alone. The synergy was also manifested by a small left-wards shift of the dose-response curve to the adenosine analogs (Fig. 13-1) or bradykinin [13]. Exactly analogous findings were obtained when ATP and adenosine were combined [11]. In this case it is important to note that ATP can be rapidly broken down to adenosine. Thus, the initial signal and its breakdown product can act synergistically, possibly leading to a prolongation of the response of the target cell. The ability of adenosine to enhance ATP response also completely explained the ability of adenosine receptor antagonists to reduce the responses to ATP. It seems likely that this type of interaction can explain part of the sometimes confusing literature regarding xanthine inhibition of ATP responses.

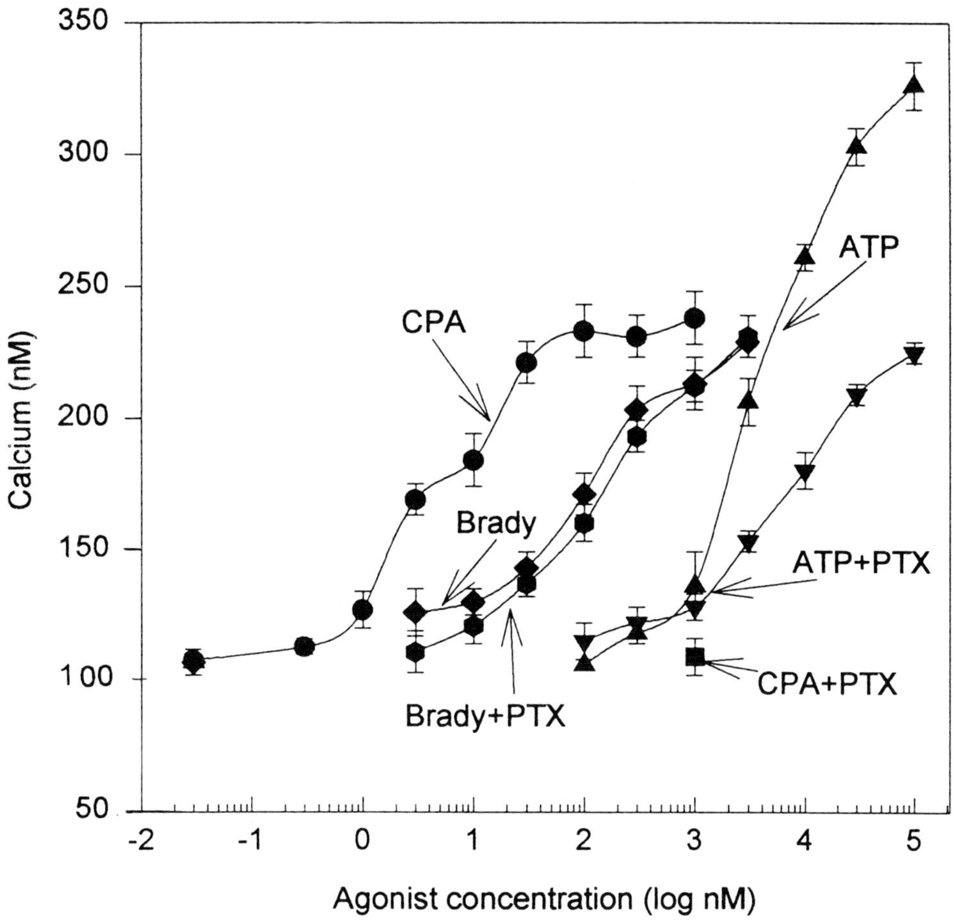

FIGURE 13-2. Effect of CPA and bradykinin on intracellular Ca^{2+} in DDT_1 MF-2 cells that were untreated or treated with pertussis toxin (200 ng/ml for 4 hours).

EFFECTS ON CALCIUM

Since $Ins(1,4,5)P_3$ is known to mobilize calcium from intracellular stores and later to cause influx of calcium from the medium, we used the Fura-2 method to determine if adenosine derivatives could also influence intracellular Ca^{2+} levels. Some of these results are summarized in Figure 13-2. It is seen that CPA, bradykinin, and ATP all increased intracellular calcium to approximately the same extent. Furthermore, in agreement with the findings on $Ins(1,4,5)P_3$, the response following pertussis toxin treatment varied very much between the agents. The response to CPA was completely blocked, the response to ATP was partially blocked, and the response to bradykinin was unaffected (Fig. 13-2). The data illustrated are the peak increases in Ca^{2+} that occurred during the first minute or so. Somewhat later there was a prolonged increase over basal that was eliminated by performing the experiments in calcium-free media.

The above-mentioned synergy between adenosine receptor agonists, on the one hand, and ATP or bradykinin, on the other, was also observed when Ca^{2+} responses were measured [11,13]. It could be argued that these synergistic responses are observed because different populations of the cells are studied. In order to rule out this possibility, experiments were also performed where Ca^{2+} responses in single cells were recorded. The synergy was equally clear-cut under these circumstances. This provides further evidence that synergistic effects are observed between receptors that signal via the G_q/G_{11} pathway and those that signal via G_i proteins.

EFFECTS ON PHOSPHOLIPASE D AND ON DIACYGLYCEROL

Phosphatidylinositol and its phosphorylated variants are not the only membrane phospholipids that are broken down in response to receptor signals. There is much recent interest in the regulation of phospholipase D that acts predominantly on phosphatidyl choline [14]. The activation of this enzyme can be monitored by taking advantage of the fact that the products of hydrolysis, in the presence of primary alcohols, including ethanol, can be converted to stable derivatives. Using this assay it was found that the adenosine analog CPA caused a concentration-dependent increase in phosphatidylethanol formation with an EC_{50} of 19 ± 2 nM. The effect was blocked by DPCPX (K_i 0.13 nM). The adenosine response was again completely blocked by pertussis toxin. In contrast, it was only slightly reduced by chelating intracellular calcium with BAPTA or by PKC inhibitors. This suggests that the activation of PLD is not only a secondary consequence of the calcium rise or of a stimulation of protein kinase C. Instead the data indicate that the receptor-mediated G-protein signal directly activates the enzyme.

Diacylglycerol is formed both as a consequence of phospholipase C (PLC) and of PLD activation. We examined the magnitude of the diacylglycerol accumulation following stimulation with CPA and bradykinin. The results are shown in Figure 13-3A. Interestingly, both agonists caused a clearly triphasic response. There was a very rapid immediate rise followed by a return to control levels after about 1 minute. Then there was a more protracted secondary rise. It seems probable that the two phases of stimulation may represent activation of PLC and PLD, respectively, but proof for this is lacking. Perhaps the most remarkable aspect of our findings is the extreme brevity of the first peak. This is clearly much more transient than the $Ins(1,4,5)P_3$ or Ca^{2+} signals. The important corollary to this is that elevations in diacylglycerol are likely to be more spatially restricted than increases in $Ins(1,4,5)P_3$ and especially in calcium.

CHANGES IN PROTEIN KINASE C

Protein kinase C is now known to comprise a family of enzymes. Three subforms α, β, and γ are known to possess calcium binding domains and are stimulated by calcium increases. Four subforms, δ, ε, η, and θ, lack the calcium-binding domains but are stimulated by diacylglycerol. The ζ and λ subforms, finally, are stimulated by neither calcium nor diacylglycerol and may be activated by membrane-associated forms of phosphatidylinositol phosphates [15]. As seen in Table 13-1, smooth muscle cells appear to express one member of each of these three classes of PKC. Interestingly, upon dedifferentiation the δ subform is lost and the ε subform is instead expressed [16].

As seen in Figure 13-3B there is a rapid, transient increase in protein kinase C activation following stimulation with bradykinin or CPA. The stimulation could be observed using two different substrates: MBP_{4-14}, which is selective for the classic enzymes including PKC α, and an ε peptide, which is a particularly good substrate for the ε subtype of PKC. There was a tendency that stimulation of the phosphorylation of the latter substrate was biphasic [Gerwins and Fredholm, unpublished]. Thus, the activation of the ε subform may show the same biphasic time course as the accumulation of diacylglycerol. By contrast, the activation of the α subform seemed to follow the level of intracellular Ca^{2+} more closely, which tallies with the fact that this subform reqires calcium for its activation.

The stimulation of PKC activation by bradykinin was unaffected by pertussis toxin pretreatment, whereas the activation by CPA was completely blocked. Furthermore, when the two stimuli were combined there was a clear-cut synergy. The activation of PKC was unaffected by changes in cAMP.

FUNCTIONAL IMPLICATIONS

The present results conclusively show what has been suggested earlier, namely, that adenosine A_1 receptor agonists can, via an action on G_i proteins, activate phospholipase C with consequent increases in intracellular calcium and diacylglycerol. In addition, there appears to be a direct activation of phospholipase D, another enzyme generating diacylglycerol. This then leads to an activation of protein kinase C that appears rather transient (Fig. 13-3B).

It must be strongly emphasized that the demonstration that PKC is activated is not a trivial consequence of the prior demonstration

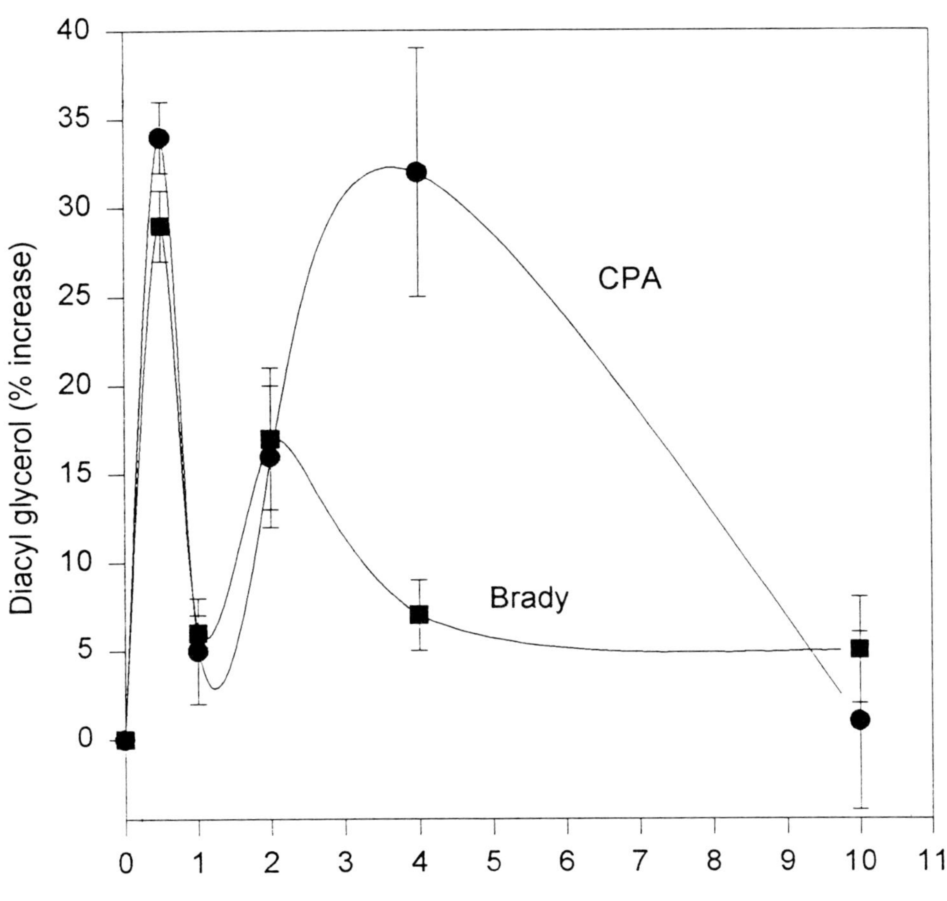

A

FIGURE 13-3. Time course of accumulation of diacylglycerol (**A**) and of activation of protein kinase C (**B**) in DDT_1 MF-2 cells after stimulation with CPA (100 nM) or bradykinin (1 μM).

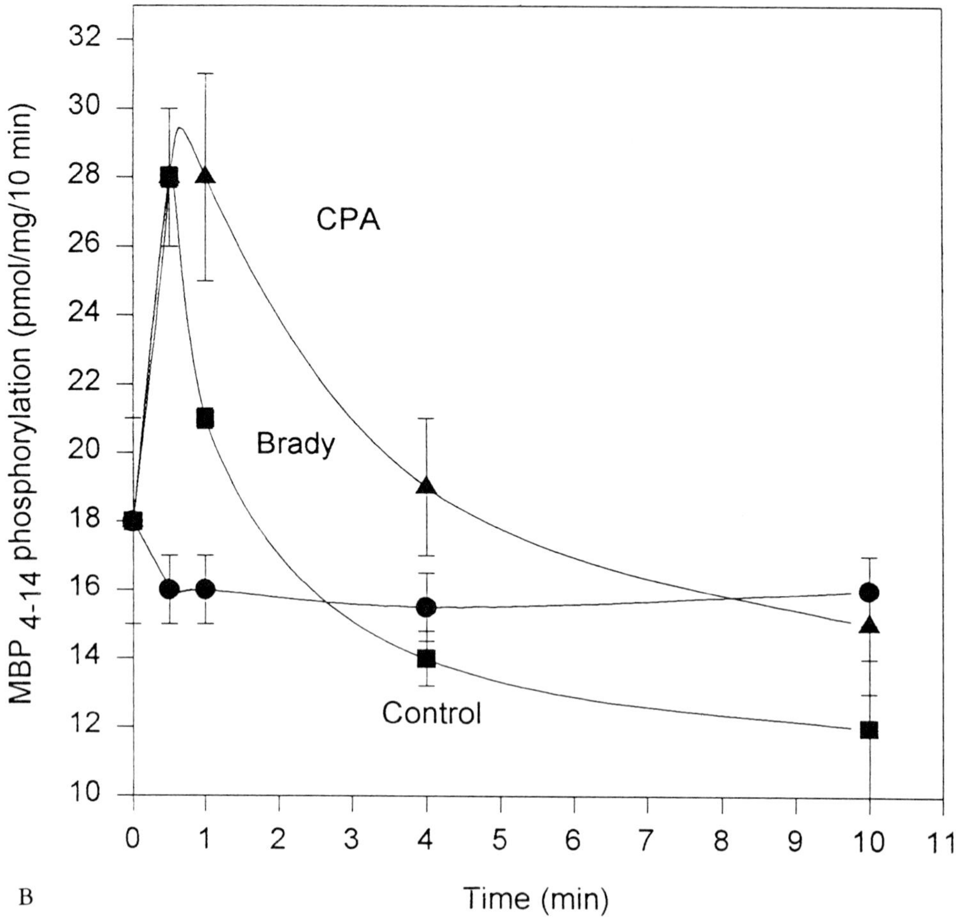

B

FIGURE 13-3 *(Continued)*. Time course of accumulation of diacylglycerol (**A**) and of activation of protein kinase C (**B**) in DDT_1 MF-2 cells after stimulation with CPA (100 nM) or bradykinin (1 μM).

TABLE 13-1. Protein kinase C subforms in smooth muscle cells

	PKC isoform											
	α		β		γ		δ		ε		ζ	
Cell type	R	P	R	P	R	P	R	P	R	P	R	P
Contractile SMC[a] (HSV, RA, VD)	+	+	−	−	−	−	+	+	−	−	(+)	(+)
Cultured SMC (6–30 passages)	+	+	−	−	−	−	+	+	+	−	(+)	(+)
Transformed SMC (DDT_1 MF-2 cells)	+	+	−	−	−	−	−	−	+	+	+	+

R = mRNA detectable by RT-PCR; P = protein detectable by Western blot; (+) = only trace amounts detectable.
[a] SMC from three sources were studied: human saphenous vein (HSV), rabbit aorta (RA), and rat vas deferens (VD). All showed the same isoform distribution, although vas deferens contained rather more PKC-ζ.

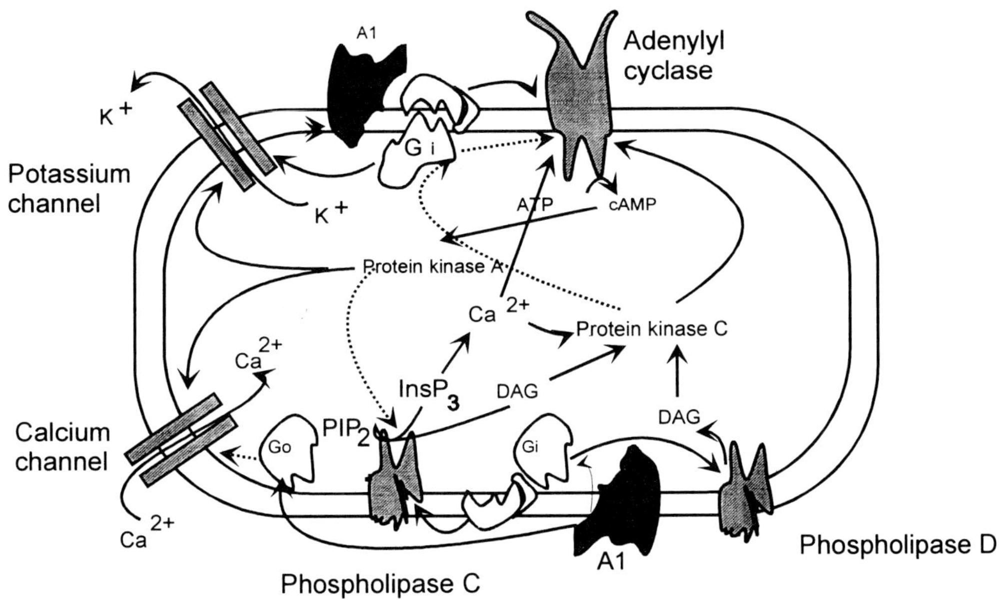

FIGURE 13-4. Schematic illustration of signaling pathways activated by adenosine A_1 receptors. Broken lines represent inhibitory pathways; intact lines represent stimulatory pathways.

that $Ins(1,4,5)P_3$ formation is stimulated. Thus, activation of P_{2u} receptors caused an increase in $Ins(1,4,5)P_3$ and of Ca^{2+} that was at least as large as that afforded by A_1 receptor stimulation. Yet there was no demonstrable PKC activation [Gerwins and Fredholm, unpublished data]. The reason for this may be that the receptor signal, particularly the accumulation of diacylglycerol, is very restricted in space due to extremely rapid degradation. Thus, we postulate that the A_1 receptors are located close to sites where PKC is present, whereas this is not the case for P_{2u} receptors.

The A_1 receptor–mediated increase in intracellular Ca^{2+} could be related to the findings that A_1 receptors may cause, for example, muscle contraction leading to vasoconstriction, at least in some vascular beds. The effect of A_1 receptor stimulation is mediated by the β,γ subunits of the G protein, and probably a high degree of G-protein activation is required [17]. One important corollary to this could be that only in cells that express A_1 receptors in high abundance, such as the DDT_1 MF-2 cell, is the response clearly seen. In other cell types, the formed β,γ subunits may not be sufficiently abundant to activate phospholipase C on their own. In these instances the synergism with the effect of receptors that activate other types of G proteins, as exemplified with bradykinin and ATP, may become very important (see Fig. 13-4). Indeed, this situation was demonstrated in the intact rat vas deferens [18].

Finally, it is interesting to speculate that the activation of PKC may be related to effects on cell proliferation or differentiation. Direct proof for this is lacking, but there are reports that stimulation of A_1 receptors can enhance cell proliferation, perhaps especially in conjunction with a submaximal activation of signaling pathways involving tyrosine kinase. One example is our previous demonstration that adenosine A_1 activation acts synergistically with low doses of PDGF to cause vascular smooth muscle cells to proliferate [19].

Acknowledgments

These studies were supported by the Swedish Medical Research Council (project no. 2553), the Swedish Cancer Fund, and Gustaf V's Tercentenary Fund. J.A. is a recepient of a British Heart Foundation fellowship.

References

1. van Calker D, Muller M, Hamprecht B (1978) Adenosine inhibits the accumulation of cyclic AMP in cultured brain cells. Nature 276: 839–841.
2. Fredholm BB, Abbracchio MP, Burnstock G, Daly JW, Harden TK, Jacobson KA, Leff P, Williams M (1994) Nomenclature and classification of purinoceptors. Pharmacol Rev 46: 143–156.
3. Trussell LO, Jackson MB (1985) Adenosine-activated potassium conductance in cultured striatal neurons. Proc Natl Acad Sci USA 82: 4857–4861.
4. Scholz KP, Miller RJ (1991) Analysis of adenosine actions on Ca^{2+} currents and synaptic transmission in cultured rat hippocampal pyramidal neurones. J Physiol (Lond) 435: 373–393.
5. Gerwins P, Nordstedt C, Fredholm BB (1990) Characterization of adenosine A_1 receptors in intact DDT_1 MF-2 smooth muscle cells. Mol Pharmacol 38:660–666.
6. Gerwins P (1993) Modification of a competitive protein binding assay for determination of inositol 1,4,5-trisphosphate. Anal Biochem 210:45–49.
7. Grynkiewicz G, Poenie M, Tsien RY (1985) A new generation of Ca^{2+} indicators with greatly improved fluorescence properties. J Biol Chem 260:3440–3450.
8. Heasly LE, Johnson GL (1989) Regulation by protein kinase C of nerve growth factor, epidermal growth factor and phorbol esters in PC12 pheochromocytoma cells. J Biol Chem 264:8646–8652.
9. Yasuda I, Kishimoto A, Tanaka S, Tominaga M, Sakurai A, Nishizuka Y (1990) A synthetic peptide substrate for selective assay of protein kinase C. Biochem Biophys Res Commun 166: 1220–1227.
10. Gerwins P, Fredholm BB (1991) Glucocorticoid receptor activation leads to up-regulation of adenosine A_1 receptors and down-regulation of adenosine A_2 responses in DDT_1 MF-2 smooth muscle cells. Mol Pharmacol 40:149–155.
11. Gerwins P, Fredholm BB (1992) ATP and its metabolite adenosine act synergistically to mobilize intracellular calcium via the formation of inositol 1,4,5-trisphosphate in a smooth muscle cell line. J Biol Chem 267: 16081–16087.
12. Camps M, Carozzi A, Schnabel P, Scheer A, Parker PJ, Gierschik P (1992) Isozyme-selective stimulation of phospholipase C-β2 by G-protein $\beta\gamma$-subunits. Nature 360:684–686.
13. Gerwins P, Fredholm BB (1992) Stimulation of adenosine A_1 receptors and bradykinin receptors, which act via different G-proteins, synergistically raises inositol 1,4,5-trisphosphate and intracellular free calcium in DDT_1 MF-2 smooth muscle cell. Proc Natl Acad Sci USA 89:7330–7334.
14. Billah MM, Anthes JC (1990) The regulation and cellular functions of phosphatidylcholine hydrolysis. Biochem J 269:281–291.
15. Nishizuka Y (1992) Intracellular signaling by hydrolysis of phopsholipids and activation of protein kinase C. Science 258:607–614.
16. Assender JW, Kontny E, Fredholm BB (1994) Expression of protein kinase C isoforms varies with state of differentiation in smooth muscle cells. FEBS Lett 342:76–80.
17. Birnbaumer L (1992) Receptor-to-effector signaling through G proteins: Roles for $\beta\gamma$ dimers as well as α subunits. Cell 71: 1069–1072.
18. Häggblad J, Fredholm BB (1987) Adenosine and neuropeptide Y enhance α_1-adrenoceptor induced accumulation of inositol phosphates and attenuate forskolin-induced accumulation of cyclic AMP in rat vas deferens. Neurosci Lett 82:211–216.
19. Jonzon B, Nilsson J, Fredholm BB (1985) Adenosine receptor-mediated changes in cyclic AMP production and DNA synthesis in cultured arterial smooth muscle cells. J Cell Physiol 124:451–456.

14. ADENOSINE DERIVED FROM NEURALLY RELEASED ATP IS THE PHYSIOLOGIC CAUSE OF SKELETAL NEUROMUSCULAR DEPRESSION

Eugene M. Silinsky
R.S. Redman

The efficiency of skeletal neuromuscular transmission is limited by the neuromuscular depression that ensues with even brief repetitive nerve stimulation [1–4]. Indeed, in patients with neuromuscular disorders such as myasthenia gravis, neuromuscular depression can be severely debilitating [5]. Neuromuscular depression is prejunctional in origin, occurring as a consequence of a reduction in the number of acetylcholine (ACh) quanta released by a nerve impulse [1–4,6]. Most attempts to quantify neuromuscular depression have been based on the assumption that the phenomenon is due to a reduction in the available vesicular store of ACh [2]. While depletion of vesicular stores of ACh contributes to depression at enormous, unphysiologic ACh outputs in the presence of very high concentrations of K^+ channel blockers [3], there is no evidence to support the depletion hypothesis at normal levels of ACh output.

In the early-mid 1970s, it was found that ATP and its hydrolysis products were released in stoichiometric amounts together with the neurotransmitter ACh from motor nerve endings [7,8]. Based upon the evidence that ATP and its hydrolysis products inhibited the release of ACh [9,10], it was suggested that ATP could indeed be the mediator of neuromuscular depression [8]. This view was reinforced in a thoughtful review on the subject [11], in which it was suggested that *Wedensky inhibition*, the term given to depression in curarized preparations [1], could be due to ATP released from nerve endings.

In the 1980s the bulk of the experimental evidence demonstrated that exogenous ATP needs to be hydrolyzed to adenosine, and it is the action of adenosine on specific adenosine receptors that mediates the effects of ATP [12, 13]. This putative purinergic model of neuromuscular depression is presented in Figure 14-1. In the 1990s, we evaluated the model shown in Figure 14-1 in more detail and found it to be an accurate depiction of the mechanism of neuromuscular depression. A description of these recent studies follows.

Methods Used to Study Neuromuscular Depression and to Minimize the Nonspecific Leakage of Adenosine

Physiologically functional ACh secretion was measured using intracellular recording of end-plate potentials (EPPs) from the innervated region of skeletal muscle (the end-plate region). Normal Ringer's solution contained 1.8 mM Ca^{2+} and sufficient concentrations of tubocurarine or α-bungarotoxin to reduce EPPs below threshold for the generation of muscle action potentials. Reagents used in this study were largely devoid of effects on postjunctional sensitivity to ACh; hence, changes in the size of the EPP in the presence of drugs that alter the actions or concentrations of adenosine or during neuromuscular depression reflect changes in the magnitude of ACh release. In some experiments, we made simultaneous measurements of perineural Ca^{2+} currents and EPPs in "Ca^{2+} current Ringer." Ca^{2+} current Ringer contained modest concentrations of the K^+ channel blockers 3,4,-diaminopyridine (DAP, 100 μM) and tetraethylammonium (TEA, 250 μM). Such K^+ channel blockade reduced the K^+ current to allow sufficiently accurate measurements of the Ca^{2+} component of the perineural current without causing the profound depletion of ACh stores that would prevent reliable measurements of EPPs. Ca^{2+} current Ringer also contained 0.9 mM Ca^{2+}, 9mM Mg^{2+}, 115 mM NaCl, 2 mM KCl, and either tubocurarine or α-bungarotoxin.

In previous studies on the action of endogenous adenosine, considerable leakage of endogenous adenosine occurred [13], leakage that could not be associated with synaptic function. To minimize this basal purine efflux and hence to evaluate better the importance of endogenous adenosine in neuromuscular depression, 1) a thin preparation, the frog cutaneous pectoris nerve-muscle preparation, was employed. This muscle is only a few cell layers thick in the region of examination; thus, the leaching of endogenous adenosine to the surface of the muscle as a consequence of deep fiber anoxia in thick neuromuscular preparations was reduced or eliminated. 2) Dissection was performed carefully to minimize cut muscle. 3) Nerve stimulation prior to experimentation was avoided. 4) Local rapid superfusion was used to apply adenosine receptor reagents. This produced a high local synaptic concentration of

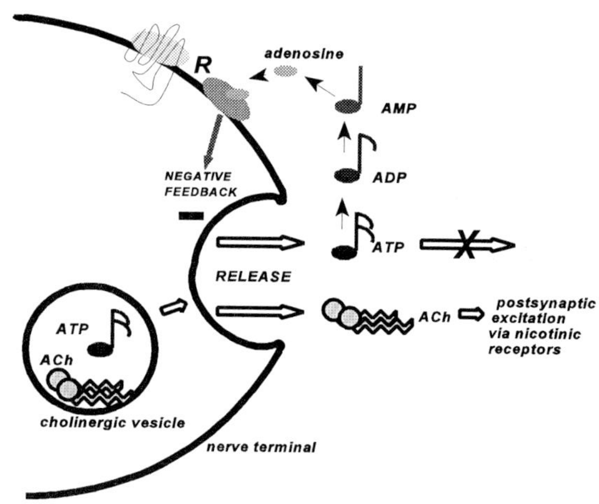

FIGURE 14-1. Diagram depiction of the putative purinergic model of neuromuscular depression. Note the storage of both ACh and ATP in cholinergic vesicles in motor nerve endings. Nerve stimulation causes the co-release of both ATP (depicted for convenience as a 16th note) and ACh. ACh produces the postjunctional excitation detected electrophysiologically as the EPP. In normal muscle (i.e., muscle that is not treated with tubocurarine or α-bungarotoxin), the EPP is so large that it produces a muscle action potential and a twitch in the muscle fiber. ATP is devoid of significant postjunctional effects on the electophysiologic correlates of neuromuscular transmission in the adult (developing myotubes possess ATP receptors linked to cation channels). However, after hydrolysis to ADP (eighth note), AMP (quarter note), and adenosine by ecto-5′-nucleotidases, adenosine is capable of binding to A_1 adenosine receptors in the plasma membrane of the nerve ending and inhibiting the subsequent release of ACh. The putative structure of the A_1 adenosine receptor (which has the conventional seven membrane-spanning domains found with other receptors linked to G proteins) is depicted as well.

these reagents, with a rapid onset of action (<50 ms). Finally, 5) highly selective adenosine antagonists, specifically, 8-cyclopentyl 1,3,dipropylxanthine (DPCPX) and 8-cyclopentyltheophylline (CPT) [14], were employed.

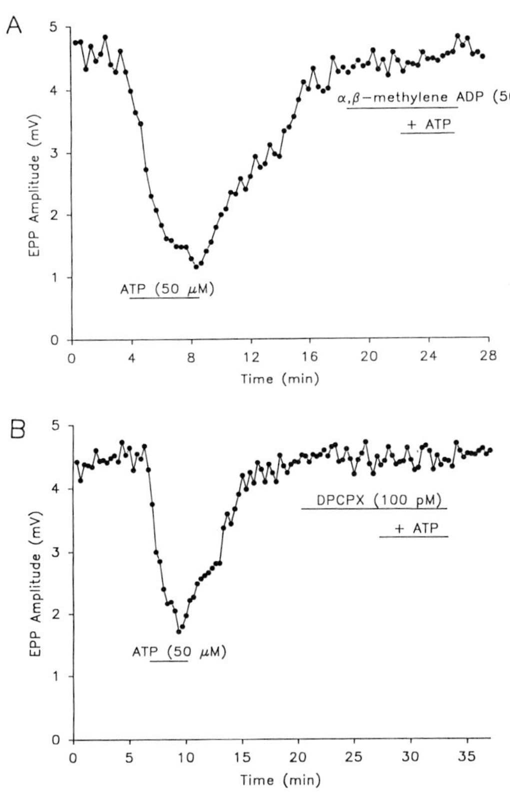

FIGURE 14-2. Evidence that exogenous ATP needs to be hydrolyzed to adenosine to inhibit ACh release. The inhibitory effect of exogenous ATP (50 μM) is prevented by α,β-methylene ADP (50 μM; **A**) and the selective adenosine receptor antagonist (8-cyclopentyl-1,3,dipropylxanthine, DPCPX, 100 pM; **B**). The nerve was stimulated at a low frequency (0.05 Hz) to prevent prejunctional neuromuscular depression. Normal Ringer's solution (1.8 mM Ca^{2+}) was employed in this experiment. The indicated drugs were superfused during the bars. The drug effects on EPP amplitudes in this and all subsequent figures represent the effects of adenosine reagents on evoked ACh release.

Exogenous ATP Must be Hydrolyzed to Adenosine to Inhibit ACh Release

At normal levels of ACh release and low frequencies of nerve stimulation, 1 every 10–20 seconds, ACh release and the size of the EPP is well maintained (Fig. 14-2A and 14-2B). These conditions were used to examine the effects of exogenous adenosine derivatives on ACh release. Figure 14-2A shows that α,β-methylene ADP (50 μM), an agent that inhibits ecto-5′-nucleotidases and therefore prevents the degradation of ATP to adenosine [15,16], prevents the inhibitory effects of ATP on ACh release (i.e., EPP amplitude). This drug did not alter the inhibitory action of adenosine on EPPs (data not shown). Likewise, the highly selective A_1 adenosine receptor antagonist (DPCPX) prevented the inhibitory effect of ATP on ACh release (Fig. 14-2B). These data suggest that, as depicted in Figure 14-1, ATP needs to be hydrolyzed to adenosine to inhibit ACh release.

Endogenous Adenosine Derived from ATP Is the Mediator of Neuromuscular Depression

In contrast to the effects at low frequencies of stimulation (1 stimulus per 10 or 20 seconds), when the frequency of nerve stimulation is increased to 0.5–1 Hz and greater, prejunctional depression of ACh ensues. Figure 14-3 (A–D) shows an experiment made under normal physiologic conditions at the frog neuromuscular junction. Note that the control-averaged EPP (Fig. 14-3A) is reduced by 43% after 20 seconds of stimulation at 0.5 Hz (Fig. 14-3B); this is the phenomenon of neuromuscular depression. Reducing the frequency of nerve stimulation allows ACh release to recover to the control level. After recovery, α,β-methylene ADP has no effect on ACh release at low frequencies of nerve stimulation (Fig. 14-3C), but when the frequency of nerve stimulation was increased again to 0.5 Hz, this inhibitor of ATP degra-

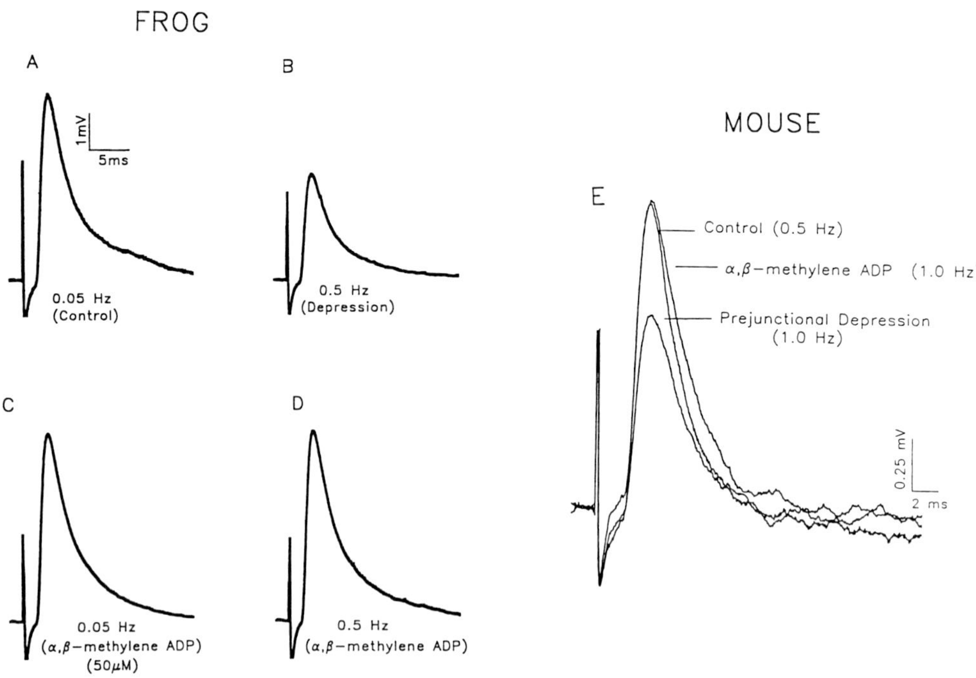

FIGURE 14-3. α,β-methylene ADP prevents neuromuscular depression at normal levels of ACh release. The figure shows the blockade of neuromuscular depression by α,β-methylene ADP in response to increasing the frequency of presynaptic stimulation in frog (A–D), and the mouse (E). Normal Ringer's solution (1.8 mM Ca^{2+}) was employed in this experiment. Each EPP is the averaged response to four consecutive stimuli. **A**: Control EPP in response to 0.05 Hz stimulation. This EPP was well maintained. **B**: Presynaptic depression of the EPP produced 20 seconds after an increase in the stimulation frequency to 0.5 Hz. **C** and **D** are similar to A and B, respectively, except for the addition of α,β-methylene ADP (50 μM). Note the absence of effect of α,β-methylene ADP on EPPs produced by low frequency stimulation (0.05 Hz, C) and the elimination of neuromuscular depression by this agent (D) during 0.5 Hz stimulation. Vertical calibration = 1 mV; horizontal calibration = 5 ms (A–D). **E** shows the averaged traces representing the control EPP (0.05 Hz), presynaptic depression (0.5 Hz), and recovery of depression by α,β-methylene ADP (0.5 Hz) nearly superimposed upon the control trace.

dation prevented neuromuscular depression (Fig. 14-3D).

Figure 14-3E shows that this effect also occurs at the mammalian neuromuscular junction at normal levels of ACh release. Note the reversal of neuromuscular depression by α,β-methylene ADP in the mouse phrenic nerve-hemidiaphragm (Fig. 14-3E).

The remaining experiments illustrate results of experiments made on frog muscle in Ca^{2+} current Ringer, which contained modest concentrations of K^+ channel blockers. This solution, in which the level of ACh release was elevated approximately 2.5-fold above that of normal solutions, had the following advantages: 1) It allowed us to study point-by-point depression at low frequencies (e.g., Fig. 14-4A, which represents the control condition) and hence to examine interimpulse regulation of evoked ACh release by adenosine, and 2) it allowed us to make simultaneous measurements of Ca^{2+} currents and evoked ACh release to inquire if neuromuscular depression or the action of endogenous adenosine is due to an action on

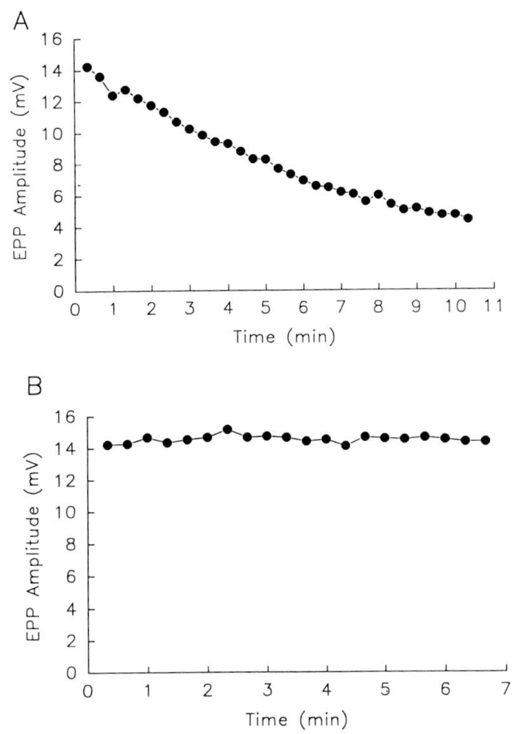

FIGURE 14-4. Presynaptic neuromuscular depression in Ca^{2+} current Ringer (**A**) and its antagonism by α,β-methylene ADP (**B**). For both A and B the motor nerve was stimulated at a frequency of 0.05 Hz. Ca^{2+} current Ringer consisted of potassium channel blockers and modified concentrations of Ca^{2+} and Mg^{2+} to allow for simultaneous measurements of EPPs and perineural Ca^{2+} currents (see Methods).

voltage-sensitive ionic channels in the plasma membrane. As Figure 14-4B shows, α,β-methylene ADP prevents neuromuscular depression in Ca^{2+} current Ringer as well.

If endogenous adenosine is the mediator of neuromuscular depression, then selective adenosine receptor antagonists such as CPT should eliminate neuromuscular depression. Furthermore, neuromuscular depression should be eliminated by the addition of adenosine deaminase to the bathing fluid. This enzyme degrades adenosine to inosine, which is inactive on adenosine recetors. Figure 14-5 (left graph) shows both of these predictions are borne out; both CPT and adenosine deaminase were able to fully reverse neuromuscular depression. In addition, during maximal neuromuscular depression the inhibitory effect of exogenous adenosine derivatives is blocked by occlusion from the endogenously released adenosine (not shown).

Neuromuscular depression and the reversal of depression by adenosine receptor antagonists or adenosine deaminase occur without effects on the Ca^{2+} components of the perineural currents measured simultaneously (Fig. 14-5, right traces, showing currents corresponding to the lettered time points in the left graph). Exogenous adenosine receptor agonists also have no effect on membrane ionic currents in frog motor nerve ending [19]. In contrast, either a reduction in the extracellular Ca^{2+} or the addition of N-type Ca^{2+} channel blockers such as Ω-conotoxin, Co^{2+}, Cd^{2+}, or aminoglycoside antibiotics (streptomycin, neomycin, or gentamicin) produced a parallel fall in EPP amplitude and Ca^{2+} currents [19,20]. For full details of results presented in this chapter, see Redman and Silinsky [17,18], and for complete details of the perineural current waveforms, see Silinsky and Solsona [19].

Summary, Speculations, and Clinical Implications

These results suggest that under the conditions of these experiments, adenosine derived from endogenously released ATP is the exclusive mediator of neuromuscular depression. This inhibitory effect of adenosine in frog occurs by a mechanism unrelated to an action on membrane ionic channels; rather, it appears that adenosine is capable of reducing the ability of Ca^{2+} to promote the secretory process [21,22]. The precise target site for the action of adenosine at the secretory apparatus is unknown. One such site could be synaptotagmin, which is the vesicular protein that serves as the Ca^{2+} sensor [23]. Indeed, synaptotagmin, which is linked via syn-

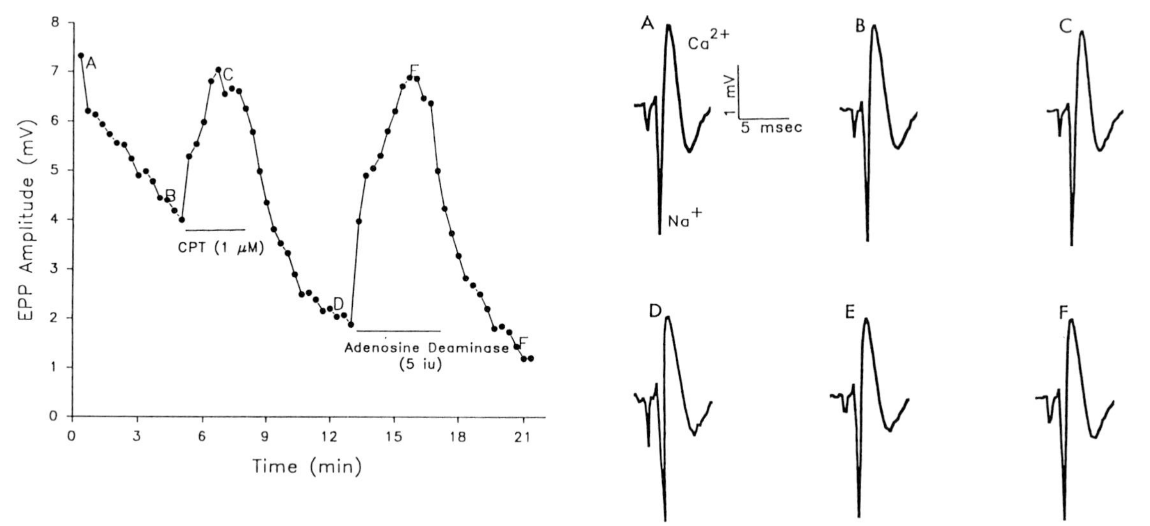

FIGURE 14-5. Neuromuscular depression is eliminated by the adenosine receptor antagonist CPT or by adenosine deaminase (**left graph**) without effects on Ca^{2+} currents recording simultaneously from motor nerve endings (**right traces**). Experiments were made in Ca^{2+} current Ringer in which a portion of the K^{+} channels were blocked. Under these conditions neuromuscular depression in produced at low frequencies of stimulation (0.05 Hz), thus eliminating the need to change stimulation parameters during the experiment. **Left Graph:** During continuous stimulation (0.05 Hz), neuromuscular depression was eliminated by the addition of 1 μM CPT; depression resumed upon washout of the drug. The application of 5 iu/ml adenosine deaminase also completely abolished neuromuscular depression. **Right Graph:** Perineural calcium currents were measured simultaneously with the EPPs. For clarity, only 6 of the 64 current traces are shown here. They correspond to the time points A–F in the graph of EPP amplitudes. It is apparent that while there are profound changes in EPP amplitude, a measure of ACh release, the perineural Ca^{2+} currents and the Na^{2+} current remain unchanged during neuromuscular depression. The calibration bar represents a voltage change of 1.5 mV produced by current flow across the perineurial resistances. (Note, only the initial component of the current traces are shown as the currents from several experiments demonstrated repetitive firing. For details of the currents, and the method for simultaneous measurements, see [19] and [32].)

taxin to the plasma membrane and to N-type Ca^{2+} channels, appears to be the most likely component for the nonplasmalemmal Ca^{2+} binding protein that controls the secretory machinery. Hence, adenosine could be acting in apparent competition with synaptotagmin or other as yet uncharacterized Ca^{2+} binding proteins for Ca^{2+} (by reducing their apparent affinity for Ca^{2+} [4]). Adenosine could also reduce the ability of such proteins, once bound with Ca^{2+}, to promote exocytosis (in which case, an uncompetitive relationship would occur). The relationship between adenosine and intracellular Ca^{2+} could also be noncompetitive, with the action of adenosine restricted to a different part of the secretory machinery than Ca^{2+} binding proteins (e.g., rab, syntaxin, unc, NSF, synapsin I, or one of the SNAP proteins) [23].

Regardless of the precise nature of the inhibitory mechanism, these results could have important clinical relevance. Specifically, if a similar mechanism of neuromuscular depression occurs in human skeletal muscle, then a selective prejunctional A_1 adenosine antagonist suitably targeted to motor nerve endings could be used to treat the debilitating neuromuscular fatigue that occurs when myasthenia gravis patients attempt to sustain repetitive neuromuscular activity [5].

Future Experiments

There is published evidence that over a prolonged time course (minutes) the release of adenosine derivatives from nerve endings occurs from stimulated rat [8,24] and frog motor nerve [25]. To provide the final support for the purinergic model of neuromuscular depression, it is necessary to show that quantal ATP release occurs over a millisecond time scale in conjunction with quantal ACh release. One ideal way to perform such experiments is to make use of the discovery that acutely dissociated celiac neurons from the guinea pig possess both nicotinic ACh receptors and highly sensitive P_2 receptors for ATP, with both nicotinic and purinergic receptors linked to a cation channel [26–28]. These excised ATP-gated cation channels have a chord conductance of approximately 22 pS (at a holding potential of −50 mV) [28]. Experiments performed in 1990 and 1991 showed that ATP is indeed the mediator of fast excitatory synaptic transmission between cultured celiac neurons [26–28], a result subsequently reproduced [29]. In this original discovery, excised outside-out patches containing ATP-gated ion channels were used to sniff the extracellular milieu for the evoked release of ATP in celiac neuronal cultures [28]. Conditions have been described to detect the release of ACh from frog motor nerve endings using outside-out patches containing ACh [30]. Indeed, the concomitant quantal release of ATP and ACh from frog motor nerve using this patch sniff method can be detected [31].

References

1. Katz B (1966) Nerve Muscle and Synapse. New York: McGraw Hill.
2. Martin AR (1966) Quantal nature of synaptic transmission. Physiol Rev 46:51–66.
3. Heuser JE, Reese TS, Dennis MJ, Jan Y, Jan L, Evans L (1979) Synaptic vesicle exocytosis captured by quick freezing and correlated with quantal transmitter release. J Cell Biol 81:275–300.
4. Silinsky EM (1985) The biophysical pharmacology of calcium-dependent acetylcholine secretion. Pharm Rev 37:81–131.
5. Lopate G, Pestronk A (1993) Autoimmune myasthenia gravis. Hosp Pract 28:109–131.
6. Otsuka M, Endo M, Nonamura Y (1962) Presynaptic nature of neuromuscular depression. Jpn J Physiol 12:573–584.
7. Silinsky EM, Hubbard JI (1973) Release of ATP from rat motor nerve terminals. Nature 243: 404–405.
8. Silinsky EM (1975) On the association between transmitter secretion and the release of adenine nucleotides from mammalian motor nerve terminals. J Physiol 247:145–162.
9. Ginsborg BL, Hirst GDS (1972) The effect of adenosine on the release of the transmitter from the phrenic nerve of the rat. J Physiol 224:629–645.
10. Ribeiro JA, Walker J (1975) The effects of adenosine triphosphate and adenosine diphosphate on transmission at the rat and frog neuromuscular junction. Br J Pharm 54:213–218.
11. Ribeiro JA (1979) Purinergic modulation of transmitter release. J Theor Biol 80:259–270.
12. Silinsky EM (1980) Evidence for specific adenosine receptors at cholinergic nerve endings. Br J Pharm 71:191–194.
13. Ribeiro JA, Sebastiao AM (1987) On the role, inactivation and origin of endogenous adenosine at the frog neuromuscular junction. J Physiol 384:571–585.
14. Lohse MJ, Klotz KN, Lindenborn-Fotinos J, Reddington M, Schwabe U, Olsson RA (1987) 8-cyclopentyl-1,3-dipropylxanthine (DPCPX)—a selective high affinity antagonist radioligand for A_1 adenosine receptors. Naunyn Schmeidebergs Arch Pharm 336:204–210.
15. Keller F, Zimmermann H (1983) Ecto-adenosine triphosphatase activity at the cholinergic nerve endings of the Torpedo electric organ. Life Sci 33:2635–2641.
16. Kreutzberg GW, Heymann D, Reddington M. 5′-Nucleotidase in the nervous system. In Kreutzberg GW, Reddington M, Zimmermann H (eds.), Cellular Biology of Ectoenzymes. Berlin: Springer Verlag, 1986, pp. 147–164.
17. Redman RS, Silinsky EM (1993) A selective

adenosine antagonist (8-cyclopentyl-1,3,-dipropylxanthine) eliminates both neuromuscular depression and the action of exogenous adenosine by an effect on A_1 receptors. Mol Pharm 45:835–840.
18. Redman RS, Silinsky EM (1994) ATP released together with acetylcholine as the mediator of neuromuscular depression at frog motor nerve endings. J Physiol 477:117–127.
19. Silinsky EM, Solsona CS (1992) Calcium currents at motor nerve endings: Absence of effects of adenosine receptor agonists in the frog. J Physiol 457:315–328.
20. Redman RS, Silinsky EM (1994) Decrease in calcium currents induced by aminoglycoside antibiotics in frog motor nerve endings. Br J Pharm 113:375–378.
21. Silinsky EM (1981) On the calcium receptor that mediates depolarization-secretion coupling at cholinergic motor nerve terminals. Br J Pharm 73:413–429.
22. Silinsky EM (1984) On the mechanism by which adenosine receptor activation inhibits the release of acetylcholine from motor nerve endings. J Physiol 346:243–256.
23. Hata Y, Slaughter CA, Sudhof TC (1993) Synaptic vesicle fusion complex contains unc- 18 homologue bound to syntaxin. Nature 366: 347–351.
24. Smith DO, Lu Z (1991) Adenosine derived from hydrolysis of presynaptically released ATP inhibits neuromuscular transmission. Neurosci Lett 122:171–173.
25. Cunha RA, Sebastiao AM (1993) Adenosine and adenine nucleotides are independently released from both the nerve terminals and the muscle fibers upon electrical stimulation of the innervated skeletal muscle of the frog. Pflügers Arch 424:503–510.
26. Gerzanich V, Matsumoto S, North RA, Silinsky EM (1991) ATP receptors in guinea-pig celiac neurons (abstr). Soci Neurosc 21:410.
27. Silinsky EM, Gerzanich V, Vanner SM (1992) ATP mediates excitatory synaptic transmission in mammalian neurones. Br J Pharm 106:762–763.
28. Silinsky EM, Gerzanich V (1993) On the excitatory effects of ATP and its role as a neurotransmitter in coeliac neurons of the guinea-pig. J Physiol 464:197–212.
29. Evans RJ, Derkach V, Surprenant A (1992) ATP mediates fast synaptic transmission in mammalian neurons. Nature 357:503–505.
30. Grinnell AD, Gunderson CB, Meriney SD, Young SH (1989) Direct measurements of ACh release from exposed frog nerve terminals: Constraints on interpretation of non-quantal release. J Physiol 419:225–251.
31. Silinsky EM, Redman RS, Watanabe M. What is the mechanism of skeletal neuromuscular depression? In Albuquerque EX (ed.), The Cholinergic Synapse: Structure, Function, and Regulation. 1994, in press.
32. Redman RS, Silinsky EM (1995) On the simultaneous electrophysiological measurements of neurotransmitter release and perineural calcium currents from frog motor nerve endings. J Neurosci Methods, in press.

15. ADENOSINE 5′-TRIPHOSPHATE RECEPTOR–MEDIATED CURRENTS IN RAT INTRACARDIAC NEURONS

T.J. Nutter
D.J. Adams

Adenosine 5′-triphosphate (ATP) activates a nonselective cation conductance in a variety of cultured peripheral neurons, including rat and bullfrog sensory neurons [1,2], rat sympathetic neurons [3], rat and guinea-pig parasympathetic neurons [4,5], and guinea-pig celiac neurons [6–9]. ATP-mediated responses have been described also in central neurons [10–12]. ATP has recently been shown to act as a fast excitatory transmitter at neuro-neuronal synapses in rat cultured celiac neurons [6,8] and in rat medial habenula slice preparations [13]. Thus, ATP has become increasingly recognized as a neurotransmitter that is active in both the central and peripheral nervous systems.

In this report, the kinetic and pharmacologic properties of ATP-evoked responses in cultured parasympathetic neurons from rat intracardiac ganglia were investigated using the patch-clamp technique. Specifically, effects of extracellular divalent cations (Ca^{2+} and Mg^{2+}) on the kinetics of ATP-evoked currents were examined in voltage-clamped neurons.

Methods

Parasympathetic neurons from rat atria were isolated and cultured as described previously [5]. Atria were isolated from neonatal (1–3 day postpartum) rats and incubated in Krebs solution containing 1 mg/ml collagenase at 37°C for 1 hour. Parasympathetic ganglia were dissected from the atria; ganglion cells were dispersed in high glucose Dulbecco's modified Eagle media (containing 10% v/v fetal bovine serum, 100 U/ml penicillin and 0.1 mg/ml streptomycin) and plated onto 18-mm laminin-coated glass coverslips. The dissociated cells were incubated at 37°C under a 95% air, 5% CO_2 atmosphere. Experiments were conducted on cells maintained in primary tissue culture for 2–3 days.

ATP-mediated responses were recorded from parasympathetic neurons under current or voltage clamp using the standard [14] and perforated-patch [15] whole-cell recording configurations of the patch-clamp technique. Plated parasympathetic neurons were transferred to a recording chamber (0.5 ml volume) and viewed at 400× magnification with a phase-contrast inverted microscope. Membrane current and voltage were monitored with a List EPC-7 patch-clamp amplifier and recorded to videotape using a digital VCR adaptor (PCM-1, Medical Systems Corp.). Whole-cell membrane current and voltage records were filtered at 4 kHz, on a chart recorder (Gould, monitored 2200S) and analyzed using a PC 486 Computer.

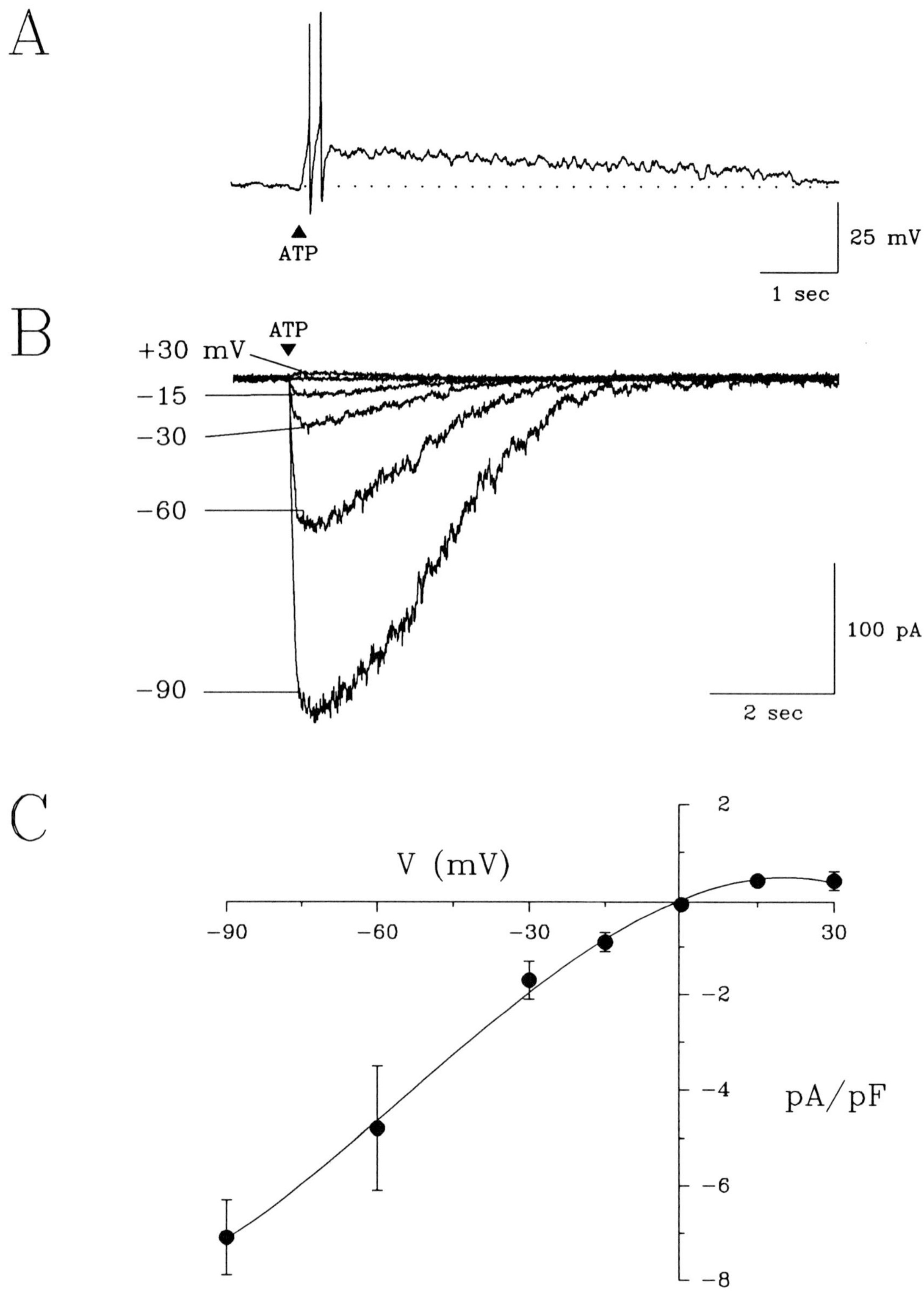
A
ATP
25 mV
1 sec
B
ATP
+30 mV
−15
−30
−60
−90
100 pA
2 sec
C
V (mV)
−90
−60
−30
30
2
−2
−4
−6
−8
pA/pF

Recording pipettes were pulled from thin-wall borosilicate glass to tip resistances of 2–4 MΩ. The intracellular pipette solution for standard whole-cell recordings contained (mM): 130 NaCl, 5 Na_4BAPTA, 2 Na_2ATP, and 10 Na-Hepes, pH 7.2. Electrodes for perforated patch experiments were dipped momentarily in a solution composed of (mM): 75 K_2SO_4, 55 KCl, 5 $MgSO_4$, 10 Hepes (pH 7.2 with KOH), and then backfilled in the same solution containing 240 μg/ml amphotericin B. The reduction in series resistance, a function of amphotericin B incorporation into the neuronal membrane, was monitored in response to 10 mV hyperpolarizing voltage steps, and experiments were begun when resistance was ≤20 MΩ (usually within 10 minutes of seal formation).

The recording chamber was continuously perfused (~2 ml/min) with a control physiological salt solution (PSS) composed of (mM): 140 NaCl, 1 $CaCl_2$, 7.7 glucose, and 10 histidine, pH 7.2. Test solutions were composed of Ca^{2+}-free PSS to which the appropriate concentration of the chloride salt of the divalent cation was added. ATP-mediated responses were evoked by focal application of Na_2ATP, dissolved in the bathing solution at the concentration indicated, via a pressure ejection device through an extracellular pipette. The ATP-receptor antagonist suramin was bath applied at a concentration of 100 μM and was also included in the pressure ejection pipette solution. All experiments were conducted at room temperature (22°C). The values reported represent the mean ± SEM.

FIGURE 15-1. Excitatory response of rat cultured parasympathetic intracardiac neurons to exogenous ATP. A: Voltage record of action potentials evoked in response to a 2-ms pulse of 100 μM ATP (arrow) applied from an extracellular pipette. The resting membrane potential was −52 mV. B: Family of ATP-evoked currents obtained in PSS at the membrane potentials indicated. Arrow indicates application of a 10-ms pulse of 100 μM ATP. C: Current-voltage (I-V) relationship for peak current amplitude evoked by ATP in normal PSS. Each point represents the mean current density (pA/pF) ±SEM from at least four cells.

Results

Under current clamp, a 2-ms pulse of 100 μM ATP applied to the soma of an intracardiac parasympathetic neuron evoked a rapid depolarization (<20 ms latency) and action potential discharge (Fig. 15-1A). Perforated-patch whole-cell currents evoked by ATP at different holding potentials are shown in Figure 15-1B. Focal application of ATP to the voltage-clamped neuron activated a transient current in >90% of the neurons examined. In PSS containing 1 mM Ca^{2+}, ATP (100 μM)-evoked currents exhibited a time to peak of ≥500 ms and a half-time of decay of 4.6 ± 0.7 seconds (n = 9), which was independent of membrane potential. The current-voltage (I-V) relationship exhibited marked inward rectification and a reversal potential of −0.1 ± 0.1 mV (n = 9) in symmetric Na^+ solutions (Fig. 15-1C). Rectification was observed in the absence of divalent cations in either the intracellular or extracellular solution, suggesting that the reduction in outward current at positive membrane potentials is unlikely to be due to divalent cation block of the ion channel. ATP-evoked current density averaged 7.1 ± 0.8 pA/pF (n = 33) at a membrane holding potential of −90 mV.

The efficacies of various purinergic receptor agonists were examined to determine the receptor subtype in rat parasympathetic intracardiac neurons. The rank order of agonist potency for purinergic-mediated responses was 2-methylthioATP = ATP > ADP > AMP = α,β-methylene ATP > β,γ-methylene ATP > adenosine, consistent with the P_2 receptor subtype [5]. The agonist sensitivity of the receptor was determined by applying various concentrations of ATP to voltage-clamped neurons. ATP-mediated current amplitude was concentration dependent, as shown in Figure 15-2A. Focal application of ATP at the concentrations indicated to the soma voltage clamped at −90 mV

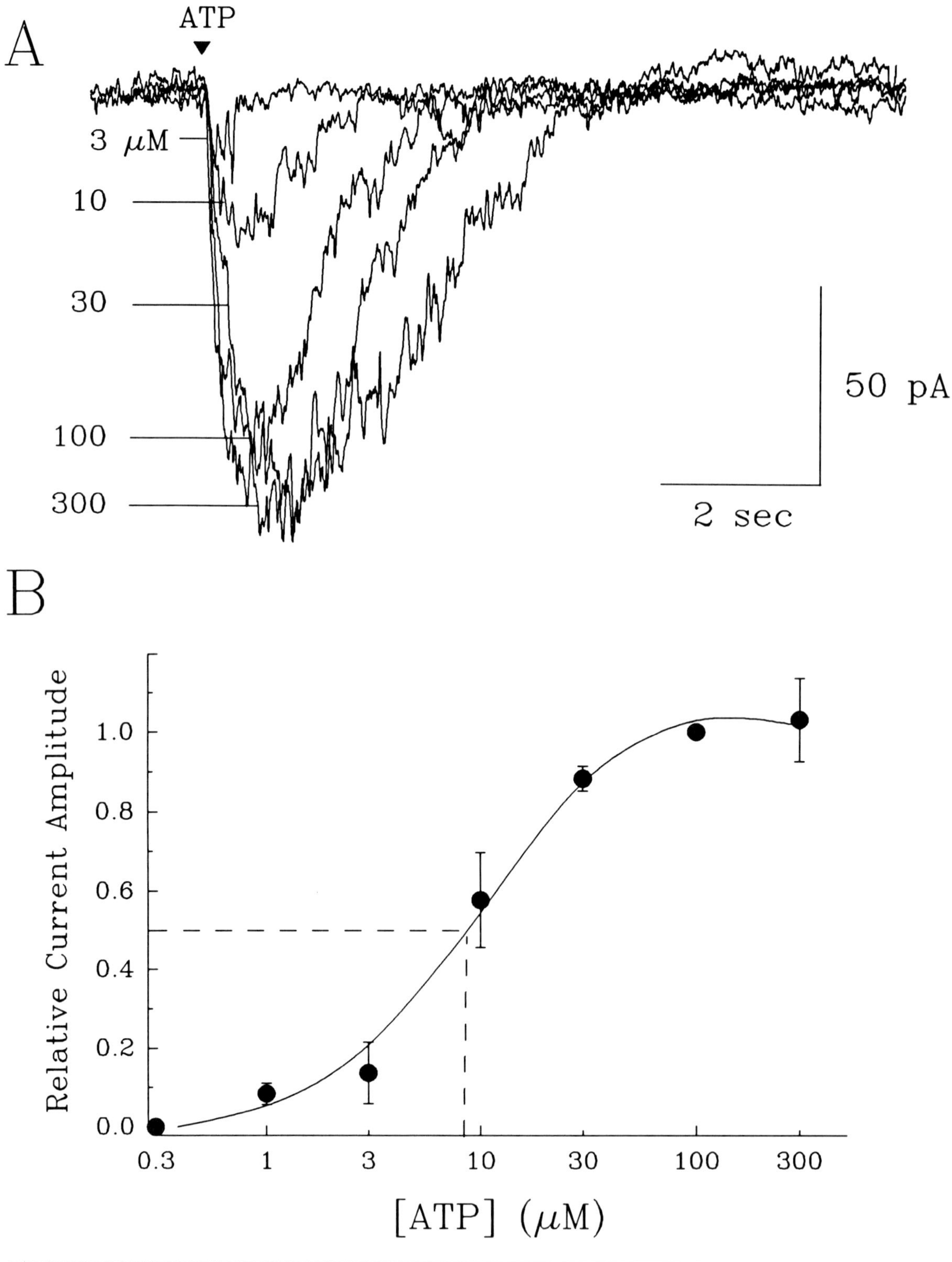
A
ATP
3 μM
10
30
100
300
50 pA
2 sec
B
Relative Current Amplitude
0.0
0.2
0.4
0.6
0.8
1.0
0.3
1
3
10
30
100
300
[ATP] (μM)

evoked an inward current, and the dose-response relationship obtained for ATP-evoked current indicated that half-maximal activation (EC_{50}) occurred with approximately 8 μM ATP under these conditions (Fig. 15-2B).

Bath application of the ATP-receptor antagonist, suramin, reduced ATP-evoked current amplitude in a dose-dependent manner (Fig. 15-3). The suramin concentration required to produce half-maximal inhibition (IC_{50}) of the ATP-activated current at a holding potential of −90 mV was approximately 6 μM (Fig. 15-3B). The reduction of ATP-mediated current amplitude by suramin was voltage-independent and reversible following bath perfusion of PSS. ATP-evoked currents in rat intracardiac neurons have also been shown to be blocked by Reactive Blue 2 (IC_{50} = 1 μM) and partially inhibited, but not completely blocked, by α, β-methylene ATP [5].

Minimal desensitization of ATP-evoked currents was observed with applications of <5 seconds duration. No attenuation of current amplitude was observed in response to 100-ms pulses of ATP (100 μM) applied at 3-second intervals (Fig. 15-4A). In the same neuron, a 3-second application of 100 μM ATP elicits a rapidly activating current that continues to increase for the duration of the pulse (Fig. 15-4B). Application of ATP for intervals longer than 5 seconds resulted in a progressive decline in current amplitude (Fig. 15-4C); the rate of current decay was independent of membrane potential (not shown).

FIGURE 15-2. Concentration dependence of ATP-activated currents in parasympathetic neurons. **A:** Currents evoked by a 50-ms focal application of ATP at the concentrations indicated. Holding potential, −90 mV. **B:** Relative peak amplitude of evoked currents as a function of ATP concentration. The curve of best fit to the data gives an EC_{50} of ~8 μM ATP. Each point represents the mean ± SEM for at least two cells.

The peak amplitude and rate of decay (desensitization) of ATP-induced currents was dependent on the species and concentrations of divalent cations in the extracellular solution (Fig. 15-5A). During a 30-second application of 100 μM ATP in PSS (1 mM Ca^{2+}), the current declined to 40.1 ± 7.3% (n = 5) of the peak amplitude. In Ca^{2+}-free PSS, the ATP-evoked current ampitude was increased by ~18% and declined to 60.0 ± 1.8% (n = 2) of the peak upon termination of agonist application. Raising external Ca^{2+} concentration ($[Ca^{2+}]$) to 10 mM reduced peak current amplitude ~40% and increased the rate of current decay such that only 26.8 ± 6.4% (n = 2) of the current remained following a 30-second exposure to 100 μM ATP. The effects of raising extracellular $[Ca^{2+}]$ on the rate of desensitization were not mimicked by Mg^{2+}. In Ca^{2+}-free PSS containing 1 mM Mg^{2+}, ATP-evoked current amplitude was only 65% of the peak attained in normal PSS. However, during a 30-second exposure to ATP, 60.4 ± 16.5% (n = 2) of the current remained upon termination of the pulse, similar to that observed in Ca^{2+}-free PSS. Increasing extracellular $[Mg^{2+}]$ to 10 mM further reduced ATP-evoked current amplitude (~20% of control in normal PSS) but had a minimal further effect on the rate of current decay (50.6 ± 19.5%; n = 2). The effects of external divalent cations on the kinetics of ATP-evoked currents are summarized in Figure 15-5B.

Discussion

The transient inward current evoked by ATP application to cultured rat parasympathetic neurons is similar to the purinergic response observed in rat sensory neurons [1] and other vertebrate neurons [16]. The agonist sensitivity sequence of 2-methylthio ATP = ATP > ADP > AMP = α,β-methylene ATP > β, γ-methylene ATP > adenosine [5] is indicative of a P_2 purinergic receptor. The short latency of current activation in response to ATP is consistent with a direct ligand action on the

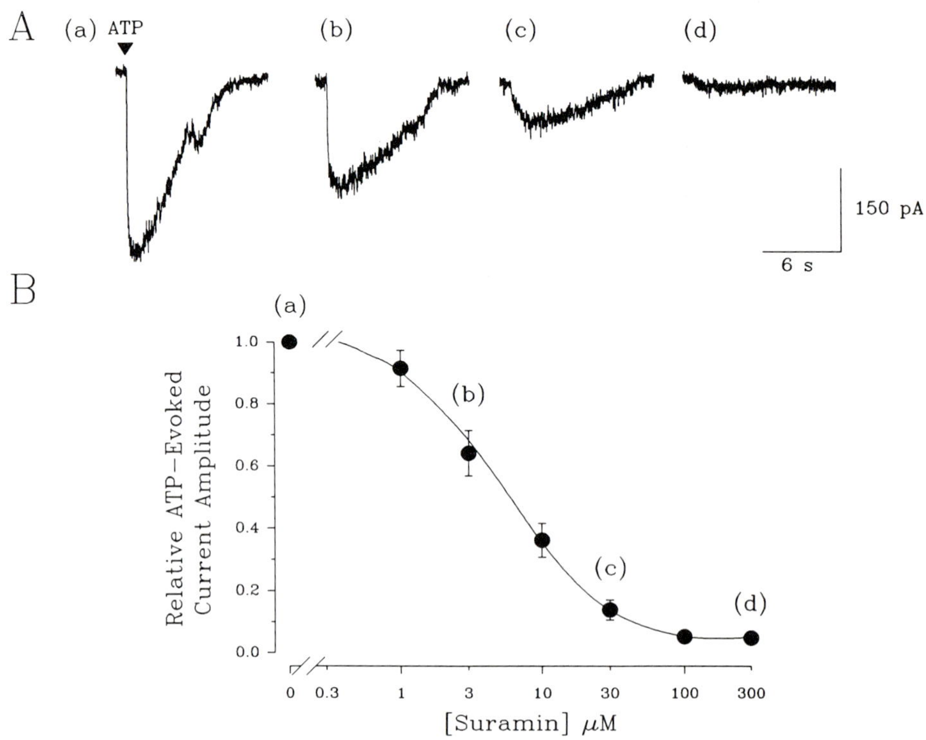

FIGURE 15-3. Dose-dependent inhibition of ATP-evoked currents by suramin. **A:** Representative whole-cell currents elicited at −90 mV by a brief pulse of ATP (100 μM, 10-ms) in the absence (a) and presence of 3 μM (b), 30 μM (c) and 300 μM (d) suramin. Recovery of ATP-evoked current amplitude was obtained upon washout. **B:** Dose-response relationship for inhibition of ATP-evoked currents by bath applied suramin. The data were fit by a single-site adsorption isotherm giving an IC_{50} of 6.3 μM and a slope parameter of 1.03. The suramin concentrations at which current traces shown in A were obtained are indicated by the small letters in parentheses (a–d).

channel; further evidence in support of direct ATP receptor–channel coupling is provided by the recording of single channel currents in excised membrane patches [5; Nutter and Adams, unpublished data]. These data are consistent with a P_{2x} purinoceptor based on a nomenclature distinguishing between a direct ligand-gated ion channel (P_{2x}) and a guanine nucleotide binding protein (G-protein)–coupled ion channel (P_{2y}) [17,18].

FIGURE 15-4. Desensitization of ATP-evoked currents in rat parasympathetic cardiac neurons. **A:** Perforated patch whole-cell current records from a neuron in response to two 100-ms pulses of ATP (100 μM) applied at a 3-second interval. Holding potential, −90 mV. **B:** Response of the same neuron to a 3-second pulse of 100 μM ATP (indicated by bar). The current amplitude continued to increase for the duration of ATP application. Holding potential, −90 mV. **C:** Desensitization of ATP-activated current during a 30-second application of 100 μM ATP. Current declined to 55% of the peak by the termination of ATP application. Holding potential, −90 mV.

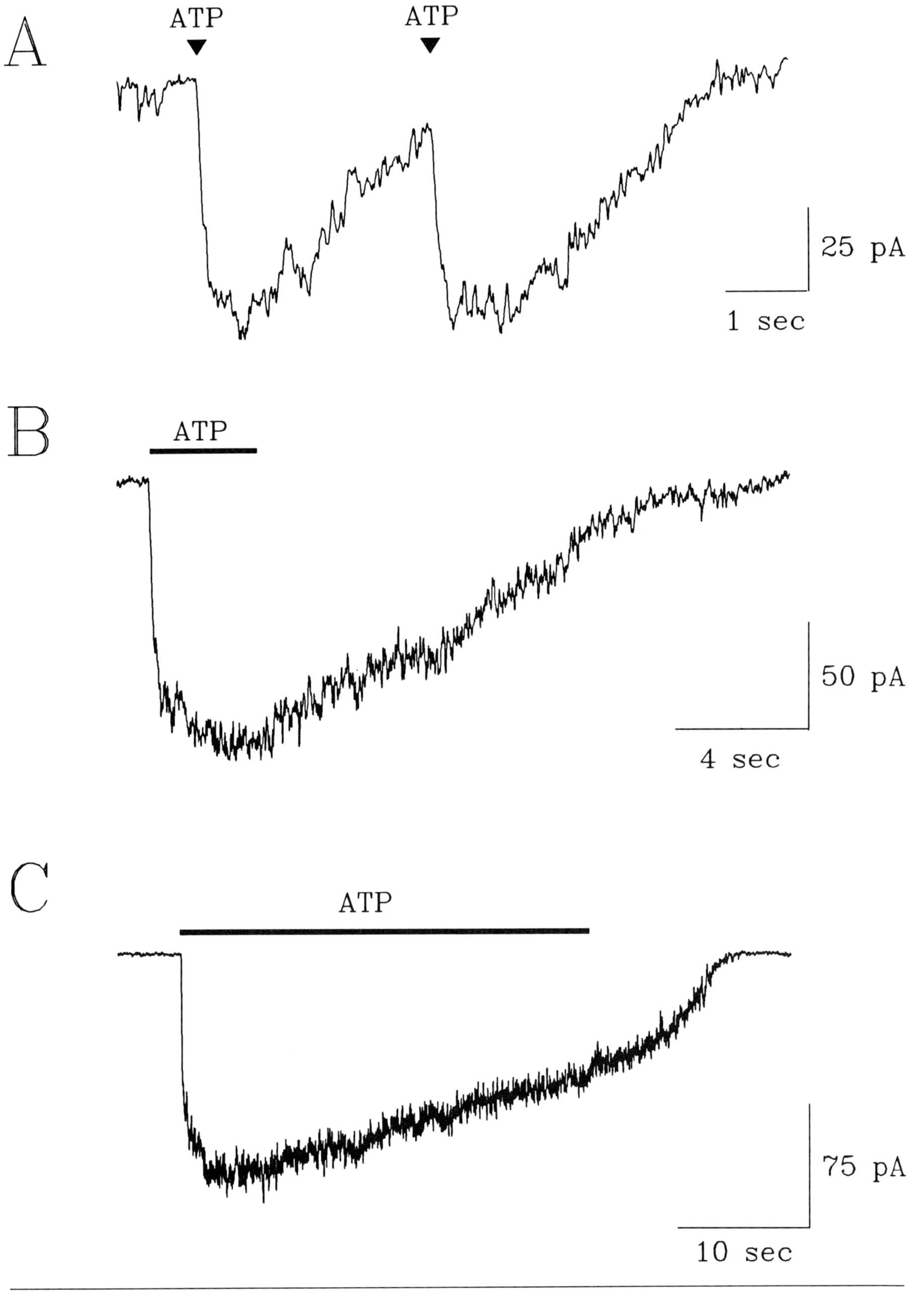
A
ATP
ATP
25 pA
1 sec
B
ATP
50 pA
4 sec
C
ATP
75 pA
10 sec

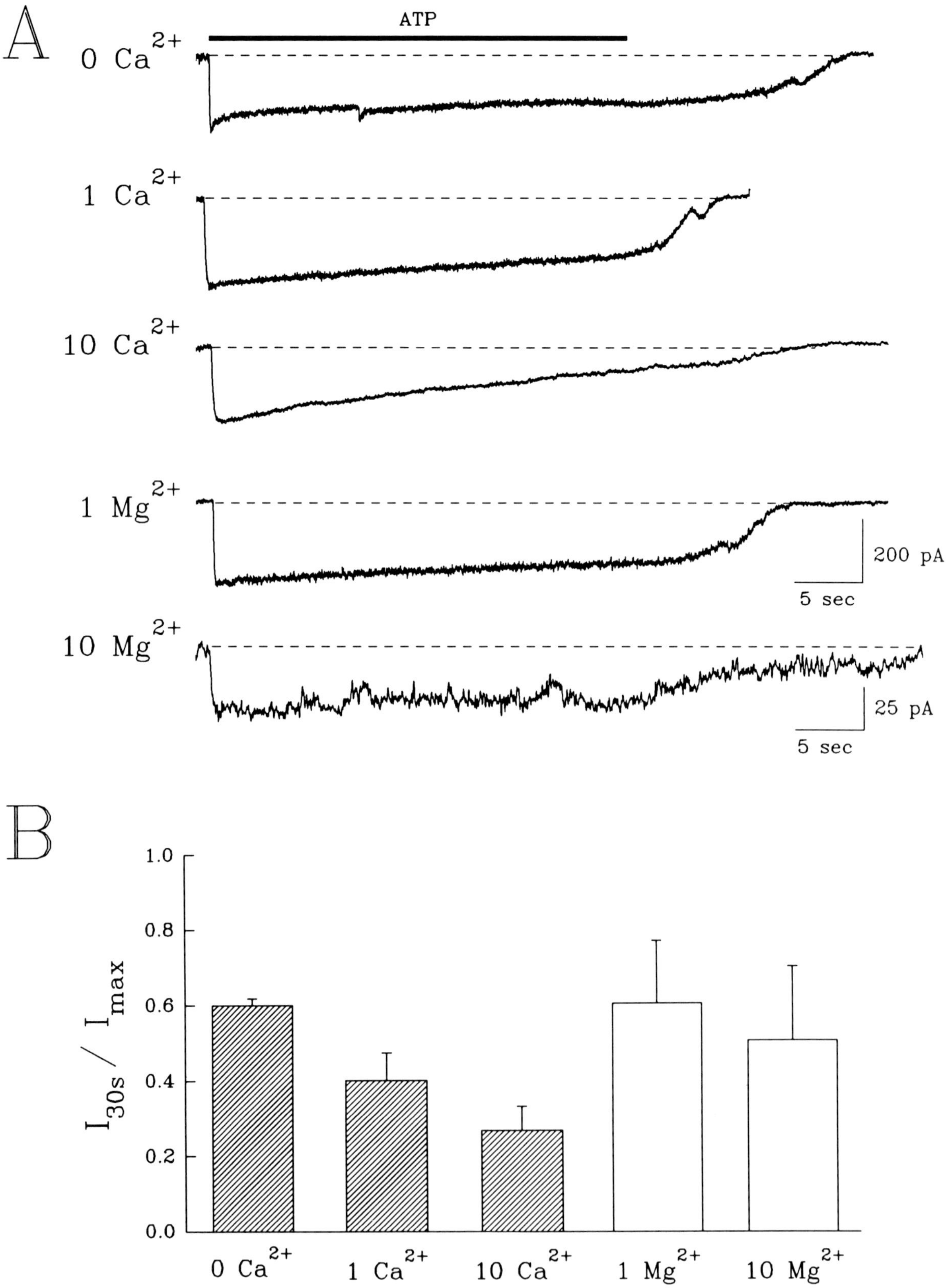

A
ATP
0 Ca2+
1 Ca2+
10 Ca2+
1 Mg2+
200 pA
5 sec
10 Mg2+
25 pA
5 sec
B
I30s / Imax
1.0
0.8
0.6
0.4
0.2
0.0
0 Ca2+
1 Ca2+
10 Ca2+
1 Mg2+
10 Mg2+

The ATP-evoked current in rat cardiac neurons is due to a nonselective cation conductance increase and the I-V relationship exhibits marked inward rectification [5]. These properties resemble those of P_{2x} receptor channels present in vertebrate neurons [8,12,19,20], pheochromocytoma (PC12) cells [21,22], and smooth muscle cells [23,24]. In rat parasympathetic neurons, rectification was observed in the absence of either internal or external divalent cations. Similar results were obtained in bullfrog sensory neurons [20], where removal of Mg^{2+} from both the intracellular and extracellular solutions did not alleviate the rectification of the I-V relationship. Unitary ATP-activated channel currents recorded in outside-out membrane patches from rat parasympathetic neurons exhibited a linear i-V relationship with a slope conductance of ~50 pS in symmetric Na^+ solutions [Nutter and Adams, unpublished data]. These results suggest that Mg^{2+} block is not responsible for the reduction in outward current, and inward rectification of the whole-cell current may be due, in part, to the voltage dependence of the channel open probability (P_o), where P_o increases with membrane hyperpolarization [5].

The purinergic receptor channel of rat intracardiac neurons exhibits a sensitivity to ATP ($EC_{50} \cong 8\,\mu M$) comparable to the concentrations of 3 μM and 10 μM required for half-maximal activation of ATP-mediated currents in bullfrog [2] or rat [1] sensory neurons, respectively. Higher ATP concentrations were required for activation of guinea-pig submucosal plexus neurons (EC_{50} = 23 μM) [9], rat nucleus solatarii neurons (K_D = 31 μM) [12], and rat superior cervical ganglion neurons (EC_{50} = 60 μM) [3].

Neuronal purinergic receptors of the P_{2x} subtype have been reported to be blocked by the trypanocidal agent, suramin. Half-maximal inhibition of ATP-evoked current amplitude in rat parasympathetic neurons was obtained with bath application of ~6 μM suramin. Higher concentrations of suramin appear to be required to inhibit ATP-activated currents in bullfrog sensory neurons (~90% inhibition by 100 μM suramin) [25] and excitatory postsynaptic currents (EPSCs) mediated by ATP in guinea pig celiac neurons (~50% inhibition by 80–230 μM suramin) [7]. ATP-activated EPSCs in neurons from rat medial habenula slices exhibit a sensitivity similar to that observed in rat parasympathetic neurons, with nearly complete block obtained with 30 μM suramin [13]. These data contrast with that reported for myenteric neurons from guinea-pig ileum, where submicromolar concentrations of suramin increased the efficacy of ATP [26].

The magnitude and kinetics of ATP-evoked currents in rat parasympathetic neurons were differentially modulated by extracellular divalent cations. Increasing the extracellular $[Ca^{2+}]$ reduced the ATP-mediated whole-cell current amplitude, as has been observed in rat [19] and bullfrog sensory neurons [2]. The inhibition of ATP-mediated current in rat sympathetic neurons [3] by elevated external $[Ca^{2+}]$ has been attributed to a saturable binding site for Ca^{2+} within the ion-conducting pore of the channel. Although ATP-activated channels in rat cardiac neurons exhibit a substantial Ca^{2+} permeability (P_{Ca}/P_{Na} = 1.48), as determined from reversal potential measurements [5], the monotonic decrease in whole-cell current amplitude upon raising the external divalent cation concentration is consistent with a saturable Ca^{2+} binding site

FIGURE 15-5. Divalent cation modulation of ATP-receptor desensitization. **A**: Representative current records obtained from a neuron in response to 30-second focal applications of 100 μM ATP. The extracellular divalent cation species and concentration is indicated to the left of each current trace. Note different current scale for 10 mM Mg^{2+}. Holding potential, −90 mV. **B**: Bar graph of the ratio of current remaining at the termination of the 30-second ATP application divided by the peak evoked current as a function of the extracellular divalent species and concentration. Ca^{2+}, but not Mg^{2+}, increases the rate of ATP-evoked current delay. Each column represents the mean ± SEM for at least two cells.

in the pore. On the other hand, the decrease in whole-cell current amplitude upon raising the divalent cation concentration in rat smooth muscle [27] and sympathetic neurons [3] involved a decrease in the free form of ATP, a possibility that cannot be disregarded. The effect of Ca^{2+} on the kinetics of current decay is similar to that observed in rat nucleus solitarii neurons [12]. The increase in the rate of current decay upon raising extracellular [Ca^{2+}] is unlikely to be due to association of ATP and Ca^{2+}, as the effects were not mimicked by external Mg^{2+}. These data suggest that Ca^{2+} acts either directly or through a Ca^{2+}-dependent second messenger to modulate ATP receptor-channel kinetics in rat parasympathetic neurons.

The existence of an ATP-activated conductance sufficient to initiate action potentials in parasympathetic neurons of rat intracardiac ganglia suggests that purinergic receptors may play a role in mediating vagal innervation of the myocardium and thus the regulation of the mammalian heart beat.

Acknowledgments

We thank Michael Parker for his participation in experiments associated with Figure 3. This work was supported by NIH grant HL35422 (DJA) and an AHA postdoctoral fellowship (TJN).

References

1. Krishtal OA, Marchenko SM, Pidoplichko VI (1983) Receptor for ATP in the membrane of mammalian sensory neurons. Neurosci Lett 35: 41–45.
2. Bean BP (1990) ATP-activated channels in rat and bullfrog sensory neurons: Concentration dependence and kinetics. J Neurosci 10:1–10.
3. Nakazawa K (1994) ATP-activated current and its interaction with acetylcholine-activated current in rat sympathetic neurons. J Neurosci 14:740–750.
4. Allen TGJ, Burnstock G (1990) The actions of adenosine 5′-triphosphate on guinea-pig intracardiac neurones in culture. Br J Pharmacol 100:269–276.
5. Fieber LA, Adams DJ (1991) Acetylcholine-evoked currents in cultured neurones dissociated from rat parasympathetic cardiac ganglia. J Physiol (Lond) 434:215–237.
6. Evans RJ, Derkach V, Surprenant A (1992) ATP mediates fast synaptic transmission in mammalian neurons. Nature 357:503–505.
7. Silinsky EM, Gerzanich V, Vanner SM (1992) ATP mediates excitatory synaptic transmission in mammalian neurones. Br J Pharmacol 106: 762–763.
8. Silinsky EM, Gerzanich V (1993) On the excitatory effects of ATP and its role as a neurotransmitter in coeliac neurons of the guinea-pig. J Physiol 464:197–212.
9. Barajas-López C, Espinosa-Luna R, Gerzanich V (1994) ATP closes a potassium and opens a cationic conductance through different receptors in neurons of guinea pig submucous plexus. J Pharmacol Exp Therap 268:1396–1402.
10. Jahr CE, Jessell TM (1983) ATP excites a subpopulation of rat dorsal horn neurons. Nature 304:730–733.
11. Salt TE, Hill RG (1983) Excitation of single sensory neurons in the rat caudal trigeminal nucleus by iontophoretically applied adenosine 5′-triphosphate. Neurosci Lett 35:53–57.
12. Ueno S, Harata N, Inoue K, Akaike N (1992) ATP-gated current in dissociated rat nucleus solitarii neurons. J Neurophysiol 68:778–785.
13. Edwards FA, Gibb AJ, Colquhoun D (1992) ATP receptor-mediated synaptic currents in the central nervous system. Nature 359:144–147.
14. Hamill OP, Marty A, Neher E, Sakmann B, Sigworth FJ (1981) Improved patch-clamp techniques for high-resolution current recording from cells and cell-free membrane patches. Pflügers Arch 391:85–100.
15. Rae J, Cooper K, Gates P, Watsky M (1991) Low access resistance perforated patch recordings using amphotericin B. J Neurosci Meth 37: 15–26.
16. Bean BP, Friel DD (1990) ATP-activated channels in excitable cells. In Narahashi T (ed.), Ion Channels, Vol. 2. New York, Plenum Press: pp 169–203.
17. Kennedy C (1990) P_1- and P_2-purinoceptor subtypes—An update. Arch Int Pharmacodyn Ther 303:30–50.
18. Edwards FA, Gibbs AJ (1993) ATP—a fast neurotransmitter. FEBS Lett 325:86–89.
19. Krishtal OA, Marchenko SM, Obukhov AG (1988) Cationic channels activated by extracellular ATP in rat sensory neurons. Neuroscience 27:995–1000.

20. Bean BP, Williams CA, Ceelen PW (1990) ATP-activated channels in rat and bullfrog sensory neurons: Current-voltage relation and single-channel behavior. J Neurosci 10:11–19.
21. Nakazawa K, Fujimori K, Takanaka A, Inoue K (1990) An ATP-activated conductance in pheochromocytoma cells and its suppression by extracellular calcium. J Physiol 428:257–272.
22. Neuhaus R, Reber BFX, Reuter H (1991) Regulation of bradykinin- and ATP-activated Ca^{2+}-permeable channels in rat pheochromocytoma (PC12) cells. J Neurosci 11:3984–3990.
23. Benham CD, Tsien RW (1987) A novel receptor-operated Ca^{2+}-permeable channel activated by ATP in smooth muscle. Nature 328:275–278.
24. Friel DD (1988) An ATP-sensitive conductance in single smooth muscle cells from the rat vas deferens. J Physiol 401:361–380.
25. Li C, Aguayo L, Peoples RW, Weight FF (1993) Ethanol inhibits a neuronal ATP-gated ion channel. Mol Pharmacol 44:871–875.
26. Barajas-López C, Barrientos M, Espinosa-Luna R (1993) Suramin increases the efficacy of ATP to activate an inward current in myenteric neurons from guinea-pig ileum. Eur J Pharmacol 250:141–145.
27. Honore H, Martin C, Mironneau C, Mironneau J (1989) An ATP-sensitive conductance in cultured smooth muscle cells from pregnant rat myometrium. Am J Physiol 257:C294–C305.

16. DESENSITIZATION OF A_1 ADENOSINE RECEPTORS

Edmund Hoppe
Martin J. Lohse

It is well known that exposure of sensitive cells to receptor agonists can cause a progressive reduction in their capacity to respond to this and related agonists. This phenomenon has been observed with cells expressing receptors from different superfamilies, and has been termed *desensitization*, *refractoriness*, *tolerance*, or *tachyphylaxis* [1]. The mechanisms responsible for this phenomenon have been studied extensively in cells that synthesize cyclic AMP in response to beta-adrenergic agonists. Exposure of such cells to isoproterenol or similar drugs may result in receptor uncoupling, receptor sequestration, and ultimately, in downregulation of receptors, resulting in a decline in the actual number of receptors [2].

Much less is known about receptors that inhibit adenylyl cyclase activity such as opiate, $alpha_2$ adrenergic, or certain types of muscarinic receptors, but desensitization of those receptors after prolonged exposure to agonist has been described in many cases [1]. Several groups have investigated the mechanisms involved in the desensitization of the A_1 adenosine receptor as another member of the family of inhibitory receptors. In analogy to other G-protein–coupled receptors, there is evidence that A_1 adenosine receptors might also undergo desensitization after prolonged or repeated incubation with agonist.

Desensitization of A_1 Adenosine Receptor–Mediated Inhibition of Adenylyl Cyclase Activity

A_1 adenosine receptor–mediated inhibition of cAMP accumulation is one of the best-described effects of this receptor and has also been used to distinguish this receptor from the A_2 receptor subtypes [3]. As a convenient system for the investigation of A_1 adenosine receptor–mediated inhibition of cAMP synthesis, adipocytes have been used frequently because they contain very few A_2 adenosine receptors [4]. Single applications of R-PIA to intact rats or to isolated adipocytes induces inhibition of cAMP accumulation in adipocytes, which can be determined either directly or indirectly by changes in lipolysis resulting in the release of free fatty acids and glycerol [5]. Prolonged administration of R-N^6-phenyl-2-propyladenosine (R-PIA) in vivo or in vitro resulted in adipocytes that did not respond fully to acute application of A_1 receptor agonists [6–8].

Isolated adipocytes that had been exposed to R-PIA had increased rates of basal and isoproterenol-stimulated glycerol release, which correlated with a similar enhancement of cAMP accumulation in these cells. Inhibition of adenylyl cyclase by prostaglandin E_1 was also attenuated in adipocytes pretreated with R-PIA.

Therefore, this type of desensitization in adipocytes can be considered to be heterologous [7]. Because after such prolonged exposure to R-PIA different stimulators of adenylyl cyclase of adipocytes, acting at different sites, could induce enhanced adenylyl cyclase activity, it was concluded that several components of the adenylyl cyclase system are regulated by such sustained A_1 adenosine receptor activation [7].

In the heart, the contractile response of myocytes can be enhanced by cAMP, and adenosine analogs can, by inhibition of cAMP accumulation, exert negative inotropic responses in these cells [9–11]. Exposure of primary atrial myocytes to R-PIA for 24 hours in vitro has been shown to induce desensitization of A_1 adenosine receptor–mediated negative inotropic effects and inhibition of cAMP accumulation [12]. Desensitization of negative inotropic responses of right ventricle strips was also observed after administration of R-PIA to chicken embryos [13]. In both systems a decrease in maximum response but not in ED_{50} values for negative inotropic effects was observed. The fact that the efficacy of R-PIA in depressing contractile development of the strips was reduced while carbachol remained fully active implies a homologous pattern of desensitization in myocytes. Thus, the A_1 receptor desensitization in heart appears to be different from the adipocyte response, which involves heterologous desensitization. The reason for this discrepancy might be due to the low density of A_1 adenosine receptors and the moderate inhibition of adenylyl cyclase mediated by these receptors in the heart.

The hamster smooth muscle cell line DDT_1MF-2 has previously been shown to contain adenosine receptors of the A_1 and A_2 type [14]. Desensitization of A_1 adenosine receptors in this cell line seems to be similar to that in myocytes. Sustained incubation of these cells with R-PIA has been shown to attenuate the A_1 adenosine receptor–mediated inhibition of cAMP accumulation without adaptive upregulation of isoproterenol-, sodium fluoride-, or forskolin-stimulated adenylyl cyclase activities [15]. Therefore, long-term R-PIA administration appears to result in a homologous pattern of desensitization in this cell line.

Mechanisms of Desensitization

Desensitization of A_1 adenosine receptors with concomitant upregulation of basal and stimulated adenylyl cyclase activity seems to be associated with changes in expression levels of G proteins. In adipocyte membranes from rats pretreated with R-PIA for 6 days, pertussis toxin–mediated ADP-ribosylation of proteins with a molecular weight of 39,000 and 41,000 Da was reduced by about 40% [7]. In addition, an increase in cholera toxin–catalyzed labeling of 42 and 47 kD proteins was observed. These effects can be interpreted as downregulation of G_i and concurrent upregulation of G_s during exposure to R-PIA. This regulation of G proteins by adenosine analogs has been analyzed in more detail with specific antibodies directed against the different alpha subunits of G proteins. After chronic A_1 adenosine receptor stimulation in adipocytes in vivo, antibodies revealed about 50% increase in immunoreactivity for $G_{s\alpha}$, about 60% reduction for $G_{i\alpha 1+2}$, and no change for $G_{i\alpha 3}$ [16]. This type of regulation of G proteins was even more pronounced in primary cultures of rat adipocytes. Antibodies specific for different G_i subtypes revealed a dramatic loss of G proteins in R-PIA–pretreated adipocytes [17]. $G_{i\alpha 1}$ and $G_{i\alpha 3}$ immunoreactivity was shown to almost completely lost, while $G_{i\alpha 2}$ was reduced to about 50% in R-PIA–treated adipocytes.

This reduction in the amount of G_i immunoreactivity in the membranes correlated with a decrease in high-affinity binding sites for agonists as detected by [^{125}I]HPIA or by R-PIA in competition experiments with radiolabeled antagonists. Furthermore, it has been suggested that the reduction of G_i might be responsible for the heterologous desensitization of other receptors employing the same types of G proteins, while the increase in G_s appears compatible with the improved capability of stimulatory effectors to activate cAMP accumulation. The mechanism for the decrease in G_i levels is most likely by control of transcription via cAMP-response elements in the $G_{i\alpha}$ genes [18,19]. In contrast, in other cell lines or tissues, such as myocytes or DDT_1MF-2, different mechanisms seem to take place. It was noted that desensitization of A_1 adenosine receptors is not always related to adaptive increases in adenylyl cyclase activity or changes in the levels of G_i or G_s [13,15]. Desensitization of A_1 adenosine receptors in these cells has been postulated to be mediated by downregulation of receptor number (see below) or to be correlated with increased phosphorylation of the receptor as shown for DDT_1MF-2 cells [15]. This is consistent with the fact that phosphorylation of A_1 adenosine receptors by the beta-adrenergic receptor kinase can uncouple receptors from G proteins, as shown by induction of low-affinity binding and reduced adenosine-stimulated GTPase activity [20].

Downregulation of A_1 Adenosine Receptors by Chronic Agonist Treatment

An additional feature of desensitization of adenosine receptors seems to be the downregulation of binding sites during chronic incubation with agonist. R-PIA infusion into rats resulted in an about 30% loss of binding sites in membranes of adipocytes as determined with [^{125}I]HPIA or [^{3}H]PIA binding [6,7]. Prolonged treatment of isolated adipocytes with A_1 adenosine receptor agonists in vitro seems to induce an even more pronounced downregulation of binding sites that can be recognized by agonists [8]. This downregulation of binding sites detected by agonists is consistent with the depletion of inhibitory G proteins and consequently an induction of receptors having a low-affinity conformation as described above. However, after exposure to R-PIA, downregulation of [^{125}I]aminobenzyl-adenosine binding sites was also observed in cell lines that did not undergo adaptive upregulation of adenylyl cyclase and where alterations of G protein levels are not likely [13].

Furthermore, downregulation of A_1 adenosine receptors after chronic R-PIA administration has also been observed using antagonist radioligands in binding studies. Long-term R-PIA infusion of rats resulted in about 50% downregulation of A_1 adenosine receptors in adipocyte membranes as determined with the antagonist [^{3}H]xanthine amine congener (XAC) [16]. Radioligand binding studies performed with [^{3}H]DPCPX or [^{3}H]XAC indicated decreases in B_{max} values of about 40% in isolated myocytes or DDT_1MF-2 cells [12,15]. Because levels of inhibitory G proteins were shown to be unchanged in these cells, downregulation of A_1 adenosine receptors after chronic exposure to agonist might be a major mechanism of desensitization in these cell lines.

Upregulation of A_1 Adenosine Receptors by Chronic Antagonist Treatment

Other investigations addressed the question of whether administration of antagonists such as theophylline or caffeine can prevent downregulation by romoval of tonic stimulation of A_1 adenosine receptors by endogenous levels of adenosine and thus cause an apparent upregulation of receptors. In several publications it has been described that chronic administration of theophylline or caffeine to animals can induce an increase in specific binding of [^{3}H]N^6-cyclohexyl-adenosine (CHA) or [^{3}H]PIA to membranes prepared from brain [21–24].

The upregulation of agonist binding seems to display a heterogeneous neuroanatomic pattern, although reports are contradictory. According to one study chronic theophylline administration significantly increased [^{3}H]PIA binding to membranes from rat cortex and cerebellum but caused only a slight increase in binding to membranes from hippocampus and diencephalon [25]. Zielke and Zielke [26] reported an increase in [^{3}H]PIA binding after chronic theophylline treatment only for cerebellar membranes but not for those prepared from cerebral cortex. Based on autoradiographic studies, Lupica et al. [27] concluded that there is a general increase of [^{3}H]CHA binding in rat brain after chronic R-PIA administration, with greatest increases in specific layers of the hippocampus, thalamus, and cerebellum. Using the same approach it has been concluded that stimulation of [^{3}H]CHA binding in brain slices from rats pretreated with theophylline was generally slight but was more easily revealed when the radioligand was used together with low concentrations of GTP [28]. The differences might be explained by species differences or by different technical approaches, such as binding to membranes versus slices, because in the latter case it was more difficult to remove endogenous adenosine.

In some studies this antagonist-induced enhancement of agonist binding seems to be due to interference with downregulation of inhibitory G proteins by endogenous adenosine and thus to a shift of low-affinity receptors to a high-affinity state. For instance, it has been reported that [^{3}H]PIA identified two affinity states of the A_1 receptors in membranes from untreated animals, but only one state that apparently corresponded to the high-affinity conformation in caffeine-pretreated animals [21]. The absolute amounts of receptors with high affinity for agonists were increased after long-term agonist treatment, suggesting a shift from low-affinity receptors to a high-affinity conformation, possibly due to improved coupling of those receptors to G proteins after pretreatment with xanthines.

Szot et al. [25] resolved data of [^{3}H]CHA binding to rat cortical membranes into two affinity states, and an increase in capacity of both high- and low-affinity receptors with no change in affinity of each conformational state was observed in theophylline-treated rats. The upregulation of high-affinity binding (72%) was more pronounced than increases of low-affinity binding (32%), indicative of an induction of high-affinity binding in addition to true upregulation of receptor number.

Further evidence for upregulation of A_1 receptor number after long-term treatment of animals with theophylline or caffeine was provided by binding experiments with antagonists that bind to receptors independently of their coupling to G proteins. Using the antagonist [^{3}H]DPX, a dose-dependent increase in the number of binding sites in membranes from mouse brain has been described after chronic caffeine administration [29,30]. However, because DPX is not specific for A_1 adenosine receptors, these data should be considered with caution. Using the A_1-specific ligand [^{3}H]-8-cyclopenty 1,3 dipropylxanthine ([^{3}H]DPCPX), a 15% increase in binding sites in rat cerebral cortical membranes was seen after 21 days of theophylline administration [31], indicating that there might be true upregulation of the amount of receptors. The mechanism of this increase in total amount of receptors is not yet clear, but it does not seem to involve enhanced expression, because mRNA levels of A_1 adenosine receptors have been shown to be unchanged during adaptation of rats to caffeine [32]. The upregulation of A_1 adenosine receptors has been shown to correlate with an improvement in signaling in terms of inhibition of adenylyl cyclase activity, noradrenaline release, or seizure generation, indicating that functionally active receptors were generated [21,25,33].

In conclusion, desensitization of A_1 adenosine receptors in adipocytes seems to involve a general heterologous adaptation with remarkable changes in G-protein expression with concomitant induction of low-affinity receptors and enhancement of cAMP accumulation. In these cells, downregulation of A_1 adenosine receptors does not seem to contribute markedly to the process of desensitization. This might be different in cell lines that do not show a general heterologous adaptation of adenylyl cyclase activity, such as cardiac myocytes or DDT_1MF-2 cells. In these cell lines desensitization of A_1 adenosine receptors might also include uncoupling of receptors by phosphorylation due to receptor specific kinases, and downregulation of total receptor number. Upregulation of receptors in response to chronic antagonist treatment indicates that this downregulation seems to occur under physiologic conditions, probably due to a tonic stimulation of receptors by endogenous adenosine. Downregulation of A_1 adenosine receptors might be even more pronounced under pathologic conditions with accompanying increases in the levels of endogenous adenosine. Therefore, it is not surprising that downregulation of receptors has been observed following brief anoxia and has been correlated with hyperexcitability of CA1 pyramidal cells, which might contribute to vulnerability of this region after anoxia [33].

References

1. Lohse MJ (1993) Molecular mechanisms of membrane receptor desensitization. Biochem Biophys Acta 1179:171–188.
2. Hausdorff WP, Caron MG, Lefkowitz RJ (1990) Turning off the signal: Desensitization of β-adrenergic receptor function. FASEB J 4: 2881–2889.
3. Van Calker D, Mueller M, Hamprecht B (1979) Adenosine regulates via two different types of receptors, the accumulation of cyclic AMP in cultured brain cells. J Neurochem 33: 999–1005.
4. Trost T, Schwabe U (1980) Adenosine receptors in fat cells. Mol Pharmacol 19:228–235.
5. Schwabe U, Ebert R, Erbler HC (1973) Adenosine release from isolated fat cells and its significance for the effects of hormones on cyclic 3′,5′-AMP levels and lipolysis. Naunyn Schmiedebergs Arch Pharmacol 276:133–148.
6. Hoffman B, Chang H, Dall'Aglio E, Reaven G (1986) Desensitization of adenosine receptor-mediated inhibition of lipolysis. The mechanism involves the development of enhanced cyclic adenosine monophosphate accumulation in tolerant adipocytes. J Clin Invest 78:185–190.
7. Parsons WJ, Stiles GL (1987) Heterologous desensitization of the inhibitory A_1 adenosine receptor adenylate cyclase system in rat adipocytes. J Biol Chem 262:841–847.
8. Green A (1987) Adenosine receptor down-regulation and insulin resistance following prolonged incubation of adipocytes with an A_1 adenosine receptor agonist. J Biol Chem 262: 15702–15707.
9. Schrader J, Baumann G, Gerlach E (1977) Adenosine as inhibitor of myocardial effects of catecholamines. Pflügers Arch 372:29–35.
10. Baumann G, Schrader J, Gerlach E (1981) Inhibitory action of adenosine on histamine- and dopamine-stimulated cardiac contactility and adenylate cyclase in guinea pigs. Circ Res 48: 259–266.
11. Martens D, Lohse MJ, Rauch B, Schwab U (1987) Pharmacological characterization of A_1 adenosine receptors in isolated rat ventricular myocytes. Naunyn-Schmiedebergs Arch Pharmacol 336:342–348.
12. Liang BT, Donovan LA (1990) Differential desensitization of A_1 adenosine receptor-mediated inhibition of cardiac myocyte contractility and adenylate cyclase activity. Relation to the regulation of receptor affinity and density. Circ Res 67:406–414.
13. Shryock J, Patel A, Belardinelli L, Linden J (1989) Down-regulation and desensitization of A_1-adenosine receptors in embryonic chicken heart. Am J Physiol 256:H321–H327.
14. Gerwins P, Nordstedt C, Fredholm BB (1990) Characterization of adenosine-A_1-receptors in intact DDT_1MF-2 smooth-muscle cells. Mol Pharmacol 38:660–666.
15. Ramkumar V, Olah ME, Jacobson KA, Stiles GL (1991) Distinct pathways of desensitization of A_1- and A_2-adenosine receptors in DDT_1MF-2 cells. Mol Pharmacol 40:639–647.
16. Longabaugh JP, Didsbury J, Spiegel A, Stiles GL (1989) Modification of the rat adipocyte-A_1 adenosine receptor-adenylate cyclase system during chronic exposure to an A_1 adenosine receptor agonist—alterations in the quantity of $G_{s\alpha}$ and $G_{i\alpha}$ are not associated with changes in their m-RNAs. Mol Pharmacol 36:681–688.
17. Green A, Johnson JL, Milligan G (1990) Down-regulation of G_i subtypes by prolonged incubation of adipocytes with an A_1 adenosine receptor agonist. J Biol Chem 265:5206–5210.
18. Weinstein LA, Kats I, Spiegel AM, Carter AD (1990) Characterization of the promoter of the human $G_{i2\alpha}$-subunit gene. Mol Endocrinol 4:958–964.
19. Eschenhagen T, Hollmann A, Friedrichsen M,

Schmitz W, Weinstein LS (1994) Regualtion of promoter activity of the human $G_{I\alpha\text{-}2}$ gene in embryonic chick cardiomyocytes. Naunyn Schmiedebergs Arch Pharmacol 349(Supplement): R11.
20. Ramkumar V, Kwatra M, Benovic JL, Stiles GL (1993) Functional consequences of A_1 adenosine-receptor phosphorylation by the β-adrenergic receptor kinase. Biochim Biophys Acta 1179: 89–97.
21. Green R, Stiles G (1986) Chronic caffeine ingestion sensitizes the A_1 adenosine receptor-adenylate cyclase system in rat cerebral cortex. J Clin Invest 77:222–227.
22. Fredholm BB (1982) Adenosine actions and adenosine receptors after 1 week treatment with caffeine. Acta Physiol Scand 115:283–286.
23. Murray TF (1982) Up-regulation of rat cortical adenosine receptors following chronic administration of theophylline. Eur J Pharmacol 82: 113–114.
24. Lupica CR, Jarvis MF, Berman RF (1991) Chronic theophylline treatment in vivo increases high affinity adenosine A_1 receptor binding and sensitivity to exogenous adenosine in the in vitro hippocampal slice. Brain Res 542:55–62.
25. Szot P, Sanders R, Murray T (1987) Theophylline-induced upregulation of A_1-adenosine receptors associated with reduced sensitivity to convulsants. Neuropharmacology 26: 1173–1180.
26. Zielke C, Zielke H (1987) Chronic exposure to subcutaneously implanted methylxanthines. Differential elevation of A_1-adenosine receptors in mouse cerebellar and cerebral cortical membranes. Biochem Pharmacol 36:2533–2538.
27. Lupica CR, Berman RF, Jarvis MF (1991) Chronic theophylline treatment increases adenosine A_1, but not A_2, receptor binding in the rat brain: An autoradiographic study. Synapse 9: 95–102.
28. Fastbom J, Fredholm BB (1990) Effects of long-term theophylline treatment on adenosine-a1-receptors in rat-brain—autoradiographic evidence for increased receptor number and altered coupling to G-proteins. Brain Res 507: 195–199.
29. Boulenger J-P, Patel J, Post RM, Parma AM, Marangos PJ (1983) Chronic caffeine consumption increases the number of brain adenosine receptors. Life Sci 32:1135–1142.
30. Marangos PJ, Boulenger J-P, Patel J (1984) Effects of chronical caffeine on brain adenosine receptors: Regional and ontogenetic studies. Life Sci 34:899–907.
31. Sanders RC, Murray TF (1988) Chronic theophylline exposure increases agonist and antagonist binding to A_1 adenosine receptors in rat brain. Neuropharmacology 27:757–760.
32. Johansson B, Ahlberg S, van der Ploeg I, Brene S, Lindefors N, Persson H, Fredholm BB (1993) Effect of long term caffeine treatment on A1 and A2 adenosine receptor binding and on mRNA levels in rat brain. Naunyn Schmiedebergs Arch Pharmacol 347:407–414.
33. Fredholm BB, Jonzon B, Lindgren E (1984) Changes in noradrenaline release and in beta receptor number in rat hippocampus following long-term treatment with theophylline or L-phenylisopropyladenosine. Acta Physiol Scand 122:55–59.
33. Lee K, Tetzlaff W, Kreutzberg G (1986) Rapid down regulation of hippocampal adenosine receptors following brief anoxia. Brain Res 380:155–158.

II. MEDICINAL CHEMISTRY

17. SELECTIVE A_2 ADENOSINE RECEPTOR AGONISTS WITH POTENT INHIBITORY ACTIVITY ON PLATELET AGGREGATION

Gloria Cristalli
Emidio Camaioni
Sauro Vittori
Rosaria Volpini

Recently, we have reported on the synthesis of the 2-hexynyl derivative of NECA, N-ethyl-1′-deoxy-1′-(6-amino-2-hexynyl-9H-purin-9-yl)-β-D-ribofuranuronamide, identified as HENECA (Fig. 17-1), which was shown to possess high affinity at A_2 adenosine receptors combined with a good A_2 versus A_1 selectivity [1]. Table 17-1 shows the ability of HENECA to inhibit the binding of radioligands to A_1 and A_2 receptors compared with that of NECA [2], 2-[[4-(2-carboxyethyl) phenethyl]amino] adenosine-5′-N-ethyluronamide (CGS 21680) [3] and of the selective A_1 agonist 2-chloro-N^6-cyclopentyl-adenosine (CCPA) [4]. In the same table the effects of these compounds on the stimulation of platelet and PC12 cell membrane adenylate cyclase are also shown [5]. The results of the cyclase assays are in good agreement with the higher affinity of HENECA of the A_2 receptors compared with the other adenosine agonists listed in Table 17-1.

In Table 17-2 is shown the negative chronotropic activity, tested in spontaneously beating rat atria, and the vasodilating activity in rat aorta of the four compounds [6], and activity in inhibiting rabbit [4] and human [5] platelet aggregation induced by ADP. The IC_{50} of NECA is 0.21 μM in the rabbit and 0.36 μM in human platelets. CCPA is a very poor inhibitor due to its A_1 selectivity. CGS 21680 is less potent than NECA, while HENECA was three to seven times more potent than the model compound.

Moreover, Dionisotti and coworkers demostrated that NECA and HENECA exhibit effective in vivo inhibitory activity on platelet function in the rabbit, whereas the selective adenosine A_1 agonist CCPA is ineffective. This antiaggregatory activity is of short duration and does not appear to be due to hemodynamic changes but rather is a result of selective drug action on platelet adenosine receptors [7].

The therapeutic potential of HENECA for treatment of cardiovascular diseases prompted us to synthesize a number of new 2-alkynyl and cycloalkynyl derivatives of NECA bearing hydroxyl, amino, chloro, and cyano groups or substituted aromatic or heteroaromatic rings in the side chain (Table 17-3) [8; Cristalli et al., submitted].

Methods

Three different approaches have been used for the synthesis of the 2-alkynyl derivatives due to the different reactivities of the side chains and separation problems during purification. Briefly, from the common intermediate, the 5′-carboxy-2′,3′-isopropyliden derivative (Fig. 17-2), the two carboxamido derivatives 4 and 5 were obtained as starting material for the three synthetic routes, depicted in Figure 17-2. The introduction of the alkynyl chain was carried out by modification of the classical palladium-catalyzed cross-coupling reaction [9].

Interaction of the new 2-alkynyl derivatives of NECA 7–50 with adenosine receptors was evaluated using both the radioligand binding technique and functional assays. Affinity for A_2 receptors was determined in competition assays in rat striatum using [^{3}H]CGS 21680 as a radioligand [10]. Affinity for A_1 receptors was determined in competition asssays in rat brain using [^{3}H]CHA [11]. NECA, HENECA [1], and CCPA [12] were included as reference compounds. Negative chronotropic activity (A_1) was tested in spontaneously beating rat atria and vasodilation (A_2) in rat aorta according to a method previously described [6].

The new derivatives were also tested as inhibitors of rabbit platelet aggregation induced by ADP [4], and the results are reported as

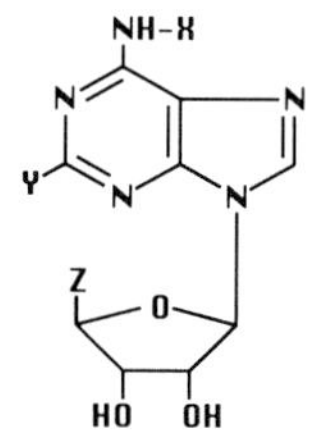

Compd.	X	Y	Z
HENECA	H	$C{\equiv}C(CH_2)_3CH_3$	CONHEt
NECA	H	H	CONHEt
CGS 21680	H	$NH(CH_2)_2Ph(CH_2)_2CO_2H$	CONHEt
CCPA	cC_5H_9	Cl	CH_2OH

FIGURE 17-1.

TABLE 17-1. Activity of HENECA on binding and cyclase assays

	Binding assay[a] Ki (nM)[b]		Adenylate cyclase[c] EC_{50} (nM)[b]	
Compound	Rat brain A_1	Rat striatum A_2	Human Platelet	PC12 cell
HENECA	130 (116–145)	2.2 (1.9–2.6)	14 (13–15)	4.1 (3.7–4.6)
NECA	10.4 (9.4–11.5)	7.8 (6.6–9.1)	110 (110–112)	99 (82–116)
CGS 21680	569 (511–634)	11.0 (9.4–12.9)	83 (78–88)	72 (66–78)
CCPA	1.3 (1.1–1.4)	650 (555–762)	1200 (1000–1400)	1200 (1170–1230)

[a] Receptor binding affinity at A_1 and A_2 receptors was determined using [^{3}H]CHA and [^{3}H]CGS21680 as radioligands, respectively [8].
[b] Data are geometrical means from at least three separate experiments; 95% confidence limits are in parenthesis.
[c] Data are from Cristalli et al. [5].

TABLE 17-2. Activity of HENECA on isolated tissues and on platelet aggregation

	Functional activity[a] EC_{50} (nM)[b]		Antiaggregatory activity[c] IC_{50} (nM)[b]	
Compound	Rat atria A_1	Rat aorta A_2	Rabbit platelet[d]	Human Platelet[e]
HENECA	>10 μM	596 (244–1460)	70 (30–150)	50 (41–60)
NECA	54.8 (30–150)	394 (209–742)	210 (160–280)	360 (350–380)
CGS 21680	>10 μM	115 (53–251)	2160 (1620–3620)	820 (640–1100)
CCPA	8.2 (4.4–15.3)	>10 μM	>10 μM	>10 μM

[a] Data are from Cristalli et al. [8].
[b] Data are means from at least three separate experiments; 95% confidence limits are in parenthesis.
[c] Platelet aggregation was induced by ADP.
[b] Data are from Dionisotti et al. [4].
[e] Data are from Cristalli et al. [5].

potency ratio versus NECA in Table 17-3. Retention of nucleosides on a reverse-phase HPLC column is reported as a measure of the relative hydrophobicity [13], and the values are listed in Table 17-3. This hydrophobicity index (k′) is calculated by the formula $k' = (t - to)/to$, where *to* represents the transit time of the solvent and *t* the retention time of each compound.

Results and Discussion

The results of binding assays and functional activity are reported elsewhere [8; Cristalli et al., submitted] and showed that there were compounds with subnanomolar activity in the rat striatal binding experiments, and some of the compounds were very potent in inducing vasorelaxation without an appreciable effect on heart rate. Moreover, from the data reported in Table 17-3 it is possible to point out some structure-activity relationships in the potency of alkynyl derivatives as platelet aggregation inhibitors.

While the presence of an aromatic or heteroaromatic ring directly conjugated to the triple bond made the compounds less active than NECA, the introduction of a hydroxyl group in the alkynyl chain led to the most potent platelet aggregation inhibitors to date. In Table 17-4 are reported only the derivatives that are more potent than HENECA itself, which showed an IC_{50} of 70 nM. All of them bear an α-hydroxyl group or, in the case of the last compound, a δ-hydroxyl group, in which, however, the presence of four methylene groups might confer a flexibility to mimic the α-position.

The importance of the hydroxyl group is stressed by comparing the three benzyl derivatives shown in Table 17-5. The hydroxyl group did not markedly influence the compounds' A_2 binding affinity, whereas its introduction into the benzyl group conferred an antiaggregatory potency 80-fold that of the parent compound and 16-fold that of NECA. Hence, the α-hydroxybenzyl derivative resulted in the most potent inhibitor in the series, with an IC_{50} of about 13 nM. However, the α-methylation markedly reduced antiplatelet activity, heart rate, and vasorelaxant activity without affecting the A_2 binding affinity, as shown by the last two componds in Table 17-5.

TABLE 17-3. Inhibition of rabbit platelet aggregation induced by ADP and hydrophobicity index of 2-alkynyl derivatives 7–50

Compound	R[a]	Antiaggr. activity[b] Potency ratio vs. NECA	k'^{c}
7	H	0.30	0.25
8	CH_2OH	2.30	0.17
9	$(CH_2)_2OH$	1.10	0.20
10	$(CH_2)_3OH$	2.20	0.22
11	$(CH_2)_4OH$	4.80	0.35
12	$CH(OH)CH_3$	4.70	0.23
13	$CH(OH)CH_2CH_3$	14.10	0.42
14	$CH_2CH(OH)CH_3$	3.20	0.29
15	$C(OH,CH_3)CH_2$-$CH(CH_3)CH_3$	0.54	1.84
16	1-Hydroxycyclopentyl	5.30	0.77
17	CH(OH)Ph	15.70	0.82
18	$C(OH,CH_3)$Ph	0.26	1.32
19	$(CH_2)_2$OPyr[d]	0.30	1.32
20	$CH(OCH_2CH_3)_2$	0.02	1.63
21	CH_2NH_2	0.40	0.15
22	$CH_2N(CH_3)_2$	2.30	0.37
23	1-Aminocyclohexyl	0.60	0.60
24	$(CH_2)_3Cl$	2.30	1.35
25	$(CH_2)_3CN$	2.10	0.32
26	1-Methylvinyl	0.66	1.27
27	1-Cyclohexen	0.03	3.90
28	Cyclohexyl	0.66	4.54
29	Ph	0.010	2.04
30	p-$PhCH_3$	0.013	3.59
31	p-$PhCh_2CN$	0.032	0.76
32	p-$PhOCH_3$	0.004	2.15
33	p-PhOH	0.006	0.66
34	p-$PhNH_2$	0.007	0.46
35	p-$PhCF_3$	0.01	5.79
36	p-PhF	0.009	2.33
37	p-$PhCONH_2$	0.003	0.32
38	p-$PhCOCH_3$	0.019	1.18
39	o-PhCHO	0.006	1.14
40	m-PhCHO	0.004	1.06
41	p-PhCHO	0.110	1.02
42	p-$Ph(CH)_2$COOtBu	0.009	2.28
43	p-$PhNO_2$	0.014	1.82
44	1-Naphtyl	0.003	5.49
45	CH_2Ph	0.01	2.40
46	$(CH_2)_2$Ph	0.19	3.16
47	$(Ch_2)_3$Ph	0.35	5.82
48	2-pyridyl	0.01	0.49
49	2-thienyl	0.023	1.63
50	2-thiazolyl	0.44	0.55

[a] The whole structure of compounds is reported in Figure 17-2.

[b] The potency ratio was calculated using the concentration of the test compound close to the IC_{50} value. In our experimental conditions, the IC_{50} value for NECA was 0.2 μM.

[c] Hydrophobicity index. Data are means from three separate experiments.

[d] Pyr = 2-tetrahydro(2H)pyrane.

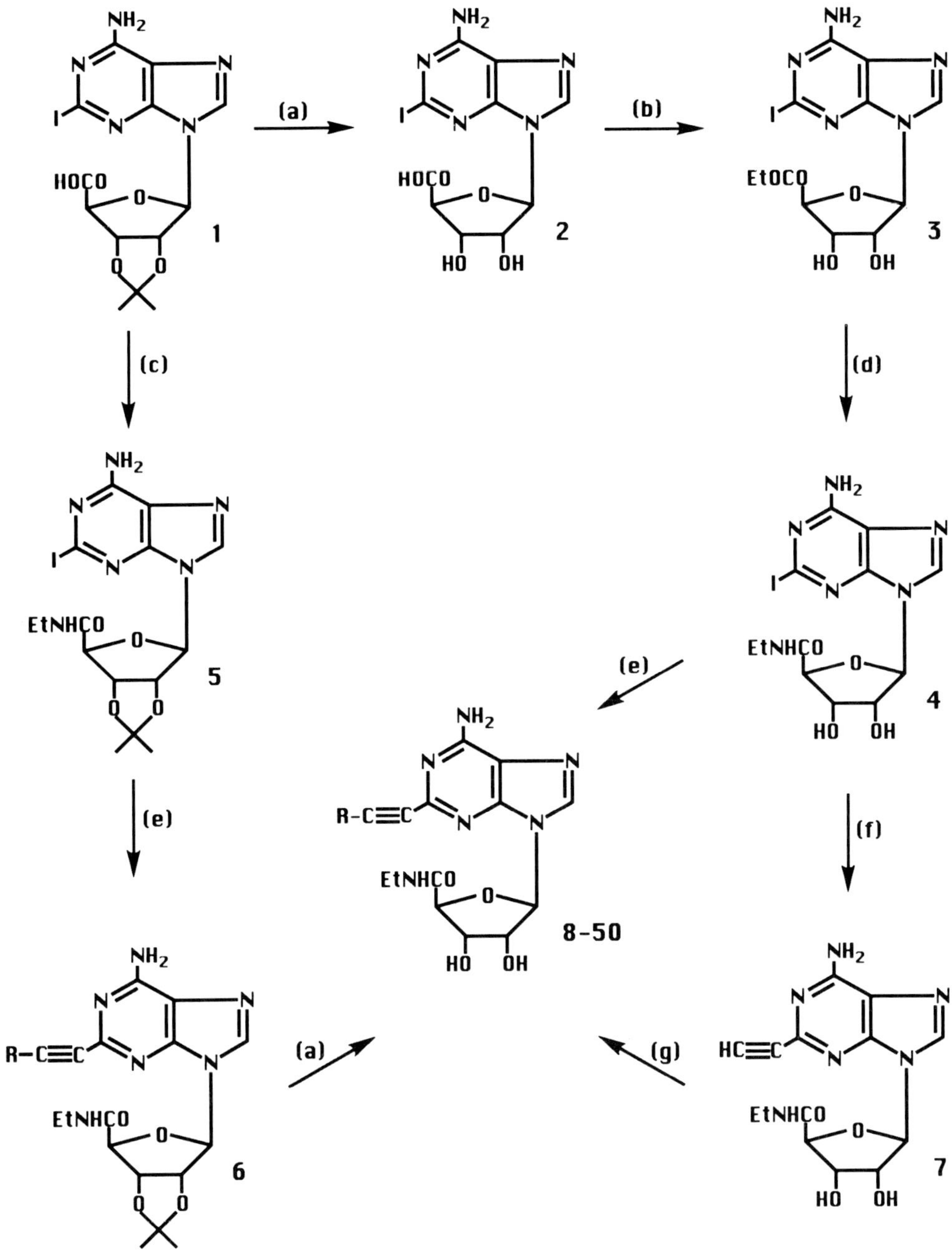

a) 50% HCOOH; b) $SOCl_2$/EtOH; c) $SOCl_2/EtNH_2$; d) $EtNH_2$;
e) R-C≡CH, $(Ph_3P)_2PdCl_2$, CuI; f) HC≡CH, $(Ph_3P)_2PdCl_2$, CuI;
g) R-X, X= I or Br.

FIGURE 17-2.

TABLE 17-4. Structure-activity relationships in α-hydroxyl derivatives.

Compd	R[a]	Antiaggr. activity IC_{50} (nM)[b]	A_2 Binding assay Ki (nM)
17	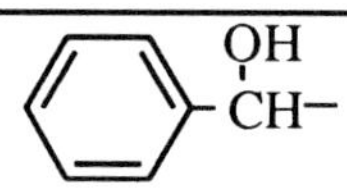	13	0.9
13	$CH_3-CH_2-CH(OH)-$	15	12.4
16	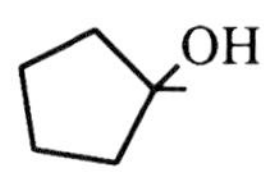	40	0.6
12	$CH_3-CH(OH)-$	45	7.6
11	$HO-CH_2-CH_2-CH_2-CH_2-$	44	6.8

[a] The whole structure of compounds is reported Figure 17-2.
[b] The IC_{50} values are calculated from the potency ratios listed Table 17-3. The IC_{50} value for NECA is 0.2 μM.

TABLE 17-5. Structure-activity relationships in benzyl derivatives.

Compd	R[a]	Antiaggr. activity IC_{50} (nM)[b]	A_2 Binding assay Ki (nM)
45	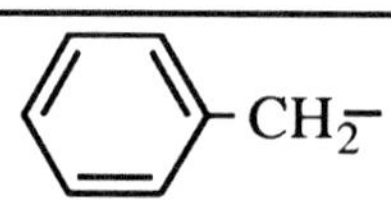	1100	1.6
17	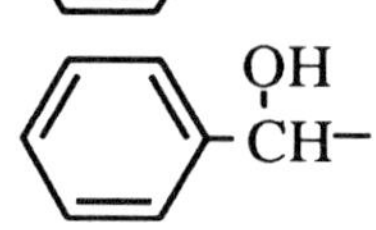	13	0.9
18	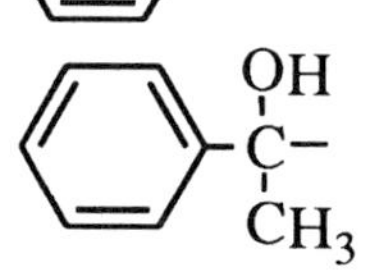	810	1.7
15	$CH_3-CH(CH_3)-CH_2-C(OH)(CH_3)-$	390	0.5

[a] The whole structure of compounds is reported in Figure 17-2.
[b] The IC_{50} values are calculated from the potency ratios listed in Table 17-3. The IC_{50} value for NECA is 0.2 μM.

The platelet receptor preference for the α-hydroxyl group is also demonstrated by comparing the homologous series reported in Table 17-6. Again, the position of the hydroxyl group and chain length are more crucial for the platelet inhibitory activity than for A_2 binding affinity.

In conclusion, introduction of an α-hydroxyl group in the alkynyl side chain increased the antiaggregatory activity in comparison with

TABLE 17-6. Structure-activity relationships in hydroxyl derivatives.

Compd	R[a]	Antiaggr. activity IC_{50} (nM)[b]	A_2 Binding assay Ki (nM)
13	$CH_3-CH_2-CH(OH)-$	15	12
14	$CH_3-CH(OH)-CH_2-$	66	56
10	$CH_2(OH)-CH_2-CH_2-$	95	11
12	$CH_3-CH(OH)-$	45	7.6
9	$CH_2(OH)-CH_2-$	190	11

[a] The whole structure of compounds is reported in Figure 17-2.
[b] The IC_{50} values are calculated from the potency ratios listed in Table 17-3. The IC_{50} value for NECA is 0.2 μM.

NECA and HENECA itself, resulting in the most potent inhibitors of platelet aggregation thus far known in the nucleoside series. However, the presence of an α-quaternary carbon markedly reduced the antiaggregatory potency without affecting the A_2 binding affinity. The introduction of a conjugated aromatic or heteroaromatic ring decreased antiplatelet activity as well as binding affinity but in some cases increased the A_2 selectivity.

The hydrophobicity index of the new nucleosides barely correlated with the binding data, whereas high k′ values were associated with increased A_2 versus A_1 selectivity but with reduced activity in all functional assays, including platelet aggregation. In conclusion, these findings lend further support to the hypothesis that the A_2 receptor on platelets is not a typical A_{2a} recognition site, as suggested in a recent study conducted in a series of 2-(ar)alkoxyadenosine derivatives [5].

Acknowledgment

We wish to thank M. Brandi, F.Lupidi, and G. Rafaiani for technical assistance.

References

1. Cristalli G, Eleuteri A, Vittori S, Volpini R, Lohse MJ, Klotz K-N (1992) 2-Alkynyl derivatives of adenosine and adenosine-5′-N-ethyluronamide as selective agonists at A_2 adenosine receptors. J Med Chem 35:2363–2368.
2. Bruns RF, Lu GH, Pugsley TA (1986) Characterization of the A_2 adenosine receptor labeled by [^{3}H]NECA in rat striatal membranes. Mol Pharmacol 29:331–346.
3. Hutchison AJ, Webb RL, Oei HH, Ghai GR, Zimmerman MB, Williams M (1989) CGS 21680C, an A_2 selective adenosine receptor agonist with preferential hypotensive activity. J Pharmacol Exp Ther 251:47–55.
4. Dionisotti S, Zocchi C, Varani K, Borea PA, Ongini E (1992) Effects of adenosine derivatives on human and rabbit platelet aggregation. Correlation of adenosine receptor affinities and antiaggregatory activity. Naunyn Schmiedebergs Arch Pharmacol 347:673–676.
5. Gristalli G, Vittori S, Thompson RD, Padgett WL, Sci D, Daly JW, Olsson RA (1994) Inhibition of platelet aggregation by adenosine receptor agonists. Naunyn Schmiedebergs Arch Pharmacol 349:644–650.
6. Conti A, Monopoli A, Gamba M, Borea PA, Ongini E (1993) Effect of selective A_1 and A_2 adenosine receptor agonists on cardiovascolar tissues. Naunyn Schmiedebergs Arch Pharmacol

347:647–676.
7. Sandoli D, Chiu PJS, Chintala M, Dionisotti S, Ongini E (1994) In vivo and ex vivo effects of adenosine A_1 and A_2 receptor agonists on platelet aggregation in the rabbit. Eur J Pharmacol, in press.
8. Cristalli G, Volpini R, Vittori S, Camaioni E, Monopoli A, Conti A, Dionisotti S, Zocchi C, Ongini E (1994) 2-Alkynyl derivatives of adenosine-5′-N-ethyluronamide (NECA): Selective A2 adenosine receptor agonists with potent inhibitory activity on platelet aggregation. J Med Chem 37:1720–1726.
9. Matsuda A, Shinozaki M, Miyasaka T, Machida H, Abiru T (1985) Palladium-catalized cross-coupling of 2-iodoadenosine with therminal alkynes: Synthesis and biological activities of 2-alkynyladenosines. Chem Pharm Bull 33: 2551–2554.
10. Jarvis MF, Schultz R, Hutchison AJ, Do UH, Sills MA, Williams M (1989) [^{3}H]CGS 21680, a selective A_2 adenosine receptor agonist directly labels A_2 receptors in rat brain. J Pharmacol Exp Ther 251:888–893.
11. Bruns RF, Daly JW, Snyder SH (1980) Adenosine receptors in brain membranes: Binding of N^6-cyclohexyl-[^{3}H]adenosine and 1,3-diethyl-8[^{3}H]phenylxanthine. Proc Natl Acad Sci USA 77:5547–5541.
12. Lohse MJ, Klotz K-N, Schwabe U, Cristalli G, Vittori S, Grifantini M (1988) 2-Chloro-N^6-cyclopentyladenosine: A highly selective agonist at A_1 adenosine receptors. Naunyn Schmiedebergs Arch Pharmacol 337:687–689.
13. Brent DA, Sabatka JJ, Minick DJ, Henry DW (1983) A simplified high-pressure liquid chromatography method for determining lipophilicity for structure-activity relationships. J Med Chem 26:1014–1020.

18. NOVEL ATP AGONISTS REVEAL RECEPTOR HETEROGENEITY WITHIN P_{2X} AND P_{2Y} SUBTYPES

Kenneth A. Jacobson
Bilha Fischer
Michel Maillard
José L. Boyer
Charles H.V. Hoyle
T. Kendall Harden
Geoffrey Burnstock

The selection of analogues of adenosine 5′-triphosphate (ATP) useful as P_2 receptor agonists is limited in number [1] relative to most other cell membrance receptors. It is our goal to establish systematically the structure-activity relationships for newly synthesized derivatives in much the same fashion as has been used for adenosine receptors [2]. Impeding this development is the current lack of high-affinity radioligands for general applicability to P_2 receptor subtypes. For example, [^{35}S]β-thio-ADP, which was shown to bind specifically to P_{2Y} receptors in turkey erythrocyte membranes [3], binds to a site in bovine brain membranes that does not resemble P_{2Y} receptors in rank order of ligand potencies [4]. Another factor that contributes to the lack of progress in the development of P_2 ligands is the incomplete knowledge of subtype classification [5,19]. The latter difficulty will be alleviated as the cloning of P_2 receptors progresses [6–8].

The structure-activity relationships for a variety of adenine nucleotide analogues at P_{2X} and P_{2Y} purinoceptors were investigated. New ATP analogues containing modifications at the triphosphate, ribose 2′ or 3′, purine C2 or C8, or at the purine N^6 positions were synthesized [9–11]. An important strategy in our investigation of the structure and function of purinoceptors has been the synthesis of new drug analogues using a functionalized congener approach [12]. By this approach, positions for attachment of chains on a pharmacophore are empirically probed, leading to knowledge of structure-activity relationships at distal sites on a molecule. The site of attachment must correspond to a region of relaxed steric requirements at or near the receptor binding site, thus offering flexibility of substitution. Reporter groups, for example, for fluorescent or radioactive labeling, may be attached at this distal site on a functionalized chain, as has been demonstrated for adenosine receptors [13]. This strategy has allowed us to target accessory sites of favorable interaction on the receptor and actually to enhance the affinity of purinoceptor ligands.

The activity of the compounds at P_{2Y} purinoceptors was established in dose-response studies of nucleotide-promoted phospholipase C activity in turkey erythrocyte membrances [14], and the relaxation of smooth muscle in three different preparations (guinea-pig taenia coli, rabbit aorta, and rabbit mesenteric artery) [15]. ATP-induced relaxation of the aorta occurs via an endothelium-dependent P_{2Y} receptor, and relaxation of the mesenteric artery occurs via an endothelium-independent P_{2Y} receptor. Activity at P_{2X} purinoceptors was established by measurement of the contraction of rabbit saphenous artery and of guinea-pig vas deferens and urinary bladder [15].

Functionalized congeners of 2-methylthioadenosine-5′-triphosphate (2-MeSATP), a potent P_{2Y}-receptor agonist, were synthesized as receptor probes [9,10]. The strategy consisted of attaching substituted alkyl thio chains at the 2-thio position. This site of chain elongation preserved high potency at P_{2Y} receptors, thus proving that this is a position on the ATP molecule having great structural tolerance in receptor binding.

Table 18-1A shows the activity profiles of selected 2-thioether analogues of ATP. Activity at P_{2Y} receptors varied somewhat, depending on the distal structural features, and activity at P_{2X} receptors varied to an even greater degree. All 11 of the synthesized [10] 2-thioethers of ATP stimulated phospholipase C in turkey erythrocyte membranes [14], leading to the production of inositol phosphates with $K_{0.5}$ values of 1.5–770 nM. In smooth muscle assay systems for activity at P_{2Y} receptors, ATP 2-thioethers displayed pD_2 values in the range of 6–8. There was a significant correlation for the 2-thioether compounds between the $pK_{0.5}$ values for inositol phosphate production in turkey erythrocyte membranes and the pD_2 values for relaxation mediated via the P_{2Y} purinoceptors in the guinea-pig taenia coli, but not for vascular P_{2Y} receptors or for P_{2X} receptors. At P_{2X} receptors, no activity was observed in the rabbit saphenous artery, but variable degrees of activity were observed in the guinea-pig was deferens and bladder, depending on distal substituents of the thioether moiety.

Figure 18-1 shows three examples of ATP 2-thioethers having unique features [10]. 2-(7-Cyanohexylthio)-ATP is more potent than 2-MeSATP at the taenia-coli P_{2Y} receptor. A *p*-aminophenethylthio ether, intended as a reporter group for radioiodination and potential crosslinking to the receptor, displayed the highest affinity of all the analogues at turkey erythrocyte P_{2Y} receptors. A *p*-nitrophenethylthio ether was relatively weak at P_{2Y} receptors but provided selectivity for a subset (vas deferens) of the P_{2X}-receptor class. 7-Aminoheptylthio- and 7-thioheptylthio- ethers (not shown), synthesized to ease the further derivatization by acylation or alkylation and to probe potential accessory binding sites on the receptor, displayed $K_{0.5}$ values of 73 and 770 nM, respectively, at erythrocyte P_{2Y} receptors.

The addition of a functionalized chain at the 2-position allowed for truncation of the triphosphate group with retention of affinity (Fig. 18-2), thus circumventing one of the major complications in interpreting ATP pharmacologic results, that is, the action of ectonucleotidases. AMP itself was inactive at P_{2Y} receptors. 2-Thioether analogues of adenosine monophosphate were full agonists at erythrocyte P_{2Y} receptors [10], although

TABLE 18-1. Activities of nucleotide analogues relative to ATP at various P_2 receptor assays

	P_{2Y} receptors			P_{2X} receptors	
A.	Taenia	Aorta	Artery	Vas	Bladder
ATP	=	=	=	=	=
2-Methylthio-ATP	++	++	++	=	=
2-Cyclohexylthio-ATP	++	+	+	+	=
2-(*n*-Hexylthio)-ATP	++	++	++	+	=
2-(5-Hexenylthio)-ATP	++	++	++	=	+
2-(7-Cyanohexylthio)-ATP	+++	++	++	++	
2-(4-Nitrophenylethylthio)-ATP	++	++	++	++	–

B.	Erythro	Taenia	Aorta	Artery	Vas	Bladder
ATP	=	=	=	=	=	=
N^6-MeATP	–	+	0	0	0	0
8-(6-Aminohexylamino)-ATP	–	–	++	0	0	0
ATP N1-oxide	–	–	++	0	0	0
2′-DeoxyATP	–	=	0	0	0	0
2′,3-DideoxyATP		=			=	+
3′-Amino-3′deoxy-ATP	+	=	++	0	+	=
3′-Acetylamino-3′deoxy-ATP	0	0	0	+	=	+
3′-HO-phenylpropionyl-NHATP	0	0	++	0	=	+
3′-Benzylamino-3′deoxy-ATP	–	0	0	0	++	++
UTP	–					=
5-F-UTP	0	–	+	0	0	++

+ through +++ increasing potency (all >ATP), = equipotent, – less potent, O inactive.
Data are from Fischer et al. [10] and Burnstock et al. [11], in which molar potency values are also found.

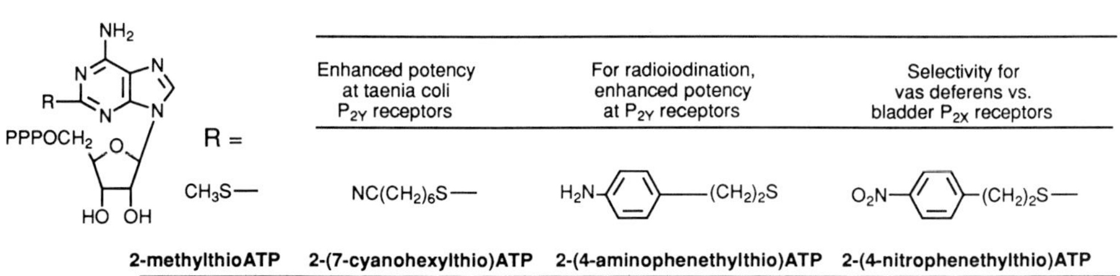

	2-methylthioATP	2-(7-cyanohexylthio)ATP	2-(4-aminophenethylthio)ATP	2-(4-nitrophenethylthio)ATP
$K_{0.5}$ erythrocytes (nM):	8	10	1.5	3000
$K_{0.5}$ taenia coli (nM):	10	2.5	-	320

FIGURE 18-1. Features of ATP 2-thioether derivatives [10]. PPP refers to a triphosphate group.

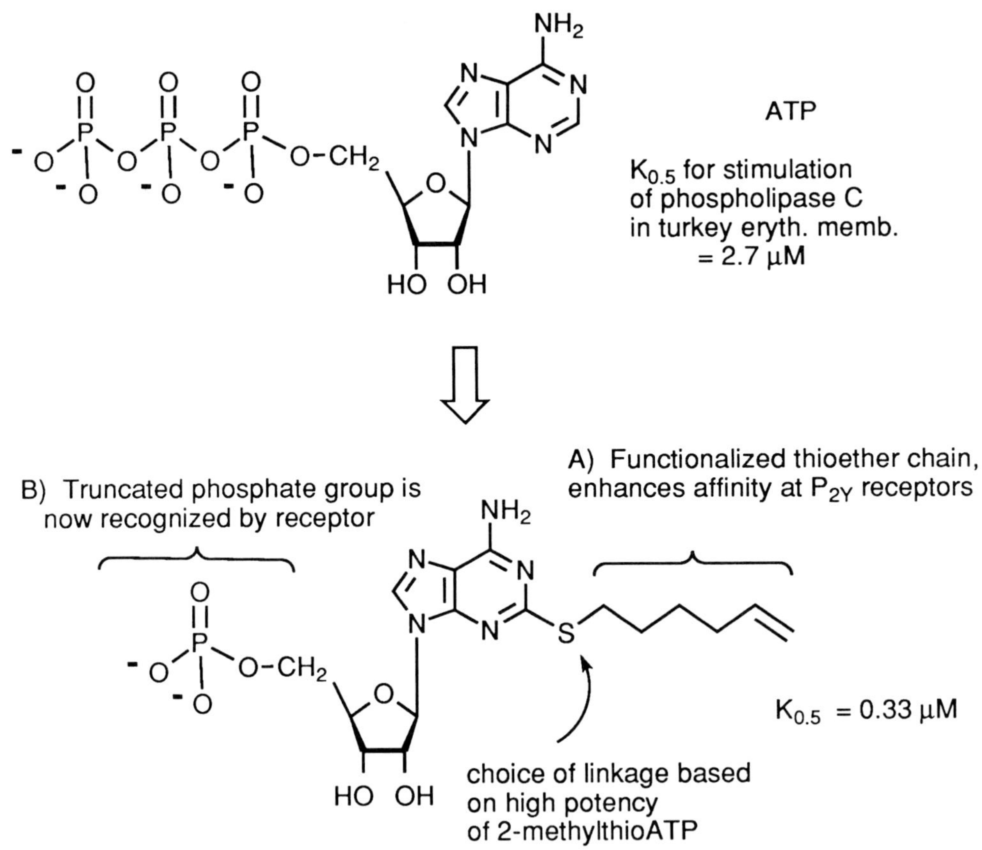

FIGURE 18-2. A functionalized congener approach applied to P_2 agonists allows for flexibility of derivatization at an extended 2-thioether chain on ATP and for removal of two phosphate groups without loss of biologic activity [10].

generally several orders of magnitude less potent than the corresponding 2-thioether triphosphate analogue. For example, a 2-hexenylthio ether of AMP was eightfold more potent than ATP itself in the stimulation of phospholipase C, but was 33-fold less potent than the corresponding triphosphate. Thus, the long chain may act as a distal anchor of the ligand at an accessory binding site on the receptor. Also, several adenosine diphosphate 2-thioether analogues proved equipotent to the corresponding ATP analogues at erythrocyte P_{2Y} receptors [10]. ADP and ATP were also nearly equipotent in this assay.

A further benefit of the presence of a long-chain thioether group at the distal 2-position was increased stability of the triphosphate group at the 5′-position (Fig. 18-3) [9]. It is likely that long chains, although at a site on the molecule distal to the triphosphate group, interfere at the binding site of nucleotidases. The apparent stability combined with high potency suggests the synthesis of radiolabeled analogues in this series.

Modifications of ATP other than 2-thioethers also resulted in unexpected selectivities [11]. Certain analogues displayed selectivity or specificity within the P_{2X}- or P_{2Y}-purinoceptor superfamilies, suggesting possible subclasses (Table 18-1B). For example, a collection of

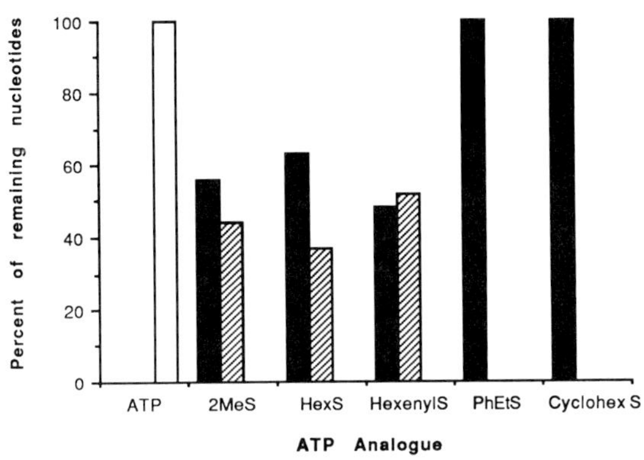

FIGURE 18-3. Stability of 2-thioether derivatives of ATP to nucleotidases, following incubation with bovine brain membranes for 40 minutes at 30°C [9]. The x-axis indicates either ATP or the 2-substituent for ATP analogues. The y-axis indicates the percentage of remaining nucleotides, as detected using HPLC, corresponding to triphosphate (solid bars), diphosphate (hatched bars), or monophosphate (open bar).

probes selective for specific receptors within the general category of P_{2Y} receptors is shown in Figure 18-4. Selectivity was achieved for P_{2Y} receptors of the mesenteric artery, aorta, or taenia coli/erythrocytes. 8-(6-Aminohexylamino)ATP and 2′,3′-isopropylidene-AMP acted selectively at endothelial rabbit aorta P_{2Y} purinoceptors versus P_{2Y} purinoceptors in the taenia coli, mesenteric artery, and erythrocytes. Both compounds were inactive at P_{2X} purinoceptors. The potent agonist N^6-methyl ATP and the somewhat less potent agonist 2′-deoxy-ATP were selective for taenia coli P_{2Y} purinoceptors versus either vascular P_{2Y} purinoceptors or all P_{2X} purinoceptors [10,11]. N^6-methyl ATP was approximately equipotent to ATP taenia-coli P_{2Y} receptors. The N^6-modification may prove to be a general means of increasing P_{2Y} selectivity, because it was compatible with other modifications. A hybrid N^6-methyl and 2-thioether ATP derivative, N^6-methyl-2-(5-hexenylthio)-ATP, was synthesized and found to be very potent at erythrocyte and taenia-coli P_{2Y} receptors and inactive at P_{2X} receptors [10].

3′-Benzylamino-3′-deoxy-ATP (Fig. 18-5) was very potent at the P_{2X} purinoceptors in the guinea-pig vas deferens (slightly less potent in the urinary bladder) but inactive at rabbit saphenous artery P_{2X} receptors and at all P_{2Y} receptors [11]. The P_{2X} potency noted was approximately an order of magnitude greater than α,β-methylene ATP (α,β-meATP), which is used widely in studies of P_{2X} receptors. We did not systematically test the compounds for desensitization, but 3′-benzylamino-3′-deoxy-ATP caused the same profile of contractile response as ATP or α,β-meATP, that is, a transient twitchlike contraction. This suggests that it too would cause rapid desensitization.

In summary, long chain 2-thioethers have been found to enhance the potency (particularly at P_{2Y} receptors) or selectivity (particularly within the P_{2X} class) of ATP analogues. The longer chain members tended to be stable to nucleotidases. The corresponding monophosphates are full agonists at P_{2Y} receptors. Groups for radiolabeling and attachment of larger reporter moieties, such as aliphatic and aryl amines, were introduced on the 2-thioether chain. A modification of the ribose 2′-position and certain purine modifications of ATP other than 2-thioethers has resulted in P_{2Y} selectivity. N^6-methyl-ATP, its congeners, and 2′-deoxy-ATP are selective agonists at taenia-coli P_{2Y} receptors. 8-(6-Aminohexylamino)ATP and ATP N-oxide are selective for endothelial P_{2Y} receptors. However, modifications at the ribose 3′-position are highly variable in receptor selectivity. For example, 3′-deoxy-ATP is a weak, but selective P_{2X} agonist, while 3′-acetamido-3′-deoxy-ATP is active at both P_{2X} receptors and mesenteric artery P_{2Y} receptors. 3′-Benzylamino-3′-deoxy-ATP is highly potent and selective for P_{2X} receptors.

The apparent heterogeneity of receptors revealed using the above ligands has given rise to a proposed new reorganized system of

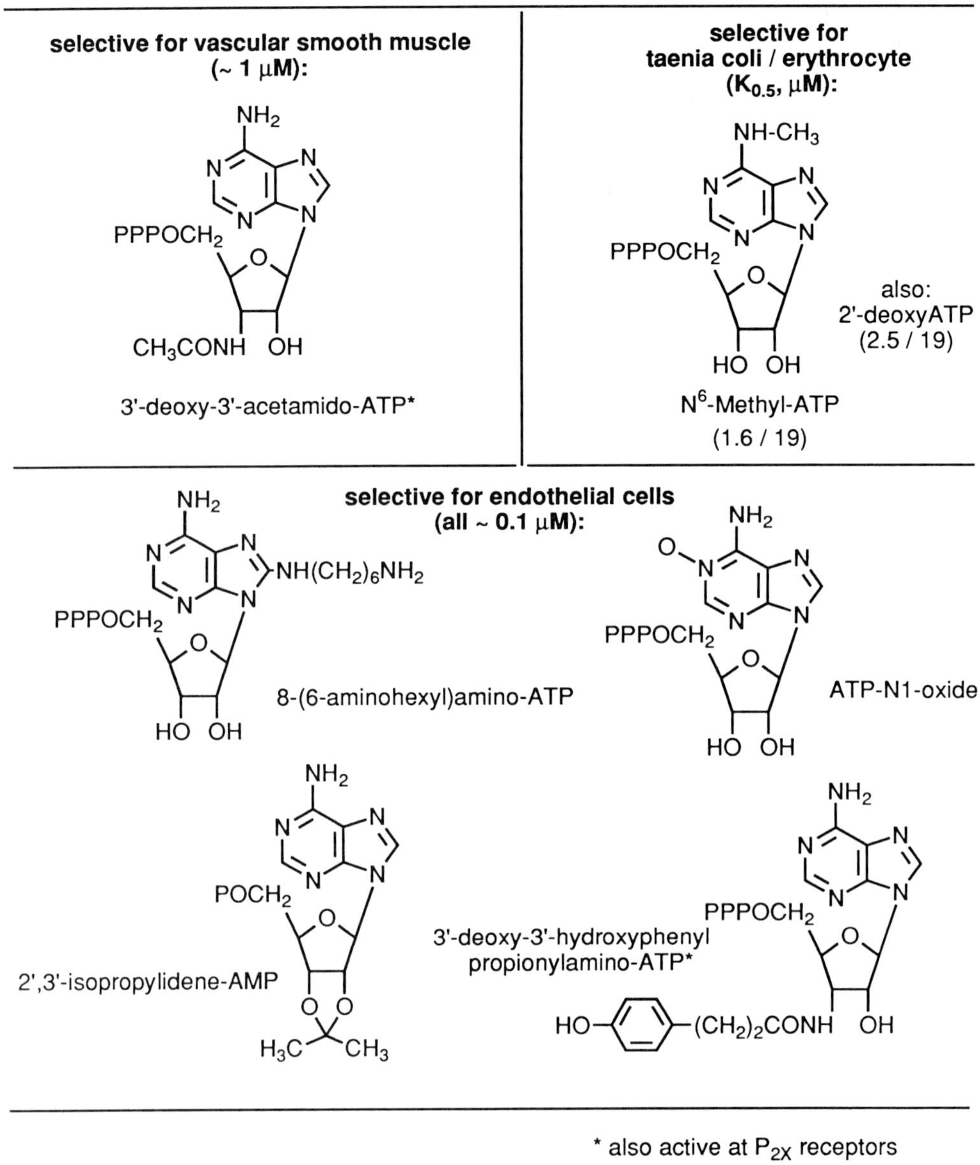

FIGURE 18-4. P_{2Y} receptor agonist probes that show selectivity for a subset of P_{2Y} receptors [10,11]. PPP refers to a triphosphate group.

nomenclature for P_2 receptors [5]. According to this proposal, the numerous P_2 receptors are divided among G-protein–coupled receptors (numbered P_{2Y1} to P_{2Y7}) and ion channels (P_{2X1} to P_{2X4}). We are currently using these ligand affinities, sequence analysis of the cloned P_{2Y} receptors [6,7], and recently reported data on site-directed mutagenesis [17] to propose a computer model for ligand binding [18]. The goal is to construct eventually a predictive model of the ligand binding site of P_{2Y} receptors.

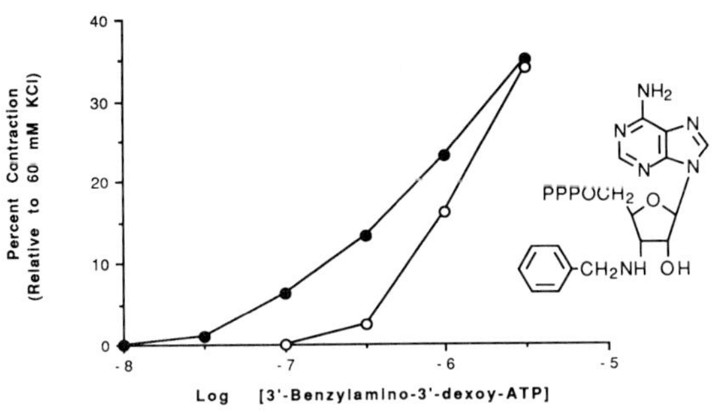

FIGURE 18-5. Smooth muscle effects of a potent, P_{2X}-selective agonist, 3′-benzylamino-3′-deoxy-ATP [11]. Contraction in the guinea-pig vas deferens (solid circles) and in the guinea-pig urinary baldder (open circles) is shown.

References

1. Cusack NJ, Hourani SMO (1990) Structure activity relationships for adenine nucleotide receptors on mast cells, human platelets, and smooth muscle. In Jacobson KA, Daly JW, Manganiello V (eds.), Purines in Cellular Signalling: Targets for New Drugs. New York: Springer, pp. 254–259.
2. Jacobson KA, van Galen PJM, Williams M, (1992) Perspective, adenosine receptors: Pharmacology, structure activity relationships and therapeutic potential. J Med Chem 35: 407–422.
3. Cooper CL, Morris AJ, Harden TK (1989) Guanine nucleotide-sensitive interaction of a radiolabeled agonist with a phospholipase C-linked P_{2Y}-purinergic receptor. J Biol Chem 264:6202–6206.
4. van Galen PJM, Melman N, IJzerman AP, Jacobson KA (1992) Characterization of [^{35}S] ADPβS binding to bovine brain membranes. Pharmacol Comm 1:279–290.
5. Abbracchio MP, Burnstock G (1994) Purinoceptors: Are there families of P_{2X} and P_{2Y} purinoceptors?, Pharmac Ther, in press.
6. Webb TE, Simon J, Krishek BJ, Bateson AN, Smart TG, King BF, Burnstock G, Barnard EA (1993) Cloning and functional expression of a brain G-protein coupled ATP receptor. FEBS Lett. 324:219–225.
7. Lustig KD, Shiau AK, Brake AJ, Julius D (1993) Expression cloning of an ATP receptor from mouse neuroblastoma cells. Proc Natl Acad Sci USA 90:5113–5117.
8. Suprenant A (1994) ATP and P_{2X} purinoceptors involved in fast synaptic transmission in neurons. Drug Dev Res 31:326.
9. Zimmet J, Järlebark L, van Galen PJM, Jacobson KA, Heilbronn E (1993) Synthesis and biological activity of novel 2-thio derivatives of ATP. Nucleosides Nucleotides 12:1–20.
10. Fischer B, Boyer JL, Hoyle CHV, Ziganshin AU, Brizzolara AL, Knight GE, Zimmet J, Burnstock G, Harden TK, Jacobson KA (1993) Identification of potent, selective P_{2Y}-purinoceptor agonists: Structure activity relationships for 2-thioether derivatives of adenosine-5′-triphosphate. J Med Chem 36: 3937–3946.
11. Burnstock G, Fischer B, Maillard M, Ziganshin A, Ralevic V, Knight G, Brizzolara A, von Isakovics A, Boyer JL, Harden TK, Jacobson KA (1994) Structure activity relationships for derivatives of adenosine-5′-triphosphate as agonists at P_{2Y} purinoceptors: Heterogeneity within P_{2X}- and P_{2Y}-subtypes. Drug Dev Res 31:206–219.
12. Jacobson KA, Daly JW (1991) Purine functionalized congeners as molecular probes for adenosine receptors. Nucleosides Nucleotides 10:1029–1038.
13. Jacobson KA, Barrington WW, Pannell LK, Jarvis MF, Ji X-D, Williams M, Hutchison AJ, Stiles GL (1989) Agonist-derived molecular probes for A_2-adenosine receptors. J Mol Recogn 2:170–178.
14. Harden TK, Hawkins PT, Stephens L, Boyer JL, Downes P (1988) Phosphoinositide hydrolysis by guanosine 5′-([gamma-thio] triphosphate)-activated phospholipase C of turkey erythrocyte membranes. Biochem J 252: 583–593.
15. Burnstock G, Kennedy C (1985) Is there a basis for distinguishing two types of P_2-purinoceptor? Gen Pharmacol 16:433–440.
16. Barrington WW, Jacobson KA, Hutchison AJ, Williams M, Stiles GL (1989) Identification of the A_2 adenosine receptor binding subunit by photoaffinity crosslinking. Proc Natl Acad Sci USA 86:6572–6576.
17. Parr CE, Sullivan DM, Paradiso AM, Lazarowski ER, Burch LH, Olsen JC, Erb L, Weisman GA, Boucher RC, Turner JT (1994) Cloning and expression of a human P_{2U} nucleotide receptor, a target for cystic fibrosis pharmacology. Proc Natl Acad Sci USA 91:3275–3279.

18. van Rhee AM, Fischer B, Jacobson KA (1994) Modelling the P_{2Y1} receptor using rhodopsin as a template. ACS National Meeting, Abstract MEDI205, August 21–25, Washington DC.

19. O'Connor SE, Dainty IA, Leff P (1991) Further subclassification of ATP receptors based on agonist studies. Trends Pharmacol Sci 12:137–141.

19. ADENOSINE RECEPTORS: SELECTIVE AGONISTS AND ANTAGONISTS

John W. Daly
Kenneth A. Jacobson

Adenosine receptors are members of the superfamily of receptors having seven transmembrane domains and coupled to guanyl nucleotide-binding protein (G proteins). The A_1, A_{2a}, A_{2d}, and A_3 adenosine receptors have been cloned from several species, including in each case rat, dog, mouse, and human. The A_1 and A_3 receptors couple via G proteins to adenylate cyclase (inhibition), phospholipase C (activation), and, in the case of A_1 receptors, to potassium channels (activation) and calcium channels (inhibition). The A_{2a} and A_{2b} receptors couple via G proteins to adenylate cyclase (activation). A putative A_4 receptor has not been cloned and will not be included in the present overview.

There are many systems that have been used to define affinity and potency and selectivity of agonists and antagonists for adenosine receptors. Because apparent selectivities will be greatly influenced by the test systems, these should be chosen carefully, that is, comparisons between intact cells and membranes, between functional assays (adenylate cyclase, phospholipase C, ion channel function, etc.) and binding assays, and between different species should be avoided. Significant differences in pharmacologic profiles exist among species for adenosine receptors, particularly the A_3 receptors. The present overview will attempt to summarize the development of selective agonists/antagonists for adenosine receptors. It is not intended to be all inclusive, since there is a plethora of agents that have been reported [1,2].

A_1-Selective Agonists

The many A_1-selective agonists have been derived by N^6-substitution of adenosine, which can markedly increase A_1 potency, while decreasing A_{2a} potency. Several such N^6-substituted adenosines are widely and successfully used as A_1-selective agonists (Fig. 19-1A). These include N^6-(R-phenylisoprophyl) adenosine (R-PIA). The diastereomeric S-PIA, by contrast, is much less potent at A_1 receptors and is not A_1 selective. N^6-phenyladenosine has not been widely used as an A_1-selective agonist, but was used as the basis for the development of ADAC and a variety of other A_1-selective probes via the functionalized congener approach [3]. N^6-p-Sulfophenyl-adenosine (SPA) provides an A_1-selective agonist that will act only peripherally [4]. N^6-cyclohexyladenosine (CHA) is the prototypic A_1-selective agonist. Further modifications have led to somewhat more selective

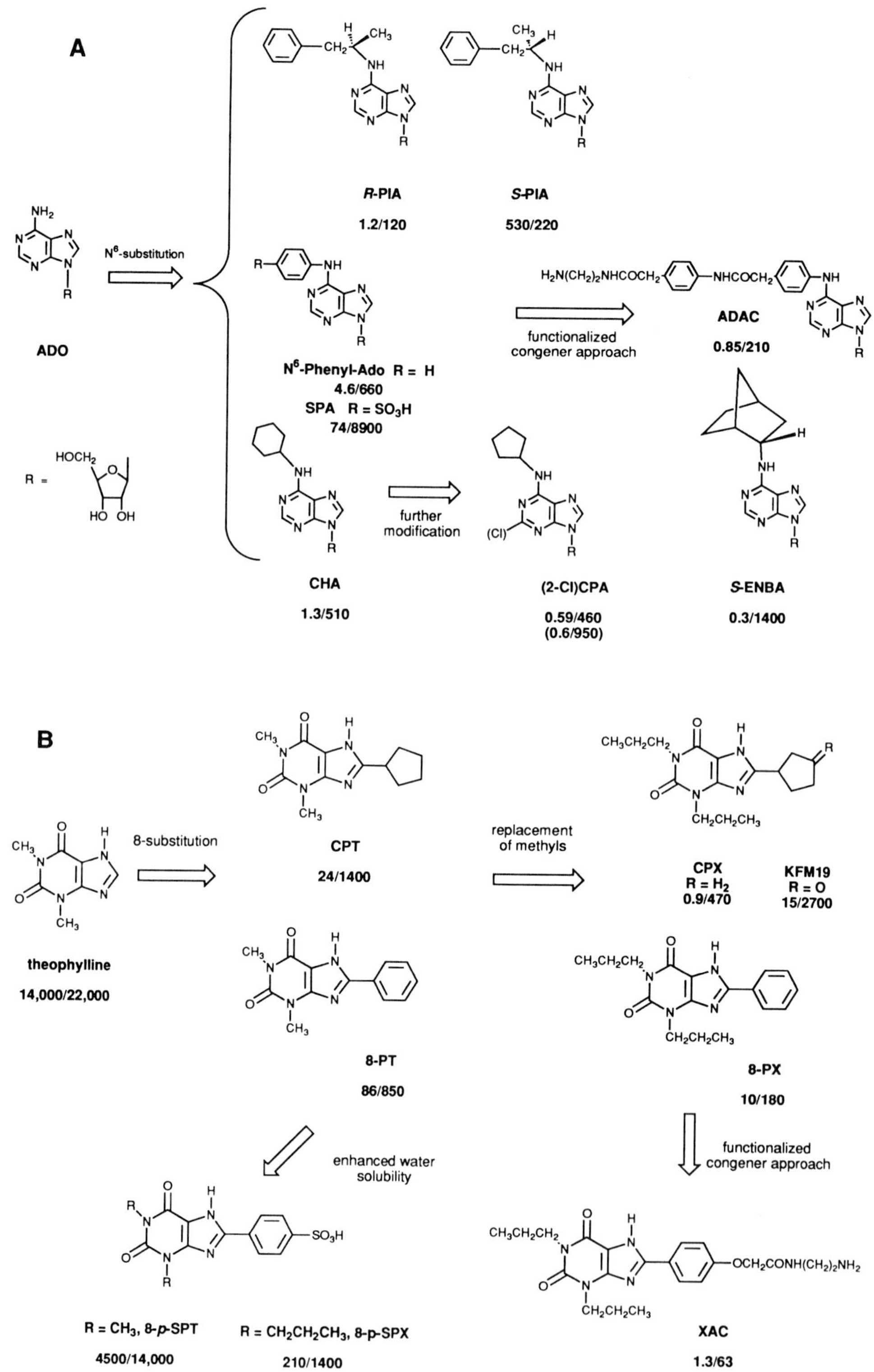
A
ADO
N6-substitution
R-PIA
1.2/120
S-PIA
530/220
N6-Phenyl-Ado R = H
4.6/660
SPA R = SO3H
74/8900
functionalized congener approach
ADAC
0.85/210
further modification
CHA
1.3/510
(2-Cl)CPA
0.59/460
(0.6/950)
S-ENBA
0.3/1400
B
8-substitution
theophylline
14,000/22,000
CPT
24/1400
replacement of methyls
CPX
R = H2
0.9/470
KFM19
R = O
15/2700
8-PT
86/850
8-PX
10/180
enhanced water solubility
functionalized congener approach
R = CH3, 8-p-SPT
4500/14,000
R = CH2CH2CH3, 8-p-SPX
210/1400
XAC
1.3/63

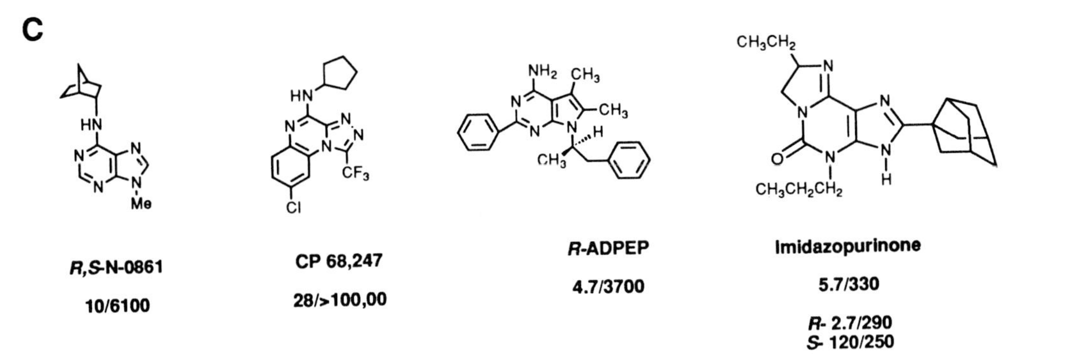

FIGURE 19-1. Evolution of selective agonists and antagonists for A_1-adenosine receptors. **A**: Agonists. **B**: Xanthine antagonists. **C**: Nonxanthine antagonists. The A_1/A_{2a} K_i values are in nanomolar and are for inhibition of binding of radioligand to rat brain A_1 receptors or rat striatal A_{2a} receptors. Data are from the literature [2,4,6].

agonists, including the N^6-cyclopentyl-(CPA and 2-Cl-CPA) and N^6-norbornyl (S-ENBA) analogs. Several A_1-selective agonists (R-PIA, CHA, CPA, ADAC, etc.) have been used as radioligands to characterize A_1 receptors. A_1-receptor assay systems include rat fat cells and membranes, hamster DDT-MF_2 smooth muscle cells and membranes for adenylate cyclase; rat FRTL thyroid cells for phospholipase C; and brain membranes for high-affinity binding.

A_1-Selective Antagonists

The many A_1-selective xanthine antagonists can be considered to have evolved from theophylline, a very weak and nonselective competitive antagonist for adenosine receptors (Fig. 19-1B). 8-Substitution with cycloalkyl groups combined with replacement of the methyl groups of theophylline with propyl groups led to the widely and successfully used A_1-selective antagonist 8-cyclopentyl-1, 3-dipropylxanthine (CPX, more widely known by the misnomer DPCPX). Replacement of the 3-propyl group of CPX with a p-aminophenethyl group yields BWA 884U, an even more selective A_1 antagonist [5]. Other A_1-selective 8-cycloalkyl-substituted xanthines [1,2] include 8-cyclopentyltheophylline (CPT), KFM19, 1,3-dipropyl-8-noradamantylxanthine (KW 3902), 8-(dicyclopropyl-methyl)-1,3-dipropylxanthine (KF 15372), and 1,3-dipropyl-8-endonorbornyl-xanthine (ENBX).

The 8-phenyl analogs (8-PT and 8-PX) are not highly A_1 selective and have low water solubility. In the case of the very potent, highly A_1-selective 8-(2-amino-3-chlorophenyl)-1,3-dipropylxanthine (PACPX), very low water solubility severely limits its usefulness. Efforts have been made to increase the limited water solubility of 8-substituted xanthines. The functionalized congener approach led to the potent, more water-soluble XAC and a variety of other selective probes for A_1 receptors [3]. The A_1 selectivity of XAC is not as great as that of the 8-cycloalkyl xanthines. The highly water-soluble 8-p-sulfophenylxanthines (8-p-SPT and 8-p-SPX) are very useful, because they act only at extracellular sites and in vivo act only peripherally. The p-sulfophenyl xanthines have only marginal A_1 selectivity. Several selective antagonists (CPX, XAC, ^{125}I-BWA 844U, ^{125}I-PAPA-XAC, etc.) have been used as radioligands to characterize A_1 receptors.

There are a number of A_1-selective non-xanthine antagonists (Fig. 19-1C) [1,2]. Low water solubility limits the usefulness of many of the nonxanthine antagonists. The N^6-*endo*-norbornyl-9-methyladenine N-0861 and the triazoloquinazoline CP 68, 247 have proved useful in spite of limited water solubility. The imidazopurines (Fig. 19-1C) are structurally closely related to 8-substituted xanthines but are very water soluble [6]. Remarkably, the R-enantiomer is highly A_1 selective, while the S-enantiomer is not. These imidazopurines should prove very useful as antagonists. The R,S-racemate has activity in an amnesia model [7].

A_{2a}-Selective Agonists

Most A_{2a}-selective agonists are based on the introduction of a large 2-substituent, which increases A_{2a} potency while markedly decreasing A_1 potency, either in an adenosine analog or in a NECA analog. CGS 21680 is now the standard A_{2a}-selective agonist, but there are a number of 2-substituted A_{2a}-selective agonists (Fig. 19-2A) [2,8,9]. The functionalized congener APEC has proved useful, because it appears to penetrate into brain to a greater extent than the anionic CGS 21680 [10]. One N^6-substituted adenosine, DMPA [11], has found use as an A_{2a}-selective agonist. [^{3}H]CGS 21680 has now supplanted [^{3}H]NECA as the preferred radioligand for A_{2a} receptors. [^{125}I]PAPA-APEC is an A_{2a}-selective radioligand with even higher affinity. Assay systems for A_{2a} receptors include rat pheochromocytoma cells/membranes and human platelet cells and membranes for adenylate cyclase and striatal membranes for high-affinity binding.

A_{2a}-Selective Antagonists

The recently developed A_{2a}-selective xanthine antagonists KF 17837 and CSC [12,13] can be considered to have evolved from caffeine, a weak and nonselective antagonist at adenosine receptors (Fig. 19-2B) The first generation of modified caffeines, DMPX, DPMX, and 8-cyclohexylcaffeine, are only slightly A_{2a} selective, while KF 17837 and CSC are highly selective and are proving useful. CSC was found to be relatively weak at A_{2b} receptors [unpublished results] (Table 19-1). [^{3}H]XAC has been employed as a species-dependent radioligand for A_{2a} receptors [14,15], as has the 8-(p-sulfoamidophenyl)xanthine PD 115, 199 [16], but an A_{2a}-selective xanthine has not been developed as a radioligand.

There are several nonxanthine antagonists that are A_{2a} selective (Fig. 19-2C), but lack of water solubility limits usefulness. The triazoloquinazoline CGS 15943 has only modest A_{2a} selectivity but has been used as a selective A_2 antagonist. In contrast, as a radioligand it labeled rat brain A_1-adenosine receptors [17]. The triazoloquinoxaline CP 66, 713 has been used in vivo as a A_{2a} selective antagonist [18].

A_{2b} Receptor

The A_{2b} receptor has a low affinity for agonists compared to A_1 and A_{2a} receptors, and hence its designation as a "low-affinity" adenosine receptor. Structure-activity relationships for agonists and antagonists at A_{2b} receptors have been studied mainly in human fibroblast membranes and intact cells using stimulation of adenylate cyclase as the assay [19–21]. Brain slices, Jurkat T cells, and astrocytes have also been used. No high-affinity binding assay is available. There are no selective agonists for this "low-affinity" receptor. NECA is more potent than 2-chloroadenosine, which is more potent than N^6-substituted adenosines (Fig. 19-3A). An N^6-substituted NECA (R-PI-NECA) is less potent than NECA. CGS 21680 and other adenosine analogs with large 2-substituents are virtually inactive at A_{2b} receptors and, there-

FIGURE 19-2. Evolution of selective agonists and antagonists for A_{2a} adenosine receptors. **A**: Agonists. **B**: Xanthine antagonists. **C**: Nonxanthine antagonists. The A_1/A_{2a} K_i values are in nanomolar for inhibition of binding to rat brain A_1 receptors or rat striatal A_{2a} receptors unless indicated otherwise. Data are from the literature [2].

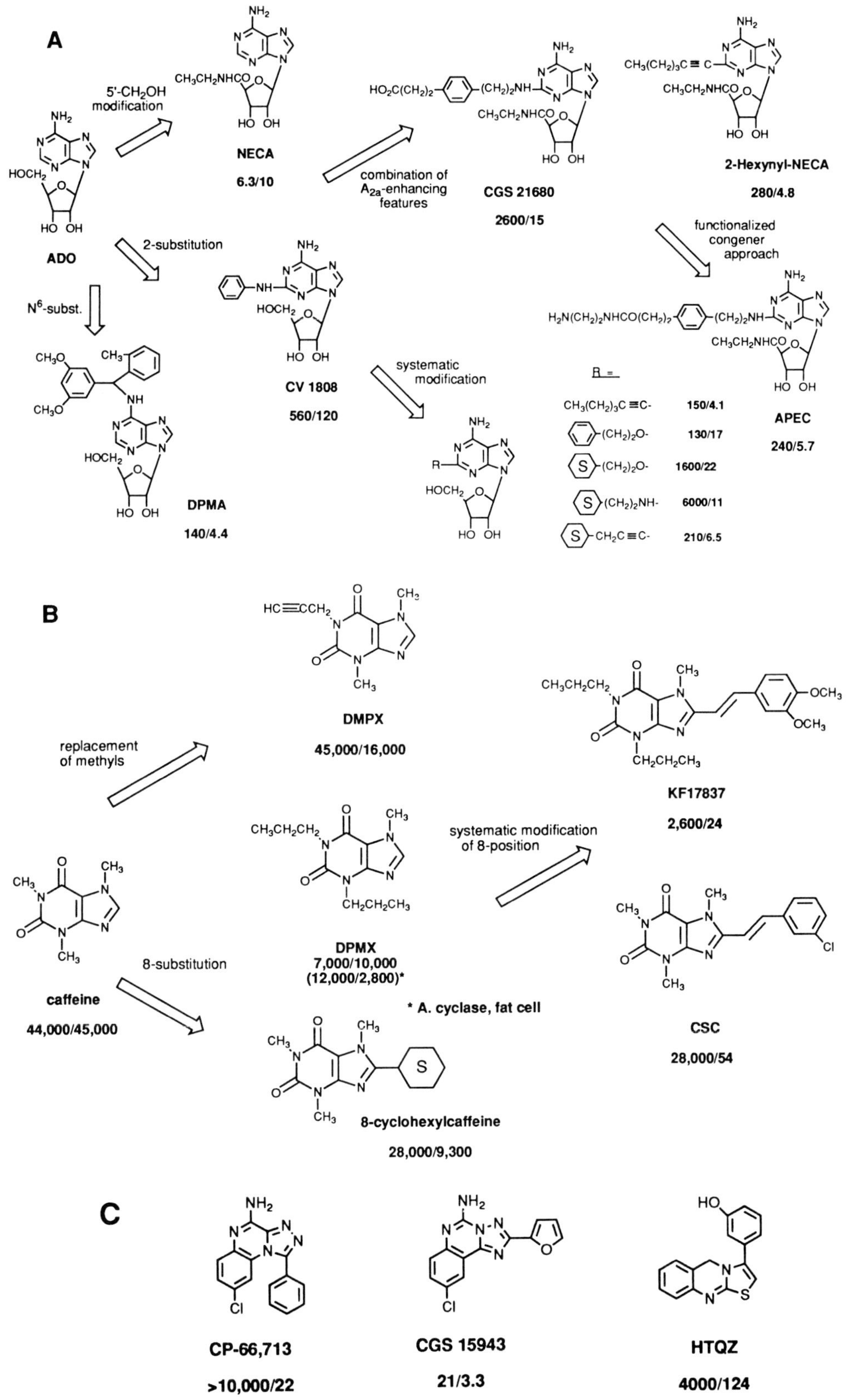
A
5'-CH2OH modification
NECA
6.3/10
combination of A2a-enhancing features
CGS 21680
2600/15
2-Hexynyl-NECA
280/4.8
ADO
2-substitution
N6-subst.
CV 1808
560/120
functionalized congener approach
APEC
240/5.7
DPMA
140/4.4
systematic modification
R =
150/4.1
130/17
1600/22
6000/11
210/6.5
B
DMPX
45,000/16,000
replacement of methyls
KF17837
2,600/24
systematic modification of 8-position
DPMX
7,000/10,000
(12,000/2,800)*
* A. cyclase, fat cell
caffeine
44,000/45,000
8-substitution
CSC
28,000/54
8-cyclohexylcaffeine
28,000/9,300
C
CP-66,713
>10,000/22
CGS 15943
21/3.3
HTQZ
4000/124

TABLE 19-1. Rank order of acivity of selected agents at adenosine receptors

A_1		A_{2a}		A_{2b}		A_3	
Agents							
CPA	0.6	DPMA	4.4	NECA	1,900	IB-MECA	1.1
R-PIA	1.2	NECA	10	R-PI-NECA	6,000	NECA	110
NECA	6.3	CGS 21680	15	2-CLADO	15,000	R-PIA	160
IB-MECA	54	IB-MECA	56	R-PIA	19,000	CPA	240
SPA	70	R-PIA	120	CHA	53,000	CGS 21680	580
DPMA	140	S-PIA	220	S-PIA	65,000	SPA	840
S-PIA	530	CPA	460	CGS 21680	Inactive	S-PIA	920
CGS 21680	2600	SPA	8900			DPMA	3600
Xanthine Antagonists							
CPX	0.9	PD 115,199	26	PD 115,119	160	I-ABOPX	1,200
XAC	1.3	CSC	54	XAC	300	CPX	5,300
8-PX	10	XAC	63	CPX	360	BWA 1433	15,000
8-PT	86	8-PX	180	8-PT	960	XAC	29,000
PD 115,119	140	CPX	470	8-p-SPT	4,500	8-p-SPX	90,000
8-p-SPX	210	8-PT	850	CSC	8,200		
8-p-SPT	4,500	8-p-SPX	1,400	DMPX	11,000		
Theophylline	14,000	CPT	1,400	Caffeine	30,000		
CSC	28,000	8-p-SPX	14,000	Theophylline	32,000		
Caffeine	44,000	CMPX	16,000				
DMPX	45,000	Theophylline	22,000				
		Caffeine	45,000				
Nonxanthine Antagonists							
N-0861	10	CGS 15943	3.3	CGS 15943	41	N-0861	41,000
CGS 15943	21	CP 66,713	22	CP 66,713	260		(sheep)
CP 66,713	270	HTQZ	124	Alloxazine	2300		
Tracaxolate	790	Tracazolate	1,500	HTQZ	4800		
HTQZ	4000	N-0861	6,100	Tracazolate	5400		
Alloxazine	9100	Alloxazine	21,000				

Structures and abbreviations are given in Figures 19-1 to 19-4. K_i values (nM) are for inhibition of binding of radioligands at rat adenosine receptors or are EC_{50} or K_i values for A_{2b}-receptor–mediated effects on adenylate cyclase in human fibroblast membranes. Data from the literature (see text) or for CSC at A_{2b} receptors of human 3T3 fibroblast membranes are unpublished (K_i 8,200 ± 3,400 nM).

fore, are useful in distinguishing between A_{2a} and A_{2b} receptors.

Based on a comparison of inhibition of an A_{2b}-receptor–mediated response in *intact* human fibroblast cells with inhibition of binding of an agonist to A_{2a} receptors in rat striatal membranes, xanthines, such as theophylline, CPT, and 8-p-SPT, appeared more potent at the A_{2b} receptors [22]. However, when adenylate cyclase data for membranes from fibroblasts (A_{2b}) and for membranes from pheochromocytoma cells (A_{2a}) are compared, none of the xanthine antagonists appear selective for the A_{2b} receptors (Fig. 19-3B) [21]. Nonxanthine antagonists, such as tracazolate, CP 66, 713, and CGS 15943, are slightly selective for A_{2a} receptors. Only alloxazine shows a slight (10-fold) selectivity for A_{2b} receptors (Fig. 19-3B).

A_3 Adenosine Receptor

Species differences complicate definition of A_3-selective agents, particularly with respect to

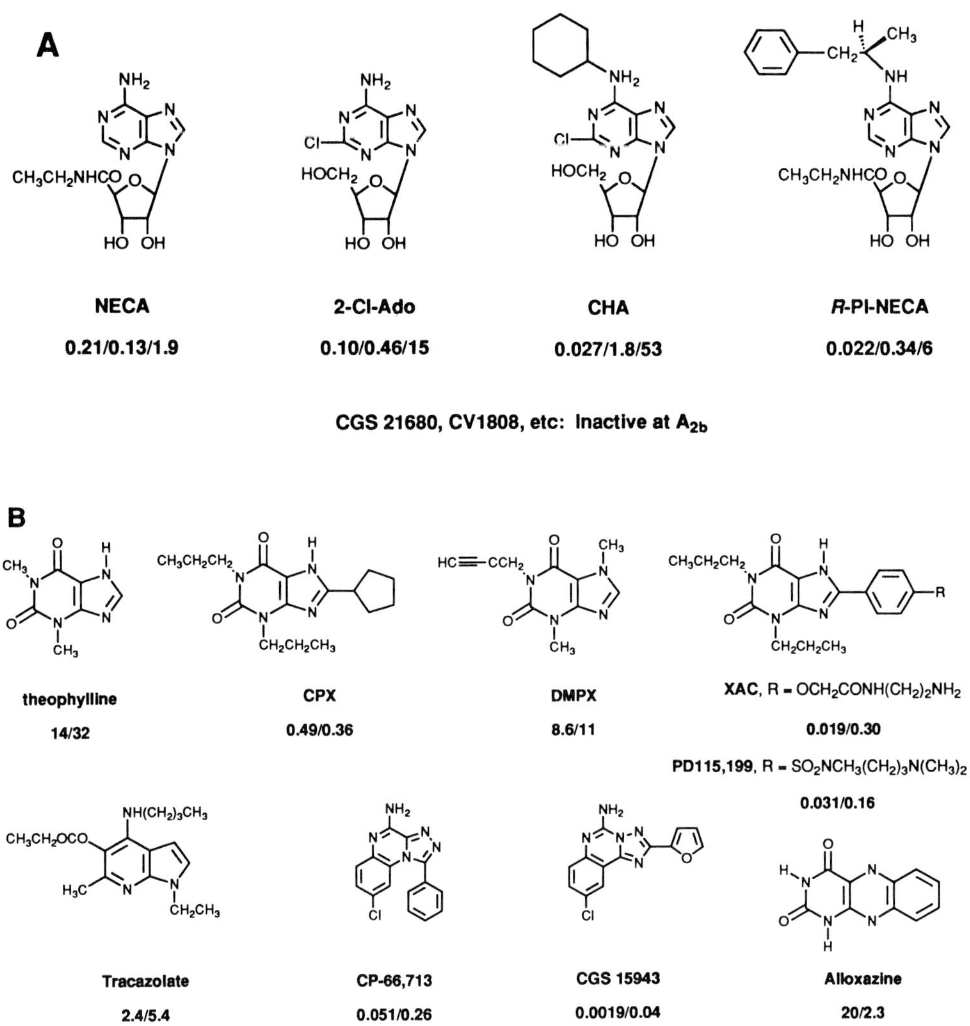

FIGURE 19-3. The A_{2b} adenosine receptor. **A**: Agonists. **B**: Antagonists. The A_1, A_{2a}, and A_{2b} EC_{50} values for agonists are in micromolar for inhibition or stimulation of adenylate cyclase in membranes from rat fat cells, rat pheochromocytoma cells, or human fibroblasts. The A_{2a}, and A_{2b} K_B values for antagonists are in micromolar for antagonism of NECA-stimulated adenylate cyclase. Data are from the literature [21].

antagonists [23–27]. Assay systems for A_3 receptors include transfected cells and membranes for adenylate cyclase, rat RBL-2H3 mast cells for phospholipase C, and transfected cells and membranes for high-affinity binding. A_3-Receptors do not appear highly conserved among species, and subtypes may need to be designated [27]. Some selective agonists have been developed through N-benzyl substitution of adenosine, NECA, and MECA (Fig. 19-4A). 3-Iodo substitution in the benzyl group enhances A_3 selectivity [26]. It should be noted that APNEA and I-ABA, which are used as iodinated radioligands (I-APNEA, I-ABA) for A_3 receptors, are not A_3 selective. IB-MECA is highly A_3 selective [26]. Xanthines, particularly those with an anionic group in the 8-substituent, are relatively potent antagonists for the human and

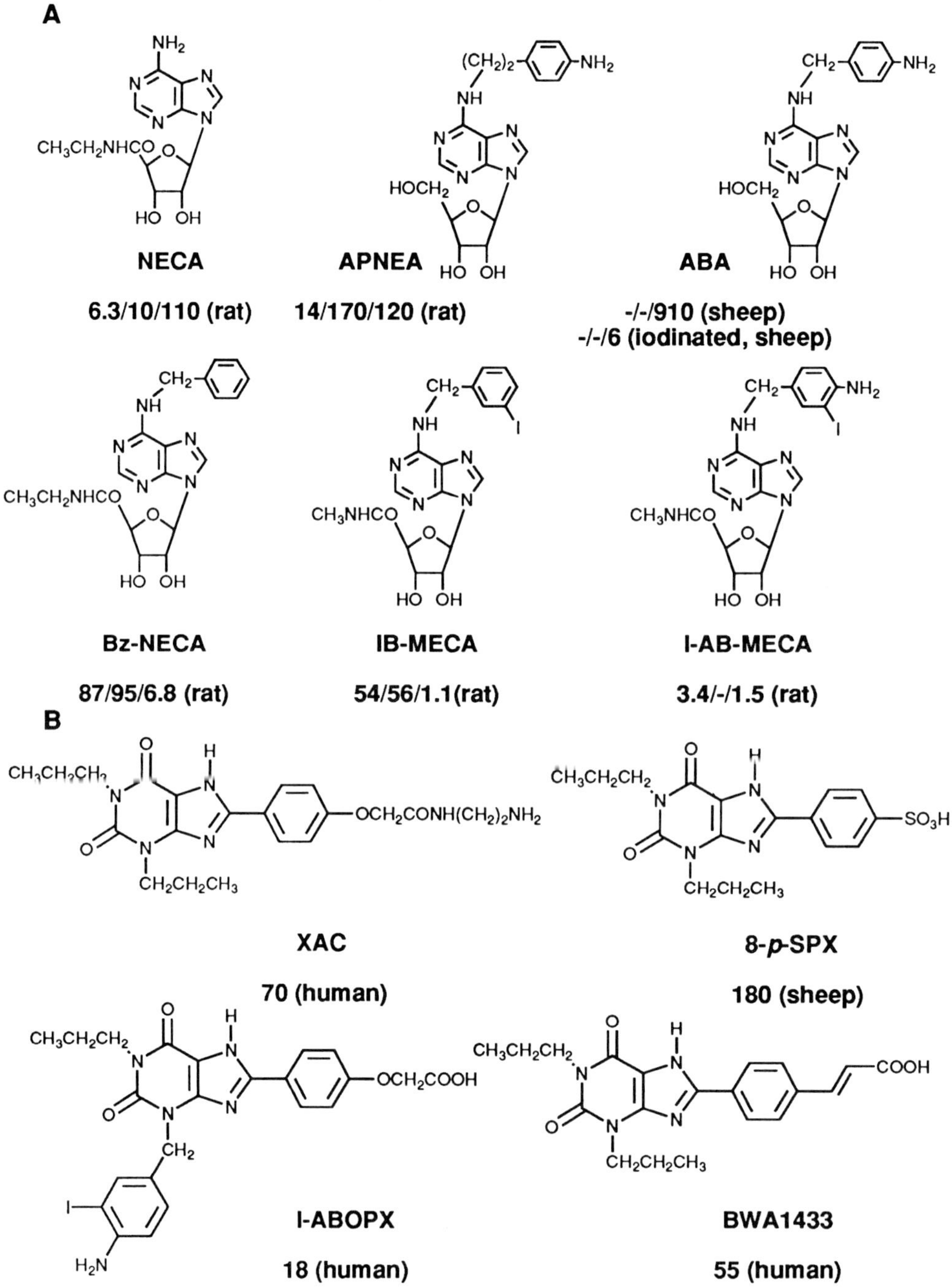

FIGURE 19-4. The A_3 adenosine receptors. **A**: Agonists. **B**: Antagonists. The A_1, A_{2a}, and A_3 K_i values for agonists are in nanomolar and are for inhibition of binding of a radioligand to brain A_1 or A_{2a} receptors or transfected A_3 receptors of rat or sheep. The A_3 K_i values for antagonists are in nanomolar and are for transfected A_3 receptors of human or sheep (see Table 19-1 for values for rat). Data are from the literature [23–27].

sheep A_3 receptors [24,25]. These include BWA 1433 and I-ABOPX (Fig. 19-4B). No potent antagonists, either xanthines or nonxanthines, for the rat A_3 receptor have been developed (see Table 19-1).

Conclusions

Rank-order profiles for agonists/antagonists have proved useful for the definition of adrenergic, adenosine, ATP, and other receptors. The rank orders for a select group of several agonists and of several antagonists for A_1, A_{2a}, A_{2b}, and A_3 receptors are presented in Table 19-1. The data are from binding assays at rat adenosine receptors, except for the A_{2b} receptors, where the data are from adenylate cyclase assays with human fibroblast membranes. Among agonists N^6-cyclopentyladenosine is highly A_1 selective, while CGS 21680 and DPMA are highly A_{2a} selective, and IB-MECA is highly A_3 selective. Among antagonists, the xanthine CPX and the adenine N-0861 are highly A_1 selective, while the xanthine CSC is highly A_{2a} selective. CSC has low potency as an antagonist at A_{2b} receptors in human 3T3 fibroblast membranes (see legend to Table 19-1). No highly selective antagonists for A_{2b} or A_3 receptors are yet available.

References

1. Jacobson KA, van Galen PJM, Williams M (1992) Adenosine receptors: Pharmacology, structure-activity relationships, and therapeutic potential. J Med Chem 35:407–422.
2. van Galen PJM, Stiles GL, Michaels G, Jacobson KA (1992) Adenosine A_1 and A_2 receptors: Structure-function relationships. Med Res Rev 12:423–471.
3. Jacobson KA, Daly JW (1991) Purine functionalized congeners as molecular probes for adenosine receptors. Nucleosides Nucleotides 10:1029–1038.
4. Jacobson KA, Nikodijević O, Ji X-D, Berkich DA, Eveleth D, Dean PL, Hiramatsu K-I, Kassell NF, van Galen PJM, Lee KS, Bartus RT, Daly JW, LaNoue KF, Maillard M (1992) Synthesis and biological activity of N^6-(p-sulfophenyl)alkyl and N^6-sulfoalkyl derivatives of adenosine: Water-soluble and peripherally selective adenosine agonists. J Med Chem 35: 4143–4149.
5. Patel A, Craig RH, Daluge SM, Linden J (1988) ^{125}I-BW-A844U, an antagonist radioligand with high affinity and selectivity for adenosine A_1 receptors and ^{125}I-azido-BW-A844U, a photoaffinity label. Mol Pharmacol 33:585–591.
6. Suzuki F, Shimada J, Nonaka H, Ishii A, Shiozaki S, Ichikawa S, Ono E (1992) 7,8-Dihydro-8-ethyl-2-(3-noradamantyl)-4-propyl-1H-imidazo[2,1-i]purin-5(4H)-one: A potent and water-soluble adenosine A_1 antagonist. J Med Chem 35:3578–3581.
7. Suzuki F, Shimada J, Shiozaki S, Ichikawa S, Ishii A, Nakamura J, Nonaka H, Kobayashi H, Fuse E (1993) Adenosine A_1 antagonists. 3. Structure-activity relationships on amelioration against scopolamine or N^6-((R)-phenylisopropyl)-adenosine-induced cognitive disturbance. J Med Chem 36:2508–2518.
8. Cristalli G, Eleuteri A, Vittori S, Velpini R, Lohse MJ, Klotz K-L (1992) 2-Alkynyl derivatives of adenosine and adenosine-5′-N-ethyluronamide as selective agonists at A_2 adenosine receptors. J Med Chem 35:2363–2368.
9. Homma H, Watanabe Y, Abiru T, Murayama T, Nomura Y, Matsuda A (1992) Nucleosides and nucleotides 112. 2-(1-hexyn-1-yl)adenosine-5′-uronamides: A new entry of selective A_2 adenosine receptor agonists with potent antihypertensive activity. J Med Chem 35: 2881–2890.
10. Nikodijević O, Daly JW, Jacobson KA (1990) Characterization of the locomotor depression produced by an A_2-selective adenosine agonist. FEBS Lett 261:67–70.
11. Bridges AJ, Bruns RF, Ortwine DF, Priebe SR, Szotek DL, Trivedi BK (1988)N^6-[2-(3,5-dimethoxy-phenyl)-2-(2-methylphenyl)ethyl] adenosine and its uronamide derivatives. Novel adenosine agonists with both high affinity and high selectivity for the adenosine A_2 receptor. J Med Chem 31:1282–1285.
12. Shimada J, Suzuki F, Nonaka H, Ishii A, Ichikawa S (1992) (E)-1,3-dialkyl-7-methyl-8-(3,4,5-trimethoxystyryl) xanthines: Potent and selective adenosine A_2 antagonists. J Med Chem 35:2342–2345.
13. Jacobson KA, Gallo-Rodriquez C, Melman N,

Fisher B, Maillard M, van Bergen A, van Galen PJM, Karton Y (1993) Structure-activity relationships of 8-styrylxanthines as A_2-selective adenosine antagonists. J Med Chem 36:1333–1342.

14. Ukena D, Jacobson KA, Kirk KL, Daly JW (1986) A [^{3}H]amine congener of 1,3-dipropyl-8-phenylxanthine: A new radioligand for A2 adenosine receptors of human platelets. FEBS 199:269–274.
15. Ji X-D, Stiles GL, Jacobson KA (1991) [^{3}H]XAC (xanthine amine congener) is a radioligand for A_2-adenosine receptors in rabbit striatum. Neurochem Int 18:207–213.
16. Bruns RF, Fergus JH, Badger EW, Bristol JA, Santay LA, Hays JJ (1987) PD 115, 199: An antagonist ligand for adenosine A_2 receptors. Naunyn Schmiedebergs Arch. Pharmacol 335: 64–69.
17. Jarvis MF, Williams M, Do UH, Sills MA (1991) Characterization of the binding of a novel nonxanthine adenosine antagonist radioligand, [^{3}H]CGS 15943, to multiple affinity states of the adenosine A_1 receptor in the rat cortex. Mol Pharmacol 39:49–54.
18. Nikodijevic O, Sarges R, Daly JW, Jacobson KA (1991) Behavioral effects of A_1 and A_2-selective adenosine agonists and antagonists: Evidence for synergism and antagonism. J Pharmacol Exp Ther 259:286–294.
19. Bruns RF (1980) Adenosine receptor activation in human fibroblasts: Nucleoside agonists and antagonists. Can J Physiol Pharmacol 58:673–691.
20. Bruns RF (1981) Adenosine antagonism by purines, pteridines and benzopteridines in human fibroblasts. Biochem Pharmacol 30: 325–333.
21. Brackett LE, Daly JW (1994) Functional characterization of the A_{2b} adenosine receptor in NIH 3T3 fibroblasts Biochem Pharmacol 47: 801–814.
22. Bruns RF, Lu GH, Pugsley TA (1986) Characterization of the A_2 adenosine receptor labeled by [^{3}H]NECA in rat striatal membranes. Mol Pharmacol 29:331–346.
23. Zhou Q-Y, Li C, Olah ME, Johnson RA, Stiles GL, Civelli O (1992) Molecular cloning and characterization of an adenosine receptor: The A3 adenosine receptor. Proc Natl Acad Sci USA 89:7432–7436.
24. Linden J, Taylor HE, Robeva AS, Tucker AL, Stehle JH, Rivkees SA, Fink JS, Reppert SM (1993) Molecular cloning and functional expression of a sheep A_3 adenosine receptor with widespread tissue distribution. Mol Pharmacol 44:524–532.
25. Salvatore CA, Jacobson MA, Taylor HE, Linden J, Johnson RG (1993) Molecular cloning and characterization of the human A_3 adenosine receptor. Proc Natl Acad Sci USA 90:10365–10369.
26. Gallo-Rodriguez C, Ji X, Melman N, Siegman BD, Sanders LH, Orlina J, Fischer B, Pu Q, Olah ME, van Galen PJM, Stiles GL, Jacobson KA (1994) Structure-activity relationships of N^6-benzyladenosine-5′-uronamides as A_3-selective adenosine agonists. J Med Chem 37: 636–646.
27. Ji X-D, von Lubitz D, Olah ME, Stiles GL, Jacobson KA (1994) Species differences in ligand affinity at central A_3-adenosine receptors Drug Dev Res, 33:51–59.

III. PURINE RELEASE AND METABOLISM

20. CARDIAC ENERGETICS AND ADENOSINE FORMATION

Jürgen Schrader
Ulrich Decking
Thomas Stumpe

Whenever cardiac work is increased, there is not only an increase in respiratory rate but also a compensatory increase in coronary flow. In the steady state ATP hydrolysis (work) and ATP production (mitochondrial oxidative phosphorylation) are precisely matched and the delivery of oxygen via the coronary circulation balances the increased respiratory rate. It thus appears that there must be two important regulatory mechanisms operating in the cardiomyocyte, one that relates to the interaction between mitochondrial oxidation and energy expenditure [13,18] and another adjusting coronary flow to changing oxygen requirements [10].

The classic model of cellular regulation of oxidative phophorylation involves feedback control by the ATP hydrolysis products, ADP and P_i [6]. In isolated mitochondria it was convincingly shown that the rate of oxidative phosphorylation is controlled by the concentration of ADP and P_i. More recent work, however, indicates that in the intact beating heart the phosphorylation potential does not change despite considerable changes in cardiac work [1,19]. One study even reports an increase in phosphocreatinine (PCr)/ATP as work load and oxygen consumption are increased with isoproterenol [31]. However, at the high work load achieved by aortic constriction and dobutamine, PCr/ATP decreased [27]. This provides evidence that during strenuous exercise, coronary reserve may become exhausted and "demand ischemia" occurs so that the increased ADP could stimulate respiration under these circumstances. Alternative mechanisms proposed in the regulation of oxidative phosphorylation include control by changes in redox potential as reflected by reduced nicotinamide adenine dinucleotide (NADH)/nicotinamide adenine dinucleotide (NAD) and/or changes in mitochondrial calcium [8,20]. Thus, control of oxidative phosphorylation is likely to be multifactorial and depends on the experimental conditions. Whatever mechanism is operative at the level of the intact heart will depend on 1) the preferred substrate that sustains the rate of mitochondrial ATP synthesis and determines NADH/NAD, 2) the cytosolic concentration of calcium that determines cardiac contractility and mitochondrial calcium, and 3) the oxygen delivery to the mitochondria via the coronary circulation, which determines tissue PO_2, and thereby ATP catabolism and free cytosolic ADP.

The fundamental link between energy turnover and coronary flow has been considered for some time to be adenosine, the dephosphorylated product of ATP [3]. Numerous studies, mostly carried out in saline-perfused isolated rodent hearts, show a close correlation between a fall in phosphorylation potential and a rise in adenosine formation (Table 20-1). The enhanced formation of adenosine could be due to an increase of tissue AMP and/or allosteric activation of cytosolic 5′-nucleotidase resulting from the fall in phosphorylation potential. Furthermore, there exists an important substrate cycle between AMP and adenosine, and small changes in free AMP may be translated into significant changes in adenosine formation [24]. It has been suggested that most interventions that increase cardiac adenosine are associated with a decrease in the oxygen supply/demand ratio or tissue PO_2 [29]. This may be particularly true for the isolated heart model, which is perfused with a medium of low oxygen carrying capacity. These hearts may be sufficiently oxygenated under control conditions, but oxygenation may become critical whenever the work load is increased, thereby leading to ATP breakdown.

Assuming equilibrium of the creatine kinase reaction, changes in the free concentration of ADP can be calculated from the equation: ADP $= (ATP \cdot Cr)/(K_{eq} \cdot PCr \cdot H^+)$, where Cr is the creatine concentration and K_{eq} is the equilibrium constant. Resting concentrations of ADP have been reported to be 50–80 μM. Assuming equilibrium of myokinase (2 ADP ⇔ ATP + AMP), the free AMP concentration will change in proportion to the square of free ADP concentration. Therefore, any increase in free ADP should lead to a parallel increase in free AMP, which, according to our present thinking, constitutes the driving force for the formation of adenosine by the action of cytosolic 5′-nucleotidase. Conversely, if there is no change in phosphorylation potential, biochemical reasons predict that ADP and AMP also do not change and that adenosine formation remains unaltered.

Table 20-1 summarizes studies on the relationship between myocardial oxygen consumption (MVO_2), phosphorylation potential, and/or adenosine. The studies are listed according to the experimental model used; first those in the isolated guinea-pig heart perfused with a saline medium, followed by studies conducted in the

TABLE 20-1. MVO_2/phosphorylation potential/adenosine

Species and intervention	Correlation (MVO_2/PP/AR)	Authors
Guinea pig heart		
Isoproterenol	Significant	Headrick et al. (1991, 1993)
Norepinephrine	Significant	He et al. (1987)
Substrates	Significant	Bünger et al. (1986, 1991)
Norepinephrine	Significant	Kang et al. (1992)
Substrate depletion	Significant	He et al. (1991)
2-deoxyglucose	Biphasic	
Dog heart in vivo		
Norepinephrine	Phasic changes in PP and AR	Gorman et al. (1992)
Norepinephrine	No change in AR	Kroll et al. (1994)
Pacing	No change in PP	Balaban et al. (1986)
Epinephrine	No change in PP	Katz et al. (1989)
Phenylephrine		
Pig heart in vivo		
Isoproterenol	Increased PP	Schwartz et al. (1994)

PP = phosphorylation potential; AR = adenosine.

blood-perfused dog or pig heart. One major insight can be gained when comparing in vito with in vivo data. All studies in the isolated heart in vitro show a significant correlation between MVO_2, phosphorylation potential, and adenosine, whereas no such correlation was found in the blood-perfused heart. In a recent study evidence was provided that in the steady state after norepinephrine stimulation, there was no change in the concentration of cytosolic adenosine [25]. Under a similar experimental condition some years ago, Balaban's group demonstrated no changes in phosphorylation potential [1,22,18]. Collectively these data support the notion that the low oxygen carrying capacity of the saline medium may have been the limiting factor when cardiac work was increased. The correlations reported in the saline-perfused heart may therefore have no direct physiologic relevance but rather reflect the limitations of the experimental model [28].

Adenosine Formation and Phosphorylation Potential in the Isolated Guinea-Pig Heart Performing Pressure Volume Work

The isolated perfused rodent heart in conjunction with nuclear magnetic resonance (NMR) spectroscopy has been used extensively in the past to study the relationship between cardiac energy metabolism and adenosine formation [15–17,21]. Due to the dimensional constraints within the superconducting magnet, hearts were perfused according to the Langendorff technique and therefore did not perform pressure volume work. To increase the cardiac metabolic rate in this model required pharmacologic stimulation with isoprotrenol or norepinephrine.

Our laboratory was interested in the preload- and afterload-induced changes in cardiac hemodynamics, energy metabolism, and adenosine formation. For this purpose we have recently developed a working heart model suited for NMR spectroscopy in which preload and afterload could be changed independently from one another [7]. The ^{31}P NMR spectra were acquired on a Bruker AMX 400 wide-bore pulsed Fourier transform NMR spectrometer equipped with a superconducting magnet (Spectrospin and Oxford Instruments) having a field strength of 9.4 Tesla and a dedicated 20-mm probehead. The magnetic field homogeneity was adjusted using the water-proton free induction decay, and a line width at half-height of the water proton of less than 20 Hz was achieved in each experiment. Each spectrum was the result of 128 acquisitions.

From the data given in Figure 20-1 it can be seen that with increasing afterload, MVO_2 increased up to 42%, while the ratio PCr/ATP, an index of the phosphorylation potential, remained unchanged. This finding is similar to the early report by Balaban [1], in which pacing-induced increases in MVO_2 of similar magnitude did not change the calculated free ADP concentration. Also, in our heart model changes in free ADP are unlikely to be causally involved in controlling the respiratory rate.

Increasing afterload paradoxically increased coronary resistance paralleled by a decrease in coronary adenosine release (Fig. 20-1). Most likely these resistance changes are due to afterload-induced changes in coronary perfusion pressure, leading to improved oxygenation despite increased oxygen consumption [2]. The observed decrease in adenosine release points in the same direction. Thus, despite improvements on the side of a NMR-suitable working heart model, oxygenation is still a major problem. Ongoing studies in our laboratory therefore use perfluorocarbons as oxygen carriers. Perfluorocarbons have an additional advantage: The ^{19}F-spin-lattice time T_1 in NMR spectroscopy is sensitive to changes in oxygen tension [26]. This property of the oxygen carrier will be used to define the extracellular PO_2 in the isolated perfused heart.

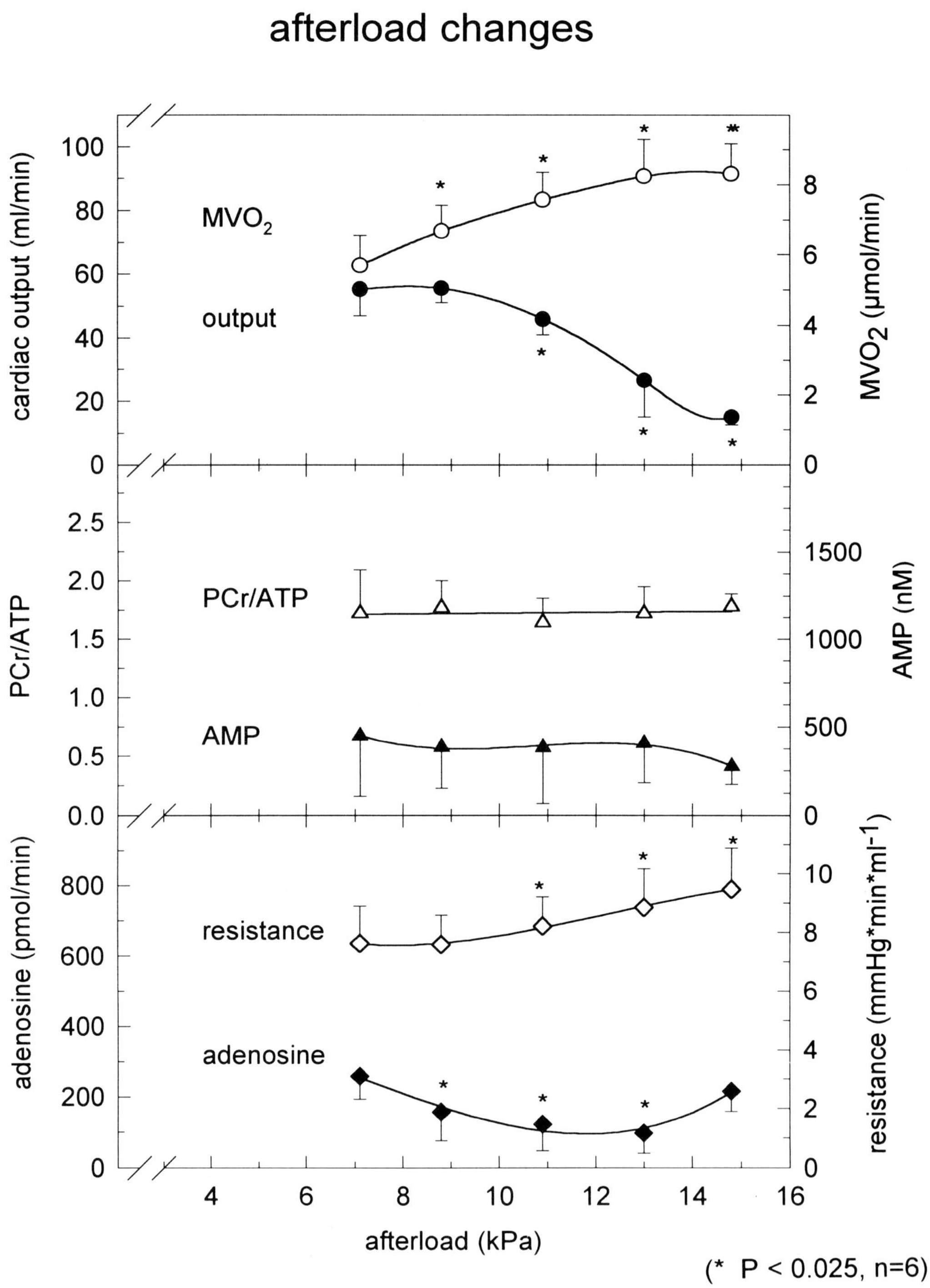

FIGURE 20-1. Cardiac output and oxygen consumption (MVO_2), PCr/ATP ratio, and calculated free AMP, coronary venous adenosine release, and coronary resistance of the isolated perfused guineapig heart at different afterloads. Cardiac output is aortic flow + coronary flow. Preload was 2 kPa.

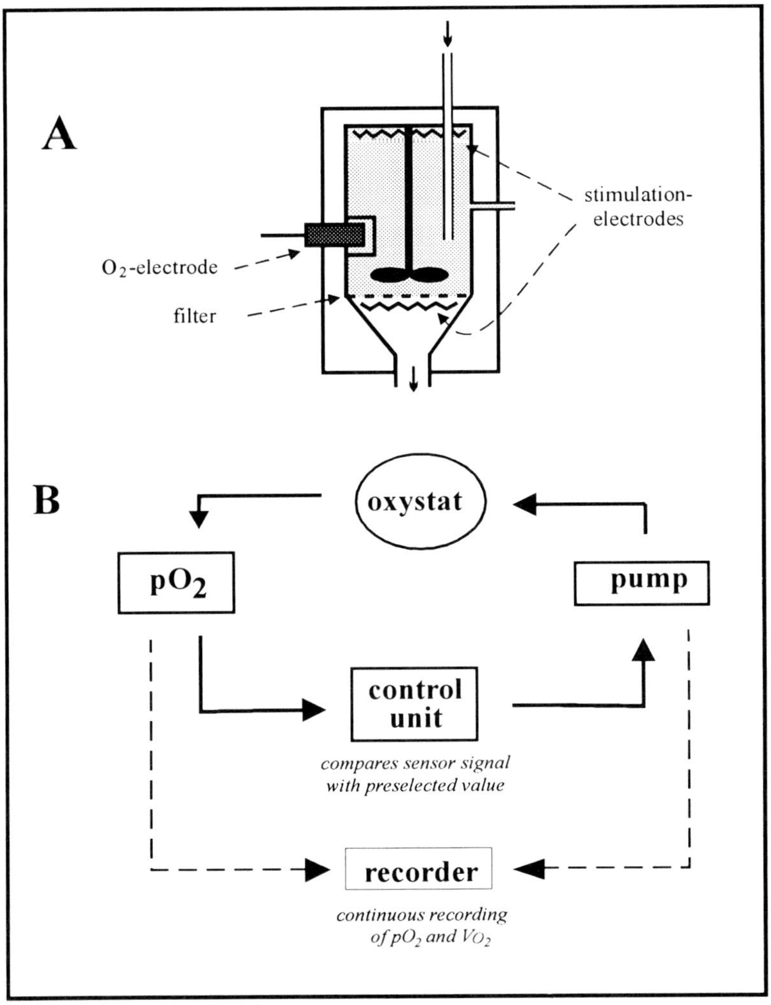

FIGURE 20-2. Schematic outline of the oxystat system.

Effect of PO_2 on Adenosine Formation and Phosphorylation Potential in Isolated Cardiomyocytes at Different Rates of Oxygen Consumption

In order to gain better insight into the regulatory mechanisms governing the formation of adenosine and control of respiratory rate by ADP, we made use of an "oxystat system" in which isolated cardiomyocytes in suspension can be incubated at any predetermined PO_2 [32]. The principal features of the oxystat system are outlined in Figure 20-2. Isolated rat cardiomyocytes were kept in a water-jacketed incubation chamber (37°C) under constant gentle stirring while PO_2 was continuously measured. In an attached "control unit" (Fig. 20-2), the actual PO_2 is then compared with the preset PO_2, and any deviation is compensated by pumping air-saturated medium to the oxystat system. From the rate of infusion at a given PO_2, the rate of cellular oxygen consumption can be easily calculated. The oxystat was also equipped with two platinum electrodes, so that the cellular rate of respiration could be increased. Stimulation at 9 Hz (540/min) with 20 μA increased oxygen consumption from a basal value of 7.9 ± 1.2 to 24.8 ± 2.6 nmol/min/mg (n = 12).

The major advantage of this system is that a selected ambient PO_2 can be maintained over extended periods during which samples of myocytes can be drawn from a side port for biochemical analyses. The following parameters were determined: 1) the free cytosolic concentration of adenosine using the SAH technique developed in our laboratory [9]; 2) ATP, ADP, phosphocreatine (PCr), and creatine (Cr) with reversed-phase HPLC [32]; 3) intracellular pH according to a modified method first described by Waddell and Butler [33]; and 4) inorganic phosphate according to Fiske and Subbarow [11]. Assuming the creatine kinase reaction to be in equilibrium, these measured values permit the calculation of the free cytosolic ADP concentration and the phosphorylation potential (PP) according to the following formulae:

$$[ADP] = \frac{[ATP] \cdot [Cr]}{[PCr] \cdot [H^+] \cdot K_{mk}} \quad (1)$$

$$PP = \frac{[PCr] \cdot [H^+] \cdot K_{ck}}{[Cr] \cdot [P_i]} \quad (2)$$

The experimental conditions were such that cardiomyocytes were incubated at two PO_2 values: 90 and 0.1 torr, with and without electrical stimulation. Data on observed changes in oxygen consumption, free cytosolic adenosine (SAH), and phosphorylation potential are summarized in Figure 20-3. At 90 torr electrical stimulation increased oxygen consumption about threefold. However, free adenosine and the phosphorylation potential did not change. These findings clearly demonstrate that changes in free ADP cannot be made responsible for the measured increase in respiratory rate. Factors other then ATP must prevail here. In addition, these findings show that changes in MVO_2 are not strictly coupled to the formation of adenosine as long as there is sufficient oxygen available. This is also true for $PO_2 > 10$ torr (data not shown). Below this range the phosphorylation potential falls and adenosine increases. This is evidenced by our experiments carried out at a PO_2 of 0.1 torr (Fig. 20-3). Compared with unstimulated controls at 90 torr, MVO_2 is suppressed, the phosphorylation potential is reduced, and the rate of SAH accumulation is drastically increased. With electric stimulation, phosphorylation potential further falls and adenosine (SAH) continues to increase.

Summary and Outlook

A major insight from recent data in the literature and from our own findings, as discussed earlier, is that there is a certain PO_2 range (>10 torr) in which a significant stimulation of cardiac respiration is not associated with any changes in phosphorylation potential and adenosine formation. This is true for the isolated heart performing pressure volume work and for isolated electrically paced cardiomyocytes. Our findings on cardiomyocytes further define the ambient critical PO_2 for respiration and support on a cellular level the early observation in the dog heart in situ showing MVO_2 and coronary flow to increase without changes in the phosphorylation potential. It therefore appears that free cytosolic ADP does not sufficiently change during moderate cardiac work to explain the observed increase in respiratory rate. As has been suggested, other factors are most likely important [18,20].

Similar reasoning also applies to adenosine. In isolated cardiomyocytes pacing increased MVO_2 about threefold; adenosine, however, remained unchanged when the ambient PO_2 was maintained at >10 torr. Only below this critical PO_2 did adenosine increase and the phosphorylation potential decrease. Therefore, energy turnover is not linked to the formation of adenosine, as was postulated from earlier experiments [30]. Over a wide range, the rate

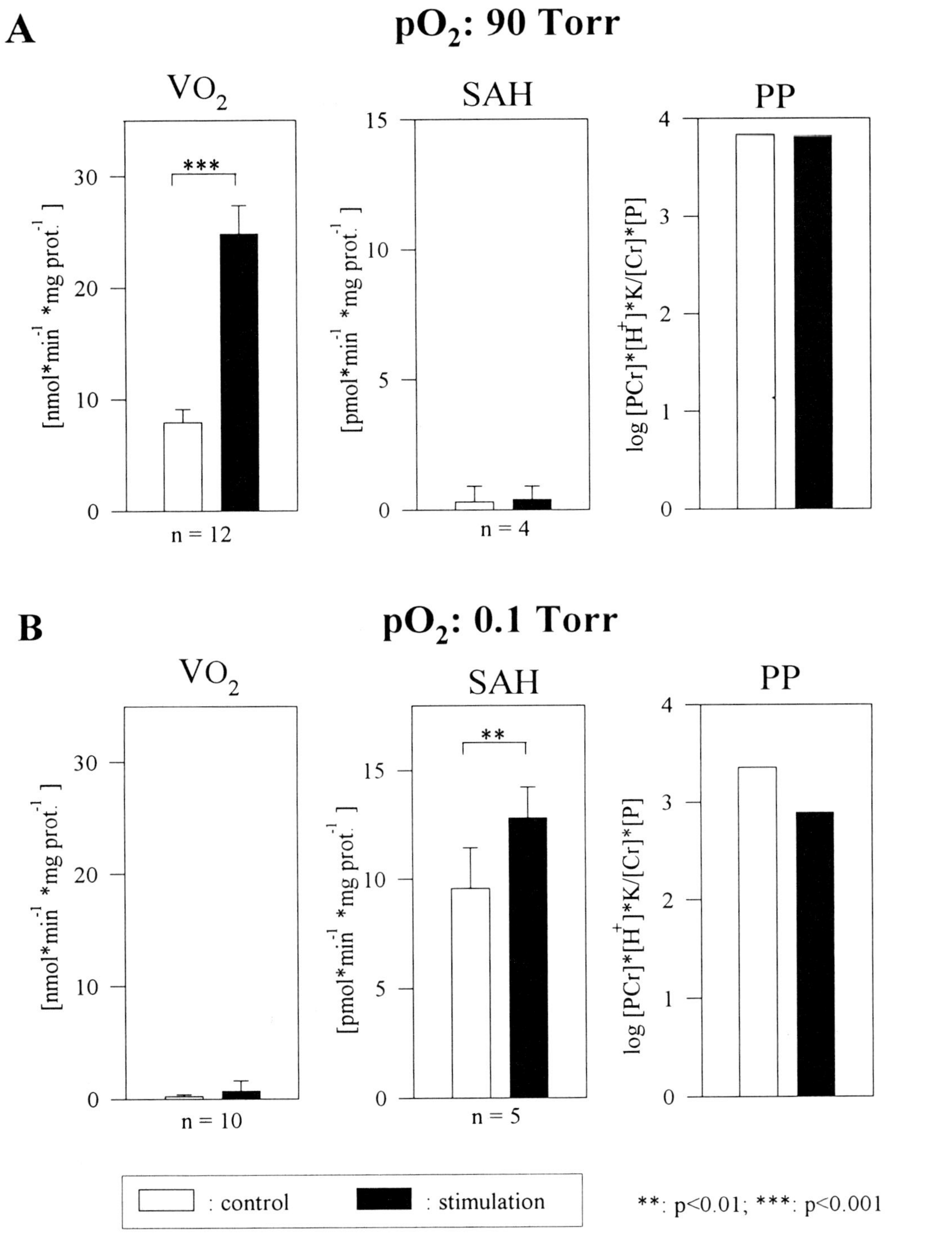

FIGURE 20-3. Effect of the electric stimulation of isolated rat cardiomyocytes on oxygen consumption (VO_2), S-adenosylhomocysteine (SAH), and phosphorylation potential (PP) at an ambient PO_2 of 90 torr (**A**) and 0.1 torr (**B**).

of rephosphorylation of ADP appears to be so efficient and fast that free ADP does not measurably increase. Since ADP and AMP can be assumed to be in equilibrium through creatine kinase, there is also no biochemical reason that free AMP, the main substrate for adenosine, should have changed. Consistent with this is that adenosine formation is linearly related to free AMP [23] and that in our experiments adenosine only increased when the phosphorylation potential decreased.

What is the physiologic relevance of the unchanged adenosine? It can safely be stated that in an adequately oxygenated heart in which tissue PO_2 is above the critical PO_2, adenosine most likely is not the key mediator linking energy turnover to coronary blood flow. In support of this conclusion is a recent study of Kroll et al. [25] showing unchanged adenosine formation in the steady state in the dog heart in situ following beta-adrenergic stimulation. Interestingly, in the same experimental model the free cellular ADP concentration was reported also not to have changed [22]. Thus, there appears to be a range of coronary blood flow regulation where factors other then adenosine are most likely important. It is, however, also conceivable that with strong cardiac stimulation, adenosine may become progressively important. Recently Massie et al. [27] have shown that in the pig myocardial energy requirements can be increased to the degree that they exceed O_2 availability in the absence of abnormalities of coronary blood flow or coronary reserve. Such demand ischemia was associated with a decrease in the phosphocreatine to ATP ratio, which is the classic condition for enhanced adenosine formation.

References

1. Balaban RS, Kantor HL, Katz LA, Briggs RW (1986) Relation between work and phosphate metabolites in the in vivo paced mammalian heart. Science 232:1121–1123.
2. Bardenheuer H, Schrader J (1986) Supply to demand ratio for oxygen determines formation of adenosine by the heart. Am J Physiol 250: H173–H180.
3. Berne RM (1980) The role of adenosine in the regulation of coronary blood flow. Circ Res 47:807–813.
4. Bünger R, Soboll S (1986) Cytosolic adenylates and adenosine release in perfused working heart: Comparison of whole tissue with cytosolic non-aqueous fractionation analyses. Eur J Biochem 159:203–213.
5. Bünger R, Mallet RT, Kang YH (1991) Guinea-pig cardiac free, bound, and interstitial adenylates. Energy-linked and energy-independent adenosine release. In Imai S, Nakazawa M (eds.), Adenosine and Adenine Nucleotides in the Biological System. Amsterdam: Elsevier, pp. 337–352.
6. Chance B, Williams CM (1956) The respiratory chain and oxidative phosphorylation. Adv Enzymol 17:65–134.
7. Decking UKM, Schulze K, Schrader J (1995) Energy metabolism and oxygen consumption in a working heart preparation adapted for NMR-spectroscopy.
8. Denton RM, McCormick JG (1990) Ca^{2+} as a second messenger within mitochondria of the heart and other tissues. Annu Rev Physiol 52: 451–466.
9. Deussen A, Borst M, Schrader J (1988) Formation of S-adenosylhomocysteine in the heart. I: An index of free intracellular adenosine. Circ Res 63:240–249.
10. Feigl EO (1983) Coronary physiology. Physiol Rev 63:1–205.
11. Fiske CH, Subbarow Y (1925) The colorimetric determination of phosphorus. J Biol Chem 66: 375–407.
12. Gorman MW, Ning XH, He MX, Portman MA, Sparks HV (1992) Adenosine release and high energy phosphates in intact dog hearts during norepinephrine infusion. Circ Res 70: 1146–1151.
13. Harris DA, Das AM (1991) Control of mitochondrial ATP synthesis in the heart. Biochem J 280:561–573.
14. He MX, Wangler RD, Dillon PF, Romig GD, Sparks HV (1987) Phosphorylation potential and adenosine release during norepinephrine infusion in guinea pig heart. Am J Physiol 253:H1184–H1191.
15. He MX, Gorman MW, Roning GD, Sparks HV (1991) Adenosine formation and energy status during hypoperfusion and 2-deoxy-glucose infu-

sion. Am J Physiol 260:H917–H926.
16. Headrick JP, Willis RJ (1990) Adenosine formation and energy metabolism: A ^{31}P-NMR study in isolated rat heart. Am J Physiol 258: H617–H624.
17. Headrick JP, Matherne GP, Berr SS, Berne RM (1991) Effects of graded perfusion and isovolumic work on epicardial and venous adenosine and cytosolic metabolism. J Mol Cell Cardiol 23:309–324.
18. Heineman FW, Balaban RS (1990) Control of mitochondrial respiration in the heart in vivo. Annu Rev Physiol 52:523–542.
19. Heineman FW, Balaban RS (1990) Phosphorus-31 nuclear magnetic resonance analysis of transient changes of canine myocardial metabolism in vivo. J Clin Invest 85:843–852.
20. Heineman FW, Balaban RS (1993) Effects of afterload and heart rate on NAD(P)H redox state in the isolated rabbit heart. Am J Physiol 264: H433–H440.
21. Kang YH, Mallet RT, Bünger R (1992) Coronary autoregulation and purine release in normoxic heart at various cytoplasmic phosphorylation potentials: Disparate effects of adenosine. Pflügers Arch 421:188–199.
22. Katz LA, Swain JA, Portman MA, Balaban RS (1989) Relation between phosphate metabolites and oxygen consumption of heart in vivo. Am J Physiol 256:H265–H274.
23. Kiviluoma KR, Peuhkurinen KJ, Hassinen IE (1986) Role of cellular energy state and adenosine in the regulation of coronary flow during variation in contraction frequency in an isolated perfused heart. J Mol Cell Cardiol 18: 1133–1142.
24. Kroll K, Decking UKM, Dreikorn K, Schrader J (1993) Rapid turnover of the AMP adenosine metabolic cycle in the guinea pig heart. Circ Res 73:846–856.
25. Kroll K, Martin GV (1994) Steady-state catecholamine stimulation does not increase cytosolic adenosine in canine hearts. Am J Physiol 265:H503–H510.
26. Mason RP, Shukla H, Antich PP (1993) In vivo oxygen tension and temperature: Simultaneous determination using ^{19}F NMR spectroscopy of perfluorocarbon. Magn Reson Med 29: 296–302.
27. Massie BM, Schwartz GG, Garcia J, Wisneski JA, Weiner MW, Owens T (1994) Myocardial metabolism during increased work states in the porcine left ventricle in vivo. Circ Res 74:64–73.
28. Nung XH, He MX, Gorman MW, Roning GD, Sparks HV (1992) Adenosine formation and energy status in isolated guinea pig hearts perfused with erythrocytes. Am J Physiol 262: H1075–H1090.
29. Schrader J, Deussen A (1988) Free cytosolic adenosine sensitively signals myocardial hypoxia in oxygen sensing in tissues. In Acker H (ed.), Berlin: Springer-Verlag, pp. 165–176.
30. Schrader J (1991) Formation and metabolism of adenosine and adenine nucleotides in cardiac tissue. In Phillis JW (ed.), Adenosine and Adenine Nucleotides as Regulators of Cellular Function. CRC Press, Boca Raton, FL, pp. 55–69.
31. Schwartz GG, Greyson CR, Wisneski JA, Garcia J, Steinman S (1994) Relation among regional O_2 consumption, high-energy phosphates, and substrate uptake in porcine right ventricle. Am J Physiol 266:H521–H530.
32. Smolenski RT, Schrader J, de Groot H, Deussen A (1991) Oxygen partial pressure and free intracellular adenosine of isolated cardiomyocytes. Am J Physiol 260:C708–C714.
33. Waddell WJ, Butler TC (1959) Calculation of intracellular pH from the distribution of 5,5-dimethyl-2,4-oxazolidinedione (DMO). Application to skeletal muscle of the dog. J Clin Invest 38:720–729.

21. EXTRACELLULAR HYDROLYSIS OF ATP AND FORMATION OF ADENOSINE IN THE NERVOUS SYSTEM

Herbert Zimmermann
Norbert Braun
Frank Nocken

Neuronal Signaling via ATP Requires Mechanisms of Rapid Inactivation

Strong evidence has recently been provided that ATP can act as a transmitter not only in smooth muscle but also in peripheral ganglia and brain [1,2]. ATP can act through ligand-gated ion channels or through receptors coupled to trimeric G proteins. At neurones from cultured or intact mammalian celiac ganglia and in slices of the medial habenula, ATP can function as a fast excitatory neurotransmitter [3]. Ca^{2+}-dependent neurogenic release of ATP can be directly measured. This was shown for fractions of synaptosomes isolated from the electric-ray electric organ and mammalian brain, and for peripheral tissue preparations such as the taenia coli, bladder, vas deferens, and rat neuromuscular junction [4,5]. ATP is costored with acetylcholine in brain and at the neuromuscular junction, with noradrenaline in sympathetic nerves and possibly also with other neurotransmitter substances [6]. In either case the vesicular nucleotide is outnumbered by its cotransmitter. Intravesicular concentrations of ATP are in the millimolar range. Molecular ratios of released acetylcholine and ATP close to those of synaptic vesicle storage have been reported [7]. K^+-induced synaptosomal ATP release was suggested to be quantal in nature [8]. However, release of ATP was not found to be directly related to its cotransmitter in all cases [9,10]. ATP can be released from cellular sites apparently lacking vesicular storage organelles. This includes the postsynaptic target cell [11,12]. In this case ATP may be released by a carrier or channel similar to the product of the multidrug resistance gene that has recently been implicated as an ATP channel [13].

Transmission using ATP as a messenger requires mechanisms for ATP inactivation. This is particularly apparent when ATP acts as a fast transmitter and impulses of high frequency need to be transmitted. Surface enzymes for hydrolysis of ATP have been found at sites of ATP release. The end product of ATP hydrolysis generally is adenosine, a messenger in its own right and a salvage product of cellular purine metabolism (Fig. 21-1).

Hydrolysis of ATP and ADP

On incubation with ATP of fractions of intact synaptosomes ADP, AMP, and adenosine sequentially appear in the medium. The ubiquitous ecto-ATPase is a surface membrane-bound enzyme that presumably occurs in neural and glial plasma membranes as well as on the surface of endothelial cells [14,15]. The activity of ecto-ATPase in synaptosomes is considerably higher than that of ecto-ADPase [15,16]. Different mechanisms of ATP hydrolysis have been suggested. There is evidence for the presence of an ecto-diphosphohydrolase (ecto-apyrase) capable of hydrolyzing both ATP and ADP.

This type of enzyme activity has been found both in non-neural tissues, such as pancreas, kidney, liver, lung, and aorta [17,18], and in preparations of synaptosomes [19,20]. Possibly an ecto-ATPase and an ecto-diphosphohydrolase coexist in the plasma membrane of a variety of cells. The enzyme activities have a broad substrate specificity for 5′-purine and pyrimidine trinucleotides. They are thus also capable of hydrolyzing GTP or UTP. Furthermore, they can be activated by either Ca^{2+} or Mg^{2+}.

Less is known about the molecular identity of these ectoenzymes. It has been suggested that members of the immunoglobulin superfamily such as the liver protein cell-CAM (Mr, 100 kD) [21] and the neural adhesion molecule N-CAM function as ecto-ATPase [22]. The ecto-ATP-dihydrophosphorylase in bovine lung and trachea has an apparent molecular mass of about 70 kD [17]. The isolation of the ecto-ATPase (or ecto-ATP-dihydrophosphorylase) from neuronal membranes has been hampered by its sensitivity to extraction by detergents and its rapid loss of activity. When solubilized in a variety of detergents, the ADPase activity of synaptic membranes from bovine brain was consistently lost [15]. In contrast, a membrane-associated

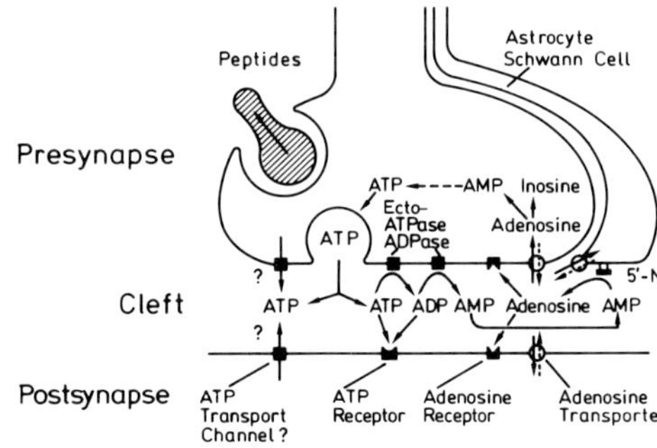

FIGURE 21-1. Life cycle and potential functions of synaptically released ATP. One single enzyme (ATP-diphosphohydrolase, apyrase) may hydrolyze both ATP and ADP. The ADP formed is also capable of activating ATP receptors. Ecto-ATPase may be localized presynaptically, postsynaptically, or on glial cells. 5′-nucleotidase may have a non-neuronal or neuronal localization. It is anchored to the cell surface by a GPI anchor. The adenosine formed can act on both presynaptic and postsynaptic adenosine receptors and is finally salvaged by the nerve terminal (or by adjacent cells) for reformation of ATP. For clarity this diagram depicts only the release of ATP, but ADP, AMP, GTP, and diadenosine polyphosphates (Ap4A, Ap5A), as well as non-nucleotide cotransmitter, may also be released in substantial amounts.

Ca^{2+}- or Mg^{2+}-activated ATPase could be solubilized with 2% polydocanol (polyoxyethylen-9-laurylether). It was purified using chromatography on Sephacryl-S300, DEAE-ion exchange chromatography, and finally glycerol density gradient centrifugation.

The enzyme had an apparent molecular mass of 50 kD and a pI of 5.3–5.9. It could be labeled by incubation of synaptosomes with [α^{32}P]8-azido-ATP or [α^{32}P]8-azido-GTP. Its pH maximum was between 7.5 and 8.5. Typical inhibitors of cytoplasmic ATPase, including the inhibitors of the mitochondrial ATPase oligomycin, NBDL-Cl, and DCCD, did not significantly inhibit enzyme activity. As expected for a surface enzyme, the ATPase was inhibited when intact synaptosomes were incubated with diazotized sulfanilic azid prior to solubilization and isolation. Diazotized sulfanilic acid is a slowly membrane-permeating and covalently binding enzyme inhibitor. Cytosolic enzymes such as lactate dehydrogenase were not inhibited under the same experimental conditions. The enzyme isolated after treatment of intact synaptosomes with diazotized sulfanilic acid revealed a small increase in molecular mass, presumably resulting from the covalent binding of sulfanilic acid molecules.

In order to obtain sequence information, the enzyme was reisolated using a modified and abbreviated procedure. This included extraction with Triton X-100 (0.5%) rather than with polydocanol and omitted the sizing column but included DEAE-ion exchange chromatography and glycerol density gradient centrifugation. A protein band of about 50 kD isolated to apparent purity was excised from the SDS polacrylamide gel and was subjected to limited proteolysis and Edman degradation. Three peptide sequences were obtained that were identical to the sequences of the F_1 complex of the mitochondrial ATP synthase: Thr219–Lys230 and Glu335–Arg347 of the α-subunit of bovine heart mitochondria and Ala387–Arg406 of the human or rat β-subunit. Theses proteins have a comparable molecular mass. At present it is not clear whether the enzyme originally isolated corresponded to the mitochondrial enzyme or whether the revised isolation procedure resulted in the isolation of the mitochondrial F_1 complex that can separate from the F_0 complex on the subcellular fractionation [23,24]. It should be noted that, like the surface ecto-ATPase, the mitochondrial proton-pumping ATPase hydrolyzes not only ATP but also a variety of 5′-purine and pyrimidine trinucleotides that can be activated by either Ca^{2+} or Mg^{2+} alone and that has an alkaline pH optimum. In this context it is of interest that the β-subunit of F1-ATPase was found to be expressed on the surface of tumor cells [25]. The molecular identity of the synaptic ecto-ATPase will be subjected to further analysis.

Ecto-5′-Nucleotidase Is a GPI-Anchored Enzyme

The primary structure of 5′-nucleotidase has now been revealed for four peripheral tissues and for the brain of the electric ray [26–30]. The enzyme is a Zn-binding sialoglycoprotein, and the disulfide-linked dimer is anchored to the plasme membrane via a glycosyl phosphatidylinositol (GPI) anchor (Fig. 21-2). There is a high degree of sequence identity between the enzymes from different sources. There are also clusters of identity with bacterial enzymes that hydrolyze not only AMP but also ADP, ATP, and nucleoside diphosphate sugars. Whereas the periplasmic bacterial 5′-nucleotidase probably serves mainly in providing a carbon source for the cell, the vertebrate enzyme has become more specialized with regard to substrate usage [31]. The two intracellular forms of 5′-nucleotidase with preferences for the hydrolysis of AMP and IMP, respectively, appear to be proteins different from the surface 5′-nucleotidase [32–34].

Ecto-5′-Nucleotidase also Occurs as a Soluble Protein

As for other GPI-anchored proteins, the GPI anchor of 5′-nucleotidase can, at least partially, be hydrolyzed in vitro by application of GPI-specific phospholipase C. This results in the formation of a soluble protein with the glycan moiety still attached and the generation of a 1,2-(cyclic)monophosphate on the inositol ring (Fig. 21-2). We demonstrated that a soluble, GPI-derived form of 5′-nucleotidase also exists in situ [35]. An antibody directed against the 1,2-(cyclic)monophosphate allows the specific identification of the soluble form of 5′-nucleotidase (Fig. 21-3). When the cyclic phosphate ring is opened by mild acid hydrolysis, this antibody no longer binds to the protein. We found that 28% of total 5′-nucleotidase activity that can be isolated from bovine brain immediately after slaughter exists in a soluble form. In contrast, in the electric organ of the electric ray the contribution of soluble 5′-nucleotidase to the total was only 1.6%. The soluble form is glycosylated, has the same molecular mass and kinetic characteristics as the membrane-bound form, and is recognized by the same antibodies.

We suggest that nervous tissue contains a pool of endogenous soluble ecto-5′-nucleotidase derived from the membrane-anchored form presumably cleaved by endogeneous ecto-phospholipase C. The physiologic implication of an endogenous solubilization is not yet understood. It should be noted that soluble forms have also been found of other GPI-anchored proteins, including cell adhesion molecules such as F3 or the receptor for ciliary neurotrophic factor (CNTFRα).

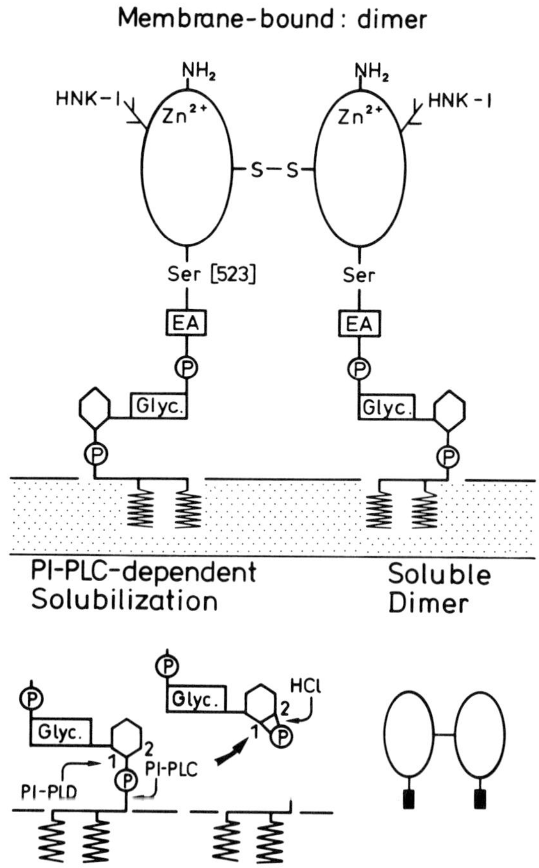

FIGURE 21-2. Molecular model of ecto-5′-nucleotidase and soluble forms derived from it. The GPI-anchored form exists as a dimer with two subunits linked by at least one disulfide bridge. In some species and under certain developmental conditions, the enzyme can carry the HNK-1 epitope. The activity of GPI-specific phospholipase C (but not of phospholipase D) leads to cleavage of the anchor with the formation of a 1,2-(cyclic)monophosphate on the inositol ring, which is susceptible to mild acid hydrolysis. This results in the formation of a soluble protein dimer with the glycan moiety still attached.

Distribution of 5′-Nucleotidase in Brain

The distribution of 5′-nucleotidase in nervous tissue has previously been analyzed with enzyme histochemical procedures and more recently also using immunocytochemistry [36]. It is puzzling that sites of extracellular phosphate formation, as revealed as lead phosphate or cerium phosphate precipitates on application of AMP to tissue sections, only partially overlap with sites of antibody binding [36,37]. Presumably glycosylation patterns of the enzyme vary between cells or tissue sources. The existence of isoforms of the enzyme or also of other surface enzymes in brain capable of hydrolyzing AMP cannot yet be excluded.

Our antibodies that are directed against 5′-nucleotides of varying tissue sources do not all reveal the same binding pattern. In mouse and rat brain the most consistently labeled structures include astrocytes and the capillary endothelial walls (Fig. 21-4E–G). Because astrocytes are in close contact with synaptic sites, glial localization may account for much of the hydrolysis of AMP derived from synaptically released ATP. It is notable that astrocyte labeling in the cerebral cortex is considerably lower than in hippopcampus and striatum. One of the antibodies labels not only astrocytes but also the CA3 region of the mouse [36] and rat hippocampus (Fig. 21-4A). The punctate labeling pattern suggests that the antibodies bind to the mossy fiber terminals. Mossy fiber terminals release not only glutamate but also ATP and opioids [38]. Possibly the high resistance of the CA3 region

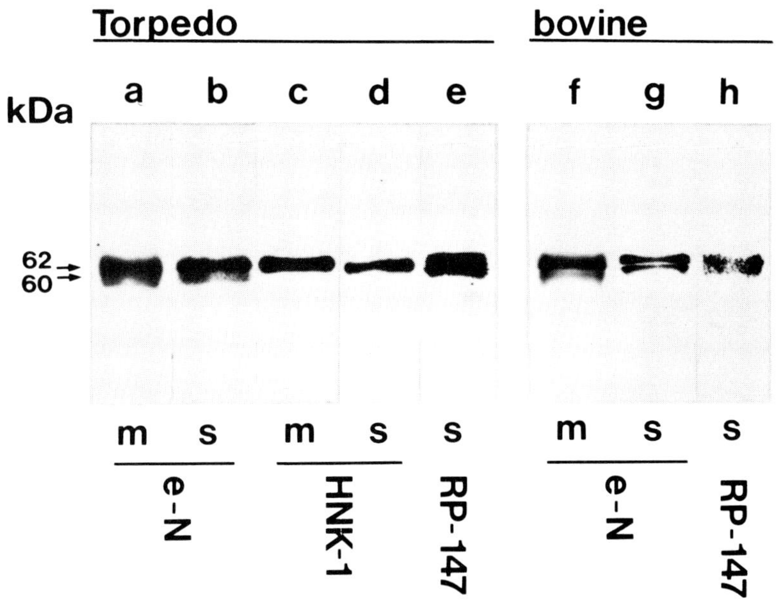

FIGURE 21-3. Demonstration of endogenous soluble forms of 5′-nucleotidase in *Torpedo* electric organ and bovine brain. Lanes a, b, f, g: Antibodies against membrane-bound (m) ecto-5′-nucleotidase (e-N) also recognize the soluble form (s). Lanes c, d: In *Torpedo* both the membrane-bound (m) and soluble (s) forms of 5′-nucleotidase carry the HNK-1 epitope. The HNK-1 epitope is absent from bovine 5′-nucleotidase. Lanes e, h: An antibody directed against the 1,2-(cyclic)monophosphate on the inositol ring (RP 147) binds to the soluble (s) form but not to the membrane-bound or the acid-treated soluble form. Acid treatment does not impair binding of the e-N or HNK-1 antibodies. (Modified from Vogel et al. [35], with permission.)

to ischemia is due to the increased extracellular concentrations of the neuroprotectant adenosine at the mossy fiber/pyramidal cell synapse. 5′-Nucleotidase antibodies also bind to the ependymal cells lining the ventricular walls (Fig. 21-4C). As expected for the distribution of GPI-anchored enzymes [39] in polarized epithelial cells, immunofluorescence is most intense at the apical surface of the cells. In addition, 5′-nucleotidase is associated with the chorioid plexus (Fig. 21-4E).

In the peripheral electromotor nerves of the electric ray, 5′-nucleotidase is associated with the Schwann-cell surface, including the presynaptic Schwann-cell fingers. The identical anti-5′-nucleotidase antibody does not bind to membranes of the myelin-forming oligodendroglia cells in the central nervous system [40].

Implication in Neural Development

As a surface-located protein, 5′-nucleotidase may interact with the extracellular matrix and also with adjoning cells. Binding of 5′-nucleotidase to both fibronectin and laminin has been reported [41,42]. Cellular expression of 5′-nucleotidase can be regulated [31]. 5′-Nucleotidase is expressed at synapses of cat visual cortex and the rat cerebellum during synapse formation and remodeling [37,43]. Enzyme cytochemistry (but not immunocytochemistry!) demonstrates a transient association of 5′-nucleotidase with synaptic membranes, including the synaptic cleft. Later in development, 5′-nucleotidase becomes restricted to an exclusively glial localization. Developing or remodeled synaptic contacts may have a parti-

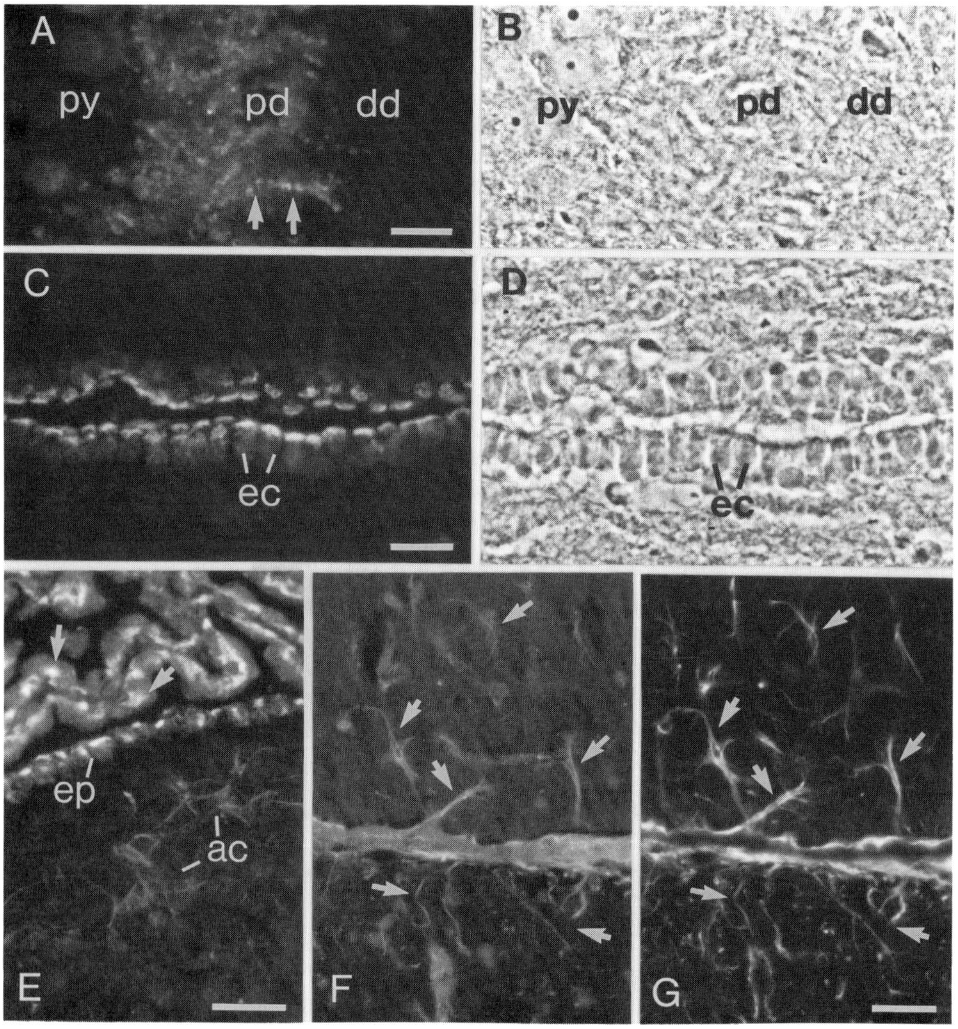

FIGURE 21-4. Distribution of 5′-nucleotidase in mammalian brain as revealed by immunocytochemistry. **A:** Immunostaining of mossy fiber terminals (arrows) at the proximal dendrites (pd) of pyramidal cells (py) in the CA3 region of the rat hippocampus using an antibody against 5′-nucleotidase from *Torpedo* electric organ (AS1). Note the absence of immunostaning at the distal dendrites (dd). **B:** Corresponding phase-contrast image. **C:** Immunostaining of the ependyma (3d ventricle of mouse brain) with strong immunofluorescence at the apical surface of ependymal cells (ec) (antibody as for A). **D:** Corresponding phase-contrast image. **E:** Chorioid plexus, ependyma (ep), and astrocytes (ac) are immunopositive (mouse brain, antibody against 5′-nucleotidase from *Torpedo*, As4). Immunopositive cells in the chorioid plexus are depicted by arrows. **F:** Astrocyte staining in the ventral hippocampus (stratum oriens) and the subjacent hypothalamus (lateral dorsal nucleus thalamus) of the mouse (antibody against 5′-nucleotidase from bovine brain, FITC-immunofluorescence). **G:** Identical image using an antibody against the astrocyte marker glial fibrillary acidic protein (GFAP) and TRITC-labeled secondary antibody. Identical astrocytic elements are labeled in F and G (arrows). Bars represent 20 μm (A–D) and 30 μm (E–G).

cular demand of extracellular nucleoside formation, but an additional functional context can be envisaged. 5′-Nucleotidase in both the electric ray electric organ and the developing cat brain contains a carbohydrate epitope that is recognized by a monoclonal antibody (HNK1, L2) [44,45]. The HNK1/L2 epitope is contained on many cell adhesion molecules, such as neural cell adhesion molecule (N-CAM), L1, J1, cytotactin, or the myelin protein myelin associated glycoprotein (MAG). The epitope may function in the formation of homophilic interactions between surface adhesion molecules and also in the adhesion of neural cells to the extracellular matrix component laminin [46]. 5′-Nucleotidase from the occipital cortex carries the HNK1/L2 epitope in kittens of 3–9 weeks age. The epitope is absent from the 5′-nucleotidase of 12-week-old kittens or adult cats (Fig. 21-5). This appearance of the HNK-1/L2 epitope on 5′-nucleotidase correlates exactly with the transient localization of the enzyme at the synapses formed by the afferents from the lateral geniculate nucleus in the visual cortex (layer IV). Possibly the neonatal ecto-5′-nucleotidase (like other HNK-1–bearing surface proteins present at that time) participates in the substrate selection of the developing neurites. Its function may thus go beyond its activity as an adenosine-producing enzyme.

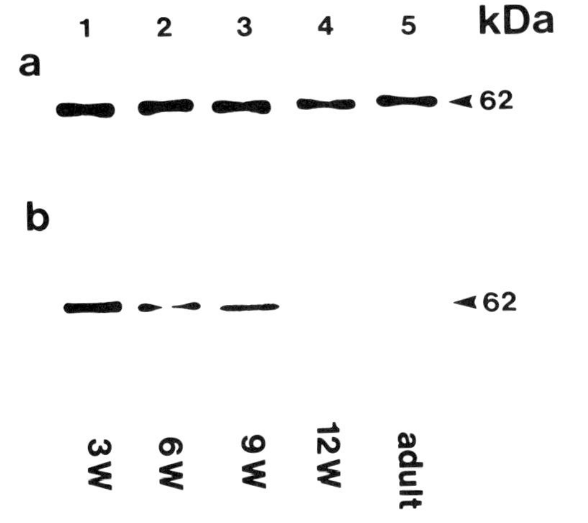

FIGURE 21-5. Demonstration of a transient association of the HNK1-epitope with 5′-nucleotidase from the cat visual cortex. Lane a: The anti-5′-nucleotidase antibody binds equally to the enzyme isolated from kitten of 3–12 weeks age and from adult cats. Lane b: The HNK-1 epitope is contained on 5′-nucleotidase only up to the age of 9 weeks. (From Vogel et al. [45], with permission.)

Acknowledgment

This work was supported by the Deutsche Forschungsgemeinschaft (SFB 269/A4).

References

1. Dubyak GR, El-Moatassim C (1993) Signal transduction via P_2-purinergic receptors for extracellular ATP and other nucleotides. Am J Physiol 265:C577–C606.
2. Salter MW, De Koninck Y, Henry JL (1993) Physiological roles for adenosine and ATP in synaptic transmission in the spinal dorsal horn. Prog Neurobiol 41:125–156.
3. Edwards FA, Gibb AJ (1993) ATP—a fast neurotransmitter. FEBS Lett 325:86–89.
4. Smith DO (1991) Sources of adenosine released during neuromuscular transmission in the rat. J Physiol 432:343–354.
5. Sawynok J, Downie JW, Reid AR, Cahill CM, White TD (1993) ATP release from dorsal spinal cord synaptosomes—characterization and neuronal origin. Brain Res 610:32–38.
6. Grondal EJM, Zimmermann H (1988) Extracellular metabolism of ATP at the cholinergic electromotor synapse, characterization and location of ectonucleotidases. In Zimmermann H (ed.), Cellular and Molecular Basis of Synaptic Transmission. Heidelberg: Springer-Verlag, pp. 395–410.
7. Richardson PJ, Brown SJ (1987) ATP release from affinity-purified rat cholinergic nerve terminals. J Neurochem 48:622–630.
8. Fiedler JL, Pollard HB, Rojas E (1992) Quantitative analysis of depolarization-induced ATP release from mouse brain synaptosomes—external calcium dependent and independent processes. J Membr Biol 127:21–33.
9. Farinas I, Solsona C, Marsal J (1992) Omega-conotoxin differentially blocks acetylcholine and adenosine triphosphate releases from *Torpedo* synaptosomes. Neuroscience 47:641–648.

10. Von Kügelgen I, Starke K (1991) Noradrenaline-ATP co-transmission in the sympathetic nervous system. Trends Pharmacol. Sci 12:319–324.
11. Vizi ES, Sperlágh B, Baranyi M (1992) Evidence that ATP, released from the postsynaptic site by noradrenaline, is involved in mechanical responses of guinea-pig vas deferens, cascade transmission. Neuroscience 50:455–465.
12. Driessen B, von Kügelgen I, Starke K (1993) Neural ATP release and its α-adrenoceptor-mediated modulation in guinea-pig vas deferens. Naunyn Schmiedebergs Arch Pharmacol 348: 358–366.
13. Abraham EH, Prat AG, Gerweck L, Seneveratne T, Arceci RJ, Kramer R, Guidotti G, Cantiello HF (1993) The multidrug resistance (mdrl) gene product functions as an ATP channel. Proc Natl Acad Sci USA 90:312–316.
14. James S, Richardson PJ (1993) Production of adenosine from extracellular ATP at the striatal cholinergic synapse. J Neurochem 60:219–227.
15. Hohmann J, Kowalewski H, Vogel M, Zimmermann H (1993) Isolation of a Ca^{2+} or Mg^{2+}-activated ATPase (ecto-ATPase) from bovine brain synaptic membranes. Biochim Biophys Acta 1152:146–154.
16. Nagy AK, Shuster TA, Delgado-Escueta AV (1986) Ecto-ATPase of mammalian synaptosomes, identification and enzymic characterization. J Neurochem 47:976–986.
17. Picher M, Coté YP, Béliveau R, Potier M, Beaudoin AR (1993) Demonstration of a novel type of ATP-diphosphohydrolase (EC3.6.1.5) in the bovine lung. J Biol Chem 268:4699–4703.
18. Picher M, Béliveau R, Potier M, Savaria D, Rousseau É, Beaudoin AR (1994) Characterization of an ecto-ATP-diphosphohydrolase in non-vascular smooth muscles of bovine trachea. Drug Dev Res 31(Suppl):308.
19. Sarkis JJF, Salto C (1991) Characterization of a synaptosomal ATP diphosphohydrolase from the electric organ of *Torpedo marmorata*. Brain Res Bull 26:871–876.
20. Müller J, João B, Rocha JBT, Battastini AMO, Sarkis JJF, Dias RD (1993) Postnatal development of ATPase-ADPase activities in synaptosomal fraction from cerebral cortex of rats. Neurochem Int 23:471–477.
21. Lin SH, Čulić O, Flanagan D, Hixson DC (1991) Immunochemical characterization of two isoforms of rat liver ecto-ATPase that show an immunological and structural identity with a glycoprotein cell-adhesion molecule with Mr-10.5000. Biochem J 278:155–161.
22. Dzhandzhugazyan K, Bock E (1993) Demonstration of (Ca^{2+}-Mg^{2+})-ATPase activity of the neural cell adhesion molecule. FEBS Lett 336: 279–283.
23. Futai M, Kanazawa H (1983) Structure and function of proton-translocating adenosine triphosphatase (F_0F_1), biochemical and molecular biological approaches. Microbiol Rev 47: 285–312.
24. Kagawa Y, Nukiwa N (1981) Conversion of stable ATPase to labile ATPase by acetylation, and the α,β and α,γ subunit complexes during its reconstitution. Biochem Biophys Res Commun 100:1370–1376.
25. Das B, Mondragon MOH, Sadeghian M, Hatcher VB, Norin AJ (1994) A novel ligand in lymphocytes-mediated cytotoxicity: Expression of the β subunit of H^+ transporting ATP synthase on the surface of tumor cell lines. J Exp Med 180:273–281.
26. Misumi Y, Ogata S, Hirose S, Ikehara Y (1990) Primary structure of rat liver 5′-nucleotidase deduced from the cDNA. Presence of the COOH-terminal hydrophobic domain for possible post-translational modification by glycophospholipid. J Biol Chem 265: 2178–2183.
27. Misumi Y, Ogata S, Ohkubo K, Hirose S, Ikehara Y (1990) Primary structure of human placental 5′-nucleotidase and identification of the glycolipid anchor in the mature form. Eur J Biochem 191:563–569.
28. Volknandt W, Vogel M, Pevsner J, Misumi Y, Ikehara Y, Zimmermann H (1991) 5′-Nucleotidase from the electric ray electric lobe: Primary structure and relation to mammalian and procaryotic enzymes. Eur J Biochem 202: 855–861.
29. Suzuki K, Furukawa Y, Tamura H, Ejiri N, Suematsu H, Taguchi R, Nakamura S, Suzuki Y, Ikezawa H (1993) Purification and cDNA cloning of bovine liver 5′-nucleotidase, a GPI-anchored protein, and its expression in COS Cells. J Biochem (Tokyo) 113:607–613.
30. Resta R, Hooker SW, Hansen KR, Laurent AB, Park JL, Balckburn MR, Knudsen TB, Thompson LF (1993) Murine ecto-5′-nucleotidase (CD73)—cDNA cloning and tissue distribution. Gene 133:171–177.
31. Zimmermann H (1992) 5′-Nucleotidase—molecular structure and functional aspects. Biochem J 285:345–365.
32. Yamazaki Y, Truong VL, Lowenstein JM (1991) 5′-Nucleotidase I from rabbit heart.

Biochemistry 30:1503–1509.

33. Darvish A, Metting PJ (1993) Purification and regulation of an AMP-specific cytosolic 5′-nucleotidase from dog heart. Am J Physiol 264:H1528–H1534.
34. Itoh R (1994) Purification and some properties of an IMP-specific 5′-nucleotidase from yeast. Biochem J 298:593–598.
35. Vogel M, Kowalewski H, Zimmermann H, Hooper NM, Turner AJ (1992) Soluble low-Km 5′-nucleotidase from electric ray (*Torpedo marmorata*) electric organ and bovine cerebral cortex derived from the glycosyl-phosphatidylinositol-anchored ectoenzyme by phospholipase-C cleavage. Biochem J 284: 621–624.
36. Zimmermann H, Vogel M, Laube U (1993) Hippocampal localization of 5′-nucleotidase as revealed by immunocytochemistry. Neuroscience 55:105–112.
37. Schoen SW, Graeber MB, Toth L, Kreutzberg GW (1991) Synaptic 5′-nucleotidase is transient and indicative of climbing fiber plasticity during the postnatal development of rat cerebellum. Dev Brain Res 61:125–138.
38. Terrian DM, Hernandez PG, Rea MA, Peters RI (1989) ATP release, adenosine formation, and modulation of dynorphin and glutamic acid release by adenosine analogues in rat hippocampal mossy fiber synaptosomes. J Neurochem 53:1390–1399.
39. Ferguson MAJ (1992) Glycosyl-phosphatidylinositol membrane anchors—the tale of a tail. Biochem Soc Trans 20:243–256.
40. Grondal EJM, Janetzko A, Zimmermann H (1988) Monospecific antiserum against 5′-nucleotidase from *Torpedo* electric organ, immunocytochemical distribution of the enzyme and its association with Schwann cell membranes. Neuroscience 24:351–363.
41. Olmo N, Turnay J, Risse G, Deutzmann R, von der Mark K, Lizarbe A (1992) Modulation of 5′-nucleotidase activity in plasma membranes and intact cells by the extracellular matrix proteins laminin and fibronectin. Biochem J 282: 181–188.
42. Mehul B, Doyennette-Moyne MA, Aubery M, Codogno P, Mannherz HG (1992) Enzymatic activity and in vivo distribution of 5′-nucleotidase, an extracellular matrix binding glycoprotein, during the development of chicken striated muscle. Exp Cell Res 203:62–71.
43. Schoen SW, Kreutzberg GW, Singer W (1993) Cytochemical redistribution of 5′-nucleotidase in the developing cat visual cortex. Eur J Neurosci 5:210–222.
44. Vogel M, Kowalewski HJ, Zimmermann H, Janetzko A, Margolis RU, Wollny HE (1991) Association of the HNK-1 epitope with 5′-nucleotidase from *Torpedo marmorata* (electric ray) electric organ. Biochem J 278:199–202.
45. Vogel M, Zimmermann H, Singer W (1993) Transient association of the HNK-1 epitope with 5′-nucleotidase during development of the cat visual cortex. Eur J Neurosci 5:1423–1425.
46. Hall H, Liu L, Schachner M, Schmitz B (1993) The L2/HNK-1 carbohydrate mediates adhesion of neural cells to laminin. Eur J Neurosci 5: 34–42.

22. P-GLYCOPROTEIN SERVES AS A TRANSPORTER OF CELLULAR ATP

Edward H. Abraham

P-glycoprotein, the gene product of the *mdr1* gene, is responsible for the interesting multidrug resistance (MDR) phenomenon. Cells that express the P-glycoprotein become resistant to a wide variety of compounds that have no obvious structural relationships other than their lipophilicity and generally cationic or zwitterionic charge distribution. It is known that exposure of tumor cells to one drug may induce cross-resistance to a variety of unrelated drugs to which the cell has not been previously exposed. The resultant overexpression of the P-glycoprotein in the plasma membrane results in a 170 kD band on membrane protein electrophoresis.

The past 10-year history of the P-glycoprotein illustrates some of the strengths and weaknesses of molecular biologic techniques at identifying protein sequences, even before the protein has been purified and its functions are fully investigated. From the sequence structure, hydrophobicity plots have predicted that P-glycoprotein may have 12 transmembrane regions and 2 nucleotide binding sites. It was initially inferred that the nucleotide binding sites also hydrolyze ATP to accomplish drug efflux from the cell. Our studies show that P-glycoprotein is associated with release of ATP [1] to the medium proportional to the concentration of the protein in the membrane and that this ATP channel activity extends to other members of the family, including cystic fibrosis transmembrane regulator (CFTR), the protein associated with cystic fibrosis [2,3].

Evolving Concept of MDR Function

The ABC (ATP binding cassette) proteins have been shown to function as transporters of drugs and membrane channels. It appears that the removal of drug is an active process that results in reduced cellular cytotoxicity. Recently the P-glycoprotein (PGP = p170 = MDR1) has been shown to function as both a volume-regulated chloride channel [4] and an ATP channel [1,5,6]. Moreover, we have found that the P-glycoprotein utilizes the transmembrane ATP electrochemical gradient to remove drug from cells (Fig. 22-1). Reversing the transmembrane ATP electrochemical gradient by depolarizing the membrane potential and reversing the ATP chemical gradient can reverse drug efflux performed by PGP. The extrusion of ATP, as first described in our recent paper [1], makes ATP available extracellularly. Besides driving the efflux of drugs and normal cell products, including some polypeptides, the ATP may modify peptide folding and function through interactions with ectoenzymes or chaperone proteins. Increase in extracellular ATP can perform a variety of functions, including phosphorylation of extracellular proteins [7] and interaction with purinergic receptors [8]. Binding to the purinergic receptors can activate intracellular enzymes, such as adenylate cyclase. In a complex multicellular organism, this response may be part of a system of communication between cells. ATP does not re-enter the cell as such but is salvaged in the form of adenosine or inosine after all phosphates have been removed. Indeed, the purine intermediates AMP, IMP, adenosine, and inosine have separate extracellular receptor interactions from ATP, generating further dimensions to this already rich set of purine signal interactions. Thus ATP released by the P-glycoprotein can participate in a transmembrane adenylate cycle (Fig. 22-2).

We have demonstrated that the P-glycoprotein functions as an ATP channel, and recent studies have shown that the transmembrane electrochemical gradient is capable of removing drugs that are lipophilic cations or neutral but zwitterionic (Fig. 22-1). We have demonstrated that reversal of the transmembrane electrochemical gradient for ATP is sufficient to reverse drug efflux through P-glycoprotein and to accomplish a net influx of drugs such as Adriamycin (doxorubicin) in resistant cells [9]. The release of ATP is capable of altering the extracellular microenvironment for a variety of the purinergic and adenosine surface receptors, the P_1, P_2, A_1, or A_2 receptors, respectively. Moreover, the extracellular ATP serves as a substrate for ectokinases, which have the potential to work in a paracrine fashion modulating cell-cell interactions. Moreover, ATP may serve to modify the function of secreted peptides such as fibroblast growth factor (FGF) [10] (Fig. 22-3).

Changes in the extracellular ATP concentration can modulate a variety of cellular responses, including apoptosis [11], regulation of the cell cycle [11], and signal transduction leading to thiol protection [12]. Zheng et al. [11] demonstrated that extracellular ATP bound to purinergic receptors led to the mobilization of intracellular calcium, presumably through the phospholipase C and inositol triphosphate pathway, leading to apoptosis. Rapaport [12] showed that treatment of human tumor cells

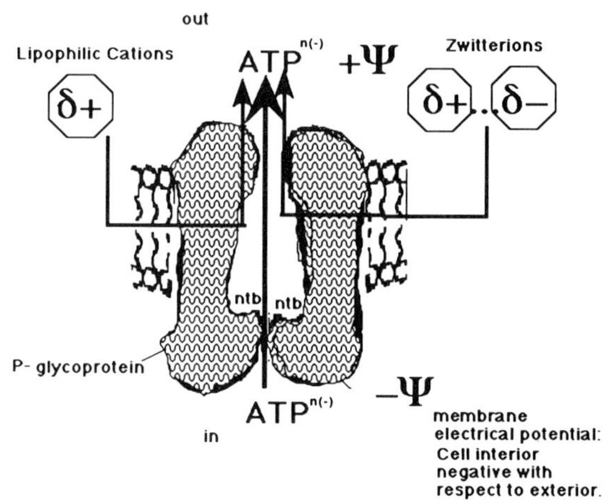

FIGURE 22-1. Model of P-glycoprotein as an ATP channel capable of cotransporting cationic and zwitterionic compounds. ntb suggests the location of the nucleotide binding domains.

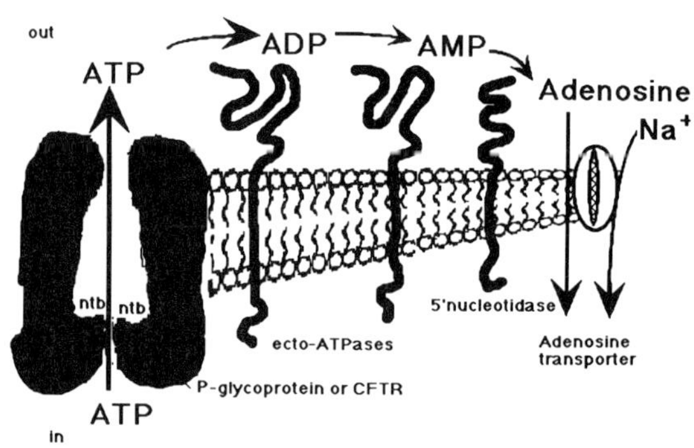

FIGURE 22-2. MDR acting as an ATP channel shown in association with membrane ectoenzymes responsible for the transmembrane purine cycle.

with ADP or ATP led to growth arrest in the S phase of the cell cycle. We have been investigating the role of extracellular ATP in modulating intracellular glutathione concentrations [13].

The role of extracellular ATP in the control of endothelial cell function has recently been reviewed [14]. The control of endothelial proliferation and modulation of endothelial function

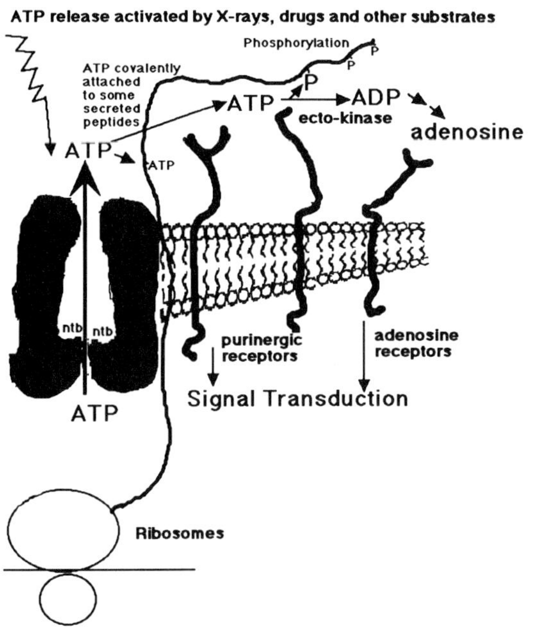

FIGURE 22-3. The effects of ionizing radiation on membrane transport. Preliminary data indicate that the effect of ionizing radiation is to alter conductance of the channel and to allow increased efflux of ATP. This can act via an autocrine loop to alter signal transduction pathways or can phosphorylate secreted peptides via the action of ecto-protein-kinases. Some peptides such as FGF may be secreted via MDR1, and the net effect of radiation may be to increase the secretion of these important modulators of the radiation response.

is crucial for tumor growth. Besides responding to extracellular ATP, endothelial cells synthesize and release two vasodilators: prostacyclin (PGI_2) and nitric oxide (NO), originally called endothelial cell relaxing factor (EDRF). PGI_2 is released rapidly following ATP stimulation of endothelial cells but returns to baseline within minutes of stimulation. Sources of ATP for this stimulation are speculated to be from platelet degranulation, red blood cell hemolysis, and endothelial cells following shear stress. Endothelial cells express the P-glycoprotein along the luminal surface and this is particularly true at the blood brain barrier where endothelial cells express high levels of P-glycoprotein [15].

Membrane Response to Ionizing Radiation

The molecular processes mediating the cellular response to ionizing radiation may in part act by way of membrane changes, which in turn activate signal transduction from cell surface receptors to the nucleus. Distinguishing whether membrane or cytosolic chemistry has the dominant role in signaling adaptive survival mechanisms depends on the identification of the initial x-ray–induced changes. The answer may be an interaction of both. The recent work of Kuo et al. [16] has demonstrated that potassium channel activation is an early response (within seconds), with irradiated cells responding to increased levels of hydrogen peroxide, an agent leading to formation of reactive oxygen intermediates. Using the technique of whole-cell voltage clamping, potassium currents increased after treatment with ionizing radiation, hydrogen peroxide, and xanthine/xanthine oxidase, but not after tritiated thymidine, suggesting a membrane effect. The effect was abolished by free radical scavengers but was maintained in the presence of agents that deplete protein kinase C. The cell membrane acts to maintain the intracellular concentration of ions. Permitting changes in ion concentration results in induction of the so-called early response genes. Ion channel proteins may have a role in initiation of the response to signal transduction, utilizing their property of rapid conformational alterations in response to changes in the membrane chemistry. A hypothesis under investigation is that opening of the MDR ATP/chloride channel in response to increased local concentrations of peroxides may be one initial response to ionizing radiation.

The acetylcholine receptor in muscle cells is directly responsible for passing an ionic signal in the form of activation of membrane sodium channels that activates the intracellular signal pathway, resulting in muscle contraction in response to acetylcholine stimulation. The MDR system may have an analogous function, through the efflux of ATP to the extracellular surface, allowing interactions with specific purinergic receptors and ectokinases. We have recently detected rapid release of ATP by cells irradiated with doses as low as 10 cGy. This suggests specificity of the MDR1 protein rather than a nonspecific membrane leak, as the magnitude of ATP release correlates with the level of MDR1 expression [9].

We continue to increase our understanding of how ATP channels and extracellular ATP fit into the cellular defense mechanisms against oxidative stress. The extensive family of MDR-related proteins may share the common feature of the ability to transport ATP. The multiple involvement of the MDR family proteins in basic biologic processes from transport and signalling to protein folding, modification, and packaging proclaim that much has yet to be discovered in this area.

References

1. Abraham EH, Prat AG, Gerweck L, Seneveratne T, Arceci RJ, Kramer R, Guidotti G, Cantiello HF (1993) The multidrug resistance (*mdrl*) gene product functions as an ATP channel. Proc Natl Acad Sci USA 90:312–316.
2. Cantiello HF, Prat AG, Reisin IL, Erole LB, Abraham EH, Amara JF, Gregory RJ, Ausiello DA (1994) External ATP and its analogs activate the cystic fibrosis transmembrane conductance regulator by a cyclic AMP-independent mechanism. J Biol Chem 269:11224–11232.
3. Reisin IL, Prat AG, Abraham EH, Amara JF, Gregory RJ, Ausiello DA, Cantiello HF (1994) The cystic fibrosis transmembrane conductance regulator is a dual ATP and chloride channel. J Biol Chem 32:20584–20591.
4. Higgins CF (1992) ABC transporters: From microorganisms to man. Annu Rev Cell Biol 8:67–113.
5. Powell SN, Abraham EH (1993) The biology of radioresistance: Similarities, differences and interactions with drug resistance. Cytotechnology 12:325–345.
6. Arias IM (1993) Is the multidrug resistance an ATP channel? Hepatology 18:216–222.
7. Skubitz KM, Goueli SA (1991) Basic fibroblast growth factor is a substrate for phosphorylation by human neutrophil ecto-protein kinase activity. Biochem Biophys Res Commun 174: 49–55.
8. Burnstock G (1990) Overview: Purinergic mechanisms. Ann NY Acad Sci 603:1–18.
9. Abraham EH, Arceci R, Steiglitz K, Herscher L, Zaidi ST, Gerweck L, Guidotti G (1994) ATP electrochemical gradient drives P-glycoprotein mediated drug fluxes. Proc Am Assoc Cancer Res p. 353 (Abst).
10. Vilgrain I, Baird A (1991) Phosphorylation of basic fibroblast growth factor by a protein kinase associated with the outer surface of a target cell. Mol Endocrinol 5:1003–1012.
11. Zheng LM, Zychlinsky A, Liu CC, Ojcius DM, Young JD (1991) Extracellular ATP as a trigger for apoptosis or programmed cell death. J Cell Biol 112:279–288.
12. Rapaport E (1983) Treatment of human tumor cells with ADP or ATP yields arrest of growth in the S-phase of the cell cycle. J Cell Physiol 114:279–283.
13. Abraham EH, Shrivastav B, Friedman N, Okunieff D (1995) Extracellular ATP increases glutathione: Role of P-glycoprotein (PGP). Proc Am Assoc Cancer Res (Abst).
14. Boeynaems JM, Pearson JD (1990) P2 purinoceptors on vascular endothelial cells: Physiological significance and transduction mechanisms. Trends Pharmacol Sci 11:34–37.
15. Shirai A, Naito M, Tatsuta T, Dong J, Hanaoka K, Mikami K, Oh-hara T, Tsuruo T (1994) Transport of clyclosporin A across the brain capillary endothelial cell monolayer by P-glycoprotein. Biochim Biophys Acta 1222: 400–404.
16. Kuo SS, Saad AH, Koong AC, Hahn GM, Giaccia AJ (1993) Potassium-channel activation in response to low doses of gamma-irradiation involves reactive oxygen intermediates in non-excitatory cells. Proc Natl Acad Sci USA 90: 908–912.

23. ON THE ORIGIN OF EXTRACELLULAR ATP IN CARDIAC BLOOD VESSELS: A DUAL ROLE FOR ENDOTHELIUM

Iain L.O. Buxton
Dennis Cheek

The exact mechanisms underlying the moment-to-moment regulation of blood flow in the heart by local mediators are not fully understood but appear to involve purines and their metabolites [1], specifically ATP, ADP, and adenosine (ADO). Determining the role of endothelial cells in the origin of each of these purines is one of the aims of our laboratory. In addition to the novel sources of ATP we propose here, ADP is released in large amounts locally when platelets aggregate [2]. ADP acts at the platelet P_2t receptor (as well as other potential nucleotide binding sites such as the fibrinogen receptor complex) to stimulate shape change, fibrinogen binding, and the platelet release reaction and thus promotion of platelet aggregation. The normal vascular response to this platelet source of purine nucleotides may be vasodilation intended to minimize the chance of clot formation. In addition, ADP would be expected to act at the endothelial cell P2y receptor, leading to relaxation of the blood vessel [3]. Our recent data [4] in cardiac endothelial cells demonstrate that the action of this extracellular ADP would include release of ATP from endothelium. This ATP could then act at the platelet P_2t receptor (ATP released from endothelium would be expected to antagonize the action of ADP at the platelet P_2t receptor), thus decreasing further platelet aggregation. Under some conditions, the effect could be locally amplified by endothelium (ATP-induced ATP release and/or ADP conversion to ATP by adjacent endothelium) so that an adherent white thrombus is not a likely event. Under other conditions, such as hypoxia, platelet aggregation might occur if these endothelial mechanisms are altered.

ATP has been found to have a dual function in the regulation of vascular tone. It acts as a cotransmitter with norepinephrine in sympathetic nerves and may cause vasoconstriction via P_{2x} purinoceptors located on vascular smooth muscle cells [5,6]. Cardiac endothelial cells are known to exert control over vascular tone by releasing relaxing factors such as EDRF (nitric oxide) and PGI_2 in response to various stimuli, including ATP. Endothelium-derived contracting factors such as thromboxane A_2 [7], which also stimulates platelets to aggregate, are also known and their release increased during hypoxia [8]. The role of hypoxia in the regulation of receptor-mediated ATP release, however, is unknown.

The notion that ATP would be released by a healthy cell is surprising to some, and we too approached the phenomenon with some skepticism. However, endothelial cells do indeed elaborate ATP into their extracellular environment in response to mechanical [9] and pharmacologic stimulation [4,10] and, as will be described here, also possess the ability to convert ADP to ATP extracellularly. It is our hypothesis that in this extracellular role, ATP is an autocrine/paracrine hormone able to 1) release vasoactive mediators such as nitric oxide from endothelial cells, 2) act in the blood vessel lumen at P_2 receptors on nearby endothelial cells downstream from the site of release to amplify the release response, 3) act in the abluminal direction on adjacent muscle cells to cause changes in blood vessel tone directly, 4) act on platelets in the bloodstream near endothelium as an antagonist (at the $P_{2}t$ receptor) to block platelet aggregation, and finally, 5) break down in the vessel lumen to adenosine downstream from the site of release, resulting in continued inhibition of platelet aggregation, vasodilation of resistance vessels, and growth maintenance and promotion of endothelium.

Here we describe studies designed to determine the role of endothelium in the extracellular enzymatic generation of ATP. The data described here, along with our previous demonstration of release of ATP from endothelial cells, support our hypothesis that a purinergic axis exists in cardiac blood vessels and participates in the moment-to-moment regulation of blood flow.

Methods

ENDOTHELIAL CELLS

Cardiac endothelial cells were isolated by collagenase (1.5 mg/ml) perfusion of guinea-pig hearts harvested and purified by Percoll density gradient centrifugation and seeded in tissue culture dishes precoated with fibronectin. Cells grown in our laboratory retain a persuasive cobblestone morphology in monolayer culture suggestive of endothelial cells in vivo. Knowing that cell morphology in tissue culture has little or no bearing as a predictor of cell function, however, we have employed a fluorescent antibody directed against Factor VIII. We have observed staining in our primary cultures, confirming that the cells we study retain this endothelial cell marker. The cells we grow readily bind the antibody, while fibroblasts (3T3) employed as a control do not. Passaged cells loose antibody staining after the eighth passage. These results are similar to those observed in a number of labs using cultured endothelial cells [11–13].

ATP DETERMINATION

We employ the luciferin-luciferase reaction to determine the presence of ATP [4]. In this procedure, the enzymatic reaction of firefly luciferase to oxidize luciferin is specific for natural ATP. ADP, added experimentally as a substrate for enzymatic production of ATP, does not support the same signal in the assay and is controlled for by using an ADP blank. The assay requires an ATP bioluminescence assay kit (Sigma) containing ATP assay mixtures and ATP standards; ATP assay mix dilution buffer is prepared according to the manufacturer's directions at pH 7.8. Samples are thawed on ice the day of the assay and are read by first diluting 1 : 1000 (or as needed) and then injecting (100 μl volume) into 100 μl assay mixture in a luminometer tube. Recordings are made in a Turner luminometer over a 10-second period, and integrations are performed under the peak of the response and averaged for duplicate or triplicate samples.

[^{3}H]ATP DETERMINATION BY HPLC

The ability of cells to convert ADP to ATP extracellularly is determined by addition of [^{3}H]ADP (1 μCi) to cell cultures incubated in Krebs buffer at 37° (pH 7.2). The buffer bathing the cells is sampled 30 seconds following the addition of label and acidified with perchloric acid (0.2%). Samples are neutralized and clarified, injected onto an anion exchange column, and separated by HPLC as described previously [4], with quantification by radioactive flow detection.

Results and Discussion

Our previous experience with guinea-pig cardiac endothelial cells grown in tissue culture led to the conclusion that several agonists that might be expected to be present in the bloodstream stimulated ATP release from these cells [4]. One of the agonists that was capable of stimulating ATP release was in fact another adenine nucleotide, ADP. The notion that adenine nucleotides induce their own release has been suggested in other systems [14] and is intriguing because it could explain the amplification of the vasoactive response to short-lived mediators such as bradykinin that are known to release ATP from endothelial cells.

While it was clear from these earlier studies that ADP did indeed act at a P_2 receptor to stimulate ATP release, addition of ADP at concentrations in the micromolar range resulted in a large component of ATP appearing in the bathing medium that could not be accounted for as the result of cellular release. This was evident because in cells prelabeled with [^{3}H]adenine to label intracellular ATP, only a fraction of the ATP generated following the addition of high concentrations of ADP was [^{3}H]-labeled (~10%). Furthermore, cells treated in this manner for up to 10 minutes were healthy, as determined by exclusion of vital dye, and if returned to the incubator for short periods of time could again release ATP upon stimulation. Thus, we reasoned that if cells were not being damaged and releasing very large amounts of cellular ATP, then it was possible that the intact cells were converting ADP to ATP extracellularly.

We tested the hypothesis that cells converted ADP to ATP extracellularly by using the

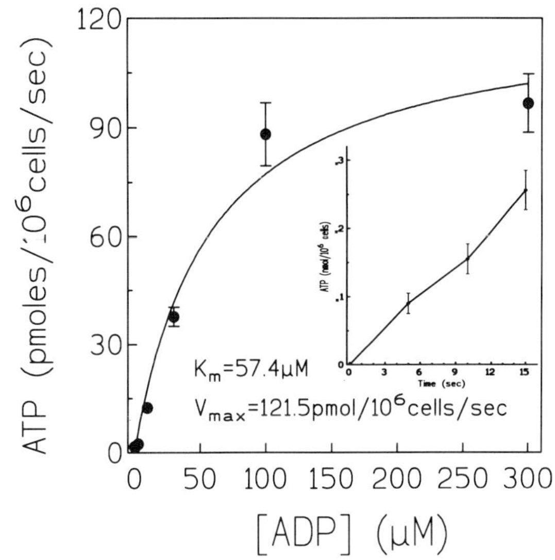

FIGURE 23-1. Cardiac endothelial cells rapidly convert ADP to ATP. Endothelial cells growing in monolayer were exposed to increasing concentrations of ADP at 37°C, and the incubation buffer was sampled for the appearance of ATP measured by the luciferase flash assay after 30 seconds. Data are the result of three experiments and are fit by nonlinear regression. The maximal rate of ATP formation and the affinity of the process for ADP are determined from the nonlinear fit of the data. **Inset:** Data showing the linear production of ATP over time following the addition of 100 μM ADP.

luciferin-luciferase assay to measure the appearance of ATP in the medium bathing cultured endothelial cells 30 seconds following the addition of ADP (Fig. 23-1). The appearance of ATP was rapid and increased linearly for the first 30 seconds following ADP addition. The apparent kinetics of ATP appearance suggested that the process was saturable with a K_m of 57.4 μM and a maximal rate of 121.5 pmol/10^6 cells/sec. This result was similar to that measured in isolated membranes and suggested that the intact cells express an extracellularly directed adenylate kinase-like activity (2ADP → ATP + AMP).

To further test the notion that the endothelial cells can generate ATP extracellularly, we added [^{3}H]ADP directly to the buffer incubating cells

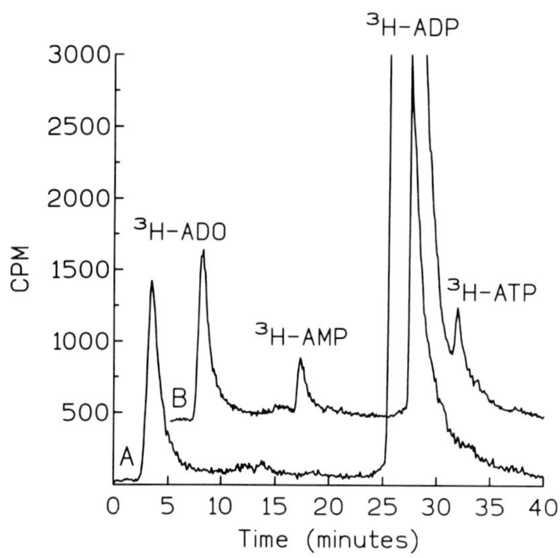

FIGURE 23-2. Conversion of ADP to ATP occurs extracellularly. Cells were exposed to [^{3}H]ADP (1 μM), and the medium bathing the cells was sampled after 30 seconds for the appearance of [^{3}H] ATP. Samples were separated over anion exchange by HPLC and radioactivity was determined continuously by flow detection. **A**: Separation of the radioactive sample from the diluted stock solution showing the absence of ATP in the radioactive source. **B**: Sample following 30-second incubation of the label with cells. The position of the adenyl purine metabolites was determined by the addition of standards. Data are representative of 10 experiments performed in duplicate.

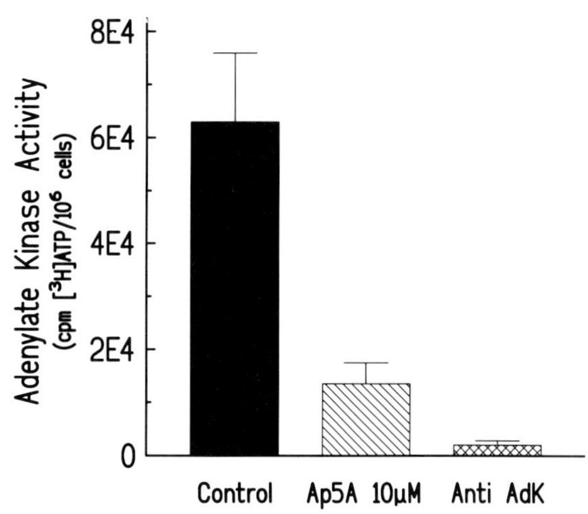

FIGURE 23-3. Inhibition of adenylate kinase blocks the formation of ATP by endothelial cells. Addition of Ap5A or anti-adenylate kinase polyclonal antibody prior to the addition of ADP resulted in the inhibition of ATP formation. Data (n = 3) are expressed as radioactivity converted following the addition of radioactivity (0.5 μM) in the form of [^{3}H]ADP (30 Ci/mmol; 37 μM).

and sampled this incubation medium 30 seconds later. Separation of the incubation buffer over strong anion-exchange high-performance liquid chomatography (HPLC) revealed that significant amounts of [^{3}H]ATP were formed (Fig. 23-2). The additional appearance of [^{3}H]AMP on the HPLC tracing, while consistent with the expected action of 5′-nucleotidase activity, is also consistent with the hypothesis that the cells express an ecto-adenylate kinase. Addition of the known adenylate kinase inhibitor Ap5A (diadenosine pentaphosphate) resulted in inhibition of the appearance of ATP, as did addition of a polyclonal antibody to adenylate kinase raised against the cytoplasmic form of the enzyme (AK1), further supporting the presence of an endothelial ecto-adenylate kinase (Fig. 23-3).

Our data presented here, taken together with other reports from our laboratory and from others, confirm the endothelial origin of extracellular ATP and suggest its importance in vascular regulation. Our data in cell culture suggest numerous questions: What is the extracellular origin of ATP in blood vessels? How is the extracellular appearance of ATP regulated? Is endothelial cell release of ATP and conversion of ADP to ATP a feature of all endothelial cells or do these mechanisms vary in disparate blood vessels? Is ATP in the region of an endothelial cell in vivo likely to remain as ATP long enough to act as ATP per se? Does the ecto-adenylate kinase we propose feed ATP to endothelial P_2 receptors in a compartmental fashion? What is the predicted result of the action of adenyl purines in a given vascular bed? Are endothelial cells polarized such that they release ATP and regenerate ATP from ADP in only the luminal direction? Answers to these questions and other basic questions we have posed are intended to lay the groundwork for the support of a purinergic hypothesis that has important implications for the regulation of cardiac blood flow in health and disease. The data presented here add to our understanding of the biochemistry of ATP actions in blood vessels by confirming an extracellular origin for luminal ATP.

References

1. Burnstock G (1990) Dual control of local blood flow by purines. Ann NY Acad Sci 603:31–45.
2. Olsson RA, Pearson JD (1990) Cardiovascular purinoceptors. Physiol Rev 70:761–846.
3. Keef KD, Pasco JS, Eckman DM (1992) Purinergic relaxtion and hyperpolarization in guinea pig and rabbit coronary artery: Role of the endothelium. J Pharmacol Exp Ther 260: 592–600.
4. Yang S, Cheek DJ, Westfall DP, Buxton ILO (1994) Purinergic axis in cardiac blood vessels. Circ Res 74:401–407.
5. White TD, Angus JA (1987) Relaxant effects of ATP and adenosine on canine large and small coronary arteries in vitro. Eur J Pharmacol 143:119–126.
6. Burnstock G, Kennedy C (1986) A dual function for adenosine 5′-triphosphate in the regulation of vascular tone excitatory cotransmitter with noradrenaline from perivascular nerves and locally released inhibatory intravascular agent. Circ Res 58:319–330.
7. Dominikzak AF, Quilley J, Bohr DF (1991) Contraction and relaxation of rat aorta in response to ATP. Am J Physiol Heart Circ Physiol 261:H243–H251.
8. Lüscher TF, Richard V, Tschudi M, Yang Z, Boulanger C (1990) Endothelial control of vascular tone in large and small coronary arteries. J Am Coll Cardiol 15:519–527.
9. Milner P, Bodin P, Loesch A, Burnstock G (1990) Rapid release of endothelin and ATP from isolated aortic endothelial cells exposed to increased flow. Biochem Biophys Res Comm 170:649–656.
10. Buxton ILO, Walther JM, Westfall DP (1990) α-adrenergic receptor-stimulated release of ATP from cardiac endothelial cells in primary culture. In Dubyak GR, Fedan JS (eds), Biological Actions of Extracellular ATP. New York: Annuals of the New York Academy of Sciences, pp. 503–506.
11. Nees S, Gerbes AL, Gerlach E (1981) Isolation, identification, and continuous culture of coronary endothelial cells from guinea pig heart. Eur J Cell Biol 24:289–297.
12. Marks RM, Czerniecki M, Penny R (1985) Human dermal microvascular endothelial cells: An improved method for tissue culture and a description of some singular properties in culture. In Vitro Cell Dev Biol 21:627–634.
13. Simionescu M, Simionescu N (1978) Isolation and characterization of endothelial cells from the heart microvasculature. Microvasc Res 16: 426–452.
14. Katsuragi T, Tokunaga T, Ogawa S, Sato C, Furukawa T (1991) Existence of ATP-evoked ATP release system in smooth muscles. J Pharmacol Exp Ther 259:513–518.

24. ALLOSTERIC MODULATION OF NUCLEOSIDE TRANSPORT BY ADENOSINE AND ATP

M. Teresa Miras-Portugal

Esmerilda G. Delicado

Raquel P. Sen

Teresa Casillas

Adenosine levels in extracellular fluids depend on a large set of physiologic or pathologic events, and there is broad evidence for a cardiovascular and neuroprotective role of this nucleoside in hypoxic or ischemic situations through its cell surface receptors [1]. Moreover, at the purinergic synapses adenosine can be considered as the last extracellular product that still maintains signaling capability.

The transport of this nucleoside ends its modulatory extracellular actions. Multiple nucleoside transport systems have been described. The sodium-dependent nucleoside transport appears to be restricted to the absorption epithelia. Nevertheless, the facilitated sodium-independent nucleoside transport has been found and characterized in most mammalian cells, and based on selectivity and affinity studies for different nucleosides and inhibitors, a wide family is now emerging [2,3]. In the central nervous system the adenosine transport has been characterized in synaptosomal preparations, but due to their heterogeneity it is difficult to demonstrate the existence of regulatory mechanisms [4,5].

The chromaffin cells from adrenal medulla, considered as the homologous of aminergic neurones, can be employed to study the adenosine transport regulation, due to their cellular homogeneity, although subpopulations containing different nucleotide receptors have recently been found [6,7]. Moreover, all their nucleoside transporters are highly sensitive to nitrobenzylthioinosine (NBTI) inhibition, thus exhibiting the necessary homogeneity to undertake the regulatory studies [8].

Extracellular signals with physiologic relevance, related to neurotransmission, neuromodulation, or the maintenance of cellular characteristics, modify adenosine transport. Figure 24-1 summarizes the current knowledge on this topic in neurochromaffin cells, which could be used for purinergic synaptic terminals.

The stimulation of acetylcholine nicotinic receptor depolarizes the plasma membrane by the entrance of extracellular Na^+ and Ca^{2+} ions, which triggers the exocytotic machinery and at the same time activation of protein kinase C (PKC) [9]. A similar decrease in the

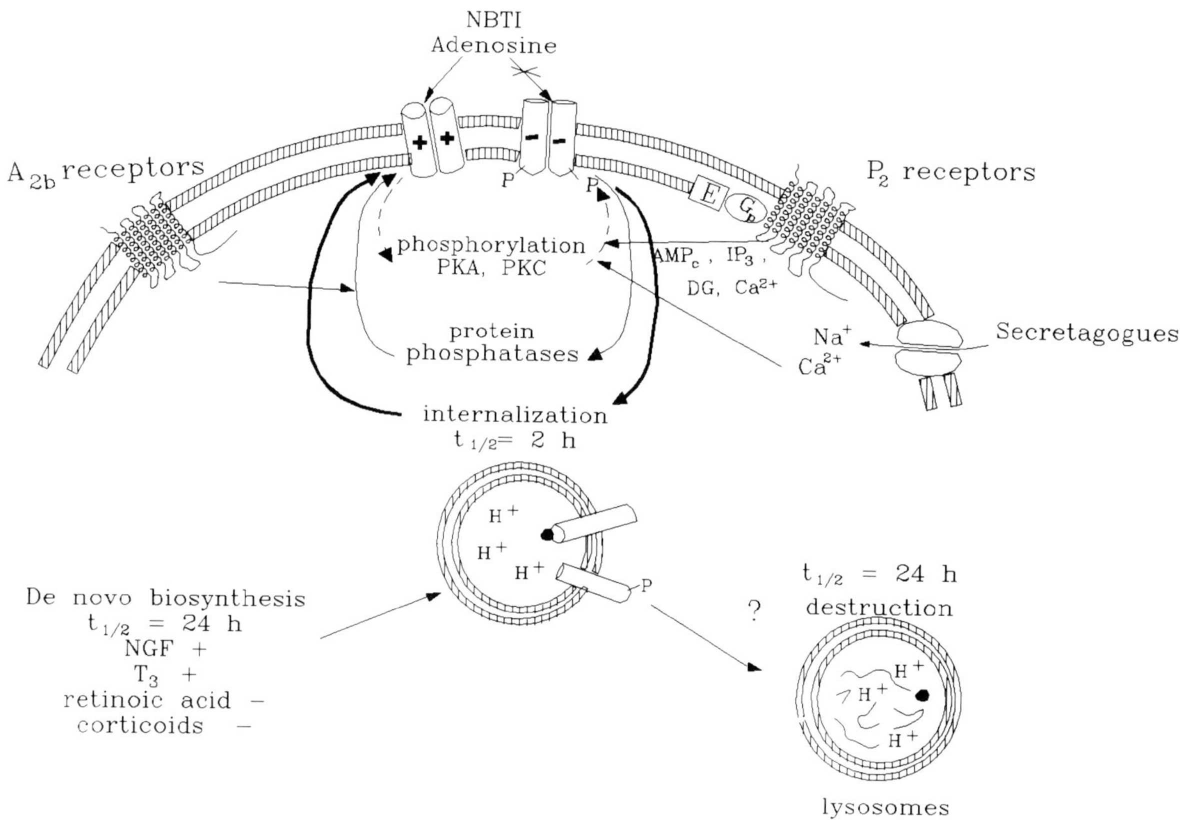

FIGURE 24-1. Adenosine transport regulation by extracellular signals in neurochromaffin cells.

transport capacity and high-affinity binding sites for NBTI is observed after secretagogue stimulation, thus maintaining the efficiency of the functional transporters that are able to internalize 12 adenosine molecules per transporter and second [10]. The same results can be obtained with direct effectors of protein kinases C and A, such as phorbol esters or chlorophenyl cyclic AMP, respectively, thus suggesting phosphorylation of the transporter or an associated protein [11]. As the loss of transport capacity and ligand binding are parallel, the binding parameter can be useful to predict modification of the adenosine transporters.

The ATP and diadenosine polyphosphates (Ap_nA) coreleased with catecholamines from chromaffin cells inhibit adenosine transport through P_2 receptors coupled to PKC activation and cytosolic calcium increase from internal stores [12]. Extracellular degradation of nucleotides produces adenosine that in chromaffin cells is able to stimulate its own transport. It is noteworthy that the A_{2b} receptors present in these cells appear to be coupled to the activation of protein phosphatases; in this case transport activation corresponds to an increase in the number of binding sites for NBTI [13,14].

From the former data a tight relationship between the secretory process and the regulation of nucleoside transport exists. This control can be considered as a short- and mid-term regulation, but data also exist in this cellular model for long-term regulation. Nerve growth factor

(NGF) and thyroid hormones can induce an increase in transport and in NBTI binding sites. On the contrary, retinoic acid and cortisol decrease adenosine transport and reverse the stimulatory effect of thyroid hormones [15,16].

Due to the relevance of adenosine transport to neural cells, for communicating between the extracellular and intracellular space in the ATP/adenosine cycle, the complexity of its regulation may be analogous to the key enzymes in metabolic pathways. The main question arising is whether this transporter can also be regulated allosterically by extracellular and intracellular metabolites. The aim of this work is to demonstrate that extracellular levels of adenosine, together with the cellular energetic state, represented by ATP levels, play a regulatory role in adenosine transport.

Kinetic and Allosteric Cooperativity in L-Adenosine Transport

No allosteric regulation for D-adenosine transport or other physiologic nucleosides has yet been reported. Nevertheless, a positive cooperativity in NBTI binding and its displacement by other inhibitors has already been described [17,18]. Recently, the transport of the nonphysiologic L-adenosine isomer has been characterized in synaptic terminals [19]. Transport of the L-isomer has been characterized in cultured chromaffin cells and plasma membrane vesicles, as previously described [20]. Nonspecific transport was determined by incubating preparations in the presence of 10 μM NBTI and subtracting the value from the assays. All data reported here correspond to specific transport. The L-[^{3}H]adenosine transport was linear during the first 15-second period in plasma membrane vesicles and during the 40-second period in cultured chromaffin cells. This compound was not metabolized, even at very long experimental times.

The concentration dependence of L-[^{3}H]adenosine transport in plasma membrane vesicles is shown in Figure 24-2A. The curve has a sigmoidal shape and the kinetic parameters obtained were $S_{0.5} = 2.5 \pm 0.2\,\mu M$ and $V_{max} = 0.26 \pm 0.01$ pmol/mg of protein · sec. The Hill coefficient was 2.8 ± 0.1, indicating a positive cooperativity.

In cultured chromaffin cells a sigmoidal plot was also obtained, as shown in Figure 24-2B. At the lowest L-adenosine concentrations, transport follows Michaelis–Menten kinetics with a K_m value of $1.1 \pm 0.1\,\mu M$ and very low capacity of transport. Above 2 μM L-adenosine concentration, transport follows a sigmoidal plot with an $S_{0.5}$ value of 6.2 ± 0.2 and a V_{max} of 19.7 ± 0.5 pmol/10^6 cells · min. The Hill plot inserted in Figure 24-2B agrees with this kinetic behavior, with a slope close to 1 at low and high substrate concentrations, and a Hill coefficient of 4.9 ± 0.3 at the intermediate concentrations of L-adenosine. This behavior can be explained only if the transporter presents kinetic cooperativity and allosteric cooperativity between subunits, similar to that reported for mnemonic enzymes.

Transport of the D-isomer of adenosine does not present cooperativity, even at the lowest experimental concentrations employed, which, due to the specific activity of the isotope, cannot be under 10 nM. Crossed inhibition of D-adenosine transport by L-adenosine and vice versa showed a much higher efficiency of the D-isomer, about four orders of magnitude [20].

Positive cooperativity is also obtained for NBTI binding to purified chromaffin cell plasma membranes, as shown in Figure 24-3A. Experiments were performed as previously described [18]. A curvilinear Scatchard plot was obtained, and to avoid erroneous interpretations it was assessed that in these experimental conditions the equilibrium had been reached. The Hill coefficient was 1.30 ± 0.03, which is much lower than that obtained in transport experiments with L-adenosine, corresponding to the cooperativity between the subunits. Moreover, a new and relevant finding was observed, that is, an increase in NBTI binding sites in the presence of ATP (Fig. 24-3A).

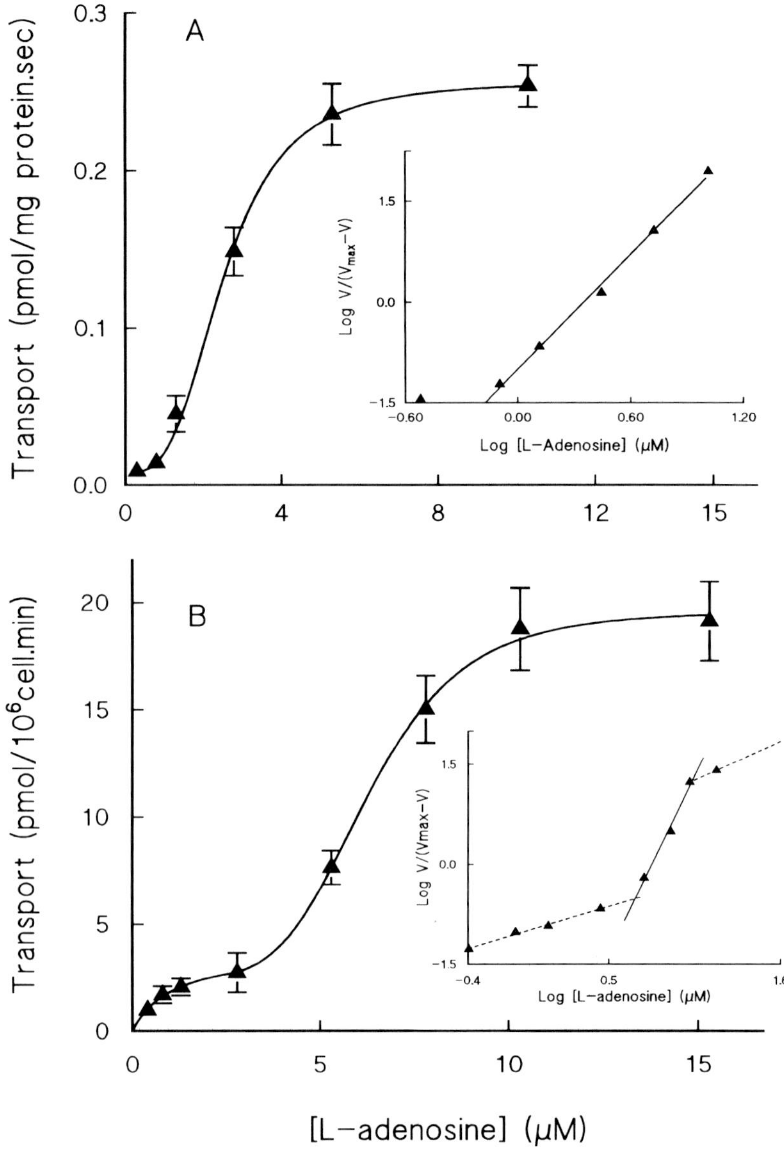

FIGURE 24-2. Concentration dependence of L-[^{3}H]adenosine transport. **A:** 0.25 mg of protein of plasma membrane vesicles were incubated for 10 seconds at 37°C at different concentrations of [^{3}H]adenosine, containing 2 μCi of labeled compound and the NBTI-sensitive transport was measured. **B:** 250,000 cultured chromaffin cells were incubated for 1 minute at 37°C with different concentrations of L-[^{3}H]adenosine (1 μCi/well) and NBTI-sensitive transport was determined. The insert corresponds to the Hill representation resulting from the data.

Modulatory Role of ATP on Nucleoside Transporters

It is currently assumed that facilitative transporters do not need from hydrolysis energy of ATP to internalize their substrates. Nevertheless, to maintain the rate of adenosine transport, ATP is necessary for phosphorylation by adenosine kinase. Similar reasoning could be

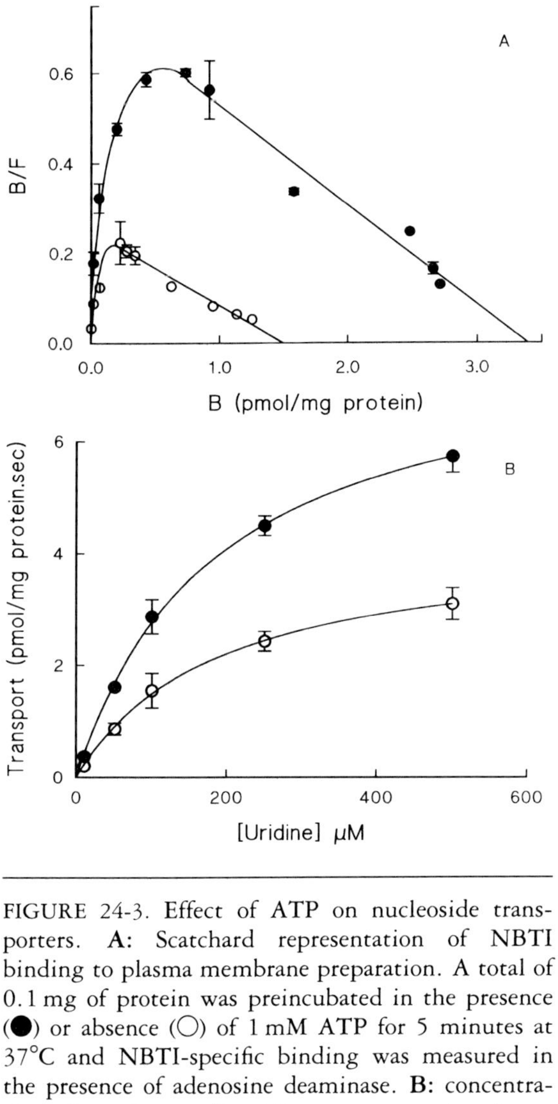

FIGURE 24-3. Effect of ATP on nucleoside transporters. **A:** Scatchard representation of NBTI binding to plasma membrane preparation. A total of 0.1 mg of protein was preincubated in the presence (●) or absence (○) of 1 mM ATP for 5 minutes at 37°C and NBTI-specific binding was measured in the presence of adenosine deaminase. **B:** concentration dependence of uridine transport in plasma membrane vesicles. A total of 0.25 mg of protein of plasma membrane vesicles were resealed in the presence (●) or absence (○) of 100 μM ATP and the specific uridine transport was determined over 10 seconds of incubating at 37°C at different extracellular concentrations of this nucleotide containing 1 μCi of labeled compound.

applied to facilitative glucose transport, in which ATP is necessary to maintain uptake by the phosphorylation process with hexokinases. In addition to the uptake process, there is a question as to whether ATP plays a direct modulatory role. In the case of facilitative glucose transporters, it has been reported that there is an ATP binding site in the cytosolic part of this protein. This site greatly modifies the transport capacity when occupied [21].

The existence of a modulatory effect of ATP on nucleoside transporters was studied by two different approaches: NBTI binding and transport of the nucleoside uridine. The effect of ATP on NBTI binding to chromaffin cell plasma membranes is shown in Figure 24-3A. The presence of the nucleotide in binding experiments increased B_{max} values from 1.62 ± 0.20 pmol/mg protein to 3.22 ± 0.31 pmol/mg of protein, which represents a 100% increase. The $K_{d'}$ values, obtained from Hill representation of the data shown in this figure were 0.52 and 0.35 nM for the control and in the presence of ATP, respectively. This parameter is the equivalent of the $S_{0.5}$ in enzymatic cooperativity and was not significantly modified

Study of nucleoside transport modulation by ATP presents difficult experimental problems, because ATP is degraded by the ATPases and ectonucleotidases present at the plasma membrane. In addition, the experiments were done in the absence of Ca^{2+} and in the presence of NaF to avoid ATP catabolism. Adenosine deaminase (ADA) was also added to destroy the adenosine that could still be formed from ATP. These experimental conditions excluded adenosine as a substrate to carry out these studies. The nucleoside uridine was used instead to circumvent the problem.

[^{3}H]uridine transport in plasma membrane vesicles was linear for at least 10 seconds at all the concentrations assayed. Uridine uptake insensitive to NBTI was subtracted from the total, and the concentration-dependent rate of transport showed typical Michaelis–Menten kinetics (Fig. 24-3B). The K_m value was 185 ± 11 μM and the V_{max} value was 4.24 ± 0.10 pmol/mg of protein · sec. The K_m value was very similar to that reported in cultured chromaffin cells.

The effect of ATP was investigated by resealing plasma membranes in the presence of this nucleotide or its analogues; the resulting vesicles showed a significant increase in uridine transport, as shown in Figure 24-3B. The V_{max} value was significantly increased (7.84 ± 0.19 pmol/mg of protein · sec), but the K_m value did not change. Similar results were obtained with the nonhydrolyzable analogue AMP-PNP. No cooperativity was observed in uridine transport.

The concentration dependence of the nucleotide effect on NBTI binding is shown in Figure 24-4. The IC_{50} values were 1 μM for ATP, 30 nM for the diadenosine tetraphosphate (Ap_4A), and 20 nM for the nonhydrolyzable analogue AMP-PNP.

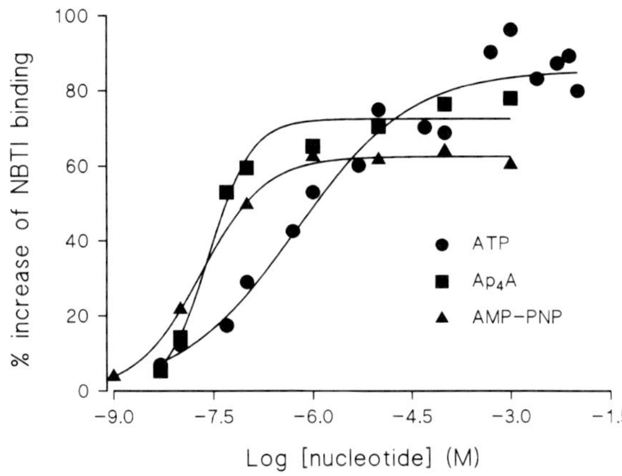

FIGURE 24-4. Effect of increasing concentrations of several nucleotides on NBTI binding to plasma membranes. A total of 0.1 mg of protein was preincubated in the presence of different concentrations of nucleotides for 5 minutes at 37°C, and the specific binding of 1 nM [^{3}H]NBTI was measured.

Discussion

Multimodulated adenosine transport in neural cells indicates the importance of this step for the proper functioning of purinergic transmission. Although no cooperativity has been observed with the D-isomer of adenosine at the lowest possible concentrations, the kinetics obtained for L-adenosine transport correlate well with those obtained for hysteretic (mnemonic) enzymes, assuming a dimeric structure [22]. This type of kinetic and allosteric cooperativity indicates that the transporter could be present in various conformations with different affinities for L-adenosine and slow transitions among them. Maintenance of the high capacity form depends on the nucleoside levels, which in the case of L-adenosine should be superior to the $S_{0.5}$ value (6.5 μM). From the transport inhibition between the L- and D-isomer it appears that a three to four order of magnitude lower D-adenosine concentration is necessary to undertake the conformational changes (about 1 nM); below this concentration the transporter has very reduced capacity [20]. At such low concentrations the mnemonic behavior of D-adenosine is not possible by direct transport measurements. The physiologic relevance of these data relies on correlation between adenosine transport and neural modulation through the A_1 receptors and ensures their continuous occupancy at the very low concentrations described in perfusion samples from the central nervous system [1].

Adenosine transport modulation by substrate at extracellular levels ensures coordination of the outside and inside of neural cells. Nevertheless, it has been shown that glucose and other transporters have a regulatory or conformational site for ATP, the ATP-binding casette family [21,23]. In the case of adenosine transporters, ATP increases both NBTI binding sites and transport of uridine in membrane preparations from chromaffin cells. This effect is mimicked by Ap_4A and nonhydrolyzable analogues of ATP. These preparations always have a small amount of ATP tightly bound, even after several washings and centrifugations [18], which could explain the levels of transport and binding in the absence of added ATP. This nucleotide or its analogues can play a conformational role, ensuring the transport rate

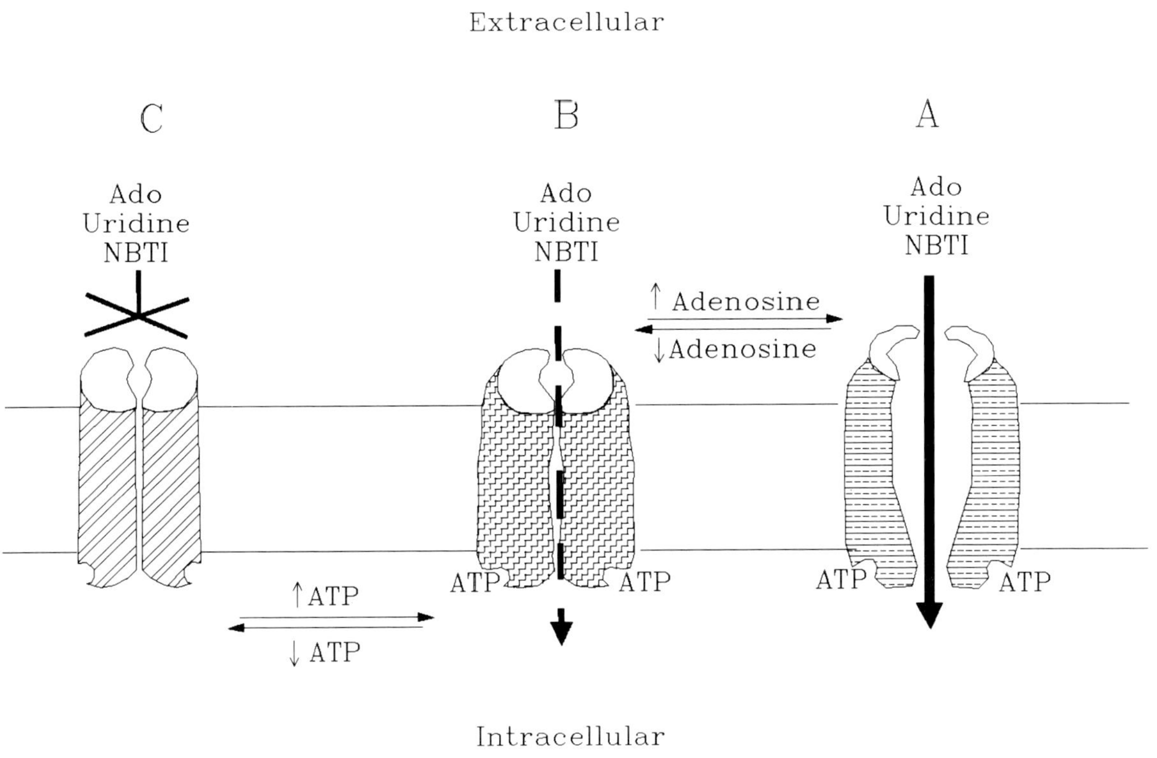

FIGURE 24-5. Schematic view of adenosine transport regulation by extracellular adenosine and intracellular ATP in neurochromaffin cells. Depending on extracellular adenosine levels and the intracellular energetic state, the nucleoside transporter can be at different conformational states. The fully active conformation (**A**) exists when extracellular levels of L-adenosine are up to 6 μM or 1 nM in the case of the physiologic D-isomer and the internal "ATP-binding site" is occupied. A decrease in the external adenosine concentrations while maintaining the occupied intracellular ATP site induces a conformational change with a lower transport capacity (**B**). When the cellular energetic state decreases, such as in hypoxic or ischemic situations, the intracellular nucleotide binding site is not occupied by ATP and the transport capacity is completely lost, even in the presence of high extracellular adenosine levels (**C**). The presence of ATP at the intracellular regulatory site is required for NBTI binding.

in normal physiologic situation. Nevertheless, the energetic cellular level, when very low, could prevent adenosine entrance, with two main objectives: first, to maintain the extracellular adenosine levels to ensure receptor occupancy and, second, to avoid expending intracellular ATP on incorporating adenosine by adenosine kinase, or on its catabolism to inosine by ADA. An illustration of the transporters based on these data is shown in Figure 24-5.

In conclusion, nucleoside transport is a multiregulated step, in which the extracellular and intracellular situations are analyzed and are integrated with the large set of extracellular signals acting via second mesengers or protein synthesis.

Acknowledgments

This work was supported by a research grant from the Spanish CICYT PB 92/0230. We thank Erik Lundin for help in preparation of the manuscript.

References

1. Rudolphi KA, Schubert P, Parkinson FE, Fredholm BB (1992) Neuroprotective role of adenosine in cerebral ischaemia. Trends Pharmacol Sci 13:439–445.
2. Plagemann PGW, Wohlhueter RM, Woffendin C (1988) Nucleoside and nucleobase transport in animal cells (review). Biochim Biophys Acta 947:405–443.
3. Lee CW, Jarvis SM (1988) Kinetic and inhibitor specificity of adenosine transport in guinea pig cerebral cortical synaptosomes. Evidence for two nucleoside transporters. Neurochem Int 12:483–492.
4. Wu PH, Phillis JW (1984) Uptake by central nervous tissues as a mechanism for the regulation of extracellular adenosine concentrations. Neurochem Int 6:613–632.
5. Shank RP, Baldy WJ (1990) Adenosine transport by rat and guinea pig synaptosomes: Basis for differential sensitivity to transport inhibitors J Neurochem 55:541–550.
6. Miras-Portugal MT, Torres M, Rotllan P, Aunis D (1986) Adenosine transport in bovine chromaffin cells in culture. J Biol Chem 261: 1712–1719.
7. Castro E, Tome AR, Miras-Portugal MT, Rosario LM (1994) Single-cell fura-2 microfluorometry reveals different purinoceptor subtypes coupled to Ca^{2+} influx and intracellular Ca^{2+} release in bovine adrenal chromaffin and endothelia cells. Eur J Physiol 426:524–533.
8. Torres M, Fideu MD, Miras-Portugal MT (1990) All nucleoside transporters in bovine chromaffin cells are nitrobenzylthioinosine sensitive. Neurosci Lett 112:343–347.
9. Terbush DR, Bittner MA, Holtz RW (1988) Ca^{2+} influx causes rapid translocation of protein kinase C to membranes. J Biol Chem 263:18873–18879.
10. Delicado EG, Sen RP, Miras-Portugal MT (1991) Effect of phorbol esters and secretagogues on nitrobenzylthioinosine binding to nucleoside transporters and nucleoside uptake in cultured chromaffin cells. Biochem J 279:651–655.
11. Sen RP, Delicado EG, Miras-Portugal MT (1990) Effect of forskolin and cyclic AMP analog on adenosine transport in cultured chromaffin cells. Neurochem Int 17:523–528.
12. Sen RP, Delicado EG, Castro E, Miras-Portugal MT (1993) Effect of P_{2y} agonists on adenosine transport in cultured chromaffin cells. J Neurochem 60:613–619.
13. Delicado EG, Rodrigues A, Sen RP, Sebastiao AM, Ribeiro JA, Miras-Portugal MT (1990) Effect of 5′-(N-ethylcarboxamido)adenosine on adenosine transport in cultured chromaffin cells. J Neurochem 54:1941–1946.
14. Mateo J, Castro E, Zwiller J, Aunis D, Miras-Portugal MT (1994) 5′-(N-ethycarboxamido)adenosine inhibits Ca^{2+} influx and activates a protein phosphatase in bovine adrenal chromaffin cells. J Neurochem, in press.
15. Fideu MD, Miras-Portugal MT (1992) Long term regulation of nucleoside transport by thyroid hormone T_3 in cultured chromaffin cells. Neurochem Res 17:1099–1104.
16. Fideu MD, Miras-Portugal MT (1993) Steroid-induced inhibition of adenosine transport in cultured chromaffin cells. Cell Mol Neurobiol 13:493–502.
17. IJzermann AP, Thedinga KH, Custers AFCM, Hoos B, Van Belle V (1989) Inhibition of nucleoside transport by a new series of compounds related to lidoflazine and mioflazine. Eur J Pharmacol 172:273–281.
18. Casillas T, Delicado EG, Miras-Portugal MT (1993) Adenosine 5′-triphosphate modulation of nitrobenzylthioinosine binding sites in plasma membranes of bovine chromaffin cells. Neurosci Lett 164:51–54.
19. Gu JG, Geiger JD (1992) Transport and metabolism of D-[^{3}H] adenosine and L-[^{3}H] adenosine in rat cerebral synaptoneurosomes. J Neurochem 58:1699–1705.
20. Casillas T, Delicado EG, Garcia-Carmona F, Miras-Portugal MT (1993) Kinetic and allosteric cooperativity in L-adenosine transport in chromaffin cells. A mnemonical transporter. Biochemistry 32:14203–14209.
21. Hebert DN, Carruthers A (1986) Direct evidence for ATP modulation of sugar transport in human erythrocyte ghosts. J Biol Chem 261:10093–10099.
22. Ricard J, Cornish-Bowden A (1987) Cooperative and allosteric enzymes: 20 years on. Eur J Biochem 166:255–272.
23. Higgins CF (1992) ABC transporters: From microorganisms to man. Ann Rev Cell Biol 8:67–113.

IV. PHYSIOLOGY AND PHARMACOLOGY OF ADENOSINE AND ATP: EFFECTS ON TISSUE AND ORGANS

A. RENAL SYSTEM

25. CHLORIDE SECRETION IN THE SHARK RECTAL GLAND: A MODEL FOR EPITHELIAL ADENOSINE RECEPTORS

John N. Forrest, Jr.
Grant G. Kelley

In the kidney, adenosine receptors have been proposed to regulate numerous physiologic functions, including hemodynamics [1,2], tubuloglomerular feedback [3,4], renin release [5,6], and tubular transport [7–9]. Adenosine receptors have been localized to several distal renal tubular segments, including rabbit cortical collecting tubule cells, medullary thick ascending limb cells in primary culture, and rat papillary collecting ducts and medullary thick ascending limb (MTAL) tubules [10,11]. However, there are very limited data on ion transport processes that may be regulated by these receptors in specific tubular segments. Spielman et al. [23] have recently provided evidence for inhibitory adenosine receptors regulating AVP-stimulated water transport in perfused rabbit cortical collecting tubules. Additionally, the metabolic regulation of specific tubular transport processes in nephron segments by endogenous adenosine has not been demonstrated. Because of the cellular diversity and anatomic complexity of the mammalian nephron, we have studied the ion transport function of A_2 and A_1 receptors in the shark rectal gland, a highly specialized elasmobranch organ with well-defined transport elements [12].

The shark rectal gland has proven to be a powerful system for investigating the physiologic functions of adenosine receptors in the coordinate regulation of cell metabolism and ion transport. Unique elements of the shark rectal gland include the following:

1. This epithelial organ consists of homogeneous branched tubules of a single cell type that are highly specialized for sodium chloride secretion. A single gland (mean weight 1.5 g) provides ample material for physiologic, biochemical, and molecular studies.
2. The organ is ideally suited for single pass in vitro perfusion with a single artery, vein, and duct that are easily canulated; the complete venous drainage of the organ can be collected readily.
3. Cells of the rectal gland contain the highest known activity of Na/K-ATPase of any tissue. This enzyme is the primary site of ATP hydrolysis and subsequent formation of adenosine, and is responsible for the high

rate of oxygen consumption observed in this tissue [13,14].

4. Basolateral membranes of the rectal gland contain the highest known activity of the Na/K/2Cl cotransporter [15]. The amount of cotransporter is at least 20 times greater in the rectal gland than in the mammalian kidney.
5. Secretion by the gland is dynamically regulated. Stimulatory hormones elicit large, rapid increases (30- to 50-fold) in sodium chloride secretion accompanied by the upregulation of both the Na/K/2Cl cotransporter and Na/K-ATPase activity.
6. Each integral membrane protein (receptors, luminal chloride channel, basolateral potassium channel, basolateral Na/K/2Cl cotransporter, and Na/K-ATPase pump) is accessible for direct study by membrane-binding or patch-clamp techniques in isolated tubules, primary culture monolayers, or perfused glands.
7. Both A_1 and A_2 adenosine receptor subtypes are present in higher density on rectal gland cells compared to renal tubular cells and are coupled to the same physiologic event, secondary active chloride transport [16,17].

In this chapter we summarize a series of studies from our laboratory on the regulatory functions of adenosine receptors in the elasmobranch rectal gland [16–20].

Methods

Rectal glands were obtained from male spiny dogfish sharks, *Squalus acanthias*, weighing 2–4 kg. Glands were removed and cannulae were placed in the single artery, vein, and duct as described previously [18]. Glands were placed in a glass perfusion chamber equilibrated to 15°C with running sea water and perfused with an elasmobranch Ringer's solution containing 270 mM NaCl, 4 mM KCl, 3 mM $MgCl_2$, 2.5 mM $CaCl_2$, 1 mM KH_2PO_4, 8 mM $NaHCO_3$, 350 mM urea, and 0.5 mM Na_2SO_4, and equilibrated to pH 7.5 by bubbling with 99% O_2 and 1% CO_2. All glands were first perfused for 30 minutes in the absence of hormones to achieve basal (unstimulated) rates of chloride secretion. Basal values are given in the text, and figures and are the last basal measurement (20–30 minute interval). Measurements of duct flow were made at 10-minute intervals in all experiments. Venous flow rates and gland weights did not vary significantly between experiments. The mean venous flow rate was 2.4 ± 0.1 ml/min, and the mean gland weight was 1.47 ± 0.08 g. Results are expressed as microequivalents of chloride secreted per hour per gram wet weight (μEq/hr/g) ± SEM.

Chloride secretion and venous effluent adenosine and inosine concentrations were measured simultaneously. Rectal glands were perfused as described above, and the venous effluent was collected on ice at 10-minute intervals for determination of adenosine and inosine concentrations. The venous effluent samples were first desalted and concentrated by passing the fluid over equilibrated C_{18} Sep-Paks (Waters Associates, #51910) at a flow rate of 4.4 ml/min using a Harvard infusion pump. The Sep-Pak cartridge was washed with 1.5 ml of 5 mM KH_2PO_4, pH 7.0, and the nucleosides were then eluted with 5 ml of 80% methanol. This solution was evaporated to dryness with a SpeedVac Concentrator (Savant Instruments, Farmingdale, NY) and then resuspended in distilled water. Adenosine and inosine were assayed by high-performance liquid chromatography (HPLC) using Rainin HPX pumps, a Gilson 231 automatic sample injector, a Gilson 112 UV detector, and a Rainin Microsorb C_{18} 250 × 4.6 mm column. (Rainin Instrument Co., Emeryville, CA). Samples (100 μl) were loaded onto the column and isocratically eluted with 5 mM KPO_4 and 14% methanol at a flow of 1 ml/min. Data were collected and peak areas were calculated by Rainin HPLC integrator software. The retention times for inosine and adenosine were routinely 5.5 and 12.0 minutes, respectively. Addition of adenosine deaminase (ADA) completely abolished the adenosine peak. Recoveries for adenosine and inosine were 95 ±

1% and 60%, respectively, and did not vary over the concentration range of the resuspended samples. Results are expressed as nanomolar concentration ± SEM. ATP, ADP, and AMP were measured using a Hydropore AX anion exchange column with a gradient of 5 mM and 750 mM NH4PO4, starting from 0% 750 mM NH4PO4 and reaching 50% 750 mM NH4PO4 in 20 minutes at a flow of 2 ml/min. The retention times for ATP, ADP, and AMP were 4.2, 8.0, and 14.2 minutes, respectively. All results are mean ± SEM. Data were analyzed for significance with Student's t test for paired or unpaired data. Forskolin and 8-phenyltheophylline were obtained from Calbiochem (San Diego, CA). C_{18} Sep-Paks were purchased from Waters Associates (Milford, MA). All other reagents were from Sigma Chemical Co. (St. Louis, MO).

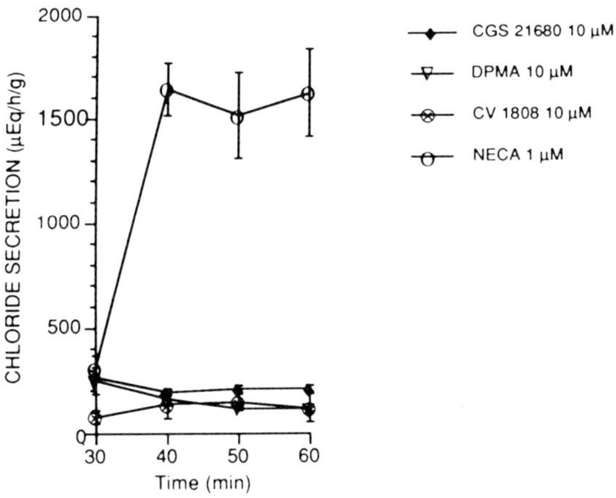

FIGURE 25-1. Stimulation of chloride secretion in the perfused rectal gland with A_2 receptor agonists. Rectal glands were perfused for 30 minutes to basal values and then perfused for 30 minutes with addition of the agonist to the perfusate. Chloride secretion rates were measured at 10-minute intervals and values are mean ± SEM.

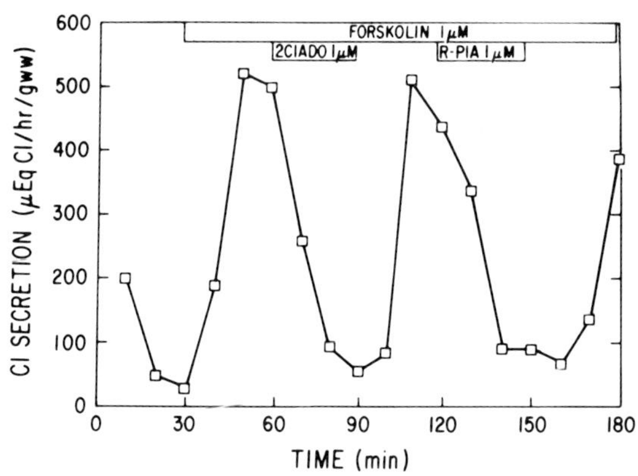

FIGURE 25-2. Inhibition of forskolin (1 µM) stimulated chloride secretion by A_1 receptor agonists (2 Cl-ADO and R-PIA) in the perfused rectal gland.

Results

Stimulation of the A_2 receptor by 5′-N′-ethylcarboxamidoadenosine (NECA), adenosine, or 2-Cl adenosine added to the perfusate markedly increases chloride secretion in the perfused rectal gland, and this effect is inhibited by both selective and nonselective A_2 receptor antagonists. However, as shown in Figure 25-1, this receptor is entirely insensitive to CGS 21680, N^6-[2-(3.5 -dimethoxyphenyl)-2-(2-methylpheny)ethyl]adenosine (DPMA), or CV 1808, indicating that the receptor is probably of the A_{2b} receptor subtype. When chloride secretion in the gland is first stimulated by forskolin (1 µM), the addition of A_1 receptor agonists (2 Cl-adenosine and R(-)N^6-phenylisopropyladenosine [R-PIA]) reversibly inhibits transport (Fig. 25-2). When A_1 agonists are added to the perfusate simultaneously with forskolin, chloride secretion is entirely inhibited to basal values, and this inhibition is reversed by addition of A_1 receptor antagonists to the perfusate (Fig. 25-3).

To correlate adenosine and inosine release with rates of chloride transport, rectal glands were perfused in vitro under varying work loads with forskolin (0.3–10 µM) and vasoactive intestinal peptide (VIP; 1–10 nM), and chloride secretion and venous effluent adenosine and inosine concentrations were determined simultaneously. Figure 25-4 illustrates the direct correlation (r = 0.98) between increasing rates of chloride secretion and venous adenosine concentrations following stimulation with both forskolin and VIP. Similar results were observed for venous inosine (data not shown). When glands are perfused with agents that prevent the interaction of endogenous adenosine with extracellular receptors (adenosine deaminaase [ADA], nitrobenzylthioinosine [NBTI], and A_1 receptor antagonists), the chloride transport response to secretagogues is increased more than twofold [18].

When tissue adenine nucleotides and tissue adenosine were measured in sections of rectal gland snap-frozen in liquid nitrogen during perfusion with 1 and 10 μM forskolin (Fig. 25-5), we found that tissue ATP was markedly depleted and the tissue content of AMP, adenosine, and inosine increased three- to five fold during maximum stimulation of chloride transport.

To demonstrate further the relationship between chloride secretion and adenosine release, and to correlate the work of ion transport with this release, two specific inhibitors of transport were employed. Bumetanide is a specific inhibitor of the Na/K/2Cl cotransporter in the gland. As shown in Figure 25-6, when chloride secretion and adenosine release are stimulated by 10 μM forskolin, the addition of bumetanide to the perfusate abruptly decreases both chloride secretion and adenosine release to basal values. Figure 25-7 illustrates that ouabain, which inhibits membrane Na/K-ATPase activity in the rectal gland, completely inhibited both chloride secretion and adenosine release.

We have examined whether hydrolyzable and nonhydrolyzable adenine nucleotides stimulate chloride secretion in this system. Similar to perfusion with 10 μM adenosine, perfusion with 10 μM ATP, ADP, or AMP increased chloride transport from basal levels (<200 μEq/hr/g) to ~1400 μEq/hr/g (Fig. 25-8A). However, nonhydrolyzable analogs of ATP do not stimulate secretion (Fig. 25-8A), and ATP, ADP, and AMP are nearly completely converted to adenosine on a single pass through the gland (Fig. 25-8B) above basal values. We conclude that the stimulatory effect of nucleotides is due to metabolism to adenosine and stimulation at the A_{2b} receptor site.

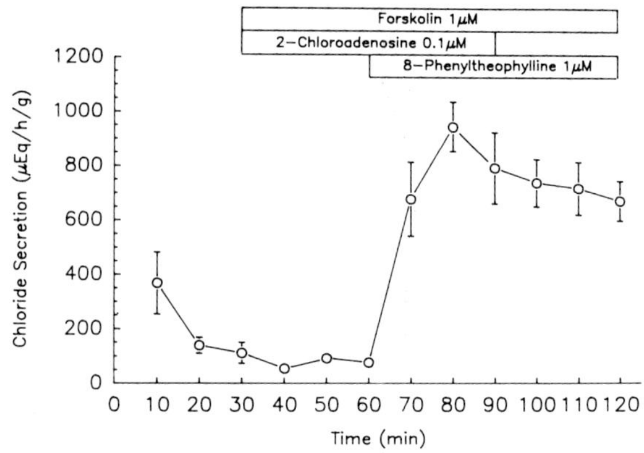

FIGURE 25-3. Effects of 0.1 μM 2 Cl-ADO to entirely inhibit the response to forskolin and reversal of this inhibition with 8-phenyltheophylline. From Kelley, et al. (1990). J Clin Invest 85:1629–1636.

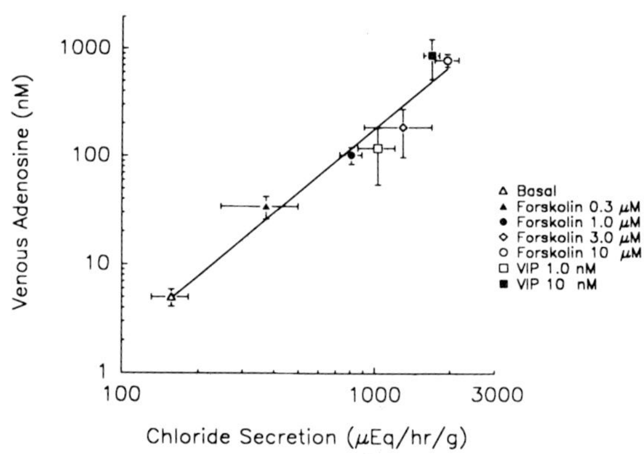

FIGURE 25-4. Relationship between venous effluent adenosine concentrations and chloride secretion rates during stimulation with increasing concentrations of forskolin and vasoactive intestinal peptide (VIP). Rectal glands were perfused for 30 minutes to basal values, and secretagogues were then added to the perfusate for an additional 30 minutes. Values are from the 50-minute period and are means ± SEM. From Kelley, et al. (1991). J Clin Invest 88: 1933–1939.

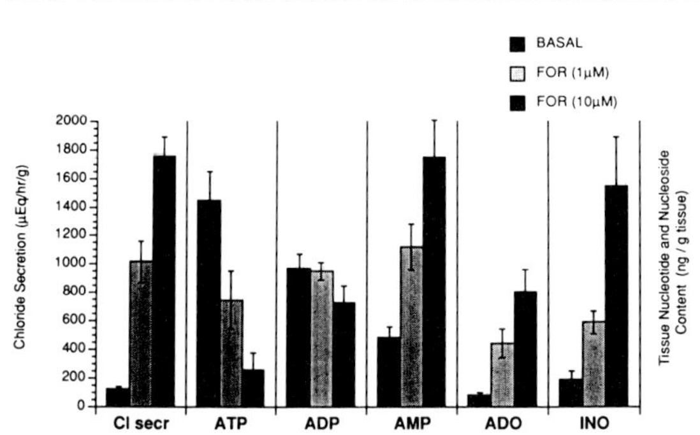

FIGURE 25-5. Tissue nucleoetide and nucleoside content and chloride secretion under basal conditions and during stimulation of chloride transport with 1 and 10 μM forskolin.

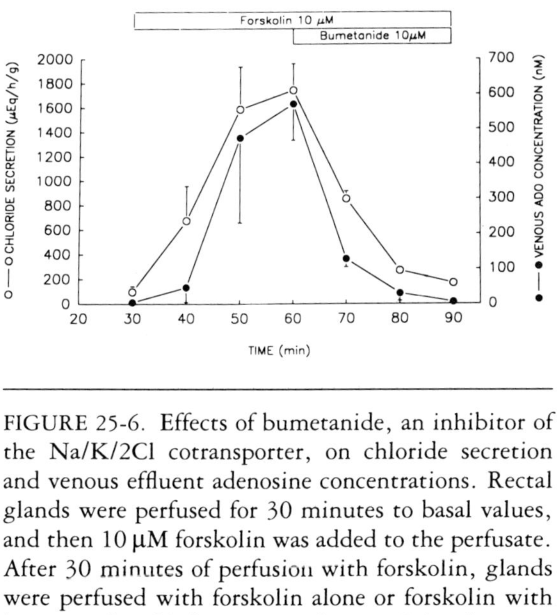

FIGURE 25-6. Effects of bumetanide, an inhibitor of the Na/K/2Cl cotransporter, on chloride secretion and venous effluent adenosine concentrations. Rectal glands were perfused for 30 minutes to basal values, and then 10 μM forskolin was added to the perfusate. After 30 minutes of perfusion with forskolin, glands were perfused with forskolin alone or forskolin with bumetanide added to the perfusate. Chloride secretion and venous adenosine concentrations were measured at 10-minute intervals, and values are the mean ± SEM.

Discussion

The rectal gland contains a high density of stimulatory and inhibitory adenosine receptors on the same cell. Our studies in this model system provide the first direct evidence that endogenous adenosine, produced by the metabolic work of ion transport, is an important autacoid regulator of ion transport in epithelia. Over a wide range of chloride transport stimulated by secretagogues, venous adenosine increases in parallel with chloride secretion. When hormone-stimulated chloride transport is inhibited with bumetanide, an inhibitor of the Na/K/2Cl cotransporter, or ouabain, an inhibitor of membrane Na/K-ATPase activity, adenosine and inosine release decrease in parallel with chloride secretion. These findings demonstrate a remarkably tight coupling between cell work and adenosine release in a tubular epithelium.

The physiologic significance of the released adenosine can be demonstrated by agents (ADA, NBTI, 8-phenyltheophylline [8-PT]) that employ diverse mechanisms to prevent the interaction of endogenous adenosine with extracellular receptors. These agents have no effect on basal secretion but markedly increased the secretory response to forskolin. The results demonstrate that the endogenous adenosine released in response to forskolin-stimulated chloride transport interacts with the A_1 adenosine receptor to inhibit secretion.

From these studies we have proposed the following model of adenosine as an inhibitory feedback regulator of chloride transport in the shark rectal gland (Fig. 25-9). Chloride secretion is stimulated by hormones acting on receptors that activate the catalytic subunit of adenylate cyclase via nucleotide binding proteins (N_S). Cyclic-AMP–dependent processes activate chloride channels in the apical plasma membrane. At the basolateral membrane chloride and potassium ions enter the cell tightly coupled to sodium via a bumetanide-sensitive Na/K/2Cl cotransporter. Potassium ions recycle across the basolateral membrane via a barium-sensitive channel. The driving force for chloride entry is an inwardly directed gradient for sodium maintained by a high specific activity of membrane-associated Na/K-ATPase. This enzyme is the primary site of ATP hydrolysis and is responsible for the high rates of oxygen consumption observed in this tissue [13,14]. Stimulation of secretion results in a marked increase in Na/K-ATPase pump activity [14, 21,22] and ATP hydrolysis, resulting in an increased tissue adenosine content. Adenosine then exits the cell down a concentration gradient via an NBTI-sensitive nucleoside transporter into the extracellular space. At this site it interacts with a high-affinity A_1 receptor coupled to an inhibitory nucleotide regulatory binding protein (N_i) to inhibit chloride transport by both cyclic-AMP–dependent and cyclic-AMP–independent processes [17]. ADA, 8PT, and NBTI block this feedback inhibition by preventing endogenous adenosine from interacting with the A_1 receptor. Thus, in the rectal gland adenosine functions as an inhibitory autacoid to regulate the metabolic work of ion transport and to link energy demand and availability.

The physiologic function of the A_{2b} receptor in the rectal gland and the nucleotide sequence of elasmobranch receptor epitopes remain to be clarified.

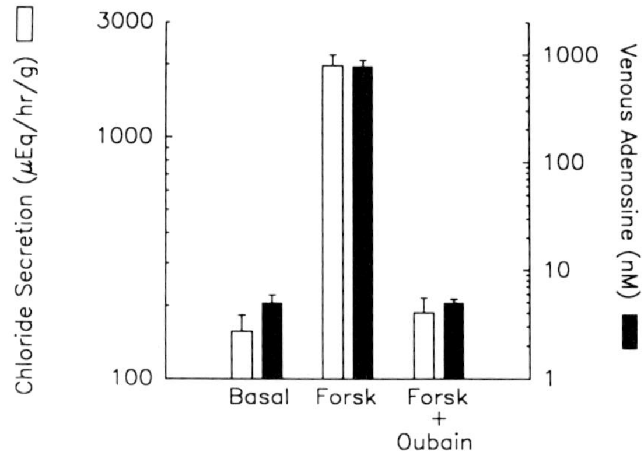

FIGURE 25-7. Effects of ouabain, an inhibitor of Na/K-ATPase activity, on chloride secretion and venous effluent adenosine concentrations. Rectal glands were perfused for 30 minutes to basal values, and then 10 μM forskolin (Forsk) was added to the perfusate. After 30 minutes of perfusion with forskolin, glands were perfused with either forskolin alone or forskolin plus ouabain. Chloride secretion and venous adenosine concentrations were measured at 10-minute intervals, and values are the mean ± SEM.

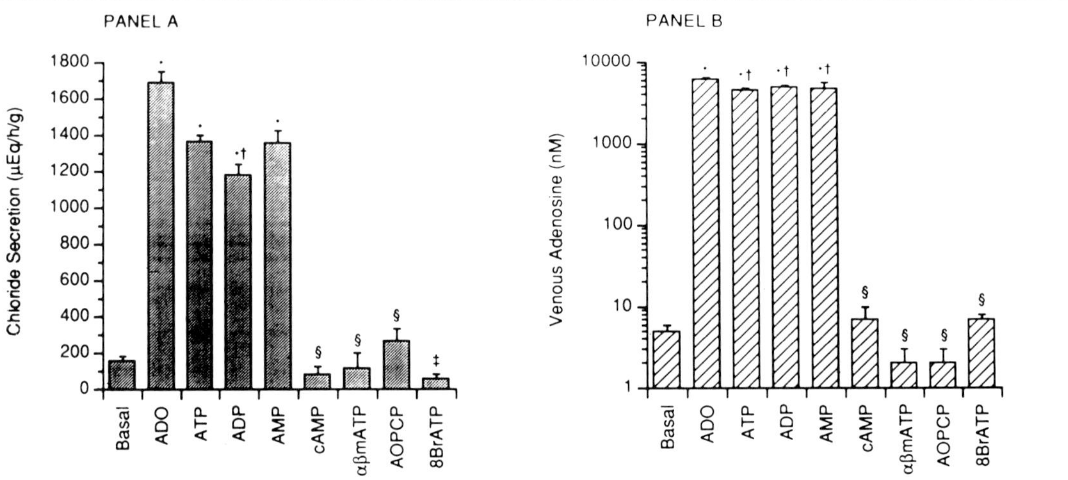

FIGURE 25-8. Effects of adenine nucleotides, nucleosides, and hydrolysis-resistant ATP receptor analogues on chloride secretion and venous adenosine concentrations in the perfused rectal gland. Rectal glands were perfused for 30 minutes, and then the agents listed were added to the perfusate for an additional 20 minutes. Chloride secretion data depicted are from the 40–50 minute period and are the mean ± SEM. **A** illustrates the effect of agents on chloride secretion (μEq/g), and **B** shows the corresponding venous adenosine concentrations (nM). *$p < 0.0001$ compared to basal; $^{\dagger}p < 0.05$ compared to ADO; $^{\#}p < 0.05$ compared to basal; $^{\S}p =$ NS compared to basal.

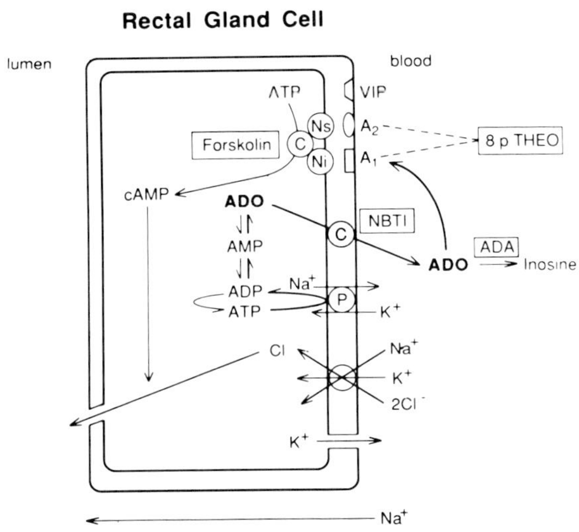

FIGURE 25-9. Model of adenosine as an inhibitory autacoid regulator of hormone-stimulated chloride transport in the rectal gland cell. Adenosine production and efflux from the cell is increased following stimulation of chloride secretion by secretagogues. Extracellular adenosine interacts with inhibitory A_1 adenosine receptors to inhibit chloride transport. ADO = adenosine; ATP = adensoine triphosphate; VIP = vasoactive intestinal peptide; A_1 = inhibitory adenosine receptor; Ns = stimulatory GTP binding protein; N_i = inhibitory GTP binding protein; Ni = inhibitory GTP binding protein; C = catalytic unit of adenylate cyclase; C = cellular nucleoside transporter; P = Na/K-ATPase; ADA = adenosine deaminase; NBTI = nitrobenzylthioinosine; 8PT = 8-phenyltheophylline. From Kelley, et al. (1991). J Clin Invest 88:1933–1939.

Acknowledgments

We thank Stephen Aller, Cindy Aller, David Opdyke, Sami Aasar, John K. Forrest, Martha Ratner, and Gwendolyn Forrest for their expert technical assistance. This work was supported by National Institutes of Health grant RO1-DK 34208 and American Heart Association Grant 92011310 to Dr. Forrest and by the Center for Membrane Toxicology (CMTS) at the Mt. Desert Island Biological Laboratory.

References

1. Osswald H, Spielman WS, Knox FG (1978) Mechanism of adenosine-mediated decreases in glomerular filtration rate in dogs. Circ Res 43: 465–469.
2. Rossi NF, Churchill PC, Amore B (1988) Mechanism of adenosine receptor induced renal vasoconstriction in the rat. Am J Physiol 255: H885–H890.
3. Osswald H, Hermes HH, Nabakowski G (1982) Role of adenosine in signal transmission of tubuloglomerular feedback. Kidney Int 22(Suppl.):S136–S142.
4. Schnerrman J, Weihprect H, Briggs J (1990) Inhibition of tubuloglomerular feedback during adenosine receptor blockade. Am J Physiol 258: F553–F561.
5. Spielman WS, Thompson CI (1982) A proposed rloe for adenosine in the regulation of renal hemodynamics and renin release. Am J Physiol 242:F423–F435.
6. Kuan CJ, Wells JN, Jackson EK (1989) Endogenous adenosine restrains renin release during sodium restriction. J Pharmacol Exp Ther 249:14659–14664.
7. Churchill PC, Churchill MC (1988) Effects of adenosine on renin secretion. In ISI Atlas of Science: Pharmacology, pp. 367–373.
8. Spielman WS, Arend LJ, Forrest JN (1987) The renal and epithelial actions of adenosine. 249–260.
9. Spielman WS, Arend LJ (1991) Adenosine receptors and signaling in the kidney. Hypertension 17:117–130.
10. Arend LJ, Sonnenberg WK, Smith WL, Spielman WS (1987) A_1 and A_2 adenosine receptors in rabbit cortical collecting tubule cells: Modulation of hormone-stimulated cAMP. J Clin Invest 79:710–714.
11. Weber RG, Brines ML, Hebert SC, Forrest JN Jr (1990) Demonstration of A_1 adenosine receptors on rat medullary thick ascending limb tubules by radioligand binding (abstr). Kidney Int 37:380.
12. Silva PJ, Stoff J, Field M, Fine L, Forrest JN, Epstein FH (1977) Mechanism of active chloride secretion by shark rectal gland: Role of Na-K-ATPase in chloride transport. Am J Physiol 233:F298–F306.
13. Silva P, Stoff JS, Solomon RJ, Rosa R, Stevens A, Epstein J (1980) Oxygen cost of chloride transport in perfused rectal gland of *Squalus acanthias*. Membr Biol 53:215–221.
14. Silva P, Stoff J, Epstein FH (1979) Indirect evidence for enhancement of Na-K-ATPase activity with stimulation of rectal gland secretion. Am J Physiol 327:F468–F472.

15. Forbush BI, Palfrey HC (1982) Bumetanide and benzmetanide binding to membranes from shark rectal gland and canine kidney (abstr). Biophysics J 37:161.
16. Forrest JN Jr, Wang F, Beyenbach KW (1983) Perfusion of isolated tubules of the shark rectal gland. Electrical characteristics and responses to hormones. J Clin Invest 72:1163–1167.
17. Kelley GG, Poeschla EM, Barron HV, Forrest, JN Jr (1990) A_1 adenosine receptors inhibit chloride transport in the shark rectal gland. J Clin Invest 85:1629–1636.
18. Kelley GG, Aassar OS, Forrest JN (1991) Endogenous adenosine is an autocoid feedback inhibitor of chloride transport in the shark rectal gland. J Clin Invest 88:1933–1939.
19. Kelley GG, Assar S, Forrest JN Jr (1987) HPLC measurements of adenosine in the rectal gland of *Squalus acanthias*: Adenosine is released at inhibitory concentrations following hormonal stimulation of transport. Bull MDIBL 27: 132–133.
20. Kelley GG, Forrest JN Jr (1994) ATP stimulates chloride transport via A_{2b} adenosine receptors in the shark rectal gland. Am J Physiol, submitted for publication.
21. Shuttleworth TJ, Thompson JL (1980) The mechanism of cyclic AMP stimulation of secretion in the dogfish rectal gland. J Comp Physiol 140:209–216.
22. Silva J, Epstein JA, Stevens A, Spokes K, Epstein FH (1983) Ouabain binding in rectal gland of *Squalus acanthias*. J Membr Biol 75: 105–1.
23. Edwards RM, Spielman WS (1994) Adenosine A_1 receptor-mediated inhibition of vasopressin action in inner medullary collecting duct. Am J Physiol 266:F791–F796.

B. CARDIOVASCULAR SYSTEM

26. MYOCARDIAL ADENOSINE RECEPTORS

Cynthia Kollias-Baker
John C. Shryock
Luiz Belardinelli

Currently there are four well-defined subclasses of adenosine receptors, namely, A_1, A_{2a}, A_{2b}, and A_3. In the heart, A_1 receptor pharmacology, regulation, and coupling to functional responses have been well characterized. The presence of an A_2 receptor that mediates adenosine-induced coronary vasodilation is also well established, although the receptor subtype (A_{2a} vs. A_{2b}) is still the subject of some debate. The A_2 receptor and its role in adenosine-mediated coronary vasodilation are reviewed in Chapter 27. Although the results of several studies provide support for the presence of an A_3 receptor in the heart, a physiologic role for the cardiac A_3 receptor has yet to be determined. Thus, the primary focus of this review will be the cardiac A_1 adenosine receptor; however physical or functional evidence for the presence of other receptor subtypes on cardiomyocytes will be presented.

Cardiac A_1 Receptor

A_1 receptor activation causes decreases in heart rate (negative chronotropic effect), impulse conduction velocity (negative dromotropic effect), and force of contraction (negative inotropic effects) that ultimately lead to a decrease in cardiac activity [1]. These effects of A_1 receptor activation are termed the *direct* or *cAMP-independent effects* of adenosine and are prominent in supraventricular tissues. Direct effects of adenosine on ventricular myocardium are less notable in most mamalian species. Adenosine also decreases cardiac activity by attenuating the stimulatory effects of catecholamines that are mediated by cAMP. This indirect or cAMP-dependent effect of adenosine is also mediated via activation of the A_1 receptor [2] and is observed in both supraventricular and ventricular tissues. The A_1 receptor is coupled to its effector units by an inhibitory guanine nucleotide regulatory binding protein or proteins (G_i or G_o proteins). The effector units of the direct (cAMP-independent) effects of the A_1 receptor are ion channels, such as the ACh and adenosine (ADO)-regulated inward-rectifying potassium channel ($I_{KACh,Ado}$) [1]. The effector units of the indirect (cAMP-dependent) effects of the A_1 receptor are not well clarified but potentially include the catalytic subunit of adenylate cyclase, cAMP phosphodiesterases, or protein phosphatases.

Regulation of Cardiac A_1 Receptor-Mediated Responses

A_1 receptor responses can be modulated to result in either an increase (sensitization) or a decrease (desensitization) in the responsiveness of the cell to agonist stimulation [3–7]. In addition, A_1-receptor–mediated responses can be enhanced by allosteric modulation of agonist binding [8–12]. While many of the studies demonstrating these modulations in cell responsiveness have been carried out with brain, adipose, and other tissues, modulation of cardiac A_1-receptor–mediated responses is well documented.

SENSITIZATION

The cardiac A_1 receptor–effector system has been reported to undergo an increase in responsiveness to agonists following prolonged exposure to the adenosine receptor antagonist theophylline [4,6]. In guinea-pig ventricular cells isolated from animals chronically treated with theophylline, an increased sensitivity to adenosine was associated with an upregulation in the number of A_1 receptors [4]. Sensitization of the heart to the anti-beta-adrenergic action of adenosine was not associated with concurrent changes in the muscarinic or adrenergic systems [4,6]. An increased sensitivity of cardiac A_1 receptor–mediated responses to adenosine was also reported in transplanted (denervated) human hearts [13]. Similar results were also obtained in syngeneic transplanted rat hearts. The IC_{50} for adenosine to cause sinus bradycardia was approximately 10-fold lower in transplanted rat hearts than in native hearts (i.e., 7.9 ± 3 vs. $82 \pm 10\,\mu M$).

DESENSITIZATION

In response to prolonged exposure to agonists, the cardiac A_1 receptor–effector system has been reported to undergo a decrease in responsiveness to further agonist stimulation. Chronic exposure of embryonic chick hearts and chick atrial myocytes to the adenosine agonist R(-)N^6-phenylisopropyladenosine (R-PIA) was associated with a downregulation of cardiac A_1 adenosine receptors and a decrease in the cardiac contractile response to R-PIA [3,5]. Pretreatment of myocytes with R-PIA also appeared to cause a conversion of A_1 receptors from a high-affinity agonist-binding state to a low-affinity state [5]. Similarly, prolonged in vivo treatment of guinea pigs with R-PIA desensitized isolated atria to the negative inotropic and chronotropic effects of adenosine [7]. This desensitization was associated with a downregulation of A_1 adenosine receptors and a decrease in the content of inhibitory G proteins [7].

ALLOSTERIC MODULATION

Recently, 2-amino-3-benzoylthiophene derivatives, such as PD 81,723, have been shown to allosterically enhance the binding of agonists to brain A_1 receptors and to potentiate the A_1 receptor–mediated inhibition of cAMP formation in FRTL-5 cells [8]. In guinea-pig isolated, perfused hearts [9] and in rat isolated atria [10], PD 81,723 was shown to specifically enhance the negative dromotropic response and the negative inotropic and chronotropic responses, respectively, to adenosine and an A_1 agonist. Recently, the potentiation by PD 81,723 of the cardiac effects of adenosine was shown to be concentration dependent, selective for A_1 and not A_{2a} receptor–mediated responses, and independent of nucleoside uptake inhibition [9,11]. Furthermore, PD 81,723 was shown to mediate this potentiation by enhancing the binding of agonists to A_1 receptors through a G-protein–dependent process [11]. PD 81,723 has also recently been shown to potentiate the effects of endogenously released adenosine in the hypoxic heart [12]. Altogether these studies indicate that allosteric enhancement of A_1 adenosine receptor–mediated responses may be a viable approach to modulate the effects of endogenously released adenosine.

Radioligand Binding for Study of A_1 Receptors

The development of very potent and highly selective adenosine agonists and antagonists allowed adenosine receptors to be studied by radioligand techniques. These techniques enable the evaluation of receptor density, drug affinity, and kinetic binding parameters. Studies of A_1 receptors of cardiac membranes have lagged behind studies of brain membranes due to the lower density of these receptors (5- to 10-fold) and the higher nonspecific binding of radioligands to cardiac tissue [14]. In one study, the rank order of potency for agonists to displace the binding of the A_1 antagonist [^{3}H]8-cyclopentyl-1,3-dipropylxanthine CPX to intact rat ventricular myocytes was R-PIA > NECA (5′-N-ethylcarboxamidoadenosine) > S-PIA [15]. A summary of the results of ligand binding studies of A_1 receptors of cardiac membranes from various species using agonist and antagonist ligands is presented in Tables 26-1 and 26-2. In general, the density of A_1 receptors in atrial tissues is reported to be higher than in ventricular tissue (Tables 26-1 and 26-2; Fig. 26-1).

Cardiac A_1 receptors have also been shown to exist in high- and low-affinity agonist-binding states (Fig. 26-2). It has been proposed that the high- and low-affinity binding states represent receptors that are coupled and uncoupled, respectively, to G proteins [19]. In intact rat ventricular cells, although 40,000 binding sites per cell were detected using the antagonist ligand [^{3}H]CPX, no high-affinity binding sites were detectable using agonist-displacement studies [15]. This lack of high-affinity agonist binding to intact cells, as opposed to membrane preparations, is consistent with the presence of high concentrations of guanine nucleotides in intact cells, which prevent the formation of stable G-protein–receptor complexes. Results of similar studies of antagonist and agonist binding to guinea-pig intact ventricular cells are shown in Figure 26-3.

TABLE 26-1. Equilibrium binding characteristics of the agonist ([^{125}I]ABA) and the antagonist ([^{3}H]CPX) ligands to atrial membranes

Species	B_{max} (fmoles/mg prot)	K_d (nM)	Reference
Porcine	16.1	1.45	Musser et al., 1993 [14]
Rat	17.2	1.8	Musser et al., 1993 [14]
Rabbit	14.5	1.6	Musser et al., 1993 [14]
Guinea pig	75.9	3.3	Musser et al., 1993 [14]
Human	26.9	2.2	Bohm et al., 1989 [16]
Rat	30	3.0	Linden et al., 1985 [17]
Rabbit	8	3.1	Froldi and Belardinelli, 1990 [18]
Guinea pig	102	9.3	Froldi and Belardinelli, 1990 [18]

TABLE 26-2. Equilibrium binding characteristics of the agonist ([^{125}I]ABA) and the antagonist ([^{3}H]CPX) ligands to ventricular membranes

Species	B_{max} (fmoles/mg prot)	K_d (nM)	Reference
Porcine[a]	11.2	0.6	Musser et al., 1993 [14]
Rat	11.5	1.5	Musser et al., 1993 [14]
Rabbit	8.5	1.9	Musser et al., 1993 [14]
Guinea pig	34.4	3.0	Musser et al., 1993 [14]
Human	15.6	1.8	Bohm et al., 1989 [16]
Rat	23	9.0	Linden et al., 1985 [17]
Guinea pig	20	4.3	Froldi and Belardinelli, 1990 [18]

[a] Denotes semipurified membranes.

Structure of A_1 Receptors

Adenosine receptors are members of a large gene family of receptors whose encoded proteins are characterized structurally by the presence of seven hydrophobic regions that are thought to be membrane-spanning domains. In addition, these receptors share the capacity to modulate the activity of effector enzymes or ion channels via guanine nucleotide regulatory binding proteins (G proteins). Photoaffinity studies have identified the A_1 receptor as a peptide with a molecular weight of 36,000 daltons [20]. This and other biochemical studies indicate that the A_1 receptor is a glycoprotein, approximately 326 amino acids in length, with complex-type carbohydrate chains, the likely presence of disulfide bonds, and histidine residues in the ligand binding site [21].

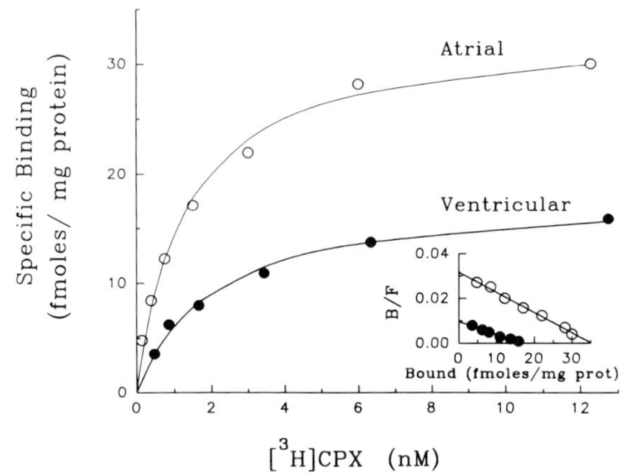

FIGURE 26-1. Equilibrium saturation binding of [^{3}H] CPX to guinea-pig atrial (●) and ventricular (○) membranes. The A_1 receptor density (B_{max}) in atrial membranes (33 fmoles/mg protein) was almost twofold greater than in ventricular membranes (18 fM/mg protein). The equilibrium dissociation constant (K_d) was the same in atrial (1.3 nM) and ventricular (1.9 nM) membranes. A Scatchard plot of the data is shown in the **inset**.

Cardiac A_2 Receptors

A_2-adenosine receptors have been identified in the heart by molecular, biochemical, pharmacologic, and functional criteria [22–24]. The regulation by A_2 receptors of coronary vascular tone and of endothelial function is discussed in Chapter 27 of this volume. Evidence for the presence and functional significance of A_2 receptors in myocardium will be discussed here. That myocardial adenosine receptors stimulate adenyl cyclase is indicated by results of experiments on heart slices [25], heart cell or tissue fragments [26,27], and isolated papillary muscles [28]. Agonists of A_2 receptors also stimulate adenyl cyclase activity and cAMP accumulation in homogeneous preparations of rat ventricular myocyte membranes [29] and guinea-pig isolated ventricular myocytes [30], respectively. Relatively high concentrations (>0.1 μM) of agonist are needed to elicit these responses [22,29,30]. Agonist-stimulated accumulation of cAMP was blocked by the A_2-adenosine receptor antagonist CGS 15943A [30]. The presence of A_2 receptors on cardiomyocytes has not been confirmed by radioligand binding.

A_2 receptor agonists caused increases of adenyl cyclase activity and contractile amplitude of embryonic chicken ventricular myocytes when the inhibitory A_1 receptor pathway had been blocked by either 8-cyclopentyl-1,3-dipropylxanthine (DPCPX) or prior exposure to pertussis toxin [23]. Under similar conditions, A_2 receptor agonists were found to increase cAMP accumulation but not the contractile amplitude of guinea-pig ventricular myocytes [30]. Indeed, some investigators have failed to observe any effect of A_2 agonists on cAMP accumulation, electric activity, or contractile function in mammalian myocytes. [31]. Cellular compartmentalization of cAMP may explain the absence of a contractile response of adult guinea-pig ventricular myocytes when cAMP content was elevated [30]. In summary, A_2 receptors that stimulate adenyl cyclase activity appear to be present in cardiomyocyte

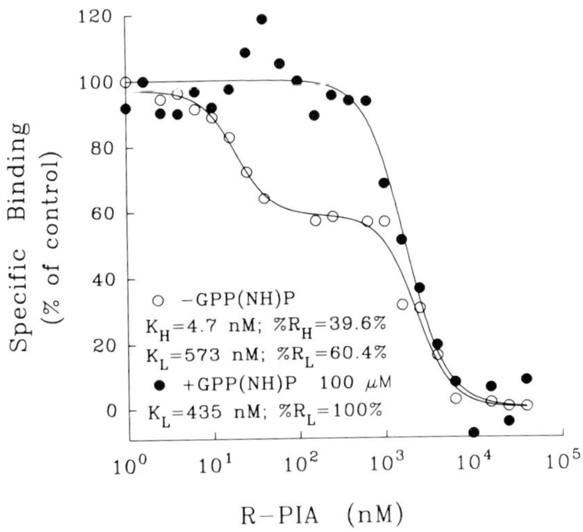

FIGURE 26-2. Competitive displacement of [^{3}H]CPX binding to guinea-pig ventricular membranes by the agonist R-PIA in the absence (○) and presence (●) of 100 μM Gpp(NH)p. K_H and K_L represent the equilibrium dissociation constants of R-PIA for the high- and low-affinity states, respectively, of the A_1 adenosine receptor. % R_H and % R_L are the fraction of A_1 receptors in the high- and low-affinity states, respectively.

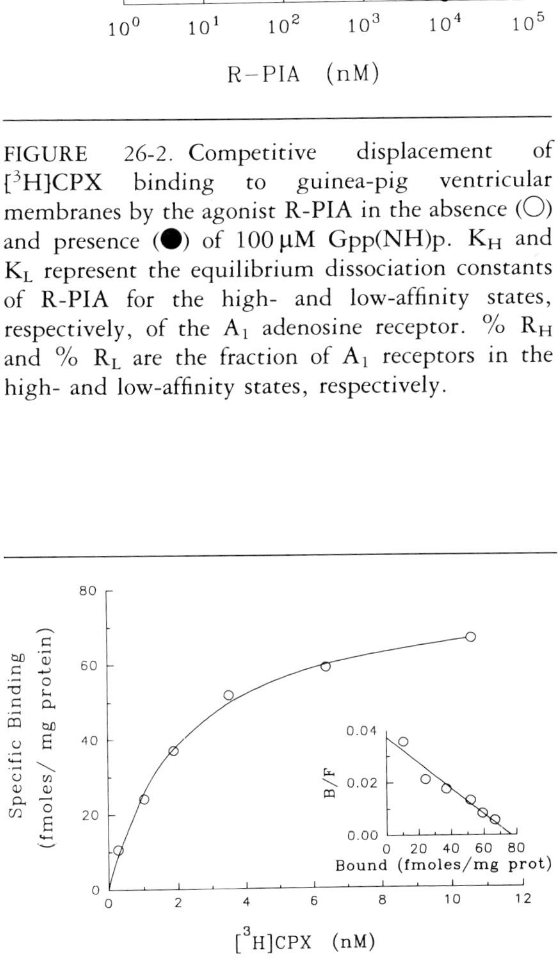

FIGURE 26-3. Equilibrium saturation binding of [^{3}H]CPX to guinea-pig intact ventricular myocytes. The A_1 receptor density (B_{max}) was 76 fmoles/mg protein, and the equilibrium dissociation constant (K_d) value was 1.8 nM.

cell membranes, but their functional significance in the mammalian heart is unclear. Activation of inhibitory A_1 receptors on cardiomyocytes may mask a functional response to activation of A_2 receptors on these cells [23,30]. Delineation of intracellular compartments of cAMP and their regulation of cellular functions may be necessary to achieve an understanding of the role of A_2 receptors on cardiomyocytes.

Cardiac A_3 Receptors

The existence of an A_3 receptor subtype was first proposed by Ribeiro and Sebastiao [32]. Their proposal was based on the results of pharmacologic studies and suggested that the A_3 receptor was present on central and peripheral neurons. Evidence for the existence of this A_3 receptor, however, is conjectural [33]. In contrast, the A_3 receptor subtype recently proposed to exist by multiple groups has been well defined by its unique sequence and pharmacologic profile and by its G-protein–mediated inhibition of adenylate cyclase [33].

The results of several studies provide support of the presence of the A_3 receptor in the heart. For example, Northern blot analysis carried out on cardiac tissue preparations from multiple species indicates the presence of small amounts of messenger RNA encoding the A_3 receptor [34–37]. Furthermore, an A_3 receptor clone was isolated from a human cDNA library of cardiac origin [38]. However, the cell type(s) (i.e., myocardial, smooth muscle, or endothelial) expressing the A_3 receptor in the heart has yet to be determined. In addition, there are few studies demonstrating a physiologic role for the A_3 receptor in the heart or in any other tissue, although a role for the A_3 receptor in mediating hypotension has been proposed [33]. The physiologic role(s) of the A_3 receptor in cardiac function remains to be determined.

Cloning of Adenosine Receptors

The first adenosine receptors were isolated from a dog thyroid cDNA library based on the similarity of sequences encoding the proteins of the G-protein–coupled receptor superfamily [39]. These clones were later shown to code for A_1 and A_{2a} receptors based on radioligand binding characteristics and in situ hybridization studies of receptor distribution. Estimates of the molecular weights of the A_1 and A_{2a} receptors based on sequence analysis were 36,000 and 45,000 daltons, respectively, which is in full agreement with previous biochemical estimates. A number of other adenosine receptor subtypes have been cloned from various species, including the A_1, A_{2a}, A_{2b}, and A_3 from human and rat tissue; the A_1 from bovine tissue; and the A_3 from sheep tissue [40,41]. To date only two receptor subtypes have been isolated from cDNA libraries derived from cardiac tissue: human and rat A_1 receptor clones, and a human A_3 receptor clone [38,40,42]. Nevertheless, there is little evidence for tissue-specific receptor subtypes [41]. The development of probes to determine the pharmacologic properties and physiologic roles of the A_3 receptor subtype is needed to corroborate the significance of the demonstration of this receptor in the heart by molecular biology methods.

Acknowledgments

The authors would like to thank Dr. Donn Dennis for assistance in mathematical analysis of A_1 adenosine receptor high- and low-affinity binding. We would also like to express our gratitude to Ms. Kathleen Duvall for her assistance in preparing this manuscript and Ms. Jackie Ruble for her technical assistance.

References

1. Belardinelli L, Linden J, Berne RM (1989) The cardiac effects of adenosine. Prog Cardiovasc Dis 32:73–97.
2. Schrader J, Baumann G, Gerlach E (1977) Adenosine as inhibitor of myocardial effects of catecholamines. Pflügers Arch 372:29–35.
3. Shryock JC, Patel A, Belardinelli L, Linden J (1989) Down regulation and desensitization of A_1-adenosine receptors in embryonic chicken heart. Am J Physiol 256:H321–H327.
4. Wu S-N, Linden J, Visentin S, Boykin M, Belardinelli L (1989) Enhanced sensitivity of heart cells to adenosine and upregulation of receptor number after treatment of guinea pigs with theophylline. Circ Res 65:1066–1077.
5. Liang BT, Donovan LA (1990) Differential desensitization of A_1 adenosine receptor-mediated inhibition of cardiac myocyte contractility and adenylate cyclase activity. Circ Res 67:406–414.
6. Lee HT, Thompson CI, Linden J, Belloni FL (1993) Differential sensitization of cardiac actions of adenosine in rats following chronic theophylline treatment. Am J Physiol 264: H1634–H1643.
7. Lee HT, Thompson CI, Hernandez A, Lewy JL, Belloni FL (1993) Cardiac desensitization to adenosine analogues after prolonged R-PIA infusion in vivo. Am J Physiol 265:H1916–H1927.
8. Bruns RF, Fergus JH (1990) Allosteric enhancement of adenosine A_1 receptor binding and function by 2-amino-3-benzoylthiophenes. Mol Pharmacol 38:939–949.
9. Amoah-Apraku B, Xu J, Lu JY, Pelleg A, Bruns RF, Belardinelli L (1993) Selective potentiation by an A_1 adenosine selective enhancer of the negative dromotropic action of adenosine in the guinea heart. J Pharmacol Exp Ther 266:611–617.
10. Mudumbi RV, Montamat SC, Bruns RF, Vestal RE (1993) Cardiac functional responses to adenosine by PD 81,723, an allosteric enhancer of the adenosine A_1 receptor. Am J Physiol 264:H1017–H1498.
11. Kollias-Baker C, Ruble J, Belardinelli L (1994) Mechanisms of allosteric enhancement of agonist binding to the A_1 adenosine receptor (abstr). Drug Dev Res 31:286.
12. Kollias-Baker C, Belardinelli (1994) Allosteric enhancer of the A_1 adenosine receptor markedly increases AV nodal conduction delay during hypoxia in guinea pig hearts (abstr). Drug Dev Res 31:287.
13. Ellenbogen KA, Thames MD, DiMarco JP, Sheehan H, Lerman BB (1990) Electrophysiological effects of adenosine in the transplanted human heart. Evidence for supersensitivity. Circulation 81:821–828.
14. Musser B, Morgan ME, Leid M, Murray TF,

Linden J, Vestal RE (1993) Species comparison of adenosine and beta-adrenoceptors in mammalian atrial and ventricular myocardium. Eur J Pharmacol-Mol Pharmacol Sec 246:105–111.
15. Martens D, Lohse MJ, Schwabe U (1988) [^{3}H]-8-Cyclopentyl-1,3-dipropylxanthine binding to A_1 adenosine receptors of intact rat ventricular myocytes. Circ Res 63:613–620.
16. Bohm M, Pieske B, Ungerer M, Erdmann E (1989) Characteristics of A_1 adenosine receptors in atrial and ventricular myocardium from diseased human hearts. Circ Res 65:1201–1211.
17. Linden J, Patel A, Sadek S (1985) ^{125}I-Aminobenzyladenosine, a new radioligand with improved specific binding to adenosine receptors in heart. Circ Res 56:279–284.
18. Froldi G, Belardinelli L (1990) Species-dependent effects of adenosine on heart rate and atrioventricular nodal conduction. Mechanism and physiological implications. Circ Res 67: 960–978.
19. Yeung SMH, Green RD (1983) Agonist and antagonist affinities for inhibitory adenosine receptors are reciprocally affected by 5′-guanylylimidodiphosphate or N-ethylmaleimide. J Biol Chem 258:2334–2339.
20. Stiles GL, Daly DT, Olsson RA (1986) Characterization of the A_1 adenosine receptor-adenylate cyclase system of cerebral cortex using an agonist photoaffinity probe. J Neurochem 47:1020–1025.
21. Jacobson KA, van Galen PJM, Ji X-D, Ramkumar V, Olah ME, Stiles GL (1993) Molecular characterization of A_1 and A_{2a} adenosine receptors. Drug Dev Res 28:226–231.
22. Stehle JH, Rivkees SA, Lee JJ, Weaver DR, Deeds JD, Reppert SM (1992) Molecular cloning and expression of the cDNA for a novel A_2-adenosine receptor subtype. Mol Endocrinol 6:384–393.
23. Xu D, Kong HY, Liang BT (1992) Expression and pharmacological characterization of a stimulatory subtype of adenosine receptor in fetal chick ventricular myocytes. Circ Res 70: 56–65.
24. Olsson RA, Pearson JD (1990) Cardiovascular purinoceptors. Physiol Rev 70:761–845.
25. Huang M, Drummond GI (1976) Effect of adenosine on cyclic AMP accumulation in ventricular myocardium. Biochem Pharmacol 25:2713–2719.
26. Webster S, Olsson RA (1981) Adenosine regulation of canine cardiac adenylate cyclase. Biochem Pharmacol 30:369–373.
27. Anand-Srivastava MB, Cantin M (1983) Regulation of adenylate cyclase in cultured cardiomyocytes from neonatal rats by adenosine and other agonists. Arch Biochem Biophys 223: 468–476.
28. Bruckner R, Fenner A, Meyer W, Nobis TM, Schmitz W, Scholz H (1985) Cardiac effects of adenosine and adenosine analogs in guinea pig atrial and ventricular preparations: Evidence against a role of cyclic AMP and cyclic GMP. J Pharmacol Exp Ther 234:766–774.
29. Romano FD, Macdonald SG, Dobson JG (1989) Adenosine receptor coupling to adenylate cyclase of rat ventricular myocyte membranes. Am J Physiol 257:H1088–H1095.
30. Stein B, Schmitz W, Scholz H, Seeland C (1994) Pharmacological characterization of A_2-adenosine receptors in guinea-pig ventricular cardiomyocytes. J Mol Cell Cardiol 26:403–414.
31. Shryock J, Song Y, Wang D, Baker SD, Olsson R, Belardinelli L (1993) Selective A_2-adenosine receptor agonists do not alter action potential duration, twitch shortening, or cAMP accumulation in guinea pig, rat, or rabbit isolated ventricular myocytes. Circ Res 72: 194–205.
32. Ribeiro JA, Sebastiao AM (1986) Adenosine receptors and calcium: Basis for proposing a third (A_3) adenosine receptor. Prog Neurobiol 26:179–209.
33. Carruthers AM, Fozard JR (1993) Adenosine A_3 receptors: Two into one won't go. Trends Pharmacol Sci 14:290–291.
34. Zhou QY, Li C, Olah ME, Johnson RA, Stiles GL, Civelli O (1992) Molecular cloning and characterization of an adenosine receptor: The A_3 adenosine receptor. Proc Natl Acad Sci 89: 7432–7436.
35. Meyerhof W, Muller-Brechlin R, Richter D (1991) Molecular cloning of a novel putative G-protein coupled receptor expressed during rat spermiogenesis. FEBS Lett 284:155–160.
36. Linden J, Taylor HE, Robeva AS, Tucker AL, Stehle JH, Rivkees SA, Fink JS, Reppert SM (1993) Molecular cloning and functional expression of a sheep A_3 adenosine receptor with widespread tissue distribution. Mol Pharmacol 44:524–532.
37. Salvatore CA, Jacobson MA, Taylor HA, Linden J, Johnson RG (1993) Molecular cloning and characterization of the human A_3 adenosine

receptor. Proc Natl Acad Sci USA 90:10365–10369.
38. Sajjadi FG, Firestein GS (1993) cDNA cloning and sequence analysis of the human A_3 adenosine receptor. Biochim Biophys Acta 1179:105–107.
39. Libert F, Parmentier M, Lefort A, Dinsart C, Van Sande J, Maenhaut C, Simons MJ, Dumont JE, Vassart G (1989) Selective amplification of four new members of the G protein-coupled receptor family. Science 244:569–572.
40. Jacobson M (1994) Cloning and expression of human adenosine receptor subtypes (abstr). Drug Dev Res 31:281.
41. Tucker AL, Linden J (1993) Cloned receptors and cardiovascular responses to adenosine. Cardiovasc Res 27:62–67.
42. Ungerer M, Obermaier-Skrobranek B, Lohse MJ (1992) Cloning and expression of a cardiac A_1 adenosine receptor from normotensive and spontaneously hypertensive rats (abstr). Circulation 86:I-304.

27. CORONARY ADENOSINE RECEPTORS: SUBTYPES, LOCALIZATION, AND FUNCTION

S. Jamal Mustafa
Ravi Marala
Worku Abebe
Neil Jeansonne
Hammed Olanrewaju
Tahir Hussain

Despite the historic observations of Drury and Szent Gyorgyi in 1929 [1] that adenosine causes coronary vasodilation, hypotension, and bradycardia, it was not until 30 years later that Berne [2] proposed the adenosine hypothesis for the metabolic regulation of coronary blood flow. A large body of literature has been accumulated on the cardiovascular actions of adenosine since that time.

Under both in vivo and in vitro conditions, adenosine has been shown to produce dilation and has been subsequently regarded as the primary mediator of coronary vasodilation in various physiologic states of the heart [2]. According to the adenosine hypothesis proposed by Berne [2], an imbalance between oxygen supply and demand leads to the breakdown of intracellular adenine nucleotides by the action of ecto-5′-nucleotidase to adenosine in the extracellular space. This nucleoside then acts on the vascular smooth muscle cells of the coronary artery through surface membrane receptors to produce relaxation [3]. The adenosine receptor has been further characterized to be of the A_2 subtype, which has been shown to be linked to adenylate cyclase through Gs protein [4–6]. The A_2-adenosine receptor has been identified in a number of species from this laboratory, including human [4–10]. However, we have reported earlier that certain adenosine analogues such as 5′-N-ethylcarboxamidoadenosine (NECA) may act on receptor through a different signaling mechanism [4]. This report [4] further suggests that the receptor signaling mechanism(s) responsible for the relaxation of vascular smooth muscle, and especially coronary smooth muscle, is more complicated than originally thought. Indeed, Mustafa [11] suggested the existence of more than one binding site for adenosine in dog and bovine coronary artery. In this article, we review the status of the coronary artery adenosine receptor and its signaling mechanisms [11,12].

A_2-*Adenosine Receptor*

BINDING STUDIES

Coronary arteries of various species have been reported to contain A_2 adenosine receptor [12]. However, characterization of such coronary adenosine receptors using specific ligand-receptor binding approach has always been a challenge. ^{3}H-labeled NECA and 2-([4-(2-carboxyethyl) phenyl]-amino)adenosine-5′-N-ethylcarboxamide (CGS 21680) have been extensively used as ligands to characterize A_2-adenosine receptors in a variety of tissues. CGS 21680, a potent vasodilator of coronary artery and a specific A_{2a} receptor agonist, did not show any specific binding in coronary artery membranes [unpublished results from this lab]. NECA, also a potent vasodilator but a nonselective ligand, binds with a K_d of approximately 1 μM and a B_{max} of 2.059 nmoles/mg protein in bovine coronary artery membranes without endothelium [unpublished data from this lab]. Since NECA has equal affinity to A_1 and A_2 receptors, N^6-cyclopentyladenosine (CPA) (50 nM) was added to the [^{3}H]-NECA binding assay to eliminate A_1 binding. Unlabeled NECA ($IC_{50} = 1.06$ μM) and 2-chloroadenosine (CAD; $IC_{50} = 2.7$ μM) were the most potent adenosine analogues in inhibiting [^{3}H]-NECA binding. A_{2a}-selective agonists (CGS 21680; N^6-[2-(3,5-dimethoxyphenyl)-2-(2-methylphenyl)ethyl]adenosine, DPMA), an A_{2a}-selective antagonist (9-chloro-2-(2-puryl)[1,2,4]-triazolo[1,5-c]quinazolin-5-amine; CGS 15943), A_1-selective agonists (CPA; R(-)N^6-(2-phenylisopropyl)adenosine, R-PIA), and A_1-selective antagonists (xanthine amine congener, XAC; 8-cyclopentyl-1, 3-dipropylxanthine, DPCPX) were found to be poor competitors ($IC_{50} > 100$ μM) of [^{3}H]-NECA binding. This potency profile suggests the presence of A_{2b} receptor.

As far as the functional response of NECA on adenylate cyclase is concerned, NECA does stimulate adenylate cyclase modestly (25%) at 10 μM only in the presence of 50 nM DPCPX. This suggests the coupling of the A_{2b} receptor to adenylate cyclase and indicates the possibility of the presence of A_1 receptor in coronary artery. As indicated earlier, CGS 21680 is a potent vasodilator, suggesting the presence of a high-affinity (A_{2a}) adenosine receptor in coronary artery. However, the fact that [^{3}H]-CGS 21680 did not show specific binding to coronary artery membranes may be due to a very low population of the receptors.

Recently, a [^{125}I]-labeled 2-substituted arylamine adenosine analogue, 2-[2-(4-aminophenethyl)ethylamino]adenosine (APE), was made available to us (synthesized by Dr. Ray Olsson and radiolabeled by Dr. Joel Linden). [^{125}I]-APE has been demonstrated to have a high affinity and selectivity to A_{2a}-adenosine receptors [13]. Membranes prepared from endothelium-denuded coronary arteries were found to contain [^{125}I]-APE binding sites with a K_d of 0.59 nM and B_{max} of 7 fmoles/mg protein. Competition with various adenosine receptor agonists suggested that 2-phenylaminoadenosine (CV 1808) and APE > NECA and CGS 21680. Furthermore, the nonxanthine antagonist CGS 15943A was found to be more potent than 8-sulphophenyltheophylline (8-SPT) in inhibiting the binding. This potency profile (CV 1808 ≫ CGS 21680) overlaps with the A_4 subtype of adenosine receptor reported in rat striatum and suggested to be present in porcine coronary artery smooth muscle cells linked to the glibenclamide-sensitive K^+ channel [14]. However, in our functional studies we were unable to block the CV-1808–induced relaxation with glibenclamide [15], thus arguing against the involvement of A_4 receptor [14].

Another characteristic common to our study and that of Cornfield et al. [14] is temperature sensitivity. The maximum binding of [^{3}H]-CV 1808 [14] and [^{125}I]-APE, as reported earlier, was detected at 4°C. Whether this is an artifact of temperature that produces changes in the binding characteristics of the competing agonists is not yet clear. However, Luthin and Linden [16] have suggested that CGS 21680 loses its affinity to A_{2a} receptor at 4°C, and hence the potency profile of the agonists on A_{2a} receptor is reversed. [^{125}I]-azido-APE indicates specific labeling of 45 kD protein, and the labeling could be displaced by unlabeled APE, CGS 15943A, and theophylline. Similar 45 kD protein was revealed by Western blot studies.

REGULATION OF CORONARY ADENOSINE RECEPTOR

Recently, we have developed an in vitro model of coronary artery in tissue culture [17] that was used to study regulation of the adenosine receptor system. In CAD-treated vascular rings, the relaxation responses to NECA, R-PIA, and CAD were attenuated compared to control arteries. This suggested the downregulation of adenosine receptor function. The stimulation of cAMP by CAD, NaF, and forskolin were significantly decreased compared to control arteries (Fig. 27-1, top) [18]. The data suggest a change in G proteins, reduced adenylate cyclase activity, and possibly decreased adenosine receptor numbers.

The α-subunits of both G proteins (G_s and G_i) in CAD-treated arteries were measured by ADP ribosylation (Fig. 27-1, bottom) [19]. Cholera and pertussis toxins catalyzed ADP ribosylation of 45- and 41-kD proteins ($G_{s\alpha}$ and $G_{i\alpha}$, respectively). $G_{s\alpha}$ labeling decreased with the increasing concentration of CAD (10^{-8} to 10^{-4} M), while $G_{i\alpha}$ labeling significantly increased with lower concentrations of CAD (10^{-8} to 10^{-6} M), followed by a decrease with a higher CAD concentration (10^{-5}, 10^{-4} M) in the treated arteries [19]. Although the changes in the ADP ribosylation do not necessarily reflect alterations in the amount of α-subunits of respective G proteins, it does suggest changes in the functional responses of both G_s and G_i proteins by CAD treatment.

To determine whether the changes in the G_s and G_i can alter the responses to other G-protein–linked receptors, beta-adrenoceptor and muscarinic receptor agonist–mediated vasoactivity was measured in CAD-treated arteries. The vasorelaxing response to isoproterenol was studied in arteries treated with 10^{-4} M CAD (wherein $G_{s\alpha}$ is maximally decreased), while the vasocontractile response to methacholine was studied in arteries treated with 10^{-6} M (wherein $G_{i\alpha}$ is maximally increased) and 10^{-4} M (wherein $G_{i\alpha}$ is maximally decreased) CAD. Surprisingly, in the treated arteries isoproterenol-mediated relaxation was not different from control [18]. Methacholine-mediated contraction in 10^{-6} M CAD-treated arteries was significantly greater than control, whereas contraction in arteries treated with 10^{-4} M CAD was unaltered [19]. Although the relaxation response to isoproterenol was unchanged in CAD-treated arteries, an increase in contractile response to methacholine at a lower concentration of CAD-treated arteries suggests a heterologous desensitization of adenosine receptor in coronary artery.

ROLE OF ENDOTHELIUM

The role of endothelium in adenosine-receptor–mediated coronary vasorelaxation has been studied by a number of investigators. Recently, we have demonstrated that the endothelium has a modulatory influence on the action of certain adenosine agonists on the coronary vasculature [20]. This effect has been suggested to be mediated via A_2-adenosine receptor(s) located on the vascular endothelium.

Endothelium-mediated coronary artery responses induced by adenosine agonists have been shown to contribute either partially or totally to the final vasorelaxations. In this regard, recently we documented that the relaxant effects of NECA and CGS 21680 on porcine coronary artery rings were partially endothelium dependent [20]. In isolated guinea-pig hearts, the vasodilatory actions of adenosine, CAD, NECA, CGS 21680, R-PIA, S(+)-N^6-(2-phenylisopropyl)adenosine (S-PIA), and CPA have been proposed to be partially endothelium dependent [21]. On the other hand, studies conducted by Newman et al. [22] reveal that coronary vasodilations elicited by adenosine and NECA in isolated guinea-pig hearts were totally dependent

A

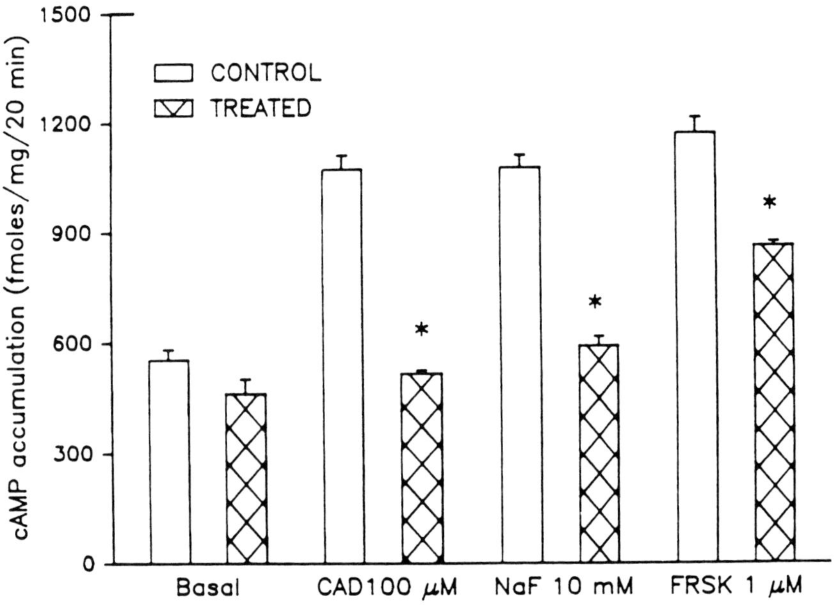

B

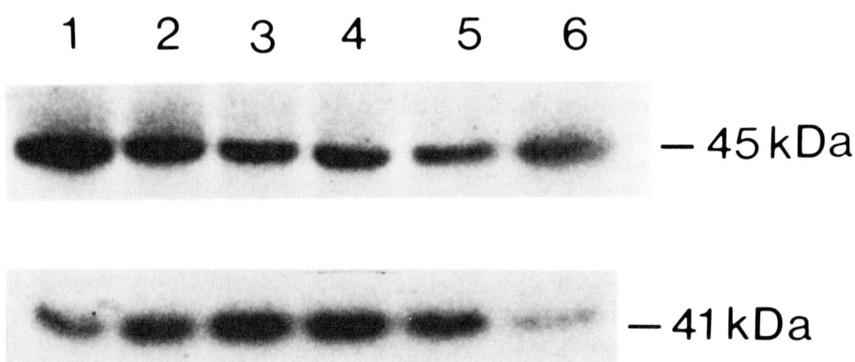

FIGURE 27-1. **A:** 2-Chloroadenosine (CAD), sodium fluoride (NaF), and forskolin stimulated cAMP accumulation in control and treated (100 μM, 3 days) arteries. Values represent mean ± SE. *Significantly different from the respective control ($p < 0.05$, Student's test). **B:** Cholera and pertussis toxin–catalyzed ADP ribosylation (45 kD and 41 kD, respectively) in membranes from control (lane 1) and 2-chloroadenosine treated (lanes 2–6, 10^{-8} to 10^{-4} M, for 3 days) arteries. Densitometrically the 45-kD band decreased from lane 1 through 6, while the 41-kD band increased from lane 1 through 4 followed by a decrease in lanes 5 and 6 (data not shown).

on the presence of the endothelium. The reason for the reported differential effects of the adenosine agonists in this respect is not apparent.

In our recent investigation, we observed that the endothelium-dependent actions of NECA and CGS 21680 in the porcine coronary artery were attenuated by the nitric oxide synthase inhibitor, N^G-monomethyl-L-arginine (L-NMMA) and by the cGMP-lowering agent, 6-anilino-5,8-quinolinedione (LY-83583) but not by the prostanoid synthesis inhibitor, indomethacin (Table 27-1). This effect was associated with augmented production of cGMP, which was also abolished by L-NMMA or LY-83583 (data not shown). The inhibitors were without effect in the absence of endothelium. These data suggest that in procine coronary artery, endothelium-dependent relaxation evoked by adenosine agonists such as NECA and CGS 21680 involve the release of nitric oxide from the endothelium and cGMP appears to be a mediator for this effect. Consistent with our observation, Vials and Burnstock [21], using isolated guinea-pig hearts, also demonstrated that reduction in perfusion pressure caused by adenosine, CPA, NECA, and CGS 21680 was inhibited by N^G-nitro-L-arginine-methyl ester (L-NAME) or N^G-nitro-L-arginine (NNLA) but not by indomethacin, further suggesting the involvement of nitric oxide in these responses.

In contrast to the above-mentioned endothelium-dependent effects, there are a number of studies that appear to indicate that the coronary vasorelaxant action of certain adenosine agonists is endothelium independent. In this respect, we have found that the relaxant effects of CAD, CPA, R-PIA, S-PIA, N^6-cyclohexyladenosine (CHA), and CV 1808 on porcine coronary artery rings were not influenced by the endothelium [20]. L-NMMA and LY-83583 were without effect on the response of the arteries to CAD, and no increase in cGMP was also observed in response to CAD (data not shown). Such a differential effect of adenosine agonists on endothelial and smooth muscle cells further indicates the existence of variations in the type of adenosine receptors located at the two sites [20].

The reason for the differences observed among the various adenosine agonists with regard to the influence of the endothelium is not well known. However, on the basis of our previous observation [20], we speculated that this could be related, at least in part, to variations in the structures of the compounds used. NECA and CGS 21680, as opposed to CAD, CHA, R-PIA, S-PIA, CV 1808, and CPA, are uronamide substituted at the 5′-carbon position of the ribose moiety. It is therefore possible that such a structural modification of these compounds may contribute to their endothelium-dependent action. However, this cannot offer a satisfactory explanation for the different results obtained using coronary artery preparations from other species. For instance, in contrast to the data obtained in porcine coronary artery [20], the effect of NECA was found to be endothelium independent in bovine coronary artery rings

TABLE 27-1. Effects of L-NMMA (30 μM) and LY-83583 (10 μM) on NECA (1 μM) and CGS 21680 (1 μM)-induced relaxation of prostaglandin $F_{2\alpha}$ (20 μM)-contracted porcine coronary artery rings in presence of endothelium

	% Relaxation		% Relaxation	
Agonists	Control	After L-NMMA	Control	After LY-83583
NECA	85.8 ± 3.0	67.6 ± 5.8[a]	84.8 ± 2.10	67.1 ± 2.8[a]
CGS 21680	92.7 ± 5.8	78.5 ± 2.7[a]	89.4 ± 1.96	64.5 ± 5.1[a]

Values are mean ± SE of five or six different observations.
[a] Significantly different from corresponding control.

[23]. Another contributory factor for the differing effects of the adenosine agonists is the nature of the contractile agents employed.

ROLE OF HYPERPOLARIZATION AND INVOLVEMENT OF K^+_{ATP} CHANNELS

Hyperpolarization of vascular smooth muscle in response to adenosine and its analogues has been reported, and the involvement of endothelial-derived factors has been demonstrated [15,24]. One such factor, called endothelium-derived hyperpolarizing factor (EDHF), increases the membrane potential by either stimulating Na^+/K^+-ATPase or activating K^+ channels of the smooth muscle membrane [25].

Glibenclamide is an antidiabetic sulfonylurea compound that appears to specifically inhibit K^+_{ATP} channels in different tissues. However, it has been reported from this laboratory that K^+_{ATP} channels are not involved in the mechanism whereby adenosine and its analogues elicit their vasorelaxant response in isolated porcine or canine epicardial vessels [15]. CV-1808 and CGS 22988 have been reported to augment whole-cell K^+ currents [14]; however, glibenclamide failed to antagonize the vasorelaxant effects of these compounds [15]. Therefore, the involvement of K^+_{ATP} channels in adenosine-receptor–mediated vasodilation is not consistent with our observations and does not involve the putative A_4 subtype, as suggested by Cornfield et al. [14].

CAD has been shown to produce both endothelium-dependent and independent hyperpolarization in human (Table 27-2), bovine, and porcine coronary arteries [24], which can be antagonized by 8-SPT (Table 27-2).

HUMAN STUDIES

Relaxation of human coronary artery by adenosine analogues was first reported in 1988 from this laboratory [7]. Further studies from this laboratory demonstrated G-protein coupling of the adenosine receptor in human coronary artery [5]. We recently compared the relaxing effects of two 5′-uronamides (CGS 21680 and NECA) with 2-cyclohexylethylamino adenosine (CGS 22492) in human blood vessels [10]. NECA possessing an unsubstituted purine heterocycle was the most potent of the three compounds tested. CGS 21680 and NECA exhibited parallel relaxation responses (Fig. 27-2). NECA achieved a significantly greater vasodilatory response than CGS 22492. Similarly, previous studies from this laboratory reported that human coronary relaxed better with NECA than with R-PIA. In vitro studies of bovine and porcine coronary artery showed adenosine-5′-uronamides to be more potent than analogues possessing an unmodified ribose ring [20,26]. More recently, we have shown the presence of A_{2a} receptor in human coronary artery using a more direct approach. We raised antibodies against polypeptide corresponding to the cloned A_{2a} receptor. By using western blot analysis with this antibody, the receptor subtype has been identified to exist in human coronary arteries with

TABLE 27-2. Effect of 2-chloroadenosine (CAD, 10^{-5} M) on resting membrane potential and influence of 8-sulfophenyltheophylline (8-SPT, 10^{-5} M) in human coronary artery branches[a]

Treatment	With endothelium	Without endothelium
	Membrane potential, mV	Membrane potential, mV
Control	-65.18 ± 1.15	-64.81 ± 1.12
CAD	-87.14 ± 1.12^c	-74.79 ± 1.15^b
CAD + 8-SPT	-66.38 ± 1.44	-65.19 ± 1.11

[a] Values are means ± SEM from two hearts; each mean represents 10 impalements in endothelium-intact and-denuded arteries.
[b] Significantly different from controls means at $p < 0.05$ and at [c] $p < 0.01$, respectively.

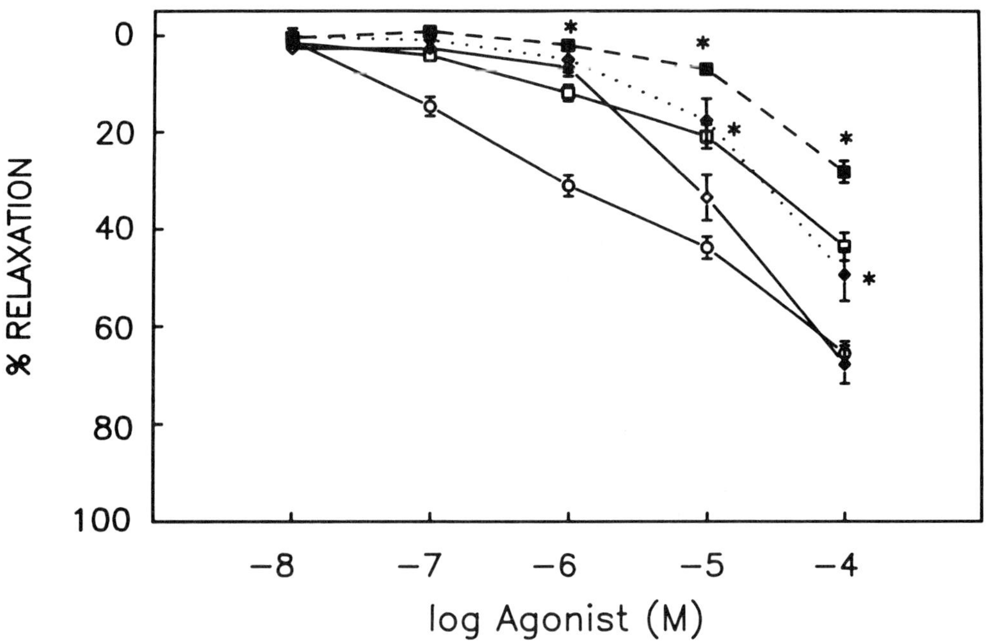

FIGURE 27-2. The antagonistic effects of CGS 15943A (◆ . . . ◆) on the relaxations produced by CGS 22492 (◇—◇); antagonistic effects of CGS 15943A (■--■) on the relaxations produced by CGS 21680 (□—□); vasodilatory effects of NECA (°—°) on 35 mM KCl-contracted human coronary rings. Each point is the mean ± SE representing 16–41 rings from four hearts. *Indicates significance from control ($p < 0.05$). With permission from Elsevier Science (Eur J Pharmacol 221:243–247, 1992).

an apparent molecular weight of 45 kD (data not shown).

A_1-Adenosine Receptor

BINDING STUDIES

Recently, several reports have suggested the existence of A_1 receptor in vascular smooth muscle, but there is controversy as to the functional role of this receptor in coronary vasculature. We have used [^{3}H]-DPCPX as a ligand for binding studies to characterize A_1 receptor in porcine coronary artery membranes. [^{3}H]-DPCPX showed a specific binding with a K_d of 0.21 nM and a B_{max} of 6.4 fmoles/mg protein. Competitors inhibited [^{3}H]-DPCPX binding with the following order of potency: (S)-N^6-[2-endo-norbornyl]adenosine (ENBA) ≈ R-PIA ≥ DPCPX ≥ XAC > NECA ≫ CGS 21680. This order of potency fits the criteria of classic A_1 receptor. ENBA (10 nM) attenuated the vasorelaxation and cAMP accumulation elicited by isoproterenol in porcine coronary rings. These effects of ENBA were found to be pertussis toxin sensitive. The data suggest the existence of a classic A_1 receptor linked to adenylate cyclase in an inhibitory manner via a pertussis toxin–sensitive G protein (most likely G_i). Despite such evidence of an inhibitory A_1 receptor, we did not observe any vasoconstriction response to A_1-selective agonists. Merkel et al, [27] have reported a modest contraction in porcine coronary rings to only CPA.

ROLE OF PROTEIN KINASE C

Protein kinase C (PKC) is believed to be an important component in the maintenance of smooth muscle tone and various other cellular signal pathways. Recent studies are generating a lot of interest in a possible role of protein kinase C in adenosine signal transduction. Other laboratories have reported that downregulation of PKC attenuated contractions produced by agents such as phorbol, 12,13-dibutyrate (PDBu), KCl, and phenylephrine in canine arterial rings [28]. Preincubation with adenosine or its analogues attenuated PDBu-induced contraction in porcine coronary arteries [29]. Phorbol esters are believed to be the direct activators of PKC, whereas endothelin-1 (ET-1) activates PKC indirectly via a receptor-mediated activation of phospholipase C. The adenosine analogues CAD, CGS 21680, and NECA produced a concentration-dependent relaxation of ET-1–contracted porcine coronary artery [29]. These studies indicate a possible role for PKC in adenosine signal transduction.

In porcine coronary artery about 75% of the PKC activity is localized in the cytosolic compartment, and the remaining 25% is membrane associated [29]. Acute exposure (20 minutes) of these arteries to PDBu translocates PKC from the cytosol to the membrane, which is a measure of its activation. Inclusion of the adenosine analogue, CAD, with PDBu completely blocked the activation of PKC (data not shown). Arteries incubated with CAD alone had about a 25% higher concentration of PKC in the cytosolic fraction with a 73% decrease in membrane-associated PKC activity, indicating disruption of PKC mobilization between the cytosol and membrane [29]. Phorbol ester is known to deplete PKC from the tissue upon chronic exposure as a consequence of the increased rate of degradation of the enzyme [30]. In accordance with the above-mentioned reports, chronic (24–48 hours) exposure of porcine coronary arteries to PDBu caused depletion of PKC activity, but, interestingly, addition of CAD significantly protected the PDBu-induced PKC depletion [29] (Fig. 27-3, inset). More importantly, chronic exposure to CAD produced an accumulation of PKC by about two-fold in the arterial tissue (Fig. 27-3, inset). This upregulation of PCK by an adenosine analogue explains its counteraction with PDBu in PKC depletion.

These effects of adenosine analogue were further substantiated by pharmacologic studies in which porcine coronary artery was chronically (24–48 hours) exposed to PDBu and the contractile responses to the activators of PKC, that is, PDBu and ET-1, were measured [29]. Chronic exposure of the arterial rings with PDBu alone significantly attenuated the contractile responses to both PDBu and ET-1. The addition of CAD with PDBu during the incubation significantly improved the contractile responses of these arterial rings when compared with PDBu alone. Incubation of arterial rings with CAD produced significantly higher contractile responses to ET-1. These effects of an adenosine analogue on contractile responses are comparable with the modulation of PKC levels in this tissue [29].

These studies provide evidence to suggest a role of adenosine in the regulation of PKC activation but do not provide any clues as to the adenosine receptor subtype(s) involved in the modulation of PKC. Adenosine receptor agonists and antagonists with high selectivity for A_1 or A_2 receptor subtypes were used instead of CAD. The A_1-receptor–selective agonist ENBA reproduced all the effects shown by CAD, that is, upregulation of PKC and protection of the PDBu-induced PKC depletion (Fig. 27-3). These responses of ENBA could be blocked by the A_1-receptor antagonist (±) N^6endo-norbornyl-9-methyl adenine (N0861) (Fig. 27-3). ENBA alone elevated PKC activity by about twofold when compared to the control (Fig. 27-3, inset). On the contrary, the A_2-receptor–selective agonist CGS 21680 failed to reproduce any of the effects of CAD and ENBA (data not shown). Incubation with CGS 21680 alone also did not significantly alter PKC activity in the arterial rings (Fig. 27-3, inset).

Summary

Considerable progress has been made in pharmacologic and biochemical characterization of the adenosine receptors of the coronary artery. However, all that is known about which ade-

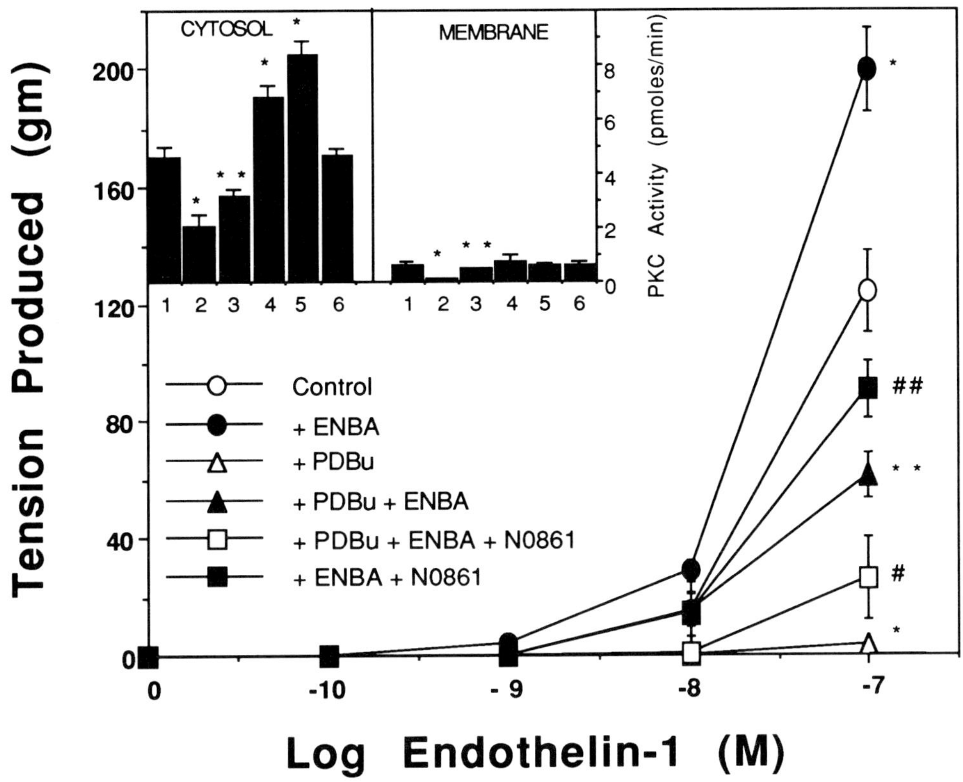

FIGURE 27-3. Modulation of protein kinase C by adenosine analogues. Endothelium-denuded porcine coronary arteries were incubated for 24 hours with PDBu (200 nM) in the presence and absence of the adenosine A_1-receptor agonist ENBA (10^{-9} M). The adenosine A_1-receptor antagonist N0861 (10^{-6} M) was added where indicated. Following incubation contractile tension was recorded in response to increasing concentrations of ET-1. **Inset:** Endothelium-denuded arteries were incubated with (1) No drugs (control); (2) PDBu (200 nM); (3) PDBu (200 nM) + CAD (10^{-4} M); (4) CAD (10^{-4} M) alone; (5) ENBA (10^{-9} M) alone; and (6) CGS 21680 (10^{-4} M) alone. Following incubation for 48 hours, tissue was homogenized, cytosol and membrane fractions were separated, and PKC activity was estimated in these fractions using a kit (RPN 77, Amersham). Values are means ± SE. Values are significantly different ($p < 0.01$) from control (*). **Significant protection of PDBu-induced PKC depletion by inclusion of ENBA or CAD ($p < 0.05$). # and ## Significant blockage of ENBA responses by N0861 ($p < 0.01$). **Inset:** The values for ENBA (5) and CGS 21680 (6) were normalized.

nosine receptors are expressed in the coronary artery is based on receptor-agonist and antagonist potency profiles. Agonist potency profiles suggest that there could be at least three (A_1, A_{2a}, A_{2b}) and possibly more subtypes of adenosine receptors expressed in the coronary artery. Another subtype, A_4, of adenosine receptors has been proposed that relaxes coronary arteries; however, we do not find much evidence for such a site.

Coronary artery apparently has A_2-adenosine receptor (A_{2a} and A_{2b}) on smooth muscle as well as on endothelium. Smooth muscle–associated A_2 receptor is linked to adenylate cyclase through G_s. Endothelium-associated A_2 receptor is involved in the release of nitric oxide, which contributes to the overall vasorelaxation. The involvement of ion channels and phospholipase C in A_2-receptor–mediated coronary vasorelaxation is also proposed. Evidence suggests the presence of an inhibitory A_1 receptor in coronary smooth muscle. The functional role of such a receptor is not clear. However, inhibition of adenylate cyclase and modulation of protein kinase C appear to be involved in A_1-receptor function. Pharmacologic studies alone are limited in their usefulness because of the complexity of the adenosine receptor/coronary artery system. We believe that it has become necessary to clone, sequence, and functionally express the adenosine receptor subtypes from coronary artery in order to better understand their involvement in the regulation of coronary artery tone.

Acknowledgment

This work was supported by National Heart, Lung and Blood Institute grant HL 27339. The authors would like to thank Dr. Joel Linden of the University of Virginia for providing ^{125}I-APE. Also, the authors would like to thank Ms. Pam Wynne for secretarial assistance.

References

1. Drury AN, Szent-Gyorgyi A (1929) The physiological activity of adenine compounds with special reference to their action upon the mammalian heart. J Physiol 68:213–237.
2. Berne RM (1963) Cardiac nucleotides in hypoxia: A possible role in regulation of coronary blood flow. Am J Physiol 204:317–322.
3. Olsson RA, Davis CJ, Khouri EM, Patterson RE (1976) Evidence for an adenosine receptor on the surface of dog coronary myocytes. Circ Res 39:93–98.
4. Cushing DJ, Brown GL, Sabouni MH, Mustafa SJ (1991) Adenosine-receptor-mediated coronary artery relaxation and cyclic nucleotide production. Am J Physiol 261:H343–H348.
5. Sabouni MH, Hussain T, Cushing DJ, Mustafa SJ (1991) G-proteins subserve relaxation mediated by adenosine receptors in human coronary artery. J Cardiovasc Pharmacol 18:696–702.
6. Hussain T, Mustafa SJ (1992) Effects of adenosine analogs on cholera toxin induced ADP-ribosylation of G-proteins in coronary artery. Am J Physiol 262:H875–H879.
7. Ramagopal MV, Chitwood R, Mustafa SJ (1988) Evidence for an A_2 adenosine receptor in human coronary arteries. Eur J Pharmacol 151: 483–486.
8. Sabouni MH, Mustafa SJ (1989) Effects of adenosine analogs on phasic contractions in human coronary artery. Eur J Pharmacol 168:271–276.
9. Sabouni MH, Ramagopal MV, Mustafa SJ (1990) Relaxation by adenosine and its analogs in potassium contracted human coronary arteries. Naunyn Schmiedbergs Arch Pharmacol 341: 388–390.
10. Makujina S, Sabouni MH, Bhatia S, Douglas FL, Mustafa SJ (1992) Vasodilatory effects of adenosine A_2 receptor agonists CGS 21680 and CGS 22492 in human vasculature. Eur J Pharm 221:243–247.
11. Mustafa SJ (1980) Cellular and molecular mechanism(s) of coronary flow regulation by adenosine. Mol Cell Biochem 31:67–87.
12. Cushing DJ, Mustafa SJ (1991) Historical perspective—Adenosine. In Phillis JW (ed.), Adenosine and Adenine Nucleotides as Regulators of Cellular Function. Boca Raton, FL: CRC Press, pp. 3–13.
13. Linden J, Prater MR, Tucker A, Thompson RD, Olsson RA (1992) ^{125}I-APE, a new and improved radioligand that binds selectively to native and recombinant adenosine A_{2a} receptors (abstr). Int J Purine Pyrimidine Res 3:86.
14. Cornfield LJ, Hu S, Hurt SD, Sills MA (1992) [^{3}H]2-phenylaminoadenosine ([^{3}H]CV-1808) labels a novel adenosine receptor in rat brain. J Pharmacol Exp Ther 263:552–561.
15. Makujina SR, Olanrewaju HA, Mustafa SJ (1994) Evidence against the involvement of ATP-sensitive K^+ channels in adenosine A_2 receptor-mediated vasorelaxation. Am J Physiol 267:H716–H724.
16. Luthin DR, Linden J (1994) Temperature-dependent binding of [^{3}H]CV 1808: Mimicry of "A4" binding in COS cells expressing only recombinant A2a adenosine receptors (abstr). Drug Dev Res 31:292.
17. Hussain T, Mustafa SJ (1993) An in vitro pharmacological model in coronary smooth muscle. J Pharmacol Toxicol Methods 30: 111–115.
18. Hussain T, Mustafa SJ (1993) Regulation of adenosine receptor system in coronary artery: Functional studies and cAMP. Am J Physiol 264 (Heart Circ Physiol 33):H441–H447.
19. Hussain T, Mustafa SJ (1994) Regulation of G proteins coupled to adenosine receptors in coronary artery. Am J Physiol 266 (Heart Circ Physiol 35):H1273–H1279.
20. Abebe W, Makujina S, Mustafa SJ (1994) Adenosine receptor-mediated relaxation of por-

cine coronary artery in presence and absence of endothelium. Am J Physiol 266:H2018–H2025.
21. Vials A, Burnstock G (1993) A_2-purinoceptor-mediated relaxation in the guinea pig coronary vascular: A role for nitric oxide. Br J Pharmacol 109:424–429.
22. Newman WH, Becker BF, Heirer H, Nees S, Gerlch E (1988) Endothelium-mediated coronary dilatation by adenosine does not depend on endothelial adenylate cyclase activation: Studies in isolated guinea pig hearts. Pflügers Arch 413:1–7.
23. Sabouni MH, Ramagopal MV, Mustafa SJ (1989) Roles of calcium and the endothelium in the relaxation produced by 5′-N-ethylcarboxamido-adenosine (NECA). Eur J Pharmacol 166:311–314.
24. Olanrewaju HA, Hargittai PT, Lieberman EM, Mustafa SJ (1994) Endothelium-dependent and independent hyperpolarization by adenosine and its analogs in porcine coronary artery. J Cardiovasc Pharmacol, in press.
25. Feleton M, Vanhoutte PM (1988) Endothelium-dependent hyperpolarization of canine coronary smooth muscle. Br J Pharmacol 93:515–524.
26. Mustafa SJ, Askar A (1985) Evidence suggesting a Ra-type adenosine receptor in bovine coronary arteries. J Pharmacol Exp Ther 232: 49–56.
27. Merkel LA, Lappe RW, Rivera LM, Cox LM, Perrone MH (1992) Demonstration of vasorelaxant activity with an A_1-selective adenosine agonist in porcine coronary artery: Involvement of potassium channels. J Pharmacol Exp Ther 266:437–443.
28. Merkel LA, Rivera LM, Colussi DJ, Perrone MH (1991) Protein kinase C and vascular smooth muscle contractility: Effects of inhibitors and down-regulation. J Pharmacol Exp Ther 257: 134–140.
29. Marala RB, Ways K, Mustafa SJ (1993) 2-Chloroadenosine prevents phorbol ester-induced depletion of protein kinase C in porcine coronary artery. Am J Physiol 264 (Heart and Circ Physiol 33):H1465–1471.
30. Young S, Parker PJ, Ullrich A, Stabel S (1987) Down-regulation of protein kinase C is due to an increased rate of degradation. Biochem J 244: 775–779.

28. BIOLOGIC ACTIVITY OF ADENOSINE A_{2a} RECEPTOR ANTAGONISTS

Ennio Ongini
Cristina Zocchi
Anna Conti
Monica Viziano
Angela Monopoli
Silvio Dionisotti

There is still little knowledge available on the function of the adenosine A_2 receptors in the mammalian organism. Characterization of their role in physiologic or pathologic states remains unexplored because thus far potent and selective A_2 antagonists have not become available [1].

Considerable research efforts made over the last decade have led to the development of the prototype A_2-selective agonist, the adenosine analog 2-[4-(2-carboxyethyl)-phenethylamino]-5′-N-ethylcarboxamidoadenosine, CGS 21680 [1], as well as a variety of other potent A_2 agonists, for example, 2-alkynyl derivatives of 5′-N-ethylcarboxamidoadenosine [2]. By using CGS 21680 as a pharmacologic probe, it has been demonstrated that A_2 receptors are distributed in specific peripheral tissues and discrete brain areas.

Classification of receptors based on the use of agonists, however, is weak, and only studies made with antagonists could lead to thorough characterization of the functional significance of the A_2 receptor. An additional complication comes from the discovery that two A_2-receptor subtypes exist in many tissues, namely, A_{2a} and A_{2b} [3]. Both these subtypes are coupled to the stimulation of adenylate cyclase. The A_{2a} subtype is the high-affinity site localized in brain striatum, platelets, and smooth muscles. This site is identified using the agonist CGS 21680. The A_{2b} subtype is the low-affinity site, which is distributed in neurons of all brain regions and occurs in human fibroblasts and in certain nonvascular tissues, such as rat intestine and bladder [3]. Drugs that are selective for the A_{2b} subtype of the receptor are not yet available.

Rapid advancements made in molecular biology have yielded useful information on the structural features of the adenosine receptors and have confirmed the heterogeneity of this G-protein–linked receptor system [1,3]. Importantly, both A_{2a}- and A_{2b}-receptor subtypes

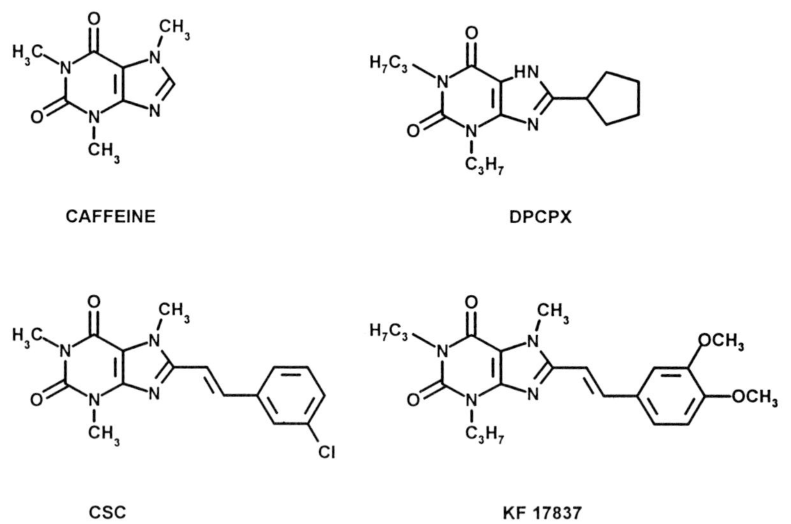

FIGURE 28-1. Structures of xanthine derivatives showing preferential affinity for A_1 (caffeine and DPCPX) or A_{2a} adenosine receptors (CSC and KF 17837): DPCPX, 8-cyclopentyl-1,3-dipropylxanthine; CSC, 8-(3-chlorostyryl)caffeine; and KF 17837, (E)7-methyl-8-(3,4-dimethoxystyryl)-1,3 dipropylxanthine.

have been cloned; however, their functional role will only become clear when selective drugs that distinguish between these sites are discovered.

With this background, it is clear that a key advance in understanding the function of the A_2 receptors, either the A_{2a} or A_{2b} subtype, could be made with the discovery of selective antagonists. Recently, some studies have revealed that important progress is being made in the development of specific A_{2a} antagonists of high pharmacologic interest. The emergence of xanthines that interact selectively with the A_{2a} receptors [4,5] and the results of our studies on new heterocycles that potently and selectively antagonize A_{2a} receptors have prompted us to review the current status of the A_{2a} antagonists.

Chemistry

While all known adenosine receptor agonists are closely related to adenosine itself, numerous structurally diverse antagonists have been identified [1]. Starting from the naturally occurring xanthines, caffeine (Fig. 28-1) and theophylline, a high number of xanthine derivatives have been studied extensively. Many of these compounds have been found to be selective for either A_1 or A_{2a} receptors. This provides the basis for distinguishing adenosine antagonists based on their chemical structures, whether xanthines or nonxanthine heterocycles.

XANTHINES

Several xanthine derivatives, most notably those substituted with 8-cycloalkyl groups, have been found to be potent and selective at the A_1 receptors [1]. Indeed, 8-cyclopentyl-1,3-dipropylxanthine, DPCPX (Fig. 28-1), having A_1 affinity in the low nanomolar range and A_1 selectivity of greater than 400-fold (Table 28-1), has become the reference A_1 antagonist for most biochemical and pharmacologic studies. More recently, it has been reported that the introduction of 8-styryl groups makes xanthines potent and selective at the A_{2a} receptor [4,5]. In particular, one compound, (E)7-methyl-8-(3,4-dimethoxystyryl)-1,3-dipropylxanthine, KF 17837 (Fig. 28-1), is reported to be a potent

TABLE 28-1. Affinity of selected adenosine antagonists in inhibiting A_1 and A_{2a} receptor binding in rat and bovine brain membranes

		Rat brain			Bovine brain		
Chemical class	Drug	A_1 Ki (nM)	A_2 Ki (nM)	A_1/A_2 Ratio	A_1 Ki (nM)	A_2 Ki (nM)	A_1/A_2 Ratio
Xanthine	DPCPX	1.5 (1.3–1.7)	706 (540–924)	0.002	0.1 (0.1–0.2)	123 (120–126)	0.001
	KF 17837	>1000	63 (47–85)	>15	237 (166–343)	74 (61–89)	3.2
Nonxanthine	CGS 15943	3.0 (2.2–4.2)	1.6 (1.2–2.1)	1.9	2.4 (2.1–2.8)	1.0 (0.8–1.1)	2.4
	8FB-PTP	3.3 (2.9–6.6)	1.2 (0.9–1.4)	2.8	2.0 (1.8–2.3)	0.07 (0.06–0.09)	29
	SCH 58261	121 (103–143)	2.3 (2.0–2.7)	53	201 (160–251)	2.0 (1.7–2.3)	101

Each value represents the geometric mean, with 95% confidence limits in parentheses, of at least four to six different experiments. Receptor affinity for A_1 binding sites was measured using ^{3}[H]-CHA in whole-brain (rat) or frontal cortex (bovine) homogenates. A_{2a} binding sites were tested using ^{3}[H]-CGS 21680 in brain striatal membranes.

A_{2a} antagonist [6]. Likewise, 8-(3-chlorostyryl) caffeine, CSC (Fig. 28-1), has been described as a selective A_{2a} antagonist of pharmacologic interest [7]. However, it is now clear that 8-styrylxanthines undergo rapid isomerization when diluted and exposed to light, the Z-isomer being less active than the E-isomer [8]. For example, the E-isomer KF 17837, which is stable in crystalline form, becomes an equilibrium mixture of E- and Z-isomers when diluted in dimethylsulphoxide and exposed to light in standard laboratory conditions [8,9]. This property, however, precludes the wide use of the 8-styrylxanthines as reference A_{2a} antagonists.

NONXANTHINE HETEROCYCLES

Among the number of chemical agents reported to antagonize A_{2a} receptors, there are a few compounds that have affinity in the low nanomolar range, namely, triazoloquinoxaline, 4-amino-8-chloro-1-phenyl-1,2,4-triazolo[4,3-a]quinoxaline, CP 66713 (Fig. 28-2), and triazoloquinazoline, 5-amino-9-chloro-2-(2-furyl)-1,2,4-triazolo[1,5-c]quinazoline, CGS 15943 (Fig. 28-2) [10]. The drug CGS 15943, which has been studied more extensively, has high affinity for both A_1 and A_{2a} receptors, and consequently it has little or no A_{2a} versus A_1 selectivity (Table 28-1). Moreover, CGS 15943 does not show competitive antagonistic properties in some biologic assays [9].

In the attempt to improve the A_{2a} selectivity of CGS 15943, a series of analogues was synthesized in which the phenyl group was replaced by a heterocyclic ring such as pyrazolo or imidazolo [11]. One of these compounds, 5-amino-8-(4-fluorobenzyl)-2-(2-furyl)-pyrazolo [4,3-e]-1,2,4-triazolo[1,5-c]-pyrimidine, 8FB-PTP (Fig. 28-2), displayed potent A_{2a} antagonistic properties in selected functional assays but still retained high affinity for the A_1 receptors (Table 28-1). Synthesis has also led to substantial optimization of A_{2a} selectivity [12]. Interestingly, one compound, namely, 7-(2-phenylethyl)-5-amino-2-(2-furyl) pyrazolo-[4,3-e]-1,2,4-triazolo [1,5-c] pyrimidine, SCH 58261, showed high affinity and good A_{2a} versus A_1 selectivity in both binding and functional assays [12].

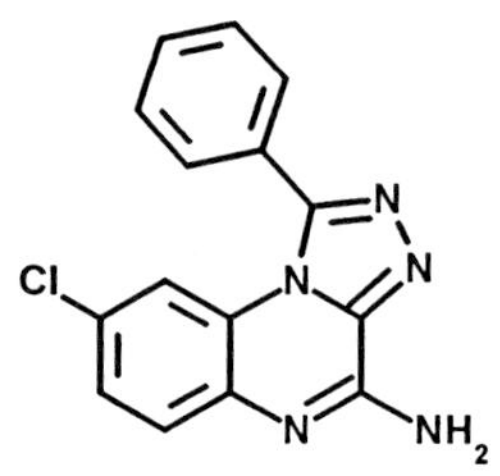

CP 66713

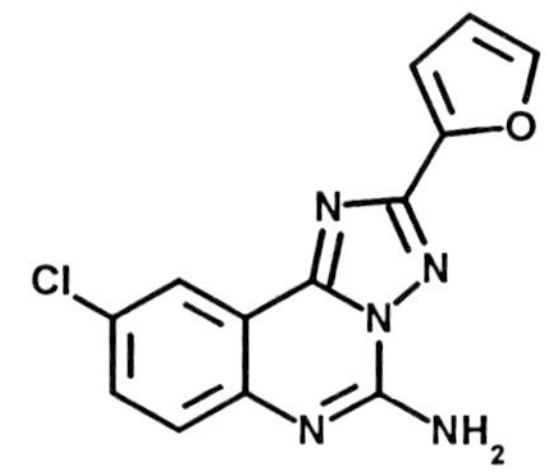

CGS 15943

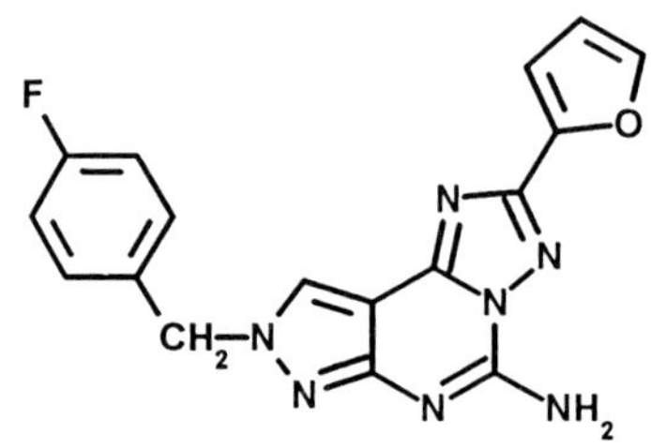

8FB-PTP

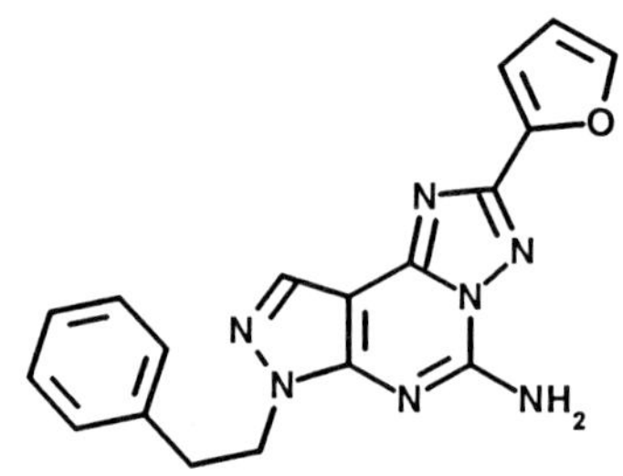

SCH 58261

FIGURE 28-2. Structures of nonxanthine heterocycles having either high affinity or some selectivity for the A_{2a} adenosine receptors. CP 66713, 4-amino-8-chloro-1-phenyl-1,2,4-triazolo[4,3-a] quinoxaline; CGS 15943, 5-amino-9-chloro-2-(2-furyl)1,2,4-triazolo[1,5-c]quinazoline; 8FB-PTP, 5-amino-8-(4-fluorobenzyl)-2-(2-furyl)-pyrazolo[4,3-e]-1,2,4-triazolo[1,5-c] pyrimidine; SCH 58261, 7-(2-phenylethyl)-5-amino-2-(2-furyl)-pyrazolo[4,3-e]-1,2,4-triazolo[1,5-c]pyrimidine.

Biology

RECEPTOR BINDING

Affinity for the A_{2a} receptors is commonly evaluated in membrane homogenates taken from brain striatum using the radioligand [^{3}H]CGS 21680. Selectivity is then calculated by comparing the A_{2a} affinity with that found for A_1 receptors, as measured in rat whole-brain or bovine frontal cortex homogenates using [^{3}H] cyclohexyladenosine ([^{3}H]CHA) as a radioligand. Since animal species show variability in terms of sensitivity to adenosine drugs and the number or distribution of adenosine receptors, it is important that drug selectivity be calculated within one animal species.

Studies made in our laboratory and shown in Table 28-1 indicate that the xanthine, KF 17837, does not have marked affinity for the A_{2a} receptor, and its selectivity is small in both rat and bovine brain membranes. These data are at variance with those originally described by Nonaka et al. [6] but in agreement with those described by Jacobson et al. [5]. Of the many 8-styrylxanthines synthesized [4,5], CSC (Fig. 28-1) was found to have a high selectivity (520-fold) in binding assays and a good affinity for the A_{2a} receptor (Ki, 54 nM).

Of the nonxanthine drugs, CGS 15943 was confirmed to possess high affinity for both A_{2a} and A_1 receptors (Table 28-1). The CGS 15943 derivative, 8FB-PTP, emerged as the most potent drug of those currently known showing an A_{2a} affinity in the subnanomolar range in bovine brain membranes (Ki, 0.07 nM). However, the drug also displayed high affinity for A_1 receptors (Table 28-1). Among the many compounds of this class [11], we have found the most interesting to be SCH 58261 (Fig. 28-2), which has high affinity for the A_{2a} receptors and a selectivity of 50- or 100-fold (Table 28-1) [12].

In general, it appears that nonxanthine heterocycles are characterized by very high affinity for the A_{2a} receptors with compounds that show a Ki around 1 nM. The weak selectivity of CGS 15943 and analogues has been overcome by the discovery of SCH 58261, whose selectivity is fairly good [12]. On the other hand, 8-styrylxanthines have certainly achieved an elevated selectivity as A_{2a} antagonists, but there are still controversial data on their potency [5]. More studies are therefore needed to define their pharmacologic profile more clearly.

Both potency and selectivity of the new A_{2a} antagonists available could be further clarified using cell lines expressing the different adenosine receptor subtypes. With such models it would also be possible to distinguish affinity at the A_{2a} versus A_{2b} receptors and to understand the affinity order profiles of the antagonists for the cloned human receptors.

FUNCTIONAL STUDIES

A variety of in vitro models indicate whether a chemical agent specifically blocks biologic responses mediated by the A_{2a} receptor system. Use of specific tissues such as platelets that contain A_{2a} receptors only [13] or vascular districts whose relaxation is modulated by A_{2a} receptors (Fig. 28-3) [14] makes it possible to examine both potency and the type of antagonism, that is, competitive or noncompetitive, of a given compound. Moreover, there are cell lines, such as PC 12 cells, which are sensitive to either stimulation or blockade of the A_{2a} receptors. Responses may vary depending on the model. For example, in platelets A_{2a} agonists inhibit aggregation [2,13] and stimulate the formation of the second messenger, cyclic AMP [7]. Either functional effect can be specifically blocked by A_{2a} antagonists. Selectivity is then explored by comparing activity in tissues that are sensitive only to A_1 receptors. For example, the beating rate of rat atria is inhibited dose dependently by A_1 agonists and is highly sensitive to A_1 receptor blockade (Fig. 28-4) [14].

In these models, the 8-styrylxanthine, KF 17837, was found to be inactive in rabbit platelet aggregation [9], bovine coronary arteries (Fig. 28-3), and rat aorta model [Conti, et al. unpublished data]. On the contrary, Nonaka et al. [6] have recently reported that KF 17837 potently antagonizes agonist-induced cyclic AMP formation in PC 12 cells. It is difficult to reconcile these contradictory findings, and further studies are necessary to elucidate the KF 17837 profile. The other xanthine, CSC, was found to reverse adenosine effects on adenylate cyclase in PC 12 cells and human platelets, and displayed selectivity when its action was compared with that produced in rat adipocytes that contain A_1 receptors [7]. More data in other functional assays are necessary in order to complete the CSC profile and to elucidate the nature of its antagonism.

The nonxanthine antagonist, CGS 15943, reversed both agonist-induced aggregation inhibition and adenylate cyclase stimulation in platelets [9,11] but was ineffective in other models, such as bovine coronary arteries (Fig. 28-3) and rat aorta [Conti, et al. unpublished data]. These data, together with other findings, explain why CGS 15943 has not become a reference A_{2a} antagonist. The drug 8FB-PTP is certainly of greater interest because it displays competitive antagonism in platelets [9], bovine coronary arteries (Fig. 28-3), and rat aorta [Conti, et al. unpublished data]. However, in keeping with its potency at A_1 receptors, the drug also antagonized the A_1-mediated response [9].

Recently a substantial advance has been achieved with the discovery of SCH 58261, which displays competitive A_{2a}-antagonistic properties (Fig. 28-3) similar to those of 8FB-PTP without interacting with A_1-mediated re-

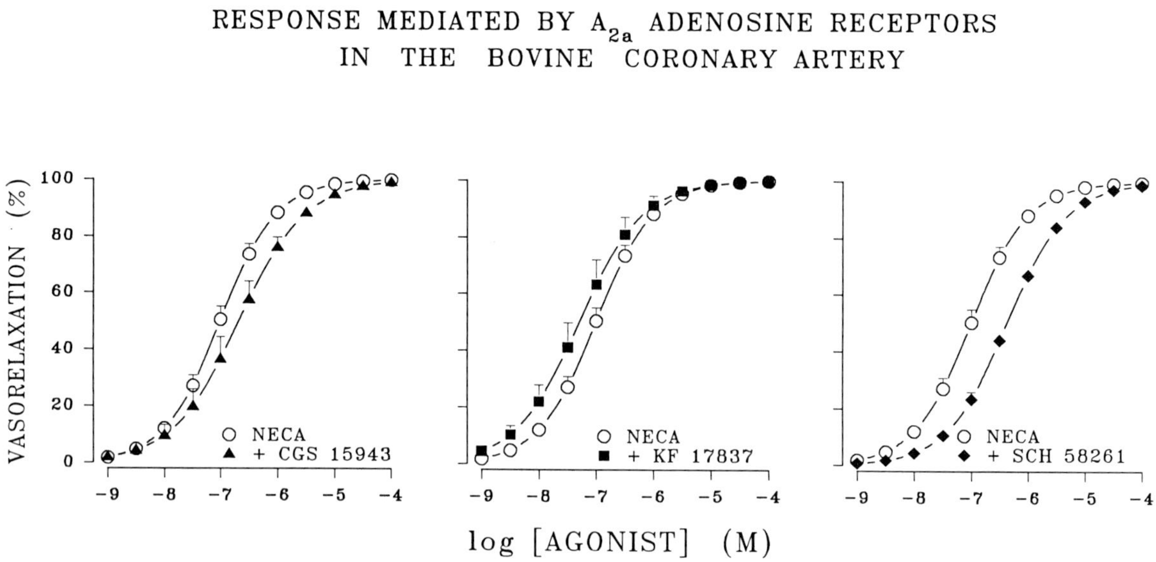

FIGURE 28-3. Effects of three different A_{2a}-adenosine antagonists on NECA-induced vasorelaxation in bovine coronary arteries. Each point represents the geometric mean, with 95% confidence limits, of at least seven replications. The dose-response curves were obtained with NECA alone (○) and in the presence of 100 nM of CGS 15943 (▲), KF 17837 (■), or SCH 58261 (◆). Note that only SCH 58261 produced parallel rightward shifts in the agonist concentration-response curve. NECA, 5′-N-ethylcarboxamidoadenosine.

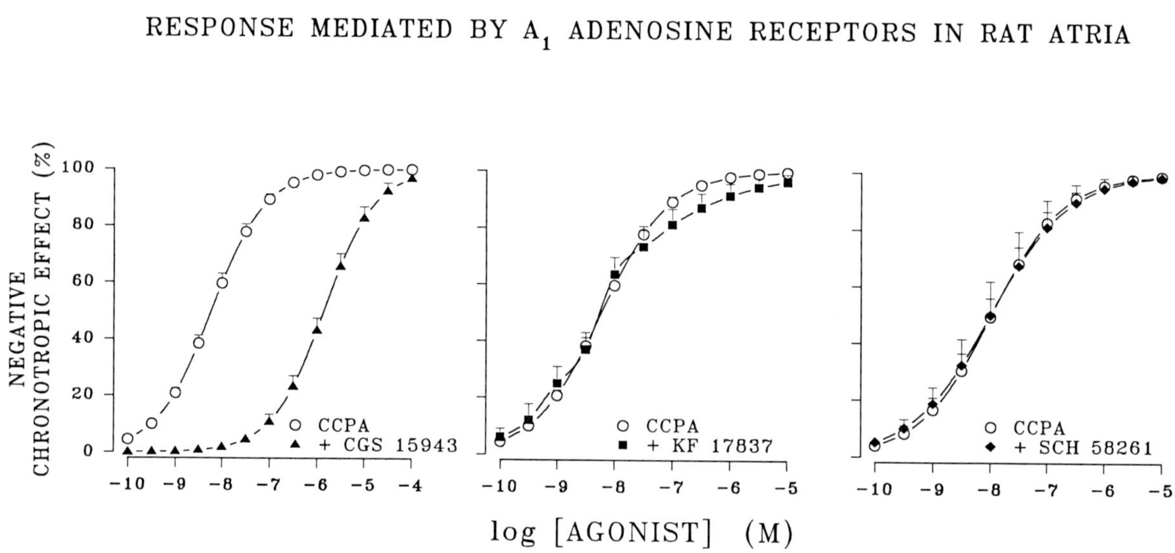

FIGURE 28-4. Effects of three different A_{2a}-adenosine antagonists on CCPA-induced negative chronotropic effects in rat atria. Each point represents the geometric mean, with 95% confidence limits, of at least seven replications. The dose-response curves were obtained with the A_1-agonist CCPA alone (○) and in the presence of 100 nM of CGS 15943 (▲), KF 17837 (■), or SCH 58261 (◆). Note that CGS 15943 potently antagonized this typical A_1-mediated response. Conversely, neither SCH 58261 nor KF 17837 affected the agonist-induced depression of spontaneously beating rate in rat atria. CCPA, 2-chloro-N^6-cyclopentyladenosine.

sponses (Fig. 28-4). Other studies now in progress will provide a more complete biologic profile of this interesting new A_{2a} antagonist.

Conclusions

In the 1980s an intensive research effort made by academia and industry led to the discovery and development of a number of selective A_{2a} agonists [1,2]. Expectations that valuable A_{2a} antagonists could also be developed, based on the early work of Williams et al. [10] with the potent but nonselective triazoloquinazoline CGS 15943, were not realized for several years. Only recently has there been a significant step forward with the discovery that xanthines having appropriate substitutions, such as the 8-styryl groups, possess the profile of selective A_{2a} antagonists [4,5]. Moreover, it has emerged that further optimization of CGS 15943 leads to interesting compounds, such as 8FB-PTP [9] or SCH 58261 [12], which combine high potency and competitive antagonism at A_{2a} receptors.

Further studies now in progress within the community of researchers on adenosine will determine whether the two classes of drugs meet all criteria of potency, competitive antagonism, and selectivity, for example, A_{2a} versus A_{2b}, which could allow one or more compounds to become reference A_{2a} antagonists. This would be a major breakthrough in the field and would stimulate the search for a functional role of the A_{2a} receptor in physiologic and pathologic states. Moreover, full characterization of a selected A_{2a} antagonist would delineate future prospects for these drugs in the treatment of neurologic disorders, such as Parkinson's disease, for which a scientific rationale exists [15].

References

1. Jacobson KA, Van Galen PJM, Williams M (1992) Adenosine receptors: Pharmacology, structure-activity relationships, and therapeutic potential. J Med Chem 35:407–422.
2. Cristalli G, Volpini R, Vittori S, Camaioni E, Monopoli A, Conti A, Dionisotti S, Zocchi C, Ongini E (1994) 2-Alkynyl derivatives of adenosine-5′-N-ethyluronamide: Selective A_2 adenosine receptor agonists with potent inhibitory activity on platelet aggregation. J Med Chem 37:1720–1726.
3. Collis MG, Hourani SMD (1993) Adenosine receptor subtypes. Trends Pharmacol Sci 14: 360–366.
4. Shimada J, Suzuki F, Nonaka H, Ishii A, Ichikawa S (1992) (E)-1,3-dialkyl-7-methyl-8-(3,4,5-trimethoxy-styryl) xanthines: Potent and selective adenosine A2 antagonists. J Med Chem 35:2342–2345.
5. Jacobson KA, Gallo-Rodriguez C, Melman N, Fisher B, Maillard M, Van Bergen A, Van Galen PJM, Karton Y (1993) Structure-activity relationships of 8-styrylxanthines as A2-selective adenosine antagonists. J Med Chem 36:1333–1342.
6. Nonaka H, Ichimura M, Takeda M, Nonaka Y, Shimada J, Suzuchi F, Yamaguchi K, Kase H (1994) KF17837 ((E)-8-(3,4-dimethoxystyryl)-1,3-dipropyl-7-methylxanthine), a potent and selective A_2 receptor antagonist. Eur J Pharmacol Mol Pharm Section 267:335–341.
7. Jacobson KA, Nikodijevic O, Padgett WL, Gallo-Rodriguez C, Maillard M, Daly JW (1993) 8-(3-Chlorostyryl)caffeine (CSC) is a selective A2-adenosine antagonist in vitro and in vivo. FEBS Lett 323:141–144.
8. Nonaka Y, Shimada J, Nonaka H, Koike N, Aoki N, Kobayashi H, Kase H, Yamaguchi K, Suzuki F (1993) Photoisomerization of a potent and selective adenosine A2 antagonist, (E)-1,3-dipropyl-8-(3,4-dimethoxystyryl)-7-methlxantine. J Med Chem 36:3731–3733.
9. Dionisotti S, Conti A, Sandoli D, Zocchi C, Gatta F, Ongini E (1994) Effects of the new A2 adenosine receptor antagonist 8FB-PTP, an 8-substituted pyrazolo-triazolo-pyrimidine, on in vitro functional models. Br J Pharmacol 112:659–665.
10. Williams M, Francis J, Ghai G, Braunwalder A, Psychoyos S, Stone GA Cash WD (1987) Biochemical characterization of the triazoloquinazoline, CGS 15943, a novel, non-xanthine adenosine antagonist. J Pharmacol Exp Ther 241:415–420.
11. Gatta F, Del Giudice MR, Borioni A, Borea PA, Dionisotti S, Ongini E (1993) Synthesis of imidazo [1,2-c]pyrazolo[4,3-e]pyrimidines, pyrazolo[4,3-e]1,2,4-triazolo [1,5-c]pyrimidines and triazolo[5,1-i]purines: New potent

A2 adenosine receptor antagonists. Eur J Med Chem 28:569–576.

12. Baraldi PG, Manfredini S, Simoni D, Zappaterra L, Zocchi C, Dionisotti S, Ongini E (1994) Synthesis and activity of a new pyrazolo [4-3-e]-1,2,4-triazolo [1,5-c] pyrimidine displaying potent and selective activity as A_{2a} adenosine receptor antagonists. Bioorg Med Chem Lett 4:2539–2544.
13. Dionisotti S, Zocchi C, Varani K, Borea PA, Ongini E (1992) Effects of adenosine derivatives on human and rabbit platelet aggregation. Correlation of adenosine receptor affinities and antiaggregatory activity. Naunyn Schmiedebergs Arch Pharmacol 346:673–676.
14. Conti A, Monopoli A, Gamba M, Borea PA, Ongini E (1993) Effects of selective A1 and A2 adenosine receptor agonists on cardiovascular tissues. Naunyn Schmiedebergs Arch Pharmacol 348:108–112.
15. Fuxe K, Ferrè S, Snaprud P, Von Euler G, Johansson B, Fredholm B (1993) Antagonistic A_{2a}/D_2 receptor interaction in the striatum as a basis for adenosine/dopamine interactions in the central nervous system. Drug Dev Res 28:374–380.

C. NERVOUS SYSTEM (CENTRAL AND PERIPHERAL)

29. ROLE OF A_{2a} RECEPTORS IN THE HIPPOCAMPUS AND MOTOR NERVE ENDINGS

Ana M. Sebastião
Rodrigo A. Cunha
Paulo Correia-de-Sá
Alexandre de Mendonça
J. Alexandre Ribeiro

The pioneering work by Ginsborg and Hirst [1] on the effects of adenosine on neurotransmitter release from motor nerve terminals was prompted by the observation of Sattin and Rall [2] that adenosine increases cyclic AMP in brain slices and that increases in intracellular cyclic AMP can facilitate neurotransmitter release [3]. In contrast to what was expected, exogenously applied adenosine *decreased*, rather than increased, neurotransmitter release from the motor nerve endings; adenine nucleotides mimic this inhibitory effect of adenosine [4,5]. Twenty years elapsed before showing that some adenosine analogues (A_2 agonists) could *increase* neurotransmitter release from motor nerve endings [6,7].

The neuromuscular junction has proved to be a useful model to study presynaptic inhibitory effects of adenosine [8]. It possesses mechanisms for the formation of extracellular adenosine, its action, and its inactivation. Thus, adenosine [9] as well as adenine nucleotides [10] are released from motor nerve endings, and extracellular catabolism of adenine nucleotides through ecto nucleotidases [11] contributes to the formation of endogenous adenosine (Fig. 29-1), which tonically inhibits neurotransmitter release [12]. Independent release of adenine nucleotides and adenosine from the muscle fiber also occurs [9]. Inactivation of adenosine occurs through uptake at the frog neuromuscular junction [12], whereas at the rat neuromuscular junction extracellular deamination of adenosine might also contribute to inactivate adenosine [13]. The inhibitory adenosine receptor at the rat motor nerve terminals belongs to the A_1 subtype, being pharmacologically similar to the inhibitory adenosine receptor at the hippocampus [14]. The "coincidence" between the way adenosine presynapticaly inhibits transmission at excitatory synapses in the hippocampus and at the neuromuscular junction was also recently emphasized [15]. In this chapter evidence will be provided demonstrating that, as occurs with the inhibitory actions of adenosine, the behavior of the neuromuscular junction and of hippocampus seem to "coincide" in relation to the excitatory effects of adenosine analogues.

Role of Excitatory A_{2a} Adenosine Receptors at Motor Nerve Terminals

The adenosine analogue, CGS 21680, at low nanomolar concentrations (2–10 nM) increases the amplitude of evoked endplate potentials recorded from phrenic-diaphragm preparations and increases the evoked release of [^{3}H]acetylcholine (ACh) from phrenic nerve terminals [7]. At these concentrations, CGS 21680 is selective for A_2 adenosine receptors [16] and does not bind to A_{2b} adenosine receptors [17]; these results suggest that the excitatory effect of CGS 21680 at the neuromuscular junction is mediated by A_{2a}-adenosine receptors. Accordingly, the excitatory effect of CGS 21680 at the rat motor nerve terminals is antagonized by the A_2/A_1 antagonist, PD 115199 [7], whereas the A_1 antagonist, 1,3-dipropyl-8-cyclopentylxanthine (DPCPX, 2.5 nM), even enhanced the excitatory effect of CGS 21680 (Fig. 29-2). Both the inhibitory effect of the A_1 agonist, R-N^6-phenylisopropyladenosine (R-PIA), and the excitatory effect of CGS 21680 can be observed in the same endplate (Fig. 29-3), indicating that excitatory and inhibitory adenosine receptors coexist in the same motor nerve terminal.

Because exogenously applied adenosine, even at very high concentrations [1,5], consistently inhibits ACh output from motor nerve endings, and because A_2 agonists facilitate ACh release, a crucial question is whether A_2 adenosine receptors have a physiologic role in neurotransmitter release. We had the opportunity to answer this question with the use of the neuropeptide, calcitonin gene-related peptide (CGRP), which is present in the mammalian nerve terminals [18] and increases the evoked release of [^{3}H]ACh from phrenic nerve terminals [19]. This excitatory effect of CGRP is prevented by adenosine deaminase [19], which indicates that the presence of endogenous extracellular adenosine is needed for the enhancement of the ACh release caused by CGRP. That this action of endogenous adenosine is a consequence of activation of A_{2a} receptors could be concluded [19] from the findings that 1) Blockade of the effect of CGRP by adenosine deaminase is no longer observed if CGS 21680 (2 nM) is added together with adenosine deaminase; and 2) PD 115199, but not DPCPX, prevented the excitatory action of CGRP. It thus appears that tonic activation of A_{2a} receptors by endogenous adenosine plays an

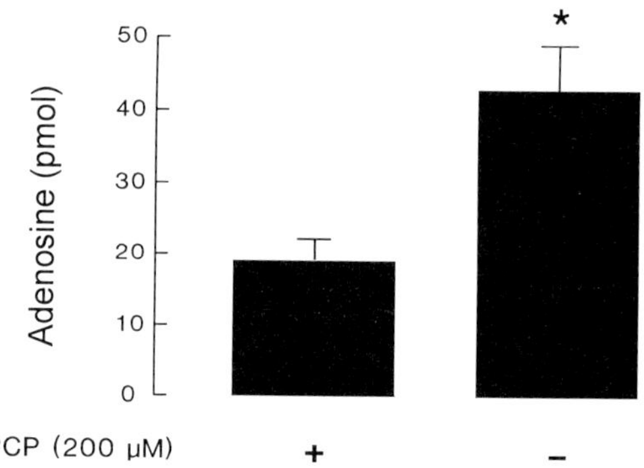

FIGURE 29-1. Released adenine nucleotides as a source of extracellular adenosine at the innervated frog sartorius muscle. The ordinates represent the adenosine levels detected in the bath (2 ml) 30 minutes after starting the electric stimulation of the nerve (0.2 Hz); the adenosine levels detected in the bath without stimulation were subtracted. Throughout the experiments, adenosine uptake was inhibited with dipyridamole (0.5 μM), and exo-AMP deaminase was blocked with coformycin (200 μM) [9]. The left column (n = 18) represents the adenosine levels when ecto-5′-nucleotidase, the enzyme responsible for the last step in the hydrolysis of adenine nucleotides into adenosine, was blocked with a supramaximal concentration (200 μM) of α,β-methylene ADP (AOPCP). The right column (n = 4) represents the adenosine levels in the absence of AOPCP. *$p < 0.05$ when compared with the left column. Note that when adenine nucleotides could be hydrolyzed into adenosine, the adenosine levels were near twice those detected upon blockade of ecto-5′-nucleotidase.

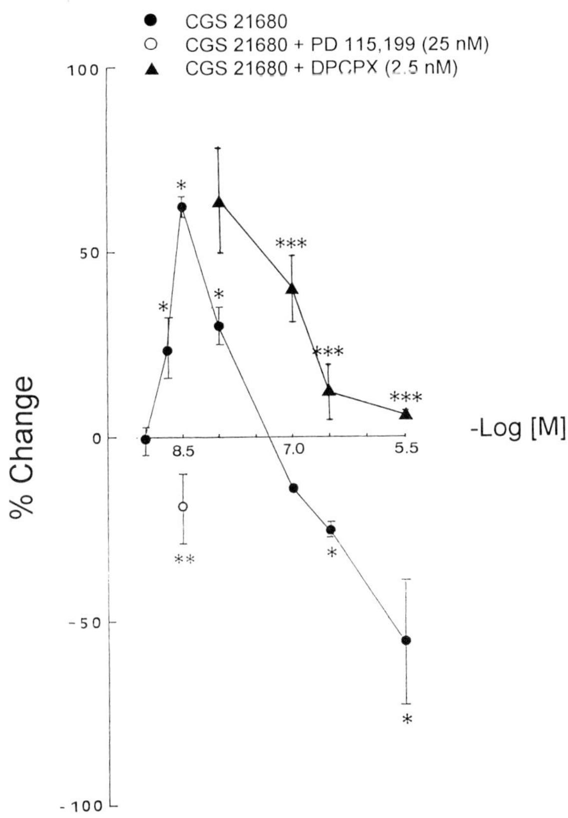

FIGURE 29-2. The excitatory effect of the A_{2a} agonist, CGS 21680, on the electrically evoked release of [^{3}H]ACh from the rat phrenic nerve endings is antagonized by the A_2/A_1 antagonist, PD 115, 199, and is enhanced by the A_1 antagonist, DPCPX. The ordinates represent the percentage change in the evoked release of [^{3}H]ACh caused by CGS 21680, and the abscissae represent the concentrations of CGS 21680 [7]. Each point is the average of two to four experiments. *p < 0.05 when compared with zero percent; **p < 0.05 when compared with the same concentration of CGS 21680 in the absence of PD 115,199; ***p < 0.05 when compared with the same concentration of CGS 21680 in the absence of DPCPX. Note that, in contrast to the excitatory effect, the inhibitory effect of high concentrations (>0.1 μM) of CGS 21680 was antagonized by DPCPX.

important role in motor nerve terminals, acting as a primer for the enhancement of ACh release by other excitatory neuromodulators such as CGRP.

Both CGRP [20] and CGS 21680 [21] increase ACh release by a cyclic-AMP–dependent pathway. These observations, taken together with the finding that the enhancement of ACh release caused by the adenylate cyclase activator, forskolin, also depends on tonic activation of A_{2a}-adenosine receptors [22], suggest that the cooperation between presynaptic A_{2a} receptors and CGRP receptors occurs at the transducing system level. One is then tempted to speculate that similar interactions might also occur between A_{2a}-adenosine receptors and other receptors positively linked to adenylate cyclase.

Other roles for the A_{2a}-adenosine receptors present in motor nerve terminals have been advanced, namely, these receptors regulate nicotinic autofacilitation of ACh release by favoring the desensitization of the presynaptic facilitatory nicotinic receptors [23]. Furthermore, activation of presynaptic A_2 receptors by endogenous adenosine seems to be particularly important when nerve terminals are strongly stimulated (1 ms pulses delivered at high frequency); under these conditions high concentrations of adenosine might accumulate around the nerve terminal, and endogenous adenosine *increases* ACh output [24]. Since the release of adenine nucleotides from nerve endings is proportional to the frequency of nerve stimulation [9], the reported [25] association between ecto-5′-nucleotidase activity and excitatory A_2-adenosine receptors in rat phrenic nerve endings may constitute an efficient mechanism for the delivery of high concentrations of adenosine near excitatory A_2 receptors.

Adenosine and the Cholinergic System in the Hippocampus

In the hippocampus, endogenous adenosine tonically inhibits the evoked ACh release by operating an A_1-adenosine receptor [26]. As at the neuromuscular junction with low frequencies of stimulation [9,12], in the hippocampus [27] the presynaptic release of adenine nucleotides, as well as the presynaptic release of adenosine as such, contribute in near equal amounts for the formation of the endogenous extracellular adenosine that inhibits ACh release.

Immunopurified cholinergic nerve terminals of the rat hippocampus are highly enriched in ecto-5′-nucleotidase activity, whereas the cholinergic nerve terminals in the cerebral cortex lack ecto-5′-nucleotidase activity [28]. It is interesting to note that in the hippocampus [27], as at the neuromuscular junction [12,29], ATP has to be hydrolyzed into adenosine before exerting its inhibitory effect on ACh release, whereas in the cerebral cortex ATP itself can inhibit the evoked release of ACh [27]. This suggests that when the enzymatic machinery to form adenosine from released adenine nucleotides is missing, ATP itself can play a role that in most cases is exclusive of adenosine. It would be interesting to know whether noradrenergic nerve terminals in the cerebral cortex also lack ecto-5′-nucleotidase activity, because ATP itself also inhibits noradrenaline release in this brain area [30]. The release of ACh [27] and noradrenaline [30] in the cerebral cortex is also inhibited by adenosine, and both the effects of ATP and of adenosine are antagonized by DPCPX. Whether the adenosine receptor is sensitive to the intact molecule of ATP or ATP acts on a putative xanthine-sensitive nucleotide receptor in this brain area [31] needs to be investigated.

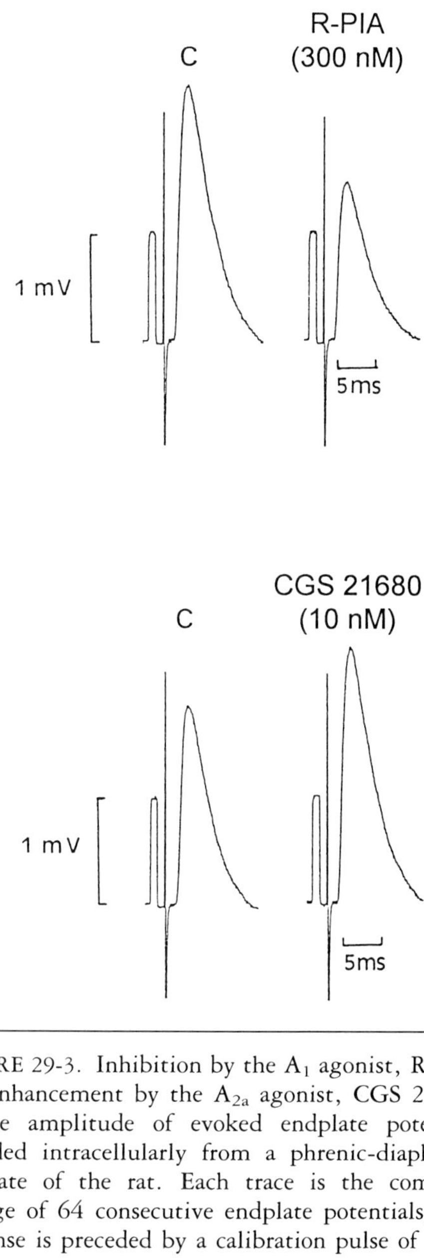

FIGURE 29-3. Inhibition by the A_1 agonist, R-PIA, and enhancement by the A_{2a} agonist, CGS 21680, of the amplitude of evoked endplate potentials recorded intracellularly from a phrenic-diaphragm endplate of the rat. Each trace is the computer average of 64 consecutive endplate potentials; each response is preceded by a calibration pulse of 1-mV amplitude and 1-ms duration and by the stimulus artifact [7]. The resting membrane potential was ≈ -83 mV. C = control response before applying R-PIA (**upper panel**) or CGS 21680 (**lower panel**). The quantal content of the evoked endplate potentials was decreased by R-PIA and increased by CGS 21680, indicating that both substances affect, in opposite ways, transmitter release from the same nerve terminal.

Enhancement of ACh Release in the Hippocampus by A_{2a} Adenosine Receptor Activation

As recently shown by Cunha et al. [32], low nanomolar concentrations of the A_{2a} agonist, CGS 21680, enhance the electrically evoked release of ACh from the CA3 and dentate gyrus areas of the rat hippocampus, but not from the CA1 area. In contrast, the mixed A_1/A_2 agonist, 2-chloroadenosine (CADO), inhibits ACh release from the three hippocampal areas. Blockade of A_1 receptors with DPCPX unmasks the ability of CADO to enhance the evoked release of ACh in the CA3 and dentate gyrus areas. As occurs with CGS 21680, CADO does not enhance ACh release in the CA1 area, even when applied in the presence of DPCPX. Taken together, these observations suggest that excitatory A_{2a}-adenosine receptors are involved

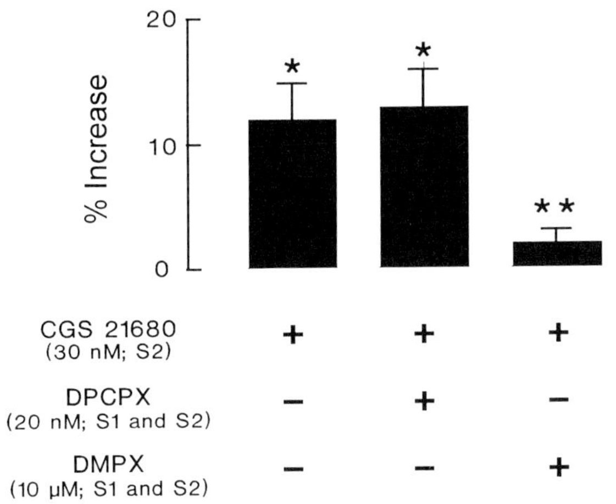

FIGURE 29-4. The excitatory effect of the A_{2a} agonist, CGS 21680, on the evoked release of $[^3H]$ACh from synaptosomes of the rat hippocampus is antagonized by the A_2 antagonist, DMPX, but not by the A_1 antagonist, DPCPX. The release of $[^3H]$ACh was evoked with veratridine (10 μM), which was applied twice (S1 and S2) (see Cunha et al. [27] for methodologic details). CGS 21680 was present during S2, and its effect on $[^3H]$ACh release was quantified as changes in the S2/S1 ratio as compared with the control; DPCPX and DMPX were present throughout the experiment. The ordinates represent the percentage enhancement of $[^3H]$ACh release caused by CGS 21680 in the absence of antagonists (left column, n = 6), in the presence of DPCPX (middle column, n = 5), or in the presence of DMPX (right column, n = 6). $^*p < 0.05$ when compared with zero percent; $^{**}p < 0.05$ when compared with the left column.

in the modulation of ACh release in the CA3 and dentate gyrus areas, but not in the CA1 area [31]. The absence of excitatory A_{2a}-adenosine receptors modulating ACh release in the CA1 area might be responsible for the inability to detect excitatory effects of CGS 21680 on ACh release in whole hippocampal slices [32,33]. However, it has been reported that 5′-N-ethylcarboxamide adenosine (NECA) in the presence of DPCPX can increase ACh release from whole hippocampal slices [34]. In cerebral cortical slices, NECA has been shown to increase ACh release, even in the absence of blockade of A_1 adenosine receptors [35].

The CGS 21680-sensitive adenosine receptors involved in enhancement of ACh release in the hippocampus are probably located in cholinergic nerve terminals, since CGS 21680 also increases the evoked release of ACh from hippocampal synaptosomes. This effect of CGS 21680 is antagonized by DMPX but not by DPCPX (Fig. 29-4), suggesting an A_{2a}-mediated effect.

In the CA3 area, endogenous adenosine might simultaneously inhibit and enhance ACh release, because removal of endogenous adenosine by adenosine deaminase causes a smaller increase in the evoked release of ACh than blockade of A_1 receptors with DPCPX [32]. In the CA1 area, adenosine deaminase and DPCPX cause a similar increase in the evoked release of ACh, indicating that in this hippocampal area endogenous adenosine only activates A_1 receptors and supporting the conclusion that A_{2a} receptors are not involved in the modulation of ACh release in the CA1 area [32]. Surprisingly, in the dentate gyrus neither DPCPX nor adenosine deaminase affect ACh release. This suggests that endogenous extracellular adenosine levels in this area are probably not enough to activate either the A_1- or the A_2-adenosine receptors.

In conclusion, these differences in the modulation of ACh release by adenosine according to the different areas of the hippocampus suggest that adenosine could in subtle ways modulate the release of transmitters in different areas of the brain. This could be of particular relevance to conceptualizing adenosine-related molecules as being of therapeutic interest.

Enhancement of Neuronal Excitability and Synaptic Transmission in the CA1 Area by A_{2a}-Adenosine Receptor Activation

Evidence for the presence of A_{2a}-adenosine receptors modulating neuronal excitability in the CA1 area of the hippocampus came from the observation that CGS 21680 (3–30 nM) increases the amplitude of pupulation spikes recorded from CA1 pyramidal cells upon stimulation of the Schaffer collaterals, an effect antagonized by DMPX (10 μM) [36]. The amplitude and slope of field excitatory post-synaptic potentials (fEPSPs) recorded from the CA1 stratum radiatum are also increased by CGS 21680 [37], which indicates that A_{2a}-adenosine receptor activation enhances synaptic transmission in the hippocampus. The recent observations that A_{2a}-mRNA is expressed in the rat hippocampus and that there are [^{3}H]CGS 21680 (30 nM) binding sites in rat hippocampal slices [37] constitute molecular evidence for the presence of A_{2a}-adenosine receptors in the hippocampus. A study on the binding of [^{3}H]CGS 21680 to rat hippocampal membranes also revealed the presence of a CGS 21680 binding site, which is pharmacologically similar to the A_{2a}-adenosine receptor in the striatum [38]. However, in this study a high-affinity CGS 21680 binding site in the hippocampus that is distinct from the typical striatal A_{2a} receptor was also detected. This atypical high-affinity CGS 21680 binding site is also present in low amounts in the striatum, and one of its most striking characteristics is its nanomolar affinity for DPCPX [38]. The characteristics and functional relevance of this binding site need to be further elucidated.

A_{2a}- and A_1-adenosine receptors are coexpressed and colocalized in the rat hippocampus [37]. The functional relevance of this colocalization is illustrated by the observation that activation of A_{2a} receptors by CGS 21680 attenuates the ability of the A_1 agonist, N^6-cyclopentyladenosine, to inhibit hippocampal excitability [37]. It thus appears that an A_{2a}-dependent blockade of A_1-mediated inhibition of neuronal excitability occurs in the hippocampus.

Adenosine, at low nanomolar concentrations, has been shown to slowly increase synaptic transmission and glutamate release in the hippocampus [39]. Whether the excitatory effects of CGS 21680 [36] and adenosine [39] are mediated by the same adenosine receptor remains to be established. CGS 21680 enhances the release of glutamate and aspartate from the cerebral cortex under ischemia [40]. That A_{2a}-receptor activation by endogenous adenosine might enhance glutamate and aspartate release and/or reverse the A_1-mediated inhibition is suggested by the observation that the A_{2a} antagonist, CGS 15943A, inhibits the release of those excitatory amino acids in the ischemic cerebral cortex [41].

Synaptic transmission in the hippocampus can be presynapticaly affected either by an increase in the release of excitatory aminoacids or by a decrease in the release of the inhibitory transmitter, gamma-aminobutyric acid (GABA). No information is available on the effect of CGS 21680 on GABA release in the hippocampus. However, in other brain areas, such as globus pallidus [42], cerebral cortex [43], or striatum [44,45], A_{2a}-adenosine receptor activation by CGS 21680 has been shown to either increase [42,45] or decrease [43,44] GABA release. Because activation of A_{2a}-adenosine receptors decreases the D_2 (dopamine)-receptor–mediated actions [46], and because activation of D_2 receptors inhibits GABA release in the striatum and globus pallidus [45], the enhancement of GABA release in these brain areas might be a consequence of the A_{2a}-mediated blockade of D_2-dopamine–induced inhibition of GABA release in slices of the globus pallidus [42] and striatum [45]. These receptor-receptor interactions might be greatly attenuated in synaptosomal preparations where CGS 21680 has been reported to decrease GABA release [44].

Besides modulating neurotransmitter release, A_2 receptors might also affect the electric properties of pyramidal neurons; in the presence of DPCPX, both R-PIA and NECA have been shown to cause membrane depolarization and to increase membrane resistance [47]. It is not known whether these excitatory non-A_1-adenosine receptors belong to the A_{2a} or A_{2b} subtypes. Excitatory, probably A_{2b}, adenosine receptors responsible for modulation of neuronal excitability might also be present in the CA3 area of the hippocampus [48]; activation of these receptors enhances P-type, but not N-type, calcium current, an effect probably mediated through activation of the cyclic-AMP–dependent protein kinase.

Modulation of Long-Term Potentiation by Activation of Inhibitory and Excitatory Adenosine Receptors

The adenosine analogue, CADO, decreases frequency-induced long-term potentiation (LTP) of the responses evoked by stimulation of the Schaffer fibers and recorded in the CA1 area of hippocampal slices [49]. The adenosine uptake blocker, nitrobenzylthioinosine, reduces LTP, whereas the A_1 antagonist, DPCPX, facilitates LTP, which suggest that endogenous adenosine by activating A_1 receptors in the hippocampus exerts a tonic inhibitory role on LTP [50]. Since induction of LTP is dependent on the activation of N-methyl-D-aspartate (NMDA) receptors [51], it is of interest that CADO inhibits the NMDA receptor-mediated component of the fEPSPs more efficiently than it inhibits the fEPSPs [52]. DPCPX enhances the NMDA-receptor–mediated component of the fEPSPs, suggesting that endogenous adenosine, by activating A_1 receptors, tonically inhibits the NMDA-receptor–mediated responses. It thus appears that the inhibitory effect of adenosine receptor agonists on LTP might be due, at least in part, to a selective inhibition of NMDA receptors, rather than being an exclusive consequence of A_1-receptor–mediated inhibition of synaptic transmission.

The inhibition of NMDA responses by adenosine, besides affecting mechanisms of synaptic plasticity, such as LTP, might also contribute to the adenosine-induced protection against hypoxia/ischemia. In fact, the hypoxia-induced inhibition of the NMDA-receptor–mediated fEPSPs is partially reversed by DPCPX [53], indicating that hypoxia-induced inhibition of NMDA responses is partially due to the release of adenosine during hypoxia.

On the other hand, the A_{2a} agonist, CGS 21680, enhances LTP of population spikes recorded in the CA1 area of hippocampal slices (Fig. 29-5), which suggests that excitatory A_{2a} adenosine receptors are involved in modulation of synaptic plasticity [50]. That endogenous adenosine might enhance LTP by activating A_2 receptors was suggested on the grounds that the A_2 antagonist CP 66713 prevents LTP of fEPSPs, though this antagonist does not affect the LTP of population spikes [54].

Conclusions

The A_{2a} adenosine receptors, as they were defined in the striatum, are receptors with low nanomolar affinity for CGS 21680. In this context, the data summarized show that A_{2a}-adenosine receptors are present in motor nerve endings and in the hippocampus. Activation of these receptors enhances ACh release, synaptic transmission, neuronal excitability, as well as phenomena of synaptic plasticity, such as LTP. There is positive cooperativity between A_{2a} adenosine receptors and other receptors involved in enhancement of neurotransmitter release; this is the case of CGRP receptors present in motor nerve terminals, which became inactive in the absence of activation of the presynaptic excitatory A_{2a} receptors, either by endogenous adenosine or by exogenous agonists. Also contributing to the enhancement of synaptic trans-

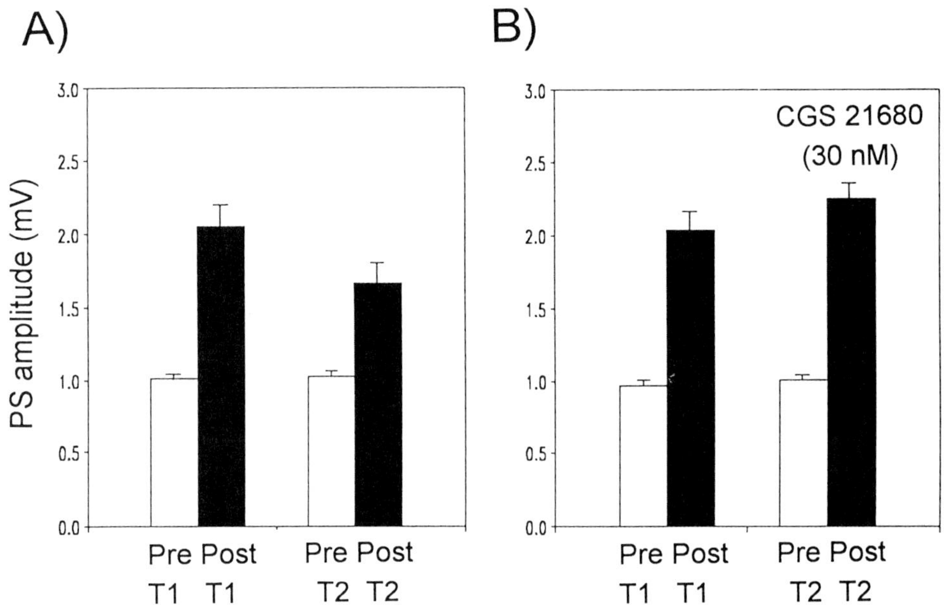

FIGURE 29-5. The A_{2a} agonist, CGS 21680, enhances long-term potentiation (LTP). The ordinates represent the averaged amplitude of population spikes (PS) recorded from the CA1 pyramidal layer of rat hippocampal slices. The open columns (Pre T) represent the averaged PS amplitude during the 8 minutes that preceded the application of a submaximal high-frequency stimulation train (100 Hz for 100 ms); the filled columns (Post T) represent the averaged amplitude of PS taken between the 20th and 28th minute after the train. Two high-frequency stimulation trains (T1 and T2) were applied to the same slice [50]. **A**: Control slices (n = 5). **B**: test slices (n = 5), in which the second train [T2] was applied in the presence of CGS 21680. Note that in the control slices LTP induced by T2 (LTP2) was smaller than that induced by T1 (LTP1), whereas in the test slices LTP2 (in the presence of CGS 21680) was greater than LTP1. The ratios LTP2/LTP1 in the test slices were significantly different ($p < 0.05$) from the ratios LTP2/LTP1 in the control slices.

mission is the ability of CGS 21680-sensitive receptors to attenuate the inhibition of neuronal excitability mediated by A_1 adenosine receptors in the hippocampus. In agreement with the idea that A_1 and A_{2a} adenosine receptors interact in the hippocampus is the finding that there is colocalization of these receptors, perhaps in the same nerve terminals as occurs in motor nerve endings, as well as coexpression of mRNA coding for A_1 and for A_{2a} adenosine receptors in the hippocampus.

In summary, information is growing that indicates CGS 21680-sensitive (A_{2a}) adenosine receptors are not exclusively located in the basal ganglia, as was originally proposed. The development of antagonists with great selectivity and affinity for A_{2a} receptors is needed to further elucidate the role of these receptors as well as the ability of endogenous adenosine to activate them.

Acknowledgments

This work was partially supported by JNICT grants (PMTC/C/SAU/792/90 and STRDA/C/SAU/292/92). R.A. Cunha (BD/148-90/ID) and P. Correia-de-Sá (BD/721-90/ID) were in receipt of JNICT studentships. The authors are involved in the concerted action ADEURO (EU, Biomed 1).

References

1. Ginsborg BL, Hirst GDS (1972) The effect of adenosine on the release of the transmitter from the phrenic nerve of the rat. J Physiol (Lond) 224:629–645.
2. Sattin A, Rall TW (1970) The effect of adenosine and adenine nucleotides on the cyclic adenosine 3′,5′-phosphate content of guinea pig cerebral cortex slices. Mol Pharmacol 6:13–23.
3. Goldberg AJ, Singer JJ (1969) Evidence for a role of cyclic AMP in neuromuscular transmission. Proc Natl Acad Sci USA 64:134–141.
4. Ribeiro JA, Walker J (1973) Action of adenosine triphosphate on endplate potentials recorded from muscle fibres of the rat-diaphragm and frog sartorius. Br J Pharmacol 49:724–725.
5. Ribeiro JA, Walker J (1975) The effects of adenosine triphosphate and adenosine diphosphate on transmission at the rat and frog neuromuscular junctions. Br J Pharmacol 54: 213–218.
6. Ribeiro JA, Correia-de-Sá P, Cunha RA, Oliveira JC, Sebastião AM (1991) Role of adenosine at the neuromuscular junction: Effects mediated by presynaptic inhibitory and excitatory receptors. In Imai S, Nakazawa M (eds.), Role of Adenosine and Adenine Nucleotides in the Biological System. Amsterdam: Elsevier Science, pp. 691–702.
7. Correia-de-Sá P, Sebastião AM, Ribeiro JA (1991) Inhibitory and excitatory effects of adenosine receptor agonists on evoked transmitter release from phrenic nerve endings of the rat. Br J Pharmacol 103:1614–1620.
8. Ribeiro JA, Sebastião AM (1991) Purinergic modulation of neurotransmitter release in the peripheral and central synapses. In Feigenbaum J, Hanani M (eds.), Presynaptic Regulation of Neurotransmitter Release, A Handbook. London: Freund Publishing House, Ltd., pp. 451–495.
9. Cunha RA, Sebastião AM (1993) Adenosine and adenine nucleotides are independently released from both the nerve terminals and the muscle fibres upon electrical stimulation of the innervated skeletal muscle of the frog. Pflügers Arch 424:503–510.
10. Silinsky EM (1975) On the association between transmitter secretion and the release of adenine nucleotides from mammalian motor nerve terminals. J Physiol (Lond) 432:343–354.
11. Cunha RA, Sebastião AM (1991) Extracellular metabolism of adenine nucleotides and adenosine in the innervated skeletal muscle of the frog. Eur J Pharmacol 197:83–92.
12. Ribeiro JA, Sebastião AM (1987) On the role, inactivation and origin of endogenous adenosine at the frog neuromuscular junction. J Physiol (Lond) 384:571–585.
13. Sebastião AM, Ribeiro JA (1988) On the adenosine receptor and adenosine inactivation at the rat diaphragm neuromuscular junction. Br J Pharmacol 94:109–120.
14. Sebastião AM, Stone TW, Ribeiro JA (1990) The inhibitory adenosine receptor at the neuromuscular junction and hippocampus of the rat: Antagonism by 1,3,8-substituted xanthines. Br J Pharmacol 101:453–459.
15. Lupica CR, Proctor WR, Dunwiddie TV (1992) Presynaptic inhibition of excitatory synaptic transmission by adenosine in rat hippocampus: Analysis of unitary EPSP variance measured by whole-cell recording. J Neurosci 12:3753–3764.
16. Jarvis MF, Schulz R, Hutchison AJ, Do UH, Sills MA, Williams M (1989) [^{3}H]CGS 21680, a selective adenosine receptor agonist directly labels A_2 receptors in rat brain. J Pharmacol Exp Ther 251:888–893.
17. Stehle JH, Rivkees SA, Lee JJ, Wesver DR, Deeds JD, Reppert SM (1992) Molecular cloning and expression of the cDNA for a novel A_2-adenosine receptor subtype. Mol Endocrinol 6:384–393.
18. Rodrigo J, Polak JM, Fernandez L, Ghatei MA, Mulderry P, Bloom SR (1985) Calcitonin gene-related peptide immunoreactive sensory and motor nerves of the rat, cat, and monkey esophagus. Gastroenterology 88:129–135.
19. Correia-de-Sá P, Ribeiro JA (1994) Potentiation by tonic A_{2a}-adenosine receptor activation of CGRP-facilitated [^{3}H]-ACh release from rat motor nerve endings. Br J Pharmacol 111:582–588.
20. Mullholland MW, Jaffer S (1990) Stimulation of acetylcholine release in myenteric plexus by calcitonin gene-related peptide. Am J Physiol 259:G934–G939.
21. Correia-de-Sá P, Ribeiro JA (1994) Evidence that the presynaptic A_{2a}-adenosine receptor of the rat motor nerve endings is positively coupled to adenylate cyclase. Naunyn Schmiedebergs Arch Pharmacol 350:514–522.
22. Correia-de-Sá P, Ribeiro JA (1993) Facilitation of [^{3}H]-ACh release by forskolin depends on A_2-adenosine receptor activation. Neurosci Lett 151:21–24.

23. Correia-de-Sá P, Timóteo, MA, Ribeiro JA (1994) The activation of tonic A_{2A} adenosine receptors receptors regulates nicotinic autofacilitation of evoked [^{3}H]-ACh release from motor nerve endings. Br J Pharmacol 112:140P.
24. Correia-de-Sá P, Timóteo MA, Ribeiro JA (1994) Tonic adenosine A_1-inhibition and tonic A_{2a}-facilitation of acetylcholine release from the rat motor nerve endings. Drug Dev Res 31:258.
25. Ribeiro JA, Correia-de-Sá P, Cunha RA, Sebastião AM (1994) Adenosine originated from released adenine nucleotide catabolism preferentially activates excitatory A_2 receptors in both the rat neuromuscular junction and hippocampal slices. Drug Dev Res 31:313.
26. Jackisch R, Strittmatter H, Kasanov L, Hertting G (1984) Endogenous adenosine as a modulator of hippocampal acetylcholine release. Naunyn Schmiedebergs Arch Pharmacol 327: 319–325.
27. Cunha RA, Ribeiro JA, Sebastião AM (1994) Purinergic modulation of the evoked release of [^{3}H]acetylcholine from the hippocampus and cerebral cortex of the rat: Role of the ectonucleotidases. Eur J Neurosci 6:33–42.
28. Cunha RA, Sebastião AM, Ribeiro JA (1992) Ecto-5′-nucleotidase is associated with cholinergic nerve terminals in the hippocampus but not in the cerebral cortex of the rat. J Neurochem 59:657–666.
29. Cascalheira JF, Sebastião AM (1992) Adenine nucleotide analogues, including γ-phosphate-substituted analogues, are metabolised extracellularly in innervated frog sartorius muscle. Eur J Pharmacol 222:49–59.
30. von Kugelgen I, Spath L, Starke K (1992) Stable adenosine nucleotides inhibit [^{3}H]-noradrenaline release in rabbit brain cortex slices by direct action at presynaptic adenosine A_1-receptors. Naunyn Schmiedebergs Arch Pharmacol 346:187–196.
31. Forsyth KM, Bjur RA, Westfall DP (1991) Nucleotide modulation of norepinephrine release from sympathetic nerves in the rat vas deferens. J Pharmacol, Exp Ther 256:821–826.
32. Cunha RA, Milusheva E, Vizi ES, Ribeiro JA, Sebastião AM (1994) Excitatory and inhibitory effects of A_1 and A_{2A} adenosine receptor activation of the electrically evoked [^{3}H]acetylcholine release from different areas of the rat hippocampus. J Neurochem 63:207–214.
33. Fredholm BB (1990) Adenosine A_1-receptor-mediated inhibition of evoked acetylcholine release in the rat hippocampus does not depend on protein kinase C. Acta Physiol Scand 140: 245–255.
34. Di Iorio P, Ballerini P, Ciccarelli R, Di Muzio M, Traversa U, Caciagli F (1991) A_1 and A_2 adenosine receptors involvement in controlling purine and acetylcholine release from rat hippocampal slices. Nucleosides Nucleotides 10: 1237–1238.
35. Spignoli G, Pedata F, Pepeu G (1984) A_1 and A_2 adenosine receptors modulate acetylcholine release from brain slices. Eur J Pharmacol 97: 341–342.
36. Sebastião AM, Ribeiro JA (1992) Evidence for the presence of excitatory A_2 adenosine receptors in the rat hippocampus. Neurosci Lett 138:41–44.
37. Cunha RA, Johansson B, van der Ploeg I, Sebastião AM, Ribeiro JA, Fredholm BB (1994) Evidence for functionally important adenosine A_{2a} receptor in the rat hippocampus. Brain Res 649:208–216.
38. Cunha RA, Johansson B, van der Ploeg I, Sebastião AM, Ribeiro JA, Fredholm BB (1994) The binding sites of the A_{2a} adenosine receptor agonist [^{3}H]CGS 21680 are different in the hippocampus and cerebral cortex and in the striatum of the rat. Drug Dev Res 31:260.
39. Okada Y, Sakurai T, Mori M (1992) Excitatory effect of adenosine on neurotransmission is due to increase in neurotransmitter release in hippocampal slices. Neurosci Lett 142:233–236.
40. O'Regan MH, Simpson RE, Perkins LM, Phillis JW (1992) The selective A_2 adenosine receptor agonist CGS 21680 enhances excitatory transmitter amino acid release from the ischemic rat cerebral cortex. Neurosci Lett 138:169–172.
41. Simpson RE, O'Reagan MH, Perkins LM, Phillis JW (1992) Excitatory transmitter amino acid release from the ischemic rat cerebral cortex: Effects of adenosine receptor agonists and antagonists. J Neurochem 58:1683–1690.
42. Mayfield RD, Suzuki F, Zahniser NR (1993) Adenosine A_{2a} receptor modulation of electrically evoked endogenous GABA release from slices of rat globus pallidus. J Neurochem 60: 2334–2337.
43. O'Regan MH, Simpson RE, Perkins LM, Phillis JW (1992) Adenosine receptor agonists inhibit the release of γ-aminobutyric acid (GABA) from the ischemic rat cerebral cortex. Brain Res 582: 22–26.

44. Kirk IP, Richardson PJ (1994) Adenosine A_{2a} receptor-mediated modulation of striatal [^{3}H]GABA and [^{3}H]acetylcholine release. J Neurochem 62:960–966.
45. Mayfield RD, Orona RA, Zahniser NR (1994) Modulation of endogenous GABA release from discrete regions of rat basal ganglia by adenosine A_{2a} and dopamine D_2 receptors. Drug Dev Res 31:297.
46. Ferré S, Fuxe K, von Euler G, Johansson B, Fredholm BB (1992) Adenosine-dopamine interactions in the brain. Neuroscience 51:501–512.
47. Ameri A, Jurna I (1991) Adenosine A_1 and non-A_1 receptors: Intracellular analysis of the actions of adenosine agonists and antagonists in rat hippocampal neurones. Brain Res 546:69–78.
48. Mogul DJ, Adams ME, Fox AP (1993) Differential activation of adenosine receptors decreases N-type but potentiates P-type Ca^{2+} current in hippocampal CA3 neurons. Neuron 10:327–334.
49. De Mendonça A, Ribeiro JA (1990) 2-Chloroadenosine decreases long-term potentiation in the hippocampal CA1 area of the rat. Neurosci Lett 118:107–111.
50. De Mendonça A, Ribeiro JA (1994) Endogenous adenosine modulates long-term potentiation in the hippocampus. Neuroscience 62:385–390.
51. Collingridge GL, Kehl SJ, McLennan H (1983) Excitatory amino acids in synaptic transmission in the rat Schaffer collateral-commissural pathway of the rat hippocampus. J Physiol (Lond) 334:33–46.
52. De Mendonça A, Ribeiro JA (1993) Adenosine inhibits the NMDA receptor-mediated excitatory postsynaptic potential in the hippocampus. Brain Res 606:351–356.
53. De Mendonça A, Canhão P, Ribeiro JA (1994) 1,3-Dipropyl-8-cyclopentylxanthine inhibits hypoxia-induced decrease of the N-methyl-D-aspartate receptor-mediated response in the hippocampus. Drug Dev Res 31:262.
54. Sekino Y, Ito K-I, Miyakawa H, Kato H, Kuroda Y (1991) Adenosine (A_2) antagonist inhibits induction of long-term potentiation of evoked synaptic potentials but not of the population spike in hippocampal CA1 neurons. Biochem Biophys Res Commun 181:1010–1014.

30. AUGMENTATION OF EXTRACELLULAR ADENOSINE FORMATION AS A POTENTIAL THERAPEUTIC STRATEGY FOR PREVENTING EXCESSIVE EXCITATORY AMINO ACID TRANSMISSION IN THE CENTRAL NERVOUS SYSTEM

Thomas D. White

In conditions such as ischemia, hypoxia, hypoglycemia, and seizures, at least some of the pathology appears to be due to the release of excitatory amino acids (EAAs), which then produce excessive EAA transmission [1,2]. There is evidence that adenosine plays a role in limiting excessive neuronal excitation by EAAs, which in extreme situations can lead to neuronal cell death [3]. We have shown that activation of ionotropic EAA receptors promotes the release of adenosine [4–9], and the latter appears to provide some inhibitory tone against excitation [7]. The adenosine can act presynaptically to diminish the release of glutamate and postsynaptically to diminish the excitatory actions of glutamate.

In the cortex, activation of N-methyl D-aspartate (NMDA) receptors releases a nucleotide that is then converted to adenosine extracellularly [8]. In contrast, activation of either kainate or AMPA receptors releases adenosine per se. In the case of NMDA receptor activation, extracellular adenosine formation is maximal when only a few of the available NMDA receptors are activated [6], so that it is unlikely that this adenosine, formed during low-level NMDA receptor activation, can provide complete protection against excessive NMDA receptor activation. The observations that the administration of exogenous adenosine and other agonists can protect against seizures [10], excitotoxicity, and cerebral ischemic damage [3] support this contention. Moreover, NMDA-evoked adenosine release does not require the formation of nitric oxide [9], whereas nitric oxide formation may be implicated in NMDA-receptor–mediated neurotoxicity [11]. On balance, it seems more likely that the extracellular adenosine formed when NMDA receptors are activated might provide an inhibitory threshold against normal NMDA-mediated processes in the cortex [7], such as memory acquisition and storage, rather than protecting against excitotoxicity. However, if one could augment the accumulation of extracellular adenosine that forms when EAA receptors are activated, this might provide protection against excitotoxicity.

Much effort has been expended to develop adenosine agonists that might be useful in treating excitotoxicity. Unfortunately these drugs usually produce numerous unacceptable side effects, reflecting the fact that adenosine receptors are located in most tissues and organs throughout the body. Another approach has been to attempt to augment the levels of endogenous extracellular adenosine in an event-specific manner. For example, AICA riboside (AICAr, acadesine) has been used to increase the extracellular levels of adenosine in ischemic regions of the heart where adenosine formation would be elevated without affecting the adenosine levels in normoxic regions [12]. The mechanism by which AICAr accomplishes this is poorly understood (inhibiting adenosine deaminase and/or adenosine kinase?), but the increased adenosine levels do appear to protect ischemic myocardium.

If one could elicit similar event-specific elevations of extracellular adenosine levels preferentially in those regions of the central nervous system (CNS) experiencing excessive EAA-induced activity without causing widespread adenosine formation, one might provide protection against excitotoxicity without causing unacceptable adverse side effects. Figure 30-1 shows the major pathways by which extracellular adenosine is formed and removed in the CNS. One can see that inhibition of adenosine kinase and/or adenosine deaminase should elevate adenosine levels; the elevated intracellular adenosine could then exit cells via the nucleoside transporter(s), thereby increasing the extracellular levels of adenosine. Adenosine formed extracellularly from the metabolism of released nucleotide(s) should accumulate there when the bidirectional nucleoside transporter is inhibited.

Adenosine antagonists are proconvulsant, particularly in promoting the spread of seizures [10]. Thus the increased extracellular adenosine formed in epileptiform tissue, presumably due to excessive EAA transmission, appears to decrease the spread of seizures. However, this protection by adenosine is incomplete, in so far as the administration of exogenous adenosine and other agonists diminishes the duration and spread of the seizures. In this regard, Zhang et al. [13] have reported some very interesting observations when studying the effects of augmenters of extracellular adenosine on seizures evoked by the administration of the $GABA_A$ antagonist bicuculline into the prepiriform cortex of the rat. They found that inhibition of adenosine kinase with 5′-amino-5′deoxyadenosine (NH_2dADO) or iodotubericidin, or block of the adenosine transporter with dilazep, were highly effective in reducing seizure activity, whereas inhibition of adenosine deaminase with 2′-deoxycoformycin (DCF) was much less efficacious. Surprisingly, AICAr did not exhibit antiseizure activity in their model.

It is important to determine whether the relative efficacies of these types of agents as anticonvulsants relate to their abilities to augment EAA-evoked formation of extracellular adenosine. In a previous study we examined the effects of nucleoside transport inhibitors on EAA-evoked adenosine release from slices of rat parietal cortex [8]. Here we examined the effects of NH_2dADO and AICAr on EAA-evoked formation of extracellular adenosine.

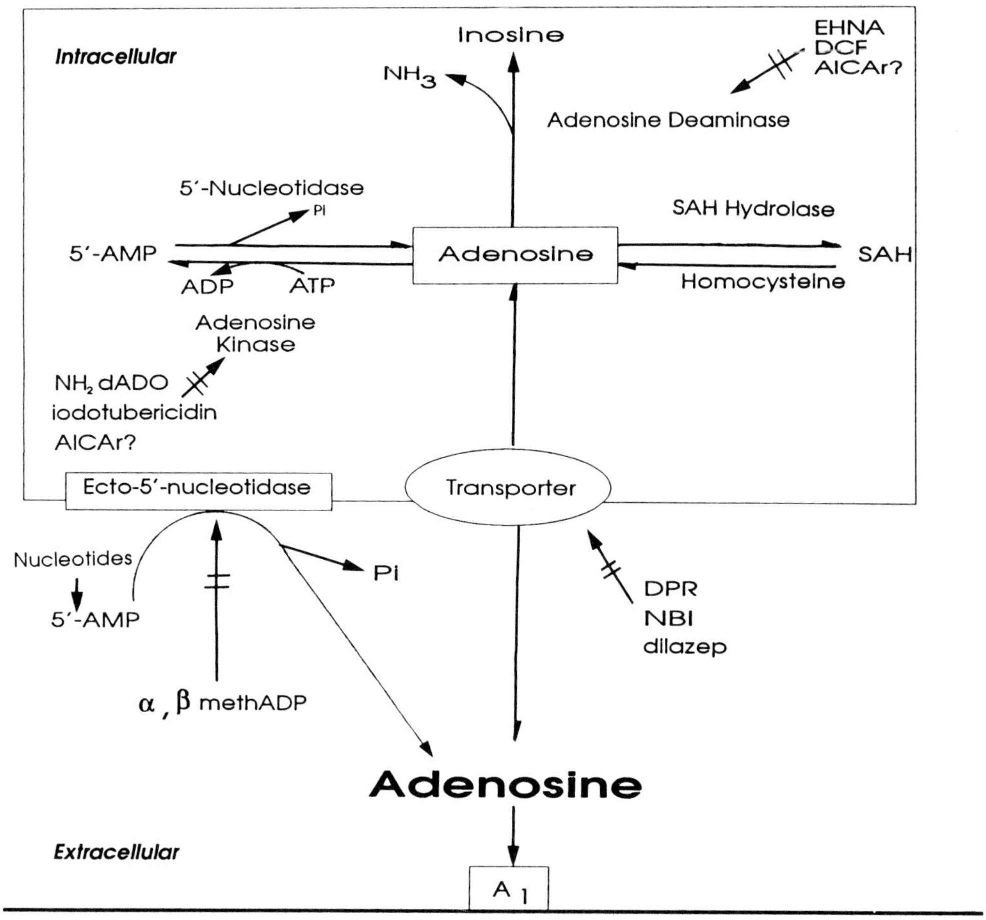

FIGURE 30-1. Routes of adenosine metabolism and the sites of action of various inhibitors (derived from Zhang et al. [13]). In the brain, metabolism by adenosine kinase is a major route of removal of adenosine. Inhibitors of adenosine kinase include NH_2dADO (amino deoxyadenosine), iodotubercidin, and possibly AICAr (AICA riboside, acadesine). Inhibitors of adenosine deaminase include EHNA, DCF (deoxycoformycin), and possibly AICAr. Inhibitors of adenosine transport include DPR (dipyridamole), NBI (nitrobenzylthioinosine), and dilazep. Ecto-5′-nucleotidase is inhibited by α,β methADP (α,β-methylene ADP).

Materials and Methods

All methods employed were identical to those described previously [8]. Rates of adenosine release at time zero immediately before exposure to EAA and at the peak rate of release in the presence of EAA were determined in the presence and absence of either NH_2dADO or AICAr. NH_2dADO and AICAr were purchased from Sigma Chemicals (St. Louis, MO).

Results

Inhibiting adenosine kinase with 20 μM NH_2dADO only slightly increased the basal rate of adenosine release, but it greatly augmented adenosine release evoked by NMDA, AMPA, or kainate (Fig. 30-2). In contrast, AICAr (20 μM) had no effect on basal adenosine release or on adenosine release evoked by NMDA or AMPA (Fig. 30-3), but it augmented kainate-evoked adenosine release from cortical slices.

Discussion

NUCLEOSIDE TRANSPORTER

We showed previously that inhibition of the nucleoside transporter with dipyridamole greatly augmented NMDA, kainate, and AMPA-evoked adenosine release from rat cortical slices [8]. One would expect that inhibition of the bidirectional nucleoside transporter would augment NMDA-evoked formation of extracellular adenosine because this adenosine is derived from a released nucleotide; in this case the uptake of the adenosine formed extracellularly would be diminished by inhibitors of the bidirectional transporter and its extracellular levels would increase. However, it was surprising that inhibition of the nucleoside transporter also augmented AMPA and kainate-evoked adenosine release, where adenosine per se is released rather than a nucleotide, which then undergoes extracellular metabolism to form adenosine. If the efflux of adenosine induced by AMPA and kainate occurs via the bidirectional dipydidamole-sensitive nucleoside transporter, release should be diminished by dipyridamole. It appears that activation of non-NMDA receptors triggers the efflux of adenosine via some other mechanism but that adenosine's removal from the extracellular space does take place on a dipyridamole-sensitive transporter. Perhaps adenosine release occurs from one cell type and removal of extracellular adenosine into another cell type.

These observations may have important therapeutic implications. If release of adenosine occurs via the bidirectional dipyridamole-

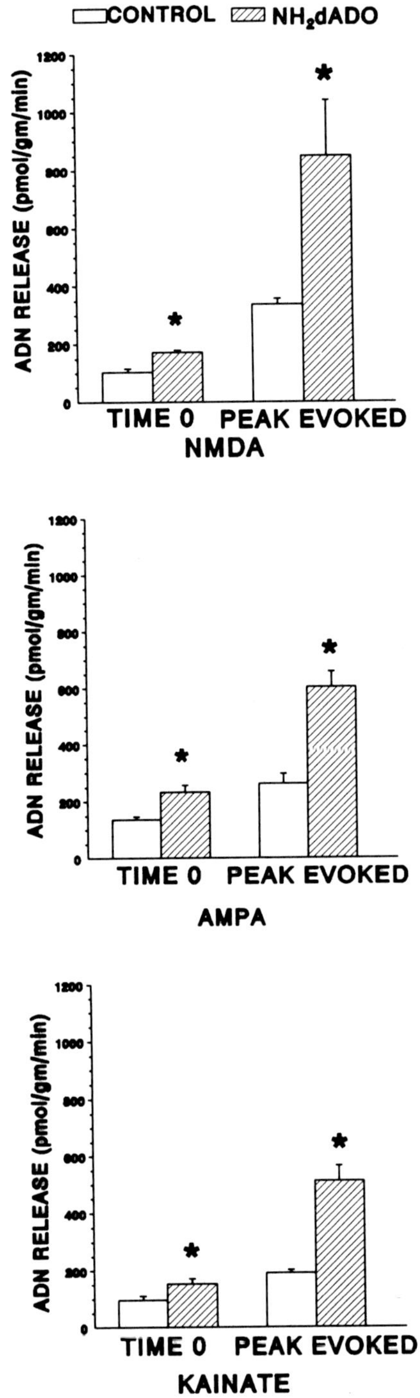

FIGURE 30-2. Effect of inhibition of adenosine kinase by NH_2dADO (20 μM) on basal and EAA-evoked adenosine (ADN) release from rat brain cortical slices. Open histograms are controls and crosshatched histograms are treated slices. Each histogram is the mean ± SEM for NMDA (100 μM, n = 5), AMPA (100 μM, n = 4), and kainate (100 μM, n = 4). *Significantly different from control ($p < 0.05$, paired t-test).

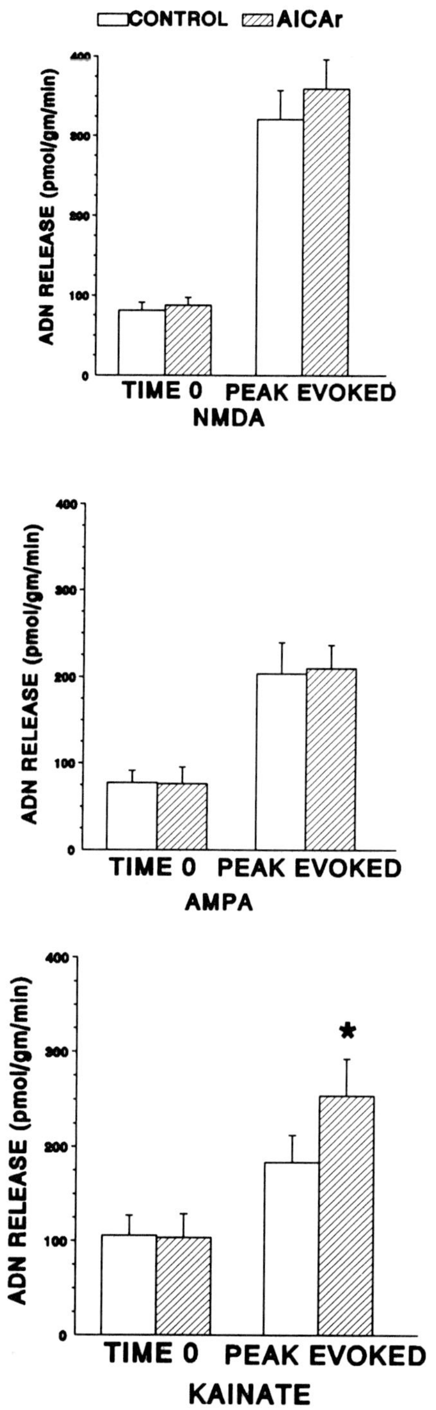

sensitive nucleoside transporter, then inhibition of this transporter might prevent its release and decrease, rather than promote, the extracellular accumulation of adenosine [14]. Assuming that extracellular adenosine is protective in the CNS, this would mean that transport inhibitors could have deleterious rather than beneficial effects on excitotoxicity produced by excessive AMPA or kainate receptor activation. However, the observation that adenosine transport inhibitors actually increase kainate and AMPA-evoked accumulation of extracellular adenosine indicates that these drugs should protect against excitotoxicity in the CNS.

ADENOSINE KINASE

In the present study inhibition of adenosine kinase with NH_2dADO slightly potentiated the basal accumulation of extracellular adenosine, but it greatly increased adenosine accumulation evoked by NMDA, AMPA, or kainate receptor activation. Using in vivo microdialysis in freely moving rats, Pedata et al. [15] have recently reported that inhibition of adenosine kinase with iodotubercidin substantially elevates striatal extracellular adenosine levels, whereas inhibition of adenosine deaminase with EHNA has much less effect on adenosine levels. These results and the findings of Zhang et al. [13] with their seizure model are consistent with the possibility that adenosine kinase inhibitors might be useful in treating excitotoxic disorders.

AICAR

In the present study AICAr had no effect on basal adenosine release from cortical slices, nor

FIGURE 30-3. Effect of AICAr (20 μM) on basal and EAA-evoked adenosine (ADN) release from rat brain cortical slices. Open histograms are controls and crosshatched histograms are treated. Each histogram is the mean ± SEM for NMDA (100 μM, n = 5), AMPA (100 μM, n = 5), and kainate (100 μM, n = 4). *Significantly different from control ($p < 0.05$, paired t-test).

did it affect NMDA- or AMPA-evoked accumulation of extracellular adenosine. However, AICAr did potentiate kainate-evoked adenosine release. This differential effect of AICAr on AMPA versus kainate-evoked adenosine release is surprising because both of these non-NMDA receptors share many common characteristics. Perhaps this difference relates to differential distributions for these two receptors; AMPA receptors are often codistributed with NMDA receptors in the outer cortical layers, whereas kainate receptors are thought to reside in deeper cortical layers [16]. The previous finding that AICAr was ineffective in preventing bicuculline-induced seizures [13] could reflect the current finding that AICAr did not potentiate the accumulation of extracellular adenosine evoked by NMDA receptor activation, assuming that NMDA receptors are involved in the generation and spread of seizures.

Certain forms of temporal lobe seizure disorders have been associated with degeneration of specific regions of the amygdala and hippocampus [17]. These regions are rich in kainate receptors and can be damaged when exposed to high levels of agonists, such as kainate and domoate. An outbreak of shellfish poisoning in Canada, later traced to the presence of high levels of domoate in mussels, resulted in several deaths and neurologic disorders, including severe anterograde memory deficits [17]. Pathologic examinations revealed neurodegeneration, especially in the amygdala and hippocampus. Although AICAr may be of limited use in treating many seizure disorders, its remarkable selectivity and specificity of action only on kainate-evoked adenosine release may make it useful in treating and/or preventing kainate receptor–mediated excitotoxicity in the CNS.

Conclusions

The patterns of augmentation of extracellular adenosine levels with AICAr and with an inhibitor of adenosine kinase, NH_2dADO, appear to parallel their anticonvulsant actions. Event-specific augmentation of extracellular adenosine formation could lead to new therapies for disorders associated with excessive EAA actions in the CNS, such as epilepsy, ischemic damage, and stroke.

Acknowledgments

The author thanks his technician, Sharon Temple, for conducting these experiments. This research was supported by the Medical Research Council of Canada.

References

1. Rothman SM, Olney JW (1987) Excitotoxicity and the NMDA receptor. Trends Neurosci 10:299–302.
2. Koh J, Goldberg MP, Hartley DM, Choi DW (1990) Non-NMDA receptor-mediated neurotoxicity in cortical culture. J Neurosci 10:693–705.
3. Marangos PJ (1991) Potential therapeutic roles for adenosine in neurologic disease. In Stone TW (ed.), Adenosine in the Nervous System. London: Academic Press, pp. 217–227.
4. Hoehn K, White TD (1990) N-methyl-D-aspartate, kainate and quisqualate release endogenous adenosine from rat cortical slices. Neuroscience 39:441–450.
5. Hoehn K, White TD (1990) Role of excitatory amino acid receptors in K^+- and glutamate-evoked release of endogenous adenosine from rat cortical slices. J Neurochem 54:256–265.
6. Hoehn K, Craig CG, White TD (1990) A comparison of N-methyl-D-aspartate-evoked release of adenosine and [^{3}H]norepinephrine from rat cortical slices. J Pharmacol Expt Ther 255:174–181.
7. Craig CG, White TD (1992) Low-level N-methyl-D-aspartate receptor activation provides a purinergic inhibitory threshold against further N-methyl-D-aspartate-mediated neurotransmission in the cortex. J Pharmacol Expt Ther 260:1278–1284.
8. Craig CG, White TD (1993) N-methyl-D-aspartate- and non-N-methyl-D-aspartate-evoked adenosine release from rat cortical slices: Distinct purinergic sources and mechanisms of action. J Neurochem 60:1073–1080.
9. Craig CG, White TD (1993) NMDA-evoked adenosine release from rat cortex does not

require the intermediate formation of nitric oxide. Neurosci Lett 158:167–169.
10. Dragunow M (1991) Adenosine and epileptic seizures. In Phillis JW (ed.), Adenosine and Adenine Nucleotides as Regulators of Cellular Function. Boca Raton, FC: CRC Press, pp. 367–379.
11. Dawson VL, Dawson TM, London ED, Bredt DS, Snyder SH (1991) Nitric oxide mediates glutamate neurotoxicity in primary cortical cultures. Proc Natl Acad Sci USA 88:6368–6371.
12. Mullane K, Young M (1993) Acadesine: Prototype adenosine regulating agent for treating myocardial ischemia-reperfusion injury. Drug Dev Res 28:336–343.
13. Zhang G, Franklin PH, Murray TF (1993) Manipulation of endogenous adenosine in the rat prepiriform cortex modulates seizure susceptibility. J Pharmacol Exp Ther 264:1415–1424.
14. White TD, MacDonald WF (1990) Neural release of ATP and adenosine. Ann NY Acad Sci 603:287–299.
15. Pedata F, Pazzagli M, Corsi C, Pepeu G (1994) Adenosine release measured "in vivo" in adult and freely moving rats. Drug Dev Res 31:306.
16. Young AB, Fagg GE (1990) Excitatory amino acid receptors in the brain: Membrane binding and receptor autoradiographic approaches. Trends Pharmacol Sci 11:126–133.
17. Teitelbaum JS, Zatorre RJ, Carpenter S, Gendron D, Evans AC, Gjedde A, Cashman NR (1990) Neurological sequelae of domoic acid intoxication due to the ingestion of contaminated mussels. N Engl J Med 322:1781–1787.

31. PURINOCEPTORS ON GLIAL CELLS OF THE CENTRAL NERVOUS SYSTEM: FUNCTIONAL AND PATHOLOGIC IMPLICATIONS

Maria P. Abbracchio
Stefania Ceruti
Geoffrey Burnstock
Flaminio Cattabeni

Glial cells make up a substantial proportion of the central nervous system (CNS) and participate in a variety of important physiologic and pathologic processes [1]. At least three types of glial cells that differ in both immunologic phenotype and functions can be identified in the mature brain: astroglia, oligodendroglia, and microglia (Table 31-1). An extensive overview of glial cell lineage and functions in the brain is not within the aims of this chapter and, in this respect, the reader is referred to authorative reviews in the field [2–4]. However, we will briefly summarize here the most widely accepted news concerning the formation of glial cells during brain development and the roles of the different glial cell types in brain, because they might represent a helpful background to correctly interpret the effects of purines on glial cell functions and could also provide useful suggestions for future research.

Results of studies employing replication-defective, recombinant retroviral vectors constructed to carry a histochemically detectable reporter gene to mark proliferative progenitor cell populations in vivo support the existence of a transient bipotential precursor cell that, at a certain step in embryonic development, can differentiate to either the neuronal or the glial lineage progenitor cell [2]. From the glial lineage-restricted progenitor cell, glial cell sublineages diverge into independent pathways of development, generating at least three different progenitors, as schematized in Figure 31-1. The radial progenitor gives rise to the radial glial cell, whose fibers attain enormous lengths (several centimeters in the midterm human fetus). This cell type is believed to support guidance to migrating neuroblasts during neurogenesis, and evidence has been produced that radial cells differentiate into mature astrocytes later during brain development, possibly as a consequence of completion of neuronal migration (Table 31-1) [2]. The type 1 astrocyte progenitor is the origin of the mature

TABLE 31-1. Types of glial cells in the brain

	Progenitor cell			
	Name	Immunologic phenotype	Immunologic phenotype	Hypothesized functions
Type 1 astrocyte	Type 1A progenitor	RAN2$^+$ GFAP$^+$ $A_2B_5^-$ GAP43$^+$ fibronectin$^-$ vimentin$^-$	GFAP$^+$ vimentin$^-$ TTX$^-$ RAN2$^+$ PDGF$^-$ CS_4^- $A_2B_5^-$ laminin$^+$ GD_3^- GAP43$^-$ fibronectin$^+$	Formation of glial limiting membrane Formation of blood-brain barrier Involved in astrogliosis
Type 2 astrocyte	O2A cell	$A_2B_5^+$ GFAP$^-$ vimentin$^+$ CS_4^+ GAP43$^+$ RAN2$^-$ GD_3^+ TTX$^+$	CS_4^- GFAP$^+$ fibronectin$^-$ GD_3^- GAP43$^+$ $A_2B_5^+$ laminin$^{+/-}$	Extend processes that associate with axons at the nodes of Ranvier? Involved in astrogliosis?
Oligodendrocyte			CS_4^- GFAP$^-$ GalC$^+$ MBP$^+$ GD_3^- GAP43$^+$ O_4^+ $A_2B_5^-$ TTX$^-$ vimentin$^-$ CNP$^+$ PLP$^+$ MAG$^+$	Formation and maintenance of myelin sheaths Involved in multiple sclerosis
Radial glia	Radial progenitor	RC1$^+$ RAN2$^+$ RAT401$^+$ $A_2B_5^-$ vimentin$^+$	RC 1$^+$ RAT401$^+$ RC2$^+$ vimentin$^-$	Offer substratum and guide for neuronal cell migration Disappear at the end of neurogenesis giving birth to tanycytes, Bergmann cells, Muller cells, and ependymal cells (RAT401$^-$; RC1$^-$) or transforming into type 1 astrocytes
Microglia	Monocytes? Neuroectodermal matrix?		MHC1$^+$ CLA$^+$ MHC2$^+$ GFAP$^-$	Major source of immunoeffector cells Involved in neurodegenerative diseases

Mature glial cells differ from each other in cell lineage, immunologic phenotype, and functions. In most cases maturation of the progenitor cell involves loss of antigenic components and acquisition of new ones. Controversy still exists on the in vivo relevance of the type-2 astrocyte [4].

A_2B_5 [5]; CLA = common leukocyte antigen; CNP = 2′3′-cyclic nucleotide 5′ phosphoidrolase; CS_4 = chondroitin 4-sulphate; GAP 43 = growth associated protein 43; GC = galactocerebroside; GD_3 [2]; GFAP = glial fibrillary acidic protein; MAG [2]; MBP = myelin basic protein; MHC 1/MHC 2 = major histocompatibility complex 1/2; O_4 = antibody binding to sulphate glycolipids; PDGF = platelet-derived growth factor; PLP [2]; RAN 2 [5]; Rat 401 [2]; RC 1/RC 2 [2]; TTX = tetanus toxin. For further details see references 2–5.

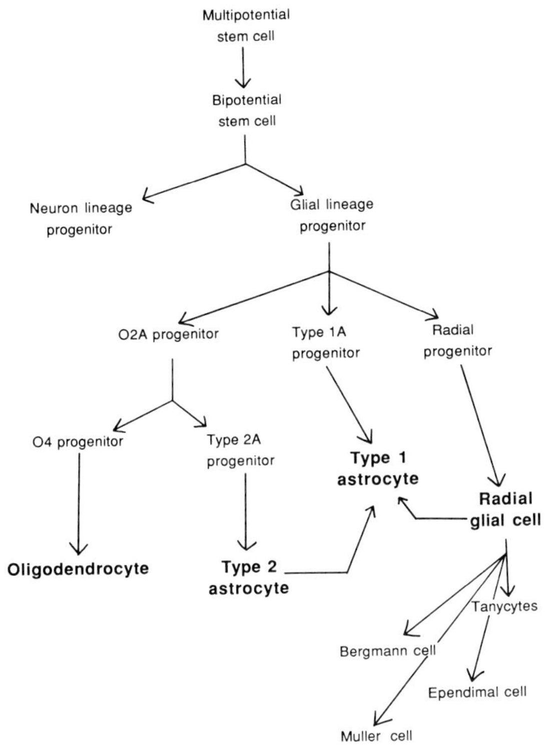

FIGURE 31-1. Glial cell lineage during the development of mammalian brain.

"fibroblast-like" type 1 astrocyte involved in the formation of the glial limiting membrane and the blood-brain barrier (Table 31-1). According to the results of in vitro studies, the so-called O2A progenitor is believed to give rise to both "neuron-like" type 2 astrocytes and to oligodendrocytes, the latter producing myelination in the central nervous system (CNS) (Table 31-1). Type 1 and type 2 astrocytes display typical morphologies when cultured in vitro (fig. 31-2), and both express the astroglial specific marker, glial fibrillary acidic protein (GFAP). Although type 2 astrocytes have been hypothesized to ensheath axonal plasma membrane at the nodes of Ranvier and to express specific neurotransmitter receptors [5], this type of astroglia has never been detected in vivo, and hence its possible roles remain unclear [4]. Astroglia is known to respond to various types of injury by rapid and vigorous astrogliosis (Table 31-1) [1,6].

In a similar manner to their daughter cells, progenitor cells also differ from each other in terms of specific immunologic phenotypes (Table 31-1). In general, it is believed that the development of differentiated cells from progenitors is an irreversible process. However, in contrast to this view, it has been recently suggested that postnatal astrocytes can be dedifferentiated to radial glial in vitro by exposure to extracts of embryonic brain [7], evidence that opens up new possibilities for the pharmacologic modulation of glial cell function. Based on this, when the factors involved in this dedifferentiation are characterized it is theoretically possible to address the phenotype of glial cells towards the cell subpopulation that most conveniently fits the brain needs under particular physiologic and pathologic conditions.

Microglial cells are resident, immunocompetent cells of the central nervous system [8]. Although there is some controversy concerning their origin, this type of glial cell is generally believed to derive from monocytes/macrophages entering the brain during embryonic development. This is also confirmed by their different immunologic profile, which more closely resembles that of macrophages (Table 31-1). Microglial cells represent a major source of immunoeffector cells in the brain, and their transition from a quiescent or resting state to an activated state occurs within hours after injury. It has long been recognized that, besides representing the first line of defense in the CNS, activated microglia can exert cytotoxicity in the brain (see also Chapter), which has implicated this type of glial cell in a number of nervous system diseases characterized by microglia activation [9]. It is now becoming evident that ameboid microglia can promote both the death of cultured neurons as well as neurite growth or regeneration depending upon different physiologic conditions [10], suggesting that microglia might also actively take part in CNS regeneration

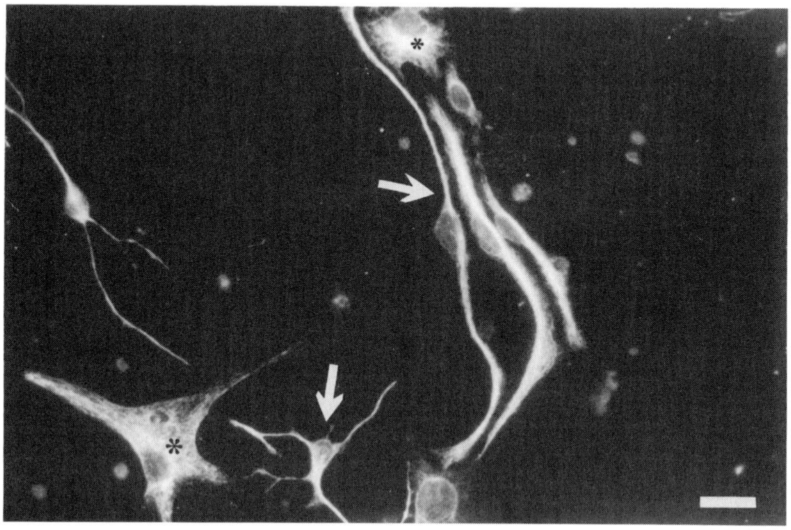

FIGURE 31-2. Immunofluorescence micrograph showing GFAP-positive control cells in primary neuron-glia cultures. Cultures obtained from 7-day-old rat corpus striatum were grown in vitro for 3 days in serum-supplemented medium, fixed, and immunostained with an anti-GFAP antibody as previously described [15]. Asterisks indicate type 1 and arrows indicate type 2 astrocytes, respectively. Scale bar = 25 μm.

following trauma and injury by influencing the formation and remodeling of neuronal networks.

Astrocytes, Reactive Astrogliosis, and Purines

Astrocytes are known to react to various types of injury by reactive astrogliosis, a reaction characterized by increased astrocyte proliferation and cellular hypertrophy (larger cellular size, presence of longer and thicker astrocytic processes, and increased staining for the GFAP marker) [1]. The exact physiologic significance of reactive astrogliosis is still under debate, but this phenomenon is believed to play a key role in the restoration of damaged brain functions following ischemia and trauma. It has been commonly assumed that astroglial cells proliferate in response to brain injury; however, more importance has been recently given to injury-associated astrocytic hypertrophy and concomitant elevation of GFAP expression, based on the fact that hypertrophy is common to every type of astrogliosis and that suppression of GFAP expression by antisense mRNA obliterates the ability of astrocytes to support neurite regrowth in vitro [6]. Increases of GFAP expression were demonstrated to occur even in noninvasive brain injury, where gliosis is not accompanied by increases in cell number [6]. Based on this, GFAP expression can be taken as a marker for the astroglial activation and differentiation known to occur in reactive astrogliosis.

Several factors have been implicated in the initiation of astrogliosis, including growth factors (EGF, FGF, PDGF), cytokines, and myelin basic proteins [11]. In addition to these factors, we and others have been exploring the hypothesis that purines might play a role in astrocytic activation following trauma and ischemia, based on 1) the fact that large amounts of ATP and adenosine are released by injured and hypoxic cells in both the cardiovascular system and in the CNS, and 2) the presence of both P_1 and P_2 purinoceptors on astroglial cells linked to modulation of cAMP formation, inositol-phosphate–dependent Ca^{2+} mobilization, and arachidonic acid production [12–14].

We have studied the effects of adenosine and ATP analogues (namely, αβ-methylene ATP [αβ-meATP] or 2-chloro-adenosine [2-CA], on the number of GFAP-positive cells in neuron-astroglia CNS primary cultures [15]. After growing cultures for 3 days in the presence of these analogues, we found that αβ-meATP increased and 2-CA decreased the number of astroglial cells in culture. For both derivatives, the effect was dose dependent. The effects of αβ-meATP were antagonized by the tripanoside suramin, indicating the involvement of a suramin-sensitive P_2 purinoceptor, whereas the effect of 2-CA was not reversed by the xanthine antagonist p-sulphonyl-phenyl-theophilline, implying the activation of a xanthine-insensitive (A_3 or A_4?) adenosine receptor subtype.

In order to investigate the molecular mechanisms underlying these effects, we have evaluated bromodeoxyuridine (BrdU) incorporation into the astrocytic DNA of GFAP-positive cells by means of indirect double-labeling immunocytochemistry. Surprisingly, both analogues could significantly increase the incorporation of BrdU into DNA, despite their different effects on astroglial cell number. It therefore seemed that whereas activation of P_2 purinoceptors on astroglial cells leads to stimulation of DNA synthesis and eventually to increases in cell number, the effects induced by 2-CA are more complex and probably involve multiple mechanisms. Due to the lack of any apparent correlation between DNA synthesis and cell number, we concluded that, besides stimulating BrdU incorporation into nuclei, 2-CA might indeed affect other parameters, such as cell adhesion and/or survival eventually leading to a reduction in cell number. An alternative explanation to this apparent discrepancy between nuclear BrdU incorporation and cell number might come from the hypothesis that the adenosine derivative is triggering apoptosis in our cell system, in keeping with previous reports of adenosine- and ATP-induced cell death. If this is true, the increased incorporation of BrdU into nuclei might reflect degradation of DNA structure rather than a proliferative event. Preliminary data obtained with flow cytometric analysis of propidium-iodide–stained cell nuclei suggest that this might be the case [Ceruti et al., unpublished]. The observation that adenosine analogues can trigger apoptosis of astroglial cells might reveal a new role for adenosine in the regulation of astrocytic cell number, which might have fascinating implications in the modulation of astroglial cell formation during brain development and in the regulation of the astrocytic response in astrogliosis.

Our results confirm and extend previous data obtained by other authors on different cell systems. Rathbone and coworkers have studied the effects of extracellular nucleotides and nucleosides on cell division in chick brain astrocytes [16]. In this system, both adenosine and ATP stimulated ^{3}H-thymidine incorporation into astrocytic DNA. The effects of ATP and several hydrolysis-resistant ATP analogues were apparently mediated by P_{2Y} purinoceptors coupled to inositol phosphate formation. In this system too, the effects induced by adenosine appeared to be complex. The increased incorporation of ^{3}H-thymidine into astrocytic nuclei demonstrated by Rathbone and coworkers following exposure to adenosine is in agreement with the increased nuclear immunoreactivity to anti-BrdU antibodies demonstrated in our study [15]. However, Rathbone and coworkers did not measure cell number, and therefore it is not possible to say whether their results really reflect increases in the number of cells and/or whether adenosine also exerts other independent effects on cultured astrocytes. Moreover, the effect observed by Rathbone and coworkers required coactivation of A_1 and A_2 receptors, and the A_2 receptor involved showed atypical responses to "classic" A_2 receptor ligands, similar to what was reported by Meininger and Granger on the purinergic stimulation of bovine endothelial cell proliferation [17].

The effects of ATP on the proliferation rate of cultured astrocytes has been also studied by

Ciccarelli and coworkers [18]. Eight hours after the addition of ATP, a dose-dependent increase in the percentage of proliferating cells was observed, whereas an opposite trend was shown at the 16th hour. The relatively stable analogue βγ-meATP only displayed a mitogenic effect, whereas the A_1 receptor analogue R-phenyl-isopropyl-adenosine (R-PIA) inhibited astrocyte proliferation. On the basis of these results, it was concluded that exogenous ATP, partly as a consequence of its rapid extracellular breakdown, exerts a dual influence on astrocyte proliferation by the involvement of both P_2 (stimulatory) and P_1 (inhibitory) purinoceptors.

Stimulatory effects of extracellular ATP on DNA synthesis have also been shown by Neary and Norenberg on primary astrocytes obtained from neonatal rat cortex [13]. These authors have looked at other parameters besides cell proliferation, such as ATP-evoked changes in astrocytic calcium signals, protein phosphorylation, and morphology [13]. Application of extracellular ATP resulted in a calcium peak followed by a sustained intracellular calcium signal, indicating that ATP stimulates mobilization of internal calcium as well as influx of calcium across the plasma membrane. These effects were associated with increased phosphate incorporation into two proteins, one of which (the 52 kDa protein band) comigrated with GFAP on sodium dodecyl-sulphate polyacrylamide gels. Both lanthanum (an inhibitor of calcium transport) and omission of calcium from the incubation medium reduced the ATP-evoked changes in phosphorylation. Consistently, similar effects were mimicked by application of a calcium ionophore. Taken together, these findings suggested that ATP-induced effects were mediated by an increase of intracellular calcium.

Since calcium is believed to play a role in astrocytic stellation, Neary and Norenberg have also studied the effect of extracellular ATP on process formation. Application of ATP resulted in marked stellation of cells [13]. This effect could be prevented by inhibiting increases of $[Ca^{2+}]i$ with BAPTA, suggesting that ATP evokes stellation by increases of intracellular Ca^{2+} levels. In an attempt to establish a correlation among ATP-evoked calcium signal, increased phosphorylation of the 52 kDa protein band, and astrocytic morphologic changes, Neary and Norenberg looked at GFAP expression after ATP application. Both ELISA and Northern blot analysis indicated that ATP can induce GFAP gene expression in astrocytes, which might be related to the observed morphologic changes.

In our system too, we have observed dramatic morphologic changes in astrocytes, including stellation and increased branching, after application of αβ-meATP (Fig. 31-3) [15]. Changes of astrocytic morphology with increased expression of GFAP-positive processes can be taken as a marker of the astroglial cell differentiation known to occur during astrocytic activation following injury [6]. Such differentiation has been demonstrated to occur following exposure to a number of putative triggers for astrogliosis, such as, for example, basic fibroblast growth factor (bFGF). Consequently, we have studied the morphologic effects induced by αβ-meATP and bFGF on the expression of GFAP-positive astrocytic processes in the same experimental model utilized for the proliferative studies. Exposure of cells to either αβ-meATP or bFGF in a serum-free medium resulted in a dose-dependent increase in the mean length of astrocytic processes per cell. Interestingly, the maximal effect induced by αβ-meATP was quantitatively comparable with the maximal effect obtained with bFGF [Abbracchio et al., in preparation].

Taken together, these results indicate that purines can affect both astroglial cell proliferation and astrocytic hyperplasia, two parameters that are indicative of astrocytic activation following injury and ischemia. In particular, our studies have demonstrated that purines can affect astroglial cell differentiation in a similar manner as classic growth factors. Moreover, the significance of the in vitro studies is confirmed by a preliminary report on the ability of ATP and

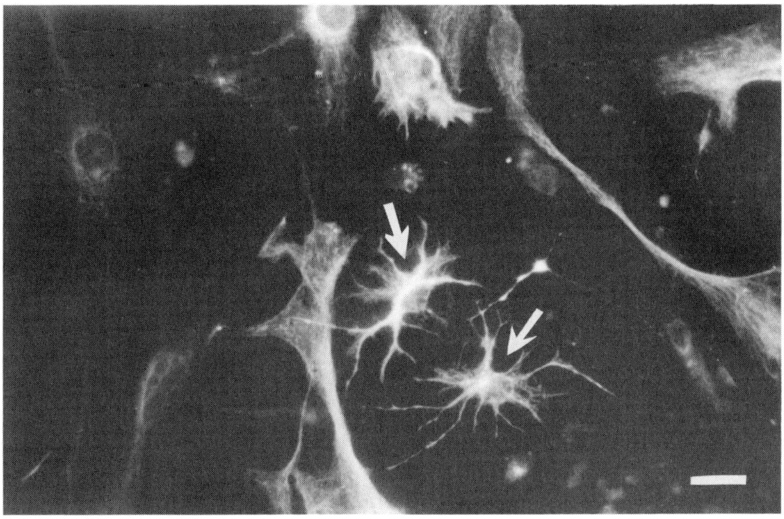

FIGURE 31-3. Immunofluorescence micrograph showing the morphologic effects induced by αβ-meATP on GFAP-positive cells in primary neuron-glia cultures. Cultures obtained from 7-day-old rat corpus striatum were grown in vitro for 3 days in medium in the presence of 10^{-5} M αβ-meATP, fixed, and immunostained with an anti-GFAP antibody as previously described [15]. Astrocytes (arrows) show increased branching and stellation. Scale bar = 25 μm.

adenosine to induce astrogliosis in vivo when infused into rat brain [19]. These findings therefore definitely implicate purines as candidates for the initiation and maintenance of reactive astrogliosis and hence for CNS repair and regeneration following injury.

Microglial Cells and Purines

Microglial cells originate from monocytes and macrophages entering the brain during early embryonic development. These resident immunocytes are capable of regressive metamorphosis in a number of pathologic conditions, such as Parkinson's and Alzheimer's diseases, HIV encephalopathy, demyelinating diseases, ischemia, nerve injury, and direct brain trauma [9]. Resting (ramified) microglia differentiate into fully activated macrophages through a number of transitional states, initially characterized by proliferation, increased or de novo expression of marker molecules (Table 31-1), migration, and eventually changes into phagocytic macrophage-like (ameboid) cells [9,20]. Microglia seem to be the first cell type to be activated in transient forebrain ischemia, well before any change in other glial cells can be detected [9]. There are two principal ways in which activated microglia can exert cytotoxicity in the brain. First, they can act as phagocytes. Second, they can release potentially cytotoxic substances, including free radicals, arachidonic acid metabolites, inflammatory cytochines, and proteolytic enzymes. The repertoire of toxic factors produced and released by microglia gives this glial element a key role in the initiation and mediation of secondary autodestructive tissue damage [9]. The endogenous factors involved in microglia activation are largely unknown; however, a possible rapid activation via contransmitters, such as ATP, has been suggested [9,21].

The electrophysiologic properties of microglial cells have been extensively characterized by Kettenmann's and Illes' groups. An inwardly rectifying current (K_{IR}) was demonstrated to be the only voltage-activated current in undifferentiated microglial cells [20]. K_{IR} was found to be inhibited by either activation of G proteins or $[Ca^{2+}]_i$ increases or exposure to ATP and ADP [22]. Although microglial cells are not considered to be elements of fast signal transfer in the nervous system (and thus to be cells expressing neurotransmitter receptors), ATP and related compounds induced a transient rapid rise in $[Ca^{2+}]_i$ in cultured microglia. ATP was more potent than ADP, while AMP and adenosine were ineffective, indicating involvement of a P_2 purinoceptor [21]. These results suggest that extracellular purines can physiologically modulate resident microglial cells. Moreover, the ATP-induced response could serve as a mediator for detecting injured neurons that may leak ATP; the purine-induced rise in intracellular Ca^{2+} could represent an early signal in the activation of microglial cells and in inducing the expression of specialized functions, such as secretion, migration, or phagocytosis.

ATP has also been shown to affect differentiated (nonproliferating) microglia. At a holding potential of −70 mV, ATP and some of its structural analogues induced a rapidly desensitizing inward current, which was blocked by P_2-purinoceptor antagonists such as suramin and reactive blue 2 [23]. At a holding potential of 0 mV, ATP caused a slowly developing and nondesensitizing outward current [23]. Hence, differentiated microglial cells express at least two P_2-purinoceptor subtypes linked to modulation of membrane currents with different kinetic characteristics (a ligand-gated channel and a slower G-protein–mediated current [24].

In addition to ATP, adenosine, via P_1 purinoceptors, also seems to play a role in modulating microglial cell function. The xanthine derivative propentofylline, which was previously shown to protect against delayed postischemic brain damage in gerbils, was demonstrated to significantly inhibit the formation of reactive oxygen intermediates in microglial cells and to interfere with the generation of phagocytic macrophagic cells [9]. This xanthine derivative is believed to favor adenosine A_2 receptor actions by blocking the adenosine membrane transporter, phosphodiesterases, and perhaps by slightly inhibiting the A_1 receptor [9]. This again underlines the involvement of both adenosine A_1 and A_2 receptor subtypes in mediating adenosine effects on glial cells, probably by a reciprocal influence on second messenger systems. McRae et al. [25] have recently shown that antibodies present in the cerebrospinal fluid (CSF) of patients with Alzheimer's disease (AD) recognize ameboid microglial cells in postischemic gerbil brain, suggesting that ischemia-associated pathologic processes and AD share common mechanisms related to the activation of microglial cells. Propentofylline treatment of gerbils after bilateral carotid occlusion was also shown to depress the immunostaining of activated microglia by AD-CSF and β-APP (the amyloid precursor protein) [26].

Taken together these findings suggest that purines are likely modulators of microglial cell function. This might be relevant both to physiologic conditions in the regulation of quiescent microglial cells and, most interestingly, to a number of diseases known to go along with an activation of microglial cells to macrophagic like elements [9]. Moreover, based on the novel roles suggested for microglial cells in CNS regeneration [10] by affecting microglia function, purines might endogenously regulate the formation and modeling of neural networks under both physiologic and pathologic conditions.

Purines, Oligodendrocytes, and Other Glial Cells

Only very limited number of studies have focused on the possible pathophysiologic roles of purines on other types of glial cells. Oligodendroglia and related cells seem likely to be affected by purines, based on the recent demonstration of an ATP-dependent inward current in mouse Schwann cells [27] and on the finding that ATP can trigger cytoplasmic Ca^{2+} release in cultured cells of the oligodendrocytic lineage [28]. In this last study, suramin-sensitive responses were demonstrated for ATP, UTP, and a number of ATP derivatives on both immature and mature oligodendrocytes. Responses were abolished by incubation with thapsigargin and intracellular administration of heparin, indicating that ATP can trigger the release of Ca^{2+} ions from inositol-phosphate–sensitive internal stores. Similarly, ATP could trigger Ca^{2+} mobilization from Bergmann glial cells [29].

These limited data point to possible effects of purines in the regulation of other glial cells besides astrocytes and microglia. The elucidation of the regulation of oligodendroglia by purines will be of particular interest in view of their possible applications to demyelinating diseases.

Conclusions

Data are accumulating that involve purines in the physiopathologic regulation of glial cells; both P_1 and P_2 purinoceptors are present. Results from in vitro and in vivo studies in experimental animals strongly implicate purines in the regulation of reactive astrogliosis and in the activation of microglial cells in the ischemic brain, and hence for CNS repair and regeneration following injury. The results reviewed here present the background for possible new therapeutic strategies for a number of neurodegenerative diseases characterized by involvement of these glial cell types. Indications from the literature point to possible additional roles of purines in brain development and myelinating processes. Future studies aimed at elucidating the effects of ATP and adenosine on glial cell maturation during nervous system development and on oligodendrocytic functions are likely to reveal additional exciting targets for potential therapeutic applications of purinoceptor ligands.

Acknowledgments

The authors wish to thank Professors P. Schubert and K. Jessen for useful discussions.

References

1. Eddleston M, Mucke L (1993) Molecular profile of reactive astrocytes—implications for their role in neurologic disease. Neuroscience 54: 15–36.
2. Cameron RS, Rakic P (1991) Glial cell lineage in the cerebral cortex: A review and synthesis. Glia 4:124–137
3. Goldman JE, Vaysse PJ-J (1991) Tracing glial cell lineages in the mammalian forebrain. Glia 4:149–156.
4. Noble M (1991) Points of controversy in the O2A lineage: Clocks and type-2 astrocytes. Glia 4:157–164.
5. Wilkin GP, Marriott DR, Cholewinski AJ (1990) Astrocyte heterogeneity. Trends Neurosci 13:43–46.
6. Hatten ME, Liem RKH, Shelanski ML, Mason CA (1991) Astroglia in CNS injury. Glia 4: 233–243.
7. Hunter KE, Hatten ME (1993) Postnatal cortical astrocytes can differentiate to the radial glia phenotype in response to embryonic signals in vivo and in vitro. Soc Neurosci Abstr 19:615.
8. Streit WJ, Graeber MB, Kreutzberg GW (1988) Functional plasticity of microglia: A review. Glia 1:301–307.
9. Banati RB, Gehrmann J, Schubert P, Kreutzberg W (1993) Cytotoxicity of microglia. Glia 7: 111–118.
10. Mallat M, Chamak B, Thery C (1994) Neurotoxic and neurite promoting effects of brain macrophages. In First European Meeting on Glial Cell Function in Health and Disease, Heidelberg, March 24–27, abstract book; p. 41.
11. Norton WT, Aquino DA, Hozumi I, Chiu FC, Brosnan CF (1992) Quantitative aspects of reactive gliosis: A review. Neurochem Res 17: 877–885.
12. Bruner G, Murphy S (1990) ATP-evoked arachidonic acid mobilization in astrocytes is via a P_{2Y}-purinergic receptor. J Neurochem 55: 1569–1575.
13. Neary JT, Norenberg MD (1992) Signaling by extracellular ATP: Physiological and pathological considerations in neuronal-astrocytic interactions. In Yu ACH, Hertz L, Norenberg MD, Sykova E, Waxman SG (eds.), Progress in Brain Research, Vol 94. Amsterdam: Elsevier, pp. 145–151.
14. Ogata T, Nakamura Y, Tsuji K, Shibata T, Kataoka K, Schubert P (1994) Adenosine enhances intracellular Ca^{2+} mobilization in conjunction with metabotropic glutamate receptor activation by t-ACPD in cultured hippocampal astrocytes. Neurosci Lett 170:5–8.

15. Abbracchio MP, Saffrey MJ, Hopker V, Burnstock G (1994) Modulation of astroglial cell proliferation by analogues of adenosine and ATP in primary cultures of rat striatum. Neuroscience 59:67–76.
16. Rathbone MP, Middlemiss PJ, Kim J-K, Gysbergs JW, DeForge SP, Smith RW, Hughes DW (1992) Adenosine and its nucleotides stimulate proliferation of chick astrocytes and human astrocytoma cells. Neurosci Res 13: 1–17.
17. Meininger CJ, Granger HJ (1990) Mechanisms leading to adenosine-stimulated proliferation of microvascular endothelial cells. Am J Physiol (Heart Circ Physiol 27) 258:H198–H206.
18. Ciccarelli R, Di Iorio P, Ballerini P, Ambrosini G, Giuliani P, Tiboni GM, Caciagli F (1994) Effects of exogenous ATP and related analogues on the proliferation rate of dissociated primary cultures of rat astrocytes. J Neurosci Res, in press.
19. Rathbone MP, Hindley S, Gysbers JW, Gabel B (1991) Purine nucleosides and nucleotides stimulate astrocyte proliferation in vitro and astrogliosis in vivo. Soc Neurosci Abstr 17: 221.3.
20. Kettenmann H, Hoppe D, Gottmann K, Banati R, Kreutzberg G (1990) Cultured microglial cells have a distinct pattern of membrane channels different from peritoneal macrophages. J Neurosci Res 26:278–287.
21. Ilschner S, Ohlemeyer C, Kettenmann H (1994) ATP induces an increase of intracellular calcium in cultured mouse microglial cells. In First European Meeting on Glial Cell Function in Health and Disease, Heidelberg, March 24–27, abstract book, p. 105.
22. Ilschner S, Kettenmann H (1994) Regulation of the inward rectifier in microglial cells by G-proteins and intracellular calcium. In First European Meeting on Glial Cell Function in Health and Disease, Heidelberg, March 24–27, abstract book, p. 105.
23. Norenberg W, Langosch JM, Gebicke-Haerter PJ, Illes P (1994) Characterization and possible function of adenosine 5′-triphosphate receptors in activated rat microglia. Br J Pharmacol 111: 942–950.
24. Abbracchio MP, Burnstock G (1994) Purinoceptors: Are there families of P_{2X} and P_{2Y} purinoceptors? Pharm Ther, in press.
25. McRae A, Rudolphi KA, Keil M (1993) Forebrain ischemia in mongolian gerbils: A model to investigate Alzheimer's microglial antibodies and amyloid. Soc Neurosci Abstr 19:260.8.
26. McRae A, Rudolphi KA, Schubert P (1994) Propentofylline depresses amyloid and Alzheimer's CSF microglial antigens after ischaemia. NeuroReport, in press.
27. Amedee T, Despeyroux S (1994) Membrane responses to externally applied ATP in mouse Schwann cells from organotypic culture of dorsal root ganglia. In First European Meeting on Glial Cell Function in Health and Disease, Heidelberg, March 24–27, abstract book, p. 51.
28. Kirischuk S, Sherer J, Kettenmann H, Verkhratsky A (1994) ATP induces cytoplasmic Ca^{2+} release via activation of $InsP_3$-sensitive channels in oligodendrocytes. In First European Meeting on Glial Cell Function in Health and Disease, Heidelberg, March 24–27, abstract book, p. 113.
29. Kirischuk S, Scherer J, Kettenmann H, Verkhratsky A (1994) ATP triggers Ca^{2+} mobilization in Bergmann glial cells via activation of P_2 purinoreceptors. In First European Meeting on Glial Cell Function in Health and Disease, Heidelberg, March 24–27, abstract book, p. 113.

32. INHIBITORY ACTION OF ADENOSINE AND ADENINE DINUCLEOTIDES ON SYNAPTIC TRANSMISSION IN THE CENTRAL NERVOUS SYSTEM

Peter Schubert
Jesus Pintor
M. Teresa Miras-Portugal

Synaptic transmission is modulated in a complex manner by adenosine, ATP, and adenine dinucleotides via distinct receptors influencing membrane ion conductances, intracellular Ca^{2+} homeostasis, and molecular signaling. All these mechansims seem to act together in modulating presynaptic transmitter release as well as the response of the postsynaptic neuron, which is further dependent on the function of inhibitory feedback loops. Effects on astrocytes that contribute to the control of the extracellular glutamate concentration add to the complexity of purinergic actions in modulating the physiologic function of a neuronal network. The predominant inhibitory influence of purines on excitatory synaptic transmission and intracellular Ca^{2+} accumulation may also provide endogenous neuroprotection against a pathologic activation of Ca^{2+}-dependent intracellular signaling, which is thought to play a pathogenic role for seizure generation and nerve cell damage [1]. The different sites (Fig. 32-1) and presumed mechanisms of the purinergic modulatory influence on synaptic transmission will be outlined, referring in particular to recent studies that have been presented at this symposium and to our own results.

Modulation of Presynaptic Transmitter Release by Adenosine

An inhibitory influence of adenosine on the presynaptic release of excitatory transmitters via G-protein–coupled A_1 receptor actions is well established [2]. There is evidence from studies on brain slices and cultured neurons that the underlying ionic mechanisms include enhancement of the membrane K^+ conductance and reduction of the presynaptic Ca^{2+} influx due to inhibition of L-type, ω conotoxin-sensitive, and probably Q-type Ca^{2+} channels [3]. However, studies performed at the neuromuscular junction indicate that adenosine may also reduce presynaptic transmitter release at a step beyond the Ca^{2+} influx by affecting the release process itself

[4]. Such an action was proposed several years ago by Silinsky, who suggested that adenosine acts presynaptically by decreasing the intracellular affinity of Ca^{2+} to a strategic component of the secretory apparatus [5]. This was also supposed to be the responsible mechanism by which adenosine prevented the recovery of synaptic transmission at reduced extracellular Ca^{2+} concentrations by K^+ channel blockers [6].

In contrast to the A_1-receptor–mediated inhibition, adenosine may enhance transmitter release through A_{2a} receptors, which are enriched in GABAergic neurons of the striatum. In the hippocampus, A_{2a} receptors are colocalized with A_1 receptors; their activation reduced the inhibitory effect of adenosine on synaptic transmission [7]. ATP has been reported to exert an excitatory effect in some central systems, such as the prepiriform cortex, where ATP may exert seizures via an P_{2x}-receptor–mediated action [8]. Since the ATP hydrolyzing enzymes, such as the astrocyte-associated 5′-nucleotidase, show a high activity in the brain, ATP is rapidly broken down to adenosine and usually acts as an inhibitor via this modulator.

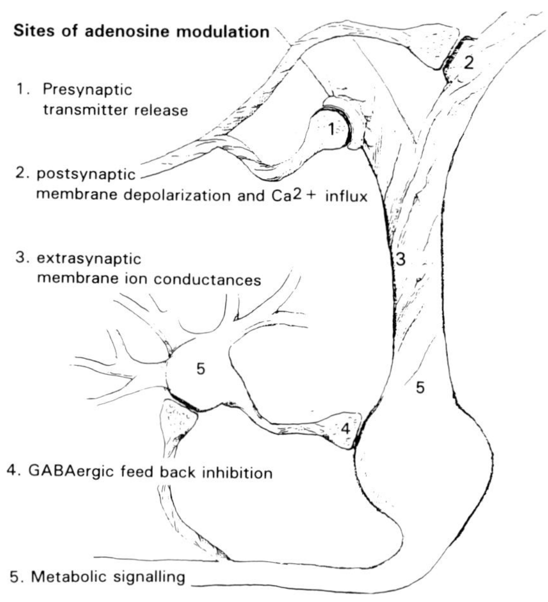

FIGURE 32-1. Presynaptic, postsynaptic, and extrasynaptic effects as well as interactions with intracellular molecular signaling contribute to modulation of synaptic transmission by adenosine, ATP, and diadenosine polyphosphates. The excitatory and inhibitory synaptic connections of a hippocampal CA1 pyramidal cell are taken as a representative neuronal model to indicate the sites of relevant purinergic actions.

Modulation of Postsynaptic Depolarization and Ca^{2+} Influx by Adenosine

Excitatory transmitters released to the extracellular space cause a postsynaptic membrane depolarization via the activation of AMPA/kainate receptors. This leads to a further depolarizing Ca^{2+} influx through voltage-dependent Ca^{2+} channels. At high synaptic input frequencies, the postsynaptic membrane depolarization may reach the critical level required to remove the Mg^{2+}-dependent voltage sensitivity of the NMDA receptor/channel complex. This initiates an additional NMDA Ca^{2+} influx known to act as an excitatory synaptic amplifier. Adenosine reduced the amount of postsynaptic membrane depolarization elicited by repetitive synaptic activation of CA1 neurons in rat hippocampal slices and depressed the evoked synaptic Ca^{2+} influx by A_1-receptor–mediated actions [9]. One micromole of adenosine completely prevented the generation of NMDA Ca^{2+} influx at 20-Hz afferent activation, whereas a blockade of the A_1 receptors by DPCPX allowed a significant synaptic Ca^{2+} influx through NMDA-receptor–operated channels, even at a reduced input frequency [10]. However, if tested in Mg-free medium, even 20 μM adenosine was ineffective, indicating that the Mg-dependent voltage sensitivity of the NMDA channel complex is a prerequisite for a significant modulation of synaptically evoked Ca^{2+} influx by adenosine. Thus, the modulation of the neuronal Ca^{2+} influx by adenosine is indirectly achieved by limiting the amount of postsynaptic membrane depolarization. The ionic mechanisms for this membrane-potential–stabilizing effect of adenosine have been studied in patch-clamp experiments in cultured hippocampal neurons. Activation of the membrane conductances for K^+ [11] and Cl^- ions [12] via G-protein–coupled A_1 receptors seems to be responsible.

Adenosine Actions at Extrasynaptic Sites and on GABA Inhibition

Since A_1 receptors are present at synaptic as well as at extrasynaptic sites [13], adenosine may not only counteract the transmitter-evoked postsynaptic depolarization at the activated synapses but also stabilize the resting membrane potential of the nonactivated parts of the postsynaptic neuron. This should antagonize dendritic signal propagation and spike generation at the soma. The relative size of the population spike and related somatic Ca^{2+} influx elicited in rat hippocampal slices by repetitive activation was depressed by 1 μM adenosine and markedly enhanced during blockade of the A_1-receptor–mediated action [9]. This could imply that adenosine counteracts the "activity-dependent disinhibition" of the soma membrane. The latter is attributed to an impairment of the feedback GABA inhibition due to intracellular Cl^- accumulation during repetitive stimulation [14]. By activating an inward-rectifying, non-GABAergic Cl^- current that is operative at the resting potential and facilitates Cl^- ourflow at the dendrites [12], adenosine could maintain a drain system that counteracts the activity-dependent Cl^- accumulation and maintains the feedback GABA inhibition of the soma membrane. Such an indirect, Cl^--related action of adenosine is supported by the finding that inhibition of the somatic Ca^{2+} influx by adenosine was largely reduced in the presence of a Cl^- channel blocker [15].

Such maintenance of the somatic (GABA) inhibition by adenosine was only seen during repetitive stimulation, when intracellular Cl^- accumulation is thought to reach a critical level. However, the paired pulse inhibition elicited in hippocampal slices by activating the feedback GABA loop with a single stimulus pulse was even reduced by 20 μM adenosine [16]. This presumably reflects the interference of increased adenosine concentrations with the activity of the inhibitory GABAergic interneurons, as supported by patch-clamp experiments on synaptically connected hippocampal nerve cell cultures [17].

In addition, these patch-clamp experiments provided evidence for regulation of the GABAergic feedback inhibition by protein kinase C (PKC). The spontaneous interneuron-evoked Cl^- currents, recorded in the synaptically connected pyramidal neurons, were increased by phorbol esters and blocked by the unselective PKC inhibitor H7. Ongoing experiments in rat hippocampal slices indicate that a transient activation of the H7-sensitive molecular signaling controls the strength of somatic GABA inhibition during repetitive activation of CA1 pyramidal neurons. Using the bicuculline-sensitive somatice Ca^{2+} influx as a measure for the power of the feedback GABA inhibition, we found that H7 led to a marked disinhibition of the somatic Ca^{2+} influx, which was no longer seen if the GABA inhibition was blocked a priori by bicuculline [17]. Therefore, repetitive synaptic activation seems to activate a H7-sensitive metabolic process (PKC-, PKA stimulation?), which maintains the feedback GABA inhibition of the somatic nerve cell membrane. The Ca^{2+}-dependent and phosphoplipase C (PLC)-linked signaling cascade that favors, for example, the activation of PKC is stimulated by a more recently recognized group of purinergic modulators, that is, by adenine dinucleotides.

Modulatory Role of Adenine Dinucleotides

Diadenosine polyphosphates (ApnA) represent a family of compounds formed by two adenosines linked by three to six phosphates (Fig. 32-2).

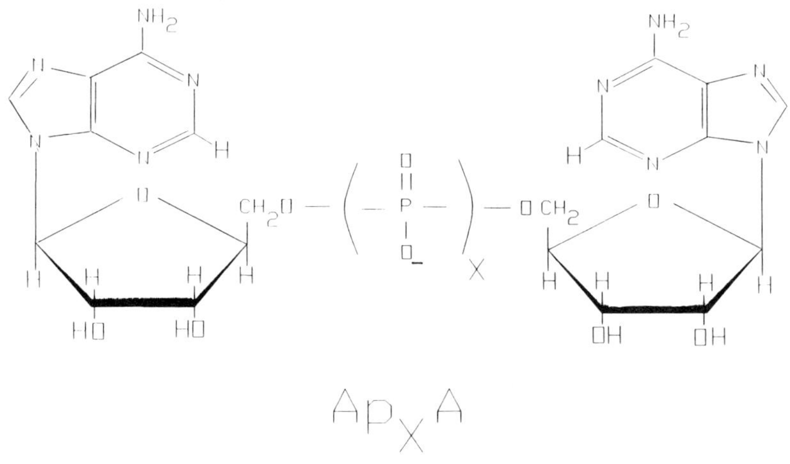

FIGURE 32-2. Structure of diadenosine polyphosphates. The members of this family of dinucleotides are characterized by a different number of phosphates bridging between the two adenosines. x indicates that the number of phosphates varies between three and six in the natural samples where these compounds have been described.

All these diadenosine phosphates are present in the dense granules of platelets [18,19]. The diadenosine polyphosphates—Ap_4A, Ap_5A, and Ap_6A—are also stored in secretory granules together with classic neurotransmitters, such as adrenaline, noradrenaline, and acetylcholine, and may act as cotransmitters. Their presence together with ATP in the chromaffin cells, a model of the peripheral monoaminergic system, has been described [20], and the coexistence of Ap_6A together with catecholamines has recently been demonstrated [21]. In synaptic vesicles of the electric ray *Torpedo marmorata*, a model of the peripheral cholinergic system, Ap_4A and Ap_5A are stored together with acetylcholine [21]. Our finding that rat brain synaptic terminals contain measurable amounts of Ap_4A and Ap_5A together with ATP demonstrates the presence of diadenosine polyphosphates in the central nervous system. Here, the concentrations of Ap_4A and Ap_5A were 169 pmol/mg and 160 pmol/mg synaptosomal proteins, respectively. In response to depolarizing agents, these dinucleotides are released together with ATP in a Ca^{2+}-dependent manner. The ratio of ATP/Ap_4A and of ATP/Ap_5A was 25, which is the same ratio observed in synaptic vesicle preparations [21].

If released to the extracellular space, diadenosine polyphosphates exert receptor-mediated physiologic actions, as shown in the chromaffin cell model [22], in *Torpedo* synaptosomes [21], in the vas deferens, and in the urinary bladder. The established purinoreceptors, P_{2X} and P_{2Y}, were responsible for mediating the actions of diadenosine polyphosphates in all these peripheral nervous systems [23].

In preparations of rat brain synaptic terminals, however, we recently found evidence for the existence of a thus far undescribed $A_{pn}A$-activated P_2 purinoreceptor, which is characterized by a distinct pharmacology and shows a higher affinity for diadenosine polyphosphates than for ATP. This new purinoceptor, named P_{2d} [24], may be responsible for the observed inhibitory influence of adenine dinucleotides in the central nervous system. It was found to depress the firing rate of cortical neurons [25] and synaptically evoked potentials in rat hippocampal slices. By addition of phospholiphase C inhibitors to the slice preparation, we could partly reverse this inhibitory effect of Ap_4A and Ap_5A [26]. This supports the notion that inhibitory influence of these adenine dinucleotides on the excitatory synaptic activation of hippocampal neurons is mediated by a metabotropic purinoceptor, probably the P_{2d} receptor.

Extended studies on rat brain synaptic terminals revealed that ApnA also trigger Ca^{2+} entry via an additional mechanism that is not sensitive to ATP (Fig. 32-3). Thus, this effect seems to be mediated by a purinoreceptor that does not belong to the P_2 family. It may therefore represent a new purinoceptor, named the P_4 receptor [22]. Activation of both the supposed P_4 receptor and the metabotropic P_{2d} receptor have to be assumed to work together to raise the intracellular Ca^{2+} concentration in the synaptic terminals. This may influence the Ca^{2+}-dependent molecular signaling and cause longer lasting changes in presynaptic transmitter release. In this context, it is interesting to note that the half-lives of diadenosine polyphosphates in the brain seem to be extremely long. Push-pull experiments performed in conscious rats revealed that Ap_4A and Ap_5A remained present in the perfusate even more than 90 minutes after amphetamine stimulation, whereas ATP and ADP were undetectable [27]. The most likely explanation is that the diadenosine polyphosphates are hydrolyzed at a much lower rate. The activity of the responsible ectonucleotidase has not been determined in the brain but may be comparable to the activity of the ecto-diadenosine polyphosphate hydrolase in neurochromaffin cells, which has been reported to be 1000 times lower than that described for ecto-ATPase [28]. These data support that diadenosine polyphosphates may act as long-term modulatory signals, leading to sustained changes in the function of a neuronal network.

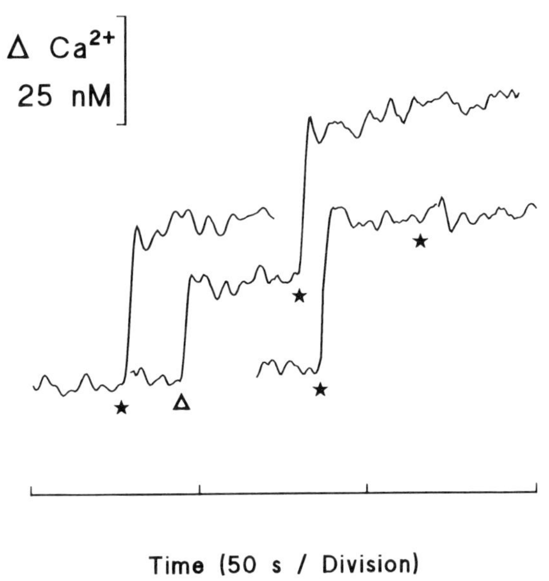

FIGURE 32-3. Evidence for an P_4-purinoceptor–mediated intracellular Ca^{2+} rise in rat brain synaptic terminals. Intracellular Ca^{2+} measurements performed with fura-2 dye in rat brain synaptic terminals indicate that the Ca^{2+} rise induced by diadenosine polyphosphates is mediated by a receptor that is not desensitized by ATP. The first record (left) shows the Ca^{2+} increase induced by 100 μM Ap_5A (solid star). The second record (middle) shows the Ca^{2+} increases induced by 100 μM ATP (open triangle), which does not reduce the further Ca^{2+} increase induced by the additional application of 100 μM Ap_5A (solid star). The third record (right) shows the Ca^{2+} increase elicited by 100 μM Ap_5A (solid star), which completely abolishes a further effect on the Ca^{2+} increase by additional application of 100 μM Ap_5A (second solid star).

Interactions with Intracellular Molecular Signaling

The ApnA-induced inhibition of synaptically evoked depolarizing potentials, observed in rat hippocampal slices, results presumably from presynaptic modulation of excitatory or inhibitory transmitter release. However, depression of the postsynaptic nerve cell response, possibly by influencing H7-sensitive feedback GABA inhibition (see earlier discussion), cannot be excluded and requires further investigation. As indicated by results from experiments on synaptic terminals, this inhibitory effect of

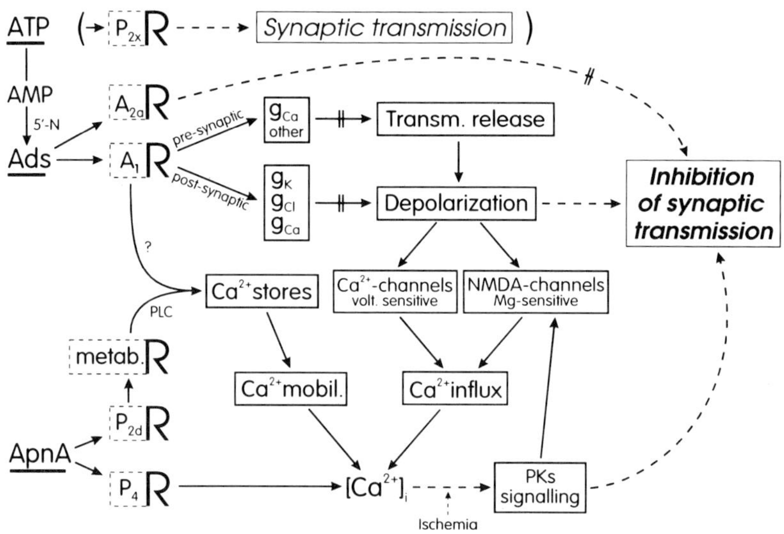

FIGURE 32-4. Flow diagram showing the presumed actions of ATP, adenosine (Ads), and diadenosine polyphosphates (ApnA) involved in purinergic modulation of synaptic transmission.

ApnA seems to be achieved by stimulation of the Ca^{2+}-dependent and PLC-linked signaling cascade, which may lead to PKC activation.

On the other hand, PKC activation has also been reported to potentiate excitatory synaptic transmission by reducing the Mg-dependent voltage sensitivity of the NMDA channel complex [29]. Accordingly, pathologic PKC activation could be responsible for the reduction of the voltage-dependent modulation of the synaptic NMDA Ca^{2+} influx by adenosine, as has been observed in gerbil hippocampal slices prepared 24 hours after preceding in vivo ischemia [17].

Such postischemic PKC activation has been described in the hippocampus [30] as well as in heart muscle [31]. This may be accompanied by pathologic activation of metabotropic receptors by the ischemia-induced rise of extracellular glutamate. Under these conditions, adenosine A_1 receptor actions may even be a "paradox": They were found to promote the ischemia-induced PKC activation and damage (?) in myocardial cells [31] and to stimulate the PLC-linked intracellular Ca^{2+} rise in muscle cells [32].

In brain cells we observed such a paradoxical adenosine A_1 receptor action, causing an increase in the intacellular Ca^{2+} level, but thus far only in astrocytes. Fluorescence imaging experiments on cultured type-1 astrocytes revealed that micromolar concentrations of adenosine A_1 agonists did not significantly influence the intracellular Ca^{2+} level if they were applied alone. But they caused a marked potentiation of the intracellular Ca^{2+} mobilization at nanomolar concentrations if the metabotropic receptors were concomitantly stimulated by t-ACPD [33]. This suggests some similarities between the paradoxical A_1-receptor–mediated actions of adenosine and the P_4/P_{2d}-receptor–mediated actions of diadenosine nucleotides (Fig. 32-4). Both tend to stimulate an intracellular Ca^{2+} rise in conjunction with metabotropic receptor activation. The resulting activation of the Ca^{2+}-dependent cascade of intracellular molecular signals may cause longer lasting changes in nerve and glial cell functions, and thus add to the purinergic modulation of synaptic transmission as summarized in Fig. 32-4.

Acknowledgments

Our studies on diadenosine polyphosphates have been supported by a research grant from the Ministry of Education and Science, DGICYT, PB 92 0230. Our studies on adenosine have been supported by a grant from the Deutsche Forschungsgemeinschaft, SFB 220. We also thank Renate Gleich for skillful technical help and Dr. Michael Knabel for producing the computer graphics.

References

1. Rudolphi KA, Schubert P, Parkinson FE, Fredholm BB (1992) Neuroprotective role of adenosine in cerebral ischemia. Trends Pharmacol Sci 13:439–445.
2. Fredholm BB, Dunwiddie TV (1988) How does adenosine inhibit transmitter release? Trends Pharmacol Sci 9:130–134.
3. Wu LG, Saggau P (1994) Adenosine inhibits evoked synaptic transmission primarily by reducing presynaptic calcium influx in area CA1 of hippocampus. Neuron 12:1139–1148.
4. Scanziani M, Capogna M, Gähwiler BH, Thompson SM (1994) Mechanism of presynaptic inhibition by adenosine. Abstract 1371, 5th. Int. Symp. on Adenosine and Adenine Nucleotides. Drug Dev Res 31:317.
5. Silinsky EM (1984) On the mechanism by which adenosine receptor activation inhibits the release of acetylcholine from motor nerve endings. J Physiol 346:243–256.
6. Schubert P, Heinemann U (1988) Adenosine antagonists combined with 4-aminopyridine cause partial recovery of synaptic transmission in low Ca media. Exp Brain Res 70:539–549.
7. Cunha RA, Johansson B, Sebastiao AM, Ribeiro A (1994) Co-expression, co-localization and functional interaction of the excitatory A_{2a} and inhibitory A_1 adenosine receptors in the hippocampus of the rat. Abstract 1124, 5th Int Symp on Adenosine and Adenine Nucleotides. Drug Dev Res 31:260.
8. Murray TE, Zhao X (1994) P_2 purinoreceptor agonists elicit seizures in the rat prepiriform cortex. Abstract 1274, 5th Int Symp on Adenosine and Adenine Nucleotides. Drug Dev Res 31:302.
9. Schubert P (1988) Physiological modulation by adenosine: Selective blockade of A_1 receptors with DPCPX enhances stimulus train-evoked neuronal Ca influx in rat hippocampal slices. Brain Res 458:162–165.
10. Schubert P, Mager R (1991) The critical input frequency for NMDA receptor-mediated neuronal Ca^{2+} influx depends on endogenous adenosine. Int J Purine Pyrimid Res 2:11–16.
11. Trussell LO, Jackson MB (1987) Dependence of an adenosine activated potassium current on GTP-binding protein in mammalian neurons. J Neurosci 7:3306–3316.
12. Mager R, Ferroni S, Schubert P (1990) Adenosine modulates a voltage-dependent chloride conductance in cultured hippocampal neurons. Brain Res 532:58–62.
13. Tetzlaff W, Schubert P, Kreutzberg GW (1987) Synaptic and extrasynaptic localization of adenosine binding sites in the rat hippocampus. Neuroscience 21:869–975.
14. Thompson SM, Gähwiler BH (1989) Activity-dependent disinhibition. II. effects of extracellular potassium, furosemide, and membrane potential on E_{Cl} in hippocampal CA3 neurons. J Neurophysiol 61:512–523.
15. Schubert P, Ferroni S, Mager R (1991) Pharmacological blockade of Cl^- pumps or Cl^- channels reduces the adenosine-mediated depression of stimulus train-evoked Ca^{2+} fluxes in rat hippocampal slices. Neurosci Lett 124: 174–177.
16. Lee K, Schubert P (1982) Modulation of an inhibitory circuit by adenosine and AMP in the hippocampus. Brain Res 246:311–314.
17. Schubert P, Ferroni S, Nakamura U, Ogata T, Rudolphi KA (1994) Modulation of GABA- and NMDA receptor-dependent synaptic transmission and intracellular Ca^{2+} by adenosine. Abstract 1374, 5th Int Symp on Adenosine and Adenine Nucleotides. Drug Dev Res 31:318.
18. Lüthje J, Ogilvie A (1983) Presence of diadenosine 5′,5′-p1,p4-triphosphate (Ap_3A) in human platelets. Biochem Biophys Res Commun 115:253–260.
19. Schlüter H, Offers E, Brüggemann G, Van der Glet M, Tepel M, Nordhoff E, Karas M, Spiker C, Witzel H, Zidek W (1994) Diadenosine polyphosphates and the physiological control of blood pressure. Nature 367:186–188.
20. Rodriguez del Castillo A, Torres M, Delicado EG, Miras-Portugal MT (1988) Subcellular distribution studies of diadenosine polyphosphates—Ap_4A and Ap_5A—in bovine adrenal medulla: Presence in chromaffin granules. J Neurochem 51:1696–1703.
21. Pintor J, Miras-Portugal MT (1993) Diadenosine polyphoshates—ApnA—as new neurotransmitters. Drug Dev Res 28:259–262.
22. Pintor J, Miras-Portugal MT (1994) P_2-puriner-

gic receptors for diadenosine polyphosphates in the nervous system. Gen Pharmacol, in press.

23. Hoyle CHV (1990) Pharmacological activity of adenine dinucleotides in the periphery. Possible receptor classes and transmitter function. Gen Pharmacol 21:827–831.
24. Pintor J, Díaz-Rey MA, Miras-Portugal MT (1993) Ap_4A and ADP-β-S binding to P_2-purinergic receptors present on rat brain synaptic terminals. Br J Pharmacol 108:1094–1099.
25. Stone TW, Perkins MN (1981) Adenine dinucleotide effect on rat cortical neurones. Brain Res 229:241–245.
26. Klishin A, Lozovaya N, Pintor J, Miras-Portugal, MT, Krishtal OA (1994) Possible functional role of diadenosine polyphosphates: Negative feedback for excitation in the hippocampus. Neuroscience 58:235–236.
27. Pintor J, Porras A, Mora F, Miras-Portugal MT (1993) In vivo induced release of diadenosine polyphosphates—Ap_4A and Ap_5A—from basal ganglia of conscious rat after amphetamine administration. Neurosci Lett 150:13–16.
28. Rodriguez-Pascual F, Torres M, Rotllán P, Miras-Portugal MT (1992) Extracellular hydrolysis of diadenosine polyphosphates, Ap_nA, by bovine chromaffin cells in culture. Arch Biochem Biophys 297:176–183.
29. Ben-Ari Y, Aniksztejn L, Bregestovski P (1992) Protein kinase C modulation of NMDA currents: An important link for LTP induction. Trends Neurol Sci 15:333–339.
30. Onodera H, Araki T, Kogure K (1989) Protein kinase C activity in the rat hippocampus after forebrain ischemia: Autoradiographic analysis by ^{3}H-phorbol 12,13-dibutyrate. Brain Res 481:1–7.
31. Strasser RH, Weinbrenner C, Röthele J, Kübler W, Marquetant R (1994) Blockade of A_1 adenosine receptors prevents senzitization of adenylcyclase and activation of protein kinase C in early myocardial ischemia. Abstract 1380, 5th Int Symp on Adenosine and Adenine Nucleotides. Drug Dev Res 31:325.
32. Gerwins P, Fredholm BB (1992) Stimulation of adenosine A_1 receptors and bradykinin receptors, which act via different G proteins, synergistically raises inositol 1,4,5-trisphosphate and intracellular free calcium in DDT_1 MF-2 smooth muscle cells. Proc Natl Acad Sci USA 89: 7330–7334.
33. Ogata T, Nakamura Y, Tsuji K, Kataoka K, Schubert P (1994) Adenosine enhances intracellular Ca^{2+} mobilization in conjunction with metabotropic glutamate receptor activation by t-ACPD in cultured hippocampal astrocytes. Neurosci Lett 170:5–8.

33. RECEPTORS FOR ATP AT PERIPHERAL NEUROEFFECTOR JUNCTIONS

Geoffrey Burnstock

Autonomic Neuroeffector Junction: Neurotransmission and Neuromodulation

There has been a major change in our understanding of autonomic control mechanisms. It is now clear that there are a multiplicity of autonomic neurotransmitters, and that cotransmission and neuromodulation are involved [1]. The autonomic neuroeffector junction consists of extensive terminal nerve fibers that contain varicosities from which transmitter is released *en passage* to reach smooth muscle cells that are in electrical contact with each other via gap junctions. The varicosities do not have a fixed relationship with particular smooth muscle cells, and the minimum distance between varicosities and muscle cells ranges from about 20 nm in densely innervated tissues, such as the vas deferens and iris, to as much as 2 μm in large elastic arteries. Muscle cells do not have postjunctional specializations. Thus, the autonomic neuromuscular junction differs from synapses (e.g., at the skeletal neuromuscular junction), where there is an established relationship with both presynaptic and postsynaptic specializations.

In addition to noradrenaline (NA) and acetylcholine (ACh), other neurotransmitters are present in autonomic nerves, including adenosine 5′-triphosphate (ATP), 5-hydroxytryptamine (5-HT), dopamine, and a number of peptides, notably neuropeptide Y (NPY), vasoactive intestinal polypeptide (VIP), enkephalin, somatostatin, cholecystokinin, substance P (SP), and calcitonin gene-related peptide (CGRP).

There is now compelling evidence that some, if not all, nerve cells store and release more than one transmitter [2,3]. Systematic studies reveal specific combinations of transmitter substances (chemical coding) for different neuron types that project to particular effector structures and have defined central connections.

A neuromodulator is defined as a substance that modifies the process of neurotransmission. It may act as a prejunctional modulator by decreasing or increasing the amount of neurotransmitter released by a nerve varicosity, or it may act as a postjunctional modulator by altering the time course or extent of action of a transmitter. There are many reports of both prejunctional and postjunctional modulation occurring at the autonomic neuromuscular junction. Neuromodulators may be circulating hormones; local agents such as prostaglandin, histamine, or bradykinin; or neurotransmitters released from other nerves nearby, or even from the same nerve varicosity.

Early History of Involvement of ATP in Autonomic Transmission and of Purinoceptors

In the early 1960s, nonadrenergic, noncholinergic (NANC) nerves [4] were shown to be strongly represented in the gastrointestinal tract of a wide range of vertebrate species and were also identified in a variety of other organs, including lung, bladder, seminal vesicles, esophagus, uterus, eye, trachea, and parts of the cardiovascular system [5]. Burnstock et al. [6] proposed that the principal active substance released from these nerves, at least those in the intestine and bladder, was the purine nucleotide, ATP; and these NANC nerves were termed *purinergic* [7]. Experiments on the CNS showed that adenosine and AMP were able to stimulate the accumulation of cAMP, an effect that was theophylline sensitive [8,9].

Early evidence for prejunctional purinergic receptors came from experiments with the isolated rat phrenic nerve-diaphragm preparation, where adenosine and adenine nucleotides were shown to reduce both the spontaneous and evoked release of ACh from motor nerve terminals [10]. Purine compounds were also shown to inhibit NA release from sympathetic nerves supplying a wide variety of tissues [11]. Prejunctional modulation of ACh release from autonomic cholinergic nerves by purines has been observed in the isolated guinea-pig ileum and the myenteric plexus longitudinal muscle preparation [12]. Purine-mediated prejunctional modulation was also demonstrated at central synapses [13]. It was not until 1979, however, that prejunctional neuromodulation of transmitter release from most nerve terminals was shown to be mediated by receptors for adenosine but not for ATP [14].

In 1978 the basis for distinguishing two types of purinergic receptor (P_1 and P_2) was proposed and relied largely on an analysis of the voluminous literature about the actions of purine nucleotides and nucleosides on a wide variety of tissue [15]. Since that time many experiments have been carried out that support and extend this proposal [16]. P_1 purinoceptors are more responsive to adenosine and AMP than to ATP and ADP; methylxanthines, such as theophylline and caffeine, are selective competitive antagonists with respect to these receptors; and occupation of these receptors leads to inhibition or activation of an adenylate cyclase system, with resultant changes in levels of intracellular cAMP. P_2 purinoceptors are more responsive to ATP and adenosine; these receptors are not antagonized by methylxanthines and do not act via an adenylate cyclase system, and their occupation may lead to prostaglandin synthesis. Subsequently, A_1 and A_2 subclasses of P_1 purinoceptor were defined partly on the basis of the relative potencies of several adenosine analogues and partly according to whether their activation stimulated or inhibited adenylate cyclase activity [17,18]. Subdivision of the P_2 purinoceptor into P_{2X} and P_{2Y} subclasses was proposed by Burnstock and Kennedy [19], and later P_{2t} and P_{2Z} purinoceptors were defined by Gordon [20].

Purinergic Cotransmission and Neuromodulation

ATP appears to be a cotransmitter in many, if not all, different nerve types [3,21].

SYMPATHETIC COTRANSMISSION

There is a substantial body of evidence to show that noradrenaline (NA) and ATP act as cotransmitters, being released from sympathetic nerves in variable proportions depending on the tissue and species [22,23]. Many of the early and more detailed studies were made on the vas deferens [24]. Evidence for purinergic cotransmission includes block of the prazosin-resistant component of the response to sympathetic nerve stimulation by the ATP antagonists arylazido amino-proprionyl-ATP ($ANAPP_3$), suramin, or by the selective desensitizer of the P_{2X} purinoceptor, α,β-methylene ATP; release of ATP during nerve stimulation, which is abolished by 6-hydroxydopamine, but is unaffected by selective depletion of NA by reserpine; selective block by ATP antagonists of excitatory junction potentials (EJPs) in response to sympathetic nerve stimulation and mimicry of EJPs by ATP but not by NA.

Studies of sympathetic cotransmission involving ATP and NA have now also been carried out on a number of different isolated blood vessels [25–28] and in whole animals [29,30]. In rabbit coronary vessels, in contrast to most other vessels, where NA and ATP cause synergistic constriction via α_1 adrenoceptors and P_{2X} purinoceptors, respectively, the predominant effect of ATP is vasodilatation via P_{2Y} purinoceptors [31]. Because in this vessel the predominant effect of NA is vasodilatation via β-adrenoceptors, this is consistent with the synergism that appears to be characteristic of cotransmission.

While early studies established adenosine as the principal modulator of transmitter release from sympathetic nerves via prejunctional P_1 purinoceptors [14,32], several authors have also identified prejunctional P_2 purinoceptors on autonomic nerve terminals in various tissues [33–36]. In the rat vas deferens and iris it has been suggested that these prejunctional ATP receptors are probably P_{2Y} purinoceptors [37,38]. Recent experiments suggest that receptors for naturally occurring dinucleotides on nerve terminals in the brain can also modulate the release of transmitters [39,40].

PARASYMPATHETIC COTRANSMISSION

ACh and ATP appear to be cotransmitters in parasympathetic neurons in the bladder [41–43]. It has been shown recently that there is plasticity in expression of cotransmitters in parasympathetic nerves supplying the human urinary bladder, in that the ratio of purinergic to cholinergic components of excitatory transmission is much increased in interstitial cystitis [44]. There is also indirect evidence for ATP being colocalized with ACh and the peptides NPY or somatostatin in subpopulations of intrinsic neurons in the heart and airways that project to small blood vessels in these organs [45,46].

SENSORY-MOTOR COTRANSMISSION

Subpopulations of sensory nerves have been claimed to use ATP as well as SP and CGRP [47–50]. By analogy with other systems, it seems likely that ATP coexists in different proportions, with these two peptides perhaps cooperating in axon reflex activity [51]. Since the role of these nerves during axon reflexes to many organs is motor rather than sensory [52], they have been termed *sensory-motor nerves* to distinguish them from the other subpopulation of afferent fibers that have a pure sensory role and whose terminals contain few vesicles and a predominance of mitochondria [53].

NONADRENERGIC, NONCHOLINERGIC (NANC) INHIBITORY COTRANSMISSION IN THE GUT

Evidence that ATP was the NANC inhibitory transmitter in nerves supplying the guinea-pig taenia coli was first presented in 1970 [54]. In the late 1970s evidence for VIP as a slow NANC inhibitory transmitter in various regions of the gastrointestinal tract was presented [55], and recently it has been proposed that nitric oxide (NO) is also a NANC inhibitory transmitter in the gut [56]. The current consensus of opinion is that ATP, VIP, and NO are probably cotransmitters mediating responses with different time courses in NANC inhibitory nerves but that their proportions vary considerably in different regions of the gut [57–60].

SOMATIC NEUROMUSCULAR TRANSMISSION

There are many reports of the coexistence and release of ATP with ACh at somatic motor nerve endings [53]. ATP acts as a neuromodulator of both the release (via adenosine) and action of ACh. There is little information available to date to support a cotransmitter role for ATP at these junctions of mature adults, but patch-clamp studies have shown that micromolar concentrations of ATP, as well as ACh, activate channels in the membrane of cultured myoblasts and myotubes [61–63].

A New Concept for P_2-Purinoceptor Subclassification

Concerns have been expressed about the recent additions to P_2-purinoceptor subclasses representing an "apparent random walk through the alphabet" with the introduction of P_{2U}, P_{2D}, and P_{2S} purinoceptors, in addition to P_{2X}, P_{2Y}, P_{2Z}, and P_{2T} purinoceptors [64,65]. It was this background that motivated Abbracchio and Burnstock [66] to propose a new broad framework for P_2-purinoceptor subclassification that will allow more rational development as more subtypes and molecular structures are recognized. Essentially, they have suggested that P_2 purinoceptors should be divided into two main families: P_{2X} purinoceptors mediating fast responses via ligand-gated ion channels and P_{2Y} purinoceptors mediating slower responses via G proteins. Further subdivisions in each of these two major families is suggested on the basis of cloned receptors [67–70] and the recent development of new selective agonists [71–73]. This basic framework of fast and slow receptors is in accordance with that developed for other neurotransmitters, including ACh (nicotinic and muscarinic receptors), GABA ($GABA_A$ and $GABA_B$), and glutamate (ionotropic and metabotropic receptors) (Table 33-1). While the precise identification and numbering of the subclasses within these two major P_2-purinoceptor families is likely to be debated in the coming period, it is hoped that the majority of readers will find this a constructive and attractive road forward.

TABLE 33-1. Comparison of the fast (via ligand-gated ion channels) and slow (via G protein) receptors for the major neurotransmitters acetylcholine (ACh), gamma-amino butyric acid (GABA), and glutamate with the newly proposed fast (P_{2X}) and slow (P_{2Y}) families of receptors for ATP

	Receptors	
Messenger	Fast (ligand-gated ion channels)	Slow (G proteins)
ACh	Nicotinic	Muscarinic
	Muscle type	M1
	Neuronal type	↓
		M5
GABA	GABA A	GABA B
Glutamate	Ionotropic	Metabotropic
	AMPA	$mGlu_1$
	Kainate	↓
	NMDA	$mGlu_7$
ATP	P2X	P2Y
	$P2X_1$	$P2Y_1$
	↓	↓
	$P2X_3$	$P2Y_7$

From Abbracchio and Burnstock [66], with permission.

References

1. Burnstock G (1986) The changing face of autonomic neurotransmission. (The First von Euler Lecture in Physiology). Acta Physiol Scand 126:67–91.
2. Burnstock G (1976) Do some nerve cells release more than one transmitter? Neuroscience 1: 239–248.
3. Burnstock G (1990) Co-transmission. The Fifth Heymans Lecture—Ghent, February 17, 1990. Arch Int Pharmacodyn Ther 304:7–33.
4. Burnstock G, Campbell G, Bennett M, Holman ME (1964) Innervation of the guinea-pig taenia coli: Are there intrinsic inhibitory nerves which are distinct from sympathetic nerves? Int J Neuropharmacol 3:163–166.
5. Burnstock G (1969) Evolution of the autonomic innervation of visceral and cardiovascular systems in vertebrates. Pharmacol Rev 21: 247–324.
6. Burnstock G, Campbell G, Satchell D, Smythe A (1970) Evidence that adenosine triphosphate or a related nucleotide is the transmitter substance released by non-adrenergic inhibitory nerves in the gut. Br J Pharmacol 40:668–688.
7. Burnstock G (1972) Purinergic nerves. Pharmacol Rev 24:509–581.
8. Sattin A, Rall TW (1970) The effect of adenosine and adenine nucleotides on the cyclic

adenosine 3′,5′-phosphate content of guinea-pig cerebral cortex slices. Mol Pharmacol 6:13–23.

9. Pull I, McIlwain H (1972) Adenine derivatives as neurohumoral agents in the brain. The quantities librated on excitation of superfused cerebral tissues. Biochem J 130:975–981.
10. Ginsborg BL, Hirst GDS (1972) The effect of adenosine on the release of the transmitter from the phrenic nerve of the rat. J Physiol (Lond) 224:629–645.
11. Hedqvist P, Fredholm BB (1976) Effects of adenosine on adrenergic neurotransmission: Prejunctional inhibition and postjunctional enhancement. Naunyn Schmiedebergs Arch Pharmacol 293:217–223.
12. Sawynok J, Jhamandas KH (1976) Inhibition of acetylcholine release from cholinergic nerves by adenosine, adenine-nucleotides and morphine antagonism by theophylline. J Pharmacol Exp Ther 197:379–390
13. Phillis JW, Edstrom JP (1976) Effects of adenosine analogs on rat cerebral cortical neurons. Life Sci 19:1041–1053.
14. De Mey J, Burnstock G, Vanhoutte PM (1979) Modulation of the evoked release of noradrenaline in canine saphenous vein via presynaptic receptors for adenosine but not ATP. Eur J Pharmacol 55:401–405.
15. Burnstock G (1978) A basis for distinguishing two types of purinergic receptor. In Straub RW, Bolis L (eds.), Cell Membrane Receptors for Drugs and Hormones: A Multidisciplinary Approach. New York: Raven Press, pp. 107–118.
16. Jacobson KA, van Galen JM, Williams M (1992) Adenosine receptors: Pharmacology, structure-activity relationships, and therapeutic potential. J Med Chem 35:407–422.
17. Von Calker D, Müller M, Hamprecht B (1979) Adenosine regulates via two different types of receptors the accumulation of cyclic AMP in cultured brain cells. J Neurochem 33: 999–1005.
18. Londos C, Cooper DMF, Wolff F (1980) Subclasses of external adenosine receptors. Proc Natl Acad Sci USA 77:2551–2554.
19. Burnstock G, Kennedy C (1985) Is there a basis for distinguishing two types of P_2-purinoceptor? Gen Pharmacol 16:433–440.
20. Gordon JL (1986) Extracellular ATP: Effects, source and fate. Biochem J 233:309–319.
21. Kupfermann I (1991) Functional studies of cotransmission. Physiol Rev 71:683–711.
22. Burnstock G (1990) Noradrenaline and ATP as cotransmitters in sympathetic nerves. Neurochem Int 17:357–368.
23. Evans RJ, Cunnane TC (1992) Relative contributions of ATP and noradrenaline to the nerve evoked contraction of the rabbit jejunal artery. Naunyn Schmiedebergs Arch Pharmacol 345:424–430.
24. Stjärne L (1989) Basic mechanisms and local modulation of nerve impulse-induced secretion of neurotransmitters from individual sympathetic nerve varicosities. Rev Physiol Biochem Pharmacol 112:1–137.
25. Burnstock G (1988) Sympathetic purinergic transmission in small blood vessels. Trends Pharmacol Sci 9:116–117.
26. Schwartz DD, Malik KU (1989) Renal periarterial nerve stimulation-induced vasoconstriction at low frequencies is primarily due to release of a purinergic transmitter in the rat. J Pharmacol Exp Ther 250:764–771.
27. Starke K, Von Kügelgen I, Bulloch JM, Illes P (1991) Nucleotides as cotransmitters in vascular sympathetic neuroeffector transmission. Blood Vessels 28:19–26.
28. Evans RJ, Surprenant A (1992) Vasoconstriction of guinea-pig submucosal arterioles following sympathetic nerve stimulation is mediated by the release of ATP. Br J Pharmacol 106: 242–249.
29. Bulloch JM, McGrath JC (1988) Selective blockade by nifedipine of "purinergic" rather than adrenergic nerve-mediated vasopressor responses in the pithed rat. Br J Pharmacol 95: 695–700.
30. Schlicker E, Urbanek E, Göthert M (1989) ATP, α,β-methylene ATP and suramin as tools for characterisation of vascular P_{2x} receptors in the pithed rat. J Auton Pharmacol 9:370–380.
31. Corr L, Burnstock G (1991) Vasodilator response of coronary smooth muscle to the sympathetic co-transmitters noradrenaline and adenosine 5′-triphosphate. Br J Pharmacol 104:337–342.
32. Sneddon P, Meldrum LA, Burnstock G (1984) Control of transmitter release in guinea-pig vas deferens by prejunctional P_1-purinoceptors. Eur J Pharmacol 105:293–299.
33. Moody CJ, Burnstock G (1982) Evidence for the presence of P_1-purinoceptors on cholinergic nerve terminals in the guinea-pig ileum. Eur J Pharmacol 77:1–9.
34. Wiklund NP, Gustafsson LE (1986) Neuromodulation by adenine nucleotides, as indicated by experiments with inhibitors of nucleotide

inactivation. Acta Physiol Scand 126:217–223.
35. Bao JX, Eriksson IE, Stjärne L (1989) On pre and/or post-junctional roles of ATP, an unknown "substance X" as sympathetic co-transmitters in rat tail artery. Acta Physiol Scand 135:65–66.
36. Sperlagh B, Vizi S (1991) Effect of presynaptic P_2 receptor stimulation on transmitter release. J Neurochem 56:1466–70.
37. Fuder H, Muth U (1993) ATP and endogenous agonists inhibit evoked [^{3}H]-noradrenaline release in rat iris via A_1 and P_{2y}-like purinoceptors. Naunyn Schmiedebergs Arch Pharmacol 348:352–357.
38. Von Kügelgen I, Kurz K, Starke K (1994) P_2-purinoceptor-mediated autoinhibition of sympathetic transmitter release in mouse and rat vas deferens. Naunyn Schmiedebergs Arch Pharmacol 349:125–132.
39. Pintor J, Díaz-Rey MA, Miras-Portugal T (1993) Ap_4A and ADP-β-S binding to P_2 purinoceptors present on rat brain synaptic terminals. Br J Pharmacol 108:1094–1099.
40. Galarreta M, Solís JM, Menéndez N, Conjero C, Martíon del Río R (1993) Nicotinamide adenine dinucleotides mimic adenosine inhibition on synaptic transmission by decreasing glutamate release in rat hippocampal slices. Neurosci Lett 159:55–58.
41. MacKenzie I, Burnstock G, Dolly JO (1982) The effects of purified botulinum neurotoxin type A on cholinergic, adrenergic and non-adrenergic, atropine-resistant autonomic neuromuscular transmission. Neuroscience 7: 997–1006.
42. Hoyle CHV, Burnstock G (1985) Atropine-resistant excitatory junction potentials in rabbit bladder are blocked by α,β-methylene ATP. Eur J Pharmacol 114:239–240.
43. Fujii K (1987) Electrophysiological evidence that adenosine triphosphate (ATP) is a co-transmitter with acetylcholine (ACh) in isolated guinea-pig, rabbit and pig urinary bladder. Proc Physiol Soc 394:26P.
44. Palea S, Artibani W, Ostardo E, Trist DG, Pietra C (1993) Evidence for purinergic neurotransmission in human urinary bladder affected by interstitial cystitis. J Urol 150:2007–2012.
45. Inoue T, Kannan MS (1988) Nonadrenergic and noncholinergic excitatory neurotransmission in rat intrapulmonary artery. Am J Physiol 254: H1142–H1148.
46. Saffrey MJ, Hassall CJS, Allen TGJ, Burnstock G (1992) Ganglia within the gut, heart, urinary bladder and airways: Studies in tissue culture. Int Rev Cytol 136:93–144.
47. Jahr CE, Jessel TM (1983) ATP excites a subpopulation of rat dorsal horn neurones. Nature 304:730–733.
48. Salt TE, Hill RG (1983) Excitation of single sensory neurones in the rat caudal trigeminal nucleus by ionotophoretically applied adenosine 5′-triphosphate. Neurosci Lett 35:53–57.
49. Fyffe REW, Perl ER (1984) Is ATP a central synaptic mediator for certain primary afferent fibers from mammalian skin? Proc Natl Acad Sci USA 81:6890–6893.
50. Krishtal OA, Marchenko SM, Obukhov AG (1988) Cationic channels activated by extracellular ATP in rat sensory neurons. Neuroscience 27:995–1000.
51. Burnstock G (1977) Autonomic neuroeffector junctions—reflex vasodilatation of the skin. J Invest Dermatol 69:47–57.
52. Maggi CA, Meli A (1988) The sensory-efferent function of capsaicin-sensitive sensory neurons. Gen Pharmacol 19:1–43.
53. Burnstock G (1986) Autonomic neuromuscular junctions: Current developments and future directions. The Third Anatomical Society Review Lecture. J Anat 146:1–30.
54. Burnstock G, Campbell G, Satchell D, Smythe A (1970) Evidence that adenosine triphosphate or a related nucleotide is the transmitter substance released by non-adrenergic inhibitory nerves in the gut. Br J Pharmacol 40:668–688.
55. Hoyle CHV, Burnstock G (1989) Neuromuscular transmission in the gastrointestinal tract. In Wood JD (ed.), Handbook of Physiology, Section 6: The Gastrointestinal System, Vol. I: Motility and Circulation. Bethesda, MD: American Physiological Society, pp. 435–464.
56. Rand MJ (1992) Nitrergic transmission: Nitric oxide as a mediator of non-adrenergic, non-cholinergic neuro-effector transmission. Clin Exp Pharmacol Physiol 19:147–169.
57. Shuttleworth CWR, Murphy R, Furness JB (1991) Evidence that nitric oxide participates in non-adrenergic inhibitory transmission to intestinal muscle in the guinea-pig. Neurosci Lett 130:77–80.
58. Sanders KM, Ward SM (1992) Nitric oxide as a mediator of non-adrenergic non-cholinergic neurotransmission. Am J Physiol 262:G379–G392.
59. Saffrey MJ, Hassall CJS, Hoyle CHV, Belai A, Moss J, Schmidt HHHW, Förstermann U, Murad F, Burnstock G (1992) Nitric oxide synthase and NADPH diaphorase activity in

cultured myenteric neurones. Neuroreport 3: 333–336.
60. Belai A, Burnstock G (1994) Evidence for coexistence of ATP and nitric oxide in non-adrenergic, non-cholinergic (NANC) inhibitory neurones in the rat ileum, colon and anococcygeus muscle. Cell Tissue Res 278:197–200.
61. Kolb H-A, Wakelam MJO (1983) Transmitter-like action of ATP on patched membranes of cultured myoblasts and myotubes. Nature 303: 621–623.
62. Hume RI, Hönig MG (1986) Excitatory action of ATP on embryonic chick muscle. J Neurosci 6:681–690.
63. Häggblad J, Heilbronn E (1988) P_2-purinoceptor-stimulated phosphoinositide turnover in chick myotubes. Calcium mobilization and the role of guanyl nucleotide-binding proteins. FEBS Lett 235:133–136.
64. Abbracchio MP, Cattabeni F, Fredholm BB, Williams M (1993) Purinoceptor nomenclature: A status report. Drug Dev Res 28:207–213.
65. Fredholm BB, Abbracchio MP, Burnstock G, Daly JW, Harden TK, Jacobson KA, Leff P, Williams M (1994) VI. Nomenclature and classification of purinoceptors. Pharmacol Rev 46:143–156.
66. Abbracchio M, Burnstock G (1994) Purinoceptors: Are there families of P2X and P2Y purinoceptors? Pharmacol Ther, in press.
67. Webb TE, Simon J, Krishek BJ, Bateson AN, Smart TG, King BF, Burnstock G, Barnard EA (1993) Cloning and functional expression of a brain G-protein coupled ATP receptor. FEBS Lett 324:219–225.
68. Lustig KD, Shiau AK, Brake AJ, Julius D (1993) Expression cloning of an ATP receptor from mouse neuroblastoma cells. Proc Natl Acad Sci USA 90:5113–5117.
69. Parr CE, Sullivan DM, Paradiso AM, Lazarowski ER, Burch LH, Olsen JC, Erb L, Weisman GA, Boucher RC, Turner JT (1994) Cloning and expression of a human P_{2U} nucleotide receptor, a target for cystic fibrosis pharmacotherapy. Proc Natl Acad Sci USA 91:3275–3279.
70. Barnard EA, Burnstock G, Webb TE (1994) G protein-coupled receptors for ATP and other nucleotides: A new receptor family. Trends Pharmacol Sci 15:67–70.
71. Fischer B, Boyer JL, Hoyle CHV, Ziganshin AU, Brizzolara AL, Knight GE, Zimmet J, Burnstock G, Kendall Harden T, Jacobson KA (1993) Identification of potent selective P_{2Y} purinoceptor agonists: Structure-activity relationships for 2-thioether derivatives of adenosine 5′-triphosphate. J Med Chem 36:3937–3946.
72. Burnstock G, Fischer B, Hoyle CHV, Maillard M, Ziganshin AU, Brizzolara AL, von Isakovics A, Boyer JL, Kendall Harden T, Jacobson KA (1994) Structure activity relationships for derivatives of adenosine 5′-triphosphate as agonists at P_2 purinoceptors: Heterogeneity within P_{2X} and P_{2Y} subtypes. Drug Dev Res 31:206–219.
73. Bo X, Fischer B, Maillard M, Jacobson A, Burnstock G (1994) Comparative studies on the affinities of ATP derivatives for P_{2X}-purinoceptors in rat urinary bladder. Br J Pharmacol 112:1151–1159.

34. ANATOMIC-FUNCTIONAL CORRELATES OF ADENOSINE-5′-TRIPHOSPHATE TRIGGERED VAGAL DEPRESSOR REFLEX

Amir Pelleg
Carl M. Hurt
Guennadi Katchanov

Adenosine-5′-triphosphate (ATP) plays a crucial role in intracellular energetics and metabolism. Once outside of the cells, however, ATP exerts pronounced effects on many organs and tissues, including the cardiovascular system [1]. Extracellular ATP exerts negative chronotropic and dromotropic actions on sinus node automaticity and atrioventricular (AV) nodal conduction, respectively [2]. A decade ago we had established the role of the vagus nerve in mediating, in part, these actions of ATP. Specifically, we have shown that ATP is more potent than adenosine in suppressing the sinus and AV nodes, and that this difference between the two compounds is due to the ability of ATP to abruptly and transiently increase vagal input to the heart [3]. Indeed, when the vagal effect is blocked by either surgical or pharmacologic perturbations, there is no difference between the negative chronotropic and dromotropic actions of ATP and adenosine [3]. The vagal component of ATP's action exhibits species variability, that is, it is manifested in cat, dog, and human but not in rat, guinea pig, and rabbit [2]. Thus, using canine models in vivo we have confirmed the vagal component in the cardiac electrophysiologic actions of ATP. Specifically, physostigmine and elevation of plasma Ca^{2+} levels enhanced the actions of ATP, while atropine and verapamil attenuated these actions [3–6]. In addition, prostaglandins, the synthesis and release of which are triggered by ATP, do not mediate the negative chronotropic and dromotropic actions of ATP [7].

In recent years we have studied the anatomic-functional correlates of the reflex-type interaction between ATP and the vagus nerve. These studies were aimed at elucidating the anatomic site of ATP's actions as well as the cascade of events culminating in vagal-dependent suppression of sinus automaticity and AV nodal conduction. These studies are briefly reviewed below.

Site of ATP Action

Early studies have indicated that the peak negative chronotropic and dromotropic effects of intra-right atrial ATP given as a rapid bolus occurred about 10–15 seconds following injection, suggesting a site other than the heart for ATP's action [3]. Indeed, when ATP and adenosine were administered directly into the sinus nodal artery, the two compounds were equipotent in the low- and mid-dosage range, and at high doses adenosine was more potent than ATP in suppressing sinus node automaticity [8]. Furthermore, the effects of both ATP and adenosine were similarly attenuated by aminophylline, an adenosine antagonist. These data were interpreted to suggest that the action of ATP was dependent on its degradation to adenosine by ectoenzymes [8]. In view of the latter and the temporal pattern of ATP's actions, we hypothesized that ATP acts on vagal nerve terminals in the lungs to trigger a depressor vagal reflex.

If this hypothesis were true, one would expect intrapulmonary ATP to be more effective in suppressing sinus node automaticity and AV nodal conduction than similar or greater doses of the nucleotide given into the right atrium. In addition, the time to peak effect of intrapulmonary ATP should be shorter than that of intra-right atrial ATP. Data obtained in recent studies support this hypothesis; specifically, intrapulmonary ATP was more potent than intra-right atrial ATP and the time to peak effect of the former was significantly shorter than that of the latter.

A typical example of these phenomena is given in Figures 34-1 and 34-2. Figure 34-1 shows the negative chronotropic action of ATP (6 μmol/kg) given (at time = 0 seconds) as a rapid bolus into the right atrium (RA) of an anesthetized dog. As can be seen, the sinus cycle length of 410 ms was prolonged maximally to 1980 ms in about 7 seconds following the administration of ATP. When 12 times less ATP, that is, 0.5 μmol/kg, was administered into the right pulmonary artery (RPA) of the same animal, a sinus cycle length of 440 ms was maximally prolonged to 1470 ms in less than 4 seconds following the administration of ATP (Fig. 34-2). Thus, the closer the site of ATP

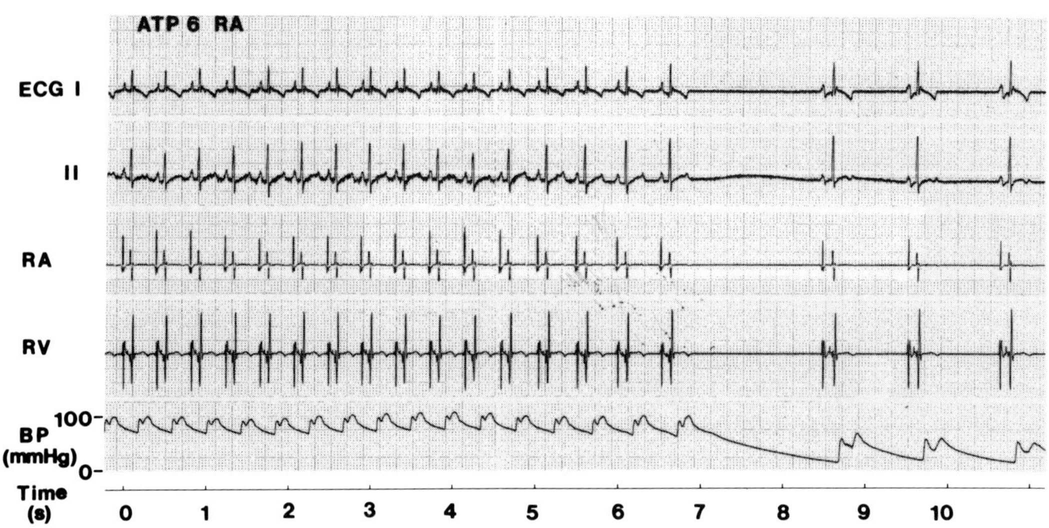

FIGURE 34-1. A typical example of the negative chronotropic action of intra-right atrial (RA) ATP (6 μmol/kg) given as a rapid (≤1 second) bolus. Tracings from top to bottom: Standard lead I and II ECG, RA electrogram, right ventricular (RV) electrogram, systemic arterial blood pressure (BP), and time marks in seconds (s). ATP was given at t = 0. See text for detalis.

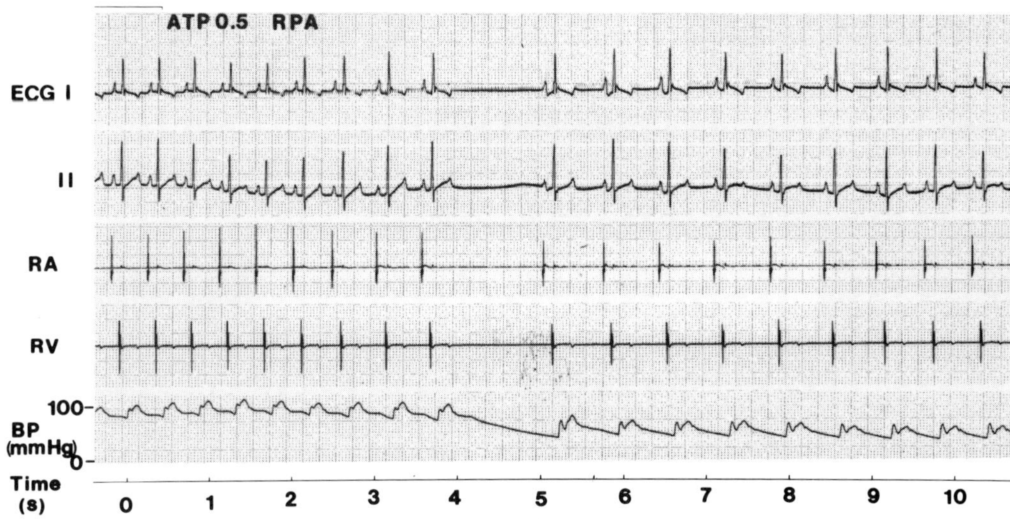

FIGURE 34-2. A typical example of the negative chronotropic action of intra-right pulmonary artery (RPA) ATP (0.5 μmol/kg) given as a rapid bolus. Same animal and tracings as in Figure 34-1. See text of details.

administration is to the site of its action, the more potent ATP is and the faster is the onset of its action. This is due to the rapid degradation of ATP by ectoenzymes [9], which reduce the amount of ATP reaching its active site during intravascular travel time.

Mechanism of ATP Action

The studies discussed above have strongly suggested that there are pulmonary sites sensitive to ATP, stimulation of which results in an abrupt and transient increase in vagal input to the heart. Thus we tested the hypothesis that ATP stimulates chemosensitive vagal afferent terminals in the lungs. For this purpose extracellular recording of the canine cervical vagal C fiber action potentials was used in combination with multiple pharmacologic perturbations. The following criteria were used for identification of pulmonary C fibers [10].

1. Stimulated by capsaicin
2. Quiescent fibers with no regular activity synchronous with respiratory and/or cardiac cycles
3. Conduction velocity <2.5 m/sec
4. Excitation by capsaicin injected upstream but not downstream from the receptors
5. Lack of excitation by capsaicine injection into bronchial circulation
6. Excitation by lung inflation to two to three times the tidal volume (V_T)
7. Excitation by gentle probing of the lung with fine forceps

Figure 34-3 is an example from one of our earlier studies of the negative chronotropic action of intra-right atrial ATP (6 μmol/kg, given as a rapid bolus) and the associated burst of neural action potentials (upper tracing) recorded from a small cervical vagal C fiber bundle. A sinus cycle length (SCL) of 407 ms was maximally prolonged to 2077 ms 10 seconds following the administration of ATP. Examples of more recent recordings are given in Figures 34-4 to 34-6. Figure 34-4 depicts a burst of neural action potentials elicited by capsaicin (10 μg/kg) given as a bolus into the

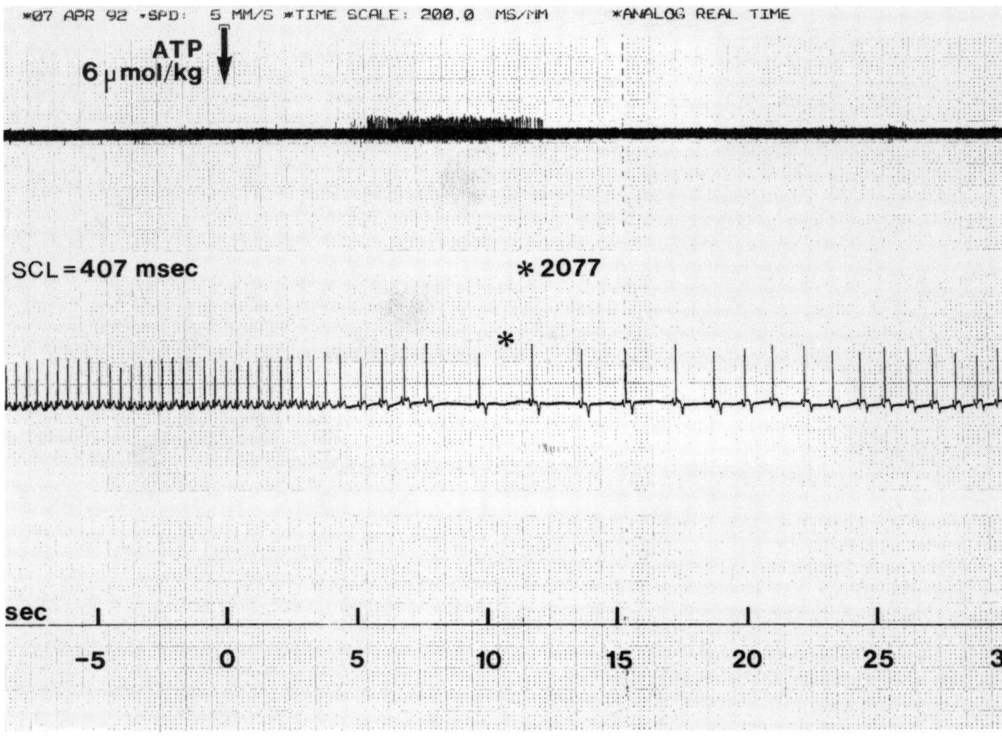

FIGURE 34-3. One of the early recordings of right vagal afferent C fiber activity associated with the negative chronotropic action of intra-right atrial ATP (6 μmol/kg) given as a rapid bolus. Tracings from top to bottom: Extracellularly recorded neuronal action potentials elicited by ATP in a small bundle of slowly conducting, otherwise quiescent, nerve fibers dissected from the right cervical vagosympathetic trunk, standard lead II ECG, and time marks in seconds (sec). ATP was given at t = 0. Note the lack of any cyclic neural activity associated with cardiac cycles. SCL = sinus cycle length. Asterisk denotes the maximal SCL following ATP (2077 ms).

right atrium (RA). Figure 34-5 depicts much shorter lasting bursts elicited in the same fibers by transient abrupt elevation and reduction of tracheal pressure (TP). Figure 34-6 depicts the effects of ATP (6 μmol/kg) given as a bolus into the right atrium. The same fibers that were stimulated by capsaicin and changes in tracheal pressure, were stimulated by ATP. Interstingly, while ATP was less potent than capsaicin in eliciting C fibers' action potentials, it was more potent in concurrently suppressing sinus node automaticity. This suggests that the quantitative relationship between afferent vagal C fiber activity and efferent vagal input to the heart is not necessarily linear. *Alternatively, ATP can affect other pulmonary and/or extrapulmonary nerve terminals that are not sensitive to capsaicin.*

Adenosine, and adenosine monophosphate and diphosphate did not elicit neural action potentials in the same fibers that were stimulated by capsaicin and ATP. This indicates that the action of ATP on pulmonary C fiber nerve terminals is mediated by a P_2 receptor. This is not surprising because a physiologic role of ATP in the autonomic nervous system has been recognized in recent years. Indeed, ATP can depolarize a subset of neurons dissociated from various sensory ganglia of rats, cats, and bull-

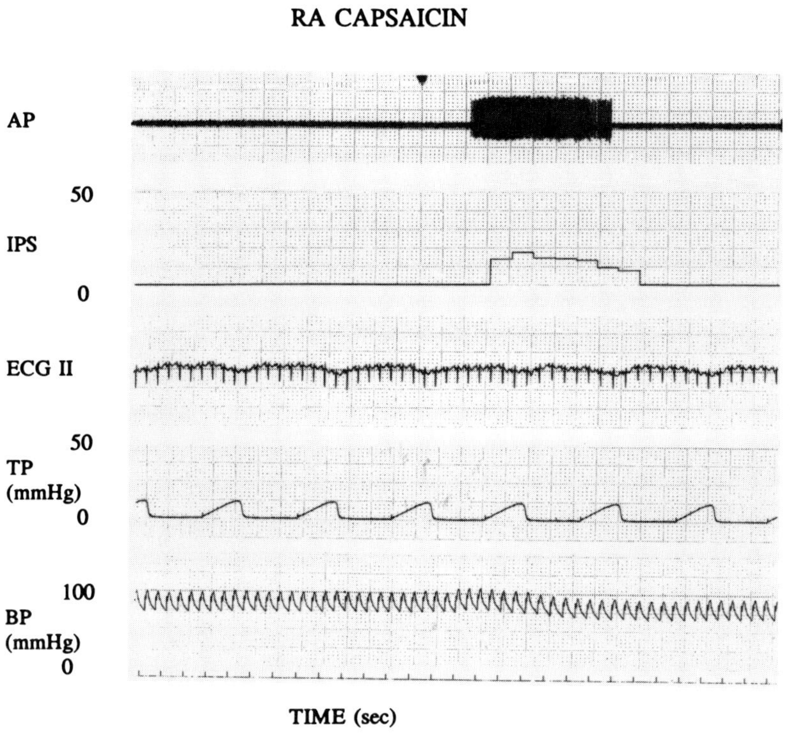

FIGURE 34-4. A burst of action potentials in cervical vagal C fibers elicited by intra-right atrium (RA) capsaicin (7 μg/kg, given as a rapid bolus at the arrowhead). Tracings from top to bottom: Extracellularly recorded neural action potentials (AP); action potentials counter output: impulses per second (IPS), standard lead II ECG, tracheal pressure (TP), systemic arterial blood pressure (BP), and time marks in seconds (sec).

frogs [11,12]. In addition, ATP can act as a fast neurotransmitter in the central nervous system [13] as well as in peripheral ganglia [14–16]. Furthermore, ATP can stimulate cutaneous chemosensitive afferent nerve terminals [17].

The precise signal transduction mechanism that mediates the action of ATP on pulmonary afferent vagal C fiber terminals is not fully known. Our preliminary data suggest that ATP acts on the P_{2x}-purinoceptor subtype and that a pertussis-toxin–sensitive G protein is not involved in this signal transduction.

Patterns of Afferent Neural Traffic Elicited by ATP

Previous studies by others on cardiopulmonary afferent traffic have suggested that the latter travel predominantly in the right vagus nerve [18]. Thus, it was interesting to determine whether the afferent traffic elicited by ATP follows this asymmetric pattern. It is well established that right vagal efferent traffic is more effective in suppressing sinus node automaticity than AV nodal conduction and vice versa, that is, left vagal efferent traffic is more effective in suppressing AV nodal conduction than sinus node automaticity [19–22]. Therefore, if ATP-elicited afferent traffic is predominantly in the right vagus nerve, then it is expected that right vagotomy would attenuate the negative chronotropic action of ATP in the sinus node much more than left vagotomy. In contrast, both right and left vagotomy should

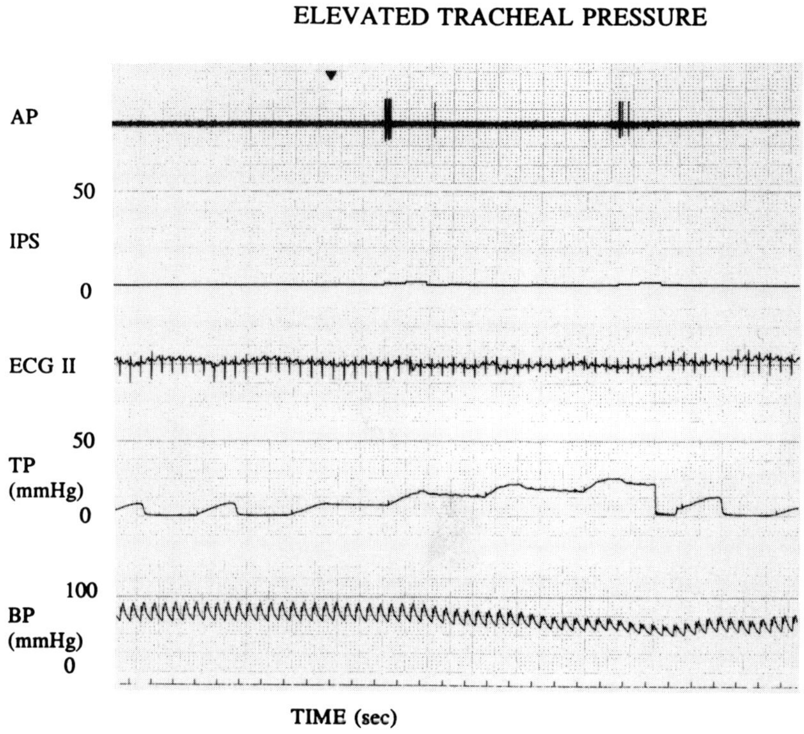

FIGURE 34-5. Bursts of neural action potentials elicited by abrupt changes in tracheal pressure (induced by transient lung inflation to two times tidal volume). Same fibers and tracings as in Figure 34-4.

similarly attenuate the negative dromotropic action of ATP in the AV node. Data obtained in recent studies in which the effect of unilateral followed by bilateral vagotomy on ATP's actions was determined support this hypothesis [23]. In addition, selective right afferent C fiber blockade alone markedly attenuated the negative chronotropic action of ATP in the sinus node [23].

While the hypothesis of right vagal afferent dominancy in ATP-elicited C fiber afferent traffic is supported by the aforementioned studies, the latter do not exclude the possibility that central processing of afferent input to the brain is responsible, at least in part, for the apparent asymmetry in a right versus left vagal role in ATP-triggered depressor reflex. Specifically, it can be hypothesized that unilateral afferent input to medullary nuclei could result in preferential ipsilateral efferent output. In this case, right vagal afferent input should result in preferential suppression of the sinus versus AV node, and vice versa, due to the asymmetry in vagal efferent innervation mentioned above. This hypothesis is congruent with findings that sensory input to medullary vagal nucleus (i.e., nucleus tractus solitarius) has only limited contralateral medullary projections.

To test this hypothesis we have studied the patterns of vagal C fiber afferent traffic elicited by unilateral intrapulmonary ATP. Results obtained in these studies have indicated that unilateral intrapulmonary ATP elicits ipsilateral vagal afferent C fiber traffic [24]. This strongly suggests that the sensory inputs from the right and left lungs are carried predominantly via the right and left vagus nerves, respectively.

According to this interpretation one would

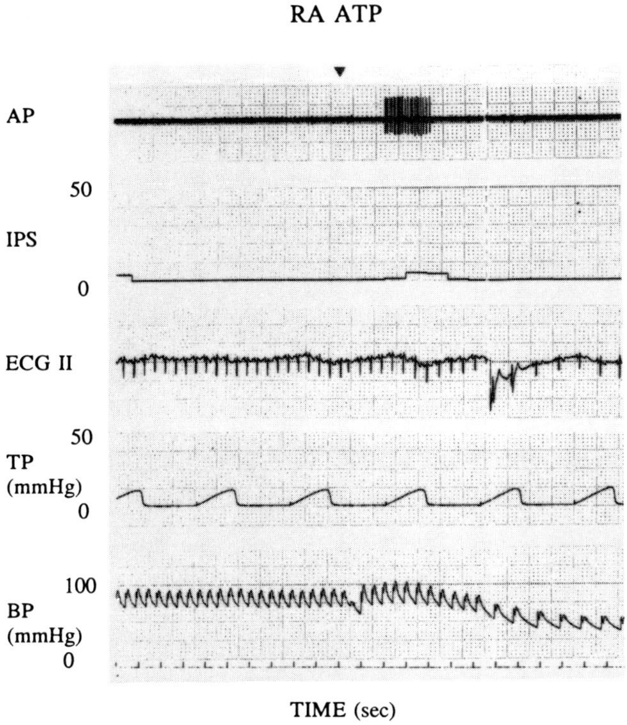

FIGURE 34-6. A burst of neural action potentials elicited by ATP. Same fibers and tracings as in Figure 34-5.

expect intra-right pulmonary ATP to suppress sinus node automaticity more than AV nodal conduction and vice versa, that is, intra-left pulmonary ATP would suppress AV nodal conduction more than sinus node automaticity. Our preliminary studies support this logical derivation. Intra-right pulmonary ATP maximally prolonged sinus cycle length (SCL) by 298 ± 45%, while the same amount of intra-left pulmonary ATP maximally prolonged SCL by 244 ± 35% ($p < 0.05$). In addition, the reverse pattern was found with regard to the negative dromotropic action of intrapulmonary ATP. Specifically, intra-right pulmonary ATP maximally prolonged the AH interval by 66 ± 8% and produced 2.8 ± 0.7 blocked atrial beats, while intra-left pulmonary ATP maximally prolonged AH interval by 80 ± 11% and produced 4.9 ± 1 blocked atrial beats ($p < 0.05$, each).

These preliminary data suggest that patterns of afferent sensory input to medullary nuclei directly affect patterns of efferent output from vagal vasomotor nuclei. Furthermore, because certain unilateral sensory input is not fully projected contralaterally, some reflex loops could have an asymmetric effect on target organs depending on the anatomic pathways employed in these loops.

Physiologic and Clinical Implications

Schematic description of the reflex loop that mediates the negative chronotropic and dromotropic actions of ATP in dogs is given in Figure 34-7. The two curved arrows emerging from the lungs depict the afferent traffic resulting from the stimulation by ATP of the chemosensitive pulmonary vagal afferent C fibers. These afferent inputs to the nuclei tractus solitarius in the medulla generate efferent traffic, which can travel in either the ipsilateral or contralateral vagus nerves. Efferent traffic in the latter, however, is dependent on intramedullary contralateral projections depicted by the two broken-line arrows within the medulla. The straight-line arrows and the dotted-line arrows denote the efferent input to the sinus node and AV node; the former arrow denotes the dominant traffic in the right and left vagi to the sinus and AV nodes, respectively. According to our hypothesis, unilateral afferent traffic resulting from unilateral stimulation of pulmonary C fiber nerve terminals will generate asymmetric sidedness-dependent efferent input to the heart.

In the studies reviewed above, exogenous ATP was given as a rapid bolus. However, endogenous ATP can be released abruptly in vivo under both physiologic and pathophysiologic conditions. ATP can be released from activated platelets, from nerve endings, as a cotransmitter, and from normoxic and ischemic muscle cells. If this release occurs in close proximity to the ATP-sensitive nerve terminals, it is highly likely that a pulmonary-cardiac depressor reflex will be elicited. Thus, ATP can be added to a long list of pulmonary autacoids that can modulate cardiovascular functions via a unique autacoid-neural axis.

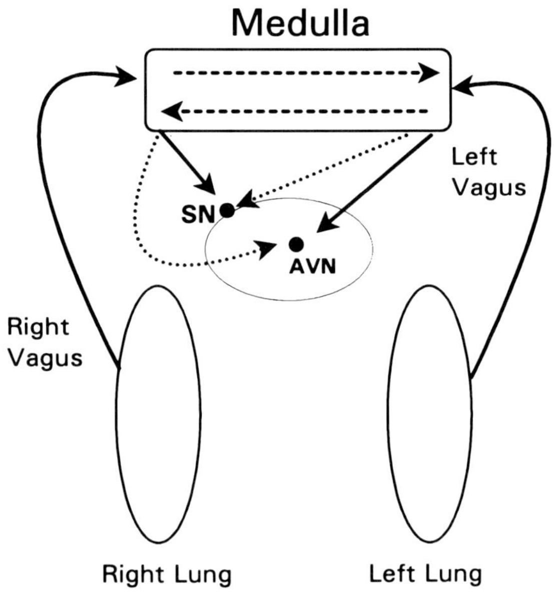

FIGURE 34-7. Schematic description of ATP-triggered pulmonary cardiac depressor reflex. The two vertical ellipsoids denote the right and left lungs. The horizontal ellipsoid denotes the heart and the rectangle denotes the medulla. The curved arrows connecting the lungs with the medulla denote the afferent traffic elicited by ATP that travels in the vagal afferent C fibers. The straight arrows and dotted-line arrows connecting the medulla with the heart denote efferent traffic to the sinus node (SN) and AV node (AVN); the former denotes the predominant efferent traffic. The broken-line arrows within the medulla denote contralateral projections of unilateral afferent medullary input. See text for details.

Acknowledgments

Drs. Belhassen, Mitamura, Mitsuoka and Miyagawa collaborated on some of the studies performed in the author's laboratory. We thank Mrs. Dwendolyn Dillard for her assistance in the preparation of this manuscript. Dr. Pelleg is supported by NIH grant HL-43006. Figures 34-4–34-6 were reproduced with permission from Am J Physiol 267:H1093, 1994.

References

1. Dubyak GR, Fedan JS (1990) Biological actions of extracellular ATP. Ann NY Acad Sci 603: 1–542.
2. Belhassen B, Pelleg A (1984) Electrophysiologic effects of adenosine triphosphate and adenosine in the mammalian heart: Clinical and experimental aspects. J Am Coll Cardiol 4:414–424.
3. Pelleg A, Belhassen B, Ilia R, Laniado S (1985) Comparative electrophysiologic effects of ATP and adenosine in the canine heart: Influence of atropine, propranolol, vagotomy, dipyridamole and aminophylline. Am J Cardiol 55:571–576.
4. Pelleg A, Mitamura H, Michelson EL (1985) Evidence for vagal involvement in the electrophysiologic actions of exogenous adenosine and adenosine triphosphate in the canine heart. J Auton Pharmacol 5:207–212.
5. Pelleg A, Michelson EL (1987) Role of the vagus in modulation by Ca^{2+} of the depressant actions of adenosine and adenosine 5′-

triphosphate on the canine sinus node in vivo. J Auton Pharmacol 7:127–34.
6. Pelleg A, Hurt CM (1990) Evidence for ATP-triggered vagal reflex in the canine heart in vivo. In Dubyak GR, Fedan JS (eds.), Biological Actions of Extracellular ATP. Ann NY Acad Sci 603:441–442.
7. Pelleg A, Mitamura H, Michelson EL, Dreifus LS (1986) Evidence against prostaglandin mediation of the electrophysiologic effects of ATP versus adenosine in the canine heart. J Cardiovasc Pharmacol 8:534–538.
8. Pelleg A, Mitsuoka T, Michelson EL (1987) Adenosine mediates the negative chronotropic action of adenosine 5′-triphosphate in the canine sinus node. J Pharmacol Exp Ther 242: 791–795.
9. Pearson JD (1985) Ectonucleotidases: Measurements of activities and use of inhibitors. Methods Pharmacol 6:83–107.
10. Coleridge JCG, Coleridge HM (1984) Afferent vagal C fiber innervation of the lungs and airways and its functional significance. Rev Physiol Biochem Pharmacol 99:1–110.
11. Illes I, Norenberg W (1993) Neuronal ATP receptors and their mechanism of action. Trends Pharmacol Sci 14:50–54.
12. Bean BP (1990) ATP-activated channels in rat and bullfrog sensory neurons: Concentration dependence and kinetics. J Neurosci 10:1–10.
13. Edwards FA, Gibb AJ, Colquhoun D (1992) ATP receptor-mediated synaptic currents in the central nervous system. Nature 359:144–147.
14. Fieber LA, Adams DJ (1991) Adenosine triphosphate-evoked currents in cultured neurons dissociated from rat parasympathetic cardiac ganglia. J Physiol 434:239–256.
15. Evans RJ, Derkach V, Surprenant A (1992) ATP mediates fast synaptic transmission in mammalian neurons. Nature 357:503–505.
16. Silinsky EM, Gerzanich V, Vanner SM (1992) ATP mediates excitatory synaptic transmission in mammalian neurons. Br J Pharmacol 106: 762–763.
17. Bleehan T (1978) The effects of adenine nucleotides on cutaneous afferent nerve activity. Br J Pharmacol 62:573–577.
18. Bishop VS, Hasser EM (1987) Physiological role of ventricular receptors. In Hainsworth R, McWilliam PN, Mary DAS (eds.), Cardiogenic Reflexes. Oxford Press, U.K., pp. 62–73.
19. Loeb JM, Dalton DP, Moran JM (1981) Sensitivity differences of SA and AV nodes to vagal stimulation: Attenuation of vagal effects of SA node. Am J Physiol 241 (Heart Circ Physiol 10):H686–H690.
20. Londegham LM, Nouton E, Stasser T, DeGeest H (1975) Additive effects of acetylcholine released by vagal nerve stimulation on atrial rate. J Appl Physiol 38:108–113.
21. Neely BH, Urthaler R (1992) Quantitative effects of sympathetic and vagal nerve stimulations or sinus and AV junctional rhythms. J Auton Nerv Syst 37:109–120.
22. Yang T, Jacobstein MD, Levy MN (1989) Synchronization of automatic cells in SA node during vagal stimulation in dogs. Am J Physiol 246 (Heart Circ Physiol 15):H585–H591.

35. EXCITATORY EFFECT OF ADENOSINE ON CARDIAC SYMPATHETIC AFFERENT FIBERS

Nicola Montano
Tomaso Gnecchi-Ruscone
Federico Lombardi
Alberto Malliani

Sensory innervation of the heart seems capable of perceiving the occurrence of transient myocardial ischemia. As a result, neural reflexes and pain can originate from the heart. However, while the link existing between myocardial ischemia and pain is elusive [1], as nonpainful episodes are a common clinical and experimental feature, complex cardiovascular reflexes seem to accompany experimental occlusion of a coronary artery [2–4] and transient myocardial ischemia in humans [5,6]. Their consequences range from simple changes in heart rate and arterial pressure to the precipitation of malignant arrhythmias [7,8]. Two hypotheses have been advanced to explain the independence of neural reflexes from pain: 1) Two sets of different afferent fibers could be responsible for mediating neural reflexes and pain, with the latter arising only concomitantly with the recruitment of high-threshold cardiac nociceptors, specifically devoted to signaling tissue damage [9]; 2) the same population of low-threshold afferent fibers would transmit different impulse codes, giving rise, respectively, to cardiovascular reflexes and pain [1]. The mechanical or chemical nature of the factors initiating these events has been repetitively debated over the years [10]. Recently, on pure clinical grounds it has been proposed that adenosine, a natural substance that is released during myocardial ischemia [11–14], might be capable of activating an algogenic code [15,16] from the heart. In this hypothesis it remained undetermined whether the peripheral mechanism of adenosine action was based upon specificity (recruitment of specific nociceptors) or intensity (massive excitation of low-threshold receptors) [17,18]. In such a framework we decided to investigate the modality through which adenosine could increase the impulse activity of afferent sympathetic fibers with left ventricular endings (i.e., the afferent pathway mediating pressor reflexes and cardiac pain).

Methods

SURGICAL PREPARATION OF ANIMALS

In eight cats anesthesia was induced with an intramuscular injection of ketamine (Ketalar, 10 mg/kg, Parke-Davis) and propriomazine (Combelen, 4 mg/kg, Bayer Leverkusen). After assessing the depth of anesthesia by the absence of behavioral reactions to a painful stimulus, such as pinching of a pinna, the animals were decerebrated by brainstem transection at the intercollicular level with a blunt spatula and the rostral brain was removed. A concomitant (20-minute) bilateral common carotid occlusion was performed to prevent hemorrhage.

In all cats the trachea was cannulated with a Y tube and artificial ventilation was performed so as to maintain an arterial PO_2 of 80–95 mmHg, PCO_2 of 25–35 mmHg, and pH of 7.33–7.42, as measured on an Astrup model RM 1304 acid-base analyzer. A polyethylene catheter was inserted into the thoracic aorta through a femoral artery and another one was inserted into a femoral vein. The heads of the second and third ribs were removed retropleurally on the left side to expose the stellate ganglion and its branches.

The thorax was opened by removing the fifth and sixth ribs, and a thread was passed around the distal portion of the descending thoracic aorta; the thread was also passed through a short, rigid polyethylene tube so that a reversible and graded constriction of the vessel could be performed by pulling. The pericardium was excised and the epicardial surface exposed. A paralyzing dose (2 mg/kg) of gallamine triethiodide was given intravenously about every 2 hours.

DATA ACQUISITION

In each animal single sympathetic afferent fiber impulse activity, surface ECG, and aortic pressure were recorded and stored on a Racall Store 7 tape recorder. From the cut central end of the left third thoracic sympathetic white ramus communicans, nerve filaments were split until impulses from a single afferent fiber were obtained [19,20]. Nerve activity was recorded with platinum electrodes connected to a Grass probe (HIP 511E) and amplified by a Grass P511 band pass amplifier. A digital neuronal spike amplifier was used to obtain an electric signal proportional to the number of neural impulses over the time and to compute the mean frequency of discharge. ECG was recorded with a Grass 7P6A ECG preamplifier. Aortic pressure was measured with Statham P 23De strain gauge. The catheter-manometer system had a flat (+5%) frequency response of 30 Hz as calculated by its response to a step increase in pressure.

MEASUREMENTS OF CONDUCTION VELOCITY OF AFFERENT FIBERS

The left inferior cardiac nerve was isolated above the aortic arch from the surrounding tissues. This portion of the nerve was stimulated through platinum electrodes with a Grass S4 stimulator through a stimulation unit. Pulses were monophasic, 0.1–1.5 ms in duration, and 10–15 V in amplitude. Conduction velocity was calculated between the proximal stimulating electrode and the closest to the recording electrode divided by the time from the stimulus artifact to the beginning of the evoked action potential.

LOCATION OF RECEPTOR ENDINGS

The immediate responsiveness of the fiber to arterial pressure changes allowed a preliminary functional localization of its sensory endings in the left side of the heart. Subsequently a gentle probing on the external surface of the beating left ventricle was performed with a cotton stick; thus, only fibers that possessed an epicardial receptive field were studied. The count of the action potentials and the assessment of the conduction velocity of the fiber was performed as previously reported [19,20]. The receptive field of each fiber was definitely localized to the left ventricle by mechanical probing performed at the end of the experiment on the nonbeating heart after the cat had been killed. Therefore, not more than one fiber could be studied in each successful experiment. Finally the destruction of the left ventricular ending always caused the disappearance of impulses in the fiber studied, confirming that the sensory fields initiating the impulse activity had been thus located.

DRUGS

Adenosine was purchased from Sigma Chemical Co. (St. Louis, MO) and dissolved in normal saline. Three different concentrations were prepared: 0.1, 1, and 10 mg/ml. The drug was applied to the epicardial surface corresponding to the fiber receptive field by means of 0.5 cm blotting paper diskettes soaked with 0.1 ml of each solution. Aminophylline (Tefamin, Recordati, Italy) was given intravenously at a dose of 5 mg/kg. The amount of P_1 blockade was not tested with specific agonists.

EXPERIMENTAL PROTOCOL

In eight animals we studied the impulse activity of left ventricular sympathetic afferent fibers during epicardial application of saline and of the three different concentrations of adenosine. The sequence of application was blind and random. The diskette was removed after 90 seconds and the epicardial surface was rinsed with saline. The time interval between each application was 20 minutes. The protocol was repeated about 20 minutes after intravenous bolus administration of aminophylline. In five animals the experimental protocol was repeated by substituting adenosine with saline, while in the remaining five cats the protocol was repeated after aminophylline administration (5 mg/kg, intravenous bolus). Furthermore, we tested the mechanical sensitivity of each fiber by means of brief (15 seconds), comparable increases in arterial pressure obtained by thoracic aortic constriction after application of saline or adenosine.

DATA ANALYSIS

After epicardial application of adenosine, we evaluated the mean peak discharge as the maximum level of excitation of the fiber during the activation period. The latency of the response was considered as the time elapsed from the application of the drug to the beginning of the activation, while the duration was considered as the time from the onset of the activation to the return to impulse activity to prestimulus levels (for these calculations 2.5-second periods were considered). Control values of impulse activity were obtained from the mean discharge of the 10-second period just preceding the epicardial application of adenosine.

STATISTICAL PROCEDURE

Results are expressed as mean ± SEM. The significance of neural responses to stimuli were tested with the Student's t test for paired observations. The contrast between responses was calculated by ANOVA for repetitive measures; $p < 0.05$ was considered significant [21].

Results

All afferent fibers displayed a spontaneous impulse activity that, under basal conditions, was largely independent from cardiac rhythm [2,6,9,17]. Six of the eight fibers studied pertained to the C group, with a conduction velocity of 0.8 ± 0.16 m/sec, while the others belonged to the A-delta group, with a conduction velocity of 2.7 ± 0.7 m/sec. All fibers exhibited the characteristic of low-threshold receptors [1], as brief and moderate increases (20 ± 8%) in aortic arterial pressure augmented their impulse activity from 1.3 ± 0.4 to 2.6 ± 0.7 impulses/sec ($p < 0.001$). In control conditions the mean RR interval was 362 ± 28 ms, while systolic and diastolic arterial pressure were, respectively, 117 ± 7 and 83 ± 7 mmHg.

Epicardial application (Fig. 35-1; Table 35-1) of adenosine at a low concentration (0.1 mg/ml) determined a significant increase of impulse

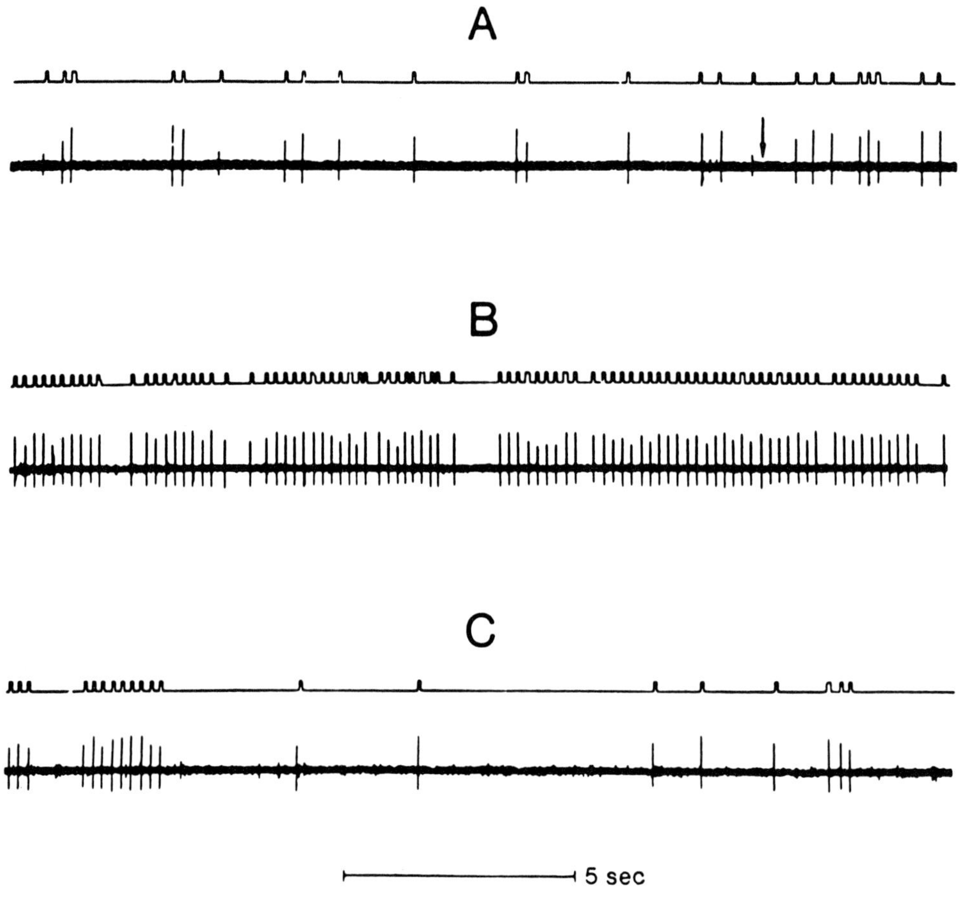

FIGURE 35-1. Impulse activity of a single sympathetic afferent fiber with left ventricular ending in control condition (**A**) and after application of adenosine (1 mg/ml). After adenosine application, indicated by the arrow, a marked activation (**B**) was observed, followed by a return to the background discharge in about 30 seconds (**C**). Only panels A and B are continuous in time. The upper tracing in each panel represents the output of a digital spike counter. (Reprinted from Montano et al. [27], with permission.)

activity (147 ± 29% from 1.5 ± 0.6 impulses/sec) with a mean latency of 17 ± 3 seconds and a duration of 35 ± 5 seconds. The medium dose (1 mg/ml) produced a greater increase in nerve activity (237 ± 42% from 1.2 ± 0.4 impulses/sec) (Fig. 35-2; Table 35-1). The activation had a mean latency of 16 ± 2 seconds and a duration of 51 ± 9 seconds. The increase in activity produced by the high dose (10 mg/ml) was greater than those observed at the lower concentrations (338 ± 39% from 1.3 ± 0.3 impules/s), with a mean latency of 14 ± 2 seconds and a duration of 63 ± 8 seconds. No significant hemodynamic changes were observed. Saline application had no effect on the afferent impulse activity. After P_1 purinergic blockade with aminophylline, adenosine application was no longer effective in exciting the impulse activity; on the contrary, the responses to aortic constrictions were unchanged by the application of adenosine (Fig. 35-3).

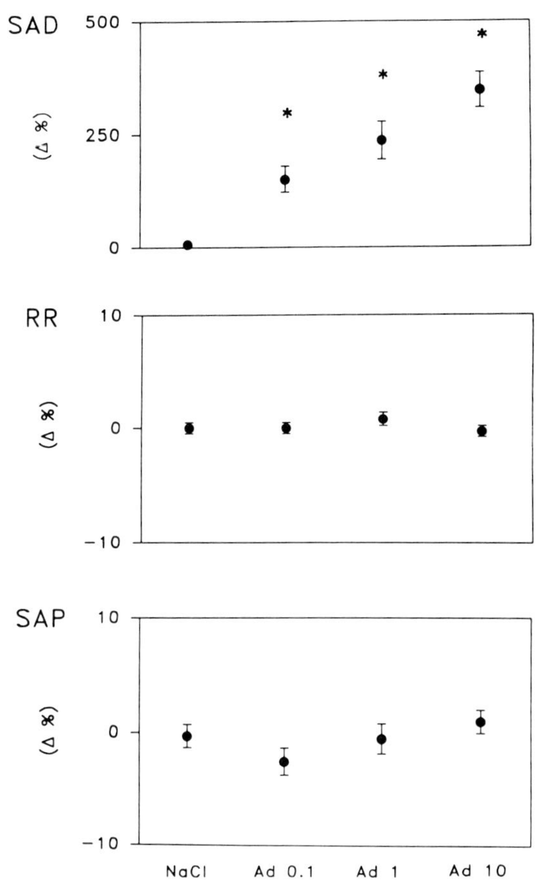

FIGURE 35-2. Percent changes in sympathetic afferent discharge (SAD), R-R interval (RR), and systolic arterial pressure (SAP) during epicardial application of saline (NaCl) or adenosine (Ad) at three different concentrations (0.1,1, and 10 mg/ml). *$p < 0.001$ versus control.

Discussion

The main result of this study is that epicardial application of adenosine induced an excitation, with a dose-related efficacy, of the impulse activity of left ventricular sympathetic afferent fibers. This responsiveness was abolished by aminophylline, a P_1 purinergic blocking agent. It is important to note that in this set of experiments, as in previous experiments conducted by our group [4,19,20], we analyzed the neural response of both myelinated or unmyelinated left ventricular sympathetic afferent fibers that displayed a spontaneous impulse activity. Moreover, epicardial application of adenosine always increased the impulse activity of these fibers and never elicited a discharge of silent fibers contained in the neural filament under study. Thus, these results seem to confirm our previous hypothesis that potential algogenic stimuli, such as coronary artery occlusion or bradykinin [20], never recruited impulses from silent afferent fibers.

This finding is crucial for interpreting the afferent code of cardiac pain [1,18], which could be based either on intensity or specificity. The intensity mechanism, the most obvious and probably the first to be formulated [18], assumes that pain results from an excessive stimulation of receptive structures already active in normal conditions, such as, in the case of cardiovascular sympathetic afferent fibers, in signalling hemodynamic events. Alternatively, pain may be conceived of as a specific sensation, that is, the product of the excitation of a well-defined nociceptor apparatus, the functional characteristics of which make it responsive only to a limited class of noxious events, such as coronary artery occlusions or increased concentrations of algogenic substances.

These present results further support our previous conclusions that an intensity mechanism seems the most likely candidate to account for the properties of the neural substrate subserving cardiac nociception [1,10,18]. Also, in the hands of other authors as well [9,22], potentially algogenic substances did not induce an appreciable recruitment of silent ventricular sympathetic afferents. This concept contrasts with the initial observations of Uchida and coworkers [23,24] that chemical substances such as bradykinin or adenosine could excite silent sympathetic afferents. However, in our opinion [1,18] the absence of spontaneous background activity in the fiber described by Uchida et al. was possibly dependent upon a low baseline arterial pressure maintaining the hemodynamic stimuli below the threshold for most of the mechanosensitive receptors rather than on a specific nociceptive attribute of the fibers. Besides these considerations, Uchida et al. [23] were the first to describe that adenosine, injected intravenously or into the left atrium, was able to excite ventricular sympathetic afferent fibers, thus suggesting a possible role for this substance in the genesis of cardiac pain. The relevance of this experimental hypothesis now has to be appreciated in light of recent human studies [15,16] suggesting an important role for adenosine in linking pain and ischemia.

The possibility that adenosine could elicit excitatory sympathetic reflexes through the activation of sympathetic afferents was recently considered by Thames and coworkers [25] in order to explain the increase in sympathetic efferent activity directed to the kidney during myocardial ischemia. A greater neural response to ischemia was observed after intravenous administration of dypiridamole, while diminished activation of renal sympathetic efferent activity followed intravenous administration of aminophylline. The present experiments described a dose-related response to an adenosine application that was similar to that observed after intracoronary administrations of bradykinin [20]. The latencies for both excitations were remarkably similar (16 ± 2 seconds after adenosine, 18 ± 2 seconds after 5 ng bradykinin) [20].

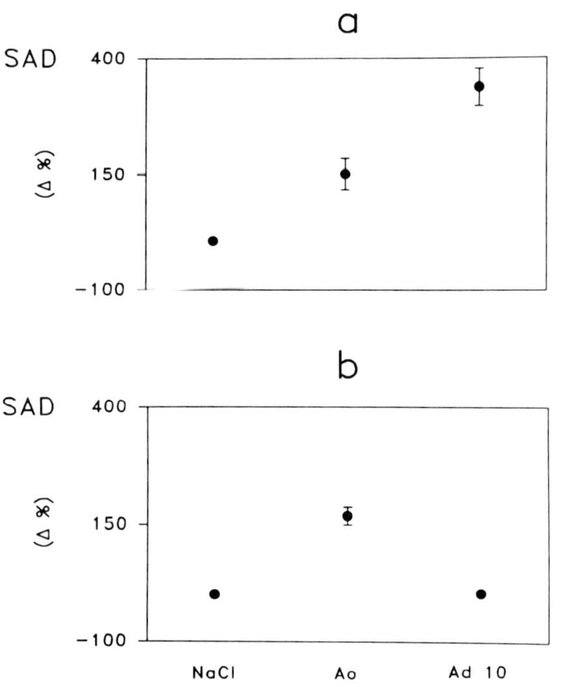

FIGURE 35-3. Percent changes in sympathetic afferent discharge following aortic occlusion (Ao) and epicardial application of saline or adenosine (1 mg/ml) before (**a**) and after (**b**) intravenous administration of aminophylline. *p < 0.001 versus control.

Although the present experiments cannot provide a direct insight into the mechanism by which adenosine produces excitation of sympathetic afferent fibers, the response seems to

TABLE 35-1. Effects of epicardial application of adenosine on sympathetic afferent discharge, RR interval, and systolic arterial pressure (n = 10)

Control				Response				
				SAD				
SAD (imp/sec)	RR (ms)	SAP (mmHg)	Stimulus Adenosine	Peak (imp/sec)	Latency (sec)	Duration (sec)	RR (ms)	SAP (mmHg)
1.5 ± 0.6	332 ± 19	115 ± 16	0.1 mg/ml	3.7 ± 1.5[a]	17 ± 3	35 ± 5	336 ± 19	115 ± 16
1.2 ± 0.4	352 ± 21	114 ± 8	1 mg/ml	3.9 ± 0.7[a]	16 ± 3	51 ± 9	358 ± 23	113 ± 8
1.3 ± 0.3	350 ± 25	114 ± 7	10 mg/ml	5.1 ± 0.9[a]	14 ± 2	63 ± 8	346 ± 26	115 ± 9

SAD = sympathetic afferent discharge; RR = interval; SAP = systolic arterial pressure.
[a]p < 0.01 versus control.

involve stimulation of P_1 purinergic receptors. These results, in addition, confirm that ventricular sympathetic afferents have functional characteristics compatible with the definition of low-threshold polymodal receptors [4,18,26], a term indicating that the receptive field is considered to be sensitive to mechanical and chemical stimuli, even though a transducer mechanism that can account for such a dual responsiveness has not yet been found.

A vast amount of experimental data by now strongly suggest that cardiovascular sympathetic afferent fibers participate in the normal regulation of circulation, with their excitatory properties interacting with the reflex inhibitory mechanisms exerted by cardiovascular supraspinal afferents [2,4]. In physiopathologic conditions, however, such as acute myocardial infarction, ventricular sympathetic afferents mediate both the algogenic code responsible for cardiac pain and the excitatory reflexes, which can exert a detrimental effect by augmenting oxygen consumption and facilitating life-threatening arrhythmias [2,4,8]. Adenosine could directly activate these fibers, thus contributing in the determination of the number of afferent sympathetic fibers involved, the intensity of their excitations, and, consequently, the characteristics of the reflexogenic and algogenic afferent impulse activity. In detail, a complex interaction of mechanical and chemical factors [10] could determine those spatially heterogeneous characteristics of the intensity mechanism (the spatio-temporal pattern) that has been hypothesized as the basis for the algogenic code from the heart [1,10].

References

1. Malliani A (1969) The elusive link between transient myocardial ischemia and pain. Circulation 73:201–204.
2. Bishop VS, Malliani A, Thorén P (1983) Cardiac mechanoreceptors. In Shepherd JT, Abboud FM, Geiger SR (eds.), Handbook of Physiology, Section 2: The Cardiovascular System: Vol. 3, Peripheral Circulation and Organ Blood Flow. Bethesda, MD: American Physiological Society, pp. 497–555.
3. Malliani A, Schwartz PJ, Zanchetti A (1969) A sympathetic reflex elicited by experimental coronary artery occlusion. Am J Physiol 217:703–709.
4. Malliani A (1982) Cardiovascular sympathetic afferent fibers. Rev Physiol Biochem Pharmacol 94:11–74.
5. Chierchia S, Mujesan L, Davies A, Balasubramian V, Gerosa S, Raftery EB (1990) Role of the sympathetic nervous system in the pathogenesis of chronic stable angina. Circulation 82(Suppl III):III71.
6. Littler WA, Honour J, Sleight P, Scott FD (1973) Direct arterial pressure and electrocardiogram in unrestricted patients with angina pectoris. Circulation 48:125–134.
7. Lown B (1979) Sudden cardiac death: The major challenge confronting contemporary cardiology. Am J Cardiol 43:313–328.
8. Malliani A, Schwartz PJ, Zanchetti A (1980) Neural mechanism in life-threatening arrhythmias. Am Heart J 100:705–715.
9. Baker DG, Coleridge HM, Coleridge JCG, Nerdrum T (1980) Search for a cardiac nociceptor: Stimulation by bradykinin of sympathetic afferent nerve endings in the heart of the cat. J Physiol (Lond) 306:519–536.
10. Malliani A, Lombardi F (1982) Consideration of the fundamental mechanism eliciting cardiac pain. Am Heart J 103:575–578.
11. Berne RM (1980) The role of adenosine in the regulation of coronary blood flow. Circ Res 47:807–813.
12. Belardinelli L, West GA, Crampton R, Berne RM (1983) Chronotropic and dromotropic actions of adenosine. In Berne RM, Rall TW (eds.), The Regulatory Function of Adenosine. The Hague: Martinus Nijhoff, pp. 378–398.
13. Bardenheuer H, Schrader J (1984) Supply-to-demand ratio for oxygen determines formation of adenosine by the heart. Am J Physiol 250:H173–H180.
14. Ely SW, Berne RM (1992) Protective effects of adenosine in myocardial ischemia. Circulation 85:893–904.
15. Sylvén C, Beermann B, Jonzon B, Brandt R (1984) Angina pectoris-like pain provoked by intravenous adenosine. Br Med J 293:227–230.
16. Crea F, Pupita G, Galassi AR, El-Tamimi H, Kaski JC, Davies G, Maseri A (1990) Role of adenosine in pathogenesis of anginal pain. Circulation 81:164–172.

17. Coleridge JCG, Hemingway A, Holmes LR, Linden RJ (1957) The location of atrial receptor in the dog: A physiological and histological study. J Physiol (Lond) 136:174–197.
18. Malliani A, Pagani M, Lombardi F (1989) In Wall PD, Melzack R (eds.), Visceral versus somatic mechanism. Textbook of Pain, 2nd ed. Edinburgh: Churchill Livingstone, pp. 128–140.
19. Casati R, Lombardi F, Malliani A (1979) Afferent sympathetic unmyelinated fibers with left ventricular endings in cats. J Physiol (Lond) 292:135–148.
20. Lombardi F, Della Bella P, Casati R, Malliani A (1981) Effects of intracoronary administration of bradykinin on the impulse activity of afferent sympathetic fibers with left ventricular endings in the cat. Circ Res 48:69–75.
21. Armitage P (1971) Statistical Methods in Medical Research. Oxford: Blackwell Scientific Publications.
22. Nishi K, Sakanashi M, Takenaka F (1977) Activation of afferent cardiac sympathetic nerve fibers of the cat by pain producing substances and by noxious heat. Pflügers Arch 372:53–61.
23. Uchida Y, Kamisaka K, Ueda H (1969) Responses of the cardiac sympathetic receptors to various substances. Jpn Heart J 10:225–242.
24. Uchida Y, Murao S (1974) Bradykinin induced excitation of afferent cardiac sympathetic fibers. Jpn Heart J 15:84–91.
25. Thames MD, Kinugawa T, Dibner-Dunlap ME (1993) Reflex sympathoexcitation by cardiac sympathetic afferents during myocardial ischemia. Role of adenosine. Circulation 87:1698–1704.
26. Burgess PR, Perl ER (1973) Cutaneous mechanoreceptors and nociceptors. In Iggo A (ed.), Handbook of Sensory Physiology—Somatosensory System, Vol. 2. Berlin: Springer Verlag, pp. 29–78.
27. Montano N, Lombardi F, Gnecchi-Ruscone T, Contini M, Guazzi M, Malliani A (1992) Effetto eccitatorio dell'adenosina sull'attività di scarica delle fibre afferenti cardiache. Cardiologia 36:953–959.

36. MECHANISM OF ANGINAL PAIN: THE KEY ROLE OF ADENOSINE

Christer Sylvén
Filippo Crea

Genesis of Anginal Pain

Anginal pain is the symptom that most often brings patients with ischemic heart disease to medical attention. However, transient myocardial ischemia and even necrosis can occur without pain, while severe anginalike pain can occur in the absence of detectable myocardial ischemia. Although the presence of cardiac ischemic pain serves the useful purpose of alerting the patient, it is unpleasant, alarming, and frequently disabling. Hence, as for other forms of response of the body to potentially noxious stimuli, pain can elicit a protective reaction. However, cardiac ischemic pain may become a major component of the disease when it is disproportionate to the severity of ischemia and bears no relation to the prognosis of the disease. A better understanding of the causes of cardiac ischemic pain and the factors that determine its occurrence and modulate its severity are important for understanding the causes of its elusive link to the severity of myocardial ischemia.

The nature of the cardiac stimuli responsible for pain has been debated for decades. It was first proposed by Colbeck about 80 years ago that ischemic cardiac pain might be due to stretching of the ventricular wall (mechanical hypothesis) [1] and about 60 years ago by Lewis that it might be due to local intramyocardial release of algogenic substances (chemical hypothesis) [2]. In the past few years a number of clinical studies have shed some new light on this topic.

By continuous monitoring of left ventricular volume during spontaneous or ergonovine-induced transient ischemic episodes, Davies et al. showed that neither the rate nor the magnitude of left ventricular dilation influence the presence or absence of anginal pain [3]. The negligible importance of mechanical distortion of myocardial fibers and the absence of specific nociceptors in the genesis of cardiac pain is supported by the observation that myocardial biopsies can be taken without any sensation of pain. These findings suggest that the mechanical hypothesis is unlikely to explain anginal pain, although they do not entirely rule out the theoretical possibility that mechanical stimuli might contribute to the modulation of its severity.

More recently, in patients with angina undergoing coronary angioplasty, Tomai et al. have shown that when the pressure used during the second balloon inflation was higher than that used during the first inflation, the severity of the cardiac pain increased even though the severity of myocardial ischemia decreased [4]. Thus, the stretching of the coronary arterial wall appears to contribute to the genesis of pain experienced by patients during coronary angioplasty. This mechanism of cardiac pain, however, might operate when myocardial ischemia is caused by coronary spasm, which is likely to cause a mechanical distortion of the arterial wall, but it cannot account for the anginal pain caused by an increase of myocardial oxygen consumption or by coronary thrombosis.

Adenosine Is a Mediator of Anginal Pain: A Proved Hypothesis

In the mid-1980s adenosine was introduced into the clinical setting 1) for the detection of critical coronary stenosis in conjunction with perfusion scintigraphy or two-dimensional echocardiography; and 2) for the diagnosis and conversion of supraventricular reentry tachycardias by giving adenosine as an intravenous bolus injection. During such use it was unexpectedly found that simultaneous with conversion of the tachycardia, patients developed short but very unpleasant sensations. These patients did not suffer from ischemic heart disease. Patient history revealed that they developed short-duration chest pain with characteristics never before experienced with complex sensory characteristics similar to those known for angina pectoris [5]. As adenosine during myocardial ischemia is released in up to a magnitude of three potencies, the hypothesis was formulated that ischemically released adenosine might be one energy-related messenger for anginal pain, centrally signaling that the tissue is in a precarious state [6].

Bleehen and Keele [7] reported in 1977 that purines could provoke pain in the blister base preparation. This is an exposed, heavily inflamed intracutaneous surface corresponding to a second-degree burn and with sensitized free nerve endings. Interestingly, in this preparation purines such as ATP, ADP, AMP, adenosine, and xantoside were reported to provoke the same degree of pain in the same μmolar range.

Sylvén et al. tested this hypothesis initially by assessing the effects of intravenous boluses of increasing doses of adenosine in healthy volunteers (Fig. 36-1). All subjects experienced anginalike chest pain. Furthermore, adenosine-induced pain was dose dependent, decreased by theophylline, and enhanced by dipyridamole [6]. The authors concluded that the pain was mediated by extracellular membrane-bound adenosine receptors. It was then observed that in anginal patients the intracoronary infusion of adenosine dose dependently and consistently caused pain exhibiting features identical to those experienced by the patients at the time of transient myocardial ischemia, but in the absence of ischemic ECG changes (Fig. 36-2) [8,9].

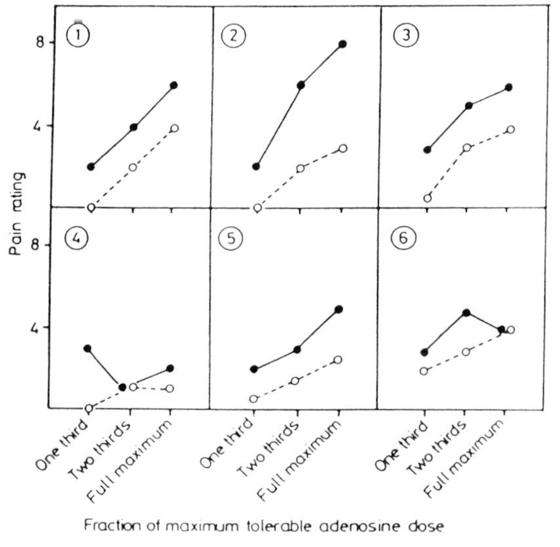

FIGURE 36-1. Dose-response relationships for adenosine-provoked chest pain in healthy volunteers following bolus injection into a peripheral vein before (●--●) and after (○--○) administration of theophyllamine. (Reproduced with permission from the British Medical Journal.)

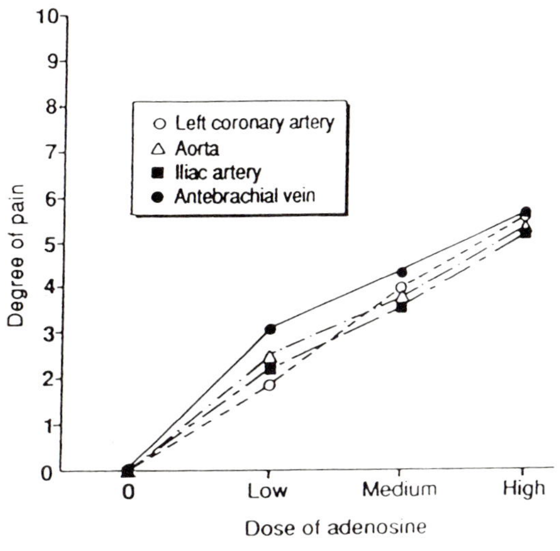

FIGURE 36-2. Dose-response relationships following intraarterial bolus injection of adenosine into the left coronary artery, the abdominal aorta, and the iliac artery, and intravenous infusion into an antebrachial vein. (Reproduced with permission from Cardiovascular Research.)

As the infusion of similar doses of adenosine into the right atrium failed to elicit symptoms, the pain elicited by the intracoronary infusion of adenosine could only originate from the heart [9]. Furthermore, the pain caused by the intracoronary infusion of adenosine appeared within 20 seconds after its administration and disappeared after another 15–20 seconds. This time course of adenosine-induced pain is very different from the anginal ischemic pain caused by coronary angioplasty, as the latter, when it occurs, begins not earlier than 30 seconds after balloon inflation. That adenosine-induced cardiac pain was not secondary to ischemia is further proved by the following observations: 1) No ischemic-like ECG changes are usually observed during adenosine-induced cardiac pain. 2) Adenosine-induced pain is observed during infusion in nonstenotic coronary branches [10]. 3) Adenosine-induced dose-dependent pain, similar in quality and time characteristics to that observed following intracoronary infusion, is also provoked in the corresponding vascular bed by infusion into the brachial, ileal, or femoral artery or in the abdominal aorta, where ischemia caused by steal cannot occur [8,11,12] (Fig. 36-2). Adenosine thus can provoke pain in different tissues. 4) The doses of intravenous or intracoronary adenosine that cause pain are higher than those that cause maximal coronary dilation (Fig. 36-3). The last finding also shows that adenosine-induced pain is not due to mechanical distention of periarteriolar nerve endings [9,13].

Theophylline reduces the severity of exercise-induced anginal pain for a similar degree of ST-segment depression (Fig. 36-4) [9]. It also reduces the ischemic pain following lower arm ischemic work (Fig. 36-5) [14], where the theophylline effect could not have been due to reduction of an adenosine-dependent steal phenomenon [15]. Thus endogenous adenosine is a mediator of pain during transient ischemia, including myocardial ischemia.

Altogether these observations appear to prove the hypothesis that adenosine is an adequate stimulus for cardiac sensory receptors and that endogenous adenosine is a mediator, the first yet identified, of cardiac ischemic pain. When adenosine is sequentially injected into the right and left coronary arteries in the same patient, about 25% of subjects experience pain in different body areas, thus suggesting the presence, at least in some patients, of a cortical representation of the myocardium [10].

Receptors Mediating the Algogenic Effects of Adenosine

The algogenic effects of adenosine can be mediated by theophylline-sensitive A_1 or A_2 receptors. This problem has been tackled in a series of clinical studies using bamiphylline. This substance is a 7,8-bisubstitute of aminophylline used to treat bronchial asthma and lung anaphylaxis with an efficiency comparable to that of theophylline but with substantially less side effects. Although functional

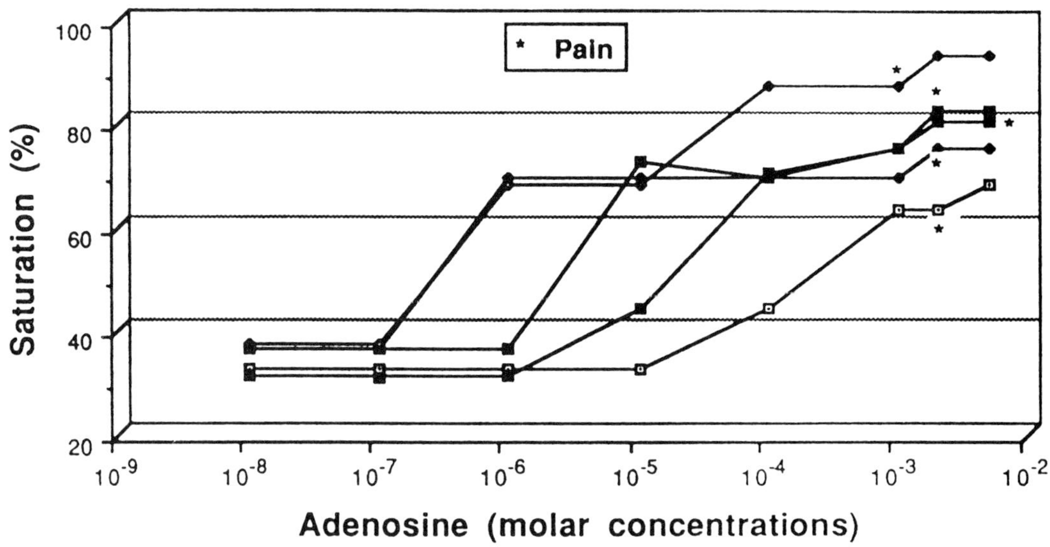

FIGURE 36-3. Plot effects on increasing intracoronary molar concentrations of adenosine (infusion rate, 1 ml/min) on blood oxygen saturation in the great cardiac vein. Massive coronary dilation, as indicated by marked increases of oxygen blood saturation, can be observed at doses that do not produce any chest pain. These findings suggest that adenosine-induced chest pain is unlikely to be caused by stretching of periarterial sensory nerve endings. (Reproduced with permission from Circulation.)

receptor specificity has not been characterized in cardiac tissues, receptor affinity has been studied in crude synaptic membranes prepared from rat brain. Bamiphylline displaces radioligands from A_1 adenosine receptors with a potency similar to that of 8-phenyl-theophylline, whereas it shows a much lower potency on A_2 adenosine receptors. This results in a high degree of A_1-receptor selectivity, indicated by a A_2/A_1 Kd ratio of about 600 [16].

Pappagallo et al. found that local pretreatment with bamiphylline reduced the severity of the pain caused by the subcutaneous administration of adenosine but not its hyperemic effect [17]. They concluded that as bamiphylline at the dose used did not prevent the vasodilatory A_2-mediated effects of adenosine, the reduction of adenosine-induced pain was due to antagonism of A_1 receptors. Similarly, Gaspardone et al. found that pretreatment with intravenous administration of bamiphylline did not affect

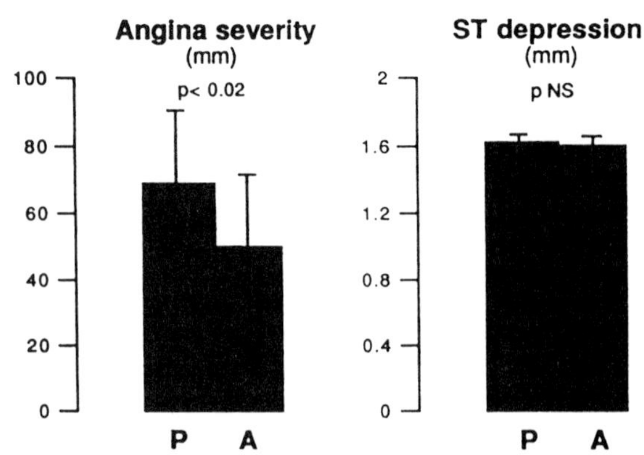

FIGURE 36-4. Bar graphs of effects of aminophylline on exercise-induced chest pain. The severity of chest pain at peak exercise was less after aminophylline (A) administration (7 mg/kg) than after placebo (P) administration, despite a comparable degree of peak ST-segment depression. (Reproduced with permission from Circulation.)

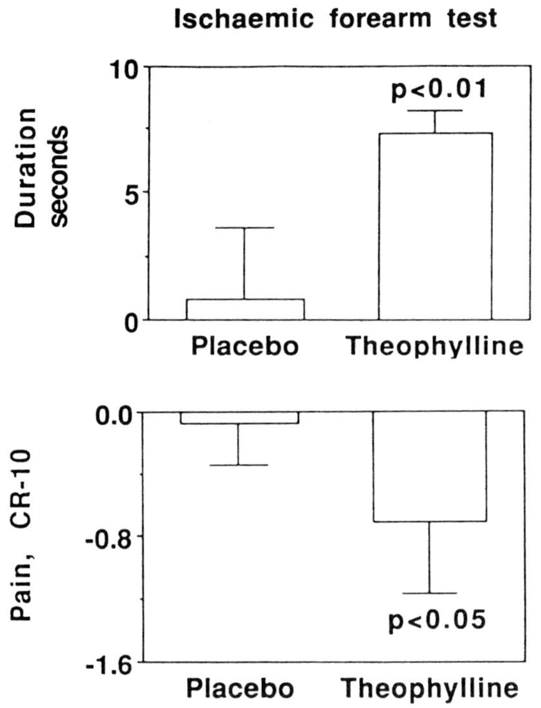

FIGURE 36-5. Effects of theophylline on the duration and intensity of pain in the ischemic forearm test in healthy volunteers. (Data derived with permission from Cardiovascular Research.)

the increase of coronary blood flow caused by intracoronary administration of adenosine, but it reduced adenosine-induced chest pain [18]. Also, bamiphylline pretreatment reduced the severity of exercise-induced anginal pain for a similar degree of ST-segment depression (Fig. 36-6). Finally, Hill et al. have recently shown that N0861, a selective A_1-adenosine antagonist, as shown both in affinity ($A_1 : A_2$ about 600) and functional ($A_1 : A_2$ about 100) studies in cardiac tissues, at a dose that blocks the A_1-mediated electrophysiologic effects, reduces the pain caused by the intravenous administration of adenosine [19]. Together these observations suggest that the algogenic effects of adenosine are, at least partially, mediated by A_1 receptors. The role of A_2 receptors has not been yet investigated in clinical trials.

Peripheral Modulation of Adenosine-Induced Pain

Adenosine may act at sensory nerves directly or indirectly [20]. Not much work has been done on the importance of indirect mechanisms of adenosine-provoked pain. By an indirect route adenosine may enhance or inhibit the release of pain-related substances such as histamine and lipidic acids from mast cells and other white corpuscular elements. While adenosine-related pain in cutaneous models may occur by both direct and indirect mechanisms [20], the short time characteristics of the pain provoked in vascular models with intravenous bolus injections suggest primarily a direct mechanism of action.

Regarding the peripheral modulation of adenosine-induced pain, metoprolol, nitroglycerin, clonidine, and cyclooxygenase fail to affect its algogenic effects, while pretreatment with nicotine enhances and β-endorphin, but not metenkephalin, reduces adenosine-provoked pain [21,22]. Furthermore, Gaspardone et al. showed that the intracoronary infusion of substance P does not cause pain per se, but it enhances adenosine-provoked pain [12]. Thus it appears that activation of nicotinergic and substance P receptors enhances, while activation of opiate receptors inhibits adenosine-provoked pain.

Failure of Adenosine to Cause Pain during Myocardial Ischemia

The results obtained in patients with angina at rest during continuous ECG and hemodynamic monitoring led to the following conclusions: 1) Transient ischemic episodes lasting less than 3 minutes and causing an increase of left ventricular pressure less than 7 mmHg are always silent. 2) Longer and/or more severe episodes can be either painful or silent, even in the same patient. 3) Milder transient ischemic episodes are more likely to be silent than severe ones, although episodes so severe as to cause a massive deficit of regional mycardial perfusion and even myocardial infarction can remain completely silent [23]. Therefore, in patients with angina at rest, a critical duration and severity of myocardial ischemia is probably needed in order to achieve an interstitial adenosine concentration adequate to stimulate cardiac sensory receptors.

The failure of myocardial ischemia of sufficient duration and severity to provoke anginal pain may result from various causes. The reasons for the variability of symptoms in patients who experience angina during some transient ischemic episodes but not during other episodes of similar severity and duration are far from being understood. The variability of symptoms may be due to 1) variability of pain perception, 2) a variable modulation in cardiac and mediastinal sympathetic ganglia of the algogenic message triggered by the interstitial release of adenosine, or 3) a variable local concentration of adenosine or of other, yet unknown, pain mediators.

In patients who have never experienced angina even during severe, prolonged ischemia and in those who have predominantly silent ischemia, the failure to develop pain could be due to psychologic factors inhibiting the transmission of nociceptive stimuli or the perception of painful stimuli. This is suggested by the generalized defective perception of painful stimuli consistently demonstrated in this subset of patients. Indeed, patients with totally silent ischemia and those with predominantly silent ischemia have a significantly higher threshold and tolerance for pain than patients with predominantly painful ischemia when challenged with painful stimuli such as forearm ischemia, cold pressor, skin electric stimulation [24,25], intravenous adenosine infusion [9], or dental pulp stimulation [26]. In addition, on a personality inventory test, patients with silent ischemia recorded significantly lower scores for nervousness and "excitability," higher scores for "masculinity," and a lower tendency to complain [24].

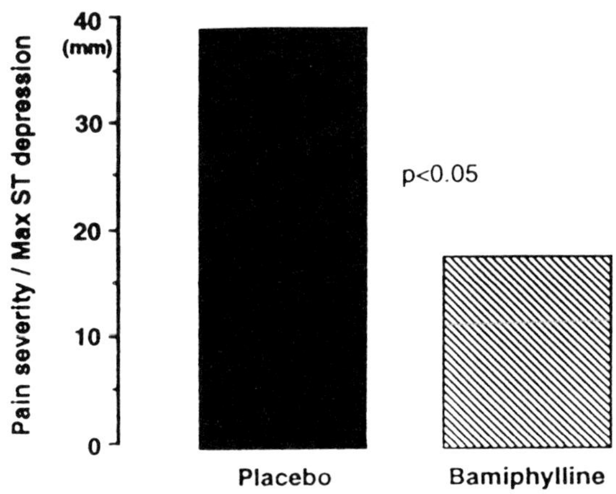

FIGURE 36-6. Bar graph of severity of exercise-induced angina normalized for maximal ST-segment depression after placebo or bamiphylline administration. Data are expressed as median values. (Reproduced with permission from Circulation.)

Despite significant differences between patients with predominantly silent ischemia and those with predominantly painful ischemia, a considerable overlap of the value of pain threshold and tolerance for the various painful stimuli has been reported in all published studies. In those patients with predominantly silent ischemia who present with a high threshold and tolerance for pain, other mechanisms, such as a lower intramyocardial concentration of adenosine or other algogenic stimuli, a reduced number of cardiac sensory receptors, or impairment of cardiac afferent fibers (for instance in diabetics), are more likely to operate.

Role of Adenosine in Genesis of Anginal pain in the Absence of Detectable Signs of Myocardial Ischemia

Patients with syndrome X are characterized by the presence of angina, normal coronary arteries at angiography, and a positive exercise test.

However, the anginal pain and the ischemialike ECG changes are not usually accompanied by detectable regional alterations of left ventricular function [27]. This is in contrast to what is observed in patients with obstructive coronary atherosclerosis in whom regional alterations of the left ventricular function typically precede both ECG changes and angina. Patients with syndrome X have been found to have a lower threshold and tolerance for pain when challenged with painful stimuli such as forearm ischemia, cold pressor, skin electric stimulation [28], or adenosine (Fig. 36-7) [29] and an abnormally high cardiac sensitivity to catheter manipulation and contrast injection [30].

It would appear, therefore, that, compared to patients with unstable, variant, and chronic stable angina who predominantly exhibit painless myocardial ischemia in spite of an obvious impairment of left ventricular perfusion and function, patients with syndrome X are at the other extreme of the spectrum of sensory manifestations of ischemic heart disease, where even myocardial ischemia not impairing ventricular function may be perceived as painful. Maseri et al. have recently proposed that, together with an enhanced perception of painful stimuli, the severe anginal pain frequently observed in patients with syndrome X in the absence or in the presence of only minor signs of myocardial ischemia might be related to a massive local release of adenosine, which on the one hand causes pain and on the other minimizes the consequences of a patchy, inappropriate prearteriolar constriction of small coronary vessels [27].

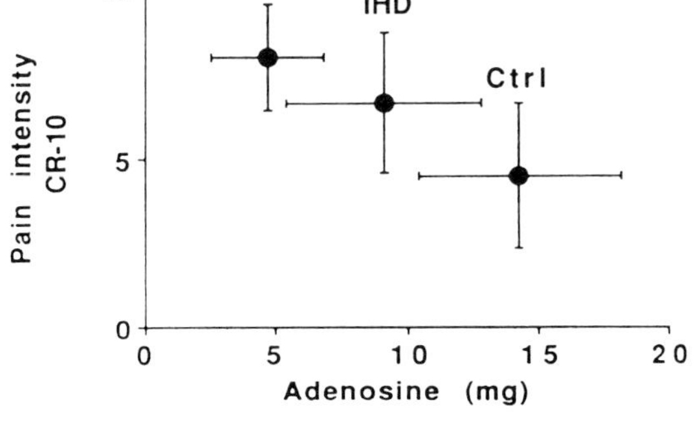

FIGURE 36-7. Dose-response relations for maximal intravenous adenosine dose in healthy volunteers, patients with coronary artery disease and angina pectoris, and patients with syndrome X. (Reproduced with permission from British Heart Journal.)

Reflexogenic Effects of Adenosine

Contrary to the known inhibitory effects of adenosine in central and peripheral efferent systems, recent studies have shown that stimulation of afferent systems by adenosine results in excitatory cardiovascular reflexes [21]. Dibner-Dunlap et al. [31] found in an animal model that intracoronary adenosine administration increases renal nerve activity; this was abolished by sympathectomy. As the A_1 agonist CPA but not the A_2 agonist CPCA enhanced renal nerve activity dose dependently, the authors suggested that these effects were mediated by adenosine A_1 receptors. Reflexogenic activation of autonomic function by adenosine is also evident from several reports in humans following both intravenous or intraarterial infusion of adenosine [21]. As beta-blockers do not entirely antagonize adenosine-induced excitatory reflexes, the latter do not appear to be mainly mediated by beta-adrenergic mechanisms. Biaggioni et al. [32] found that in humans the intravenous infusion of adenosine increased heart rate, systolic arterial pressure, minute ventilation, and muscle sympathetic efferent activity. As adenosine increased nerve activity much more than nitroprusside at doses that caused an equivalent reduction in arterial pressure, the authors concluded that adenosine increases sympathetic tone by activating afferent nerves, including arterial chemoreceptors [33].

Cox et al. assessed the local coronary excitatory effects of adenosine in a human model. They found that the intracoronary injection of adenosine, at a dose that did not cause systemic effects, resulted in a small but consistent increase of arterial blood pressure, which was not observed in transplanted patients. Thus, the authors proposed that stimulation of cardiac afferent fibers results in a reflex increase of arterial blood pressure [34]. Marzilli et al. also reported a reflex increase in blood pressure following intracoronary infusion of adenosine [35]. However, in another study Crea et al. found an increase of blood pressure during intracoronary adenosine administration only at doses similar or higher than those causing pain; they suggested that the increase of arterial blood pressure observed during intracoronary adenosine administration might be simply secondary to the anxiety caused by the pain [10].

Direct Neural Excitatory Effects of Adenosine

In the last few years the local excitatory effects of adenosine on afferent systems have been studied extensively in a variety of experimental models. Katholi et al. [36] in 1983 reported that renal nerve afferents were excited by the application of adenosine in the renal pelvis. In 1981 McQueen and Ribeiro [37] reported that carotid sinus afferent nerve activity was enhanced by adenosine; this effect was mediated by A_2-subtype receptors [38]. Also, dorsal root ganglion cells can be excited by adenosine [39,40]. Finally, adenosine and ATP as well as their analogues can enhance or depress neural activity in cardiac intrinsic ganglia [41]. Montano et al. [42] were the first to directly study cardiac primary afferents. They found that in a cat model the epicardial application of 10 μM adenosine resulted in a consistent activation of cardiac sympathetic afferent dorsal root neurons. Armour et al. [43] showed that following epicardial application of adenosine in a dog model, both 1 μM adenosine and ATP could enhance or depress neural activity in mediastinal ganglia, while coronary occlusion in part of the preparations increased activity (Fig. 36-8).

When activity of dorsal root ganglion was studied, adenosine in the 1 μM range, and to a lesser extent ATP, enhanced neural activity [44]. A_1, A_2, and substance P antagonists decreased baseline dorsal root ganglion neural activity, suggesting that all these receptors located at the same afferent neuron are tonically active. Furthermore, application of adenosine enhanced dorsal root ganglionic neural activity; the effect was partially inhibited by epicardial application of A_1 or A_2 (DPCPX, KF 17387) antagonists. Total blockade was obtained only after combined A_1 and A_2 antagonism (CGS). Substance P enhanced dorsal root ganglionic neural activity; the latter was inhibited by spantide, a substance P antagonist. The same neuron could thereafter be excited by adenosine. After occlusion of the left anterior descending artery, dorsal root ganglionic neural activity increased. This increase was prevented by application of CGS and substance P. Finally, following application of EGTA or nifedipine at the same neuron, neural activity increased to at least the levels observed following adenosine application, while following application of the Ca channel agonist Bay K 8644 the neural activity decreased [44].

Organization and Significance of the Cardiac Sympathetic Afferent System

The cardiac sympathetic afferent system has a complex structure. Afferent nerve cell bodies are present in intrinsic cardiac, mediastinal, and dorsal root ganglia. Application of adenosine or ATP at intrinsic cardiac ganglia or mediastinal ganglia can enhance or decrease neural activity [38,40]. The effect of adenosine and ATP on cardiac intrinsic ganglia may be of clinical interest, as these ganglia receive their blood supply from the epicardial coronary arteries. As the metabolic needs of these ganglia are similar to those of the myocardium, coronary occlusion may cause ischemia of these ganglia followed by the release of purines that can modulate afferent neural traffic and cause pain. While both

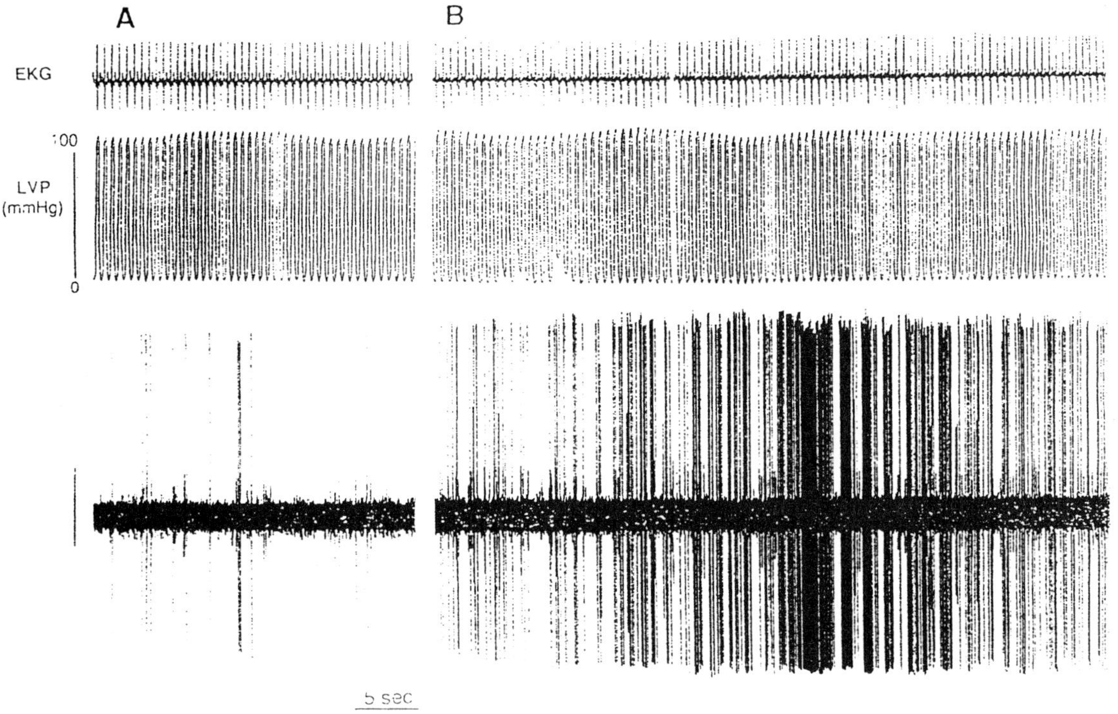

FIGURE 36-8. Epicardial application of CPA. N6-cyclopentyl-adenosine, 1.5 mM (between panels **A** and **B**) to the ventral left ventricle resulted in activity being increased without alterations in the cardiac variables monitored. Note that the activity generated by this neuron was irregular with respect to time. Panel B was obtained 12 minutes after the application of CPA had terminated. (Reprinted with permission from Cardiovascular Drugs and Therapy.)

adenosine and ATP appear to enhance or decrease neural activity in the intrinsic cardiac and mediastinal ganglia, enhancement of neural activity only was observed in dorsal root ganglia following epicardial purine application [41]. In addition, in power spectral analysis of activity generated by dorsal root ganglia cardiac neurons, at least one adenosine-specific peak occur at about 40 Hz [Armour, personal communication]. Purinergic neural modulation of cardiac sympathetic fibers may be of importance for the beat-to-beat regulation of cardiac performance. During an ischemic insult, at least part of this activity is the basis for the appearance of cardiac pain. As both adenosine and ATP can also decrease cardiac afferent neural activity, at least at the cardiac and mediastinal level, neural modulation during an ischemic insult may result in the absence of pain. Although ATP has the potential to be a pain messenger, it has not yet been shown to fulfill the criteria for a pain messenger, as suggested by Lewis [2].

Conclusions

Adenosine is so far the only substance that fulfills Lewis criteria as a pain messenger of myocardial ischemia. Indeed, adenosine is released during ischemia, and its administration provokes pain and activation of afferent nerves. Other substances, such as ATP, may also be involved, but this has not yet been established. Adenosine-provoked pain has been demonstrated in healthy volunteers and in patients with ischemic heart disease. Patients with silent ischemia and those with syndrome X are hyporeactive and hyper-reactive to adenosine, respectively. Antagonism of endogenous adenosine decreases ischemic pain. The adenosine-provoked pain is exerted at extracellular membrane-bound adenosine receptors. Additional stimuli, such as substance P and nicotine, appear to enhance adenosine-induced pain. Experimental studies have shown that adenosine A_1 and A_2 and substance P receptors are present at the same afferent dorsal root ganglion neuron. Clinical studies, however, have hitherto tested A_1 receptors only, showing that the algogenic effects of adenosine are, at least partially, mediated by the stimulation of adenosine A_1 receptors. The adenosine effect on the cardiac afferent nerve appears to involve calcium-dependent mechanisms.

References

1. Colbeck EH (1903) Angina pectoris: A criticism and a hypothesis. Lancet 1:793–795.
2. Lewis T (1932) Pain in muscular ischemia—its relation to anginal pain. Arch Intern Med 49: 713–727.
3. Davies EJ, Bencivelli W, Chierchia S, Fragasso G, Crea F, Crow J, Crean P, Pratt T, Morgan M, Maseri A (1988) Sequence and magnitude of ventricular volume changes in painful and painless myocardial ischemia. Circulation 78: 310–319.
4. Tomai F, Crea F, Gaspardone A, Versaci F, Esposito C, Chiariello L, Gioffré PA (1993) Mechanisms of cardiac pain during coronary angioplasty. J Am Coll Cardiol 22:1892–1896.
5. Eriksson B, Vuorisalo D, Sylvén C (1994) Quality and intensity of chest pains in patients treated in coronary care. J Intern Med 235: 473–478.
6. Sylvén C, Beermann B, Jonzon B, Brandt R (1986) Angina pectoris-like pain provoked by IV adenosine in healthy volunteers. Br Med J 293:227–230.
7. Bleehen T, Keele CA (1977) Observations of the algogenic actions of adenosine compounds on the human blister base preparation. Pain 3: 367–377.
8. Lagerqvist B, Sylvén C, Beermann B, Helmius G, Waldenström A (1990) Intracoronary adenosine causes angina pectoris like pain—an inquiry into the nature of visceral pain. Cardiovasc Res 24:609–613.
9. Crea F, Pupita G, Galassi AR, el-Tamimi H, Kaski JC, Davies G, Maseri A, et al. (1990) Role of adenosine in pathogenesis of anginal pain. Circulation 81:165–173.
10. Crea F, Gaspardone A, Kaski JC, Davies G, Maseri A (1992) Relation between stimulation site of cardiac afferent nerves by adenosine and location of cardiac pain in patients with stable angina. J Am Coll Cardiol 20:1498.
11. Sylvén C, Jonzon B, Borg G, Fredholm BB, Kaijser L (1988) Adenosine injection into the brachial artery produces ischemia-like pain or discomfort in the forearm. Cardiovasc Res 22: 674–678.
12. Gaspardone A, Crea F, Tomai F, Iamele M, Crossman DC, Pappagallo M, Versaci F, Chiariello L, Gioffré PA (1994) Substance P potentiates the algogenic effects of the intra-arterial infusion of adenosine. J Am Coll Cardiol 24:477–482.
13. Sylvén C, Jonzon B, Edlund A (1989) Adenosine pectoris-like pain provoked by iv bolus of adenosine: Relationship to coronary sinus blood flow, heart rate and blood pressure in healthy volunteers. Eur Heart J 10:48–54.
14. Jonzon B, Sylvén C, Kaijser L (1989) Theophylline decreases pain in the ischemic forearm test. Cardiovasc Res 23:807–809.
15. Crea F, Pupita G, Galassi AR, el-Tamimi H, Kaski JC, Davies G, Maseri A (1989) Effect of theophylline on exercise-induced myocardial ischaemia. Lancet 8640:683–686.
16. Abbracchio MP, Cattabeni F (1987) Selective activity of bamiphylline on adenosine A1-receptors in rat brain. Pharmacol Res Commun 19:537–545.
17. Pappagallo M, Gaspardone A, Tomai F, Iamele M, Crea F, Gioffré PA (1993) Analgesic effect of bamiphylline on pain induced by intradermal injection of adenosine. Pain 53:199–204.
18. Gaspardone A, Crea F, Iamele M, Tomai F, Versaci F, Pellegrino A, Chiariello L, Gioffré PA (1993) Bamiphylline improves exercise-induced myocardial ischemia through a novel mechanism of action. Circulation 88:502–508.
19. Hill JA, Bertolet B, Kerenski N, Nicholos W, Cowley S, Miller J, Belardinelli L (1994) Attenuation of adenosine A_1-receptor mediated effects by N-0861 in humans. Drug Dev Res 4:278.
20. Sawynok J, Yaksh TL (1993) Caffeine as an

analgesic adjuvant: A review of pharmacology and mechanisms of action. Pharmacol Rev 45: 43–85.

21. Sylvén C (1993) Mechanisms of pain in angina pectoris—a critical review of the adenosine hypothesis. Cardiovasc Drugs Ther 7:745–59.
22. Eriksson B, Maixner W, Ballenger K, Sheps D, Sylvén C (1994) Intravenous infusion of β-endorphin but not of metenkephalin counteracts angina pectoris-like pain provoked by intravenous bolus injection of adenosine. N Engl J Med, submitted.
23. Chierchia S, Lazzari M, Freedman B, et al. (1983) Impairment of myocardial perfusion and function during painless myocardial ischemia. J Am Coll Cadiol 1:924–930.
24. Droste C, Roskamm H (1983) Experimental pain measurements in patients with asymptomatic myocardial ischemia. J Am Coll Cardiol 1:940–945.
25. Glazier JJ, Chierchia S, Brown M, Maseri A (1986) Importance of generalized defective perception of painful stimuli as a cause of silent myocardial ischemia in chronic stable angina pectoris. Am J Cardiol 58:667–672.
26. Falcone C, Sconocchia R, Guasti L, Codega S, Montemartini C, Specchia G (1988) Dental pain threshold and angina pectoris in patients with coronary artery disease. J Am Coll Cardiol 12:348–352.
27. Maseri A, Crea F, Kaski JC, Crake T (1991) Mechanisms of angina pectoris in syndrome X. J Am Coll Cardiol 17:499.
28. Turiel M, Galassi AR, Glazier JJ, Kaski JC, Maseri A (1987) Pain threshold and tolerance in women with Syndrome X and women with stable angina pectoris. Am J Cardiol 60: 503–508.
29. Lagerqvist B, Sylvén C, Waldenström A (1992) Low threshold for adenosine induced chest pain in patients with angina pectoris and normal coronary angiogram. Br Heart J 68:282–283.
30. Cannon RO, Quyyumi A, Schenke WH, et al. (1990) Abnormal cardiac sensitivity in patients with chest pain and normal coronary arteries. J Am Coll Cardiol 16:1359–1366.
31. Dibner-Dunlap ME, Kinugawa T, Thames MD (1993) Activation of cardiac sympathetic afferents: Effects of exogenous adenosine and adenosine analogues. Am J Physiol 265:H395–400.
32. Biaggioni I, Killian T, Mosqueda-Garcia R, Robertson RM, Robertson D (1991) Adenosine increases sympathetic nerve traffic in humans. Circulation 83:1668–1675.
33. Costa F, Biaggioni I (1993) Adenosine activates forearm afferents eliciting a pressor reflex. Circulation 88:361.
34. Cox DA, Vita JA, Treasure CB, Fish RD, Selwyn AP, Ganz P (1989) Reflex increase in blood pressure during the intracoronary administration of adenosine in man. J Clin Invest 84:592–596.
35. Marzilli M, Klassen GA, Marraccini P, Camici P, Trivella MG, L'Abbate A (1989) Coronary effects of adenosine in conscious man. Eur Heart J 10(Suppl F):78–81.
36. Katholi RE, Hageman GR, Whinlow PL, Woods WT (1983) Hemodynamics and afferent renal nerve response to intrarenal adenosine in the dog. Hypertension 5(Suppl 1):1149–1154.
37. McQueen DS, Riberio JA (1981) Effects of adenosine on carotid chemoreceptor activity in the cat. Br J Pharmacol 74:129–136.
38. Monteiro EC, Ribeiro JA (1987) Ventilatory effects of adenosine mediated by carotid body chemoreceptors in the rat. Arch Pharmacol 335: 143–148.
39. MacDonald RL, Skerritt JH, Werz MA (1986) Adenosine agonists reduce voltage-dependent calcium conductance of mouse sensory neurons in cell culture. J Physiol 370:75–90.
40. Dolphin AC, Forda SR, Scott RH (1986) Calcium-dependent currents in cultured rat dorsal root ganglion neurons are inhibited by an adenosine analogue. J Physiol 373:47–61.
41. Huang MH, Sylvén C, Smith FM, Pelleg A, Armour JA (1993) Adenosine and adenosine 5-triphosphate modulate intrinsic cardiac neuronal activity in the dog. Am J Physiol 265:R914–922.
42. Montano N, Lombardi F, Ruscone TC, Contini M, Guazzi M, Malliani A (1991) The excitatory effect of adenosine on the discharge activity of the afferent cardiac sympathetic fibers. Cardiologia 36:953–959.
43. Armour JA, Huang MH, Pelleg A, Sylvén C (1994) Responsiveness of in situ canine nodose ganglion afferent neurons to epicardial mechanical or chemical stimuli. Cardiovasc Res 28: 1218–1225.
44. Huang MH, Sylvén C, Horackova M, Armour JA (1995) Cardiac sensory neurons tonically activated by endogenous adenosine and substance P. Am J Physiol, in press.

D. OTHER SYSTEMS

37. THE P_{2Z} RECEPTOR AND ITS REGULATION OF MACROPHAGE FUNCTION

Francesco Di Virgilio
Davide Ferrari
Maddalena Munerati
Simonetta Falzoni
Martin Villalba
Vincenzo Bronte
Annalisa Zambon
Paola Zanovello
Thomas H. Steinberg

In 1967 the late Zanvil A. Cohn reported that adenine nucleotides had striking effects on the morphology and pinocytic activity of mouse macrophages [1]. These early studies were followed by the demonstration by Sung et al. that extracellular ATP (ATP_e) caused plasma membrane depolarization, increased intracellular Ca^{2+} ($[Ca^{2+}]_i$), and inhibition of Fc receptor-mediated phagocytosis [2]. These intriguing effects remained unexplained until the work by Steinberg and Silverstein, who started a thorough investigation of the mechanism of macrophage responses to extracellular nucleotides, ATP in particular [3,4].

Most of the studies have been performed in mouse macrophages, but very recently human monocyte/macrophage cells have also been investigated [5]. We will first describe the P_{2Z} receptor in mouse macrophages and then report very recent data obtained in human cells.

How Many Macrophage Receptors for Extracellular ATP?

Steinberg et al. [3] discovered that extracellular ATP (ATP_e) opened a pore in the plasma membrane of mouse macrophages sufficiently large to allow aqueous molecules <900 Da to enter or leave the cells. This striking response had been previously seen in rat mast cells by Gomperts and colleagues [6] and in some transformed mouse fibroblasts by Heppel and coworkers [7]. It seemed likely that this finding could explain all the effects of ATP_e reported by Sung et al. [2]; however, the increase in $[Ca^{2+}]_i$ occurred at lower concentrations of ATP_e than did plasma membrane depolarization or pore formation (Figs. 37-1 and 37-2).

This observation led Greenberg et al. [8] to further examine ATP_e-induced $[Ca^{2+}]_i$ transients. They found that mouse macrophages exhibited a clear biphasic response to stimulation with ATP_e. At low nucleotide concentrations ($<300\,\mu M$) in the presence of 1 mM Ca^{2+} and Mg^{2+} and pH 7.4, the most remarkable effect was an increase in the cytoplasmic Ca^{2+} concentration ($[Ca^{2+}]_i$) as a result of both influx across the plasma membrane and release from intracellular stores [8]. At ATP_e concentrations $>500\,\mu M$, the macrophage response changed dramatically and was dominated by a nonselective increase in plasma membrane permeability to hydrophylic solutes of molecular weight <900 Da [4].

Three lines of evidence demonstrated that ATP_e-induced Ca^{2+} mobilization and ATP_e-mediated pore formation (and membrane depolarization) were mediated by different ATP receptors. First, the dose-response curves were different, as noted above. Second, Ca^{2+} mobilization could be elicited by many other nucleotides; in fact, extracellular UTP (UTP_e) was a more potent agonist than ATP_e, while, on the contrary, plasma membrane depolarization could only be induced by ATP_e and a few ATP analogs. Third, ATP_e-resistant (ATPR) cell lines that were selected for the ability to survive in medium containing high concentrations of ATP_e and did not exhibit pore formation in the presence of ATP_e, nevertheless retained the $[Ca^{2+}]_i$-mobilizing response [8]. Thus, mouse macrophages express purinergic receptors of both the P_{2Y} and P_{2Z} subtypes (Fig. 37-3).

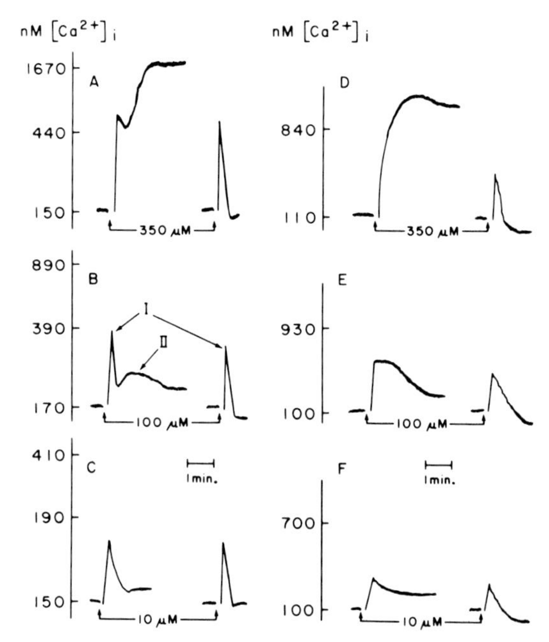

FIGURE 37-1. Kinetics of $[Ca^{2+}]_i$ increases triggered by extracellular ATP in mouse macrophages. The tracings on the left represent experiments performed in saline solution containing 1 mM Mg^{2+} and 1 mm Ca^{2+}; the tracings on the right represent experiments performed in the presence of 5 mM EGTA. A–C: Indo-1-loaded cells, D–F: Quin-2-loaded cells. Reproduced from Greenberg et al. [8], with permission.

P_{2Z} Receptor of Mouse Macrophages

Steinberg and colleagues have shown that ATP_e causes a generalized increase in plasma membrane permeability to a variety of membrane-impermeant fluorescent dyes smaller than 900 Da [4]. Opening of the pore is almost immediate upon ATP_e addition, as shown by the recording of membrane potential in the whole-cell patch-clamp configuration [9]. Likewise, closure of the pore begins almost immediately after ATP_e is removed from the bathing solution. The absence of a lag in the activation of the pore is also suggested by kinetic measurement of ethidium bromide uptake by ATP_e-pulsed macrophages [10]. Thus, we think that the P_{2Z} receptor/pore is directly activated by ATP_e and is not modulated by known intracellular second messengers. This hypothesis is also supported by the fact that ATP pore formation occurs normally in the whole-cell patch-clamp configuration, because in this cir-

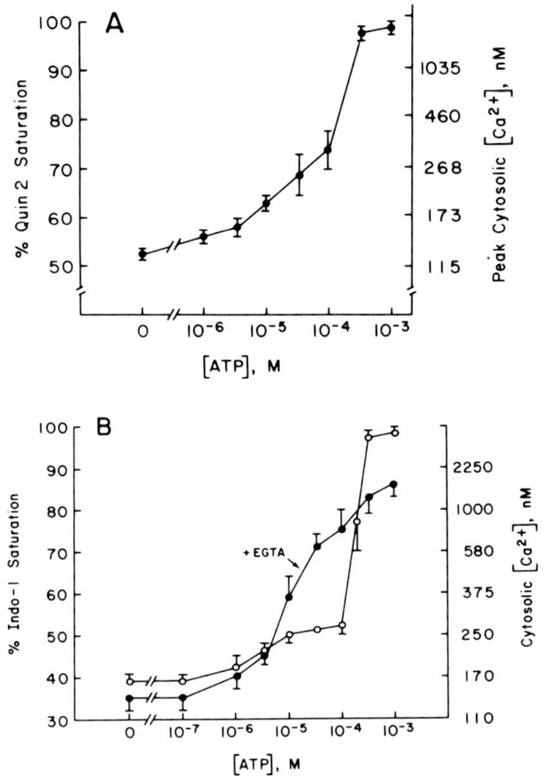

FIGURE 37-2. ATP dose-response curves for quin2 and indo-1-loaded mouse macrophages. (**A**) quin2-loaded and (**B**) indo-1-loaded cells stimulated with ATP in saline medium containing 1 mM Ca^{2+} and 1 mM Mg^{2+}. Reproduced from Greenberg, et al. [8] with permission.

cumstance small molecules in the cytosol are removed by equilibration with the medium in the patch pipette.

Similar to P_{2Z} receptors described in rat mast cells, transformed mouse fibroblasts, and mouse neuroblastoma cells [11], in mouse macrophage also the active ligand is ATP^{4-} and, among a few active ATP analogues, benzoylbenzoic ATP (BzATP) is a more potent agonist than ATP itself [Di Virgilio, unpublished observations].

Opening of the ATP_e-gated pore causes a dramatic perturbation of macrophage intracellular ion homeostasis and morphology. Release of intracellular monovalent cations (monitored as ^{86}RB efflux) starts almost immediately, as well as Na^+ and Ca^{2+} influx and plasma membrane potential collapse [3,9]. These changes in intracellular ion concentration are followed by cell swelling after a brief lag of about 5–10 seconds. Swelling is easily detected by microscopic examination, but can also be kinetically monitored by measuring the light-scattering increase of the macrophage suspension [10] (Fig. 37-4). Removal of ATP_e within 15–30 minutes fully reverses the morphologic changes and allows macrophages to resume their physiologic functions, for example, phagocytosis [12]. However, if incubation is prolonged beyond this time, irreversible damage and macrophage death occur.

In contrast to other cell types in which ATP_e causes death by apoptosis, macrophage cell death is due to colloido-osmotic lysis [10]. Rather intriguingly, death is Ca^{2+} and Na^+ independent and is completely prevented by performing ATP_e permeabilization in Ca^{2+}-free, sucrose-containing medium. The protective effect of sucrose is very surprising, because under these conditions macrophages were normally permeabilized by ATP_e. Time-course experiments show that irreversible plasma membrane damage very probably starts after 1 hour of incubation in the presence of an at least 1 mM ATP_e concentration (Ca^{2+}-free, 1 mM Mg^{2+}-containing saline) because the release of high molecular weight cytoplasmic constituents (lactic dehydrogenase) is first detectable at this time.

An important proof of the key role of the P_{2Z} receptor in ATP_e-mediated cytotoxicity has been recently offered by experiments showing that macrophage pretreatment with oxidized ATP (oATP), a covalent and specific reagent of ATP binding sites, completely prevented plasma membrane permeabilization and the cytotoxic effect of several millimolar ATP_e concentrations without decreasing $[Ca^{2+}]_i$ mobilization due to activation of $P_{2Y/P2U}$ receptor [13].

P_{2Z} Receptor of Human Monocytes/Macrophages

Early experiments performed in monocyte-derived human macrophages [T.H. Steinberg, unpublished] failed to reveal ATP_e-triggered plasma membrane permeabilization, the unequivocal hallmark of the P_{2Z} receptor, thus suggesting that human cells might lack this kind of receptor. However, in 1991 Djeu and coworkers showed that human macrophages were sensitive to the cytotoxic effect of ATP_e and that gamma-interferon greatly potentiated this effect [5]. This response, however, was not further characterized, thus leaving open the question of the specificity of ATP_e cytotoxic effect and the purinergic receptor subtype involved.

We are currently re-examining this issue, and preliminary evidence indicates an expression of P_{2Z} receptor by human macrophage but not monocyte cells [Falzoni et al., submitted]. Peripheral blood monocytes (PBM) stimulated with ATP_e soon after their isolation show an increase in $[Ca^{2+}]_i$ typical of activation of $P_{2Y/P2U}$ receptor. The fast early transient is followed by a second shoulder of $[Ca^{2+}]_i$ increase, typical of receptors that open plasma membrane Ca^{2+} channels in addition to releasing Ca^{2+} from intracellular stores. In monocytes, neither ATP_e nor other nucleotides trigger uptake of ethidium bromide or lucifer yellow, and even plasma membrane depolarization is negligible, thus indicating a lack of any membrane-permeabilizing effect.

However, the pattern of responses changes dramatically after 5 days of in vitro culture, a stage at which monocytes have acquired many of the typical differentiation features of macrophages. In these cells ATP_e stimulation triggers a massive influx of Ca^{2+}, which is soon followed by the efflux of fura-2 into the extracellular medium. Ethidium bromide uptake is also induced, and these cells become permeable to lucifer yellow. As can be anticipated, they also become sensitive to ATP_e-dependent cytotoxicity.

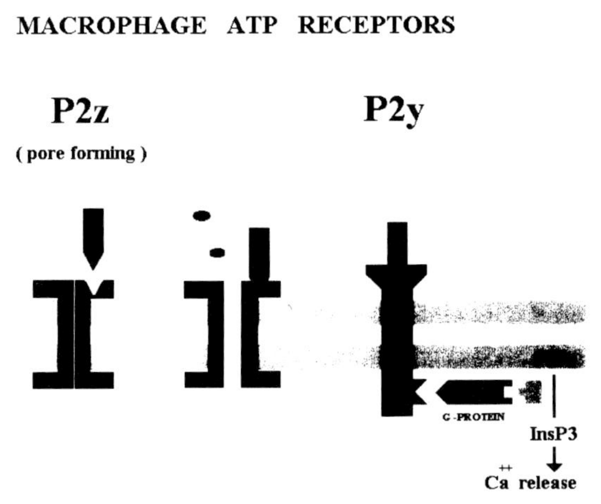

FIGURE 37-3. Macrophage P_2 receptor subtypes.

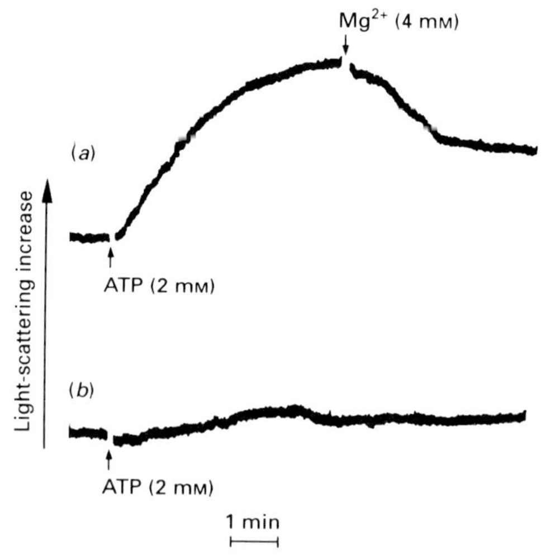

FIGURE 37-4. Swelling of J774 mouse macrophages trigered by extracellular ATP. (**a**) Wild-type J774 cells; (**b**) ATP-resistant ATPRB2 cells. Reproduced from Murgia, et al. [10] with permission.

Gamma-interferon dramatically potentiates these responses. Conversely, pretreatment with oATP fully abrogates membrane permeabilization as well as cytotoxicity.

What is the Role of P_{2Z} Receptor in Macrophage Physiology?

The P_{2Z} receptor was initially viewed at best as a tool to allow manipulation of the cytoplasm, if not as a mere curiosity without any real biologic relevance. We think that this was at least in part due to the fact that most of the studies were until very recently performed in mouse and rat cells and the few observations reported in human cells were either negative or inconclusive, thus, incorrectly, suggesting that the P_{2Z} receptor was not expressed by cells relevant to human pathology. A second reason for lack of interest in the P_{2Z} receptor paradoxically stems from its very peculiar properties. The only response thus far linked to this receptor is reversible plasma membrane permeabilization, something that most investigators found difficult to find of biologic relevance.

Interest towards this receptor, and all purinergic receptors in general, is rapidly increasing, as it is becoming clear that whether or not a given cell expresses this receptor could be a life-or-death question [14]. That stimulation of some cell types with ATP_e could lead to cell death was a common but often neglected observation, until Steinberg and Silverstein exploited this approach to select J774 macrophage clones resistant to ATP_e-dependent permeabilization [4]. These investigators exposed J774 macrophage cultures to repeated rounds of incubation in the presence of high (10 mM) ATP_e concentrations until cells were no longer killed by ATP_e. Resulting cell clones did not take up lucifer yellow or ethidium bromide, and did not depolarize when stimulated with ATP_e. The ATP-resistant mutants (ATPR) were a clear indication that injurious effects of ATP_e were more than a mere nonspecific perturbation of the plasma membrane but rather involved a specific receptor-mediated pathway.

Interest in ATP_e-dependent cytotoxicity grew even further when we and others showed that other cells besides macrophages were also susceptible to ATP_e-mediated killing and that within the same populations cell subtypes inherently resistant to ATP_e were present [15–18]. The most striking example is provided by mouse lymphocytes. Resting B and T lymphocytes and several lymphoma cell lines are readily killed by ATP_e, while cytotoxic T lymphocytes (CTL) and lymphokine-activated killer (LAK) cells are eminently resistant [15,16]. Stimulation with IL-2 seems to have a crucial role in the acquisition of ATP_e resistance.

Insensitivity to ATP_e of cytotoxic cells, the apoptotic effect of ATP_e in many cell types, and the reported ability to release ATP of some CTL clones suggested the appealing hypothesis that ATP_e might be one of the soluble mediator of T-lymphocyte–mediated cytotoxicity and the P_{2Z} receptor might he the target molecule [11]. We have thoroughly tested this hypothesis and have come to the conclusion that, at least in the mouse, the P_{2Z} receptor is unlikely to be involved in T-cell–mediated cytotoxicity [19].

It can be anticipated that were the P_{2Z} receptor a principal target of cytotoxic mediator molecules, cells lacking it (e.g., ATPR macrophages) and oATP-treated wild-type J774 macrophages should be protected against CTL-dependent killing. In fact, this is not the case, as ATPR cells are fully sensitive and oATP pretreatment does not protect from CTL-mediated killing.

These results do not rule out ATP_e as a cytotoxic molecule, because cell death is a common event in many in vivo conditions in which ATP is released by activated or injured cells. For example, ATP_e might be a soluble factor responsible for spreading a necrotic injury to neighboring cells. However, we think that more in vivo experiments are needed before a role for the P_{2Z} receptor in accidental or developmentally regulated cell death can be put forward.

An alternative view would be that cytotoxicity is a side effect of activation of the P_{2Z} receptor/pore elicited under extreme situations (exposure to high ATP_e concentrations for prolonged times) that only rarely occurs under normal physiologic conditions (e.g., extensive tissue damage). The size of the ATP_e-gated pore appears to be similar to that of the gap junction connexin43, a membrane complex that is expressed by J774 and bone-marrow–derived mouse macrophages but absent is in some ATPR macrophage clones [20]. Gap junctions couple adjacent cells, allowing the diffusion of electric and biochemical signals and the operation of single cells like a functional syncitium. Macrophages perform numerous gamma-interferon–modulated tasks that involve cell-to-cell contact, such as antigen presentation, cytotoxicity, and the generation of multinucleated giant cells. It is possible that the ATP_e-gated pore, which, as described earlier, is upmodulated by gamma-interferon and during monocyte to macrophage differentiation, participates in these specialized responses and therefore has a central role in the immune/inflammatory reaction.

Acknowledgments

This work was made possible by grants from the National Research Council of Italy (target projects Biotechnology & Bioinstrumentation and Clinical Application of Cancer Research, and special project Biology & Pathology of Calcium), the Ministry of Scientific Research, the Italian Association for Cancer Research, the AIDS project, and Telethon of Italy.

References

1. Cohn ZA, Parks E (1967) The regulation of pinocytosis in mouse macrophages. J Exp Med 125:457–466.
2. Sung SS, Young JD-E, Origlio AM, Heiple JM, Kaback HR, Silverstein SC (1985) Extracellular ATP perturbs ion fluxes, elevates cytosolic Ca^{2+} and inhibits phagocytosis in mouse macrophages. J Biol Chem 260:13442–13449.
3. Steinberg TH, Silverstein SC (1987) Extracellular ATP^{4-} promotes cation fluxes in the J774 mouse macrophage cell line. J Biol Chem 262:3118–3122.
4. Steinberg TH, Newman AS, Swanson JA, Silverstein SC (1987) ATP^{4-} permeabilizes the plasma membrane of mouse macrophages to fluorescent dyes. J Biol Chem 262:8884–8888.
5. Blanchard DK, McMillen S Djeu JY (1991) IFN-gamma enhances sensitivity of human macrophages to extracellular ATP-mediated lysis. J Immunol 147:2579–2585.
6. Cockcroft S, Gomperts BD (1979) ATP induced permeability in rat mast cells. Nature 279: 541–542.
7. Rozengurt E, Heppel LA, Friedberg I (1977) Effect of exogenous ATP on the permeability properties of transformed cultures of mouse cell lines. J Biol Chem 252:4584–4590.
8. Greenberg S, Di Virgilio F, Steinberg TH, Silverstein SC (1988) Extracellular nucleotides mediate Ca^{2+} fluxes in J774 macrophages by two distinct mechanisms. J Biol Chem 263: 10337–10343.
9. Buisman HP, Steinberg TH, Fischbarg J, Silverstein SC, Vogelzang SA, Ince C, Ypey DL, Leijh PCJ (1988) Extracellular ATP induces a large nonselective conductance in macrophage plasma membranes. Proc Natl Acad Sci USA 85:7988–7992.
10. Murgia M, Pizzo P, Steinberg TH, Di Virgilio F (1992) Characterization of the cytotoxic effect of extracellular ATP in J774 mouse macrophages. Biochem J 288:897–901.
11. Di Virgilio F, Steinberg TH (1993) ATP permeabilization. Wiley (London) Cell Tissue Cultures: Laboratory Procedures 5B:4.1–4.8.
12. Di Virgilio F, Meyer BC, Greenberg S, Silverstein SC (1988) Fc receptor-mediated phagocytosis occurs in macrophages at exceedingly low cytosolic Ca^{2+} levels. J Cell Biol 106:657–666.
13. Murgia M, Hanau S, Pizzo P, Rippa M, Di Virgilio F (1993) Oxidized ATP: An irreversible inhibitor of the macrophage purinergic P_{2Z} receptor. J Biol Chem 268:8199–8203.
14. Murgia M, Pizzo P, Zanovello P, Zambon A, Di Virgilio F (1992) In vitro cytotoxic effects of extracellular ATP. Altern Lab Anim 20: 66–70.
15. Di Virgilio F, Bronte V, Collavo D, Zanovello P (1989) Responses of mouse lymphocytes to extracellular ATP. J Immunol 143:1955–1960.
16. Filippini A, Taffs RE, Agui T, Sitkovsky MV (1990) Ecto-ATPase activity in cytolytic T-lymphocytes. Protection from the cytolytic effects of extracellular ATP. J Biol Chem 265: 334–340.
17. Zanovello P, Bronte V, Rosato A, Pizzo P, Di Virgilio F (1990) Responses of mouse lymphocytes to extracellular ATP. II. Extracellular ATP causes call type-dependent lysis and DNA

fragmentation. J Immunol 145:1545–1550.
18. Pizzo P, Murgia M, Zambon A, Zanovello P, Bronte V, Pietrobon D, Di Virgilio F (1992) Role of purinergic P_{2Z} receptors in ATP-mediated killing of tumor necrosis factor-sensitive and tumor necrosis factor-resistant L929 fibroblasts. J Immunol 149:3372–3378.
19. Zambon A, Bronte V, Di Virgilio F, Hanau S, Steinberg TH, Collavo D, Zanovello P (1994) Role of extracellular ATP in cell-mediated cytotoxicity: A study with ATP-sensitive and ATP-resistant macrophages. Cell Immunol 156: 458–467.
20. Beyer EC, Steinberg TH (1991) Evidence that the gap junction protein connexin43 is the ATP-induced pore of mouse macrophages. J Biol Chem 266:7971–7974.

38. EFFECTS OF PURINOCEPTOR AGONISTS ON INSULIN SECRETION

Marie-Madeleine Loubatières-Mariani
Pierre Petit
Jeannie Chapal
Dominique Hillaire-Buys
Gyslaine Bertrand
Gérard Ribes

It is well known that adenosine triphosphate plays an important intracellular role in the insulin-secreting pancreatic B cell. For many years ATP was only considered as being an intracellular fuel; but more recently its key role in the regulation of insulin secretion through ATP-dependent K^+ channels has been recognized and is now well documented. However, in addition to these intracellular effects, ATP is also able to act on the extracellular side of the B-cell membrane to increase insulin secretion. Furthermore, a nonphosphorylated adenine derivative, adenosine, can decrease this secretion. These two opposite effects are mediated through different types of purinoceptors for adenine nucleotides and nucleoside. In this paper we present a brief overview of the effects of ATP, adenosine, and structural analogues on insulin secretion. We discuss the pharmacologic characterization of B-cell purinoceptors, the proposed mechanisms involved, and the possible implication in physiologic and/or pathophysiologic states.

P_2 Purinoceptors Involved in Stimulation of Insulin Secretion

STIMULATING EFFECT OF ATP ON INSULIN SECRETION

As early as 1963, Candela and Garcia-Fernandez [1] reported that exogenous ATP stimulated insulin release in vivo in the rat. Later, most of the studies were performed on in vitro preparations. Loubatières et al. [2], using isolated perfused rat pancreas, reported that exogenously applied ATP induced a stimulation of insulin secretion higher than that elicited by cAMP. Then it was shown that glucose was required for the nucleotide action. Indeed, in the absence of glucose ATP was ineffective on insulin secretion; in the presence of a nonstimulating glucose concentration, it induced a weak and transient stimulation; in contrast, in the presence of a slightly stimulating glucose concentration, ATP elicited a marked and long-lasting stimulation of insulin secretion [3]. Therefore, ATP is able to potentiate insulin secretion in the presence of glucose but, unlike the sugar, cannot initiate release on its own. From the kinetic study of insulin release carried out on the isolated perfused rat pancreas, it can be seen that the insulin response to ATP displayed a biphasic pattern: A first phase in a peak form was followed by a long-lasting second phase (Fig. 38–1). This insulin-stimulating effect was concentration dependent between 10^{-6} and 10^{-4} M [4,5].

PHARMACOLOGIC CHARACTERIZATION OF A P_2 PURINOCEPTOR ON THE B CELL

The pharmacologic characterization of a P_2 purinoceptor on the B cell was performed in our laboratory using natural adenosine derivatives as well as structural analogues [5]. From the study of natural adenine derivatives it could be observed that ATP and ADP displayed similar insulin-stimulating effects, while AMP and adenosine were much less effective (Fig. 38–2). The study of triphosphate nucleotides other than ATP showed that only purine nucleotides were effective, with a rank order of relative potency ATP > GTP > ITP, while pyrimidine nucleotides, CTP and UTP, had little or no activity on insulin secretion. Thus, the purine base and the presence of diphosphate or triphosphate groups seemed to be essential for nucleotide-induced insulin secretion; this structure-activity relationship supported the view of a purinoceptor of the P_2 type on the B cell membrane.

In order to rule out a metabolic role for ATP and ADP, structural analogues known to be resistant to hydrolysis were studied. We first used analogues modified on the polyphosphate chain between $_{\beta}P$ and $_{\gamma}P$: adenyl-imido-diphosphate or AMP-PNP and β,γ-methylene ATP or AMP-PCP. Whereas the former was 10 times less potent than ATP, the latter was unable to stimulate insulin release. This dis-

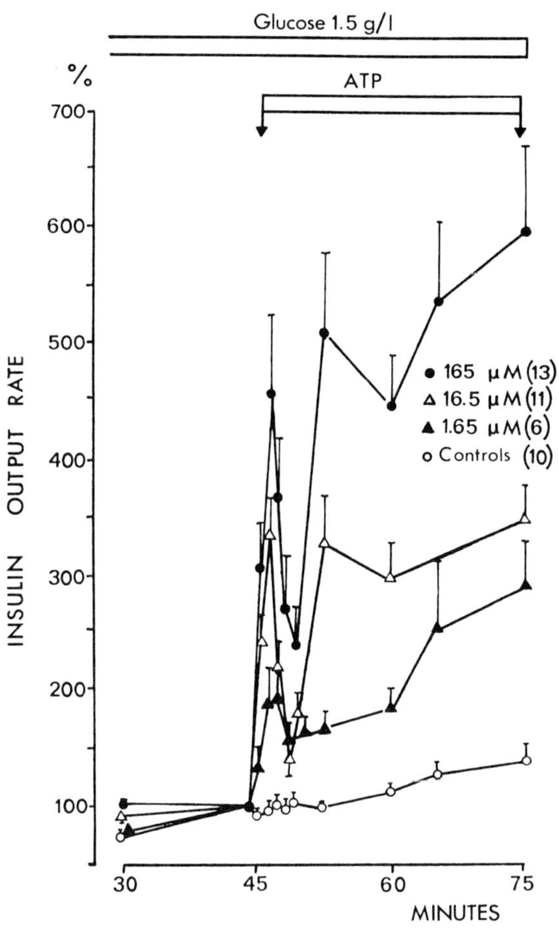

FIGURE 38-1. Kinetic studies of ATP on insulin secretion from isolated rat pancreas.

crepancy in the insulin-stimulating effects of these analogues could be related to the differences in electronic and steric characteristics of the polyphosphate chain, particularly between ATP and AMP-PCP [5,6]. However, when the polyphosphate chain was modified by a methylene residue between $_{\alpha}P$ and $_{\beta}P$ (α, β-methylene ATP and α, β-methylene ADP), the insulin-secreting activity was not altered and both substances were as potent as ATP (Fig. 38-2). All these results support the view of an extracellular nonmetabolic effect of ATP and emphasize the importance of steric and electronic characteristics of the polyphosphate

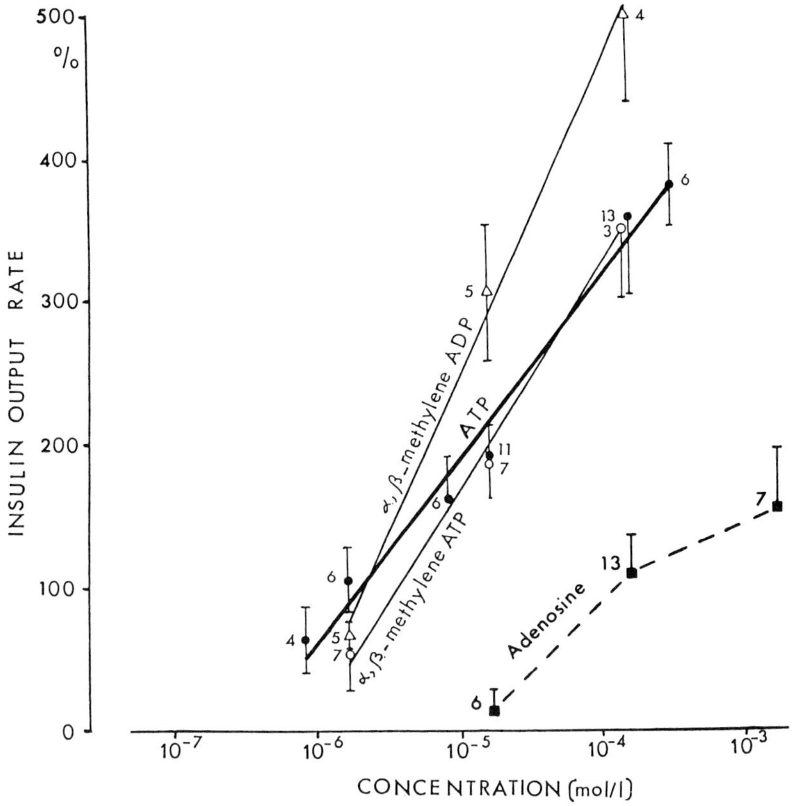

FIGURE 38-2. Concentration-response curves for ATP, structural analogues, and adenosine.

chain, particularly at the $_{\beta}P\ _{\gamma}O$ level, for binding to B-cell purinoceptors [6].

Studies performed with a purinoceptor antagonist, 2–2′ pyridylisatogene tosylate (PIT), provided another argument in favor of an extracellular effect of ATP. This substance, which was a specific ATP antagonist on some preparations, counteracted the insulin-stimulating effect of ATP. This effect was selective because , in the same conditions, PIT did not modify the stimulating effect of glucose or acetylcholine [7].

Later, when P_2 purinoceptors were subdivided into P_{2x} and P_{2y} types, experiments were performed with an ATP analogue more specific for P_{2y} receptors, 2-methylthio ATP. The relative potency of this compound compared with ATP made it possible to characterize the P_2 purinoceptor of the B cell as being of the P_{2y} subtype [8]. More recently, this pharmacologic characterization was confirmed by a study showing that a stable analogue of ADP modified on the $_{\beta}P$ and more specific for the P_{2y} purinoceptor, adenosine 5′-O-(2-thiodiphosphate) (ADPβS), was 100 times more potent than ATP or ADP in stimulating insulin secretion [9]. Thus, from all the results obtained in isolated perfused rat pancreas, it can be concluded that pancreatic B cell is provided with P_2 purinoceptors of the P_{2y} subtype, the activation of which stimulates insulin secretion.

MECHANISMS OF ACTION

The mechanisms of action that operate during B-cell P_2 purinoceptor activation remain to be fully elucidated (Fig. 38-3). Indeed, if an increase in cytosolic calcium is generally admitted, some authors focus on increased Ca^{2+} uptake, whereas others focus on Ca^{2+} mobilization involving the phosphatidyl inositol pathway. Thus, Petit et al. [10] observed that the stimulation of insulin secretion induced by α, β-methylene ADP was concomitant with an increased Ca^{2+} uptake by rat isolated islets. These results, as well as those reported by Geschwind et al. [11] on the clonal B cell line HIT, suggest the involvement of Ca^{2+} channel opening. For other authors, Ca^{2+} mobilization from intracellular stores may be involved in the insulin secretory effect of purinergic agonists; this is the case for Gylfe and Hellman [12] using pancreatic islets from ob/ob mice and for Blachier and Malaisse [13] using rat isolated islets. The latter authors also reported the involvement of polyphosphoinositide hydrolysis in the stimulus-secretion coupling of P_2 purinoceptors. However, in other experiments this pathway did not seem to account for P_2-purinoceptor–induced insulin release [14]. These discrepancies may result from differences in the structural analogues tested, their concentrations, and the experimental preparations used. Furthermore, species variations regarding the adenine nucleotide effects on B cells have also been described. Thus, in mice, α,β-methylene ADP inhibited insulin release from perifused islets and transiently reduced electric activity of the B-cell membrane [15]. Recently Li et al. [16] showed that extracellular ATP depolarized the membrane of rat insulinoma cells (RINm5F) by closing ATP-dependent K^+ channels and that insulin secretion induced by ATP involved phospholipase C-dependent and -independent mechanisms.

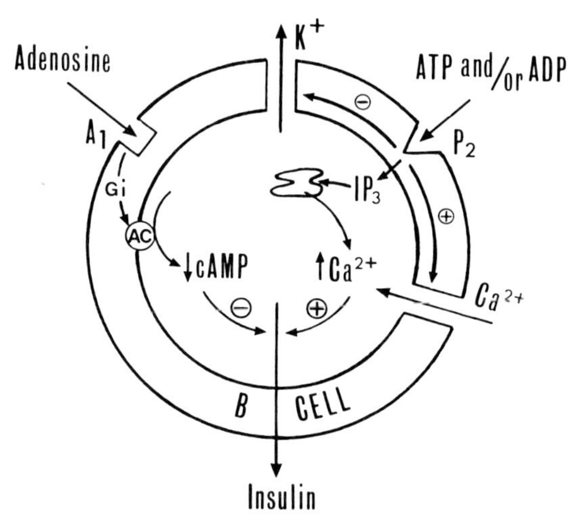

FIGURE 38-3. Transduction mechanisms proposed to be involved in the purinergic control of insulin secretion.

POSSIBLE PHYSIOLOGIC IMPLICATIONS

With regard to their physiologic relevance, P_2 purinoceptors could modulate insulin secretion. Extracellular ATP may interact with the cholinergic stimulation of insulin secretion. Indeed, ATP and its derivatives are considered as neurotransmitters, cotransmitters, or neuromodulators in the central and peripheral nervous systems; there is evidence that ATP is stored and coreleased with acetylcholine. Accordingly, ATP released with acetylcholine could potentiate the effect of this neurotransmitter on insulin secretion. In fact, ATP or ADP and acetylcholine act in potentiating synergism on B cells [17]. This potentiation may play a part in the parasympathetic control of insulin secretion, particularly during the early prandial period (cephalic phase).

Another and perhaps more important source of ATP and ADP are the B cells themselves. Indeed, these nucleotides are present in high concentrations in B-cell secretory granules and are released during glucose-induced insulin secretion. Thus, ATP and ADP may exert a positive feedback and amplify glucose-induced insulin secretion. In fact, in isolated perfused rat pancreas the activation of P_2 purinoceptors by ATP or ADP potentiates the biphasic response to a physiologic glucose increment [18]. Thus, the activation of P_2 purinoceptors may be important for amplification of the insulin secretory response; consequently, it can be suggested that activation of these receptors may participate in the control of glucose homeostasis.

P_1 Purinoceptors Involved in Inhibition of Insulin Secretion

PHARMACOLOGIC CHARACTERIZATION OF A P_1 PURINOCEPTOR ON THE B CELL

Depending on the concentration used, adenosine has been successively reported to enhance (millimolar concentrations) or to decrease (micromolar concentrations) insulin secretion. However, on the isolated perfused rat pancreas the stimulation of insulin secretion induced by adenosine displayed a time course different from that elicited by ATP; it was delayed, occurring after a first phase of transient inhibition [5]. It is known that adenosine and its deaminated derivative, inosine, have a similar stimulating effect on insulin secretion; adenosine and inosine effects are both abolished by an inhibitor of nucleoside transport, p-nitrobenzylthioguanosine, whereas only the adenosine effect is suppressed by an inhibitor of adenosine deaminase. Thus adenosine, after its intracellular breakdown into inosine, is metabolized as this latter in the B cell and, when used at high concentration, stimulates insulin secretion by its metabolic degradation.

Concerning the inhibitory effect of adenosine, Campbell and Taylor [19], using a structural analogue very pooly taken up by cells, N^6-phenylisopropyladenosine (PIA), showed that this compound decreased insulin secretion. These data supported the view of an inhibitory P_1 purinoceptor on the B cell. Later, when P_1 purinoceptors were subdivided into the A_1 and A_2 subtypes, the receptor was characterized on the isolated perfused rat pancreas as being of the A_1 subtype [20]. This characterization was supported by the great stereoselectivity for R-PIA, which was effective in the nanomolar range and was about 100 times more potent than S-PIA. In conclusion, these data show that the pancreatic B cell is provided with A_1-subtype adenosine receptors, the activation of which inhibits insulin secretion.

MECHANISMS OF ACTION

It was shown that adenosine and its analogues inhibited adenylyl cyclase activity [19]. In isolated perfused rat pancreas, the inhibiting effect of R-PIA on insulin secretion involves a pertussis toxin-sensitive G protein [21], presumably of the G_i type, coupled to adenylyl cyclase. It is known that modification of intracellular cAMP levels is not the unique transduction mechanism involved in A_1 purinoceptor activation: Adenosine can also directly affect Ca^{2+} and K^+ channels via G proteins in some tissues. However, it has been shown that adenosine and R-PIA only slightly inhibited $^{45}Ca^{2+}$ and $^{86}Rb^+$ efflux in rat islet cells [22] and electric activity in mouse islet cells [23]. Furthermore, the inhibition of insulin release by adenosine or R-PIA was observed in the presence of forskolin (a direct activator of adenylyl cyclase) but not in the presence of exogenous cAMP [22,23]. Thus, it appears that inhibition of adenylyl cyclase, rather than direct changes in membrane Ca^{2+} and K^+ conductances, is involved in the inhibition of insulin secretion by adenosine.

POSSIBLE PHYSIOLOGIC IMPLICATIONS

Extracellular adenosine could play a role in the regulation of insulin secretion. It is well known that ATP, released in various physiologic conditions, is rapidly broken down into adenosine. It can be speculated that insulin secretion is modulated via the activation of the two types of purinoceptors, P_1 and P_2, with adenosine and ATP having opposite effects. Thus, ATP and/or ADP released from B cells during insulin exocytosis may amplify insulin response (positive feedback), and this effect may be limited by adenosine which, due to its ability to reduce insulin secretion, may constitute a negative regulatory feedback. Similarly, adenosine may negatively regulate insulin secretion induced by acetylcholine and ATP coreleased from the vagal nerve. Thus, it can be suggested that ATP and adenosine, via their opposite effects on insulin secretion, may constitute a fine and local purinergic regulatory system.

Search for a New Class of Antidiabetic Drugs

B-cell dysfunction, particularly an impairment of the insulin response to glucose, is involved in the pathophysiology of non-insulin-dependent diabetes mellitus. Therefore, P_{2y} agonists may be considered as potential new substances in the treatment of this form of diabetes due to their ability to stimulate insulin release. Consequently, it was necessary to establish that in vivo–administered P_{2y} agonists were effective on insulin secretion and glycemia. This in vivo study raised the problem of the metabolic stability of P_{2y} purinoceptor agonists. Indeed, the first structural analogue used in vivo in dogs, 2-methylthio ATP, had to be injected directly in the pancreatico-duodenal artery to

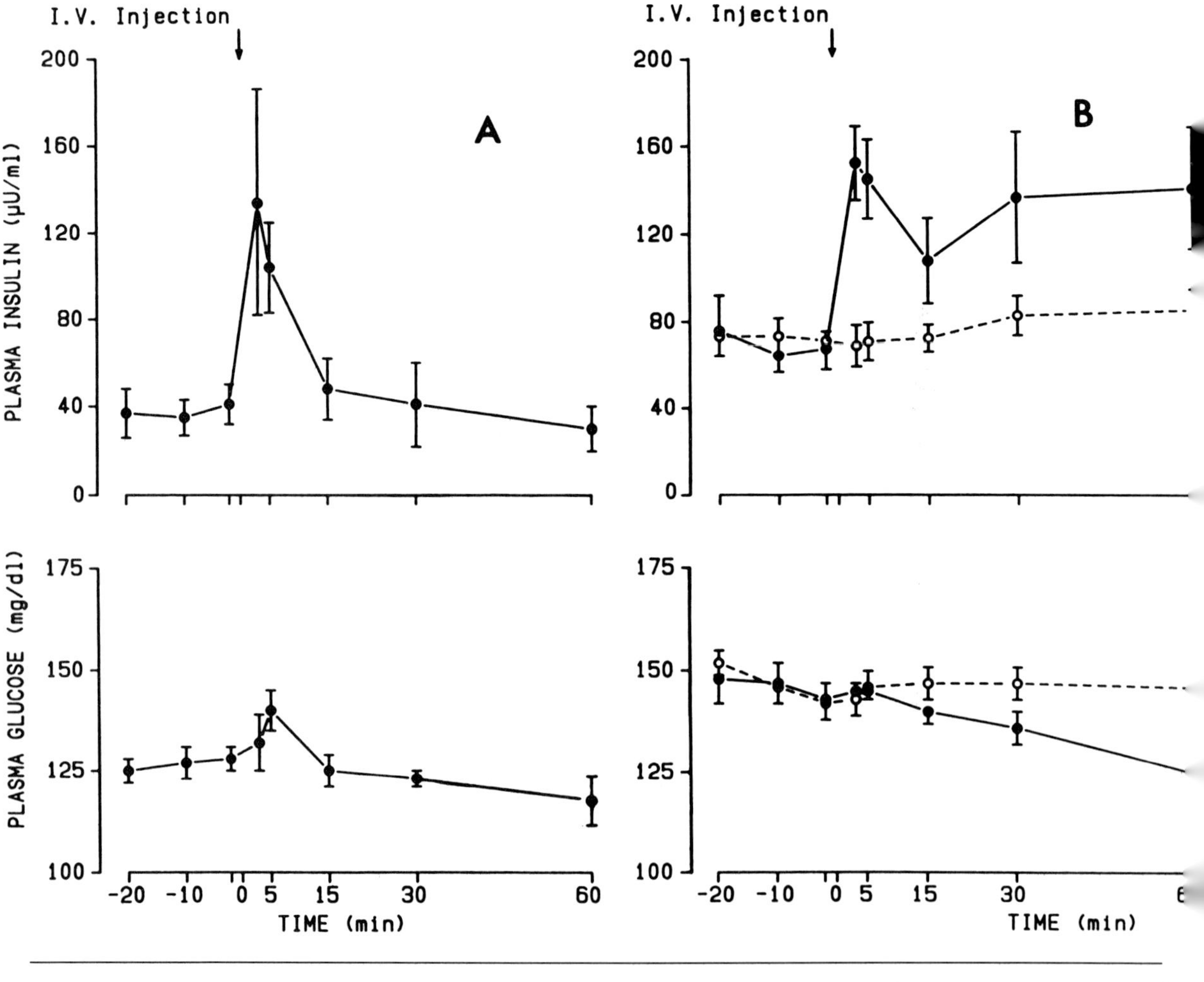

FIGURE 38-4. Effects of intravenous injection of ADPβS (0.2 mg/kg) on plasma insulin and blood glucose levels in anesthetized rats. (**A**) Overnight-fasted rats; (**B**) fed rats.

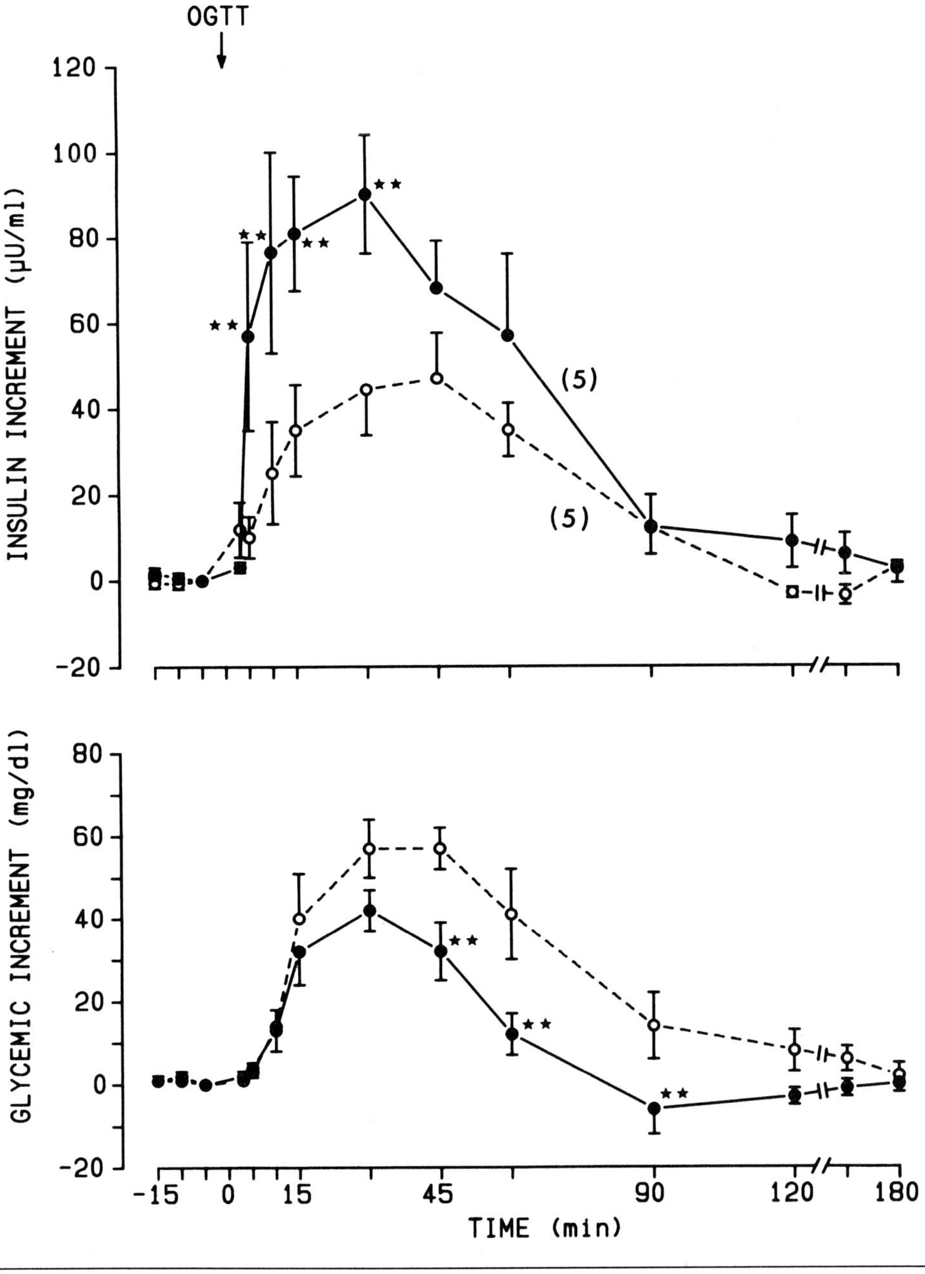

FIGURE 38-5. Increments in plasma insulin and blood glucose levels after oral glucose tolerance test (OGTT) in fasted conscious dogs in response to glucose alone (1 g/kg) (○) or glucose (1 g/kg) and ADPβS (0.1 mg/kg) (●).

avoid its rapid breakdown into adenosine. Under these conditions it was shown to stimulate insulin release and to decrease glycemia slightly [24].

To continue our investigations it was necessary to have a stable structural analogue that could be administered peripherally. The P_{2y}-selective agonist ADPβS, which was found to be 100 times more potent than ATP in vitro, was used for in vivo investigations in rats and dogs. Experiments were performed using ADPβS in basal conditions as well as during glucose tolerance tests [25]. In anesthetized rats, ADPβS injected into a peripheral vein increased plasma insulin levels. However, this insulin-stimulating effect appeared to be dependent on the nutritional state of the animals (Fig. 38-4). Indeed, in overnight-fasted rats, ADPβS elicited only a transient insulin response without significant modification in glycemia; in fed animals, the same dose of ADPβS induced a sustained insulin response with a significant reduction in glycemia. However, the basal glycemia was slightly higher in fed rats than in fasted animals and ADPβS decreased glycemia to a similar level as the baseline in fasted animals. Thus, the lack of an ADPβS effect under fasted conditions may limit the risk of hypoglycemia. Another advantage of ADPβS could be its ability to improve glucose tolerance. When glucose was administered intravenously in anesthetized rats, the simultaneous administration of ADPβS strongly increased insulin secretion and accelerated the return of glycemia to baseline values.

However, because it is known that purine nucleotides are metabolized in the gastrointestinal tract, it was essential to investigate the effects of ADPβS after oral administration. In vivo experiments in conscious dogs have shown that this substance was effective after oral administration, transiently increasing plasma insulin and punctually reducing glycemia. Moreover, during an oral glucose tolerance test, ADPβS also markedly enhanced insulin secretion and reduced hyperglycemia (Fig. 38-5). From these results, it can be summarized that the P_{2y} purinoceptor agonist ADPβS, which is particularly potent in vitro, is also a potent insulin secretagogue in vivo, improves glucose tolerance, and is effective after oral administration.

Conclusions

From these pharmacologic investigations it can be suggested that 1) the pancreatic B cell is submitted to fine and local purinergic control (through activation of P_{2y} and A_1 purinoceptors, respectively), which could participate in the regulation of insulin release; 2) the P_{2y} purinoceptors of this cell may be potential targets for a new class of oral antidiabetic drugs.

Acknowledgments

The authors thank M. Tournier and V. Montesinos for technical assistance in the preparation of this manuscript.

References

1. Candela JLR, Garcia-Fernandez MC (1963) Stimulation of secretion of insulin by adenosine triphosphate. Nature 197:A1210.
2. Loubatières A, Loubatières-Mariani MM, Chapal J (1972) Adénosine triphosphate (ATP), adénosine 3′5′ monophosphate cyclique (3′5′ AMPc) et sécrétion d'insuline. C R Soc Biol 166:1742–1746.
3. Loubatières-Mariani MM, Loubatières A, Chapal J, Valette G (1976) Adénosine triphosphate (ATP) et glucose. Action sur les sécrétions d'insuline et de glucagon. C R Soc Biol 170:833–836.
4. Loubatières-Mariani MM, Chapal J, Valette G (1977) Adénosine triphophate (ATP) et insulino-sécrétion: Effet de différentes concentrations et influence de la température. C R Soc Biol 171:864–869.
5. Loubatières-Mariani MM, Chapal J, Lignon F, Valette G (1979) Structural specificity of nucleotides for insulin secretory action from the isolated perfused rat pancreas. Eur J Pharmacol 59:277–286.
6. Chapal J, Loubatières-Mariani MM (1981) Effects of phosphate-modified adenine nucleotide analogues on insulin secretion from perfused rat pancreas. Br J Pharmacol 73:105–110.
7. Chapal J, Loubatières-Mariani MM (1981) Attempt to antagonize the stimulatory effect of ATP on insulin secretion. Eur J Pharmacol 74:127–134.
8. Bertrand G, Chapal J, Loubatières-Mariani MM, Roye M (1987) Evidence for two different P_2-purinoceptors on β cell and pancreatic

vascular bed. Br J Pharmacol 102:783–787.
9. Bertrand G, Chapal J, Puech R, Loubatières-Mariani MM (1991) Adenosine-5′-O-(2-thiodiphosphate) is a potent agonist at P_2 purinoceptors mediating insulin secretion from perfused rat pancreas. Br J Pharmacol 102: 627–630.
10. Petit P, Manteghetti M, Puech R, Loubatières-Mariani MM (1987) ATP and phosphate-modified adenine nucleotide analogues: Effects on insulin secretion and calcium uptake. Biochem Pharmacol 36:377–380.
11. Geschwind JF, Hiriart M, Glennon MC, Najafi H, Corkey BE, Matschinsky FM, Prentki M (1989) Selective activation of Ca^{2+} influx by extracellular ATP in a pancreatic β-cell line (HIT). Biochim Biophys Acta 1012: 107–115.
12. Gylfe E, Hellman B (1987) External ATP mimics carbachol in initiating calcium mobilization from pancreatic β-cells conditioned by previous exposure to glucose. Br J Pharmacol 92:281–289.
13. Blachier F, Malaisse WJ (1988) Effect of exogenous ATP upon inositol phosphate production, cationic fluxes and insulin release in pancreatic islet cells. Biochim Biophys Acta 970:222–229.
14. Petit P, Manteghetti M, Loubatières-Mariani MM (1988) Differential effects of purinergic and cholinergic activation on the hydrolysis of membrane polyphosphoinositides in rat pancreatic islets. Biochem Pharmacol 37: 1213–1217.
15. Petit P, Bertrand G, Schmeer W, Henquin JC (1989) Effects of extracellular nuleotides on the electrical, ionic and secretory events in mouse pancreatic β-cells. Br J Pharmacol 98:875–882.
16. Li G, Milani D, Dunne MJ, Pralong WF, Theler JM, Petersen OH, Wollheim CB (1991) Extracellular ATP causes Ca^{2+}-dependent and -independent insulin secretion in RINm5F cells. J Biol Chem 266:3449–3457.
17. Bertrand G, Chapal J, Loubatières-Mariani MM (1986) Potentiating synergism between adenosine diphosphate or triphosphate and acetylcholine on insulin secretion. Am J Physiol 251:E416–E421.
18. Bertrand G, Gross R, Chapal J, Loubatières-Mariani MM (1989) Difference in the potentiating effect of adenosine triphosphate and α,β-methylene ATP on the biphasic insulin response to glucose. Br J Pharmacol 98:998–1004.
19. Campbell IL, Taylor KW (1982) Effects of adenosine, 2-deoxyadenosine and N^6-phenylisopropyladenosine on rat islet function and metabolism. Biochem J 204:689–696.
20. Hillaire-Buys D, Bertrand G, Gross R, Loubatières-Mariani MM (1987) Evidence for an inhibitory A_1 subtype adenosine receptor on pancreatic insulin-secreting cells. Eur J Pharmacol 136:109–112.
21. Hillaire-Buys D, Gross R, Loubatières-Mariani MM, Ribes G (1989) Effect of pertussis toxin on A_1-receptor mediated inhibition of insulin secretion. Br J Pharmacol 96:3–4.
22. Bertrand G, Nenquin M, Heanquin JC (1989) Comparison of the inhibition of insulin release by activation of adenosine and α_2-adrenergic receptors in rat β-cells. Biochem J 259:223–228.
23. Bertrand G, Petit P, Bozem M, Henquin JC (1989) Membrane and intracellular effects of adenosine in mouse pancreatic β-cells. Am J Physiol 257:E473–E478.
24. Ribes G, Bertrand G, Petit P, Loubatières-Mariani MM (1988) Effects of 2-methylthio ATP on insulin secretion in the dog in vivo. Eur J Pharmacol 155:171–174.
25. Hillaire-Buys D, Bertrand G, Chapal J, Puech R, Ribes G, Loubatières-Mariani MM (1993) Stimulation of insulin secretion and improvementof glucose tolerance in rat and dog by the P_{2y} purinoceptor agonist, adenosine-5′-O-(2-thiodiphosphate). Br J Pharmacol 109:183–187.

V. CYTOPROTECTIVE ACTIONS OF ADENOSINE

A. CARDIOPROTECTION

39. RECEPTOR-MEDIATED AND METABOLIC EFFECTS OF ADENOSINE IN ISCHEMIC AND POSTISCHEMIC MYOCARDIUM

Robert D. Lasley
Rolf Bünger
Robert M. Mentzer, Jr.

Myocardial ischemia is associated with rapid decreases in cardiac contractility and the high-energy phosphates creatine phosphate (CrP) and adenosine triphosphate (ATP). Ischemic durations of 20 minutes or less induce no cell necrosis, but upon reperfusion the myocardium at risk exhibits prolonged contractile dysfunction, which may persist for hours to days. This postischemic depressed contractility, despite restoration of normal coronary blood flow, is defined as stunned myocardium. Longer durations of ischemia are associated with cell death, that is, myocardial infarction. Although the exact mechanisms underlying ischemic/reperfusion injury are not known, there is substantial evidence that the purine nucleoside adenosine protects the heart against both myocardial stunning and infarction, in part via activation of sarcolemmal adenosine A_1 receptors located on the cardiac myocytes [1,2]. This review will focus on the receptor-mediated and metabolic effects of adenosine in the ischemic heart and the attenuation of postischemic ventricular dysfunction by adenosine.

Receptor-Mediated Effects of Adenosine in Ischemic and Postischemic Myocardium

With the recognition that adenosine exerts its effects in nonischemic myocardium by acting on specific adenosine receptor subtypes [3], studies of the mechanism(s) underlying the cardioprotective effect of adenosine have also focused on the role of adenosine receptors. This has been accomplished with the use of adenosine receptor agonists and antagonists relatively selective for the A_1 or A_2 receptor, and the use of techniques to estimate interstitial fluid (ISF) adenosine levels. Evidence obtained with adenosine analogues suggests that the beneficial effect of adenosine on postischemic function is mediated via adenosine A_1-receptor activation [4,5]. Since the A_1 receptor is located primarily on cardiac myocytes that are bathed by the ISF, this hypothesis requires that ISF adenosine accumulate in a sufficient concentration to activate this receptor. Using the cardiac microdialysis technique we have shown that augmentation of ISF adenosine levels prior to ischemia is associated with attenuation of myocardial stunning [6,7].

ADENOSINE RECEPTOR STUDIES IN ISOLATED RAT HEARTS

The results of isolated rat heart studies in our laboratory with adenosine agonists and antagonists suggest that the beneficial effect of adenosine receptor activation in the ischemic/reperfused myocardium is mediated via the adenosine A_1 receptor. Initial studies showed that the adenosine-induced delay in the onset of ischemic contracture could be mimicked by the adenosine A_1-receptor agonist R-phenylisopropyladenosine (PIA), but the A_2-receptor agonist phenylaminoadenosine (PAA) exerted no beneficial effect [8]. The results of two subsequent studies are summarized in Figure 39-1. Brief treatment of hearts immediately prior to 30 minutes of global no flow ischemia (37°C) with adenosine (100 μM) and the A_1-receptor agonist cyclohexyladenosine (CHA, 0.25 μM), but not PAA (0.25 μM), attenuated postischemic contractile dysfunction [4]. The cardioprotective effect of adenosine itself was blocked by the selective adenosine A_1-receptor antagonist 8-cyclopentyl-1,3-dipropylxanthine (DPCPX, 5 μM).

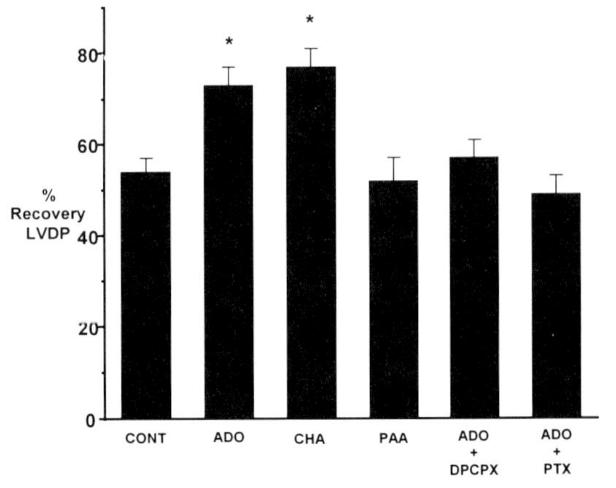

FIGURE 39-1. The effects of adenosine (ADO), adenosine receptor agonists cyclohexyladenosine (CHA), and phenylaminoadenosine (PAA), the adenosine A_1 receptor blocker 8-cyclopentyl-1,3-dipropylxanthine (DPCPX), and pertussis toxin (PTX) on recovery of left ventricular developed pressure (LVDP) after global ischemia in isolated perfused rat hearts. $^*p < 0.05$ versus control (CONT).

Further evidence for an A_1-receptor–mediated cardioprotective effect of adenosine was obtained in rats pretreated with pertussis toxin [9]. Ventricular myocyte adenosine A_1 receptors are coupled to inhibitory guanine nucleotide (G_i) binding proteins, which are disabled by pertussis toxin, thus blocking the effects of A_1 receptor stimulation [3]. Pertussis toxin pretreatment (25 μg/kg, i.p., 48 hours prior to isolation) alone had no effect on recovery of function but did block the beneficial effects of adenosine and CHA. These results indicate that adenosine-induced cardioprotection is mediated in part via an A_1 receptor, pertussis-toxin–sensitive G_i protein mechanism.

Although the antiadrenergic actions of adenosine remain the only well-characterized effect of adenosine A_1-receptor activation in ventricular myocardium [3], recent work from our laboratory suggests that adenosine may modulate other signal transduction pathways, specifically protein kinase C (PKC) [10]. As shown in Figure 39-2, treatment of constant flow-perfused rat hearts with the phorbol ester phorbol 12-myristate 13-acetate (PMA), 10 nM for 30 minutes), which activates PKC, resulted in a prolonged negative inotropic effect. The negative inotropic effect of PMA was blocked by the PKC inhibitor bisindolylmaleimide (0.5 μM), and the inactive phorbol ester 4α phorbol 12,13-didecanoate had no effect on ventricular function, suggesting that the PMA effect was indeed due to activation of PKC.

Treatment with adenosine (100 μM) and the adenosine A_1-receptor agonist 2-chloro-N^6-cyclopentyladenosine (CCPA, 0.1 μM) blocked the negative inotropic effect of PMA, whereas the adenosine A_2-receptor

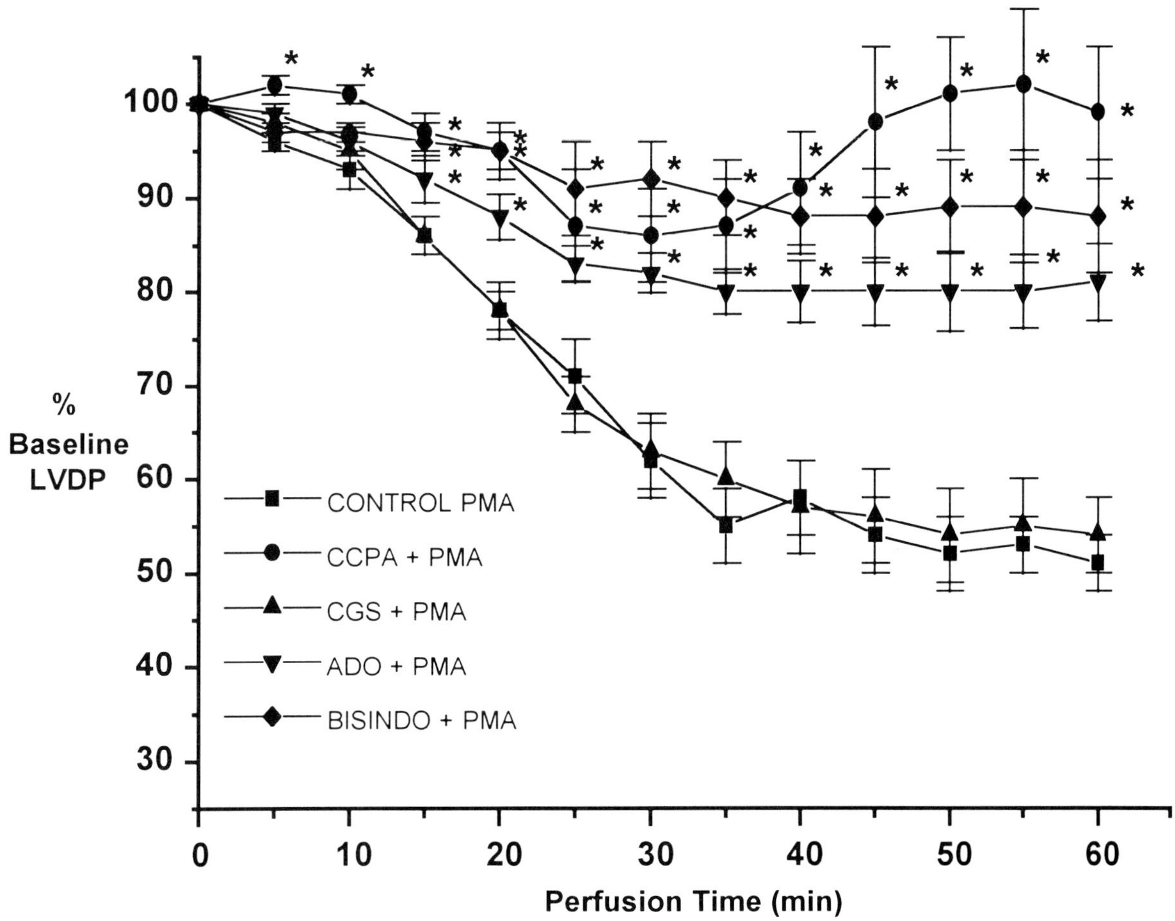

FIGURE 39-2. Effects of adenosine (ADO), adenosine receptor agonists 2-chloro-N^6-cyclopentyladenosine (CCPA) and 2-p-(2-carboxy-ethyl) phenethylamino-5′-N-ethylcarboxamido adenosine hydrochloride (CGS 21680), and the PKC inhibitor bisindolylmaleimide (BISINDO) on the negative inotropic effect of the phorbol ester phorbol 12-myristate 13-acetate (PMA). PMA was infused for 30 minutes and then allowed 30 minutes of recovery. *$p < 0.05$ versus control PMA hearts.

agonist 2-p-(2-carboxy-ethyl) phenethylamino-5′-N-ethylcarboxamido adenosine hydrochloride (CGS 21680, 0.1 μM) had no effect. Treatment with 1,2-dioleoyl-sn-glycerol (DIOG, 0.4 μM), an analog of the endogenous PKC activator diacylglycerol, produced a negative inotropic effect similar to PMA, an effect that was also blocked by CCPA. These results suggest that adenosine A_1 receptor activation may modulate PKC activity itself or attenuate the effects of increased PKC activity. It remains to be determined whether this effect of adenosine plays any role in adenosine-mediated protection of the ischemic heart.

INTERSTITIAL FLUID ADENOSINE AND POSTISCHEMIC VENTRICULAR DYSFUNCTION

Our laboratory has utilized the cardiac microdialysis technique to assess changes in in vivo interstitial fluid (ISF) adenosine levels during interventions known to increase adenosine, such as myocardial ischemia. The microdialysis

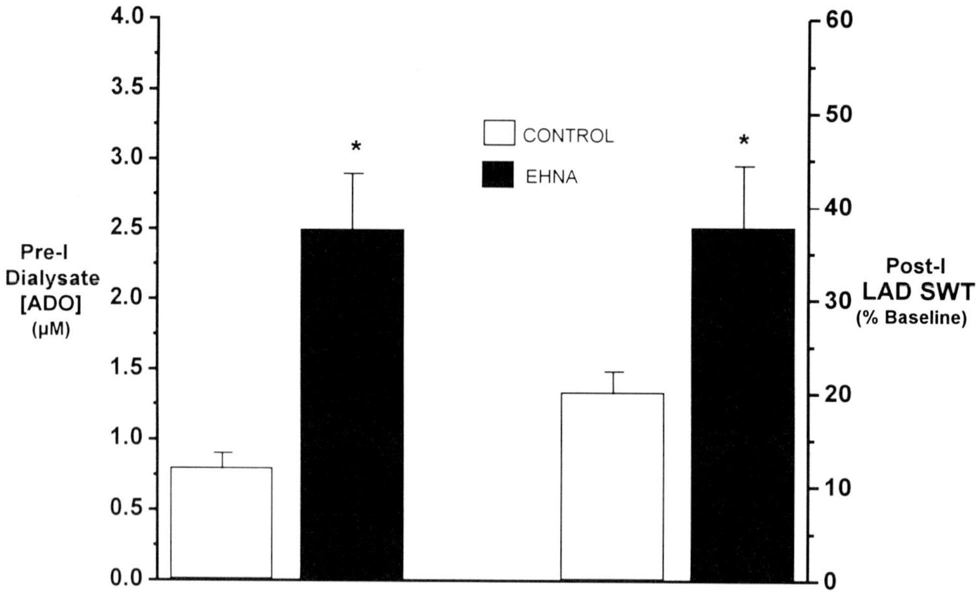

FIGURE 39-3. Effects of the adenosine (ADO) deaminase inhibitor EHNA on preischemic dialysate adenosine concentration and recovery of postischemic systolic wall thickening (SWT) in in vivo canine myocardium subjected to 15 minutes of LAD occlusion and 60 minutes of reperfusion. $^*p < 0.05$ versus control.

technique, which has been described in detail previously [11,12], involves the implantation of a hollow dialysis fiber (200 μm i.d.) into the wall of the left ventricular myocardium. The dialysis fiber is perfused with Krebs–Henseleit buffer, and as the fluid passes through the fiber diffusion occurs between the fluid within the fiber and the ISF surrounding it. The concentration of a compound in the effluent (termed the *dialysate concentration*) is thus representative of the ISF concentration of that substance.

Initial studies in an in vivo canine preparation revealed that although adenosine concentration in the ISF rapidly increased during regional myocardial ischemia, the adenosine metabolites inosine and hypoxanthine increased to a much greater extent. We thus tested the effects of the adenosine deaminase inhibitor erythro-9-(2-hydroxy-3-nonyl) adenine HCl (EHNA) on ISF adenosine levels in the ischemic heart to determine whether augmentation of ISF adenosine would enhance postischemic function [6]. Regional myocardial stunning was induced in open-chest dogs by 15-minutes LAD occlusion, and regional ventricular function was assessed by measuring systolic wall thickening via sonomicrometry. Control animals were compared to those treated with an intraatrial infusion of EHNA (5 mg/kg/hr) initiated 20 minutes prior to occlusion and maintained throughout the 60-minute reperfusion period. Although EHNA increased coronary blood flow prior to ischemia, it had no effect on mean arterial blood pressure, heart rate, or ventricular function.

Figure 39-3 shows that EHNA treatment was associated with an increase in preischemic dialysate adenosine concentration (from 0.8 ± 0.1 μM to 2.5 ± 0.4 μM). After 15 minutes ischemia EHNA-treated hearts exhibited 25-fold greater dialysate adenosine levels than control hearts, and three- and two-fold decreases in dialysate inosine and hypoxanthine levels, respectively. This increase in ISF adenosine levels was associated with an almost twofold increase in recovery of LAD systolic wall thickening (Fig. 39-3; 38 ± 7% of preischemic values) compared to untreated control hearts (20 ± 2%). These results suggest that preischemic augmentation of endogenous ISF adenosine attenuates myocardial stunning.

The next series of experiments was designed to determine whether exogenous adenosine administered prior to ischemia attenuated postischemic dysfunction [7]. In a similar in vivo canine preparation of regional myocardial stunning, adenosine (50 μg/kg/min) was infused into an LAD catheter immediately prior to 15-minutes LAD occlusion. Intracoronary adenosine infusion increased LAD blood flow sevenfold but had no effect on LAD systolic wall thickening or mean arterial blood pressure. Dialysate adenosine levels increased from 0.6 ± 0.1 μM to 1.2 ± 0.3 μM, but dialysate inosine and hypoxanthine levels increased eight- and fourfold, respectively, providing in vivo evidence of the coronary endothelial metabolic barrier to adenosine [13].

Adenosine-pretreated hearts showed significantly greater systolic wall thickening throughout reperfusion. After 90-minute reperfusion recovery of LAD systolic wall thickening in adenosine-pretreated animals was 26 ± 10% of baseline ($p < 0.05$ versus control), whereas untreated hearts continued to exhibit negative wall thickening (−14 ± 10%). When the same dose of adenosine was infused for the first 30 minutes of reperfusion, regional function increased from −11 ± 13% of baseline to 3.6 ± 17% ($p < 0.05$). However, function rapidly deteriorated to pre-adenosine infusion levels upon termination of the adenosine infusion in parallel with the decrease in adenosine-induced hyperemia. These results suggest that 1) pretreatment with adenosine, in a concentration sufficient to accumulate in the ISF but with little or no systemic effects can reduce postischemic dysfunction; and 2) adenosine must be administered prior to ischemia to provide a sustained attenuation of myocardial stunning.

The majority of adenosine studies have been performed in the normothermic ischemic heart, but hypothermia is used during clinical cardioplegia and during heart preservation for transplantation. The University of Wisconsin (UW) preservation solution, which contains 5 mM adenosine in addition to numerous other constituents, has been shown to provide excellent prolonged hypothermic (4°C) myocardial preservation. We thus tested whether adenosine played a beneficial role in UW solution during these profound hypothermic conditions [14]. Isolated rabbit hearts were flushed with UW solution ± adenosine and stored for 18 hours at 4°C. In one series of hearts dialysis fibers were placed for continuous monitoring of ISF purines during the period of cold storage. The second series of hearts was reperfused with Krebs–Henseleit buffer to assess postpreservation function. In hearts flushed with normal UW solution (i.e., containing 5 mM adenosine) dialysate adenosine levels during cold storage were 20- to 40-fold higher than in hearts flushed with UW solution without adenosine. After 45-minute reperfusion hearts flushed with UW solution exhibited a ≈40% greater rate-pressure product than hearts preserved with UW solution without adenosine. These results suggest that adenosine exerts a beneficial effect even during prolonged hypothermic ischemia in a clinically relevant setting.

Metabolic Effects of Adenosine in Ischemic and Postischemic Myocardium

As recently as 5 years ago it was thought that adenosine improved postischemic ventricular function by stimulating the purine salvage pathway for ATP resynthesis [15]. However, subsequent studies dissociated adenosine-enhanced postischemic function from increased myocardial ATP levels [4,16], and other studies reported little correlation between the degree of postischemic dysfunction and ATP content [17,18]. Nevertheless, treatment of the ischemic heart with adenosine has been shown to decrease the rate of ATP breakdown during the initial minutes of myocardial ischemia both in vitro and in vivo. This effect occurs incependent of adenosine's coronary vasodilatory and negative chronotropic effects, and can be mimicked with adenosine A_1-receptor agonists PIA and CHA [4,8,15]. In addition, adenosine receptor blockade during both zero flow and low-flow ischemia is associated with a more rapid rate of ATP catabolism [8,19]. It remains to be determined what role, if any, this ATP-sparing effect of adenosine in the initial period of myocardial ischemia plays in the cardioprotective effect of adenosine.

ADENOSINE AND MYOCARDIAL PHOSPHORYLATION POTENTIAL

Although there is little correlation between myocardial ATP levels and postischemic ventricular function, the energy available for energy-dependent cellular processes such as the Na/K-ATPase, the SR Ca/Mg-ATPase, the sarcolemmal Ca-ATPase, and the myofibrillar contractile apparatus (actomyosin ATPase) is not determined by tissue ATP content but instead by the energy derived from ATP hydrolysis. This energy of hydrolysis is determined by the following equation:

$$\Delta G_p = \frac{\Delta G^0 + RT \ln[ADP] \times [P_i]}{[ATP]},$$

where [ ] refers to the free cytosolic concentrations of ATP, ADP, and P_i, and R and T are the usual constants. ΔG_p is called the cytosolic phosphorylation potential (expressed as kJ/mole), and since ΔG^0 (the standard free energy) varies relatively little, changes in ATP hydrolysis energy can be monitored by measuring the concentration ratio $[ATP]/([ADP] \times [P_i])$. This phosphorylation ratio is in equilibrium with the creatine kinase reaction, and the following relationship holds for steady-state conditions:

$$\frac{[ATP]}{([ADP] \times [P_i])} = \frac{[CrP]}{[Cr] \times [P_i]} \times \frac{H^+}{K_{ck}},$$

where H^+ is the cytosolic hydrogen ion concentration, and K_{ck} is the pH- and Mg^{2+}-dependent creatine kinase equilibrium constant. The rapid decrease in contractility with the onset of ischemia correlates with the decrease in CrP and CrP/P_i ratio and the increase in P_i [20]. Decreased contractility during reperfusion has also been reported to correlate well with low CrP/P_i ratios and reduced cytosolic phosphorylation potential values [17,18,20].

We have obtained evidence that adenosine treatment increases postischemic myocardial phosphorylation potential both in vitro and in vivo [1,21]. Adenosine treatment (100 μM) of isolated Krebs-perfused guinea-pig hearts had no effect on cytosolic phosphorylation potential during normoxia, but during reperfusion after 15-minute low-flow ischemia (1 ml/min/g wet mass) adenosine-treated hearts exhibited a greater $CrP/Cr{*}P_i$ ratio, indicative of a greater phosphorylation potential [1]. Attenuation of in vivo myocardial stunning with adenosine pretreatment is also associated with increased myocardial phosphorylation potential [21]. Regional myocardial stunning was induced in open-chest pigs by 10 minutes of left anterior descending coronary artery (LAD) occlusion. In one group of animals adenosine (50μg/kg/min) was infused into the LAD for 10 minutes immediately prior to occlusion.

Figure 39-4 illustrates that although adenosine infusion had little effect on regional function prior to ischemia, this adenosine pretreatment resulted in significantly improved recovery of postischemic function in the stunned LAD-perfused myocardium. After 90 minutes of reperfusion rapid-frozen transmural biopsy samples were obtained with an air-driven biopsy drill from the stunned LAD bed and the normal left circumflex perfused bed. As shown in Figure 39-5 preischemic adenosine treatment had no effect on reperfusion ATP levels but significantly increased the postischemic $CrP/Cr{*}P_i$ ratio. In a separate series of animals, it was observed that intracoronary adenosine infusion in nonischemic myocardium

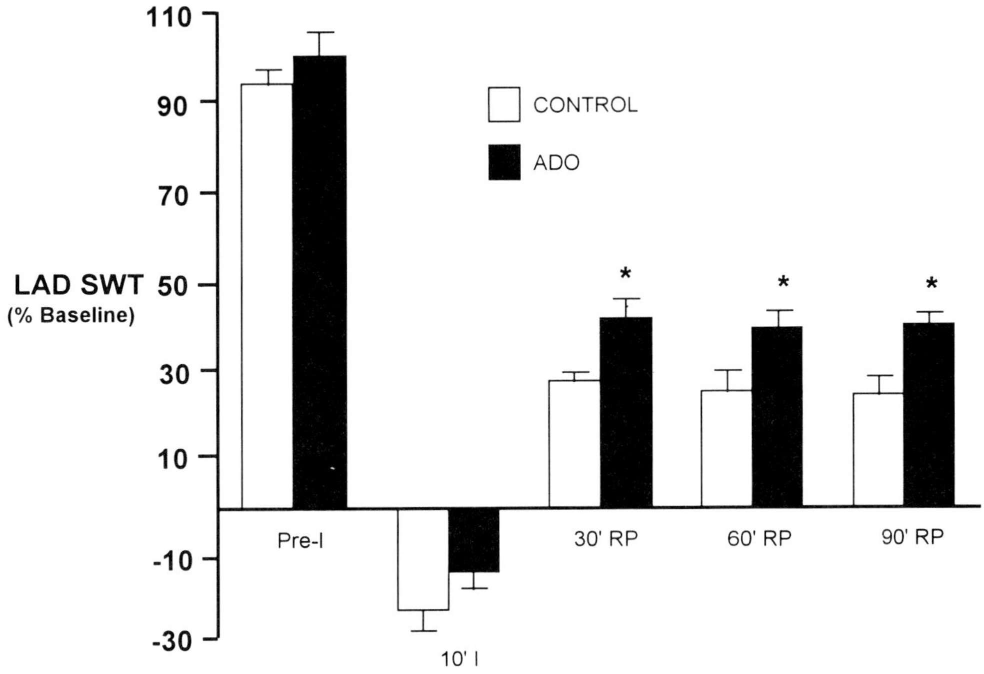

FIGURE 39-4. Attenuation of in vivo porcine myocardial stunning with adenosine (ADO) pretreatment (50 μg/kg/min, intracoronary). Adenosine infusion prior (Pre-I) to 10 minutes of LAD occlusion (10′I) had no effect on regional systolic wall thickening (SWT) but significantly improved SWT during reperfusion (RP). *$p < 0.05$ versus control.

did not increase myocardial phosphorylation potential, and there was also no effect of adenosine on regional function in normal myocardium. It is not clear what role increased phosphorylation potential plays in adenosine-mediated attenuation of myocardial stunning, but it is plausible that increased phosphorylation potential may provide more energy for intracellular sodium and hence calcium homeostasis. It also remains to be determined whether adenosine-mediated enhancement of the reperfusion phosphorylation potential is receptor mediated.

ADENOSINE AND GLUCOSE METABOLISM

There is also evidence that adenosine modulates glucose metabolism both in normal and ischemic myocardium. Adenosine receptor blockade with 8-(p-sulfophenyl)-theophylline has been shown to attenuate hypoxia-induced increases in glycolytic flux in isolated perfused rat hearts [22]. Adenosine infusion has also been reported to increase glucose uptake in constant-flow perfused rat hearts [23] and in constant-pressure perfused guinea-pig hearts [1]. In the latter experiments, adenosine treatment prior to and following low-flow ischemia was associated with increased intracellular levels of glycolytic metabolites, including fructose-6-phosphate, the substrate of the rate-controlling enzyme phosphofructokinase, and 3-phosphoglycerate, the product of the powerful glyceradehyde 3-phosphate dehydrogenase/phosphoglycerate kinase (GAPDH/PGK) system. The increase in the levels of these glycolytic intermediates may be a mechanism by which accumulation of inorganic phosphate (P_i) during ischemia/reperfusion could be reduced, thus stabilizing, if not increasing, the $(ATP)/[(ADP) \times (P_i)]$ ratio.

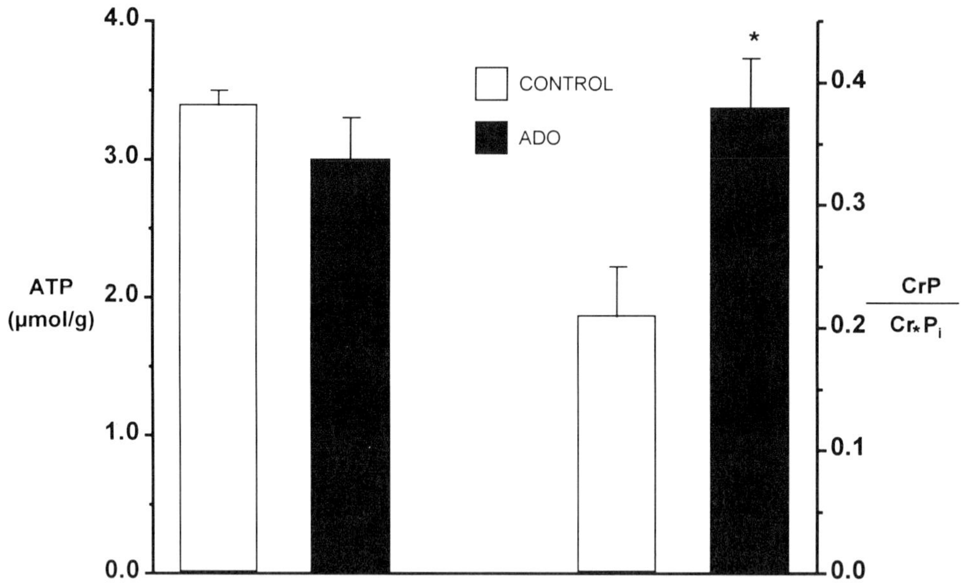

FIGURE 39-5. Effect of adenosine (ADO) pretreatment on postischemic ATP content (μmol/g wet wt) and (CrP)/(Cr) · (P_i) ratio (μmol^{-1}) in LAD stunned myocardium after 90 minutes of reperfusion. See Figure 39-4 and text for details. *$p < 0.05$ versus control.

Other evidence indicates that the effects of adenosine on glucose metabolism may play a role in adenosine-mediated cardioprotection. During low-flow ischemia (0.5 ml/min) in isolated perfused rat hearts, adenosine treatment (100 μM) significantly increased lactate release and delayed the time to onset and rate of ischemic contracture development [19]. Coinfusion of the adenosine deaminase inhibitor EHNA (50 μM) decreased adenosine metabolism and potentiated lactate release. Treatment with the adenosine receptor antagonist BW A1433U (5 μM), a para-phenyl carboxyl-substituted derivative of 1,3-dipropyl-8-phenylxanthine, significantly decreased lactate release and accelerated the onset and rate of contracture development. When glucose was omitted from the perfusate, adenosine + EHNA treatment failed to alter lactate release or delay contracture. These results suggest that adenosine may protect the heart during low-flow ischemia via modulation of glucose metabolism and/or H^+ handling. The latter mechanism could be a result of increased monocarboxylate, that is, mainly lactate, release because the export of glycolytic monocarboxylates is likely to be coupled stoichiometrically to H^+ export [24].

Summary

Substantial experimental evidence now exists that indicates the purine nucleoside adenosine enhances myocardial tolerance to ischemia and attenuates postischemic myocardial stunning. Although the exact mechanism of adenosine's beneficial effect is not known, it appears that it is mediated via activation of adenosine A_1 receptors and modulation of myocardial metabolism, both of which may limit intracellular calcium overload. Adenosine's beneficial effects on metabolism include enhanced phosphorylation potential, increased glycolytic intermediates, and improved H^+ handling. In order to provide sustained attenuation of postischemic ventricular dysfunction, adenosine must be administered prior to ischemia and in a concentration sufficient to overcome coronary endothelial cell and red blood cell metabolism. Adenosine also has cardioprotective properties during hypothermic ischemia, indicating its utility in clinical settings of cardiac surgery and heart preservation for transplantation.

References

1. Mentzer RM, Bünger R, Lasley RD (1993) Adenosine enhanced preservation of myocardial function and energetics. Possible involvement of the adenosine A_1 receptor system. Cardiovasc Res 27:28–35.
2. Thornton JD, Liu GS, Olsson RA, Downey JM (1992) Intravenous pretreatment with A_1-selective adenosine analogues protects the heart against infarction. Circulation 85:659–665.
3. Belardinelli L, Linden J, Berne RM (1989) The cardiac effects of adenosine. Prog Cardiovasc Dis 22:73–97.
4. Lasley RD, Mentzer RM Jr (1992) Adenosine improves the recovery of postischemic myocardial function via an adenosine A_1 receptor mechanism. Am J Physiol 263:H1460–H1465.
5. Yao Z, Gross GJ (1993) Glibenclamide antagonizes adenosine A_1 receptor-mediated cardioprotection in stunned canine myocardium. Circulation 88:235–244.
6. Dorheim TA, Hoffman A, Van Wylen DGL, Mentzer RM Jr (1991) Enhanced interstitial fluid adenosine attenuates myocardial stunning. J Surg 110:136–145.
7. Randhawa MPS, Lasley RD, Anderson GM, Hegge J, Mentzer RM Jr (1992) Adenosine pretreatment attenuates myocardial stunning. Circulation 86(Suppl I):I613.
8. Lasley RD, Rhee JW, Van Wylen DGL, Mentzer RM Jr (1990) Adenosine A_1 receptor mediated protection of the globally ischemic isolated rat heart. J Mol Cell Cardiol 22:39–47.
9. Lasley RD, Mentzer RM Jr (1993) Pertussis toxin blocks adenosine A_1 receptor mediated protection of the ischemic rat heart. J Mol Cell Cardiol 25:815–821.
10. Lasley RD, Noble MA, Paulsen KL, Mentzer RM Jr (1994) Adenosine attenuates phorbol ester-induced negative inotropic and vasoconstrictive effects in the isolated rat heart. Am J Physiol 266:H2159–H2166.
11. Van Wylen DGL, Willis J, Sohdi J, Weiss RJ, Lasley RD, Mentzer RM Jr (1990) Cardiac microdialysis to estimate interstitial adenosine and coronary blood flow. Am J Physiol 258: H1642–H1649.
12. Van Wylen DGL, Schmit TJ, Lasley RD, Gingell RL, Mentzer RM Jr (1992) Cardiac microdialysis in isolated rat hearts: Interstitial purine metabolites during ischemia. Am J Physiol 262:H1934–1938.
13. Nees S, Herzog V, Becker BF, Böck M, Des Rosiers C, Gerlach E (1985) The coronary endothelium: A highly active metabolic barrier for adenosine. Basic Res Cardiol 80:515–529.
14. Lasley RD, Mentzer RM Jr (1994) The role of adenosine in extended myocardial preservation with the University of Wisconsin solution. J Thorac Cardiovasc Surg 107:1356–1363.
15. Wyatt DA, Ely SW, Lasley RD, Walsh R, Mainwaring R, Berne RM, Mentzer RM (1989) Purine-enhanced asanguineous cardioplegia retards adenosine triphosphate degradation during ischemia and improves postischemic ventricular function. J Thorac Cardiovasc Surg 97:771–778.
16. Ambrosio G, Jacobus WE, Mitchell MC, Litt MR, Becker LC (1989) Effects of ATP precursors on ATP and free ADP content and functional recovery of postischemic hearts. Am J Physiol 256:H560–H566.
17. Bünger R, Mallet RT, Hartman DA (1989) Pyruvate-enhanced phosphorylation potential and inotropism in normoxic and postischemic isolated working heart. Near-complete prevention of reperfusion contractile failure. Eur J Biochem 180:221–233.
18. Mallet, RT, Bünger, R (1993) Metabolic protection of post-ischemic phosphorylation potential and ventricular performance. In Sideman S, Beyar R (eds.), Interactive Phenomena in the Cardiac System. New York: Plenum Press, pp. 233–241.
19. Lasley RD, Mentzer RM Jr (1993) Adenosine increases lactate release and delays the onset of contracture during global low flow ischemia. Cardiovasc Res 27:96–101.
20. Schwartz GG, Schaefer S, Meyerhoff DJ, Gober J, Fochler P, Massie B, Weiner MW (1990) Dynamic relation between myocardial contractility and energy metabolism during and following brief coronary occlusion in the pig. Circ Res 67:490–500.
21. Zhou Z, Bünger R, Lasley RD, Hegge JO, Mentzer RM Jr (1993) Adenosine pretreatment

increases cytosolic phosphorylation potential and attenuates postischemic cardiac dysfunction in swine. Surg Forum 44:249–52.

22. Wyatt DA, Edmunds MC, Rubio R, Berne RM, Lasley RD, Mentzer RM (1989) Adenosine stimulates glycolytic flux in isolated perfused rat hearts by A_1-adenosine receptors. Am J Physiol 257:H1952–H1957.

23. Mainwaring RD, Lasley R, Rubio R, Wyatt D, Mentzer R Jr (1988) Adenosine stimulates glucose uptake in the isolated rat heart. Surgery 103:445–449.

24. Büinger R, Mallet RT (1993) Mitochondrial pyruvate transport in working guinea-pig heart. Work-related versus carrier-mediated control of pyruvate oxidation. Biochim Biophys Acta 1151:223–236.

40. ROLE OF ACTIVATION OF ECTO-5′-NUCLEOTIDASE FOR CARDIOPROTECTION IN ISCHEMIC PRECONDITIONING

Masafumi Kitakaze
Tetsuo Minamino
Koichi Node
Kazuo Komamura
Toshinao Kurihara
Michitoshi Inoue
Masatsugu Hori
Takenobu Kamada

When brief periods of ischemia precede sustained ischemia, infarct size is markedly limited, a phenomenon known as *ischemic preconditioning* [1,2]. The precise mechanisms underlying this phenomenon have been investigated [3,4] because identification of the primary mediator of ischemic preconditioning may contribute to the development of the potential treatment of acute myocardial infarction. Several lines of evidence suggest that beneficial effects of ischemic preconditioning are observed in clinical settings [5,6]. Recently, Liu et al. [7] experimentally demonstrated that an exposure to 8-sulfophenyltheophylline blunts the infarct-size–limiting effect of ischemic preconditioning and that brief periods of exposures to adenosine, instead of transient ischemia, mimic ischemic preconditioning. Thornton et al. [8] showed that adenosine A_1-receptor activation is responsible for the infarct-size–limiting effect of ischemic preconditioning. Several lines of evidence support this observation [9,10]. Indeed, it has been clarified that adenosine can limit infarct size caused by ischemia and reperfusion [11]. There are two different possibilities by which adenosine links with cardioprotection in ischemic preconditioning. One possibility is that release of adenosine during ischemia and reperfusion is enhanced by ischemic preconditioning, and the other possibility is that released adenosine makes myocardium resistant to ischemia and reperfusion. Here we discuss the possibility that ischemic preconditioning activates the enzymes responsible for adenosine release, that is,

5′-nucleotidase, and test the cause-effect relationship between activation of 5′-nucleotidase and the infarct-size–limiting effect in ischemic preconditioning.

Adenosine Release and 5′-Nucleotidase Activity in Ischemic Preconditioning

First, we tested whether the ischemic preconditioning procedure increases ecto-5′-nucleotidase activity and adenosine release during reperfusion [12]. In anesthetized open-chest dogs, after intravenous administration of heparin (500 U/kg), we cannulated and perfused the left anterior descending (LAD) coronary artery with blood from the left carotid artery through an extracorporeal bypass tube. Coronary blood flow and coronary perfusion pressure in the perfused area were measured. A small-caliber (1 mm), short (7 cm) collecting tube was introduced into a small coronary vein near the center of the perfused area to sample coronary venous blood for measurements of adenosine. The left atrium was catheterized for microsphere injection to determine the collateral flow during ischemia.

After the bypass tube to the LAD coronary artery was occluded four times for 5 minutes, both ecto- and cytosolic 5′-nucleotidase activity was increased (Fig. 40-1, lower panel). Ecto- and cytosolic 5′-nucleotidase activity was defined as the 5′-nucleotidase activity of membrane and cytosolic fractions [12]. The upper panel of Figure 40-1 illustrates the adenosine concentration of coronary arterial and venous blood during reperfusion following 40 minutes of ischemia. Adenosine concentration in the coronary venous blood was higher in the group subjected to ischemic preconditioning than in the untreated control group.

Does activation of ecto-5′-nucleotidase directly link with enhanced release of adenosine during ischemia and reperfusion? Adenosine is produced simply by the following enzymatic reactions: The dephosphorylation of 5′-AMP by 5′-nucleotidase and the hydrolysis of S-adenosylhomocysteine (SAH) by SAH-hydrolase. Although adenosine is produced through the latter pathway in normoxic hearts [13,14], 5′-nucleotidase is related to adenosine production during ischemia and hypoxia [15–18]. This idea is supported by the fact that α, β-methylene adenosine 5′-diphosphate (AOPCP, an inhibitor of ecto-5′-nucleotidase; 80 μg/kg/min) potently reduces adenosine production in the ischemic myocardium [19]. Furthermore, the extent of decreases in reactive hyperemic flow following a brief period of coronary occlusion is comparably reduced under treatments with AOPCP and 8-phenyltheophylline (an adenosine receptor antagonist), suggesting that adenosine production during ischemia is attributable to the activity of 5′-nucleotidase [19].

However, accumulation of 5′-AMP seems to be the other factor that regulates adenosine production. Cytosolic 5′-AMP concentration crucially depends on the duration and severity of ischemia and culminates in 1×10^{-2} to 10^{-4} M. Considering that production of adenosine is 1×10^{-6} to 10^{-8} M, even considerable changes in AMP concentration would not affect adenosine production during ischemia.

FIGURE 40-1. The **upper panel** indicates the graph showing adenosine release during reperfusion after 40 minutes of coronary occlusion with an without ischemic preconditioning. Adenosine release in the group that underwent ischemic preconditioning is enhanced ($p < 0.01$) for 40 minutes compared with the untreated control condition. The **lower panel** indicates bar graphs showing ecto- and cytosolic 5′-nucleotidase activity in the control and ischemic preconditioned myocardium before the 40-minute coronary occlusion. Both ecto- and cytosolic 5′-nucleotidase activity was augmented by ischemic preconditioning. (From Kitakaze et al. [12], with permission.)

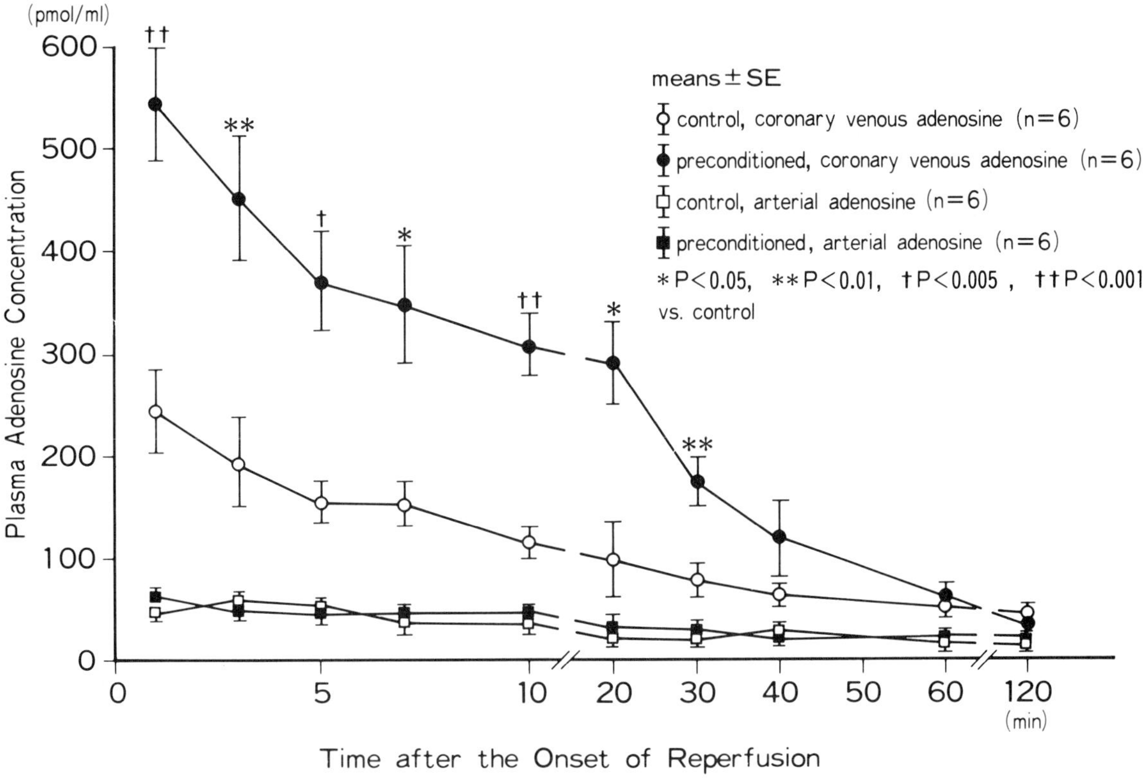
(pmol/ml)
Plasma Adenosine Concentration
means ± SE
control, coronary venous adenosine (n=6)
preconditioned, coronary venous adenosine (n=6)
control, arterial adenosine (n=6)
preconditioned, arterial adenosine (n=6)
* P<0.05, ** P<0.01, † P<0.005, †† P<0.001
vs. control
Time after the Onset of Reperfusion
(min)

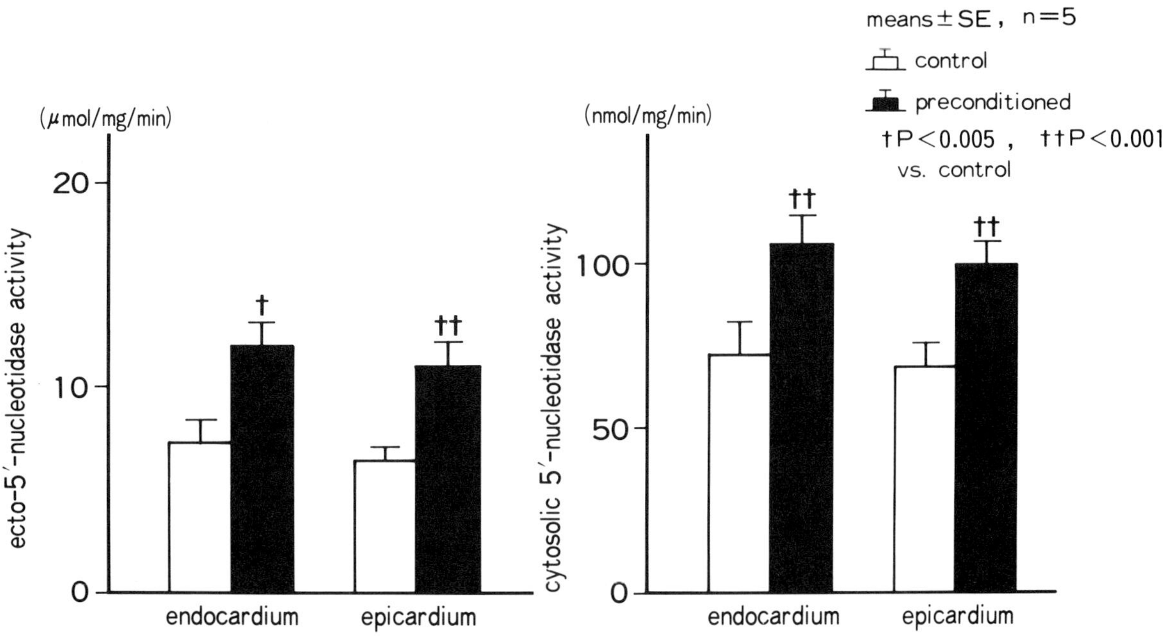
means ± SE, n=5
control
preconditioned
† P<0.005, †† P<0.001
vs. control
(μmol/mg/min)
ecto-5′-nucleotidase activity
(nmol/mg/min)
cytosolic 5′-nucleotidase activity
endocardium
epicardium

Role of Activation of 5′-Nucleotidase and Enhanced Adenosine Release in Salvage of Myocardial Necrosis in Ischemic Preconditioning

To test the cause-and-effect relationship between activation of 5′-nucleotidase and the infarct-size–limiting effect in ischemic preconditioning, we examined whether AOPCP blunted the infarct-size–limiting effect of ischemic preconditioning [20]. We occluded the coronary artery four times for 5 minutes each with intracoronary administration of AOPCP. AOPCP was administered into the LAD coronary artery 5 minutes prior to the ischemic preconditioning (IP) procedure and was continued for 60 minutes of reperfusion, except for the coronary occlusion period (the AOPCP treatment with IP group). In the other dogs, AOPCP was administered into the LAD coronary artery 40 minutes prior to ischemia and was continued for 60 minutes of reperfusion, except for the coronary occlusion period (the AOPCP treatment group).

To discriminate between the role of increases in 5′-nucleotidase activity during the ischemic preconditioning procedure (the AOPCP pretreatment with IP group) or during reperfusion (the AOPCP during reperfusion with IP group) on the infarct-size–limiting effect, we infused AOPCP only during the ischemic preconditioning procedure or only during up to 60 minutes of reperfusion in the ischemic-preconditioned dogs. In our preliminary study this dose of AOPCP was sufficient to attenuate ecto-5′-nucleotidase activity of the canine myocardium: At 5 minutes of exposure to AOPCP, ecto-5′-nucleotidase activity was reduced to 3.2 ± 4.9 nmol/mg protein/min from 37.1 ± 4.1 nmol/mg protein/min ($p < 0.001$; $n = 5$). There are no significant differences in risk area and collateral flow during ischemia between the six groups. Figure 40-2 illustrates the infarct size in the six groups. Ischemic preconditioning markedly attenuated infarct size. AOPCP completely abolished the infarct-size–limiting effect of ischemic preconditioning. In the AOPCP pretreatment with IP and the AOPCP during reperfusion with IP groups, infarct size was partially attenuated compared with the AOPCP treatment with IP group and the AOPCP treatment group. The infarct sizes of these two groups were smaller than the control and AOPCP groups, and larger than that in the ischemic preconditioning group.

Figure 40-3 depicts the relation between infarct size normalized by risk area and collateral flow to the inner half of the LAD-dependent myocardium during the sustained ischemia period. Figure 40-3 clearly indicates that 1) ischemic preconditioning attenuates infarct size, 2) AOPCP treatment blunts the infarct-size–limiting effect of ischemic preconditioning, and 3) administration of AOPCP during ischemic preconditioning procedure and during the reperfusion period partially blunts the infarct-size–limiting effect. These results indicate that both increased 5′-nucleotidase activity and adenosine release during ischemic preconditioning procedure and during early reperfusion synergistically contribute to the infarct-size–limiting effect of ischemic preconditioning.

Despite the activation of 5′-nucleotidase and enhanced release of adenosine during reperfusion, these observations do not necessarily indicate that adenosine release is augmented during ischemia. Indeed, we did not measure adenosine release during the sustained period of ischemia. Moreover, there is a report showing that increases in adenosine concentration in the interstitial space are not augmented during sustained ischemia in ischemia preconditioning [21].

One possible explanation for the differences between this result and our observation [12] is that the activity of the other enzymes, which also regulates adenosine metabolism during ischemia, for example, adenosine kinase and adenosine deaminase [14,15], may be augmented by ischemic events, for example, acidosis or energy levels. If this is the case, adenosine production due to increased 5′-nucleotidase activity during sustained ischemia may be at-

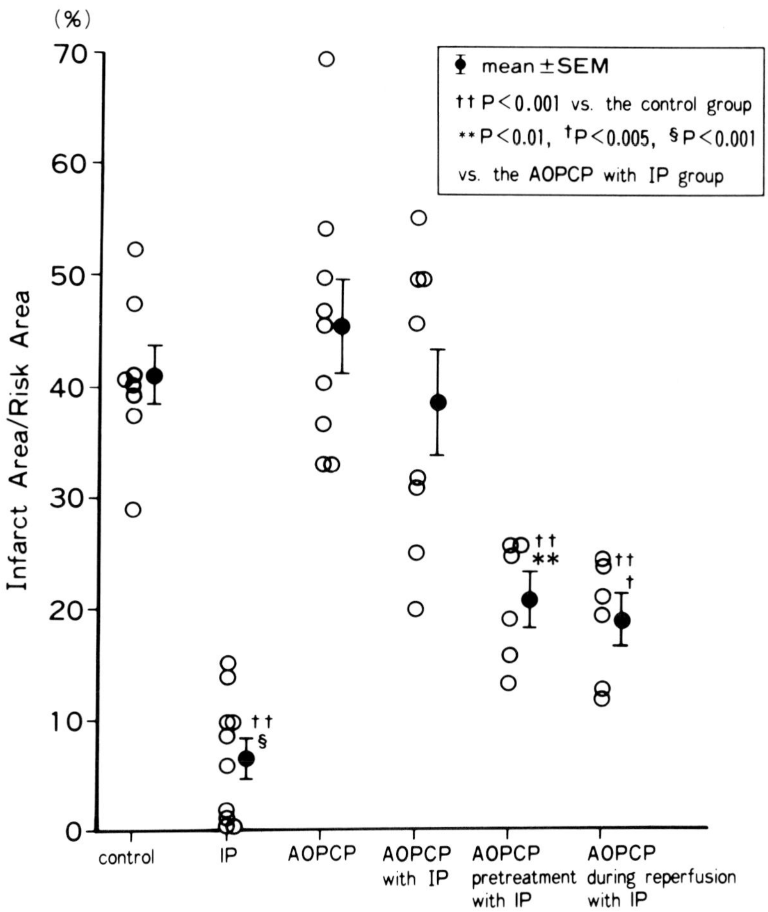

FIGURE 40-2. Infarct size in the control group, the ischemic preconditioning group, the AOPCP treatment group, the AOPCP treatment with IP group, the AOPCP pretreatment with IP group, and the AOPCP during reperfusion with IP group. Infarct size was markedly decreased in ischemic preconditioning. The infarct-size–limiting effect of ischemic preconditioning was completely abolished by administration of AOPCP. Administration of AOPCP during the ischemic preconditioning procedure or during reperfusion following sustained ischemia attenuated ($p < 0.001$) infarct size compared with infarct size in the control group. On the other hand, the infarct size of AOPCP prereatment with the IP group and the AOPCP during reperfusion with IP group was larger ($p < 0.01$) than that in the ischemic preconditioning group. (From Kitakaze et al. [20], with permission.)

tenuated. Another possibility is that adenosine uptake into the myocytes is enhanced during ischemia in the preconditioned group. Adenosine taken into the cardiomyocytes may be converted to ATP via 5′-AMP, which may contribute to the preservation of ATP in ischemic preconditioning. Since ecto-5′-nucleotidase produces adenosine outside the cardiomyocytes and endothelial cells, the advantage of exerting the action of adenosine on its A_1 receptors may primarily contribute to attenuate infarct size. Furthermore, if ecto-5′-nucleotidase in coronary endothelium cells is predominantly activated compared with ecto-5′-nucleotidase in cardiomyocytes, the release of adenosine into coronary venous blood may be increased despite no

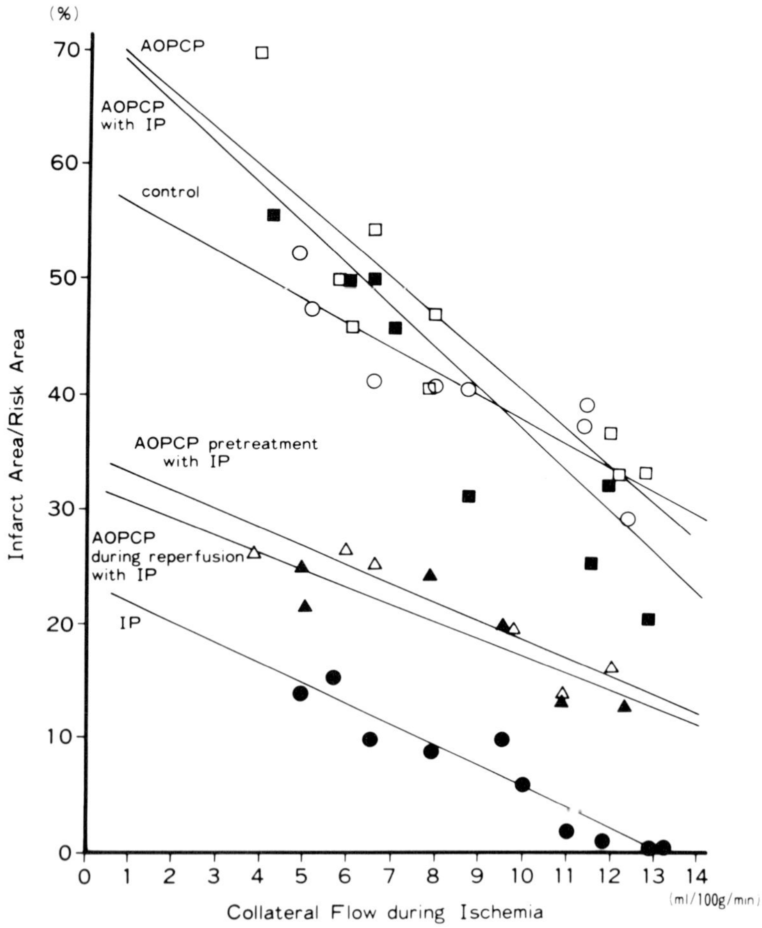

FIGURE 40-3. Plot of infarct size expressed as a percentage of the risk area and regional collateral flow during ischemia. In the control group (open circles), the ischemic preconditioning group (closed circles), the AOPCP treatment with IP group (closed squares), the AOPCP treatment group (open squares), the AOPCP pretreatment with IP group (open triangles), and the AOPCP during reperfusion with IP group (closed triangles), there are inverse relations between normalized infarct area and collateral flow. Infarct size in the preconditioned hearts was smaller ($p < 0.001$) with respect to any given value for collateral blood flow than that in the control hearts. Infarct size was markedly increased ($p < 0.001$) in the AOPCP treatment with IP group compared with the ischemic preconditioning group. Infarct size in the AOPCP pretreatment with IP group and the AOPCP during reperfusion with IP group were not different; however, these values are larger ($p < 0.001$) than that in the ischemic preconditioning group and smaller ($p < 0.001$) than the control and AOPCP treatment groups. (From Kitakaze et al. [20], with permission.)

changes in adenosine from the cardiomyocytes, as is detected in interstitial fluid.

Role of Alpha$_1$-Adrenoceptor Activity in Salvage of Myocardial Necrosis in Ischemic Preconditioning

To test the role of alpha$_1$-adrenoceptor activation in cardioprotection in ischemic preconditioning [23], constant infusion of prazosin (4 µg/kg/min) into the LAD coronary artery was begun 5 minutes prior to ischemic preconditioning and continued through the first 60 minutes of the reperfusion period, except during coronary occlusion (the prazosin with IP group). In other dogs, prazosin was infused into the LAD coronary artery beginning 40 minutes prior to ischemia without ischemic preconditioning and was continued for 60 minutes of reperfusion, except during coronary occlusion (the prazosin group). To test whether alpha$_1$-adrenoceptor stimulation mimics the infarct-size–limiting effect of ischemic preconditioning, we administered methoxamine into the LAD (40 µg/kg/min, 4 cycles for 5 minutes each with 5-minute intervals; methoxamine group). Following methoxamine exposure, 90 minutes of coronary occlusion and 6 hours of reperfusion were imposed. Furthermore, to examine the role of increased ecto-5′-nucleotidase activity due to exposure to methoxamine, we concomitantly infused AOPCP 5 minutes prior to ischemic preconditioning and continued 60 minutes of reperfusion, except during 90 minutes of coronary occlusion in animals treated with methoxamine (the methoxamine with AOPCP group). In another group, we determined the effect of AOPCP on infarct size (the AOPCP group). In this group, AOPCP was administered prior to 40 minutes of coronary occlusion and during 1 hour of reperfusion following 90 minutes of coronary occlusion.

Ischemic preconditioning increased both ecto- (Fig. 40-4) and cytosolic 5′-nucleotidase activity in the myocardium. Prazosin administration without ischemic preconditioning decreased ecto- and cytosolic 5′-nucleotidase activity and blunted the increases in ecto- and cytosolic 5′-nucleotidase activity due to ischemic preconditioning. Methoxamine increased both ecto- and cytosolic 5′-nucleotidase activity to the levels obtained by ischemic preconditioning. Figure 40-5 shows that ischemic preconditioning markedly attenuates infarct size compared with controls, although collateral flow is not different. Figure 40-5 also shows that prazosin completely abolishes the infarct-size–limiting effect of ischemic preconditioning. With methoxamine administration, infarct size was attenuated to the level seen with ischemic preconditioning (Fig. 40-5). However, the infarct-size–limiting effect due to exposure of methoxamine was blunted by AOPCP, and no difference in infarct size existed between the methoxamine with AOPCP and the AOPCP groups.

We also checked the regression plots of infarct size, as a percentage of the area at risk against collateral flow and obtained comparable data in Figure 40-5 [22]. This observation is consistent with previous studies. It is reported that alpha$_1$-adrenoceptor activation is intimately involved in attenuation of the severity of ischemia and reperfusion [17,23] and ischemic preconditioning [24]. The present results strongly suggest that alpha$_1$-adrenoceptor stimulation mediates cardioprotection seen in ischemic preconditioning, which is attributable to activation of ecto-5′-nucleotidase.

According to the data of Ytrehus et al. [25], protein kinase C is tightly linked to ischemic preconditioning. We also have reported that activation of protein kinase C increases ecto- and cytosolic 5′-uncleotidase activity in vitro [26]. Since protein kinase C is activated by ischemia per se [27] and by norepinephrine released during brief periods of ischemia, ischemic preconditioning may increase ecto-5′-nucleotidase activity via protein kinase C activation. Although we have not elucidated the mechanisms whereby protein kinase C increases 5′-nucleotidase activity, we speculate that protein kinase C may

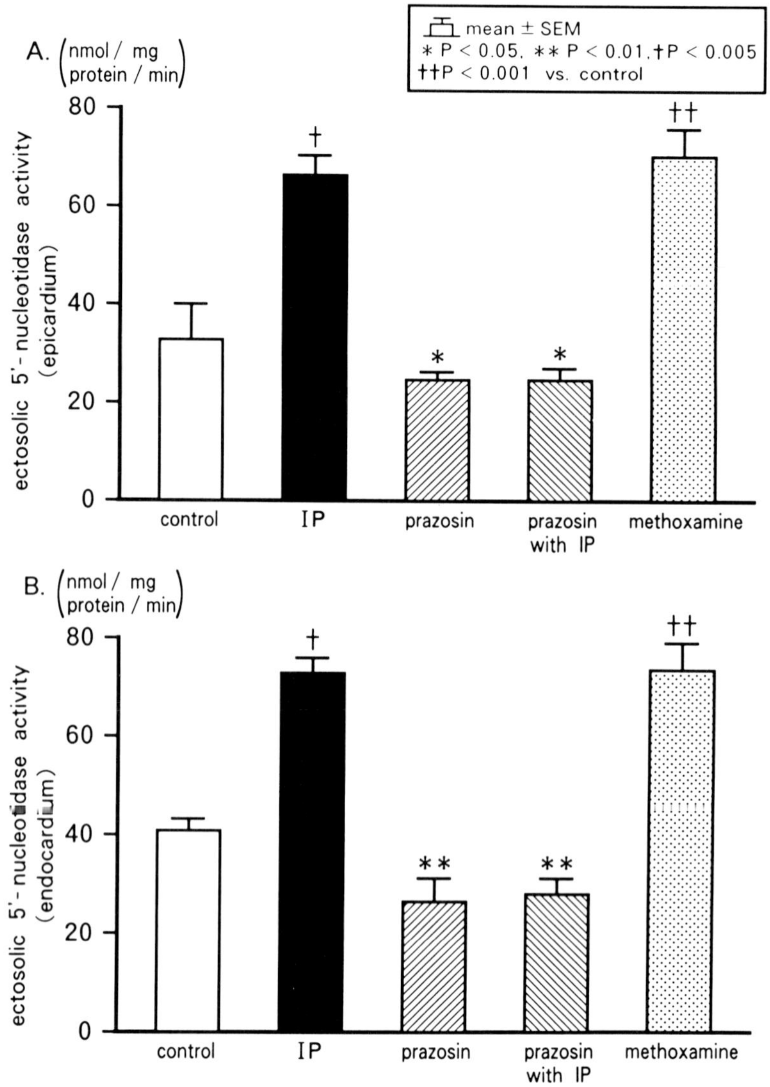

FIGURE 40-4. Comparison of canine ecto-5′-nucleotidase activity in the epicardium (**A**) and endocardium (**B**) following 40 minutes of steady-state observation (control group), 40 minutes of prazosin administration (n = 5), ischemic preconditioning with and without prazosin administration, and intermittent exposures to methoxamine. Ischemic preconditioning increased ecto-5′-nucleotidase activity, which was blunted by prazosin and mimicked by methoxamine. (From Kitakaze et al. [22], with perimission.)

change the characteristics of the active site of ecto-5′-nucleotidase or induce a conformational change in the structure of 5′-nucleotidase.

Clinical Relevance

The finding in the present study suggests two clinical applications for treatment of acute myocardial infarction. One strategy to limit infarct size is to find a method to increase 5′-nucleotidase activity. As we observed,

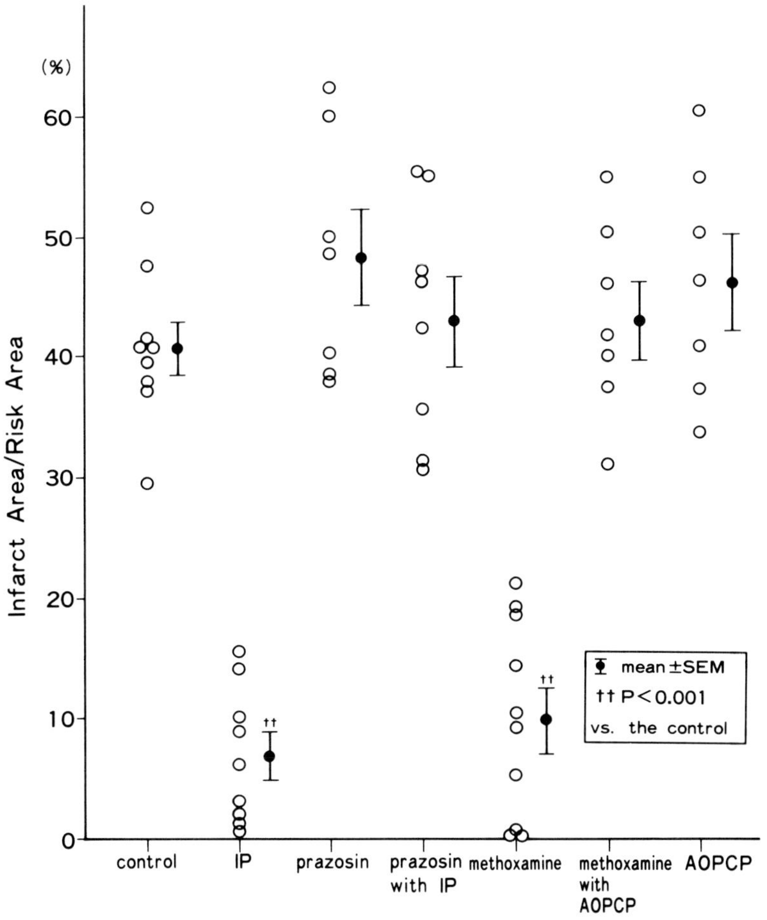

FIGURE 40-5. Infarct size in the control group, the ischemic preconditioning group, the prazosin with IP group, the prazosin group, the methoxamine group, the methoxamine with AOPCP group, and the AOPCP group. Infarct size was decreased in the ischemic preconditioning group. The infarct-size–limiting effect of ischemic preconditioning was abolished by prazosin. On the other hand, methoxamine mimicked the infarct-size–limiting effect of ischemic preconditioning, which was blunted by AOPCP. (From Kitakaze et al. [22], with permission.)

protein kinase C activation may increase 5′-nucleotidase activity and may attenuate contractile dysfunction and infarct size. Another possibility is to enhance adenosine release. Administration of adenosine and potentiators of adenosine production, for example, acadesine, dilazep, or dipyridamole, may limit infarct size. Both strategies merit clinical investigation, although enhanced understanding of the basic process involved in ischemic preconditioning is necessary.

References

1. Murry CE, Jennings RB, Reimer KA (1986) Preconditioning with ischemia: A delay of lethal cell injury in ischemic myocardium. Circulation 74:1124–1136.
2. Li GC, Vasquez JA, Gallagher KP, Lucchesi BR (1990) Myocardial protection with preconditioning. Circulation 82:609–619.
3. Murry CE, Richard VJ, Reimer KA, Jennings RB (1990) Ischemic preconditioning slows

energy metabolism and delays ultrastructural damage during a sustained ischemic episode. Circ Res 66:913–931.
4. Steenbergen C, Fralix TA, Perlman M, Gabel SA, London RE, Murphy E (1991) Na^+/H^+ and Na^+/Ca^{2+} exchange during ischemia in preconditioned rat hearts in vitro (abstr). Circulation 84:II210.
5. Deutsch E, Berger M, Kussmaul WG, Hirshfeld KW Jr, Herrmann HC, Laskey W (1990) Adaptation in ischemia during percutaneous transmural coronary angioplasty. Clinical, hemodynamic, and metabolic features. Circulation 82:2044–2051.
6. Okazaki Y, Kodama K, Sato H, Kitakaze M, Hirayama A, Mishima M, Hori M, Inoue M (1993) Attenuation of increased regional oxygen consumption during exercise as a major cause of warm-up phenomenon. J Am Coll Cardiol 21: 1597–1604.
7. Liu GS, Thornton J, Van Winkle DM, Stanley AWH, Olsson RA, Downey JM (1991) Protection against infarction afforded by preconditioning is mediated by A_1 adenosine receptors in rabbit heart. Circulation 84:350–356.
8. Thornton JD, Liu GS, Olsson RA, Downey JM (1992) Intravenous pretreatment with A_1-selective adenosine analogue protects the heart against infarction. Circulation 85:659–665.
9. Miura T, Ogawa T, Iwamoto T, Shimamoto K, Iimura O (1992) Dipyridamole potentiates the myocardial infarct size-limiting effect of ischemic preconditioning. Circulation 86: 979–985.
10. Van Winke DM, Chien GL, Wolff RA, Soifer BE, Davis RF (1992) Intracoronary infusion of R-phenylisopropyl adenosine prior to ischemia-reperfusion reduces myocardial infarct size in swine (abstr). Circulation 86:I213.
11. Olafsson B, Forman MB, Puett DW, Pou A, Cates CU, Friesinger GC, Virmani R (1987) Reduction of reperfusion injury in the canine preparation by intracoronary adenosine: Importance of the endothelium and the no-reflow phenomenon. Circulation 76:1135–1145.
12. Kitakaze M, Hori M, Takashima S, Sato H, Inoue M, Kamada T (1993) Ischemic preconditioning increases adenosine release and 5′-nucleotidase activity during myocardial ischemia and reperfusion in dogs. Implication for myocardial salvage. Circulation 87:208–215.
13. Berne RM (1980) The role of adenosine in the regulation of coronary blood flow. Circ Res 47:807–813.
14. Llyod HGE, Schrader J (1987) The importance of the transmethylation pathway for adenosine metabolism in the hearts. In Gerlach E, Becker BF (eds.), Topics and Perspectives in Adenosine Research. Berlin: Springer-Verlag, pp. 199–207.
15. Sparks NV Jr, Bardengeuer H (1986) Regulation of adenosine formation by the heart. Circ Res 58:193–201.
16. Hori M, Kitakaze M (1991) Adenosine, the heart, and coronary circulation. Hypertension 18:565–574.
17. Kitakaze M, Hori M, Kamada T (1993) Role of adenosine and its interaction with alpha adrenoceptor activity in ischemic and reperfusion injury of the myocardium. Cardiovasc Res 27:18–27.
18. Imai S, Nakazawa M, Imai M, Jin H (1986) 5′-Nucleotidase inhibitors and the myocardial reactive hyperemia and adenosine content. In Gerlach E, Becker BF (eds.), Topics and Perspectives in Adenosine Research. Berlin: Springer-Verlag, pp. 416–424.
19. Kitakaze M, Hori M, Takashima S, Iwai K, Sato H, Inoue M, Kitabatake A, Kamada T (1992) Superoxide dismutase enhances ischemia-induced reactive hyperemic flow and adenosine release in dogs: a role of 5′-nucleotidase activity. Circ Res 71:558–566.
20. Kitakaze M, Hori M, Morioka T, Minamino T, Takashima S, Sato H, Shinozaki Y, Chujo M, Mori H, Inoue M, Kamada T (1994) The infarct size limiting effect of ischemic preconditioning is blunted by inhibition of 5′-nucleotidase activity and attenuation of adenosine release. Circulation 89:1237–1246.
21. Dorheim TA, Menzer RM, Van Wylen DGL (1991) Preconditioning reduces interstitial fluid purine metabolites during prolonged myocardial ischemia. Circulation 84(Suppl II):II191.
22. Kitakaze M, Hori M, Morioka T, Minamino T, Takashima S, Sato H, Shinozaki Y, Chujo M, Mori H, Inoue M, Kamada T (1994) $Alpha_1$-adrenoceptor activation mediates the infarct size-limiting effect of ischemic preconditioning through augmentation of 5′-nucleotidase activity. J Clin Invest 93:2197–2205.
23. Kitakaze M, Hori M, Sato H, Iwakura K, Gotoh K, Inoue M, Kitabatake A, Kamada T (1991) Beneficial effects of $alpha_1$-adrenoceptor activity on myocardial stunning in dogs. Circ Res 68:1322–1339.
24. Banerjee A, Locke-Winter C, Rogers KB, Mitchell MB, Brew EC, Cairns CB, Bensard DD, Harken AH (1993) Preconditioning against

myocardial dysfunction after ischemia and reperfusion by alpha-adrenergic mechanism. Circ Res 73:656–670.

25. Ytrehus K, Liu Y, Downey J (1993) Preconditioning protects the ischemic rabbit heart by protein kinase activation (abstr). FASEB J 7:A418.
26. Kitakaze M, Kitabatake A (1991) Increased 5′-nucleotidase activity caused by protein kinase C enhances adenosine production in hypoxic cardiomyocytes of rats (abstr). Circulation 84: II620.
27. Strasser RH, Walendzik H, Marquetant R (1991) Prolonged myocardial ischemia provokes an induction of protein kinase C activity (abstr). Circulation 84:II620.

41. ADENOSINE UPTAKE BLOCKERS FOR CARDIOPROTECTION

Herman Van Belle

The ever increasing number of communications at the International Symposia on Adenosine (the first held in Banff, 1978) illustrates the fascination this molecule now has for biochemists, biologists, and pharmacologists. This is in sharp contrast to the almost 30 years of virtual silence following the description, in 1929 [1], of adenosine's effect on the mammalian heart as "remarkable". Undoubtedly, it is the progress in bioanalytical and biophysical technics that has contributed most to a better understanding of the "adenosinergic" system. In addition, I believe, the hope of getting interesting medicines from the many tools that were, and are, created to intervene with this system has also provided the matrix for much research in the field. How and where we stand now with potential therapeutics is briefly discussed in this chapter.

Dipyridamole, a Prime Mover

After a long-lasting hibernation, an outbreak of adenosine research emerged in the early 1960s when in Germany (Kübler, Bretschneider, and Gerlach) and the United States (Berne and Rubio) the mechanisms behind adenosine formation in the ischemic myocardium were unraveled. At that time dipyridamole, known as a potent coronary vasodilator, was found to inhibit nucleoside (adenosine) transport [2]. It is an open question whether this finding has been the consequence of or the impetus for linking up adenosine formation with metabolic regulation of coronary flow. However, in the following years dipyridamole came to challenge the original hypothesis because if adenosine formation occurred in the myocyte and the transporter remained situated at the myocyte membrane, how could inhibition of the efflux be reconciled with vasodilation? Until it was discovered in the 1980s that the endothelial layer largely determines adenosine's fate (see later discussion), the debate about the validity of Berne's hypothesis remained largely confined to the biochemical aspects of adenosine formation, and dipyridamole became almost buried as a tool to manipulate adenosine levels.

Receptor Agonists and Antagonists: A Breakthrough in our Knowledge

Burnstock's proposal of purinergic nerves [3] and subsequent classification of purinergic receptors in the 1970s initiated a second burst of studies aimed at gathering a more complete picture of adenosine's pharmacology. Adenosine as such, because of rapid and extensive participation in several metabolic pathways, turned out to be a poor tool for characterization of its receptors. Hence many analogs were synthesized that were less prone to metabolism and exhibited more appropriate binding properties. While their availability was thought to be of great help, scientifically, the hope was to get compounds sufficiently potent and specific for one or another receptor in a particular organ, so that valuable therapeutic agents could be developed. The first goal has indeed been achieved, and the use of agonists and antagonists, covering a whole spectrum of potencies and affinities for various receptors, has contributed, and still does contribute, to fill up a broad spectrum of activities of adenosine [4].

An important spin-off is the growing awareness that adenosine is more than a coronary vasodilator in that it may limit the release of and excessive stimulation by exitatory neurotransmitters; it may interfere with energy substrate utilization (antilipolytic; enhancement of glucose utilization); it may prevent platelet as well as leukocyte activation; and it may preserve ion homeostasis (explaining antiarrhythmic activity) [5,6]. An equally impressive list of activities of adenosine on the central nervous system is known, most of which are aimed at suppressing energy demand and/or improving energy supply, exactly as holds true for the myocardium. The rapid restoration of the supply/demand balance may well be a prerequisite for improved functional recovery from ischemia (adenosine as a "retaliatory metabolite" [7]).

Unfortunately, the existing adenosine agonists have not been successful thus far in therapy [8], and several pharmaceutical companies have even terminated research projects on them [9]. Failure is partly due to unfavorable pharmacokinetics, but the major problem remains selectivity in hitting a specific target. Indeed, despite fairly high specificity having been achieved for receptors in vitro, discrimination in vivo is far less apparent [10]. The same, of course, applies to exogenously administered adenosine, with which continuous infusion of high doses is required to compensate for the very short half-life. In this respect, the marked cardioprotection by adenosine, as documented in a series of in vivo experiments in dogs [11], indicates the validity of the concept of (cardio)protection by adenosine but may well be restricted to the canine species, in which adenosine removal in the systemic circulation proceeds much more slowly than in human blood.

Back to Nucleoside Transport Inhibition and Inhibitors

ROLE OF THE TRANSPORTER

In the mid-1980s, attention was focused on the endothelial cell as a major partner in handling adenosine [12,13]. This cell was found to have a dual role. Being well equipped intracellularly with the enzymes catabolizing adenosine, it may be responsible, to a major extent, for the breakdown of the nucleoside locally at its site of formation (the ischemic area). Of utmost importance for this function is the nucleoside transporter, which within the myocardium is located almost exclusively at the surface of the endothelial cells [14], and facilitates diffusion according to the gradient. The second role of endothelial cells is their function as a physical barrier, as a continuous layer around the (micro) vessels that prevents passage of many products between the interstitial space and the vasculature. This barrier is bypassed *for the nucleosides* by the transporter, which thus enhances the efflux (= washout) via the vessels draining the (ischemic) area. The striking biochemical effects from inhibition of the nucleoside transporter fit well with this concept.

As we have shown in isolated cat and rabbit hearts (where any interference with the removal of nucleosides through red blood cells is excluded), nucleoside transport inhibition markedly prevents tissue catabolism of adenosine *during ischemia* and delays nucleoside efflux *upon reperfusion* [15]. The net result is a prolonged local accumulation of adenosine within the interstitial space of the ischemic and the reperfused area of the myocardium. (It thus no longer matters whether adenosine is produced inside the myocyte or from nucleotides present on the outside and dephosphorylated by the ectonucleotidase, which is abundantly available on the surface of the microvessels.) This adenosine can then trigger its many receptors located in the vasculature (vasodilation), fat cells (inhibition lipolysis), or nerve endings (antiadrenergic effects). In addition, and paradoxically [6], more unchanged adenosine is released into the vessels draining the ischemic/reperfused area to prevent platelet aggregation (antithrombotic effect) and leukocyte activation (tissue damage). The biochemical effect of nucleoside transport inhibition is not only documented in cat and rabbit myocardium but has also been reported in guinea-pig and canine hearts, and more recently in isolated human myocardium [16]. Rat hearts are the exception because of the well-known insensitivity of the rat nucleoside transporter to existing drugs.

TRANSPORT INHIBITORS

Concomitant with the progress in analytical techniques enabling detailed assessment of the biochemical effect from nucleoside transport inhibition and a better understanding of how these marked changes are achieved came the availability of more efficacious drugs. For a long time, dipyridamole and, to a much lesser extent, dilazep, have been the only choice. Both, however, have several drawbacks.

The therapeutic use of dipyridamole as an antithrombotic has been ascribed to inhibition of phosphodiesterase(s) [17] so that nucleoside transport inhibition was almost forgotten for a few decades. Revival of interest in dipyridamole as a nucleoside transport inhibitor came with its application in pharmacologic stress testing to screen for myocardial perfusion defects in patients. Is it not remarkable that nucleoside transport inhibition has been neglected in the effect of dipyridamole on platelets (where dilazep is equally effective but has no effect on phosphodiesterase) although phosphodiesterase inhibition (and stimulation of prostacyclin production) may definitely contribute to dipyridamole's "steal" (especially at the high doses used in stress testing), but nucleoside transport inhibition is regarded as the *sole* culprit [18]? Unfavorable pharmacokinetics, such as a high affinity for plasma proteins (heavily competing for the effect on the transporter) and poor, unpredictable oral bioavailability (almost excluding the achievement of statistical significance for efficacy in large trials using oral treatment) may explain the general failure of dipyridamole as a therapy for cardioprotection. The major problem with dilazep, which is a potent and very specific nucleoside transport inhibitor, for therapy is its very short duration of action and almost absent oral activity, both of which are attributable to its being a diester.

As discussed in more detail elsewhere, with the development of draflazine, the (−)-enantiomer of R075231 (= racemate), we now possess a much more appealing drug. It has a very strong affinity for the transporter, is barely hindered by the presence of plasma proteins (studies in human erythrocyte), has a long duration of action (as evidenced from ex vivo transport inhibition in circulating erythrocytes), has a very high specificity (no known interaction with a multitude of receptors or enzymes; no effect in an array of pharmacologic tests in vivo [19]), and has excellent oral bioavailability (in animals and humans) [6].

TRANSPORT INHIBITION FOR THERAPY

The hope of finding useful therapeutics within the adenosine agonists and the shortcomings of the existing nucleoside transport inhibitors may explain the paucity in the 1980s of studies on cardioprotection by nucleoside transport inhibition. More recently, however, by using draflazine or its racemate, several experimental studies have evidenced major benefit in isolated as well as in in situ hearts under various conditions of ischemia and reperfusion and, importantly, in many species (Table 41-1).

At first glance efficacy of one drug in all these various experimental conditions may look like too much of a good thing. However, it should be kept in mind that it is not the drug itself that

TABLE 41-1. Experimental evidence for cardioprotection: benefit from nucleoside transport inhibition

Species	Benefit
Dog	Heart transplantation
	Normothermic global ischemia, in situ
	Stunning
Pig	Ventricular fibrillation and death during LAD occlusion
	Stunning during repetitive ischemia
Cat	Postischemic recovery function (isolated)
Rabbit	Normothermic global ischemia (isolated)
	Preconditioning
	Catecholamine cardiotoxicity
Guinea pig	Adenosine-sensitive tachycardia (isolated) [20]
Sheep	Cardiovascular function and survival after severe asphyxia in fetal lambs [21]

LAD, left anterior descending.
Most studies are referenced in Van Belle [6].

is the active principle. Benefit, indeed, stems from the many interesting properties of accumulating adenosine, acting on different targets *in concert* and *in time*, that is, locally, when and where needed. In this respect, the marked reduction in mortality from catecholamine cardiotoxicity [19] is illustrative. This model has been known for decades, is simple, and may be relevant for human pathology. Nevertheless, no single drug has been found that offers protection, most probably because other drugs may handle one single problem arising during ischemia and reperfusion, whereas adenosine may handle all of them. It is also evident that if the degree of transport inhibition can be adjusted to yield *local* accumulation of adenosine without raising the concentration of adenosine in the circulation, the antithrombotic effect [22], for example, may be expressed in the ischemic/reperfused area without hampering platelet (and leukocyte) functions in the periphery.

In view of the remarkable prevention of catecholamine cardiotoxicity, draflazine may well delay progressive deterioration of the failing heart if it is caused by excessive neurohormonal stimulation [23]. Other clinical applications may include heart preservation and subsequent transplantation, angioplasty, and thrombolysis, in which the *local*, adenosine-mediated, antithrombotic effect in addition to that of aspirin [6] may attenuate the low-flow phenomenon. In view of the known modulatory effect of endogenous adenosine on renin release [24], nucleoside transport inhibition can be expected to limit reperfusion injury and to afford protection much like the angiotensin-converting enzyme inhibitors. A dual effect may be envisaged on atrial fibrillation: The antiarrhythmic effect may shorten the fibrillatory episode [20], and if stroke is a consequence brain protection may be equally achieved [25]. Assessment of potential benefit from draflazine in cardiac surgery is the main goal of ongoing phase II trials.

Other Interventions

A few other possibilities for intervening with the adenosinergic system should be briefly discussed [26]. The most straightforward procedure to raise adenosine is to bring on ischemia. The success of preconditioning in reducing myocardial infarcts is most probably based on this principle, but its applicability for human therapy is strongly limited because of the problem of getting preconditioned *in time*. Inhibition of adenosine deaminase, which is the second step in adenosine catabolism (the first being the transport into cells containing the enzyme), has hardly resulted in any benefit for the ischemic myocardium thus far, except when combined with nucleoside transport inhibition. Failure may be due to rapid removal of adenosine during reperfusion (with no hindrance of washout, in contrast to inhibition of the transporter). Apart from their toxicity, doses of existing drugs such as EHNA or deoxycoformycin, which may block all deaminase activity, may yield increases in systemic adenosine and result in unwanted effects (as mentioned for the adenosine agonists).

AICA-riboside, or acadesine, is a specific case of an "adenosinergic" drug. While originally it was hoped to act as a purine precursor to replenish myocardial ATP (lost during ischemia and reperfusion), it did not do so in vivo, at least not in the myocardium, although the hyperuricemica seen when high doses are infused for long periods in humans may be indicative of purine formation elsewhere (in the liver?). In one study and in one species (the dog), infusion of acadesine was shown to raise adenosine in blood from the ischemic area, concomitant with reduced arrhythmias, reduced granulocyte accumulation, and a better endocardial flow to the ischemic myocardium (at 5 minutes of ischemia). No data indicating improved functional recovery were shown in this study. Further attempts to substantiate the claim of raising adenosine in myocardium from other species have failed. Nevertheless, clinical trials in humans have been initiated in the setting of coronary bypass surgery. Because the results of the first two trials were rather equivocal, a third study is ongoing.

It is still too early to judge the therapeutic potential of adenosine binding enhancement (prototype drug: PD 81723). Exploitation of endogenous adenosine is attractive, much like nucleoside transport inhibition. The enhancement seems to be confined to A_1 receptors, which limits the full expression of adenosine's beneficial effects.

In conclusion, the possibility of useful therapy for the ischemic and reperfused myocardium (and brain) based on intervention with the "adenosinergic" system resides entirely with nucleoside transport inhibition. If this approach fails, one may question the real benefit of adenosine as a "natural" defense system.

References

1. Drury AN, Szent-Györgi A (1929) The physiological activity of adenine compounds with special reference to their action upon the mammalian heart. J Physiol (Lond) 68:213–237.
2. Koss FW, Beisenherz G, Maerkisch R (1962) Die Eliminierung von Adenosin aus dem Blut under dem Einfluss von Dipyridamole und Papaverin. Arzneimittelforschung 12:1130–1131.
3. Burnstock G (1972) Purinergic nerves. Pharmacol Rev 24:509–581.
4. Collis MG, Hourani SMO (1993) Adenosine receptor subtypes. Trends Pharmacol Sci 14: 360–366.
5. Ely SW, Berne RM (1992) Protective effects of adenosine in myocardial ischemia. Circulation 85:893–904.
6. Van Belle H (1993) Specific metabolically active antiischemic agents: Adenosine and nuleoside transport inhibitors. In Singh B, Dzau V, Vanhoutte P, Woosley R (eds.), Cardiovascular Pharmacology and Therapeutics. New York: Churchill Livingstone, pp. 217–235.
7. Newby AC (1984) Adenosine and the concept of "retaliatory metabolites." Trends Biochem Sci 9:42–44.
8. Jacobson KA, Van Galen PJM, Williams M (1992) Adenosine receptors: Pharmacology, structure-activity relationships and therapeutic potential. J Med Chem 35:407–422.
9. Deckert J, Gleiter CH (1990) Adenosinergic psychopharmaceuticals: Just another cup of coffee? J Psychopharmacol 4:183–187.
10. Gerencer RZ, Finegan BA, Clanachan AS (1992) Cardiovascular selectivity of adenosine receptor agonists in anaesthetized dogs. Br J Pharmacol 107:1048–1056.
11. Forman MB, Velasco CE, Jackson EK (1993) Adenosine attenuates reperfusion injury following regional myocardial ischemia. Cardiovasc Res 27:9–17.
12. Nees S (1989) The adenosine hypothesis of metabolic regulation of coronary flow in the light of newly recognized properties of the coronary endothelium. Z Kardiol 78(Suppl 6): 42–49.
13. Bassingthwaighte JB, Wang CY, Chan IS (1989) Blood-tissue exchange via transport and transformation by capillary endothelial cells. Circ Res 65:997–1020.
14. Parkinson FE, Clanachan AS (1991) Adenosine

receptors and nucleoside transport sites in cardiac cells. Br J Pharmacol 104:399–405.
15. Van Belle H (1993) Nucleoside transport inhibition: A therapeutic approach to cardioprotection via adenosine? Cardiovasc Res 27:68–76.
16. Masuda M, Sukehiro S, Möllhoff T, Van Belle H, Flameng W (1993) Effect of nucleoside transport inhibition on adenosine and hypoxanthine accumulation in the ischemic human myocardium. Arch Int Pharmacol Pharmadyn 322:45–54.
17. FitzGerald GA (1987) Dipyridamole. N Engl J Med 316:1247–1257.
18. Van Belle H (1993) Nucleoside transport inhibition for cardioprotection: New perspectives for the clinic. Drug Dev Res 28:344–348.
19. Van Belle H, Verheyen W, Ver Donck K, Janssen PAJ, Robertson JIS (1992) Prevention of catecholamine-induced cardiac damage and death with a nucleoside transport inhibitor. J Cardiovasc Pharmacol 20:173–178.
20. Dennis D, Martens J, Bellardinelli L (1994) Frequency-dependent negative dromotropic effect of adenosine metabolism inhibitors in guinea pig heart (abstr). Drug Dev Res 31:263.
21. De Haan HH, De Haan J, Van Reempts JLH, Van Belle H, Hasaart THM (1993) The effect of adenosine transport inhibition on cardiovascular function and survival after severe asphyxia in fetal lambs. Pediatric Res 33:185–189.
22. Beukers MW, Kerkhof CJM, Yzerman AP, Soudijn W (1994) Nucleoside transport inhibition and platelet aggregation in human blood: R75381 and its enantiomers, draflazine and R88016. Eur J Pharmacol 266:57–62.
23. Packer M (1992) The neurohormonal hypothesis: A theory to explain the mechanism of disease progression in heart failure. J Am Coll Cardiol 20:248–254.
24. Brown NJ, Porter J, Ryder D, Branch RA (1991) Caffeine potentiates the renin response to diazoxide in man. Evidence for a regulatory role of endogenous adenosine. J Pharmacol Expt Ther 256:56–61.
25. Rudolphi KA, Schubert P, Parkinson FE, Fredholm BB (1992) Adenosine and brain ischemia. Cerebrovasc Brain Metab Rev 4: 346–369.
26. Van Belle H (1993) Adenosine promotors: An overview of existing strategies. Curr Opin Invest Drugs 2:1191–1199.

42. ENDOTHELIAL ADENOSINE RECEPTORS MODULATE LEUKOCYTE ADHESION: RELEVANCE FOR MYOCARDIAL REPERFUSION DAMAGE

Bernhard F. Becker
Stefan Zahler
Gert Richardt

The contribution of polymorphonuclear neutrophilic granulocytes (PMN) to acute postischemic reperfusion damage in the heart is a controversial issue but, due to their being equipped with proteolytic enzymes and potent mechanisms for generating reactive oxygen species, these cells certainly have the potential to rapidly cause functional derangement. In experiments conducted on isolated perfused hearts of the guinea pig, PMN of both guinea pig or human origin were able to significantly suppress the pumping ability after ischemia by 50% when applied in small numbers (2–40 million cells) [1,2]. Antioxidants protected from this cardiodepressant action, and the PMN effect was dependent on prior myocardial ischemia [1].

In continuing with this work, the possible involvement of adenosine released in the coronary effluent of the partially ischemic tissue at levels of 0.1–0.2 μM attracted our attention [3]. The ability of adenosine to suppress radical production and adhesion of PMN to the vascular endothelium via leukocyte A_2 receptors is well documented [4,5]. Based on this familiar concept, we expected to find myocardial damage by PMN aggravated if endogenously formed adenosine was removed or inhibited. To this end, experiments were performed in the presence of adenosine deaminase or theophylline. Surprisingly, both interventions completely abolished the action of PMN in the working-heart model. Moreover, the A_1-receptor antagonist DPCPX (1,3-dipropyl-8-cyclopentylxanthine, 0.1 μM) was equally cardioprotective [3]. A possible explanation for this rather consternating result seemed to be proffered by a report of Cronstein et al. that the chemotaxis and adhesivity of human PMN were enhanced via A_1-receptor stimulation of these cells [5]. However, we could find no influence of adenosine at submicromolar levels on chemotaxis induced by the peptide stimulus N-formyl-L-methionyl-L-leucyl-L-phenylalanine (FMLP) or on aggregation and the FMLP-induced oxidative burst reaction of human and guinea-pig PMN [6].

There is growing awareness that cell adhesion

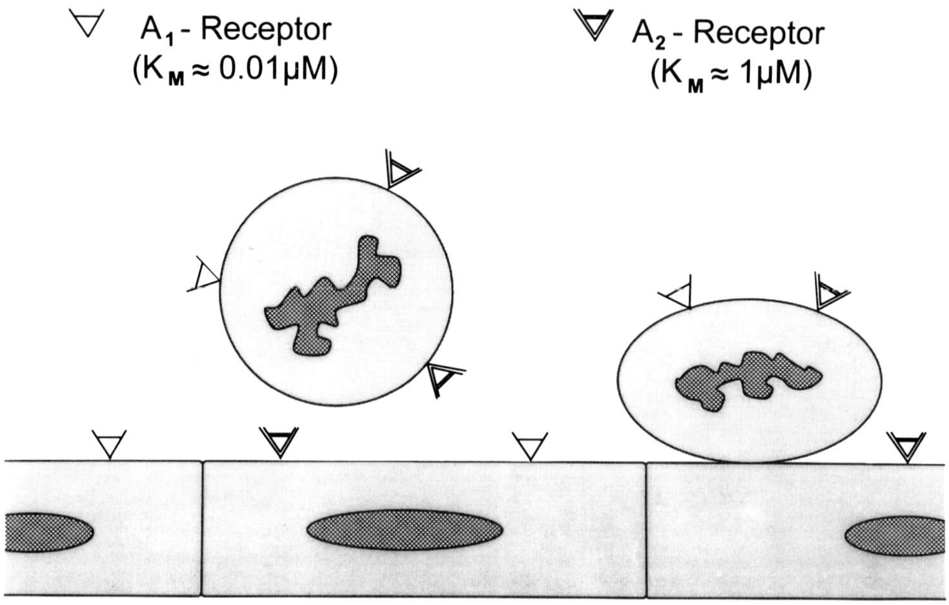

FIGURE 42-1. Scheme of the alternative sites and modes of action of adenosine with respect to the interaction of granulocytes with endothelial cells.

to the endothelium is an absolute requirement in the sequence of events leading to tissue damage by granulocytes [7]. Whether adenosine could perhaps influence PMN endothelial interactions by way of receptors localized on the endothelial cells has, to our knowledge, not been seriously considered. This is surprising because not only PMN but also coronary endothelial cells, at least in the guinea pig, possess both A_1 and A_2 receptors [8]. A quantitative assay for determining PMN adhesion in the intact coronary system was thus developed, with the specific aim of testing for receptor- and site-specific effects of adenosine (Fig. 42-1). Furthermore, it was deemed essential to establish concentration-response relationships, beginning at the level of about 0.1 µM. As indicated in Figure 42-1, A_1 receptor–mediated actions commence at submicromolar adenosine concentrations. However, effects evolving from activation of A_2 receptors should become increasingly manifest at micromolar levels of adenosine, as A_2 receptors have lower affinity for the physiologic agonist [9]. In addition, 0.1 µM adenosine was encountered in our postischemic heart model. Submicromolar levels also pertain to the human heart, in which coronary sinus adenosine increases from about 0.04–0.08 µM to 0.3–0.8 µM after balloon dilatation of increasing duration of occlusion [2,10]. Since these latter values do not relate to acute reperfusion after more prolonged ischemia, new data were sought for patients undergoing coronary reopening after myocardial infarction.

Methods

ISOLATED HEART PREPARATION

Hearts of male guinea pigs (approximately 250 g body weight) were isolated and retrogradely perfused with modified Krebs–Henseleit buffer at constant flow of 5 ml/min as nonworking preparation [3,6]. The upper and lower caval veins and the azygos vein were ligated, with coronary venous effluent appearing quantitatively through the cannulated pulmonary artery. Coronary perfusion pressure was monitored with a Statham P23Db force transducer.

PREPARATION OF PMN

Granulocytes were freshly prepared from arterial blood of guinea pigs or venous blood obtained from healthy human volunteers, with EDTA (0.1%) serving as an anticoagulant. The blood was immediately centrifuged (15 minutes at 350 × g), and the platelet-rich plasma was discarded. The cellular portion was sedimented in a 6% solution of hydroxyethyl starch (75 minutes), and the supernatant was then centrifuged at 350 × g for 10 minutes. The pellet was next resuspended in phosphate-buffered saline containing 0.1% EDTA, layered over Percoll, and sedimented at 400 × g for 25 minutes. After lysis of the remaining red blood cells, the PMN were washed, resuspended in Tris-buffered Tyrode's solution, and the cell count was adjusted to 600,000/ml (Coulter Counter ZM, Coulter Electronics, UK). As shown previously, cells prepared in this manner gave strong chemotactic, oxidative-burst, and aggregatory responses when stimulated with FMLP [1,2,6].

ADHESION EXPERIMENTS

A standardized bolus of 600,000 PMN in 1 ml of Tyrode's solution was infused steadily over 1 minute into the intact coronary system of isolated guinea pig hearts. Cells emerging with the coronary effluent were collected and counted. Control experiments revealed that a sampling period of 100 seconds, commencing with the onset of cell infusion, sufficed to accumulate those cells not interacting with the endothelium (>98% of all cells leaving the heart within about 5 minutes appeared within these 100 seconds). The numerical difference between the applied cells and those recovered in the effluent then yielded the number of cells retained in the heart, that is, the percentage of cells undergoing at least primary adhesion [6,11]. To localize adherent cells in the tissue, some hearts were immersion-fixed in formalin after the 100-second washout period and then stained for leukocyte esterase (Naphthol AS-D chloroacetate kit; Sigma, St Louis, MO). The red PMN could then be easily visualized by light microscopy.

ADENOSINE MEASUREMENTS IN CORONARY VENOUS AND ARTERIAL HUMAN BLOOD

In a preliminary study approved by the university ethics committee, five patients with acute anterior myocardial infarction undergoing coronary reopening with the aid of percutaneous transluminal coronary angioplasty (PTCA) catheters consented to having blood samples taken prior to, immediately after, and 5–10 minutes after reperfusion. At these time points, paired samples of whole blood were taken from the aorta and the coronary sinus. Blood (5 ml) was drawn through appropriately positioned catheters into an equal volume of ice-cold isotonic saline without any stopping solution or additives. After mixing, the samples were immediately added to 1 ml of cold perchloric acid (3 M), mixed thoroughly, and kept on ice until further processing. Two 1-ml aliquots were placed in Eppendorf cups, centrifuged, and 0.5-ml portions of the clear supernatant were stored at −30°C to await chromatography. As described previously [12], adenine nucleotides were removed by high-pressure chromatography (HPLC) over an NH_2-nucleosil column (Macherey & Nagel, Düren, FRG) before the purine nucleoside and base fraction were applied to a C18-nucleosil column. Using a gradient of 5 mM $NH_4K_2PO_4$ (pH 5.0) against 60% methanol in water (0–40% in 20 minutes), adenosine was prefractionated (retention time approximately 18 minutes at 1 ml/min), shifted to inosine with adenosine deaminase, and rechromatographed under the above-mentioned conditions. Quantification occurred via UV absorption (254 nm) and comparison with inosine standards (retention time approximately 11 minutes). Corrections were made for all dilutions of blood.

Results and Discussion

ADHESION EXPERIMENTS

Under control conditions (perfusion with Krebs–Henseleit buffer at physiologic shear stress) about 20% of the applied PMN did not reemerge from the coronary system within 100 seconds, irrespective of whether guinea pig or human cells were introduced. As indicated in Table 42-1, preperfusion of the hearts with 0.1 μM adenosine for 5 or 30 minutes increased the percentage of primary adherent cells significantly to about 30%, with identical values being found for both times. It is pertinent that this adenosine effect could be blocked by coinfusion of the specific A_1 receptor antagonist DPCPX and mimicked by the A_1 receptor agonist CPA (Table 42-1). Thus, adenosine-based enhancement of adhesion of PMN in the guinea pig coronary system obviously occurred via an A_1 receptor–mediated action. Furthermore, the onset of action was fast and was not subject to rapid tachyphylaxis. Histologic examination of adenosine-stimulated hearts revealed PMN adherent to the intraluminal surface of small vessels bearing characteristics of venules and small veins (not shown).

Investigation of the concentration dependency of the adhesive response to adenosine gave a surprising result (Fig. 42-2). Adhesion attained a plateau, after increasing from the control level without adenosine, at adenosine concentrations in the submicromolar range of 0.1–0.3 μM. Instead of climbing further, at 1 μM adenosine

TABLE 42-1. Adhesion of guinea pig and human PMN in the coronary system of isolated guinea pig hearts

Heart	PMN	Adhesion (% of 600,000 cells)
Control	g.p.	22 ± 2.0
ADO (0.1 μM, 5 min)	g.p.	30 ± 1.4[a]
ADO (0.1 μM, 30 min)	g.p.	30 ± 7[a]
ADO (0.1 μM, 5 min) + DPCPX (0.1 μM)	g.p.	22 ± 1.8
CPA (0.01 μM, 5 min)	g.p.	32 ± 1.8[a]
ADO (0.1 μM, 5 min + 90 sec washout)	g.p.	35 ± 3.6[a]
ADO (0.1 μM, 5 min)	g.p. (ADO 1 μM)	32 ± 2.0[a]
Postischemic (1 min after 15 min ischemia)	g.p.	35 ± 3.0[a]
Postischemic + DPCPX (0.1 μM)	g.p.	23 ± 1.0
Thrombin (0.3 U/ml, 5 min)	g.p.	34 ± 2.4[a]
Iloprost (0.01 μM, 5 min)	g.p.	21 ±2.3
Control	Human	21 ± 2.9
ADO (0.1 μM, 5 min)	Human	32 ± 2.3[a]
WEB 2086 (100 μM)	Human	19 ± 1.8
ADO (0.1 μM, 5 min) + WEB 2086(100 μM)	Human	24 ± 4.2
Thrombin (0.3 U/ml, 5 min)	Human	31 ± 1.4[a]

Values represent the percent of cells retained in the coronary system 100 seconds after application of a bolus of 600,000 PMN at an effective flow rate of 6 ml/min and washed out at 5 ml/min. Means (± SEM) of five experiments each, except for the g.p. controls (n = 10) are given.

[a] $p < 0.05$ versus control (Student–Newman–Keul's test).

ADO = adenosine; DPCPX = dipropylcyclopentylxanthine; CPA = cyclopentyladenosine; WEB 2086 is a PAF antagonist (Boehringer Ingelheim, FRG); g.p. = guinea pig.

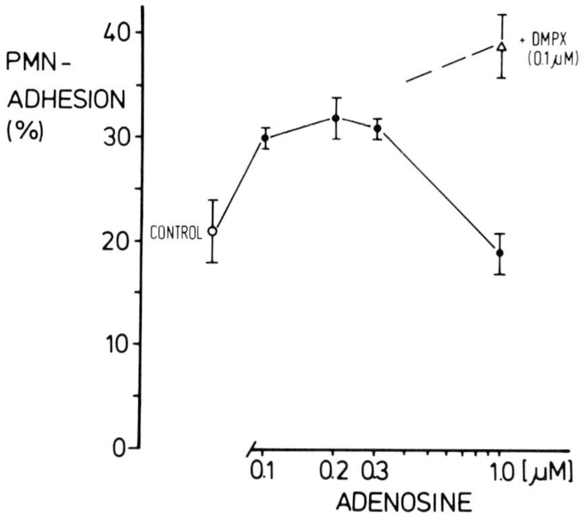

FIGURE 42-2. Concentration-response curve for the effect of adenosine on adhesion of polymorphonuclear neutrophilic leukocytes (PMN) of the guinea pig in the intact coronary system of isolated guinea pig hearts. Adhesion is expressed as the percent of cells retained of an applied bolus of 600,000 PMN (mean values ± SEM of five experiments at each concentration). DMPX = experiments conducted in the presence of the A_2 receptor antagonist dimethylpropargylxanthine.

adherence had returned to the control level. However, as illustrated in Figure 42-2, in the presence of the A_2 receptor antagonist DMPX (3,7-dimethyl-1-propargylxanthine) 1 μM adenosine approximately doubled the degree of adherence of guinea pig PMN. The biphasic response curve can therefore be readily explained on the basis of masking of the proadhesive A_1 response by an opposing A_2 receptor–mediated inhibitory action of adenosine. In accordance with the known lower affinity of the A_2 receptor toward adenosine (Fig. 42-1), such an inhibitory action would become increasingly manifest with adenosine rising into the micromolar concentration range.

In experiments designed to establish the cell-specific localization of the A_1 receptors responsible for enhancing PMN adhesion to the coronary endothelium, adenosine infusion (0.1 μM, 5 minutes) was terminated 90 seconds before PMN application. Separate studies with radiolabeled adenosine had already demonstrated that such a washout interval suffices to lower the intracoronary level of the infused adenosine to about 0.1% [11]. From the data in Table 42-1 it is clear that adhesion of PMN was just as high as if adenosine had still been present during PMN application. In other words, because only the endothelial cells had experienced elevated adenosine levels, the A_1 receptors enhancing adhesion must be localized on these cells and not on the PMN. An endothelial "memory" for A_1 stimulation must also be construed.

As a kind of "reversal" of this setup, PMN were preincubated with high adenosine (1 μM) for 15 minutes, before application to hearts that, in turn, had been preexposed to low adenosine (0.1 μM) for 5 minutes (the concentration of adenosine rose to a calculated 0.27 μM during the bolus infusion). Despite the prestimulation of inhibitory A_2 receptors of the PMN, PMN adhesion remained elevated versus control (Table 42-1). This result befits more the usual effect of low concentrations at the endothelial A_1 sites (Fig. 42-2). Thus, in the present model, inhibition of PMN adhesion via the A_2 mechanism seems to depend on exposure of the endothelial cells to micromolar adenosine.

Whereas the familiar proadhesive agent thrombin also increased retention of guinea pig and human PMN in the heart, the prostacyclin analogue iloprost, an adenylyl cyclase activator like micromolar adenosine, did not (Table 42-1). Interestingly, 0.01 μM iloprost and 0.1 μM adenosine led to identical reductions in coronary perfusion pressure of the hearts (Fig. 42-3), indicative of equal coronary vascular dilatation and, hence, of similar increases in perfused surface area as well as reductions in shear stress. This result illustrates that the dilatory effect of adenosine cannot per se be made responsible for the increases in adhesion. This conclusion is further supported by the behavior of CPA: Adhesion was enhanced (Table 42-1), while perfusion pressure remained practically unaltered (Fig. 42-3).

Table 42-1 also presents data relating to the action of endogenous adenosine on PMN adhesion. Guinea pig PMN applied to hearts in the second minute of reperfusion following a 15-minute period of global ischemia displayed markedly enhanced adhesion, an effect that was fully antagonized by supplementing the perfusate with DPCPX (Table 42-1). The levels of adenosine measured in the coronary effluent increased from preischemic values of 4–6 nM to about 0.8 μM during the first minutes of reperfusion, fell to 0.4 μM at the time of PMN application, and then declined rapidly to baseline within the next 10 minutes. Interventions such as preconditioning have since been found to elevate postischemic adenosine levels into the micromolar range in these heart preparations, with concomitant inhibition of postischemic adhesion [11]. Appropriately, for individual hearts the degree of adhesion was negatively correlated with the coronary venous adenosine concentration prevailing immediately after ischemia. This phenomenon is highly reminiscent of the biphasic concentration-response relationship described earlier.

With respect to the nature of the membrane adhesion molecule(s) possibly involved in the adenosine-stimulated interaction between endothelium and PMN, we have very little information to date. In the case of thrombin, acute effects are known to be mediated, at least in part, via rapid externalization of the adhesion molecule P selectin by the endothelium [13]. Because thrombin and adenosine initiated quantitatively similar increases in adhesion, both for guinea pig and human PMN (Table 42-1), P selectin may be a candidate. However, there are certain doubts as to whether human PMN can at all interact with guinea pig P selectin [14]. On the other hand, thrombin can also generate platelet activating factor (PAF) in endothelial cells. This molecule remains largely associated with the cell membrane and, in turn, enhances PMN adhesion [13]. In experiments conducted with human PMN we were able to demonstrate that adenosine-induced adherence was in fact attenuated by the PAF antagonist WEB 2086 (Table 42-1).

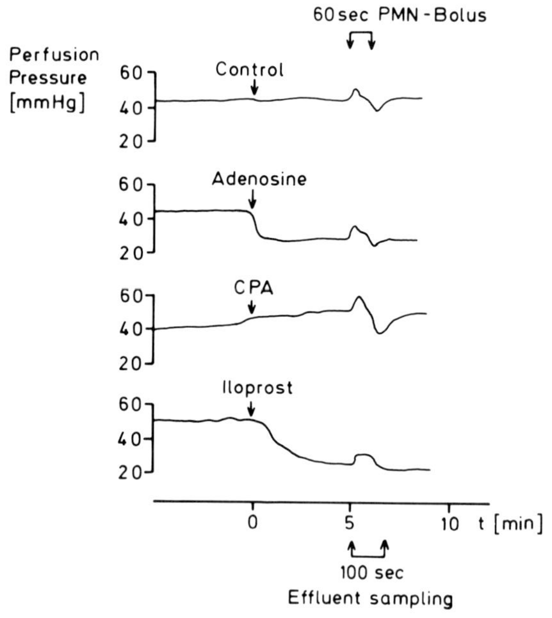

FIGURE 42-3. Coronary perfusion pressure of hearts perfused at constant flow (5 ml/min, except during the bolus when flow was 6 ml/min) in the absence and presence of adenosine (0.1 μM), the A_1 agonist cyclopentyladenosine (CPA, 0.01 μM), or iloprost (0.01 μM). The periods of application of PMN and of sampling of coronary effluent are indicated, and the autoregulatory responses of coronary tone in conjunction with the bolus infusion are apparent (exemplary tracings of five experiments each).

Finally, in preliminary experiments on cultured human umbilical vein endothelial cells (HUVEC), it was possible to show an enhancement of adhesion of human PMN by preincubation of the cultures with 0.1 μM adenosine, whilst 1 μM adenosine left adhesion at the control level. Despite this encouraging result, further studies with specific antagonists and agonists of adenosine and cells of other vascular beds will be required to substantiate the presence of A_1 and A_2 receptors on the endothelium of various organs in humans and to

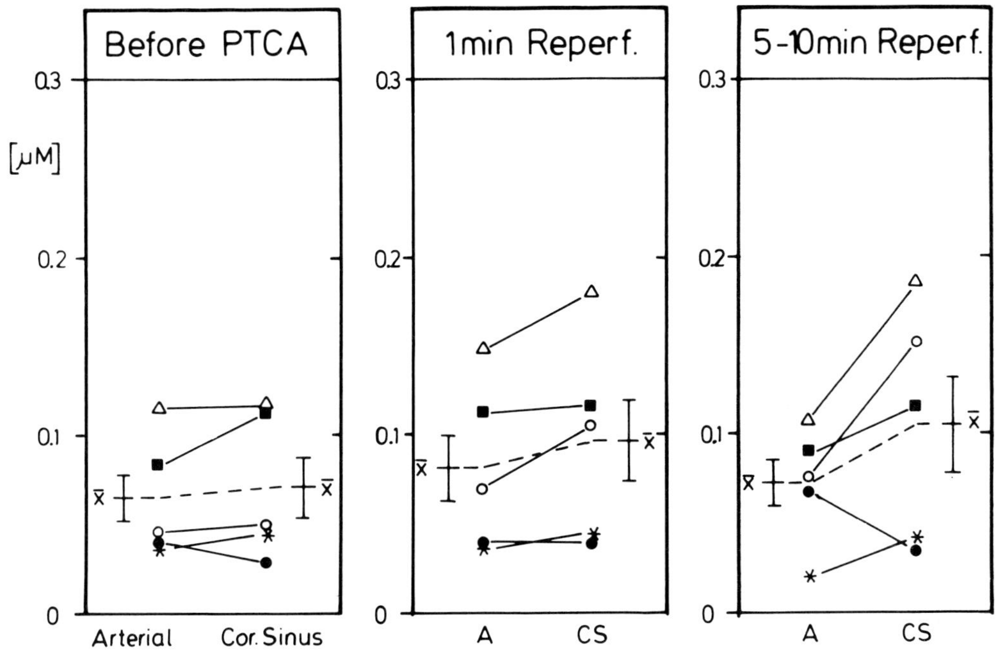

FIGURE 42-4. Coronary adenosine levels of five patients with fresh anterior myocardial infarction undergoing acute reperfusion (coronary reopening with PTCA catheter). A = arterial; CS = coronary sinus blood sample. Symbols represent individual patients; dashed lines connect the mean values; bars indicate SEM.

establish their action on leukocyte adhesion in the state of reperfusion.

CORONARY ADENOSINE IN HUMANS

Assuming the human coronary endothelium to react in the same manner toward adenosine as did guinea pig endothelial cells in situ and HUVEC in culture, a biphasic concentration-response profile of PMN adhesion is anticipated. The basal concentration of adenosine in coronary blood (arterial and venous) of resting human subjects not taking adenosine uptake blockers is on the order of 0.04–0.08 μM [10,12,15,16]. In hearts subjected to regional ischemia of 30–90 second duration (PTCA), coronary sinus levels of adenosine after reperfusion increase briefly up to maximum values of 0.4–0.8 μM [10]. From the available data enhancement of PMN adhesion is to be expected after PTCA.

Reperfusion damage of the heart, to which PMN may well contribute, is, however, more pronounced after periods of ischemia far longer than those encountered during balloon dilatation. Preliminary results in an attempt to determine how much coronary adenosine increases immediately after reperfusion of infarcted myocardium are depicted in Figure 42-4, which reports on five patients whose coronary occlusion was acutely reopened. Although the arterial and coronary sinus blood values showed considerable interindividual variability before reperfusion, the mean values were below 0.1 μM and there were practically no arteriovenous differences [12]. Blood drawn during the first minute of reperfusion revealed marked increases in coronary sinus adenosine in some instances but not in others, with the maximum increase rising to about 0.2 μM. (Retrospectively it was possible to link these results to the individual time lapse between the onset of symptoms and the catheter intervention and/or to the severity of infarction.) The arterial levels also tended to increase. The individual changes observed after 1 minute were generally maintained after 5–10 minutes of reperfusion (Fig. 42-4, right-hand side), with coronary AV differences becoming more pronounced in some instances. However, in no case did the measured content of adenosine in whole blood taken from the aorta or coronary sinus extend into the micromolar range.

On the basis of these results, endogenously formed adenosine must be expected to enhance PMN adhesion in reperfused human myocardium. Due to the large discrepancy in the concentration found (maximally 0.2 μM) and the level presumably required to mask the proadhesive effect (about 1 μM), this conclusion is not altered by considerations of possible adenosine compartmentalization between plasma and red blood cells (RBC) or transport into these cells. Should the RBC contain no adenosine at all, the plasma value would have been underestimated by about 60–80% (hematocrit 40–45%). Transport into the RBC (already preequilibrated with arterial adenosine) would make no substantial difference, because metabolism of adenosine to AMP or inosine is relatively slow and is the rate-limiting aspect of adenosine removal by RBC [17]. At normal coronary flow, an estimated maximum of 10 seconds elapse between the appearance of adenosine in the coronary microvascular lumen and entry of the blood samples into the ice-cold saline of the sampling syringe. The chosen method of sample processing—extraction of total nucleoside into perchloric acid—allows for determination of any sequestered adenosine.

One time-honored uncertainty remains linked with the present protocol, namely, the level of adenosine in vessels localized directly in the postischemic tissue. However, even if local adenosine is high, at some place in the course of washout intravascular adenosine must nevertheless fall into the proadhesive range.

Conclusions

According to numerous studies adenosine exhibits A_1 receptor–mediated cardioprotective effects not necessarily linked to the presence of PMN [18,19]. In view of the present results, however, intracoronary application of adenosine A_1 receptor agonists may augment the contribution of PMN to myocardial reperfusion damage, via stimulation of PMN adhesion to coronary endothelial cells. Conversely, A_2 receptor agonists could act on the endothelium as well as directly on blood cells to blunt the adhesion of PMN (and platelets!). With respect to endogenously generated and exogenously applied adenosine and considering the biphasic concentration-response relationship, stimulation of PMN adhesion must be expected unless micromolar levels of adenosine can be attained. Interventions designed to elevate blood and tissue levels of adenosine (transport blockers, methotrexate, acadesine, etc.) may fall short of this goal. The prognosis of the response in any given human individual seems difficult. Although application of adenosine A_1 blockers, but certainly not of A_2 receptor antagonists, would appear advantageous to mitigate enhancement of PMN adhesion by adenosine, the absolute benefit remains to be established.

Acknowledgments

These studies were supported in part by the Friedrich-Baur Foundation of the University of Munich. The expert technical assistance of Mr. W. Schrödl and Ms. D. Kiesl is gratefully acknowledged. Particular thanks are due to Dr. Götz Münch of the Cath Lab, Technical University of Munich, for aiding in the sampling of blood, and to the five patients who participated in the study.

References

1. Raschke P, Becker BF, Leipert B, Schwartz LM, Zahler S, Gerlach E (1993) Postischemic dysfunction of the heart induced by small numbers of neutrophils via formation of hypochlorous acid. Basic Res Cardiol 88:321–339.
2. Becker BF, Raschke P, Bardenheuer H, Gerlach E (1993) Granulocytes, reactive oxygen species

and uric acid: Their role in postischemic reperfusion damage of the heart. Pharm Pharmacol Lett 3:139–142.

3. Schwartz LM, Raschke P, Becker BF, Gerlach E (1993) Adenosine contributes to neutrophil-mediated loss of myocardial function in post-ischemic guinea pig hearts. J Mol Cell Cardiol 25:927–938.
4. Cronstein BN, Daguma L, Nichols D, Hutchison AJ, Williams M (1990) The adenosine/neutrophil paradox resolved: Human neutrophils possess both A_1 and A_2 receptors that promote chemotaxis and inhibit O_2^- generation, respectively. J Clin Invest 85:1150–1157.
5. Cronstein BN, Levin RI, Philips M, Hirschhorn R, Abramson SB, Weissmann G (1992) Neutrophil adherence to endothelium is enhanced via adenosine A_1 receptors and inhibited via adenosine A_2 receptors. J Immunol 148:2201–2206.
6. Becker BF, Zahler S, Raschke P, Schwartz LM, Beblo S (1992) Adenosine enhances neutrophil sticking in the coronary system: A novel mechanism contributing to cardiac reperfusion damage. Pharm Pharmacol Lett 2:8–11.
7. Ley K (1992) Leukocyte adhesion—molecular basis and physiological consequences. Clin Hemorheol 12:93–108.
8. Parkinson FE, Clanachan AS (1991) Adenosine receptors and nucleoside transport sites in cardiac cells. Br J Pharmacol 104:399–405.
9. Olsson RA, Pearson JD (1990) Cardiovascular purinoceptors. Physiol Rev 70:761–845.
10. Bardenheuer HJ, Fabry A, Höfling B, Peter K (1994) Adenosine: A sensitive marker of myocardial ischemia in man. Cardiovasc Res 28: 656–662.
11. Zahler S, Becker BF, Raschke P, Gerlach E (1994) Stimulation of endothelial adenosine A_1-receptors enhances adhesion of neutrophils in the intact guinea pig coronary system. Cardiovasc Res 28:1366–1372.
12. Becker BF, Leipert B, Raschke P, Gerlach E, Permanetter B (1991) Formation, release and scavenger function of uric acid derived from adenine nucleotides in heart and lung. In Imai S, Nakazawa M (eds.), Role of Adenosine and Adenine Nucleotides in the Biological System. Amsterdam: Elsevier, pp. 321–336.
13. Lorant DE, Patel KD, McIntyre TM, McEver RP, Prescott SM, Zimmerman GA (1991) Coexpression of GMP-140 and PAF by endothelium stimulated by histamine or thrombin: A juxtacrine system for adhesion and activation of neutrophils. J Cell Biol 115:223–234.
14. Zahler S, Becker BF, Raschke P, Beblo S, Gerlach E (1993) Thrombin-induced adhesion of human granulocytes to guinea pig coronary endothelium: Contribution of platelet activating factor (PAF) but not of P-selectin under conditions of shear stress. Pflügers Arch 422(Suppl 1):R111.
15. Utterback DB, Staples ED, White SE, Hill JA, Belardinelli L (1994) Basis for the selective reduction of pulmonary vascular resistance in humans during infusion of adenosine. J Appl Physiol 76:724–730.
16. Hamm CW, Kupper W, Bredehorst R, Hilz H, Bleifeld W (1988) Quantitation of coronary venous adenosine in patients: Limitations evaluated by radioimmunoassay. Cardiovasc Res 22:236–243.
17. Decking UKM, Alves C, Spahr R, Schrader J (1994) 2-Fluoroadenosine uptake by erythrocytes and endothelial cells studied by ^{19}F-NMR. Am J Physiol 266:H1596–H1603.
18. Thornton JD, Liu GS, Olsson RA, Downey JM (1992) Intravenous pretreatment with A_1-selective adenosine analogues protects the heart against infarction. Circulation 85:659–665.
19. Lasley RD, Mentzer RM Jr (1992) Adenosine improves recovery of postischemic myocardial function via an adenosine A_1 receptor mechanism. Am J Physiol 263:H1460–H1465.

B. NEUROPROTECTION

43. ADENOSINE AND BRAIN ISCHEMIA

Karl A. Rudolphi
Peter Schubert

Stroke is the second or third major cause of death in the "industrialized countries." About 30% of patients die within a year after the first insult, and 20% remain dependent on care. So far there is no effective therapy available but one is desperately needed. At present worldwide many pharmacotherapeutic approaches are under preclinical and clinical investigation in which the focus is on agents that counteract the deleterious effects of excessively released excitatory neurotransmitters such as glutamate. Recently the endogenous nucleoside adenosine or its reinforcement has been demonstrated to exert a variety of actions that counteract ischemia-induced brain damage. In this chapter we briefly review and discuss the experimental evidence for this hypothesis in the framework of the pathophysiology of brain ischemia. The corresponding literature has been extensively reviewed in a recent publication [1]. We therefore only include newer references.

Mechanisms Underlying the Neuroprotective Effects of Adenosine

The physiologic function of adenosine as a cell modulator is achieved by a mosaic of actions that influence membrane ion conductances and intracellular metabolic signaling in neurons as well as in glial cells. The same interconnected mechanisms that tend to maintain the physiologic "resting state" in a neuronal network also seems to be responsible for the neuroprotective effect of adenosine counteracting a pathologic "activation" of nerve and glial cells. The complex cascade of pathologic events induced by brain ischemia is shown in a rather simplified flow diagram in Figure 43-1, indicating the multiple sites at which adenosine may interfere with the pathogenic process leading to nerve cell death. As illustrated in the figure, the thromboembolic occlusion of an artery supplying the brain—in more than 90% of stroke cases it is the middle cerebral artery—triggers a cascade of hemodynamic, electrophysiologic, and neurochemical events, which include many interwoven vicious circles and finally result in excessive intracellular accumulation of free ionic calcium and in the neuronal necrosis [2,3]. In a stroke patient the final infarct size (tissue with irreversible neuronal necrosis) is determined by the degree and efficiency of collateralization. The densely ischemic core is surrounded by an expanding ring of tissue that does not receive enough oxygen and glucose to maintain electric activity but that is essentially salvagable, provided treatment starts early enough. This zone, called the penumbra, which has a cerebral blood flow (CBF) above the critical level for neuronal necrosis, is the target for any therapeutic inter-

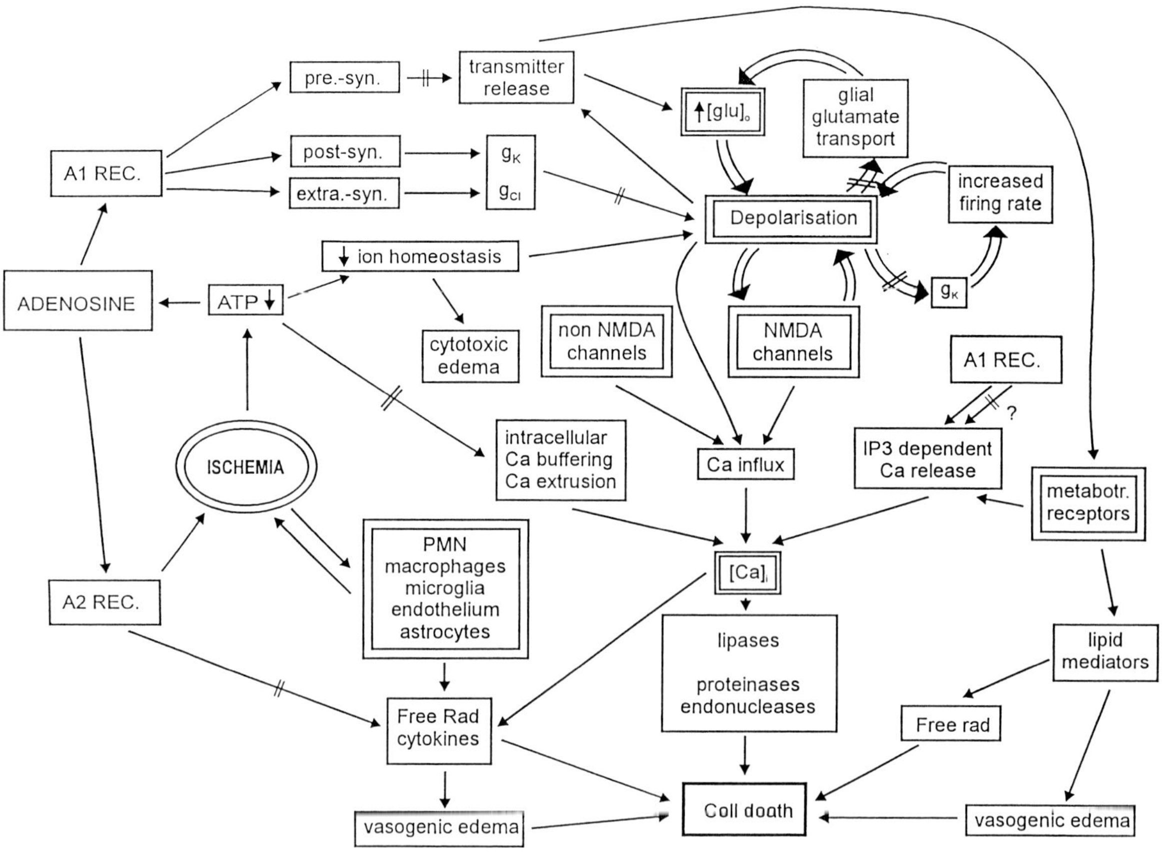

FIGURE 43-1. Schematic diagram of the pathophysiology of brain ischemia and the presumed neuroprotective effects of adenosine, (Modified from Rudolphi et al. [1], with permission).

vention. The decrease of CBF beyond a critical threshold leads to a rapid breakdown of oxidative energy metabolism of the nervous tissue, a fall in ATP and phosphocreatine content, and tissue acidosis. As a consequence of ATP depletion and a failure of ATP-dependent ion pumps, especially the Na/K-ATPase pump, there is a rapid derangement of ion homeostasis with an increase of intracellular Na^+, Cl^-, Ca^{2+}, and extracellular potassium. The influx of Na and Cl ions is accompanied by the intracellular accumulation of osmotically obligated water, not only in neurons but also in astrocytes leading to cytotoxic edema, which further compromises blood supply at the level of the microcirculation by swollen astrocytes arround the cerebral capillaries. This underlines the significance of ischemia-induced changes of astrocyte properties. The loss of ionic homeostasis leads to a sustained and strong membrane depolarization and a subsequent Ca^{2+} influx, which presynaptically triggers the release of neurotransmitters, including the potentially deleterious excitatory amino acids glutamate and aspartate. The following activation of ionotropic and metabotropic receptors leads to an opening of receptor-operated Ca^{2+} ion channels in the postsynaptic membrane and may also cause the release of Ca^{2+} from intracellular compartments via IP3. This, together with a disturbance of the Ca^{2+}

buffering mechanisms, seems to be responsible for the excessive and irreversible accumulation of free intracellular Ca^{2+} in the neuron. Metabotropic receptor stimulation also causes an activation of protein kinase C (PKC) and enhances the N-methyl-D-aspartate (NMDA)-type glutamate receptor–mediated Ca^{2+} influx by over-running the Mg^{2+} block of this channel. The reduction of the Mg^{2+}-dependent voltage sensitivity by a pathologic PKC activation during the reperfusion phase may further increase the deleterious NMDA Ca^{2+} influx. Ischemic depolarization also disturbs the Na^+-dependent electrogenic uptake mechanisms for glutamate in presynaptic nerve terminals and astrocytes, which may largely contribute to the massive and sustained increase of extracellular glutamate. Ca^{2+} accumulation is one of the key factors for irreversible brain damage because it activates lipid peroxidation, free radical generation, and proteolysis, and may stimulate pathologic gene expression. Secondary disturbances of blood flow in the ischemic brain area, such as breakdown of cerebral blood flow (CBF) autoregulation, exacerbation of ischemia by vasospasms, and compromised microcirculation (enhanced platelet aggregation and capillary plugging with activated neutrophils) contribute to tissue damage. They are triggered by the enhanced formation and release of vasoconstrictive, pro-aggregatory, and proinflammatory substances such as, for example, serotonin and endothelin, lipid peroxidation products (arachidonic acid, thromboxane, platelet activating factor, prostaglandins, leuktrienes), bradykinin, substance P, and cytokines. Many of these agents are also involved in the generation of vasogenic edema, which is one of the most serious complications in stroke patients and is very often responsible for a fatal outcome.

Cytokines may also be involved in the activation process of microglial cells during brain ischemia. These immunocompetent cells, when pathologically activated, seem to be an important source of free oxygen radicals and other potentially neurotoxic agents, which may play a pathogenic role in the development of neurodegenerative diseases including stroke [4].

How Much Endogenous Adenosine Limits Ischemic Brain Damage?

Several studies applying microdialysis or the cortical cup technique showed a rapid and dramatic increase of extracellular adenosine during ischemia. In the ischemic brain tissue adenosine is formed from AMP after ATP catabolism and transported into the extracellular space by specific transport systems [1]. Recently, the CBF threshold for the ischemia-induced increase of the extracellular adenosine and glutamate concentration was measured in cats during permanent middle cerebral artery occlusion, which is so far one of the clinically most relevant experimental models of stroke. Adenosine started to rise at a higher CBF threshold (25 ml/100 g/min) than glutamate (20 ml/100 g/min), possibly indicating an inherent inhibitory effect of endogenous adenosine on glutamate release. In contrast to the sustained glutamate increase, the rise of extracellular adenosine concentration was transient, reflecting the rapid exhaustion of its source (ATP). At later time points when the extracellular adenosine concentration had returned to nanomolar levels the CBF threshold for glutamate increase was markedly enhanced (to about 30 ml/100 g/min), again indicating that endogenous adenosine may temporarily inhibit glutamate release during ischemia [5].

Figure 43-1 shows schematically that there are several mechanisms for the neuroprotective actions of adenosine against ischemic brain damage. Presynaptically adenosine inhibits the release of glutamate. This seems to result from an A_1-receptor–mediated inhibition of the Ca^{2+} influx, a membrane hyperpotarization by opening K^+ ion channels and interference with the vesicular secretary apparatus itself. Indeed, the topical application of the A_1-selective adenosine agonist cyclopentyladenosine (CPA) and the mixed A_1/A_2 agonist N-ethylcarboxamidoadenosine in nanomolar concentrations significantly depressed ischemia-induced glutamate and aspartate release in a model of global temporary forebrain ischemia in rats [6]. In the same study, the A_2-specific agonist CGS 21680 significantly enhanced the release of glutamate and aspartate. The pathophysiologic relevance of this effect remains to be clarified.

Membrane hyperpolarization by adenosine via A_1 receptors may also counteract the release of glutamate from astrocytes by improving the function of the electrogenic glutamate uptake mechanism or by inhibiting its reversal. Postsynaptically, adenosine tonically depresses the neuronal Ca^{2+} influx by limiting the degree of the actvity-evoked membrane depolarization, thereby reducing the opening of voltage-dependent Ca^{2+} channels. By increasing the critical synaptic input frequency required to overcome the voltage-dependent Mg^{2+} block of the NMDA receptor ion channel complex adenosine prevents the NMDA Ca^{2+} influx at lower frequency synaptic activation. As a consequence, the threshold to initiate the ischemia-induced cascade of pathologic events is raised by this adenosine action exerted already at low physiologic concentrations. However, the NMDA Ca^{2+} influx could no longer be influenced at the later postischemic phase by even increased adenosine concentrations [7], possibly because the NMDA channel has lost its voltage sensitivity (due to pathologic PKC activation?). This points to a restriction of the initially powerful adenosine-mediated protection by a time window that presumably is not yet passed in the "penumbra" of a focal ischemic insult where the pathogenic process is just beginning.

The primary adenosine action that limits an excessive membrane depolarization is achieved by presynaptic, postsynaptic, and nonsynaptic enhancement of K^+ and Cl^- conductances. By controlling this important initial pathomechanism, adenosine counteracts intracellular Ca^{2+} accumulation and may influence the Ca^{2+}-dependent metabolic processes that are responsible for ischemic nerve cell death.

Effects of Adenosine on Glial Cells

The amount of nerve cell damage seems to be also dependent on the ischemia-induced alterations of the intracellular metabolic signaling and consecutive functional changes of the glial cells. There is increasing evidence that adenosine influences the pathologic reactions of astrocytes as well as of microglial cells, and this may contribute to neuroprotection (for a review see Chapter 31 and the references therein). Under pathologic conditions, if the metabotropic receptors of astrocytes are activated, for example, by excessive extracellular glutamate, the concomitant activation of A_1 adenosine receptors may lead to an even "paradox" effect. Then, A_1-receptor agonists were found to initiate in cultured astrocytes a rise of Ca^{2+} by potentiating its mobilization from intracellular stores [8]. This may favor the postischemically observed astrocyte reaction and concomitant functional changes such as the release of trophic factors. Treatment of cultured astrocytes with propentofylline, which is thought to favor A_2-receptor–mediated adenosine actions, has been shown to increase the release of nerve growth factor (NGF) in vitro [9] and in vivo [10]. On the other hand, an intracellular pathologic Ca^{2+} rise by a "paradox" A_1 receptor action may favor the swelling of astrocytes, which has been shown to be Ca^{2+} dependent [11]. Thus, it remains to be determined whether the potentially neuroprotective or deleterious effects of adenosine are prevalent on astrocytes. The ischemia-induced pathologic activation of microglial cells, which is assumed to be mainly neurotoxic (see earlier discussions), was inhibited by propentofylline. This drug largely reduced or even prevented the free oxygen radical formation in peripheral (peritoneal) and microglia-derived rat macrophages [12]. In gerbils, the post-treatment with propentofylline, started 24 hours after 5 minutes of transient forebrain ischemia largely depressed the expression of several surface antigens in the previously ischemia-activated microglia [13]. The finding that their recognition by cerebrospinal fluid (CSF) antibodies from Alzheimer patients was also suppressed by propentofylline suggests that a pharmacologic favoring of adenosine A_2 actions may be used for a treatment of such neurodegenerative diseases in which an ongoing microglial cell activation is thought to play a pathogenic role.

Neuroprotective effects of adenosine and some of its synthetic analogues have also been demonstrated in experimental in vitro models of brain ischemia (oxygen and glucose deprivation in brain slices and neuronal cell cultures) that lack a vascular system. Moreover, besides its modulatory actions on neurons and glial cells, adenosine may improve CBF and counteract brain ischemia itself by its vasodilatory, antithrombotic effects and by inhibiting neutrophil activation and capillary plugging [1].

Pharmacologic Manipulation of the Adenosine System to Influence Ischemic Brain Damage

Pharmacologic studies with administration of adenosine receptor antagonists, adenosine analogues, or agents that inhibit the physiologic mode of adenosine inactivation have been used to either investigate the role of adenosine in brain ischemia and/or to find ways of drug therapy for stroke patients [1]. In the majority of studies, using different in vivo and in vitro models of ischemic brain damage, the single-dose pretreatment with adenosine antagonists enhanced brain damage.

However, up until now the effect of adenosine antagonists has surprisingly not been studied in permanent middle cerebral artery occlusion, which is one of the clinically most relevant experimental models of stroke. When gerbils were pretreated with the adenosine antagonist theophylline and then subjected to a short period (2 minutes) of global forebrain ischemia, which by itself does not produce substantial degree of damage of hippocampal CA1 pyramidal neurons, the neuronal necrosis was significantly enhanced [14,15]. These studies clearly indicated an inherent neuroprotective effect of endogenous adenosine. They were confirmed by using other experimental models of brain ischemia and administrations of caffeine and the A_1-receptor–specific compound DPCPX [16,17]. Contrarily, the chronic pretreatment with DPCPX or caffeine led to a significant reduction of ischemic brain damage [17,18]. In one of these studies [18] this was accompanied by an increase of A_1-binding sites in the hippocampus and cortex as shown by quantitative [^{3}H]CHA autoradiography, whereas such a receptor upregulation could not be observed in the study with DPCPX in a membrane binding assay [17]. In contrast to antagonists, adenosine itself and some of its metabolically stable synthetic analogues, especially A_1-selective compounds such as R-PIA, CHA, and CPA, reduced ischemic brain damage when given in single doses before or after ischemia, whereas CPA has recently been shown to enhance the damage following chronic pretreatment and withdrawal [17].

Adenosine actions can be reinforced by inhibitors of its physiologic inactivation mechanisms such as cellular reuptake by nucleoside transporters or enzymatic degradation by adenosine deaminase (ADA). The effects of these compounds would be largely restricted to compromised brain areas where the extracellular adenosine concentration is increased and where unwanted systemic side effects, such as those of adenosine agonists (sedation, hypotension, cardiodepression, renal vasoconstriction), are less likely to occur. The experimental results with adenosine transport blockers and the ADA inhibitor pentostatin are equivocal. The xanthine derivative propentofylline is a combined inhibitor of adenosine transporters and cyclic nucleotide phosphodiesterases which favors A_2-receptor–mediated actions. It is in advanced clinical investigation in patients with stroke and vascular dementia. Propentofylline effectively reduces ischemic brain damage with pretreatment and post-treatment in models of global and focal brain ischemia. It also depresses ischemia-induced astrocytic hypertrophy and the formation of free oxygen radicals by activated microglial cells, inhibits their transformation into brain macrophages, and blocks the ischemia-induced expression of cell antigens which are indicative for a potential neurotoxic function of activated microglia [13]. Because many of its in vitro and in vivo actions, including neuroprotection, cannot be diminished by the adenosine antagonist theophylline or ADA, it is still unclear whether reinforcement of adenosine actions is the predominant mechanism for the neuroprotective effects of propentofylline.

Open Questions and Conclusions

Some of the recently discovered actions of adenosine seem to be paradoxical with respect to neuroprotection. Adenosine agonists increased ischemia-induced glutamate release [6] and electrophysiologic studies in brain slices revealed excitatory effects [19]. Adenosine may also induce astrocytic swelling [20]. A massive increase of free intracellular Ca^{2+} has recently been observed in astrocytes treated with 2-chloradenosine together with stimulation of metabotropic glutamate receptors with t-ACPD [8]. Adenosine is a potential source for free radical production during reperfusion. There also seems to be a rather narrow therapeutic time window for the neuroprotective effects of adenosine because of rapid desensitization of adenosine A_1 receptors following ischemia. Agonist-induced receptor desensitization and downregulation may also limit the efficacy of sustained treatment with adenosine analogues. The following observations underline the existence of the limited neuroprotective capacity of adenosine: In transient global ischemia of longer duration (5–10 minutes), blockade of endogenous adenosine actions by theophylline no longer elicited enhanced brain damage [14, 15] and the adenosine-induced depression of the activity-evoked synaptic Ca^{2+} influx could not be observed in gerbil hippocampal slices prepared 18–24 hours after 5-minute ischemia [7]. The ischemia-induced increase of endogenous extracellular adenosine is transient because of the limited availability of its precursor (ATP), and therefore, the pharmacologic reinforcement of endogenous adenosine by transport blockers or ADA inhibitors may be inefficient. This is probably not relevant in focal ischemia (stroke), because there may be transients of adenosine increases in the advancing penumbra, which could be reinforced by these drugs if treatment starts early enough. Despite the limitations, the overall impression from pharmacologic and pathophysiologic studies is that adenosine has powerful neuroprotective effects and that adenosine-agonistic pharmacotherapy can reduce ischemic brain damage. However, many of the underlying neuromodulatory actions of adenosine, that we have discussed in this chapter, are deduced from in vitro studies under "physiologic" study conditions and their pathophysiologic relevance has to be confirmed in experimental models of brain ischemia.

References

1. Rudolphi KA, Schubert P, Parkinson FE, Fredholm BB (1992) Adenosine and brain ischemia. Cerebrovasc Brain Metab Rev 4: 346–369.
2. Siesjö B (1992) Pathophysiology and treatment of focal cerebral ischemia. Part I: Pathophysiology. J Neurosurg 77:169–184.
3. Siesjö B (1992) Pathophysiology and treatment of focal cerebral ischemia. Part II: Mechanisms of damage and treatment. J Neurosurg 77: 337–354.
4. Banati RB, Gehrmann J, Schubert P, Kreutzberg GW (1993) Cytotoxicity of microglia. Glia 7:111–118.
5. Matsumoto K, Graf R, Rosner G, Shimada N, Heiss WD (1992) Flow thresholds for extracellular purine catabolite elevation in cat focal ischemia. Brain Res 579:309–314.
6. O'Reagan MH, Simpson RE, Perkins LM, Phillis JW (1992) The selective A_2 adenosine receptor agonist CGS 21680 enhances excitatory amino acid release from the ischemic rat cerebral

cortex. Neurosci Lett 138:169–172.
7. Schubert P, Keller F, Rudolphi KA (1993) Depression of synaptic transmission and evoked NMDA Ca^{2+} influx in hippocampal neurons by adenosine and its blockade by LTP or ischemia. Drug Dev Res 28:399–405.
8. Ogata T, Nakamura Y, Tsuji, Kataoka K, Schubert P (1994) Adenosine enhances intracellular Ca^{2+} mobilization in conjunction with metabotropic glutamate receptor activation by t-ACPD in cultured hippocampal astrocytes. Neurosci Lett 170:5–8.
9. Shinoda I, Furukawa Y, Furukawa S (1990) Stimulation of nerve growth factor synthesis/secretion by propentofylline in cultured mouse astroglial cells. Biochem Pharmacol 39: 1813–1816.
10. Nabeshima T, Nitta A, Hasegawa T (1993) Impairment of learning and memory and the accessory symptom in aged rat as senile dementia model (3): Oral administration of propentofylline produces recovery of reduced NGF content in the brain of aged rats. Jpn J Psychopharmacol 13:89–95.
11. O'Connor ER, Kimelberg H (1993) Role of calcium in astrocyte volume regulation and in the release of ions and amino acids. J Neurosci 13:2638–2650.
12. Banati RB, Schubert P, Rothe G, Gehrmann J, Rudolphi K, Valet G, Kreutzberg GW (1994) Modulation of intracellular formation of reactive oxygen intermediates in peritoneal macrophages and microglia/brain macrophages by propentofylline. J Cereb Blood Flow Metab 14:145–149.
13. McRae A, Rudolphi KA, Schubert P (1994) Propentofylline depresses amyloid and Alzheimer's CSF microglial antigens after ischaemia. NeuroReport 5:1193–1196.
14. Rudolphi KA, Keil M, Hinze HJ (1987) Effect of theophylline on ischemically induced hippocampal damage in mongolian gerbils: A behavioral and histopathological study. J Cereb Blood Flow Metab 7:74–81.
15. DeLeo J, Schubert P, Kreutzberg GW (1988) Propentofylline (HWA 285) protects hippocampal neurons of mongolian gerbils against ischemic brain damage in the presence of an adenosine antagonist. Neurosci Lett 84: 307–311.
16. Rudolphi KA, Keil M, Fastbom J, Grome JJ, Fredholm BB (1990) Contrary effects of acute versus chronic caffeine administration on ischemically induced hippocampal damage in the gerbil—involvement of adenosine receptors? In Jacobson KA, Daly JW, Manganiello V (eds.), Purines in Cellular Signalling. Targets for New Drugs. New York: Springer, pp. 411–412.
17. Von Lubitz DKJE, Lin RCS, Melman N, Ji XD, Carter MF, Jacobson KA (1994) Chronic administration of selective adenosine A_1 receptor agonist or antagonist in cerebral ischemia. Eur J Pharmacol 256:161–167.
18. Rudolphi KA, Keil M, Fastbom J, Fredholm BB (1989) Ischemic damage in gerbil hippocampus is reduced following upregulation of adenosine (A_1) receptors by caffeine treatment. Neurosci Lett 103:275–280.
19. Okada Y, Sakurai T, Mori M (1992) Excitatory effect of adenosine on neurotransmission is due to increase of transmitter release in the hippocampal slice. Neurosci Lett 142:233–236.
20. Bourke RS, Kimelberg HK, Daze MA (1978) Effects of inhibitors and adenosine on (HCO_3-/CO_2) stimulated swelling and Cl^- in brain slices and cultured astrocytes. Brain Res 154: 195–202.

44. ADENOSINE METABOLITES ARE A SOURCE OF OXYGEN FREE RADICALS IN THE ISCHEMIC/ REPERFUSED RAT BRAIN

John W. Phillis
Michael H. O'Regan

The pathophysiology of brain ischemia/ reperfusion injury is characterized by a complex sequence of events, including biochemical, hemodynamic, and electrophysiologic processes, which frequently lead to neuronal death. Decreases in cerebral blood flow (CBF) below a critical threshold result in energy failure, tissue acidosis, disturbed ion homeostasis with cellular Na^+ and Ca^{2+} influx and K^+ efflux, membrane depolarization, and cytotoxic edema. There is a massive release of the excitatory amino acid neurotransmitters, glutamate and aspartate, which can trigger further cell depolarization and intracellular calcium accumulation. The increased level of free cytosolic Ca^{2+} appears to play a pivotal role in the progression of events that lead to irreversible neuronal injury, in that it precipitates a series of neurotoxic processes, including free radical generation, lipid peroxidation, activation of proteolytic enzymes, and the induction of potentially injurious protooncogene expression.

During brain ischemia the tissue stores of adenosine triphosphate (ATP) are rapidly depleted, resulting in a further loss of ability to restore the normal transmembrane gradients of Na^+ and Ca^{2+}, as well as in increases in the levels of adenosine and its metabolites, inosine, hypoxanthine, xanthine, and uric acid in the brain. Two major hypotheses have been developed to account for the phenomenon of ischemia/reperfusion-induced neuronal death. The excitotoxic amino acid neurotransmitter hypothesis [1] is largely aimed at events during the actual period of ischemia, whereas the free radical hypothesis [2,3] is directed largely at events during reperfusion. There is convincing evidence both for glutamate and aspartate release during cerebral ischemia and that excessive activation of NMDA and AMPA receptors by these amino acids promotes depolarization and the accumulation of intracellular Ca^{2+}, leading to neuronal injury and death.

The free radical hypothesis of cerebral ischemia/reperfusion injury was formulated to provide a conceptual basis for the sequence of events that lead to delayed neuronal death in brains subjected to a period of ischemia followed by reperfusion. The brain appears to be particularly susceptible to oxidative damage in that it contains high levels of readily peroxidizable fatty acids. Furthermore, the brain is relatively poorly endowed with protective antioxidant enzymes or antioxidant compounds. It is possible that neurons that are selectively vulnerable to ischemia/reperfusion injury either lack adequate antioxidant defences or have a high endogenous content of iron, which possesses an ability to catalyze the production of damaging oxygen free radical species. Free radicals are able to initiate damage to many cellular elements, such as lipids, proteins, and nucleic acids. The damage to membrane lipids can result in changes in fluidity and permeability as well as in the function of membrane-bound receptors, ion channels, and enzymes.

Because they are largely concerned with two different phases of injury (i.e., ischemia versus reperfusion) the excitotoxic amino acid and free radical hypotheses are not mutually exclusive, and the two mechanisms may operate in tandem to initiate delayed neuronal death. Indeed, Ca^{2+} influx through NMDA-gated channels may lead to the formation of reactive oxygen species (ROS). Conversely, free radical–derived damage to plasma membranes can result in the efflux of aspartate and glutamate.

Reactive Oxygen Species and Lipid Peroxidation

Oxygen-derived free radicals arise in sequential fashion from molecular oxygen by successive single-electron reduction reactions. ROS are themselves chemically reactive because in order to regain thermodynamic stability they attempt to remove electrons or hydrogen atoms from neighboring molecules, thus radicalizing these molecules in turn, creating a so-called cascade effect. The initial radical in the sequence is the superoxide radical ($O_2^{\cdot -}$), which can be converted to hydrogen peroxide (H_2O_2) by the enzyme superoxide dismutase. H_2O_2, in the presence of ferrous iron, is converted into the highly reactive hydroxyl radical (·OH, Fenton reaction). Hydrogen peroxide can be removed by either of two enzymes, glutathione peroxidase or catalase, which reduce it to water. The ·OH is extremely reactive and capable of interacting with almost every type of molecule found in living cells, including DNA bases and organic acids, such as polyunsaturated fatty acids [4].

A significant consequence of ROS formation is lipid peroxidation of fatty acids in cellular membranes by ·OH or other radicals. Lipid peroxidation is important because it is a chain reaction, in which the initiating free radical can precipitate the destruction of numerous adjacent molecules. Peroxidation of a polyunsaturated fatty acid (PUFA) results from the removal of a methylene hydrogen atom, whereupon the resultant fatty acid converts to a conjugated diene. This moiety is itself peroxidized by the addition of O_2 to yield a peroxy radical, which can initiate a new cycle by repeating the hydrogen atom sequestration from another PUFA molecule, yielding a lipid hydroperoxide and a lipid alkyl radical. In the presence of a transition metal, the lipid hydroperoxide can be converted into a lipid alkoxyl radical, which can initiate further lipid peroxidation, with further expansion of chain reaction. Termination of the chain reaction occurs when the propagating radicals are neutralized by organic scavenger compounds (vitamin E, ascorbic acid) or by enzymic reduction of the lipid hydroperoxides and hydrogen peroxide by glutathione peroxidase.

Sources of Reactive Oxygen Species

Reactive oxygen radicals may be generated by a number of pathways [3], including the oxidation of ubisemiquinone radical, the diffusable electron transfer agent in the mitochondrial respiratory chain, or as a result of the metabolism of arachidonic acid, itself a product of membrane lipolysis by Ca^{2+}-activated phospholipases. Arachidonic acid can be metabolized by either cyclooxygenase or lipoxygenase to form prostaglandins, prostacyclins, thromboxanes, and leukotrienes, with the simultaneous production of superoxide and hydroxyl radicals. Free radical generation may also occur as a result of neutrophil and microglial cell activity or as a result of the interaction of nitric oxide

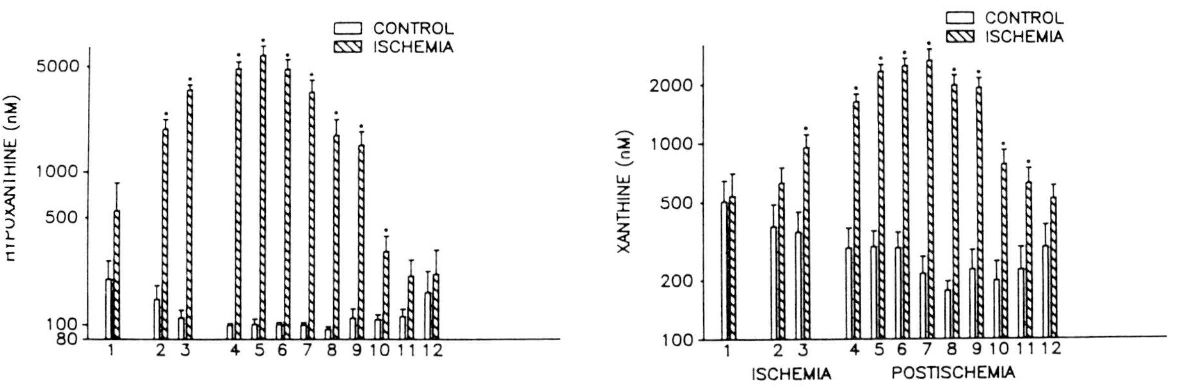

FIGURE 44-1. Time course of changes in levels of hypoxanthine and xanthine sampled from perfusates of the cerebral cortices of sham-operated control nonischemic and ischemic rats. Samples were collected at 10-minute intervals and analyzed by HPLC. The ischemia was of 20 minutes duration. Data are presented as mean ± SEM. The bars in 1 represent preischemic superfusate purine concentrations; 2–3, those during ischemia; 4–9, those during six consecutive 10-minute collections; 10–12 were collected at 30-minute intervals.

with superoxide anion to form peroxynitrite ($ONOO^{-\cdot}$).

Another potential source of ROS is as a consequence of purine metabolism. During ischemia, ATP is utilized but cannot be resynthesized due to the cessation of mitochondrial oxidative phosphorylation. Its metabolites—adenosine, inosine, hypoxanthine, and xanthine—accumulate in the extracellular space [5] (Fig. 44-1). In the normoxic brain, hypoxanthine is metabolized by xanthine dehydrogenase to xanthine and ultimately to uric acid. With the onset of ischemia, the rise in intracellular Ca^{2+} results in a conversion of xanthine dehydrogenase to xanthine oxidase (XO), which uses molecular oxygen instead of the nucleotide radical of NAD as its electron acceptor, catalyzing the formation of $O_2^{\cdot -}$ during reperfusion. In rat brain, XO is concentrated in the capillary endothelium [6], and during ischemia there is a substantial conversion of xanthine dehydrogenase to XO [7]. Generation of hydroxyl radical formation during an anoxia/reoxygenation sequence has been demonstrated in isolated rat brain microvessels [8].

Inhibition of Xanthine Oxidase

Release of purines from the normoxic and hypoxic/ischemic rat cerebral cortex was studied using a cortical cup technique. Allopurinol and its metabolite, oxypurinol, had comparable effects, reducing the release of uric acid to 10% of predrug levels, indicating an effective block of XO activity. Basal and hypoxia/ischemia-evoked xanthine and hypoxanthine levels in the cortical superfusates increased following allopurinol or oxypurinol administration [9]. Although basal levels of adenosine and inosine were not affected, the hypoxia/ischemia-evoked release of both purines was enhanced.

Oxypurinol was then tested for cerebroprotective activity in a gerbil forebrain global ischemia model and a rat middle cerebral artery occlusion (MCA) focal ischemia model. When administered either prior to or 30 minute after-ischemia, oxypurinol (40 mg/kg) significantly reduced both the hypermotility and the extent of CA1 hippocampal pyramidal cell damage in comparison with that observed in control ischemic gerbils [10]. Oxypurinol was also protective when administered either prior to or 1

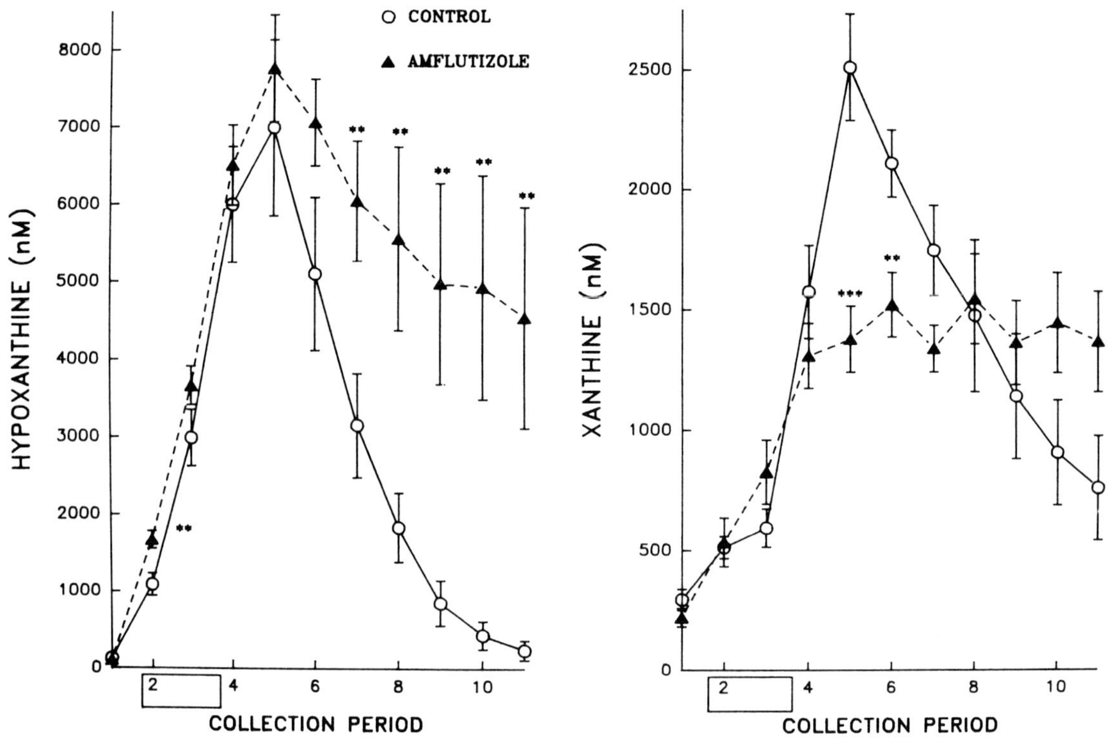

FIGURE 44-2. Effects of the xanthine oxidase inhibitor, amflutizole, on rat cortical cup superfusate hypoxanthine and xanthine concentrations prior to, during, and following a 20-minute period of four-vessel occlusion (4 VO) cerebral ischemia. Amflutizole (10 μM) significantly increased hypoxanthine levels and decreased xanthine concentrations in comparison with those of control 4 VO rats.

hour after-ischemia in the rat MCA occlusion model. It attenuated the degree of neurologic deficit observed in control animals and significantly reduced both the extent of brain swelling and infarct volume [11].

Various potential mechanisms can be proposed to account for oxypurinol's cerebroprotective actions, and several of these may operate simultaneously. They include enhancement of extracellular levels of adenosine, which itself has cerebroprotective actions; promotion of purine salvage by conserving hypoxanthine for resynthesis into inosine monophosphate and ATP; and the prevention of free radical formation by xanthine oxidase.

Amflutizole (4-amino-3[3-trifluoromethyl phenyl]5-isothiazole carboxylic acid); [12], a potent noncompetitive inhibitor of xanthine oxidase, has been tested by topical application (10 μM) in cerebral cortical superfusates of a four-vessel occlusion rat cerebral ischemia model [13]. This compound also enhanced ischemia-evoked hypoxanthine release while suppressing xanthine efflux (Fig. 44-2).

Adenosine Metabolites and Free Radical Generation in the Ischemia/Reperfused Rat Brain

Direct evidence for a role of XO in the generation of free radicals in the ischemia/reperfused rat brain has been obtained with the application of spin-trapping and electron paramagnetic

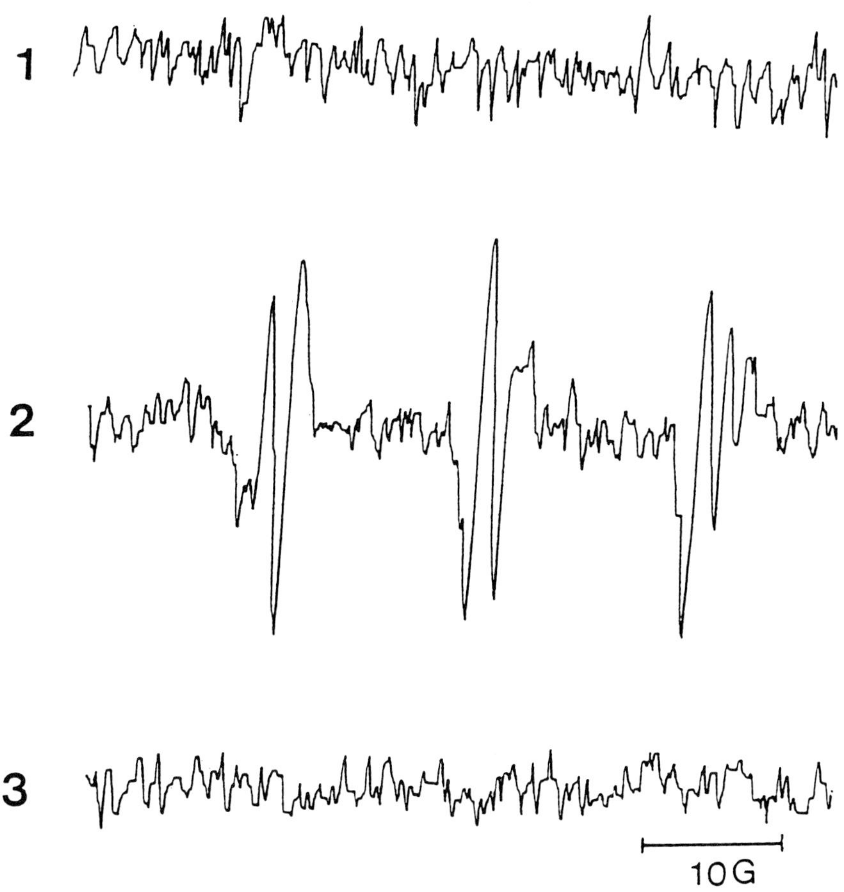

FIGURE 44-3. ESR spectra obtained from cerebral cortical superfusates samples of a rat 4 VO ischemia/reperfusion model, using POBN as a spin-trapping agent. Samples were collected: 1, prior to ischemia; 2, at the onset of reperfusion; 3, after 60 minutes of reperfusion.

resonance (EPR, ESR) spectroscopy. Short-lived free radicals are "trapped" with nitrone or nitroso spin-trapping agents, upon which the resulting adduct gives rise to a stable nitroxide free radical, which can be measured with high sensitivity and specificity by EPR.

Hydroxyl radical production in vivo during ischemia/reperfusion of the rat cerebral cortex was measured by EPR using the cortical cup technique with pyridyl-N-oxide-*tert*-butyl nitrone (POBN) as a trapping agent [14,15]. Serial samples of POBN-containing artificial cerebrospinal fluid were collected from the cortices of four-vessel occlusion rats during ischemia and following reperfusion. Preischemic samples did not elicit an EPR signal. During and following a 30-minute period of ischemia; ·OH adduct signals (characterized by six-line EPR spectra; $\alpha_N = 15.4G\alpha_H^\beta = 2.3G$; Fig. 44-3) were detected in the superfusate samples. These signals achieved their maximal amplitude during the initial 20 minutes of reperfusion and were no longer detectable after 90 minutes of reperfusion. The formation of oxygen-derived free radicals during a period of ischemia (= oxygen deprivation) may initially appear to constitute an anomaly. Studies in several laboratories have, however, shown that four-vessel occlusion does not result in a complete cessation of cerebral blood flow [16–18], and it is likely that the residual delivery of oxygen was adequate to support the generation of free radicals.

Evidence that ·OH was generated by XO activity was obtained by administration of oxypurinol (40 mg/kg) or amflutizole (30 mg/kg). Both inhibitors greatly attenuated the appearance of ·OH adduct in cerebral cortical superfusates during ischemia and reperfusion, confirming that XO is likely to be involved in ROS formation [15,19]. It may be significant that XO is concentrated in rat capillary endothelial cells, as these will inevitably be an important site of free radical formation. The resultant destruction of the blood-brain barrier may make a major contribution to the edema formation in brains that have suffered a stroke.

Conclusions

The formation of oxygen-derived free radicals in the ischemic/reperfused rat brain, as a result of the metabolism of hypoxanthine and xanthine by XO, may play a critical role in initiating the phenomenon of delayed neuronal death. Inhibition of this enzyme by oxypurinol or amflutizole essentially abolishes ·OH formation and release during the initial stages of ischemia and reperfusion. Oxypurinol possesses demonstrable cerebroprotective activity in various ischemic models, probably as a consequence of its ability to attenuate free radical formation.

References

1. Rothman SM, Olney JW (1986) Ann Neurol 19:105–111.
2. Kontos HA (1989) Chem Biol Interact 72:229–255.
3. Phillis JW (1994) Progr Neurobiol 42:441–448.
4. Halliwell B (1992) J Neurochem 59:1609–1623.
5. Phillis JW, Walter GA, Simpson RE (1991) Int J Purine Pyrimid Res 2:41–48.
6. Betz AL (1985) J Neurochem 44:574–579.
7. Grammas P, Liu G-J, Wood K, Floyd RA (1993) Free Rad Biol Med 14:553–557.
8. Kinuta Y, Kikuchi H, Ishikawa M, Kimura M, Itokawa Y (1989) J Neurosurg 71:421–429.
9. O'Regan MH, Phillis JW, Walter GA (1989) Neurochem Int 14:91–99.
10. Phillis JW, Clough-Helfman C (1990) Int J Purine Pyrimid Res 1:31–35.
11. Lin Y, Phillis JW (1992) Brain Res 571:272–280.
12. Wortman RL, Ridolfo AS, Lightfoot RW, Fox IH (1985) J Rheumatol 12:540–543.
13. O'Regan MH, Smith-Barbour M, Perkins LM, Cao X, Phillis JW (1994) Neuropharmacology, in press.
14. Sen S, Phillis JW (1993) Free Rad Res Commun 19:255–265.
15. Phillis JW, Sen S (1993) Brain Res 628:309–312.
16. Furlow TW (1982) Stroke 13:852–855.
17. Kameyama M, Suzuki J, Shirane R, Ogawa A (1985) Stroke 16:489–493.
18. Shirane R, Shimizu H, Kameyama M, Weinstein PR (1991) J Cereb Blood Flow Metab 11:949–956.
19. Phillis JW, Sen S, Cao X (1994) Neurosci Lett 169:188–190.

45. HYPERGLYCEMIC ISCHEMIA AND REPERFUSION: EFFECTS ON ADENOSINE AND ADENINE NUCLEOTIDES

Sean S.-F. Hsu
Joseph Meno
Jie-Gang Zhou
Ellen Gordon
H. Richard Winn

Elevated blood glucose adversely impacts neurologic outcome in brain ischemia and reperfusion [1]. Myers and Yamaguchi [2] first showed that infusion of intravenous glucose into monkeys before cardiac arrest worsened brain injury. This deleterious effect of hyperglycemia has been confirmed by multiple basic investigators [3,4]. Recently, adverse effects of hyperglycemia have been reported in stroke, as well as head injury [5], and following aneurysmal subarachnoid hemorrhage [6].

The mechanism underlying hyperglycemia-enhanced cerebral ischemic injury is, however, unclear. One explanation for the increased brain damage in hyperglycemic ischemia is the rise in lactate production and concomitant tissue acidosis [3]. Nevertheless, it remains controversial whether acidosis plays a significant role in the pathogenesis of ischemic brain injury [7]. Additionally, recent data also question the link between elevations in brain lactate and subsequent tissue injury or neurologic deficits [8].

Because some investigators [8] have reported preservation of brain ATP during ischemic hyperglycemic conditions, we hypothesized that ischemic brain adenosine production would consequently be attenuated. Because adenosine is derived from the adenine nucleotides, in general its concentration varies inversely with the ATP concentration in a variety of tissues, including the brain [9,10]. Adenosine, a purine nucleoside, is a potent vasodilator [9,11] and has been proposed as a regulator of cerebral blood flow [12]. Concentrations of adenosine in brain tissue, cerebrospinal fluid (CSF), and interstitial fluid increase during ischemia [9,10], hypotension [13], hypoxia [14], and seizures [15]. In addition to its role as a vasodilator, increased adenosine concentrations can decrease neuronal activity [16], suppress the release of excitatory amino acids [17], and inhibit neutrophil-endothelium interactions [18]. All of these cerebroprotective actions of adenosine would potentially decrease the extent of ischemic injury, and, conversely, attenuation of ischemic brain adenosine levels could increase the degree of injury.

This chapter reviews the changes in brain adenine nucleosides and nucleotides during hyperglycemic and normoglycemic ischemia and reperfusion [19,20]. In addition, we will review the impact of alteration of adenosine availability and activity on the extent of neuronal damage in hyperglycemic ischemia and reperfusion [21].

Methods

Three sets of experiments were performed in male adult Sprague–Dawley rats:

1. Analysis of changes in CSF and brain adenosine, inosine, hypoxanthine, ATP, ADP, and AMP before, during, and after 15 minutes of ischemia and 60 minutes of reperfusion [19]. In these studies CSF was obtained from a closed cranial window and brain tissue by means of freeze-blow technique. We compared normoglycemic and hyperglycemic animals during 15 minutes of ischemia and 60 minutes of reperfusion. Ischemia was achieved by the two-vessel hypotensive model [22] in anesthetized animals.
2. Analysis of adenine nucleotides, phosphocreatine, Pi, and intracellular pH by ^{31}P nuclear magnetic resonance spectroscopy [20]. In these studies we used the same two-vessel hypotensive model of ischemia but utilized a different technique (i.e., NMR spectroscopy) to analyze adenine nucleotides and pH during 15 minutes of hyperglycemic ischemia.
3. Effects of alteration in adenosine activity on neuronal injury following normoglycemic and hyperglycemic ischemia [21]. In these studies we pretreated animals with either an adenosine receptor blocker, theophylline (0.20 μmol/g, i.p.) or an adenosine agonist, cyclohexyladenosine (CHA). CHA (6.25 nmol) was administered into the lateral ventricle. Control animals were administered saline in the percutaneum (theophylline series) or in the cerebral ventricle (CHA series). We studied both normoglycemic and hyperglycemic ischemia and reperfusion using the four-vessel occlusion model of ischemia. After a 5-day survival, the animals were sacrificed and the extent of neuronal injury was quantified by an observer blinded to the experimental conditions.

In all studies moderate hyperglycemia was achieved by administrating 17% D-glucose (3 g/kg, i.p.) 15 minutes prior to ischemia. The normoglycemic animals received an equivalent volume of saline. In all groups, animals were carefully monitored to maintain constant skull and body temperature. Arterial blood gases and pH did not differ between the hyperglycemic and normoglycemic animals during or after ischemia.

Results

Changes in CSF and brain adenine nucleosides and nucleotides were as follows.

TISSUE ADENOSINE, INOSINE, AND HYPOXANTHINE

There were no significant differences in adenosine, inosine, or hypoxanthine levels in nonischemic saline-treated and glucose-treated animals. After 15 minutes of ischemia there was a 150-fold increase in brain adenosine levels in the saline-treated animals (Fig. 45-1). In contrast, in the hyperglycemic animals the increase in adenosine was significantly ($p < 0.01$) attenuated, rising only 50-fold. Ischemic levels of inosine and hypoxanthine were also significantly lower ($p < 0.01$) in the glucose-treated rats compared with the animals that received saline. The reduced levels of adenosine, inosine, and hypoxanthine associated with hyperglycemia persisted after 5 minutes of reperfusion ($p < 0.01$) as levels in both normoglycemic and hyperglycemic groups were decreasing from ischemic values.

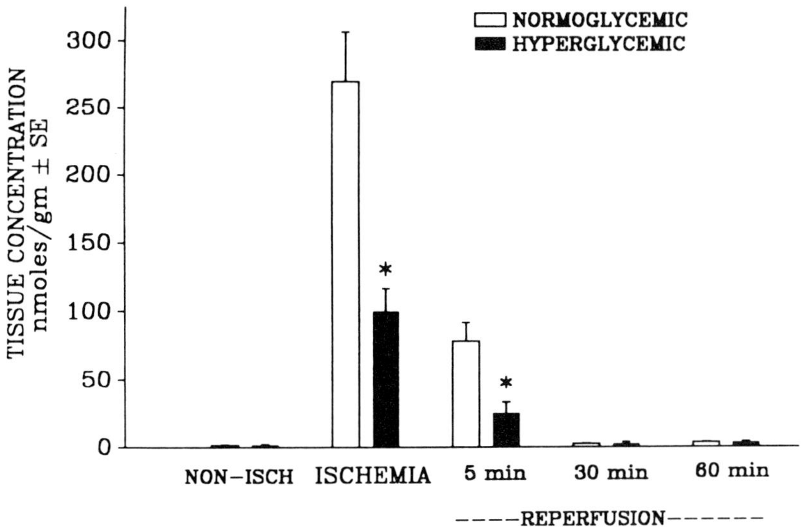

FIGURE 45-1. Brain adenosine concentrations in nonischemic (Non-Isch), 15-minute ischemic, and reperfused normoglycemic and hyperglycemic rats. Values are mean ± SE with six animals in each data group (total = 60). Adenosine concentrations in hyperglycemic rats were significantly (*$p < 0.01$) lower than in corresponding normoglycemic rats during ischemia and after 5 minutes of reperfusion.

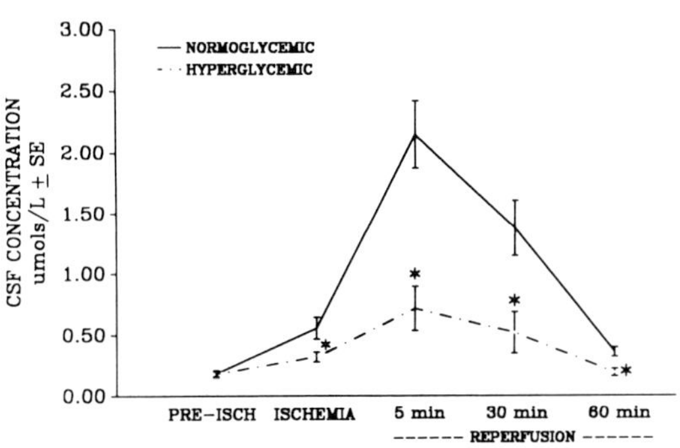

FIGURE 45-2. Cerebrospinal fluid (CSF) adenosine concentrations before ischemia (Pre-Isch) after 15 minutes of ischemia and during reperfusion in six normoglycemic and five hyperglycemic rats. Values are means ±SE. Adenosine concentrations were significantly (*$p < 0.05$) lower during ischemia, as well as after 5, 30, and 60 minutes reperfusion in hyperglycemic rats compared with the corresponding values in normoglycemic rats.

CSF ADENOSINE, INOSINE, AND HYPOXANTHINE

CSF adenosine concentrations also increased during ischemia, and this increase was attenuated in the hyperglycemic rats ($p < 0.01$; Fig. 45-2). Additionally, as with tissue levels, after the first 5 minutes of reperfusion adenosine levels in CSF were increased significantly ($p < 0.05$) more in normoglycemic compared with hyperglycemic animals. The elevated adenosine concentration in normoglycemic rats persisted throughout 60 minutes of reperfusion, and the levels in the hyperglycemic animals remained significantly lower. This persistent elevation of adenosine in CSF contrasts with tissue levels, which return close to normal by 30 minutes of reperfusion. In contrast to normoglycemia, hyperglycemia also reduced CSF inosine concentrations during ischemia and after 5 minutes of reperfusion, as well as reducing hypoxanthine concentrations after 5 minutes of reperfusion.

TISSUE ADENINE NUCLEOTIDES

Fifteen minutes of severe forebrain ischemia caused a dramatic reduction in brain ATP levels in saline-treated animals. Brain ATP levels of the glucose-treated animals, although also diminished after ischemia, were, however, sig-

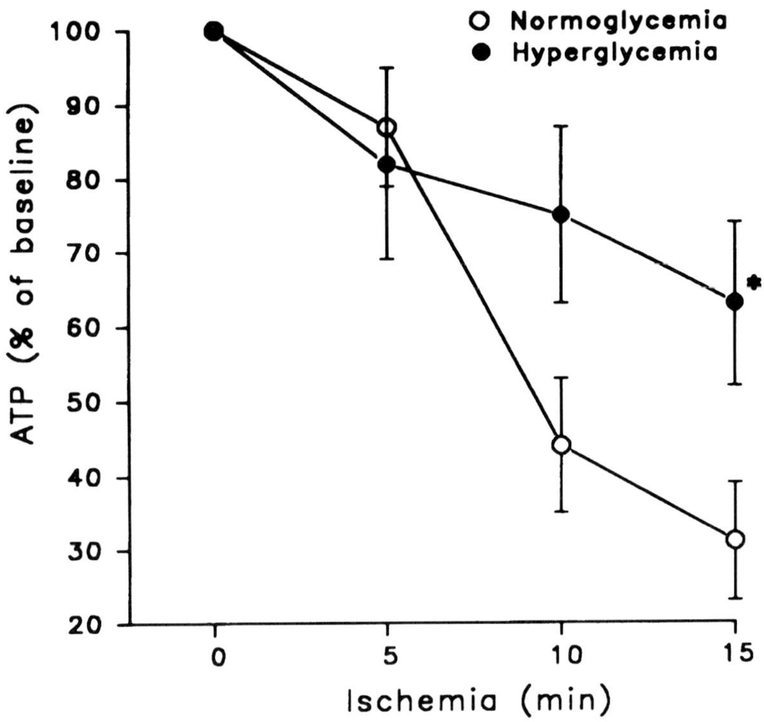

FIGURE 45-3. ATP concentration determined from peak area measurements by ^{31}P NMR spectroscopy. Percentages were calculated from baseline values and plotted against time of ischemia for normoglycemic (open circles, n = 7) and hyperglycemic (solid circles, n = 6) rats. Values are means ± SE. ATP concentration was significantly different after 15 minutes of ischemia; *$p < 0.05$.

nificantly higher ($p < 0.01$) than in comparable saline-treated animals. During ischemia, ADP levels and energy charge were also significantly higher ($p < 0.01$) in glucose-treated animals, whereas AMP levels were significantly lower ($p < 0.01$). Hyperglycemia did not alter adenine nucleotide values in the nonischemic saline- or glucose-treated animals.

NMR spectroscopic analysis of adenine nucleotides and pH during ischemia in normoglycemic and hyperglycemia animals: ATP pools, as measured by ^{31}PNMR spectroscopy, decayed with ischemia in both normoglycemic and hyperglycemic animals (Fig. 45-3). After 15 minutes of ischemia, the residual ATP was significantly greater in hyperglycemic (63 ± 11%) versus normoglycemic (31 ± 8%) rats ($p < 0.05$). In contrast, although pH declined significantly ($p < 0.01$), the decrease in pH was similar for both hyperglycemic and normoglycemic animals.

EFFECTS OF ALTERED ADENOSINE ACTIVITY AND AVAILABILITY ON NORMOGLYCEMIC VERSUS HYPERGLYCEMIC ISCHEMIA AND REPERFUSION

Histopathologic results for animals treated with either saline or theophylline prior to normoglycemic and hyperglycemic ischemia are depicted in Figure 45-4. Three brain regions—CA_1, cortex, and caudate—were analyzed for ischemic injury. Hyperglycemic cerebral ischemia resulted in a significantly greater percentage of dead neurons in all three brain regions than normoglycemia. In addition, the percentage

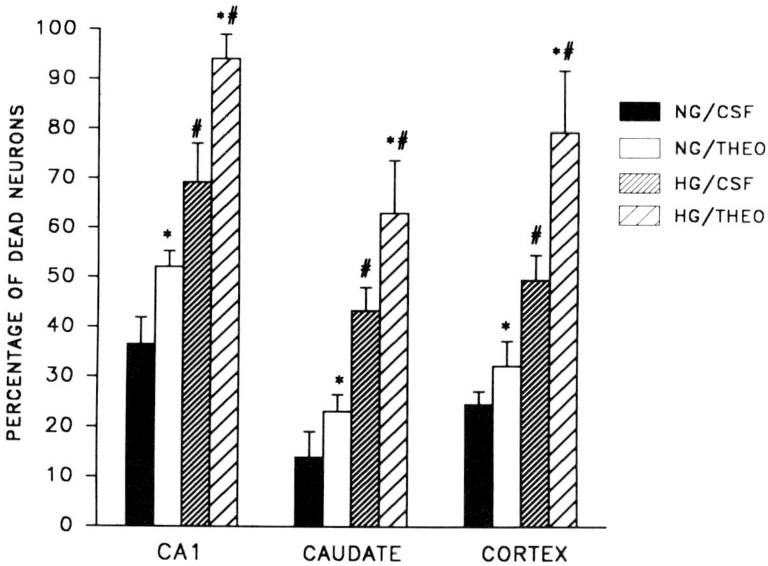

FIGURE 45-4. The effect of theophylline on normoglycemic and hyperglycemic ischemic injury. Data are expressed as means ± SD (n = 6). THEO = theophylline (0.20 μmol/g, i.p.); NG = normoglycemia; HG = hyperglycemia. *$p < 0.01$ versus corresponding saline values. #$p < 0.01$ versus corresponding normoglycemic values.

of dead neurons in each of these brain regions was significantly ($p < 0.01$) greater in theophylline-treated rats.

Figure 45-5 depicts the results for animals treated with either intraventricular (12.5 μl) CSF or CHA (6.25 nmol) prior to normoglycemic and hyperglycemic ischemia. Again, hyperglycemia significantly enhanced ischemic injury in the hippocampus, cortex, and caudate. The animals treated with CHA had significantly ($p < 0.01$) less injury in the three brain regions analyzed.

Discussion

In the present chapter we reviewed studies [19–21] that demonstrated adenosine concentrations in CSF and brain are attenuated in hyperglycemic ischemia and reperfusion compared to normoglycemic ischemia and reperfusion. Furthermore, during ischemia tissue concentrations of ATP, as measured by two different techniques (i.e., freeze-blow and NMR spectroscopy) are better preserved in hyperglycemia. In contrast, the changes in pH are similar in both hyperglycemia and normoglycemia. Lastly, in animals pretreated with theophylline, an adenosine receptor blocker, tissue injury was greater in than those treated with saline in both the normoglycemic and hyperglycemic groups. In contrast, pretreatment with CHA, an adenosine agonist, resulted in an attenuation of neuronal injury. Our studies in normoglycemic ischemia and reperfusion confirm the findings of Rudolphi et al. [23] and von Lubitz et al. [24]. The hyperglycemic data suggest that the depressed levels of adenosine, a known neuroprotector during hyperglycemic ischemia and reperfusion, may play an etiologic role in the adverse outcome in hyperglycemic ischemia.

Adenosine concentrations could influence outcome by multiple known adenosine mechanisms, such as alteration of vessel tone and cerebral blood flow, depression of neuronal activity and metabolic demand, blocking release

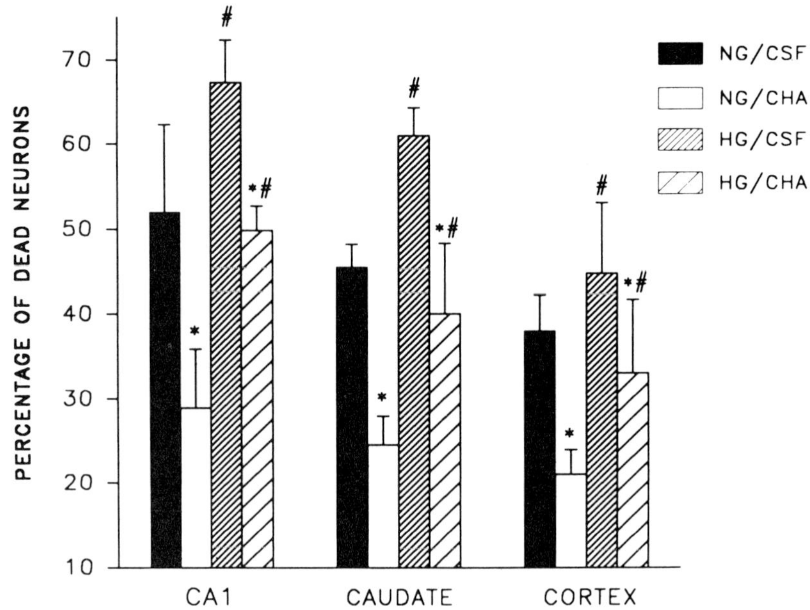

FIGURE 45-5. The effect of cyclohexyladenosine on normoglycemic and hyperglycemic ischemic injury. Data are expressed as means ± SD (n = 6). CHA = cyclohexyladenosine (6.5 mmol intraventricularly); NG = normoglycemia; HG = hyperglycemia. *p < 0.01 versus corresponding CSF values. #p < 0.01 versus corresponding normoglycemic values.

of excitatory amino acids and inhibiting leukocyte-endothelial adherence, and the generation of oxygen-derived free radicals. With a relative decrease in adenosine concentration (i.e., hyperglycemia), one or more of these adenosine-related mechanisms could lead to increased neuronal injury. In contrast, with increased adenosine concentration (e.g., normoglycemia), there would be increased vasodilation, decreased neuronal and metabolic activity, blockage of free radical generation, inhibition of excitatory amino acid release, and leukocyte endothelial adherence.

In summary, the data are supportive of the hypothesis that adenosine is involved mechanistically with the adverse outcome observed following hyperglycemic ischemia.

References

1. Plum F (1983) What causes infarction in ischemic brain? The Robert Wartenberg Lecture. Neurology 33:222–233.
2. Myers RE, Yamaguchi S (1977) Nervous system effects of cardiac arrest in monkeys. Arch Neurol 34:65–74.
3. Pulsinelli WA, Waldman S, Rawlinson D, Plum F (1983) Moderate hyperglycemia augments ischemic brain damage: A neuropathologic study in the rat. Neurology 32:1239–1246.
4. Siemkowicz E, Hansen AJ (1978) Clinical restitution following cerebral ischemia in hypo-, normo- and hyperglycemic rats. Acta Neurol Scand 58:1–8.
5. Lam AM, Winn HR, Cullen BF, Sundling N (1991) Hyperglycemia and neurologic outcome in patients with head injury. J Neurosurg 75: 545–551.
6. Lanzino G, Kassell NF, Germanson T, Truskowski L, Alves W (1993) Plasma glucose levels and outcome after subarachnoid hemorrhage. J Neurosurg 79:885–891.
7. Paljarvi L, Soderfeldt B, Kalimo H, Olsson Y, Siesjo BK (1982) The brain in extreme respiratory acidosis. Acta Neuropathol 58: 87–94.
8. Hoffman WE, Braucher E, Pelligrino DA, Thomas C, Albrecht RA, Miletich DJ (1990)

Brain lactate and neurologic outcome following incomplete ischemia in fasted, nonfasted and glucose-loaded rats. Anesthesiology 72: 1045–1050.
9. Berne RM, Rubio R, Curnish RR (1974) Release of adenosine from ischemic brain. Circ Res 35: 262–271.
10. Winn HR, Rubio R, Berne RM (1979) Brain adenosine production in the rat during 60 seconds of ischemia. Circ Res 45:486–492.
11. Morii SA, Ngai AC, Winn HR (1986) Reactivity of rat pial arterioles and venules to adenosine and carbon dioxide: With detailed description of the closed cranial window technique in rats. J Cereb Blood Flow Metab 6:34–41.
12. Winn HR, Rubio R, Berne RM (1981) The role of adenosine in the regulation of cerebral blood flow. J Cereb Blood Flow Metab 1:239–244.
13. Van Wylen DGL, Park TS, Rubio R, Berne RM (1988) Cerebral blood flow and interstitial fluid adenosine during hemorrhagic hypotension. Am J Physiol 255 (Heart Circ Physiol 24): H211–H218.
14. Winn HR, Rubio R, Berne RM (1981) Brain adenosine concentration during hypoxia in rat. Am J Physiol 241:H235–H2421.
15. Winn HR, Welsh J, Rubio R, Berne RM (1980) Changes in brain adenosine durin bicuculline-induced seizures in rats: Effects of hypoxia and altered systemic blood pressure. Circ Res 47:568–577.
16. Phillis JW, Kostopoulos GK, Limacher JJ (1975) A potent depressant action of adenine derivatives on cerebral cortical neurones. Eur J Pharmacol 30:125–129.
17. Coradetti R, LoConte G, Mornoi F, Passani MB, Pepeu G (1984) Adenosine decreases aspartate and glutamate release from rat hippocampal slices. Eur J Pharmacol 104:19–26.
18. Cronstein BN, Rosenstein ED, Kramer SB, Weiss-Mann G, Hirschhorn R (1985) Adenosine: A physiologic modulator of superoxide anion generation by human neutrophils. Adenosine acts via an A_2 receptor on human neutrophils. J Immunol 135:1366–1371.
19. Hsu SS-F, Meno JR, Zhou J-G, Gordon EL, Winn HR (1991) Influence of hyperglycemia on cerebral adenosine production during ischemia and reperfusion. Am J Physiol 262 (Heart Circ Physiol 30):H398–H403.
20. Hsu SS-F, Meno JR, Gronka R, Kushmerick M, Winn HR (1994) Moderate hyperglycemia affects ischemic brain ATP levels but not intracellular pH. Am J Physiol 266 (Heart Circ Physiol 35):H258–H262.
21. Zhou J-G, Meno JR, Hsu SS-F, Winn HR (1994) Effects of theophylline and cyclohexyladenosine on brain injury following normo- and hyperglycemic ischemia: A histophathologic study in the rat. J Cereb Blood Flow Metab 14:166–173.
22. Nordstrom CH, Siesjo BK (1978) Effects of phenobarbital in cerebral ischemia. Part 1. Cerebral energy metabolism during incomplete ischemia. Stroke 9:327–335.
23. Rudolphi KA, Keil M, Hinze H-J (1987) Effect of theophylline on ischemically induced hippocampal damage in Mongolian gerbils: A behavioral and histopathological study. J Cereb Blood Flow Metab 7:74–81.
24. von Lubitz DKJE, Dambrosia JM, Kempski O, Redmond J (1988) Cyclohexyladenosine protects against neuronal death following ischemia in the CA1 region of the hippocampus. Stroke 19: 1133–1139.

C. RENAL PROTECTION

46. STUDIES ON THE MECHANISM BY WHICH ADENOSINE RECEPTOR ANTAGONISTS ATTENUATE ACUTE RENAL FAILURE

Edwin K. Jackson
Zaichuan Mi
William A. Herzer

Acute renal failure (ARF) is an extremely serious medical condition characterized by an abrupt reduction in glomerular filtration rate (GFR), which leads to the retention of nitrogenous wastes. Five percent of all hospitalized patients develop acute renal failure, and the mortality rate in such patients is high [1]. Therefore, development of effective therapeutic strategies for ARF would be a significant medical advance.

It is important to realize that ARF is a complicated disorder of multiple etiologies. Causes may be either extrinsic (i.e., prerenal and postrenal failure) or intrinsic to the kidney. Intrinsic causes of ARF include acute tubular necrosis (i.e., injury to renal epithelial cells), cortical necrosis, interstitial nephritis, precipitation of material in renal tubules, renovascular disease, and glomerulonephritis. Although many types of ARF exist, acute tubular necrosis (ATN) accounts for two thirds of all cases of intrinsic ARF, and approximately 50% of patients with ATN die [2].

Because of the seriousness and relatively high incidence of ATN-induced ARF, it is noteworthy that blockade of adenosine receptors, particularly A_1 adenosine receptors, significantly attenuates the reduction of GFR and renal blood flow (RBF) that accompanies experimental ATN. At least eight separate studies utilizing six different adenosine receptor antagonists have reported significant protection against the decline in renal function that accompanies glycerol-induced ATN [3–10]. Furthermore, blockade of adenosine receptors appears to preserve GFR and RBF during experimental ATN induced by gentamicin [11], amphotericin B [12], and severe renal ischemia [13]. Likewise, blockade of adenosine receptors was demonstrated to preserve renal function in four separate studies using cisplatin as the offending nephrotoxin [10,11,14,15].

Since the efficacy of adenosine receptor antagonists in experimental ATN-induced ARF is securely established, it is appropriate to inquire as to the mechanism by which this class of drugs affords renoprotection. In this regard, the current model of ATN-induced ARF is that epithelial cell damage leads to a reduction in GFR by three mechanisms: 1) obstruction of tubules, 2) backleak of ultrafiltrate into the renal interstitium, and 3) functional alterations in the renal microcirculation [16]. Most likely in ATN-induced ARF, adenosine participates in the functional alterations of the renal microcirculation.

Figure 46-1 describes in more detail possible explanations regarding the role of adenosine in ATN. Injury to epithelial cells may alter energy production, leading to depletion of ATP and ADP [17] with increased levels of AMP and adenosine. Also, a reduction in proximal reabsorption increases distal delivery of solutes, which causes the macula densa to activate tubuloglomerular feedback via an adenosine-mediated process [18–20]. Adenosine is well known to increase the resistance of the afferent arteriole via A_1 receptors and thereby decreases GFR and increases renal vascular resistance (RVR) [21–25].

In most respects, however, the model shown in Figure 46-1 regarding how adenosine receptor antagonists preserve renal function during ATN is untested. Accordingly, the purpose of the present study was to test the hypothesis that adenosine receptor antagonists attenuate ATN induced ARF by blocking the functional renovascular effects of increased levels of endogenous adenosine.

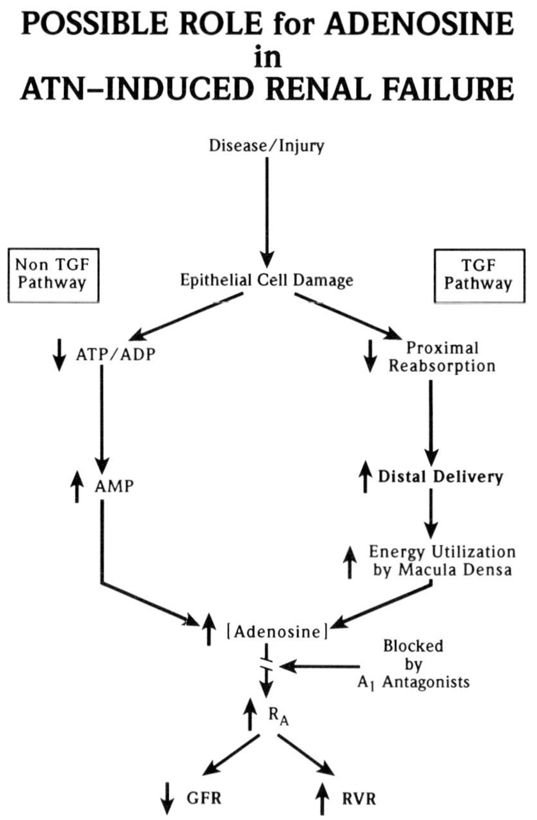

FIGURE 46-1. Hypothetical mechanism by which endogenous adenosine may participate in ATN-induced acute renal failure. TGF = tubuloglomerular feedback; R_A = afferent arteriolar resistance; GFR = glomerular filtration rate; RVR = renal vascular resistance.

Methods

Male Sprague-Dawley rats (387 ± 59 gr, mean ± SD) were obtained from Charles River (Wilmington, MA) and were administered intraperitoneal injections of either cisplatin (4 mg/kg) or the vehicle for cisplatin (0.9% saline). Six days later, rats were anesthetized with thiobutabarbital sodium (100 mg/kg, i.p.). The animals were placed on a Deltaphase isothermal pad (Braintree Scientific, Braintree, MA), and body temperature was monitored with a rectal temperature probe (Physiotemp, Clifton, NJ). A heat lamp was used to maintain body temperature at 37°C by adjusting the height of the lamp above the animal.

The trachea was cannulated with polyethylene (PE)-240 tubing to facilitate respiration, and catheters (PE-10 and PE-50) were inserted into the jugular vein. Also, a PE-50 catheter was placed in the left carotid artery and connected to a digital blood pressure analyzer (Micro-Med, Inc., Louisville, KY).

A laparotomy was performed, and a PE-10 catheter was placed in the left ureter for urine collections. A transit-time flow probe (1 mm) was placed around the left renal artery and connected to a flowmeter (model T206, Transonic Systems, Inc., Ithaca, NY). A microdialysis probe with a 20,000 Dal membrane cutoff and a membrane outer diameter of 0.5 mm (CMA/20 microdialysis probe 4 mm, Bioanalytical Systems, Inc., West Lafayette, IN) was introduced into the renal cortex of the left kidney. The microdialysis probe was perfused at 2 μl/min with 0.9% saline containing heparin (10 U heparin/ml) with a Braintree infusion pump (model BSP 99).

Next, a 32-gauge needle attached to PE-10 tubing was carefully inserted into the left renal artery, and an infusion of 0.9% saline was begun at 20 μl/min directly into the renal artery. Then, rats were administered intravenous ^{14}C-carboxyl-inulin dissolved in 0.9% saline (0.75 μCi bolus followed by a continuous infusion at 0.05 μCi/80 μl/min).

After the surgery was completed, animals were allowed to stabilize for 1 hour before the protocol was begun. The first half of the protocol consisted of four 30-minute periods. In some animals only saline (20 μl/min) was infused into the renal artery throughout the four periods (time-control groups), whereas other animals received increasing doses of the A_1 agonist N^6-cyclopentyladenosine (CPA; 0, 10, 30, and 100 ng/min) into the renal artery during the four experimental periods. Mean arterial blood pressure (MABP) was time-averaged over each 30-minute experimental period, and RBF for each 30-minute experimental period was calculated as the average of three measurements taken every 10 minutes. During the last 20 minutes of each 30-minute period, urine was collected and a midpoint arterial blood sample (150 μl) was taken.

The intrarenal infusion of CPA was replaced with saline, and the animal was allowed to stabilize for an additional 30 minutes. Then dialysate from the renal cortex was collected for 2 hours by allowing the dialysate exiting from the probe to drip directly into a freezing apparatus. During the first hour, dimethylsulfoxide (DMSO) was infused into the PE-10 jugular catheter (0.7 μl/min), and during the second hour rats received either DMSO (time/vehicle control) or 1,3-dipropyl-8-cyclopentylxanthine (DPCPX, 3 μg/min). DPCPX is an effective and selective A_1 antagonist when administered intravenously to rats at the dose employed in this study [26]. MABP was time-averaged across each 1-hour period, and a 20-minute urine collection and midpoint arterial blood sample was taken during the last 20 minutes of each 1-hour period. RBF for each 1-hour period was calculated as the average of six measurements taken every 10 minutes.

The dialysate was thawed and analyzed for purines with an Isco (Lincoln, NE) high-pressure liquid chromatographic system (pump model 2350, gradient programmer model 2360, V^4 absorbance detector, 4.6×250 mm C_{18} column with 5 μm particle size; ChemResearch Data Management System) using a previously published method [27]. Radioactivity was determined in the urine and midpoint plasma samples by liquid scintillation counting, and GFR was calculated as the clearance of ^{14}C-carboxyl-inulin. RVR was calculated by dividing MABP by RBF. Statistical analyses were conducted using the Number Crunchers Statistical System (Kaysville, UT), and the criterion of significance was $p < 0.05$.

Results

As shown in Figures 46-2 and 46-3 (top panels), in cisplatin-treated rats GFR was significantly reduced by approximately 50%, and RVR was significantly elevated by approximately 100% compared with control rats that did not receive cisplatin. Furthermore, GFR and RVR were reasonable stable over the four consecutive 30-minute experimental periods. CPA significantly reduced GFR by approximately 20% in both control and cisplatin-treated rats (Fig. 46-2, bottom panel); however, CPA did not significantly affect RVR in either group (Fig. 46-3, bottom panel).

In cisplatin-treated rats, when only the vehicle for DPCPX (i.e., DMSO) was infused for 2 hours, urine volume, RVR and GFR were stable over the 2-hour period (Fig. 46-4). When

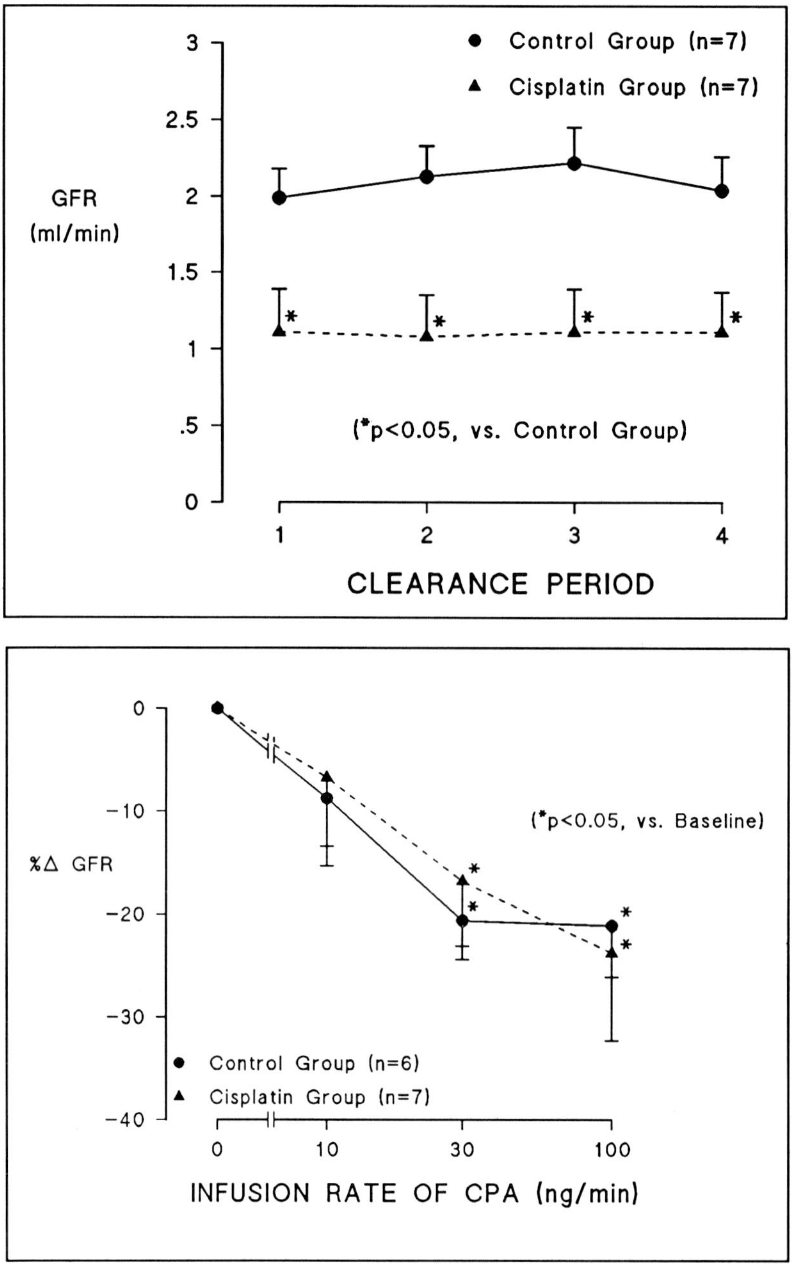

FIGURE 46-2. **Top panel:** Glomerular filtration rate (GFR) in control and cisplatin-treated rats during four consecutive clearance periods. **Bottom panel:** Percentage change in GFR induced by increasing doses of N^6-cyclopentyladenosine (CPA) infused directly into the renal artery of control and cisplatin-treated rats.

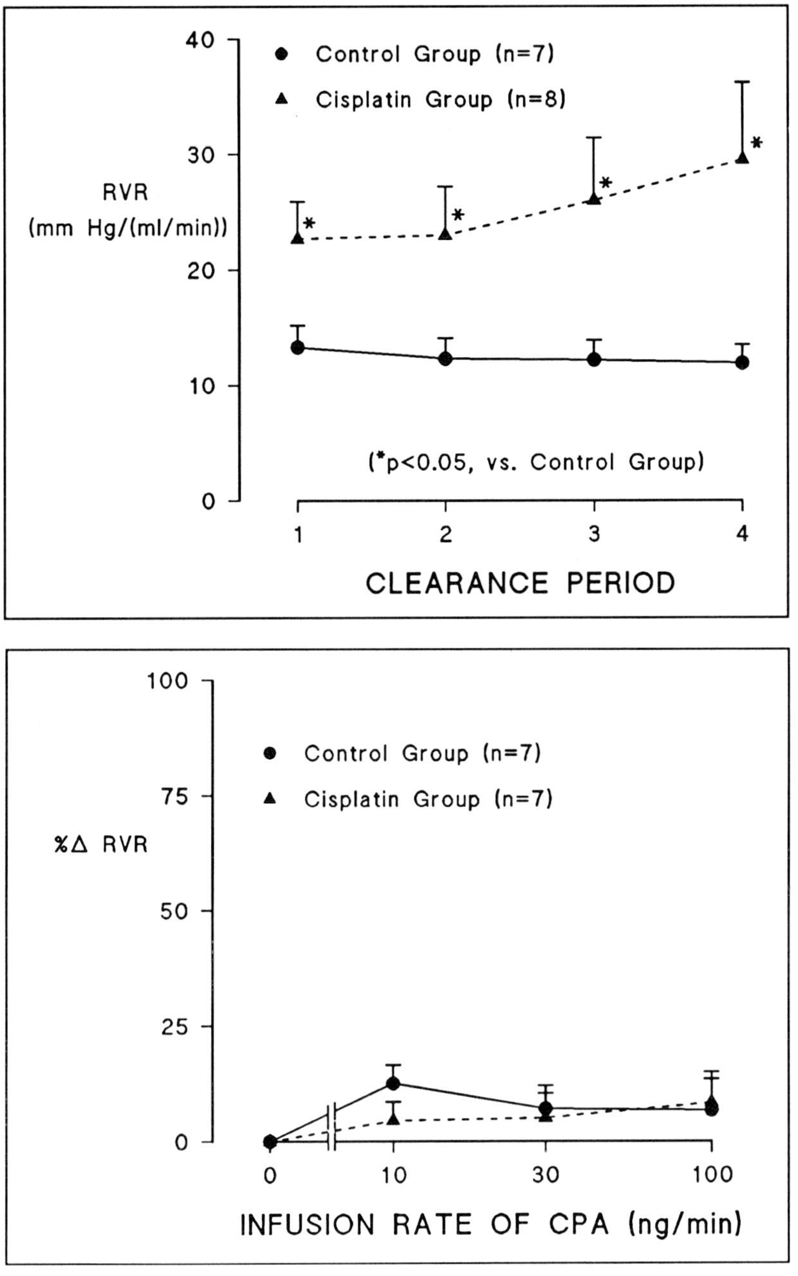

FIGURE 46-3. **Top panel:** Renal vascular resistance (RVR) in control and cisplatin-treated rats during four consecutive clearance periods. **Bottom panel:** Percentage change in RVR induced by increasing doses of N^6-cyclopentyladenosine (CPA) infused directly into the renal artery of control and cisplatin-treated rats.

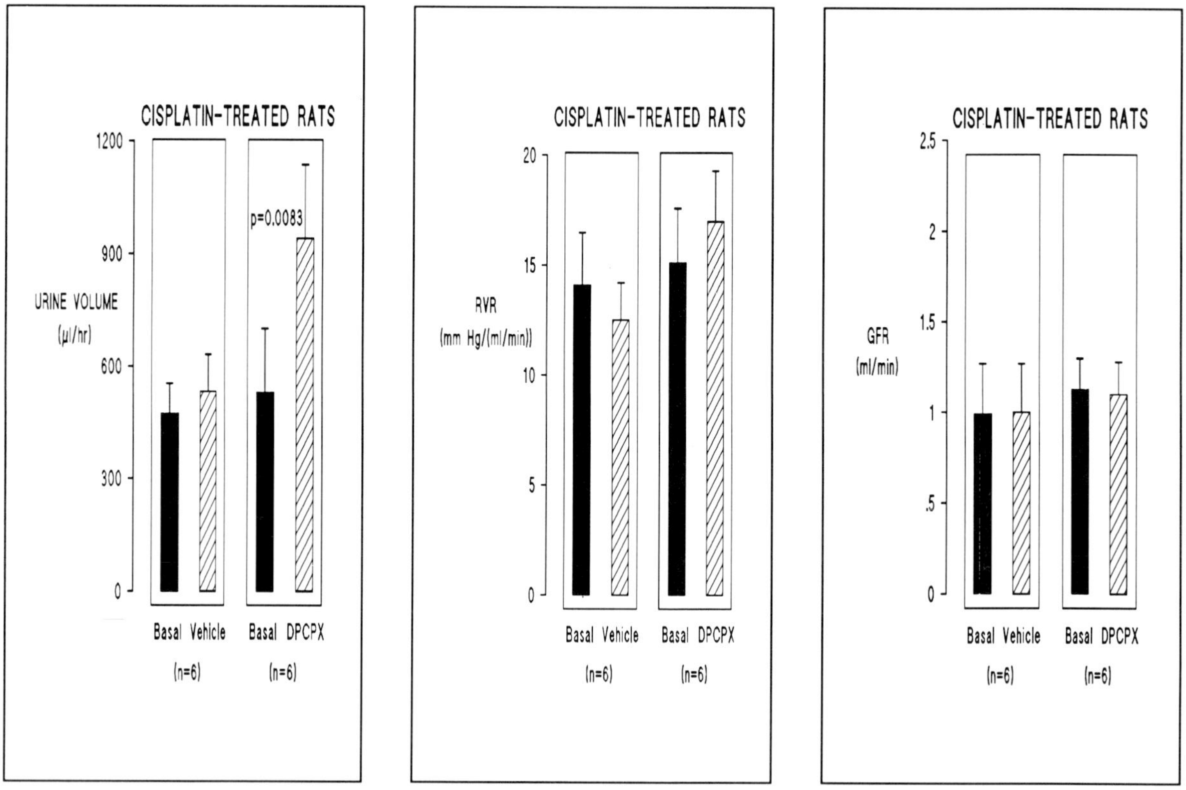

FIGURE 46-4. Glomerular filtration rate (GFR, **right panel**), renal vascular resistance (RVR, **middle panel**), and urine volume (**left panel**) were measured in two groups of cisplatin-treated rats. In both groups, baseline values (Basal) were determined in the presence of DMSO. Rats then received either 1,3-dipropyl-8-cyclopentylxanthine (DPCPX) dissolved in DMSO or continued to receive only DMSO (Vehicle).

DPCPX was infused during the second hour, urine volume was significantly elevated compared with the first hour. However, DPCPX did not influence RVR or GFR in cisplatin-treated rats (Fig. 46-4). In control rats not treated with cisplatin, DPCPX also increased urine volume but did not alter RVR or GFR (data not shown). The renal cortical interstitial levels of adenosine, inosine, xanthine, and hypoxanthine were not increased in cisplatin-treated rats compared with control rats (Fig. 46-5). In fact, hypoxanthine levels were significantly lower in cisplatin-treated compared with control rats.

Discussion

Regarding the effectiveness of adenosine receptor antagonists in ATN-induced ARF, if the model illustrated in Figure 46-1 is correct, four predictions should be fulfilled: 1) An intrarenal infusion of an A_1 agonist should mimic the renal hemodynamic consequences of ATN-induced ARF. 2) In animals with established ATN-induced ARF, the renal responses to an exogenous A_1 agonist should be diminished or absent due to saturation of A_1 receptors by elevated levels of endogenous adenosine. 3) Renal cortical interstitial levels of adenosine and/or its metabolites should be elevated in ATN-induced ARF. 4) Acute blockade of A_1 receptors should rapidly improve renal function

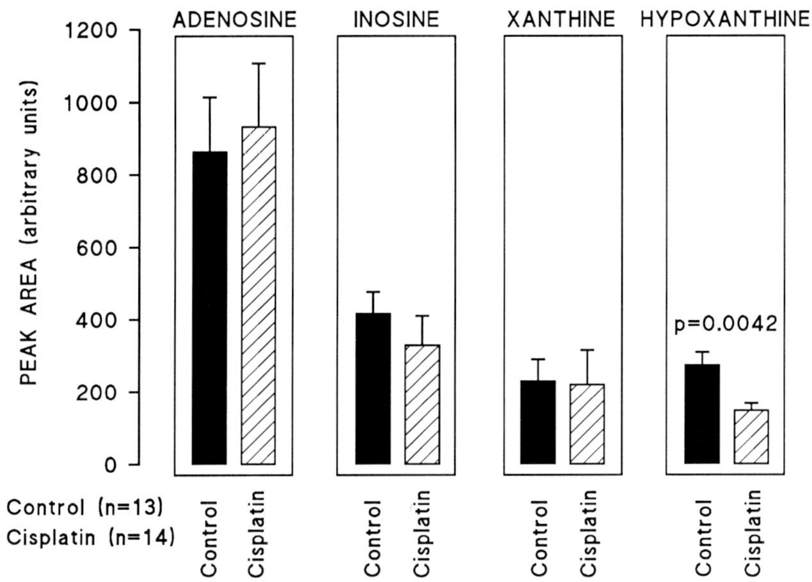

FIGURE 46-5. Amount of adenosine, inosine, xanthine, and hypoxanthine measured by high-pressure liquid chromatography in the dialysate recovered from a microdialysis probe inserted into the renal cortex of control and cisplatin-treated rats. Values are expressed as area under the chromatographic peak obtained when 80 μl of dialysate were analyzed.

even after the development of ATN-induced ARF. In this regard it is important to note that prior studies [3–15] were performed with coadministration of an adenosine antagonist with the nephrotoxin, and administration of the antagonist was then continued for one or more days after the renal insult.

In the current study these predictions were tested by treating rats with either cisplatin or vehicle. After 6 days, animals were anesthetized, and dose-response curves to 30-minute intrarenal infusions of CPA were obtained with respect to GFR and RVR. GFR and RVR were also determined before and 1 hour into an infusion of either DPCPX or its vehicle. In addition, a microdialysis probe was inserted into the renal cortex, and dialysate was collected directly into a freezing apparatus and analyzed for adenosine and its metabolites using reverse-phase high-performance liquid chromatography in the gradient elution mode.

Cisplatin, a compound known to injure renal epithelial cells, reduced GFR approximately 50% and doubled RVR, indicating that cisplatin had indeed caused ATN-induced ARF. However, CPA reduced GFR similarly in both the control and cisplatin-treated rats, and the maximum effect of CPA in control rats was only a 20% reduction in GFR, much less than the 50% reduction in GFR that was achieved even with the modest dose of cisplatin used in this study. Furthermore, intrarenal infusions of CPA did not significantly increase RVR in either control or cisplatin-treated rats. In both control rats and cisplatin-treated rats, a 1-hour infusion of DPCPX did not alter either GFR or RVR, whereas DPCPX did increase urine volume, indicating that a pharmacologically active dose of DPCPX had indeed been administered. Finally, cisplatin treatment was not associated with an increase in the renal cortical interstitial levels of adenosine, inosine, xanthine, or hypoxanthine.

The results obtained in this study are not consistent with the hypothesis outlined in

Figure 46-1. One important caveat, however, is that these studies were conducted 6 days after the administration of cisplatin. It is possible that A_1-receptor–mediated renal vasoconstriction occurred early after cisplatin administration and induced irreversible renal damage due to renal ischemia. The present study would have missed such an occurrence. However, because intrarenal administration of CPA to control rats decreased GFR without increasing RVR, CPA appears to decrease, not increase, energy requirements of the tubules. It is possible, though, that intrarenal administration of adenosine agonists does not mimic endogenously generated adenosine. In this regard, it is important to note that interstitial administration of adenosine appears to increase RVR much more than does intrarenal administration of adenosine [25,28].

To summarize, we found that 1) the renal hemodynamic consequences of cisplatin-induced ARF cannot be mimicked with an intrarenal infusion of an A_1 agonist, 2) cisplatin-induced ARF is not associated with a diminished GFR response to an A_1 agonist, 3) acute blockade of A_1 receptors does not improve renal function in animals with established cisplatin-induced ARF, and 4) cisplatin-induced ARF is not associated with increased renal cortical levels of adenosine or its metabolites. Therefore, the hypothesis shown in Figure 46-1 regarding how adenosine receptor antagonists attenuate ATN-induced ARF when coadministered with a nephrotoxin is incorrect. However, another mechanism by which A_1 antagonists might protect the kidney from nephrotoxins is through inhibition of epithelial cell transport. Inhibition of epithelial cell transport would decrease the workload on injured epithelial cells and also would dilute the nephrotoxin in the tubular fluid. This hypothesis is under investigation.

Acknowledgments

This work was supported by grants from the National Institutes of Health (HL40319, HL35909, and HL14192).

References

1. Hou SH, Bushinsky DA, Wish JB, Cohen JJ, Harrington JT (1983) Hospital-acquired renal insufficiency: A prospective study. Am J Med 74:243–248.
2. Lunding M, Steiness I, Thaysen JH (1964) Acute renal failure due to tubular necrosis. Immediate prognosis and complications. Acta Med Scand 176:103–119.
3. Bidani AK, Churchill PC (1983) Aminophylline ameliorates glycerol-induced acute renal failure in rats. Can J Physiol Pharmacol 61: 567–571.
4. Bowmer CJ, Collis MG, Yates MS (1986) Effect of the adenosine antagonist 8-phenyltheophylline on glycerol-induced acute renal failure in the rat. Br J Pharmacol 88:205–212.
5. Bidani AK, Churchill PC, Packer W (1987) Theophylline-induced protection in myoglobinuric acute renal failure: Further characterization. Can J Physiol Pharmacol 65:42–45.
6. Kellett R, Bowmer CJ, Collis MG, Yates MS (1989) Amelioration of glycerol-induced acute renal failure in the rat with 8-cyclopentyl-1, 3-dipropylxanthine. Br J Pharmacol 98: 1066–1074.
7. Ishikawa I, Shikura N, Takada K (1993) Amelioration of glycerol-induced acute renal failure in rats by an adenosine A_1 receptor antagonist (FR-113453). Ren Fail 15:1–5.
8. Shimada J, Suzuki F, Nonaka H, Karasawa A, Mizumoto H, Ohno T, Kubo K, Ishii A (1991) 8-(Dicyclopropylmethyl)-1,3-dipropylxanthine: A potent and selective adenosine A_1 antagonist with renal protective and diuretic activities. J Med Chem 34:466–469.
9. Panjehshahin MR, Munsey TS, Collis MG, Bowmer CJ, Yates MS (1992) Further characterization of the protective effect of 8-cyclopentyl-1,3-dipropylxanthine on glycerol-induced acute renal failure in the rat. J Pharm Pharmacol 44:109–113.
10. Suzuki F (1992): KW-3902. Drugs Fut 17: 876–878.
11. Andoh TF, Terai T, Nakano K, Horiai H, Mori J, Kohsaka M (1991) Protective effects of FR113453, an adenosine Al-receptor antagonist, on experimental acute renal failure in rats. In International Symposium on Acute Renal Failure (October 2–4, 1991), University of North Carolina, Chapel Hill, NC.
12. Heidemann HT, Bolten M, Inselmann G (1991) Effect of chronic theophylline administration on

amphotericin B nephrotoxicity in rats. Nephron 59:294–298.
13. Lin JJ, Churchill PC, Bidani AK (1988) Theophylline in rats during maintenance phase of post-ischemic acute renal failure. Kidney Int 33:24–28.
14. Heidemann HT, Muller S, Mertins L, Stepan G, Hoffmann K, Ohnhaus EE (1989) Effect of aminophylline on cisplatin nephrotoxicity in the rat. Br J Pharmacol 97:313–318.
15. Knight RJ, Collis MG, Yates MS, Bowmer CJ (1991) Amelioration of cisplatin-induced acute renal failure with 8-cyclopentyl-1,3-dipropylxanthine. Br J Pharmacol 104:1062–1068.
16. Lieberthal W, Levinsky NG (1992) Acute clinical renal failure. In Seldin DW, Giebisch G (eds.), The Kidney: Physiology and Pathophysiology, 2nd ed. New York: Raven Press, pp. 3181–3225.
17. Beck F, Thurau K, Gstraunthaler G (1992) Pathophysiology and pathobiochemistry of acute renal failure. In Seldin DW, Giebisch G (eds.), The Kidney: Physiology and Pathophysiology, 2nd ed. New York: Raven Press, pp. 3157–3225.
18. Osswald H, Hermes HH, Nabakowski G (1982) Role of adenosine in signal transmission of tubuloglomerular feedback. Kidney Int Suppl 12:S136–S142.
19. Schnermann J, Weihprecht H, Briggs JP (1990) Inhibition of tubuloglomerular feedback during adenosine$_1$ receptor blockade. Am J Physiol 258:F553–F561.
20. Franco M, Bell PD, Navar LG (1989) Effect of adenosine A_1 analogue on tubuloglomerular feedback mechanism. Am J Physiol 257: F231–F236.
21. Murray RD, Churchill PC (1985) Concentration dependency of the renal vascular and renin secretory responses to adenosine receptor agonists. J Pharmacol Exp Ther 232:189–193.
22. Nies AS, Beckmann ML, Gerber JG (1991) Contrasting effects of changes in salt balance on the renovascular response to A_1-adenosine receptor stimulation in vivo and in vitro in the rat. J Pharmacol Exp Ther 256:542–546.
23. Inscho EW, Carmines PK, Navar LG (1991) Juxtamedullary afferent arteriolar responses to P_1 and P_2 purinergic stimulation. Hypertension 17:1033–1037.
24. Barrett RJ, Droppleman DA (1993) Interactions of adenosine A_1 receptor-mediated renal vasoconstriction with endogenous nitric oxide and ANG II. Am J Physiol 265:F651–F659.
25. Yagil Y (1994) The effects of adenosine on water and sodium excretion. J Pharmacol Exp Ther 268:826–835.
26. Kuan CJ, Herzer WA, Jackson EK (1992) An experimental paradigm for investigating the role of endogenous adenosine/A_1 receptor interactions in vivo. J Pharmacol Exp Ther 263:657–662.
27. Mi Z, Herzer WA, Zhang Y, Jackson EK (1994) 3-Isobutyl-1-methylxanthine decreases renal cortical interstitial levels of adenosine and inosine. Life Sci 54:PL277–PL282.
28. Agmon Y, Dinour D, Brezis M (1993) Disparate effects of adenosine A_1- and A_2-receptor agonists on intrarenal blood flow. Am J Physiol 265: F802–F806.

VI. THERAPEUTIC APPLICATIONS

A. CARDIOVASCULAR

47. CLINICAL ELECTROPHYSIOLOGY OF ADENOSINE

Anne B. Curtis

Adenosine is an endogenous nucleoside that has become an important agent in the treatment of cardiac arrhythmias. Its major electrophysiologic effects include slowing of the heart rate (negative chronotropic effect) and depression of conduction through the atrioventricular (AV) node (negative dromotropic effect). It is the latter effect that has made adenosine one of the more commonly used drugs for the acute termination of supraventricular tachycardias in which the AV node is involved in a reentrant circuit. The purpose of this chapter is to review the clinical cardiac electrophysiology of adenosine, including its role in the diagnosis and termination of tachycardias.

Physiology

Adenosine is an intermediate metabolite as well as an important physiologic mediator. Of the two cardiac adenosine receptors that have been identified, the A_1 receptor is found on cardiac myocytes and is responsible for the electrophysiologic actions, depression of cardiac contractility, and antiadrenergic effects. The A_2 receptor is found on smooth muscle and endothelial cells, and is responsible for the vascular effects of adenosine, such as coronary vasodilation.

Response to activation of the A_1 receptor is cAMP independent and mediated by a guanine nucleotide regulatory protein (G_i). The most important result of activation of the A_1 receptor is an increase in the steady-state outward potassium current, $I_{K+Ach,Ado}$ [1]. The potassium channels carrying this current are found mostly in the atria, sinus node, and atrioventricular (AV) node. Activation of this current in the atria leads to shortening of action potential duration and hyperpolarization of the membrane. There is also a minor depression of the inward calcium current (I_{Ca}). In the sinus and AV nodes, adenosine directly depresses the action potential and slows conduction [2]. The indirect actions of adenosine are cAMP dependent and mediated by beta-adrenergic inhibition. Antagonism of catecholamine-stimulated adenyl cyclase lowers cAMP levels, leading to decreases in I_{Ca}, the pacemaker current I_f, and I_{Ti}.

Electrophysiologic Effects

When given as a bolus injection, adenosine provokes a biphasic response in heart rate. The initial effect, seen within 15–30 seconds of a peripheral venous injection, is sinus bradycardia that lasts for approximately 5–10 seconds. This is followed by a reflex sinus tachycardia, probably caused by activation of carotid body chemoreceptors, resulting in respiratory stimulation and secondary activation of pulmonary stretch receptors [3].

In addition to its negative chronotropic effect, adenosine has a negative dromotropic effect on the human AV node. When administered to humans, it causes transient (seconds) block of AV conduction [4,5]. This explains the efficacy of adenosine in terminating supraventricular tachycardias in which the AV node is part of a reentrant circuit. There is indirect evidence that the negative dromotropic effect of adenosine in the human heart is frequency dependent. For example, the dose of adenosine required to prolong the atrium-to-His bundle (AH) interval and to cause AV block is greater during normal sinus rhythm or atrial pacing at long cycle lengths than during supraventricular tachycardia [4]. Frequency dependence, in which a drug has a greater effect at more rapid heart rates, has been demonstrated for the type 1 antiarrhythmic drugs as well as verapamil, and may be a key feature of drugs that are effective for tachycardias.

Adenosine has limited effects on other tissues in the heart. It shortens atrial action potential duration and refractoriness [6]. There is no effect on the His Purkinje system. Most accessory atrioventricular pathways are insensitive to adenosine. However, there are occasional patients with adenosine-sensitive accessory pathways; conduction in these pathways is normally decremental [7].

Effects on Specific Arrhythmias

Adenosine will cause transient slowing of sinus tachycardia because of membrane hyperpolarization caused by activation of $I_{KAch,Ado}$ and the antiadrenergic effects on I_f. There are few reports on the effect of adenosine on sinus-node reentrant tachycardia. Adenosine has been reported to have no effect on the tachycardia [8], but in another report transient slowing or termination was demonstrated in two patients [9]. On the other hand, a recent study demonstrated termination of the arrhythmia in all six patients given adenosine [10].

Early reports suggested that atrial tachycardias were insensitive to adenosine [4,8,11]. Haines and DiMarco reported on 14 patients with intraatrial reentrant tachycardia in whom adenosine caused AV block without effect on the atrial rhythm in 13 patients [10]. More recent data and the experience in our own laboratory confirm that it is possible for atrial tachycardias to terminate with adenosine. It is still not clear what precisely determines whether atrial tachycardia will terminate with adenosine. It has been proposed that sensitivity to adenosine is governed by proximity of the arrhythmia focus to the crista terminalis [12]. Adenosine terminated 95% of ectopic atrial tachycardias arising from the crista terminalis but only 29% of those arising elsewhere; the mechanism of the tachycardias was not specified. The response to adenosine may be dictated by the mechanism of the tachycardia. In a recent study, adenosine terminated atrial tachycardia due to triggered activity in two patients, transiently suppressed automatic atrial tachycardia in seven out of seven patients, and had no effect in five patients with intraatrial reentrant tachycardia [10]. The electrophysiologic effects of adenosine may explain the different responses to these tachycardias. Adenosine does not prolong refractoriness in the atrium and thus should have little effect on intraatrial reentrant tachycardia. Transient suppression of automatic atrial tachycardia might be due to activation of $I_{KAch,Ado}$. Finally, termination of atrial tachycardia due to triggered activity could be explained by the antiadrenergic action of adenosine on cellular cAMP levels.

Adenosine has no effect on atrial flutter or fibrillation. Administration of adenosine will cause AV block, which may be helpful in unmasking flutter waves and hence identifying the mechanism of the tachycardia in some patients with atrial flutter [13]. In addition, rather than terminating these arrhythmias, adenosine may actually precipitate atrial flutter or fibrillation because of its ability to shorten

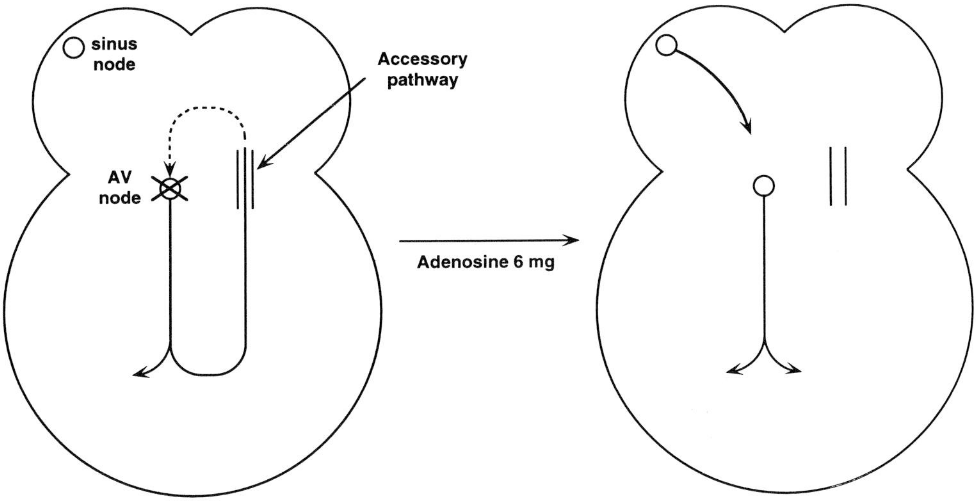

FIGURE 47-1. Diagram showing the reentrant circuit in AV reentrant tachycardia and the effect of adenosine. On the **left**, the circuit is shown to utilize the AV node and His-Purkinje system in the antegrade direction and the accessory pathway in the retrograde direction. Adenosine blocks conduction in the AV node, restoring sinus rhythm (**right**).

atrial action potential duration and refractoriness [8,14].

Adenosine has an important therapeutic role in the termination of two clinically significant tachycardias, AV nodal reentrant tachycardia and AV reentrant tachycardia. Both tachycardias are reentrant circuits in which the AV node is a necessary link in the completion of the electric circuit. The most common form of AV nodal rentrant tachycardia uses a "slow" AV nodal pathway (i.e., a long AH interval) in the antegrade direction and a "fast" pathway (a short HA interval) in the retrograde direction. The depressant effect of adenosine on conduction through the AV node causes termination of AV nodal reentrant tachycardia in well over 90% of patients when given as a rapid bolus injection [4,8]. Indeed, failure of the tachycardia to terminate after a 12-mg bolus of adenosine should call into question the accuracy of the diagnosis, unless the patient is on a medication such as theophylline that antagonizes the effects of adenosine.

The most common tachycardia seen in the Wolff–Parkinson–White syndrome is AV reentrant tachycardia. This tachycardia is caused by a macro-reentrant circuit, involving the AV node in the antegrade direction and an accessory pathway in the retrograde direction. Impulses travel through the atrium and ventricle to complete the circuit (Fig. 47-1). Termination of the tachycardia requires only transient block in any limb of the reentrant circuit. Given the vulnerability of the AV node to pharmacologic manipulation, this is the usual site for termination of these tachycardias. In patients with concealed accessory pathways, any AV nodal blocking agent could theoretically terminate the tachycardia, and the usual choices are verapamil and adenosine [4,8]. As the tachycardia terminates, adenosine sometimes has the added advantage of unmasking transient preexcitation (Fig. 47-2), which may help to confirm a diagnosis of AV reentrant tachycardia.

Adenosine has no significant direct effect on ventricular myocardium, and thus it is not surprising that most forms of ventricular tachycardia are insensitive to the drug [15]. The most common type of ventricular tachy-

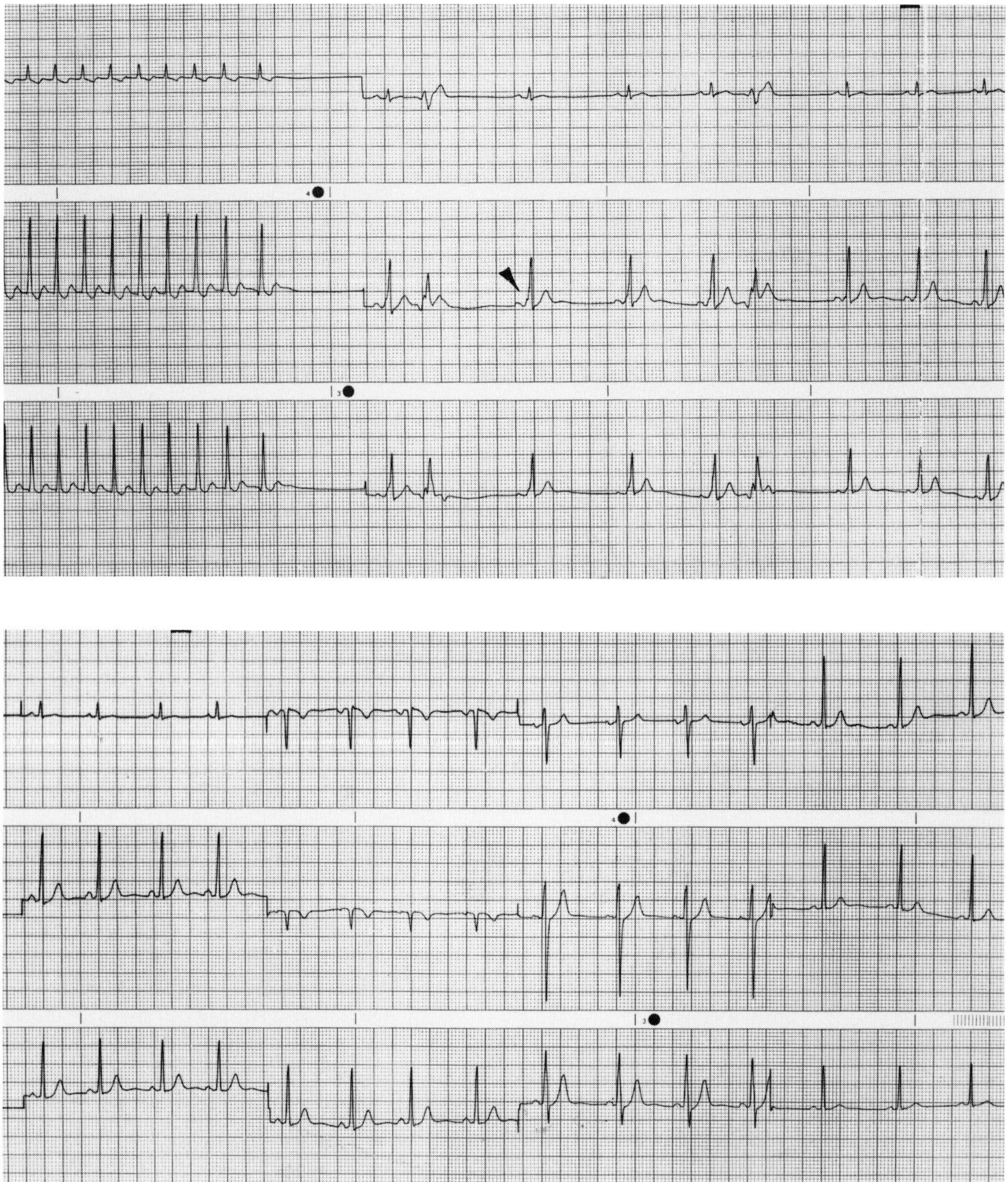

A

B

FIGURE 47-2. **A:** Rhythm strip of electrocardiographic leads I, II, and III during administration of 6 mg of adenosine in a patient with AV reentrant tachycardia. As the tachycardia terminates, a delta wave is evident (arrowhead), confirming that an accessory pathway is present and is likely to be a participant in the tachycardia. **B:** The patient's electrocardiogram during sinus rhythm showed no evidence of Wolff–Parkinson–White syndrome.

cardia, which is due to reentry, is most often found in patients with structural heart disease, usually coronary artery disease. This tachycardia is completely insensitive to adenosine. On the other hand, there is a specific form of ventricular tachycardia characterized by a left bundle branch block, inferior axis morphology that is most commonly seen in younger patients without structural heart disease. The mechanism of this tachycardia has been debated, but there is strong support for the hypothesis that the arrhythmia is due to cAMP-triggered activity [16]. This tachycardia originates in the right ventricular outflow tract, commonly terminates with adenosine, and is curable with radiofrequency ablation.

Adenosine as a Diagnostic Agent

The consistent block of AV nodal conduction afforded by adenosine has created interest in the use of the drug for diagnostic purposes in patients with undiagnosed tachycardias. Adenosine can be used in both narrow and wide complex tachycardias. A rapid 6-mg intravenous bolus is given, followed by 12 mg in 2 minutes if there is no effect from the first bolus. If the AV node is a critical part of a reentrant circuit sustaining the tachycardia, such as in AV nodal reentrant tachycardia or AV reentrant tachycardia, then adenosine will terminate the tachycardia in almost all cases. As explained above, certain other narrow complex tachycardias, such as sinus node reentrant tachycardia and some atrial tachycardias, will also terminate with adenosine (Table 47-1). On the other hand, sinus tachycardia will slow down transiently but not break. Atrial fibrillation, atrial flutter, and multifocal atrial tachycardia will not terminate, but higher degrees of AV block induced transiently by adenosine may make it easier to see atrial activity and thus diagnose the tachycardia. In one recent study, the accuracy of a group of internal medicine housestaff in diagnosing unknown tachycardias was improved significantly when they were shown rhythm strips during the administration of adenosine [17].

TABLE 47-1. Effects of adenosine on tachycardias

Tachycardias that terminate
AV nodal reentrant tachycardia
AV reentrant tachycardia
Atrial tachycardia (some)
Sinus node reentrant tachycardia
cAMP-mediated ventricular tachycardia
Tachycardias that are suppressed transiently
Sinus tachycardia
Atrial tachycardia (some)
Tachycardias that are unaffected (except for provocation of AV block)
Atrial fibrillation
Atrial flutter
Atrial tachycardia (some)
Ventricular tachycardia

Adenosine may also be used in the differential diagnosis of wide-complex tachycardias. The overwhelming majority of patients with coronary artery disease and wide-complex tachycardias have ventricular tachycardia. However, there are situations in which supraventricular tachycardia with aberration is considered possible, and a diagnostic agent to discriminate the two would be useful. It is well known that administration of verapamil to such patients can have disastrous hemodynamic consequences [18]. Adenosine is a much safer alternative for diagnostic or therapeutic purposes in such patients, since the effects last only a few seconds. If adenosine actually terminates the tachycardia, then the differential diagnosis includes right ventricular outflow tract ventricular tachycardia, AV or AV nodal reentrant tachycardia or atrial tachycardia with bundle branch block aberration, and antidromic tachycardia.

Adenosine has also been advocated as an aid in the assessment of the success of radiofrequency or surgical ablation of accessory pathways. If adenosine is to be used in this

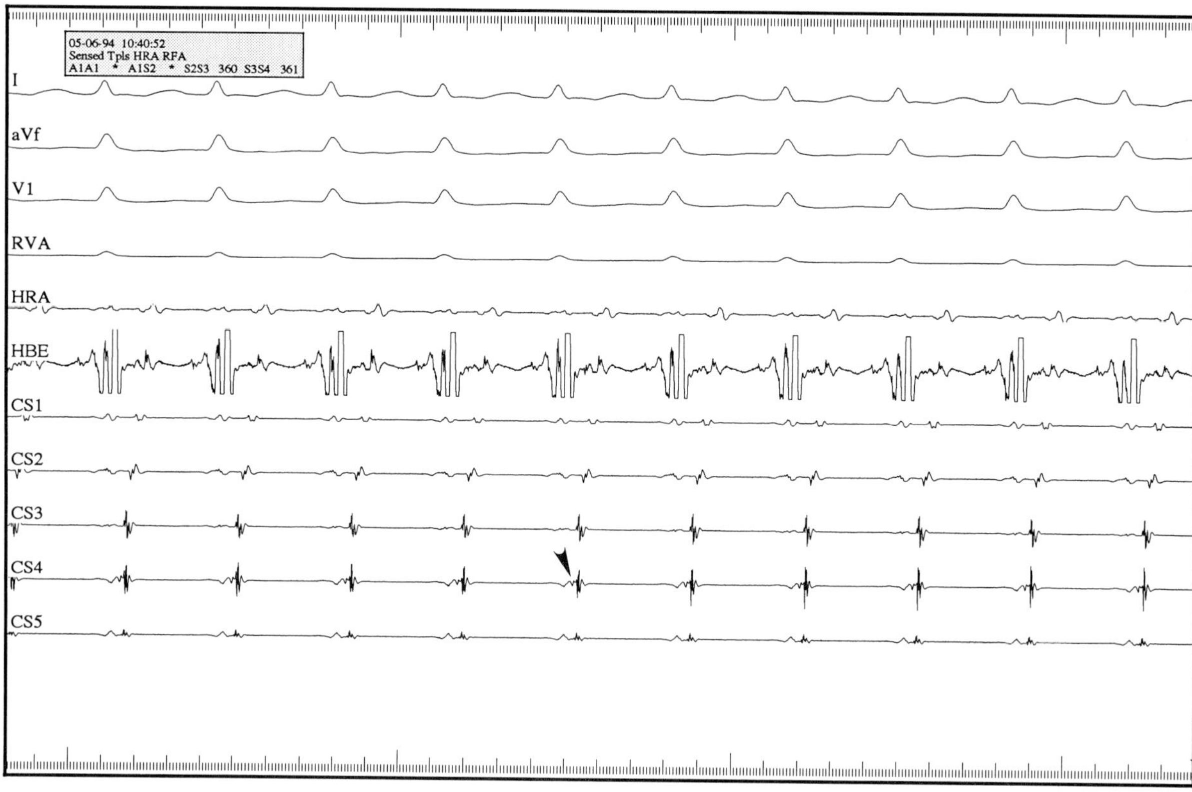

05-06-94 10:40:52
Sensed Tpls HRA RFA
A1A1 * A1S2 * S2S3 360 S3S4 361
I
aVf
V1
RVA
HRA
HBE
CS1
CS2
CS3
CS4
CS5

A

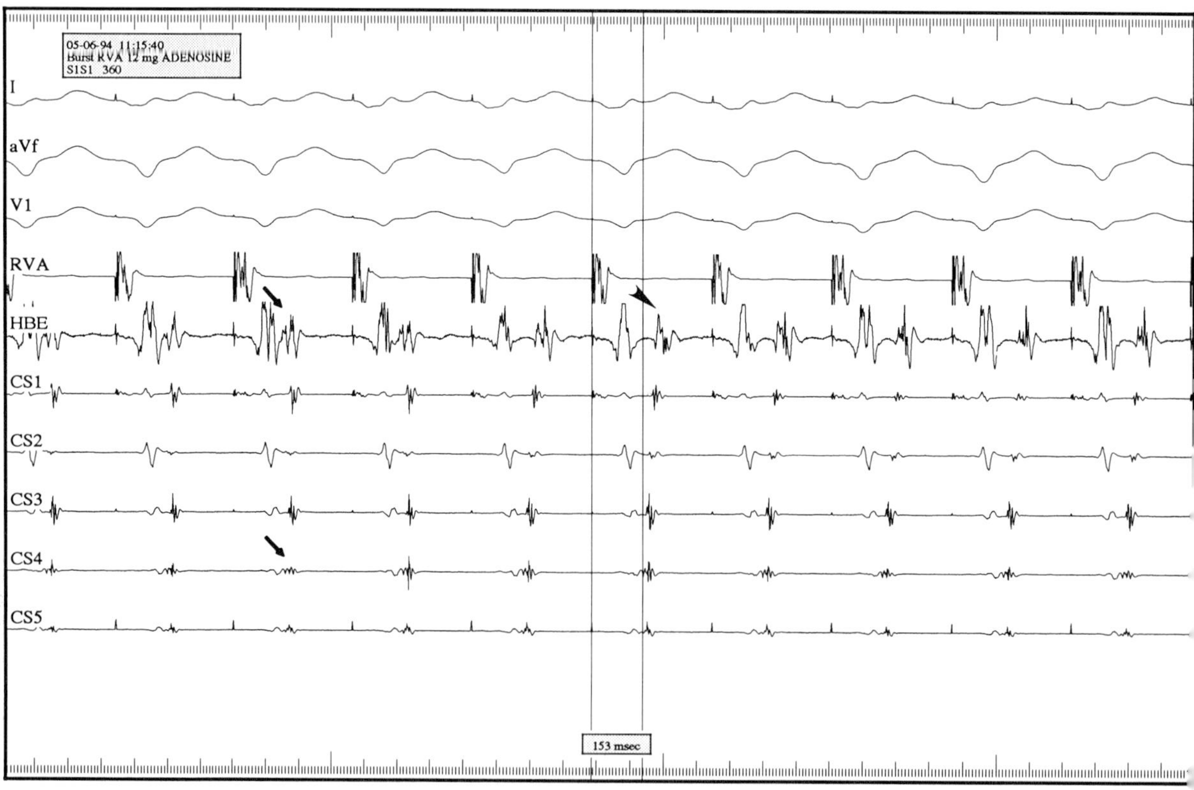

05-06-94 11:15:40
Burst RVA 12 mg ADENOSINE
S1S1 360
I
aVf
V1
RVA
HBE
CS1
CS2
CS3
CS4
CS5
153 msec

B

manner, it is important to administer the drug prior to ablation in order to be certain that the accessory pathway itself is not sensitive to adenosine, which sometimes occurs. If the accessory pathway is insensitive to adenosine, then adenosine may be given after ablation to assess the results. If all communication between the atria and the ventricles has been severed except for the AV node, then administration of adenosine should lead to transient AV block. If residual accessory pathway conduction remains, preexcitation or VA conduction over the pathway may become manifest transiently. In one case in our own series, this technique was helpful in achieving successful ablation of a posteroseptal pathway [19]. There have been occasional reports of exceptions [20], but in general adenosine can be a very useful tool in assessing the presence and persistence of accessory pathways during ablation procedures (Fig. 47-3).

Sick Sinus Syndrome

Patients with sick sinus syndrome may have a constellation of problems, including sinus bradycardia and pauses, atrial dysrhythmias, atrial fibrillation, and AV block. Symptomatic sinus bradycardia and sinus pauses may require implantation of a permanent pacemaker. In order to attempt to avoid pacemaker implantation in some patients, there has been interest in the possibility of using theophylline as an adenosine antagonist to raise the sinus rate sufficiently to avoid the need for pacemaker implantation. The hypothesis with this approach is that the bradycardia is caused by or accentuated by endogenous adenosine; however, there is no proof of this as yet. Nevertheless, there have been at least two studies published so far attesting to the efficacy of this therapeutic approach [21,22]. Theophylline has also been reported to be effective in increasing the ventricular response in patients with atrial fibrillation and a slow ventricular response [23].

FIGURE 47-3. **A:** Orthodromic AV reentrant tachycardia utilizing a left freewall accessory pathway. Shown are surface electrocardiographic leads I, a Vf, and V1 as well as intracardiac recordings from the right ventricular apex (RVA), high right atrium (HRA), His bundle (HBE), and coronary sinus (CS) from proximal (CS1) to distal (CS5). There is a probable accessory pathway potential on CS4 (arrowhead). **B:** During right ventricular pacing, atrial activation is nearly simultaneous in the AV node and in the coronary sinus (arrows). Adenosine, 12 mg intravenously, delays activation through the AV node (arrowhead), thus clarifying the location of the accessory pathway at CS4. Radiofrequency current delivered at this site successfully ablated the accessory pathway.

Role of Adenosine in AV Block Associated with Inferior Myocardial Infarction

Adenosine appears to be a mediator of the sinus bradycardia and AV block that may occur in patients with acute inferior infarction. Adenosine mimics the effects of hypoxia on AV nodal conduction, and during hypoxia and ischemia adenosine is released in amounts sufficient to account for the observed effects on AV conduction. The use of aminophylline, a nonspecific adenosine antagonist, was demonstrated to reverse atropine-resistant AV block in a patient with an acute inferior infarction [24]. However, because aminophylline can cause sinus tachycardia, which is undesirable in patients with acute myocardial infarction, its use should be restricted to very selected patients with acute inferior infarction and symptomatic AV block that is resistant to atropine.

Heart Transplantation

Recipients of orthotopic heart transplants are supersensitive to the negative chronotropic and dromotropic effects of adenosine [25]. Inflammatory cells are potent releasors of adenosine, which may contribute to the depression of sinus node function that may be observed in these patients [26]. In one recent study, administration of theophylline to 15 transplant recipients with bradyarrhythmias restored sinus rhythm in 14 patients; only one required permanent pacemaker implantation [27]. Although the efficacy of theophylline suggests that adenosine may be responsible for these bradyarrhythmias, further work is needed in this area to elucidate the mechanism of post-transplantation bradyarrhythmias.

Drug Interactions

Dipyridamole blocks the reuptake of adenosine into cells and thus enhances its action. It is therefore important to reduce the dose of adenosine in any patient taking dipyridamole, or possibly to avoid it entirely. Theophylline is a nonspecific adenosine antagonist that may be used to reverse any adverse effects of adenosine that occur. Patients who are receiving theophylline may have little or no response to administration of adenosine. It is important to recognize in such patients that lack of termination of the tachycardia may simply be due to antagonism by theophylline and not because the tachycardia mechanism itself is necessarily insensitive to adenosine.

Adverse Effects of Adenosine

Minor side effects, such as dyspnea, chest pain, and flushing, are observed in approximately 30% of patients who receive adenosine [28]. It should be used with caution in patients with asthma because of the potential for bronchoconstriction. Nausea, headache, and lightheadedness may occur infrequently. Precipitation of atrial fibrillation is possible, because of shortening of the action potential duration. There has been a case reported of a patient with long-QT syndrome who developed torsade de pointes after receiving adenosine [29]. Another investigator reported two patients, one who developed prolonged sinus arrest and syncope, and the other who developed syncope with bradycardia and hypotension following administration of adenosine [30]. Administration of adenosine can be associated with more rapid ventricular responses to atrial flutter and fibrillation in patients with Wolff–Parkinson–White syndrome, possibly due to shortening of antegrade refractoriness of accessory connections [31].

Despite occasional reports of adverse reactions to adenosine, overall the drug is quite safe and well tolerated. There is only one report of its use during pregnancy; adenosine successfully terminated supraventricular tachycardia without any effect on fetal heart rate [32]. Adenosine is preferred over verapamil in neonates and infants in whom verapamil can cause hemodynamic collapse [33].

Adenosine Agonists as Antiarrhythmic Agents

The depressant effect of adenosine on AV nodal conduction could make it a useful drug for the chronic management of supraventricular tachycardias, except that its half-life is far too short. Adenosine agonists with longer durations of action could therefore potentially be useful as antiarrhythmic agents. We have studied the adenosine uptake blocker dipyridamole for its antiarrhythmic effects [34]. We measured coronary sinus adenosine levels in patients undergoing electrophysiologic studies for standard indications. Dipyridamole consistently elevated adenosine levels, predominantly during rapid atrial pacing, and caused an increase in the AV Wenckebach cycle length as well as an increase in the functional refractory period of the AV node. When dipyridamole was given to 10 patients with supraventricular tachycardia (AV nodal reentrant tachycardia or AV reentrant tachycardia with a concealed accessory pathway), tachycardia cycle length prolonged in all and the tachycardia terminated in 4 of 10 patients. Adenosine levels were significantly elevated, and the effects of dipyridamole could be reversed by administration of theophylline. These results suggest that dipyridamole enhanced the effects of endogenous adenosine. The development of long-acting adenosine agonists might thus have real potential for a new approach to antiarrhythmic treatment of supraventricular tachycardias.

Conclusions

Adenosine has important electrophysiologic effects on the sinus and AV nodes, as well as on the atria. It is useful for the acute management of supraventricular tachycardia, and it has diagnostic utility as well. Longer acting adenosine agonists may have potential as antiarrhythmic agents, but they are still in the early stages of clinical investigation.

References

1. Lerman BB, Belardinelli L (1991) Cardiac electrophysiology of adenosine: Basic and clinical concepts. Circulation 83:1499–1509.
2. Clemo HF, Belardinelli L (1986) Effect of adenosine on atrioventricular conduction. I: Site and characterization of adenosine action in the guinea pig atrioventricular node. Circ Res 59:427–436.
3. Watt AH, Routledge PA (1986) Transient bradycardia and subsequent sinus tachycardia produced by intravenous adenosine in healthy adult subjects. Br J Clin Pharmacol 21:533–536.
4. DiMarco JP, Sellers TD, Berne RM, West GA, Belardinelli L (1983) Adenosine: Electrophysiologic effects and therapeutic use for terminating paroxysmal supraventricular tachycardia. Circulation 68:1254–1263.
5. Favale S, DiBiase M, Rizzo U, Belardinelli L, Rizzon P (1985) Effect of adenosine and adenosine-5′-triphosphate on atrioventricular conduction in patients. J Am Coll Cardiol 5:1212–1219.
6. Haines DE, Lerman BB, DiMarco JP (1988) Intravenous adenosine shortens atrial refractoriness in man. Circulation 78(Suppl II):II153.
7. Lerman BB, Greenberg M, Overholt ED, Swerdlow CD, Smith RT, Sellers TD, DiMarco JP (1987) Differential electrophysiologic properties of decremental retrograde pathways in long RP′ tachycardia. Circulation 76:21–31.
8. DiMarco JP, Sellers TD, Lerman BB, Greenberg ML, Berne RM, Belardinelli L (1985) Diagnostic and therapeutic use of adenosine in patients with supraventricular tachyarrhythmias. J Am Coll Cardiol 6:417–425.
9. Griffith MJ, Garrat CJ, Ward DE, Camm AJ (1989) The effects of adenosine on sinus node reentrant tachycardia. Clin Cardiol 12:409–411.
10. Engelstein ED, Lippman N, Stein KM, Lerman BB (1994) Mechanism-specific effects of adenosine on atrial tachycardia. Circulation 89:2645–2654.
11. Haines DE, DiMarco JP (1990) Sustained intraatrial reentrant tachycardia: Clinical, electrocardiographic and electrophysiologic characteristics and long-term follow-up. J Am Coll Cardiol 15:1345–1354.
12. Shenasa H, Cooper RA, Pressley J, Sorrentino RA, Greenfield RA, Merrill JJ, Wharton JM (1994) Site of origin of ectopic atrial tachycardia predicts its response to adenosine. J Am Coll Cardiol 23:250.
13. Rankin AC, Oldroyd KG, Chong E, Rae AP, Cobbe SM (1989) Value and limitations of adenosine in the diagnosis and treatment of narrow and broad complex tachycardias. Br Heart J 162:195–203.
14. Lerman BB, Wesley RC, Belardinelli L (1989) Electrophysiologic effects of dipyridamole on atrioventricular nodal conduction and supraventricular tachycardia: Role of endogenous adenosine. Circulation 80:1536–1543.
15. Griffith MJ, Linker NJ, Garratt CJ, Ward DE, Camm AJ (1990) Relative efficacy and safety of intravenous drugs for termination of sustained ventricular tachycardia. Lancet 336:670–673.
16. Lerman BB, Belardinelli L, West GA, Berne RM, DiMarco JP (1986) Adenosine-sensitive ventricular tachycardia: Evidence suggesting cyclic AMP-mediated triggered activity. Circulation 74:270–280.
17. Conti JB, Curtis AB, Belardinelli L (1992) The value of adenosine in the diagnosis of tachyarrhythmias. PACE 15:595.
18. Stewart RB, Bardy GH, Greene HL (1986) Wide complex tachycardia: Misdiagnosis and outcome after emergent therapy. Ann Intern Med 104:766–771.
19. Keim S, Curtis AB, Epstein ML, Staples ED, Belardinelli L (1992) Adenosine-induced atrioventricular dissociation: A method to assess surgical ablation of accessory atrioventricular

pathways. J Am Coll Cardiol 19:1005–1012.
20. Engelstein E, Wilber D, Wadas M, Stein K, Lippman N, Lerman B (1994) Limitations of adenosine in assessing the efficacy of radiofrequency catheter ablation of accessory pathways. Am J Cardiol 73:774–779.
21. Alboni P, Ratto B, Cappato R, Rossi P, Gatto E, Antonioli GE (1991) Clinical effects of oral theophylline in sick sinus syndrome. Am Heart J 122:1361–1367.
22. Saito D, Katashi M, Yamanari H, Obayashi N, Uchida S, Maekawa K, Sato T, Mizuo K, Kobayashi H, Haraoka S (1993) Effects of oral theophylline on sick sinus syndrome. J Am Coll Cardiol 21:1199–1204.
23. Alboni P, Paparella N, Cappato R, Pirani R, Yiannacopulu P, Antonioli GE (1993) Long-term effects of theophylline in atrial fibrillation with a slow ventricular response. Am J Cardiol 72:1142–1145.
24. Wesley RC, Lerman BB, DiMarco JP, Berne RM, Belardinelli L (1986) Mechanism of atropine-resistant atrioventricular block during inferior myocardial infarction: Possible role of adenosine. J Am Coll Cardiol 8:1232–1234.
25. Ellenbogen KA, Thames MD, DiMarco JP, Sheehan H, Lerman BB (1990) Electrophysiological effects of adenosine in the transplanted human heart: Evidence of supersensitivity. Circulation 81:821–828.
26. Madara JL, Patapoff TW, Gillece-Castro B, Colgan SP, Parkos CA, Delp KC, Mrany RJ (1993) 5′-Adenosine monosphosphate is the neutrophil-derived paracrine factor that elicits chloride secretion from T84 intestinal epithelial cell monolayers. J Clin Invest 91:2320–2325.
27. Redmond JM, Zehr KJ, Gillinov MA, Baughman KL, Augustine SM, Cameron DE, Stuart RS, Acker MA, Gardner TJ, Reitz BA, Baumgartner WA (1993) Use of theophylline for treatment of prolonged sinus node dysfunction in human orthotopic heart transplantation. J Heart Lung Transplant 12:133–139.
28. DiMarco JP, Miles W, Akhtar M, Milstein S, Sharma AD, Platia E, McGovern B, Scheinman MM, Govier WC, The Adenosine for PSVT Study Group (1990) Adenosine for paroxysmal supraventricular tachycardia: Dose ranging and comparison with verapamil. Ann Intern Med 113:104–110.
29. Wesley RC, Turnquest P (1992) Torsades de pointe after intravenous adenosine in the presence of prolonged QT syndrome. Am Heart J 123:794–796.
30. Reed R, Falk JL, O'Brien J (1991) Untoward reaction to adenosine therapy for supraventricular tachycardia. Am J Emerg Med 9:566–570.
31. Garratt CJ, Griffith MJ, O'Nunain S, Ward DE, Camm AJ (1991) Effects of intravenous adenosine on antegrade refractoriness of accessory atrioventricular connections. Circulation 84:1962–1968.
32. Harrison JK, Greenfield RA, Wharton JM (1992) Acute termination of supraventricular tachycardia by adenosine during pregnancy. Am Heart J 123:1386–1388.
33. DiMarco JP (1994) Adenosine: Diagnostic and therapeutic uses in cardiac arrhythmias. Cardiol Rev 2:33–41.
34. Conti JB, Utterback D, Belardinelli L, Curtis AB (1995) Endogenous adenosine is an antiarrhythmic agent. Circulation, in press.

48. MYOCARDIAL IMAGING DURING ADENOSINE INFUSION

Mario S. Verani
John J. Mahmarian

It is ironic that adenosine has only been used as an adjunct to cardiac imaging in the last 4 years, in view of the potent coronary vasodilatory actions of adenosine, which have been known since at least 1929 [1]. Adenosine has been also widely used to elicit changes in coronary blood flow in animal experimentation. Dipyridamole, which acts by increasing the blood and tissue levels of adenosine, has also been widely used as an adjunct to cardiac imaging since 1978 [2].

The very first attempt to use pharmacologically induced coronary vasodilation to detect coronary stenosis by cardiac imaging was that of Strauss and Pitt [5]. These investigators used ethyladenosine in dogs with experimentally produced subcritical coronary stenosis. These investigators demonstrated that the myocardial thallium-201 uptake ratio between the myocardium perfused by a stenotic artery and that perfused by normal arteries was 0.56 after administration of ethyladenosine, compared to a ratio of near unity in the baseline state. This reduced thallium uptake ratio was well matched by a similar decrease in absolute flow ratio measured by radioactive microspheres. Moreover, Strauss and Pitt performed thallium-201 imaging in a few dogs who received a thallium injection after establishment of the coronary stenosis and following the infusion of ethyladenosine. In these animals, these investigators documented the presence of perfusion defects.

A few years after Strauss and Pitts' report, Leppo et al. [6] used direct infusions of adenosine in combination with thallium perfusion scintigraphy in dogs with experimental coronary stenosis. These investigators demonstrated a large increase in coronary blood flow, averaging four times the baseline values, in normal coronary arteries after adenosine, with a concomitant lack of increase in flow through the stenotic artery. These authors demonstrated it was possible to readily identify perfusion abnormalities on their images.

Nearly a decade went by since the above-mentioned experiments before adenosine was finally considered for cardiac imaging in patients with coronary artery disease. It is interesting to speculate on the reasons for such a long interval between the animal experiments and the clinical investigations. Investigators may have been concerned with the ultrashort half-life of exogenous adenosine (2–10 seconds), which would raise the question of how much adenosine would arrive in the coronary arteries from a peripheral arm vein infusion. Second, there may have been concerns regarding the safety of large intravenous doses of a drug with potent coronary vasodilatory actions, which, under appropriate circumstances, could lead to a coronary steal phenomenon and potentially trigger severe myocardial ischemia in patients with coronary artery disease. Another concern might have been the potential for slowing the atrioventricular conduction, which for years had been known to occur after intravenous or intracoronary bolus injections of adenosine.

These concerns notwithstanding, adenosine was finally used as an adjunct to myocardial perfusion imaging. Siffring et al. in 1989 reported the use of adenosine during thallium-201 scintigraphy in healthy volunteers. Earlier that year Verani's group presented their initial experience with adenosine thallium scintigraphy for the diagnosis of coronary artery disease [7]. This was rapidly followed by full reports from Verani et al. [8] and Iskandrian and colleagues [9]. Other reports followed under diagnostic uses of adenosine thallium scintigraphy [10–21].

Mechanism of Perfusion Defects during Adenosine Scintigraphy

Coronary artery disease is identified by adenosine perfusion scintigraphy when one sees a perfusion defect in the images obtained shortly after the tracer injection. Gould [2] masterfully established the physiologic basis of such perfusion defects, namely, a large (three- to sixfold) increase in coronary blood flow in the normal arteries and a lack of increase, a lesser increase, or even a flow decrease in the vascular territory perfused by the stenotic artery under the effect of a potent coronary vasodilator. In most cases, the flow in the territory with the perfusion defect is similar to the flow at baseline, and because the myocardial oxygen demands are only slightly increased, if at all, during coronary vasodilation, there is no true myocardial ischemia but rather a state of heterogeneous myocardial perfusion. However, there is often a redistribution of flow across the ventricular wall in the territory perfused by the stenotic vessel after adenosine infusion, that is, the epicardial flow increases somewhat whereas the subendocardial flow decreases. If the reduction in the endocardial flow is severe, then a state of true myocardial ischemia may ensue due to this transmural or endocardial to epicardium coronary steal.

On the basis of the above-described mechanism, one would expect that myocardial perfusion scintigraphy would be a sensitive technique to demonstrate the presence of coronary stenosis based on the visualization of heterogeneous myocardial perfusion, whereas techniques based on the demonstration of deterioration of wall motion during administration of adenosine, which requires the presence of ischemia, should be less sensitive. These predictions have indeed been borne out by the available data.

Practical Considerations during Adenosine Infusion

The initial protocol proposed by Verani et al. [8] consisted of an intravenous infusion of adenosine through a pump, titrated from an initial dose of 50 μg/kg/min to a maximal of 140 μg/kg/min, with dose increases every minute. Thallium-201 was injected after 1 minute at the highest dose, which was then maintained for two additional minutes. Single-photon emission tomography images were obtained 5 minutes after the thallium administration and redistribution images 4 hours later. Nearly all patients tolerated the 140 μg/kg/min dose. On that basis, the protocol was then simplified to a continuous infusion of 140 μg/kg/min, with the tracer injection taking place midway through the infusion.

A few technical details warrant discussion. First, it is absolutely necessary that adenosine be injected through pump infusion to maintain a steady, accurate infusion rate. Second, the constancy of the infusion rate should not be interrupted with maneuvers such as inflating and deflating a blood pressure cuff above the site of the infusion, or administering thallium-201 and flushing it in the same line where adenosine is being infused, which could have detrimental consequences if the rate of adenosine infusion is suddenly accelerated. In order to avoid problems with flow steadiness, it was proposed to inject adenosine and thallium in two separate veins in the upper extremities. Subsequently, we have demonstrated that with the appropriate setup it is possible to inject adenosine and thallium on the same line, provided that thallium is injected slowly on a side branch near the venous entry and is pushed in by the dripping adenosine solution, rather than with a separate flush injection. Third, it is important to ascertain that patients are not receiving dipyridamole in the 24 hours preceding an adenosine test, because dipyridamole would markedly raise the blood and tissue levels of adenosine by inhibiting its transport across the cell membrane. It is equally important to ascertain that patients are not taking medications containing xanthines (theophylline, caffeine), because these drugs antagonize the coronary vasodilator effect of adenosine by binding with the adenosine cell-wall receptors.

Hemodynamic Effects of an Intravenous Adenosine Infusion

Adenosine-induced vasodilation is much more prominent in the coronary arteries than in the systemic arteries. Consequently, large increases in coronary flow are accompanied by relatively minor changes in the systemic circulation. Accordingly, the coronary flow velocity measured by coronary Doppler catheter increases an average of 4.3-fold above the resting values [22]. The flow velocity by the Doppler method is linearly related to flow; hence, a flow increase of the same magnitude is expected. The changes in systemic hemodynamics include a modest decrease in systolic and diastolic blood pressure and heart rate [8,11]. The cardiac output increases by an average of 56%, and the pulmonary arterial wedge pressure increases slightly in normal subjects and more markedly so in patients with coronary artery disease [23]. The increase in cardiac output is almost exclusively due to the increase in heart rate, with little change in stroke volume. In normal individuals, a slight increase in ejection fraction may occur. If and when myocardial ischemia occurs during adenosine administration, the ejection fraction may fall, with concomitant deterioration in ventricular wall motion.

The cerebral blood flow also increases during administration of adenosine, an observation that is important with respect to the testing of patients with cerebral vascular disease. The test is usually well tolerated in these patients, despite the fall in systolic blood pressure, possibly because of the cerebral vasodilation.

Tolerance and Side Effects of Adenosine

Adenosine administration produces side effects in most patients. In the large series reported by Abreu et al. [15], side effects occurred in 79% of the patients but were generally well tolerated. The side effects usually start within 1–2 minutes of the infusion at the highest dose and subside within 1 or 2 minutes after discontinuing the infusion. A similarly high frequency of side effects has been observed by other investigators [9–12,24]. The most frequent side effects are facial flushing (35%), chest pain (34%), headache (21%), and dyspnea (19%) [15]. While the mechanism of flushing and headache is related to the vasodilatory effect of adenosine and dyspnea is due in most patients to direct stimulation by adenosine of the carotid respiratory chemoreceptors, the etiology of chest pain is not clear. In patients with coronary artery disease, chest pain may mimic their usual angina and may be due to a coronary steal. However, many normal individuals also experience chest pain during adenosine infusion, and in fact the frequency of chest pain appears to be dose related. Crea et al. [25] have suggested that adenosine stimulation of the A_1 adenosine receptors is the reason for chest pain. Bronchospasm has occurred only occasionally, but adenosine has the potential to induce bronchospasm in patients with asthma or severe chronic obstructive pulmonary disease, and thus should not routinely be administered to these patients.

Electrocardiographic changes of ischemia (≥1 mm ST-segment depression) occurs in 8–25% of the patients and may be more frequent in patients who have transient perfusion defects. The ST-segment depression is a good indication of coronary artery disease and is associated with high-grade coronary stenosis and the presence of collateral circulation [16].

First-degree atrioventricular block occurs in approximately 10% of all patients receiving adenosine, and second-degree AV block occurs in about 4%. Third-degree AV block occurs in less than 1% [26]. In a large series from the National Registry on Adenosine Scintigraphy, including over 9000 patients studied, the test had a remarkable safety record with no deaths and only one case of acute myocardial infarction and one of syncope. However, both patient screening as well as the performance of the test itself need to be meticulously done because there is the potential for severe complications, just like any other stress test applied to patients with severe coronary artery disease [27].

Sensitivity and Specificity of Adenosine Scintigraphy for Coronary Artery Disease Detection

The diagnostic value of adenosine scintigraphy has been reported by several investigators in a total of over 800 patients. In these studies, thallium-201 has been used as the usual tracer, although Marwick et al. [28] have instead used Tc-99 m sestamibi. In the reported studies, the sensitivity has ranged from 83% to 97% and specificity ranged from 38% to 94%. The somewhat low specificity in some series may be ascribed to referral bias and/or problems with the SPECT technique. In most reports, however, the specificity has been closer to 85–90%. In patients with coronary artery disease without a previous myocardial infarction, perfusion defects elicited by adenosine administration typically fill in during redistribution or at rest, in a similar fashion as with exercise or dipyridamole scintigraphy.

Several studies have also compared adenosine scintigraphy to exercise scintigraphy in the same patients. In general, similar sensitivity and specificity have been found between exercise and adenosine scintigraphy [29,30]. In the multicenter trial reported by Nishimura et al. [19], myocardial perfusion defect sizes during adenosine scintigraphy correlated well with defect sizes measured by exercise scintigraphy ($r = 0.80$); however, the defect sizes were significantly larger during adenosine than during exercise scintigraphy; thus, adenosine may actually be more sensitive than exercise for assessing the total extent of jeopardized myocardium.

Adenosine Scintigraphy in the Preoperative Risk Stratification for Vascular Surgeries

We assessed the use of adenosine thallium tomography to risk stratify 107 patients prior to vascular arterial reconstructive surgery. Abnormal adenosine thallium tomograms occurred in 57 patients (33%). There were three postoperative cardiac deaths, all in patients undergoing aortic aneurysmectomy. Another patient with an aortic aneurysm developed unstable angina and one patient with lower limb arterial surgery had postoperative pulmonary edema. No patients without transient defects had an event. Cardiac events only occurred in patients with transient thallium defects. However, only 5 of 47 such patients had events. The perfusion defects and the ischemic defect sizes were nearly threefold larger in patients with than those without events. Thus, absence of reversible hypoperfusion during adenosine thallium tomography ensures the virtual absence of postoperative cardiac events [29].

RISK STRATIFICATION BEFORE MAJOR OPERATIONS

Shaw et al. [17] have recently reported their experience with adenosine thallium scintigraphy for risk stratifying patients under consideration for vascular, orthopedic, or general surgeries. In this study mixed perfusion effects (those with partial redistribution at rest, probably indicative of a combination of scar and ischemia) were the strongest predictor of subsequent cardiac events; all patients who subsequently developed cardiac events had either partial or totally reversible perfusion defects during adenosine scintigraphy. Patients without perfusion defects had no events.

RISK STRATIFICATION AFTER ACUTE MYOCARDIAL INFARCTION

Our group has recently reported the use of adenosine thallium tomography performed an average of 5 days after myocardial infarction in 120 stable patients. The test was well tolerated without any serious untoward effects and showed a high sensitivity (95%) for detection of the infarct-related artery and moderate sensitivity (70%) for stenosis in the non-infarct-related artery. The total extent of the perfusion defects, assessed by computerized maps, and the amount of defect reversibility were the best predictors of short-term events [21].

ADENOSINE AS AN ADJUNCT TO POSITRON EMISSION COMPUTER TOMOGRAPHY

Gupta et al. [31] have used adenosine in combination with nitrogen-13 ammonia and positron emission tomography. These authors found a high accuracy (97%) for detection of coronary artery disease with this technique.

ADENOSINE TWO-DIMENSIONAL ECHOCARDIOGRAPHY

Zoghbi et al. [32] were the first to report the combination of adenosine with two-dimensional (2D) echocardiography in patients with coronary artery disease. These investigators found an overall sensitivity of 85% for detection of $\geq$75% coronary stenosis, with a specificity of 92%. However, in patients with normal baseline electrocardiograms the sensitivity was only 60%. Other investigators have assessed the use of 2D echocardiography in combination with adenosine [9,28,33]. In these studies, adenosine has had a relatively low sensitivity, similar to the sensitivity previously reported for dipyridamole echocardiography.

Thus, in selected patients adenosine 2D echocardiography may allow the documentation of transient wall motion deterioration induced by adenosine, which is an indirect marker of significant coronary stenosis. However, because in many patients true myocardial ischemia does not occur during adenosine infusion, a deterioration of wall motion cannot be depended upon as a marker of coronary artery disease. On the other hand, because perfusion scintigraphy only depends on the production of heterogeneous myocardial blood flow distribution, it has good sensitivity for coronary artery disease detection. Marwick et al. [28] have compared perfusion tomography using Tc-99m sestamibi with 2D echocardiography during adenosine infusion and, as expected, found higher sensitivity for adenosine perfusion tomography.

Future Perspectives

Adenosine perfusion scintigraphy has been shown to be a very effective technique to demonstrate the presence of coronary artery disease by noninvasive means. However, the price to pay is a high frequency of side effects, which although very transient, can be quite annoying to some patients. Many of the side effects are believed to be due to stimulation of the adenosine A_1 receptors. Thus, current research is focusing on the possible combination of adenosine A_1 antagonists with adenosine to minimize the side effects during infusion. A second alternative would be the use of selective adenosine A_2 agonists. Although these modalities have not been tested as yet in human beings, it is expected they will be soon.

References

1. Drury AN, Szent-Gyorgyi A (1929) The physiological activity of adenine compounds with

special reference to their action upon the mammalian heart. J Physiol 68:213–237.
2. Gould KL (1978) Noninvasive assessment of coronary stenoses by myocardial perfusion imaging during pharmacologic coronary vasodilation: I. Physiologic basis and experimental validation. Am J Cardiol 41:267–278.
3. Gould KL, Westcott RJ, Albro PC, et al. (1978) Noninvasive assessment of coronary stensosis by myocardial imaging during pharmacologic coronary vasodilatation: II. Clinical methodology and feasibility. Am J Cardiol 41: 279–287.
4. Albro PC, Gould KL, Westcott RJ, et al. (1978) Noninvasive assessment of coronary stenoses by myocardial imaging during pharmacologic coronary vasodilation III. Clinical trial. Am J Cardiol 42:751–760.
5. Strauss HW, Pitt B (1977) Noninvasive detection of subcritical coronary arterial narrowing with a coronary vasodilator and myocardial perfusion imaging. Am J Cardiol 39:403–406.
6. Leppo J. Rosenkrantz J, Rosenthal R, et al. (1981) Quantitative thallium-201 redistribution with a fixed coronary stenosis in dogs. Circulation 63:632–639.
7. Staudacher RA, Mahmarian JJ, Hixson JB, Boyce TM, Pacifico A, Kugiyama K, Verani MS (1989) Adenosine thallium-201 scintigraphy: Feasibility, safety and initial results in man. J Am Coll Cardiol 13:161 (Abst).
8. Verani MS, Mahmarian JJ, Hixson JB, et al. (1990) Diagnosis of coronary artery disease by controlled coronary vasodilation with adenosine and thallium-201 scintigraphy in patients unable to exercise. Circulation 82:80–87.
9. Nguyen T, Heo J, Ogilby JD, et al. (1990) Single-photon emission computed tomography with thallium-201 during adenosine-induced coronary hyperemia correlation with coronary arteriography, exercise thallium imaging and two-dimensional echocardiography. J Am Coll Cardiol 16:1375–1383.
10. Coyne EP, Belvedere DA, Vande-Streek PR, et al. (1991) Thallium-201 scintigraphy after intravenous infusion of adenosine compared with exercise thallium testing in the diagnosis of coronary artery disease. J Am Coll Cardiol 17: 1289–1294.
11. Nishimura S, Mahmarian JJ, Boyce TM, et al. (1991) Quantitative thallium-201 single-photon emission computed tomography during maximal pharmacologic coronary vasodilation with adenosine for assessing coronary artery disease. J Am Coll Cardiol 18:736–745.
12. Gupta NC, Esterbrooks DJ, Hilleman DE, et al. (1992) Comparison of adenosine and exercise thallium-201 single-photon emission computed tomography (SPECT) myocardial perfusion imaging. J Am Coll Cardiol 19:248–257.
13. O'Keefe JH Jr, Bateman TM, Silvestri R, et al. (1992) Safety and diagnostic accuracy of adenosine thallium-201 scintigraphy in patients unable to exercise and those with left bundle branch block. Am Heart J 124:614–621.
14. Iskandrian As, Heo J, Nguyen T, et al. (1992) Tomographic myocardial perfusion imaging with technetium-99 m teboroxime during adenosine-induced coronary hyperemia: Correlation with thallium-201 imaging. J Am Coll Cardiol 19:307–312.
15. Abreu A, Mahmarian JJ, Nishimura S, et al. (1991) Tolerance and safety of pharmacologic coronary vasodilation with adenosine in association with thallium-201 scintigraphy in patients with suspected coronary artery disease. J Am Coll Cardiol 18:730–735.
16. Nishimura S, Kimball KT, Mahmarian JJ, et al. (1993) Angiographic and hemodynamic determinants of myocardial ischemia during adenosine thallium-201 scintigraphy in coronary artery disease. Circulation 87:1211–1219.
17. Shaw L, Miller DD, Kong BA, et al. (1992) Determination of preoperative cardiac risk by adenosine thallium-201 myocardial imaging. Am Heart J 124:861–869.
18. Nishimura S, Mahmarian JJ, Verani MS. (1992) Significance of increased lung thallium uptake during adenosine thallium-201 scintigraphy. J Nucl Med 33:1600–1607.
19. Nishimura S, Mahmarian JJ, Boyce TM, et al. (1992) Equivalence between adenosine and exercise thallium-201 myocardial tomography: A multicenter, prospective, crossover trial. J Am Coll Cardiol 20:265–275.
20. Iskandrian AS, Heo J, Nguyen T, et al. (1990) Left ventricular dilation and pulmonary thallium uptake after single-photon emission computed tomography using thallium-201 during adenosine-induced coronary hyperemia. Am J Cardiol 66:807–811.
21. Mahmarian JJ, Pratt CM, Nishimura S, et al. (1993) Quantitative adenosine thallium-201 single-photon emission computed tomography for the early assessment of patients surviving acute myocardial infarction. Circulation 87: 1197–1210.
22. Wilson RF, Wyche K, Christensen BV, et al.

(1990) Effects of adenosine on human coronary arterial circulation. Circulation 82:1595–1606.
23. Ogilby JD, Iskandrian AS, Untereker WJ, et al. (1992) Effect of intravenous adenosine infusion on myocardial perfusion and function: Hemodynamic/angiographic and scintigraphic study. Circulation 86:887–895.
24. Iskandrian AS, Heo J, Nguyen T, et al. (1991) Assessment of coronary artery disease using single-photon emission tomography with thallium-201 during adenosine-induced coronary hyperemia. Am J Cardiol 67:1190–1194.
25. Crea F, Pupita G, Gallassi AR, et al. (1990) Role of adenosine in pathogenesis of anginal pain. Circulation 81:164–172.
26. Lee J, Heo J, Ogilby JD, et al. (1992) Atrioventricular block during adenosine thallium imaging. Am Heart J 123:1569–1574.
27. Cerqueira MD, Verani MS, Schwaiger M, Heo J, Iskandrian AS, and the Investigators of the Multicenter Adenoscan Trial (1994) Safety profile of adenosine stress perfusion imaging: Results from the adenoscan multicenter trial registry. J Am Coll Cardiol 23:384–389.
28. Marwick T, Willemart B, D'Hondt AM, et al. (1993) Selection of the optimal nonexercise stress for the evaluation of ischemic regional myocardial dysfunction and malperfusion: Comparison of dobutamine and adenosine using echocardiography and ^{99m}Tc MIBI single-photon emission computed tomography. Circulation 87:345–354.
29. Verani MS (1993) Pharmacologic stress myocardial perfusion imaging. Curr Probl Cardiol 8:481–528.
30. Iskandrian AS, Verani MS, Heo J (1994) Pharmacologic stress testing: Mechanism of action, hemodynamic responses, and results in detection of coronary artery disease. J Nucl Cardiol 1:94–111.
31. Gupta NC, Esterbrooks D, Mohiuddin S, et al. (1991) Adenosine in myocardial perfusion imaging using positron emission tomography. Am Heart J 122:293–301.
32. Zoghbi WA, Cheirif J, Kleiman NS, et al. (1991) Diagnosis of ischemic heart disease with adenosine echocardiography. J Am Coll Cardiol 18:1271–1279.
33. Martin TW, Seaworth JF, Johns JP, et al. (1992) Comparison of adenosine, dipyridamole, and dobutamine in stress echocardiography. An Intern Med 116:190–196.

49. INTRAVENOUS ADENOSINE AND LIDOCAINE IN THE LIMITATION OF INFARCT SIZE: PRELIMINARY SAFETY DATA

Kirk N. Garratt
Raymond J. Gibbons
Guy S. Reeder
Dennis A. Laudon
Joseph K. Lobl
David R. Holmes, Jr.

Reperfusion injury has been defined as "those metabolic, functional and structural consequences of restoring coronary arterial flow . . . that can be avoided or reversed by modification of the conditions of reperfusion" [1]. The adverse consequences of reperfusion with unmodified whole blood may be manifest as reversible injury (an injury that reduces the functional efficiency of myocardium without causing cell death) or lethal reperfusion injury (an injury that results in "the death of myocytes, alive at the time of reperfusion, as a direct result of some aspect of the process of reperfusion" [2]. Although controversial, the existence of reperfusion injury is suggested by several lines of animal investigation.

In humans, studies of reperfusion injury have been primarily of interest to cardiothoracic surgeons. Animal laboratory observations formed the basis for modifications made in cardioplegia solutions and methods of reperfusion used following circulatory arrest and cardiopulmonary bypass support [3–12]. In this setting, the myocardial ischemia that develops during a period of iatrogenic cardiac arrest, necessary to complete a surgical procedure, may lead to myocyte necrosis if the ischemia is prolonged and/or reperfusion is established with a perfusate of warm, whole blood. Use of cardioplegic and reperfusion solutions that contain excess buffer, reduced concentrations of ionized calcium, increased osmolality, and compounds that support substrate-level phosphorylation (such as glutamate) results in significantly enhanced myocardial performance [1,13].

Unfortunately, human myocardial ischemia is most frequently encountered in a much less controlled setting. Clinical syndromes related to myocardial ischemia develop when thrombus forms over an ulcerated atherosclerotic plaque within a coronary artery, severely reducing anterograde blood flow. Partial occlusion of a coronary artery may cause unstable angina pectoris, while an obstructive thrombus, in the absence of significant collateral blood vessels, may result in prolonged myocardial ischemia, culminating in myocyte necrosis. More than a million myocardial infarctions occur annually in the United States [14]. In the past, nearly 40% of the 700,000 cardiac-related deaths occurring each year have been due to acute myocardial infarction [14,15]. Mortality related to myocardial infarction has been sharply reduced, however, through the use of thrombolytic agents or direct balloon angioplasty techniques (use of an inflatable balloon catheter instead of thrombolytic drugs to relieve coronary obstruction) that can restore coronary patency.

The benefit of reperfusion therapy may be limited by many variables, the most important of which is likely to be the time from the onset of symptoms to the administration of effective therapy. However, on the basis of observations made in animal studies of reperfusion and in patient studies of cardiopulmonary bypass support, the conditions of reperfusion may also be of importance. Currently, all patients successfully treated with thrombolytic drugs or direct angioplasty undergo reperfusion with unmodified whole blood. Since modification of the perfusate following *controlled* myocardial ischemia has been shown to preserve or augment myocyte viability, it is possible that modification of the perfusate following *uncontrolled* myocardial ischemia may also be of benefit. Although the intracoronary or systemic application of some agents could theoretically attenuate reperfusion injury in the setting of human acute myocardial infarction, there are no clinical data demonstrating efficacy for any of the agents that have been shown to be effective in animal studies. From a clinical standpoint, the most useful agents would be those that could be conveniently administered to patients immediately prior to reperfusion therapy. Before clinical trials with any drug or drug combination are possible, preliminary safety data must be obtained to ensure that agents given during acute myocardial infarction do not provoke intolerable adverse events among the recipients.

We have begun a pilot study designed to test the efficacy of adenosine plus lidocaine in limiting reperfusion injury among patients with acute myocardial infarction who are treated with direct balloon angioplasty. This pilot study was designed to achieve three goals: first, to assess the safety of intravenous adenosine plus lidocaine among humans with acute myocardial infarction; second, to assess myocardial salvage when intravenous adenosine plus lidocaine is given prior to balloon angioplasty for acute myocardial infarction, using serial myocardial perfusion imaging with sestamibi to measure myocardial salvage; and third, to compare myocardial salvage after adenosine plus lidocaine to a historical control group.

Preliminary Work

ADENOSINE

Extensive animal research [3,16–44] has demonstrated that reperfusion with oxygenated whole blood after a period of prolonged ischemia can result in accelerated necrosis of myocytes (Fig. 49-1). Although the mediators of reperfusion injury are not well understood, the activity of mononuclear leukocytes [16,41], platelets [45,46], and endothelial cells [41,47–49] all probably contribute. Several compounds can attenuate the amount of myocardial injury occurring during reperfusion when administered directly into the coronary artery or into the systemic circulation [18,42,49–56]. Adenosine, an endogenous purine nucleoside, antagonizes many of the biochemical and physiologic functions that have been implicated as mediators of reperfusion injury and has been demonstrated to reduce postischemic and/or reperfusion ventricular dysfunction and myocyte necrosis in several animal species [37,55,57]. Coronary sinus effluent levels of adenosine are elevated

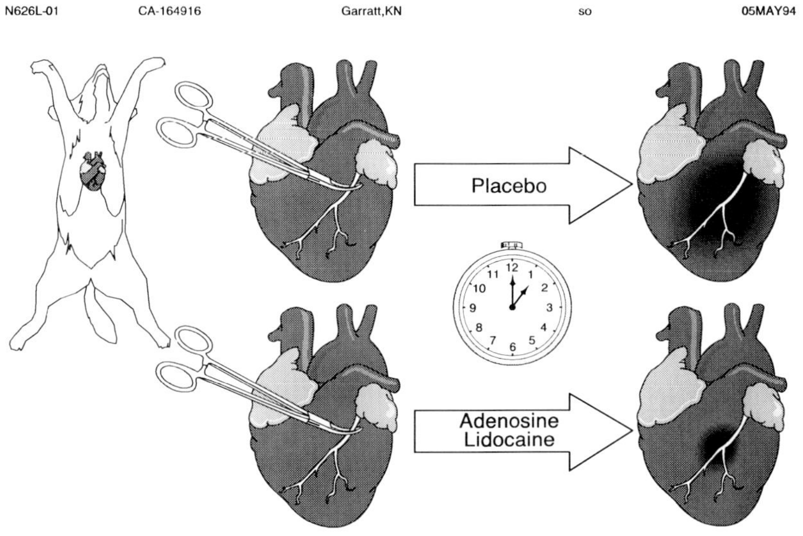

FIGURE 49-1. Using a canine model of myocardial ischemia, investigators at Vanderbilt University demonstrated that the area of myocardial infarction (expressed either as a percentage of the total left ventricular mass or as the proportions of area at risk) following prolonged ischemia can be reduced by as much as 75–80% if the conditions of reperfusion are modified by the addition of adenosine and lidocaine [49].

following myocardial ischemia, a reflection of increased endogenous adenosine production that occurs as a consequence of the consumption and degradation of adenosine triphosphate [58]. Adenosine can reduce granulocyte adhesion and production of radical oxygen species [59,60], which are postulated to cause the endothelial membrane injury that is central to lethal reperfusion injury. In addition, adenosine enhances local collateral blood flow, reduces calcium influx, and may act as a substrate for reformation of high-energy phosphate compounds [61].

LIDOCAINE

Lidocaine is a class I antiarrhythmic agent that enhances the electric stability of myocyte membranes by suppressing the fast inward sodium current phase 0 of the action potential [62]. Perhaps due to its ability to "stabilize membranes," or perhaps because it is able to suppress granulocyte adhesion and the formation of oxygen radical species, lidocaine alone can provide some protection against reperfusion injury in animal studies [63,64]. Concomitant lidocaine is required for the full benefit of adenosine to be realized in some animal models [54,61].

SESTAMIBI

Technetium-99m-hexakis-2-methoxyisobutyl isonitrile, also known as RP-30a or ^{99m}Tc sestamibi, is a radionuclide compound suitable for the assessment of myocardial perfusion [65]. Sestamibi is taken up by myocytes in proportion to the myocardial flow and binds avidly; consequently, minimal washout and no significant redistribution occurs with this agent. Because of these properties, sestamibi imaging conducted after the patient has received reperfusion therapy will produce an image that reflects the perfusion pattern prior to therapy; thus, the myocardial area "at risk" can be reliably measured without delaying care of the patient [65–72]. ^{99m}Tc decay produces relatively high-energy photons (about 140 keV) with a half-life of 6 hours; consequently, high-quality tomographic images can be obtained up to 8 hours after a bolus injection of 30 mCi of sestamibi. Repeat sestamibi studies conducted days or weeks after treatment will reflect the perfusion pattern following reperfusion therapy (in the absence of additional ischemic events). The difference between the initial and late perfusion defects can be quantified and serves as a measure of myocardial mass salvaged by therapy [65,66,68]. The relatively short half-life of sestamibi helps to minimize the radiation exposure to the patient, which enhances the safety of serial studies.

The left ventricular ejection fraction can be estimated using sestamibi by means of a "first-pass" technique [73–75]. When given as a bolus intravenous injection, the initial passage of sestamibi through the left ventricle can be imaged. The dimensions of the left ventricle can be imaged through several cardiac cycles before the sestamibi has passed through the left ventricle. Images can be selected that represent the radioactivity within the left ventricle at the end of diastole and the end of systole; the difference between these two values, divided by the diastolic value, represents the left ventricular ejection fraction. The reliability of first-pass left ventriculography with sestamibi is equivalent or superior to that achieved with other ^{99m}Tc radionuclides [73,74].

Study Design

Patients recruited into this study must qualify by meeting clinical and angiographic entry criteria.

CLINICAL QUALIFICATION

Study Patients. Men and women who are not more than 80 years old are eligible for inclusion; unsterilized women who are of child-bearing age are excluded. All patients must be suitable for immediate coronary angiography and possible percutaneous transluminal coronary angioplasty.

Although early reperfusion is unquestionably desirable, late reperfusion therapy with thrombolysis or direct angioplasty may be of clinical benefit compared with persistent coronary occlusion [76–78]; late reperfusion may confer a benefit by mechanisms other than myocardial cell salvage. Many patients experience a "stuttering" course of waxing and waning chest pain, which may reflect cyclic platelet and fibrinogen aggregation, with alternating complete and incomplete coronary occlusion [79, 80]. Furthermore, the relationship (if any) between the timing of reperfusion and benefit

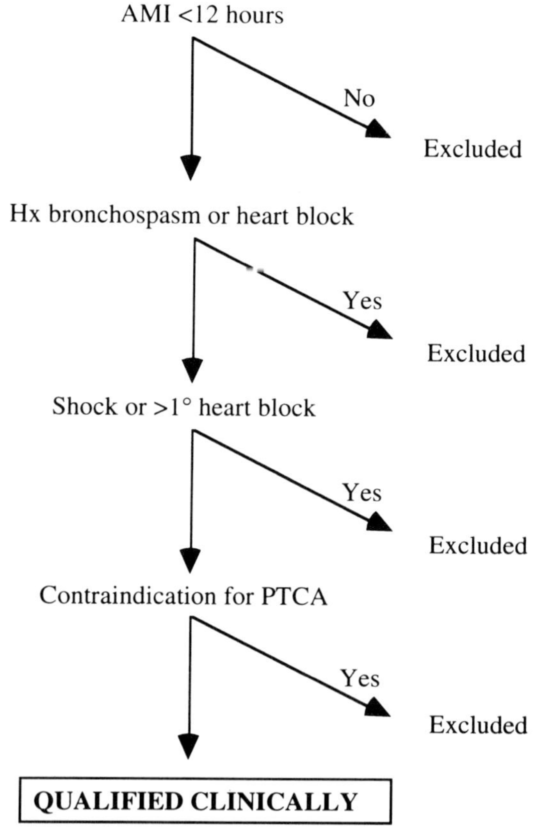

FIGURE 49-2. Flow diagram indicating the inclusion and exclusion criteria used to identify patients who are qualified clinically for study participation. AMI = acute myocardial infarction; Hx = history of; PTCA = percutaneous transluminal angioplasty.

when modifying the conditions of reperfusion in humans is unknown. For these reasons, patients are eligible for enrollment into this study if they have had chest pain typical of ischemia for up to 12 hours (Fig. 49-2).

Exogenous adenosine may produce significant, even fatal, bronchospasm among susceptible patients [81,82]. Therefore, patients with any history of bronchospastic disorder, or who have ever been treated with bronchodilator medications, are excluded.

The ability of exogenous adenosine to slow conduction through the atrioventricular node (AV node) forms the basis for its utility in the management of supraventricular arrhythmias [83]. However, patients with acute myocardial infarction may develop AV node conduction delays as a manifestation of focal ischemia near the AV node or, more commonly, as a consequence of heightened vagal tone. In these patients, the additional AV nodal effects of adenosine may precipitate high-grade AV node block and the need for temporary transvenous pacing support. Patients with a history of untreated heart block or who developed greater than first-degree AV node block as a consequence of myocardial infarction are excluded.

The vasodilator actions of adenosine are well known and may account for some of the benefit of exogenous adenosine during reperfusion [58, 61,76]. Patients with acute myocardial infarction may develop systemic hypotension due to impaired ventricular performance or reflex vasodilatation. In these patients, additional peripheral vasodilatation may further reduce coronary blood flow and so would not likely be well tolerated. For these reasons, patients whose systemic blood pressure is less than 90 mmHg or who require pressor agents or mechanical assist devices for support of their systemic blood pressure are excluded.

Electrocardiographic Criteria. Study patients must manifest electrocardiographic changes consistent with acute myocardial infarction: At least 1 mm of ST-segment elevation in two or more contiguous electrocardiographic leads, or 2+ mm of ST-segment depression in leads V_1 through V_3 (to allow inclusion of true posterior wall myocardial infarctions), are accepted as electrocardiographic evidence of acute myocardial infarction.

Patients who do not meet the exclusion criteria are considered qualified clinically for the study. Once they provide informed consent, they receive a bolus injection of lidocaine, 1 mg/kg intravenously, followed by a continuous infusion of lidocaine, 2 mg/min intravenously. A bolus injection of at least 20–30 mCi sestamibi is given intravenously. Use of aspirin, beta-blocking drugs, oxygen, nitrates, and other drugs appropriate in the management of acute myocardial infarction are administered as deemed appropriate by the attending physician. With a minimum of delay, patients are taken to the adult cardiac laboratory for direct balloon angioplasty of the infarct-related artery.

Angiographic Qualification. Patients are prepared for angiography in the usual manner. A sterile field is established, and a vascular sheath is inserted into the right or left femoral artery using local anesthesia (10–20 ml of 1% lidocaine). Mild sedation is achieved with low-dose intravenous fentanyl (a narcotic analgesic with few cardiovascular effects) with or without diazepam or lorazepam (benzodiazepine sedative hypnotics). Diagnostic angiography is performed using conventional techniques. Patients are considered qualified angiographically to continue in the study if the following conditions are met: 1) The infarct-related artery is identifiable. Under some circumstances, extensive multivessel disease may be present, which makes identification of the artery responsible for the acute event difficult or impossible. 2) Reperfusion has not already taken place. Reperfusion is defined as normal or nearly normal anterograde blood flow through the infarct-related artery. Blood flow is graded according to the scale developed by the Thrombolysis in Myocardial Infarction (TIMI) investigators. TIMI grade 0 (absent) or 1 (severely reduced) blood flow is required for participation in the study. 3) Coronary anatomy is suitable for balloon angioplasty. In the presence of extensive diffuse disease or severe three-vessel disease, or significant left main coronary artery disease, the culprit lesion may be deemed unsuitable for direct angioplasty by the operator, and the patient is disqualified from further participation.

If a patient qualifies angiographically for continued participation (Fig. 49-3), a continuous intravenous infusion of adenosine 70 mcg/kg/min is begun. Adenosine is allowed to infuse for at least 10 minutes prior to the first inflation of the angioplasty balloon and continues for at least 1 hour following restoration of normal (TIMI grade 3) blood flow in the infarct-related artery.

Sestamibi Imaging

To determine the myocardial mass at risk during the acute myocardial event, an acute-phase sestamibi image is obtained within 8 hours of the initial sestamibi injection. A second image is obtained prior to hospital discharge (generally between 5 and 7 days after the hospital admission date). A third sestamibi image is obtained 6 weeks after the initial hospital admission date. The second and third sestamibi studies permit measurement of the myocardial mass having a persistent severe perfusion defect after therapy, which is a measure of the amount of myocardium that is infarcted. The difference in perfusion defects observed between the acute image and the 6-week image is used to determine the myocardial mass salvaged with the therapies applied.

As an adjunct to the myocardial perfusion data, left ventricular ejection fraction is also measured in the study patients. Acute (initial) estimates of left ventricular ejection fraction cannot be obtained using radionuclide techniques, because (unlike sestamibi perfusion imaging) the images must be obtained immediately; the delays in patient care that this would entail are not acceptable. Whenever possible, contrast left ventriculography is performed just before or after balloon angioplasty and is used to estimate the initial left ventricular ejection fraction. Pre-hospital discharge and 6-week follow-up estimates of left ventricular

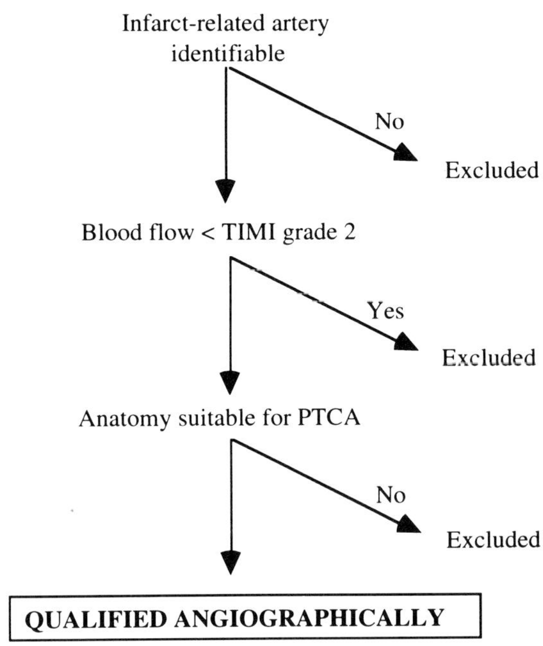

FIGURE 49-3. Flow diagram indicating the inclusion and exclusion criteria used to identify patients who are qualified angiographically for study participation. PTCA = percutanceous transluminal angioplasty; TIMI = Thrombolysis In Myocardial Infarction.

TABLE 49-1. Patient enrollment into pilot study

Characteristics	No. patients
Enrolled	31
Retained in study	21
Reason for withdrawal from study	
Arrest before adenosine adminstered	3
Arrest after adenosine administered	1
TIMI grade 3 blood flow	1
Inappropriate anatomy for angioplasty	3
Aortic stenosis	1
Unsuccessful angioplasty attempt	1

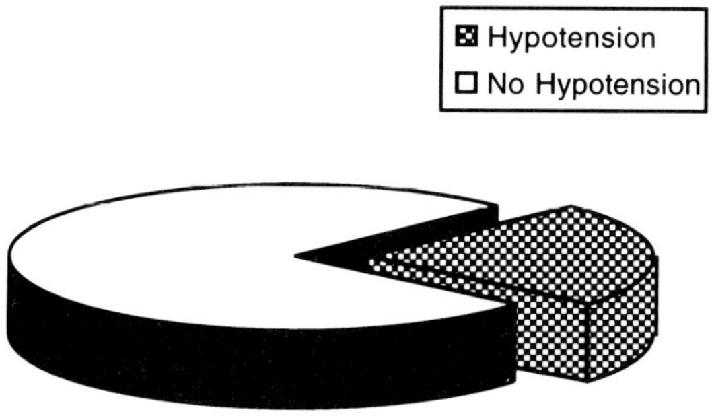

FIGURE 49-4. Pie chart illustrating the proportion of patients (5/21) experiencing significant systemic hypotension after infusion of adenosine.

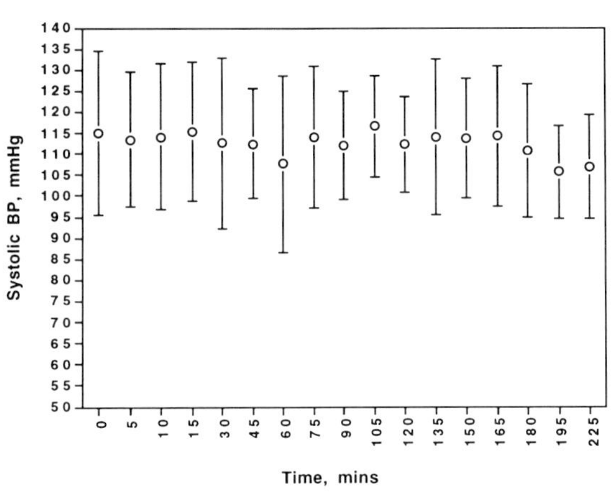

FIGURE 49-5. Mean systolic blood pressure measurements (±SD) obtained during and after adenosine infusion. Zero time indicates central blood pressure reading immediately prior to initiation of adenosine infusion. Pressures reflect central arterial measurements made during angiography (generally the first 60–120 minutes) and peripheral arterial measurements made after completion of angiography.

ejection fraction are obtained using the first-pass sestamibi technique or resting gated radionuclide angiography.

Results

The pilot study is designed to evaluate 50 patients, and at the time of this writing 31 patients have been identified as clinically appropriate candidates for study participation and have been enrolled. Of this group, 10 patients were withdrawn prior to obtaining the initial sestamibi image. The reasons for patient withdrawal from the protocol are displayed in Table 49-1. Most patients were excluded because of an anatomy that was not felt to be suitable for balloon angioplasty or because of cardiopulmonary arrest or significant ventricular arrhythmias developing prior to the initial sestamibi imaging; these latter patients were felt to be too unstable to be transported to the radionuclear imaging suites for acute imaging and were therefore disqualified from further participation. Unexpected, severe aortic stenosis was identified in one patient; because systemic hypotension is poorly tolerated in such patients, this individual was disqualified from further participation. In one additional patient the attempted balloon angioplasty procedure was unsuccessful.

HYPOTENSION

Systemic hypotension was observed in 24% of patients following administration of intravenous adenosine (Fig. 49-4). Significant hypotension that developed after beginning the adenosine infusion resolved, in all cases, with a reduction in dose or a temporary interruption of adenosine. When adenosine infusion was interrupted, it was restarted at a lower infusion rate and was well tolerated. The mean systolic blood pressure values obtained during the adenosine infusion period are displayed in Figure 49-5.

ATRIOVENTRICULAR BLOCK

The occurrence of AV block after lidocaine and adenosine administration is displayed in Figure 49-6. Transient second-degree AV block developed in two patients. The AV block was well tolerated in both patients and a dose reduction of adenosine was not necessary. No sustained AV block or 3° AV block was observed, and transvenous temporary pacing was not required in any study patient.

OTHER COMPLICATIONS

A variety of other adverse clinical events have been observed among study patients. All observed significant adverse events are displayed in Table 49-2. Ventricular tachyarrhythmias (sustained ventricular tachycardia or ventricular fibrillation) were associated with cardiac arrest. Congestive heart failure symptoms appeared, in all cases, to be the result of impaired left ventricular performance (usually assessed by transthoracic echocardiography) rather than pulmonary bronchospasm. Profound hypotension and bradycardia related to stimulation of the Bezold–Jarisch reflex [84] were observed in one patient after successful balloon dilatation of the right coronary occlusion and in one patient after removal of the vascular access hemostatic sheath.

Other complications unlikely to be related to the infusion of adenosine and lidocaine are listed in Table 49-3. Bleeding at the femoral artery that was punctured to gain arterial access for the conducting of angioplasty was observed in three patients, and one additional patient developed gastrointestinal bleeding. This frequency of bleeding complications is not unusual for acute myocardial infarction patients treated with balloon angioplasty, because they are treated with high doses of heparin, which may be continued as an intravenous infusion for several hours after angioplasty, as well as oral aspirin.

Although there is a possibility that the electrophysiologic properties of adenosine may contribute to the development of atrial arrhythmias [85,86], the development of atrial fibrillation frequently complicates acute myocardial infarction. This rhythm disturbance may develop as a consequence of concomitant atrial infarction, or as a result of increased atrial wall stress that develops in response to the elevated left ventricular diastolic pressures that occur during myocardial ischemia or as a result of heightened circulating catecholamines [21,84,85].

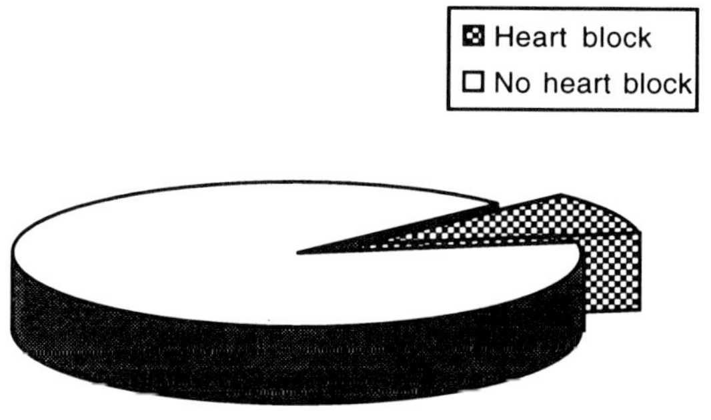

FIGURE 49-6. Pie chart illustrating the proportion of patients (2/21) experiencing greater than first-degree atrioventricular block.

TABLE 49-2. Major complications observed in study patients

Complication	No. patients
Systemic hypotension	5
$>1°$ atrioventricular block	2
Ventricular tachyarrhythmia/cardiac arrest	3
Congestive heart failure	4
Abnormal liver function tests	1
Bezold–Jarisch reflex/vasovagal reaction	2
Fever without infection	1

TABLE 49-3. Complications observed in study patients that are not likely to be related to use of adenosine plus lidocaine

Complication	No. patients
Vascular access site bleeding	3
Other bleeding	1
Atrial fibrillation	1
Infection	2

Conclusions

A sound theoretical framework exists to support the investigation of adenosine as an adjunct to reperfusion therapy among humans with acute myocardial infarction. We have designed a pilot study that will provide preliminary data regarding the safety and efficacy of intravenous adenosine plus lidocaine in patients with acute myocardial infarction who undergo direct balloon angioplasty to achieve reperfusion.

The initial safety data from this pilot study suggest that intravenous use of adenosine plus lidocaine in patients with acute myocardial infarction is associated with a low incidence of significant atrioventricular block and systemic hypotension. These adverse events, which respond to adjustments in the adenosine infusion rate, may be observed in about one fourth of patients. Additionally, a low frequency of other adverse events may be observed among patients receiving these drugs. These data suggest that the use of these drugs in patients with acute myocardial infarction, in the dosages and manner of administration described, is sufficiently safe that continued study of their potential merit in reducing myocardial infarction size is justified. Completion of this pilot study will also provide data regarding the mass of myocardium salvaged by the therapies applied that, when compared with an appropriate historical control population, will provide information regarding the potential efficacy of this treatment strategy. The preliminary safety and efficacy data collected will be useful in the construction of a randomized trial of direct balloon angioplasty alone versus direct balloon angioplasty with adjunctive adenosine plus lidocaine.

References

1. Rosenkranz E, Buckberg G (1983) Myocardial protection during surgical coronary reperfusion. J Am Coll Cardiol 1:1235–1246.
2. Jennings R, Yellon D (1992) Reperfusion injury: Definitions and historical background. In Jennings R, Yellon D (eds.), Myocardial Protection: The Pathophysiology of Reperfusion and Reperfusion Injury. New York: Raven Press, pp. 1–11.
3. Acar C, Partington MT, Buckberg GD (1991) Studies of controlled reperfusion after ischemia. XIX. Reperfusate composition: Benefits of blood cardioplegia over fluosol DA cardioplegia during regional reperfusion—importance of including blood components in the initial reperfusate. J Thorac Cardiovasc Surg 101:284–293.
4. Boehm DH, Human PA, von Oppell U, et al. (1991) Adenosine cardioplegia: Reducing reperfusion injury of the ischaemic myocardium? Eur J Cardiothorac Surg 5:542–545.
5. Coetzee A, Roussouw G, Fourie P, Lochner A (1990) Preservation of myocardial function and biochemistry after blood and oxygenated crystalloid cardioplegia during cardiac arrest. Ann Thorac Surg 50:230–237.
6. Gunnes S, Ytrehus K, Sorlie D, Helgesen KG, Mjos OD (1987) Improved energy preservation following gentle reperfusion after hypothermic, ischemic cardioplegia in infarcted rat hearts. Eur J Cardiothorac Surg 1:139–143.
7. Julia PL, Buckberg GD, Acar C, Partington MT, Sherman MP (1991) Studies of controlled reperfusion after ischemia. XXI. Reperfusate composition: Superiority of blood cardioplegia over crystalloid cardioplegia in limiting reperfusion damage—importance of endogenous oxygen free radical scavengers in red blood cells. J Thorac Cardiovasc Surg 101:303–313.
8. Landymore RW, Marble AE, Eng P, MacAulay MA, Trillo A, Li QY (1991) Effect of high-volume cardioplegia on small-amplitude electrical activity during cardioplegia arrest. Eur J Cardiothorac Surg 5:395–399.
9. Liu FM, Xu SC (1993) Effect of captopril cardioplegia on renin-angiotensin system, prostaglandins, free radicals and electrolytes in the isolated hypothermic ischemic and reperfusion rabbit hearts. Chin Med J 106:903–906.
10. Takahashi A, Chambers DJ, Braimbridge MV, Hearse DJ (1988) Optimal myocardial protection during crystalloid cardioplegia. Interrelationship between volume and duration of infusion [see comments]. J Thorac Cardiovasc Surg 96:730–740.
11. Tian GH, Mainwood GW, Biro GP, et al. (1991) The effect of high buffer cardioplegia and secondary cardioplegia on cardiac preservation and postischemic functional recovery: A ^{31}P NMR and functional study in Langendorff perfused pig hearts. Can J Physiol Pharmacol 69: 1760–1768.
12. Wikman CJ, Wagner S, Wu S, Parmley W (1991) Alcohol and pyruvate cardioplegia. Twenty-four-hour in situ preservation of hamster hearts. J Thorac Cardiovasc Surg 101: 509–516.

13. Follette D, Key K, Buckberg G (1981) Reducing postischemic damage by temporary modification of reperfusate calcium, potassium, pH, and osmolarity. J Thorac Cardiovasc Surg 82:221–238.
14. Gersh B, Clements I, Chesebro J (1991) Acute myocardial infarction. In Guiliani ER, Fuster V, Gersh BJ, McGoon MD, McGoon DC (eds.), Cardiology: Fundamentals and Practice, Vol. 2. St. Louis: Mosby Year Book, pp. 1318–1361.
15. Spiekerman R, Brandenberg J, Achor R, et al. (1962) The spectrum of coronary heart disease in a community of 30,000: A clinicopathologic study. Circulation 25:57–65.
16. Zimmerman BJ, Granger DN (1994) Mechanisms of reperfusion injury. Am J Med Sci 307:284–292.
17. Abd EAS, Jessen ME, Lekven J, Doherty NE3, Brunsting LA, Wechsler AS (1988) Myocardial reperfusion injury. Role of myocardial hypoxanthine and xanthine in free radical-mediated reperfusion injury. Circulation III: 224–235.
18. Babbitt DG, Virmani R, Forman MB (1989) Intracoronary adenosine administered after reperfusion limits vascular injury after prolonged ischemia in the canine model. Circulation 80:1388–1399.
19. Bagchi D, Das DK, Engelman RM, Prasad MR, Subramanian R (1990) Polymorphonuclear leucocytes as potential source of free radicals in the ischaemic-reperfused myocardium. Eur Heart J 11:800–813.
20. Bhimji S, Godin DV, McNeill JH (1987) Coronary artery ligation and reperfusion in rabbits made diabetic with alloxan. J Endocrinol 112:43–49.
21. Bril A, Rochette L, Verry A, Maupoil V, Man RY, Opie LH (1990). Effects of the free radical generating system $FeCl_3$/ADP on reperfusion arrhythmias of rat hearts and electrical activity of canine Purkinje fibres. Cardiovasc Res 24: 669–675.
22. Darley UVM, Smith DR, O'Leary VJ, Stone D, Hardy DL, Clark JB (1990) Hypoxia-reoxygenation induced damage in the myocardium: The role of mitochondria. Biochem Soc Trans 18:526–528.
23. Darley UVM, Stone D, Smith D, Martin JF (1991) Mitochondria, oxygen and reperfusion damage. Ann Med 23:583–588.
24. Das DK, Engelman RM, Rousou JA, Breyer RH, Otani H, Lemeshow S (1986) Role of membrane phospholipids in myocardial injury induced by ischemia and reperfusion. Am J Physiol 251:H71–H79.
25. Diederichs F, Wittenberg H, Sommerfeld U (1990) Myocardial cell damage and breakdown of cation homeostasis during conditions of ischaemia and reperfusion, the oxygen paradox, and reduced extracellular calcium. J Clin Chem Clin Biochem 289:139–148.
26. Ferrero ME, Marni A, Parise M, Salari PC, Gaja G (1991) Protection of rat heart from damage due to ischemia-reperfusion during procurement and grafting by defibrotide. Transplantation 52:611–615.
27. Forman MB, Virmani R, Puett DW (1990) Mechanisms and therapy of myocardial reperfusion injury. Circulation 81:IV69–IV78.
28. Ganote CE, Vander HRS (1986) Morphologic aspects of reperfusion injury. J Mol Cell Cardiol 4:23–26.
29. Godin DV, Garnett ME (1989) Altered antioxidant status in the ischemic/reperfused rabbit myocardium: Effects of allopurinol. Can J Cardiol 5:365–371.
30. Hearse D (1977) Reperfusion injury of the ischemic myocardium. J Mol Cell Cardiol 9: 605–616.
31. Hearse D (1990) Ischemia, reperfusion and the determinants of tissue injury. Cardiovasc Drugs Ther 4:767–776.
32. Hearse D. Myocardial injury during ischemia and reperfusion: Concepts and controversies (1992) In Jennings R, Yellon D (eds.), Myocardial Protection: The Pathophysiology of Reperfusion and Reperfusion Injury. New York: Raven Press, pp. 13–34.
33. Kim MS, Akera T (1987) O_2 free readicals: Cause of ischemia-reperfusion injury to cardiac Na^+-K^+-ATPase. Am J Physiol 252:H252–H257.
34. Masini E, Giannella E, Palmerani B, Pistelli A, Gambassi F, Mannaioni PF (1989) Free radicals induce ischemia-reperfusion injury and histamine release in the isolated guinea pig heart. Int Arch Allergy Appl Immunol 88: 132–133.
35. Nayler WG, Elz JS (1986) Reperfusion injury: Laboratory artifact or clinical dilemma? Circulation 74:215–221.
36. Nayler WG (1991) Basic mechanisms involved in the protection of the ischaemic myocardium. The role of calcium antagonists. Drugs 2:21–27.
37. Opie L (1989) Reperfusion injury and its pharmacologic modification. Circulation 80: 1049–1062.

38. Otani H, Umemoto M, Kagawa K, et al. (1986) Protection against oxygen-induced reperfusion injury of the isolated canine heart by superoxide dismutase and catalase. J Surg Res 41:126–133.
39. Quaife RA, Kohmoto O, Barry WH (1991) Mechanisms of reoxygenation injury in cultured ventricular myocytes. Circulation 83:566–577.
40. Sellevold OF, Jynge P (1988) Modification of myocardial ischemic injury: A concentration response study of glucocorticoid supplementation during reperfusion. J Cardiothorac Anesth 2:45–55.
41. Sheridan FM, Dauber IM, McMurtry IF, Lesnefsky EJ, Horwitz LD (1991) Role of leukocytes in coronary vascular endothelial injury due to ischemia and reperusion. Circ Res 69:1566–1574.
42. Velasco CE, Turner M, Cobb MA, Virmani R, Forman MB (1991) Myocardial reperfusion injury in the canine model after 40 minutes of ischemia: Effect of intracoronary adenosine. Am Heart J 122:1561–1570.
43. Virmani R, Forman MB, Kolodgie FD (1990) Myocardial reperfusion injury. Histopathological effects of perfluorochemical. Circulation 81:IV57–IV68.
44. Wroblewski LC, Aisen AM, Swanson SD, Buda AJ (1990) Evaluation of myocardial viability following ischemic and reperfusion injury using phosphorus 31 nuclear magnetic resonance spectroscopy in vivo. Am Heart J 120:31–39.
45. Squadrito F, Ioculano M, Altavilla D, et al. (1993) Platelet activating factor interaction with tumor necrosis factor in myocardial ischaemia-reperfusion injury. J Lipid Mediat 8:53–65.
46. Zhu B, Sievers R, Sun Y, Morse-Fisher N, Parmley W, Wolfe C (1994) Is the reduction of myocardial infarct size by dietary fish oil the result of altered platelet function? Am Heart J 127:744–755.
47. Liu JJ, Gu XH, Casley DJ, Nayler WG (1990) Reoxygenation, but neither hypoxia nor intermittent ischemia, increases [^{125}I] endothelin-1 binding to rat cardiac membranes. J Cardiovasc Pharmacol 15:436–443.
48. Hashimoto K, Pearson PJ, Schaff HV, Cartier R (1991) Endothelial cell dysfunction after ischemic arrest and reperfusion: A possible mechanism of myocardial injury during reflow. J Thorac Cardiovasc Surg 102:688–694.
49. Olafsson B, Forman MB, Puett DW, et al. (1987) Reduction of reperfusion injury in the canine preparation by intracoronary adenosine: Importance of the endothelium and the no-reflow phenomenon. Circulation 76:1135–1145.
50. Thelin S, Hultman J, Ronquist G (1991) Effects of adenosine infusion on the pig heart during normothermic ischemia and reperfusion. Scand J Thorac Cardiovasc Surg 25: 207–213.
51. Norton ED, Jackson EK, Virmani R, Forman MB (1991) Effect of intravenous adenosine on myocardial reperfusion injury in a model with low myocardial collateral blood flow. Am Heart J 122:1283–1291.
52. Masuda M, Demeulemeester A, Chen CC, Hendrikx M, Van BH, Flameng W (1991) Cardioprotective effects of nucleoside transport inhibition in rabbit hearts. Ann Thorac Surg 52:1300–1305.
53. Koke JR, Fu LM, Sun D, Vaughan DM, Bittar N (1989) Inhibitors of adenosine catabolism improve recovery of dog myocardium after ischemia. Mol Cell Biochem 86:107–113.
54. Homeister JW, Hoff PT, Fletcher DD, Lucchesi BR (1990) Combined adenosine and lidocaine administration limits myocardial reperfusion injury. Circulation 82:595–608.
55. Ely SW, Berne RM (1992) Protective effects of adenosine in myocardial ischemia. Circulation 85:893–904.
56. Babbitt DG, Virmani R, Vildibill HDJ, Norton ED, Forman MB (1990) Intracoronary adenosine administration during reperfusion following 3 hours of ischemia: Effects on infarct size, ventricular function, and regional myocardial blood flow. Am Heart J 120:808–818.
57. Hori M, Kitakaze M (1991) Adenosine, the heart, and coronary circulation. Hypertension 18:565–574.
58. Berne R (1980) The role of adenosine in the regulation of coronary blood flow. Circ Res 47:807–813.
59. Cronstein B, Levin R, Belanoff J, Weissmann G, Hirschorn R (1986) Adenosine: An endogenous inhibitor of neutrophil-mediated injury to endothelial cells. J Clin Invest 78: 760–770.
60. Cronstein B, Kramer S, Weissmann G, Hirschorn R (1983) Adenosine: A physiologic modulator of superoxide anion generation by human neutrophils. J Exp Med 158:1160–1177.

61. Forman M, Velasco C, Jackson E (1993) Adenosine attenuates reperfusion injury following regional myocardial ischemia. Cardiovasc Res 27:9–17.
62. Hammill S (1991) Antiarrhythmic drugs. In Guiliani ER, Fuster V, Gersh BJ, McGoon MD, McGoon DC (eds.), Cardiology: Fundamentals and Practice, Vol. 2. St. Louis: Mosby Year Book, pp. 694–729.
63. Lesnefsky E, VanBenthuysen K, McMurtry I, Shikes R, Johnston RJ, Horwitz L (1989) Lidocaine reduces canine infarct size and decreases release of a lipid peroxidation product. J Cardiovasc Pharm 13:895–901.
64. Baron D, Sunamori M, Harrison C (1983) Preservation of oxidative phosphorylation by lidocaine in ischemic and reperfused myocardium. Adv Myocardiol 4:567–573.
65. Gibbons R. Nuclear Cardiology (1991) In Guiliani ER, Fuster V, Gersh BJ, McGoon MD, McGoon DC (eds.), Cardiology: Fundamentals and Practice, Vol. 2. St. Louis: Mosby Year Book, pp. 459–491.
66. Gibson W, Christian T, Pellikka P, Behrenbeck T, Gibbons R (1992) Serial tomographic imaging with technetium-99m-sestamibi for the assessment of infarct-related arterial patency following reperfusion therapy. J Nucl Med 2080–2085.
67. Glover D, Okada R (1993) Myocardial technetium 99m sestamibi kinetics after reperfusion in a canine model. Am Heart J 125:657–666.
68. Sinusas A, Trautman K, Bergin J, et al. (1990) Quantification of area at risk during coronary occlusion and degree of myocardial salvage after reperfusion with technetium-99m methoxyisobutyl isonitrile. Circulation 82: 1424–1437.
69. De Coster P, Merlin J, Detry J, Brasseur L, Beckers C, Col J (1985) Coronary artery reperfusion in acute myocardial infarction: Assessment by per- and postintervention thallium-201 myocardial perfusion imaging. Am J Cardiol 55:889–895.
70. Beller G, Glover D, Edwards N, Ruiz M, Simanis J, Watson D (1993) 99mTc-sestamibi uptake and retention during myocardial ischemia and reperfusion. Circulation 87: 2033–2042.
71. Mousa S, Cooney J, Stevens S (1992) Kinetics of technetium-99m-sestamibi and thallium-201 in a transient ischemic myocardium animal model: Insight into the 'redistribution' phenomenon. Cardiology 81:157–163.
72. Mortelmans L, Nuyts J, Vanhaecke J, et al. (1993) Experimental validation of a new quantitative method for the analysis of infarct size by cardiac perfusion tomography (SPECT). Int J Card Imag 9:201–212.
73. Iskandrian A, Heo J (1992) Technetium-labeled myocardial imaging agents. Int J Card Imag 8:277–287.
74. Williams K, Taillon L, Draho J, Foisy M (1993) First-pass radionuclide angiographic studies of left ventricular function with technetium-99m-teboroxime, technetium-99m-sestamibi and technetium-99m-DTPA. J Nucl Med 34:394–399.
75. Iskandrian A, Kegel J, Tecce M, Wasserleben V, Cave V, Heo J (1993) Simultaneous assessment of left ventraicular perfusion and function with technetium-99 m sestamibi after coronary artery bypass grafting. Am Heart J 126:1199–1203.
76. Ross A (1993) The current controversies regarding reperfusion therapy for acute myocardial infarction. Zeit Kardiol 82:113–117.
77. Topol E, Califf R, Vandormael M, et al. (1992) A randomized trial of late reperfusion therapy for acute myocardial infarction. Thrombolysis and Angioplasty in Myocardial Infarction-6 Study Group. Circulation 2090–2099.
78. Gulba D (1993) The value of late thrombolytic perfusion in acute myocardial infarct. Zeit Kardiol 82:183–185.
79. Ikeda H, Koga Y, Kuwano K, et al. (1993) Cyclic flow variations in a conscious dog model of coronary artery stenosis and endothelial injury correlate with acute ischemic heart disease syndromes in humans. J Am Coll Cardiol 21:1008–1017.
80. Maseri A, Crea F (1991) The elusive cause of instability in unstable angina. Am J Cardiol 68:16B–21B.
81. Cerqueira M, Verani M, Schwaiger M, Heo J, Iskandrian A (1994) Safety profile of adenosine stress perfusion imaging: Results from the Adenoscan Multicenter Trial Registry. J Am Coll Cardiol 384–389.
82. Burkhart K (1993) Respiratory failure following adenosine administration. Am J Emerg Med 11:249–250.
83. Pelleg A (1993) Mechanisms of action and

therapeutic potential of adenosine and its analogues in the treatment of cardiac arrhythmias. Cor Art Dis 4:109–115.
84. Gersh BJ (1991) Arrhythmias. In Guiliani ER, Fuster V, Gersh BJ, McGoon MD, McGoon DC (eds.), Cardiology: Fundamentals and Practice, Vol. 2. St. Louis: Mosby Year Book, pp. 1453–1478.
85. Curtis MJ (1991) The pathophysiological basis of arrhythmogenesis in myocardial ischemia and reperfusion: Possible target for intervention. Bratisl Lek Listy 92:91–101
86. Podzuweit T, Binz KH, Nennstiel P, Flaig W (1989) The anti-arrhythmic effects of myocardial ischaemia. Relation to reperfusion arrhythmias? Cardiovasc Res 23:81–90.

50. ROLE OF ADENOSINE IN THE PHENOMENON OF ISCHEMIC PRECONDITIONING

James M. Downey
Michael V. Cohen

The major complication in the patient with acute myocardial infarction is loss of ventricular mass through necrosis with resulting heart failure. While it is now possible with thrombolytic therapy to restore perfusion to ischemic regions, reperfusion can seldom be accomplished before a substantial amount of myocardial necrosis has occurred. For several decades there has been a concerted research effort to identify interventions that could make the cardiomyocytes more tolerant of a transient period of ischemia, which would in turn reduce infarct size and thus preserve pump function. To date no such therapy is available to the clinician. Antioxidants, calcium antagonists, and beta blockers have all been tested but have failed to provide a consistent antiinfarct effect in preclinical trials [1–3].

Recent studies with ischemic preconditioning (PC), however, reveal that such protection is theoretically possible. *Murry and colleagues in 1986 reported that exposing the heart to four brief periods of ischemia, each 5 minutes in duration, paradoxically caused the heart to tolerate a subsequent 45-minute ischemic insult with only a fraction of the infarction realized in non-PC hearts [4]*. PC is the first intervention that has achieved widespread consensus concerning its antiinfarct efficacy. If the mechanism of this adaptive response can be understood, it should be possible to institute this form of protection pharmacologically in the coronary artery disease patient.

PC with a single 5-minute period of ischemia was found to be as effective as multiple cycles [5,6]. Also the effects are fairly short lived, lasting for only about 1 hour [6–8]. In most models of ischemic PC, the protection is equivalent to a shortening of the ischemic period by about 20 minutes.

Adenosine Appears to Trigger PC

There is mounting evidence that adenosine triggers this phenomenon in a number of species. Adenosine has been the focus of studies in the heart for many years but only since the studies of Ely and colleagues [9] has adenosine been proposed to have direct cardioprotective effects. Three years ago we found that nonselective adenosine-receptor–blocking agents abolished protection from PC [10]. In addition, intracoronary infusion for 5 minutes of adenosine or an A_1-selective adenosine receptor agonist followed by 10 minutes of washout mimicked PC [10]. While the direct effects of adenosine, delayed conduction and coronary dilation were quickly reversed during the washout period, the protected state persisted. Those observations have generally been suported by other laboratories and in a variety of species [11–14]. *We, therefore, proposed that adenosine released during the PC ischemia triggers a protective change in the heart via the A_1 receptor and this change lasts for about 1 hour*. The one known exception to this paradigm seems to be the rat. Adenosine receptor blockers have no effect on the myocardial protection from ischemic PC in the rat heart [15,16].

Adenosine Receptor Subtypes in PC

Our original study implicated the A_1 adenosine receptor because the moderately A_1-selective agonist N^6-1-(phenyl-2R-isopropyl) adenosine (PIA) given at a dose that did not dilate the coronary arteries (an A_2 effect) was able to mimic PC's protection [10]. Also the highly A_2-selective agonist CGS 21680 failed to protect the heart [17]. In a subsequent study, however, we found that the highly A_1-selective antagonist, 8-cyclopentyl-1,3-dipropylxanthine (DPCPX), would not block protection from PC in our rabbit model [18]. We knew that the rabbit's A_1 receptors were blocked by DPCPX because it completely blocked the A_1-induced bradycardia from adenosine.

One possible explanation of these seemingly contradictory observations was that the newly discovered A_3 receptor may be involved [19]. We found that the known A_3 agonist, N^6-[2-(4-aminophenyl)ethyl]adenosine (APNEA), could also mimic PC [18]. While DPCPX could not block protection from PC, adenosine, or APNEA, the A_3 antagonist BW 1433 did block protection from ischemic preconditioning. Based on these observations we believe that the enosine A_3 receptor participates in ischemic preconditioning, at least in the rabbit. Because A_3 receptor activation has minimal hemodynamic side effects, triggering PC in the ischemic patient with an A_3 agonist has promise. Recent studies reveal that human ventricle expresses the message for A_3 receptors and thus likely has them [20]. Whether adenosine A_1 receptors constitute a parallel pathway to A_3 receptors on the heart or whether A_3 receptors act alone remains unclear.

Role Played by Catecholamines Is Uncertain

While adenosine receptors may trigger PC, little is known of the effector that accomplishes the protection. One theory is that PC slows the rate of ATP loss during ischemia [21]. Jennings and coworkers have suggested that norepinephrine released during deep ischemia augments ATP utilization and that adenosine may block that release. Indeed, dogs depleted of norepinephrine by reserpine failed to exhibit the adenosine-induced slowing of ATP depletion [22]. Additionally, Wiltse et al. found that reserpine blocks PC's protection in the rabbit model [23]. However, another explanation may be possible. With reduced beta-adrenergic stimulation, metabolism may be low, which would reduce purine formation during the PC ischemia. Suzuki et al. [24] also failed to see protection from PC in reserpinized rabbits. Data not in the abstract but presented at the meeting revealed that raising oxygen consumption with an infusion of isoproterenol during PC restored the ability of PC to protect the reserpinized hearts. Their interpretation was that in reserpinized rabbits too little adenosine was released during a 5-minute occlusion to trigger the protective response. Thus catecholamines may be modulating adenosine release. It should be noted that if norepinephrine accelerates ATP usage during ischemia, then sympathectomy with reserpine should mimic PC. This has never been observed, however [23–25]. It is interesting to note that most investigators precondition with a single 5-minute period of ischemia, which is probably very close to the threshold for protection. A slight fall in adenosine release during that period could easily result in a subthreshold amount of receptor stimulation.

ATP-Sensitive Potassium Channels May Be the End Effector of Protection

Gross and Auchampach have proposed that ATP-sensitive potassium channels (K_{ATP}) are opened in PC myocardium and protect by reducing calcium influx [26]. Blocking those channels with glibenclamide selectively inhibits the salutary effects of ischemic PC in dogs [26] and adenosine-induced protection in pigs [27]. We, however, failed to see this effect in pentobarbital-anesthetized rabbits [28] or rats [15]. More recently, Toombs et al. [29] were able to block PC's protection with the K_{ATP}-channel blocker glibenclamide in a ketamine/xylazine-anesthetized rabbit. We have subsequently confirmed that switching anesthetics does allow glibenclamide to block PC's protection in rabbits [30]. Thus, it is possible that adenosine receptors may act to open K_{ATP} channels, but it is unclear why this effect is species- and anesthesia-dependent. We proposed that glibenclamide might modulate interstitial adenosine concentration during ischemia in the ketamine/xylazine-anesthetized rabbit, but direct measurements of interstitial adenosine with microdialysis disproved that possibility [31].

Opening K_{ATP} channels might reduce calcium influx by shortening the action potential. Another possibility is that open channels may help to maintain volume regulation. One discrepant feature of this hypothesis is that the opening of K_{ATP} channels is usually proarrhythmic, while ischemic PC is very antiarrhythmic [32]. Adenosine A_1 receptors reportedly open K_{ATP} channels in rat myocytes by direct G_i coupling [33], which is also difficult to reconcile with the protein kinase C data presented below. On the other hand, in some cell types, including atrial myocytes, activation of PKC has been seen to open K_{ATP} [34,35].

Other Possible Effectors of Protection

Many theories of PC have been put to rest, including those based on mitochondrial ATPase [36], augmented collateral flow [4], and increased antioxidants [37]. Omar et al. [38] found that substituting pyruvate for glucose at reperfusion mimicked PC's protection in their recovery-of-function model, suggesting that PC might prevent a defect in glucose metabolism. We, however, failed to confirm that theory when infarction in rabbit heart was used as the end point [39]. Cyclooxygenase products also do not appear to be involved [40,41]. Heat-shock proteins were considered as a possible mediator, but Schoot et al. could not cause regional tissue heating to salvage ischemic dog myocardium [42]. Also, total inhibition of protein synthesis could not block protection from ischemic PC in rabbits [43], which also argues against expression of any new protein being part of the protection process. Furthermore, protection has completely worn off 2 hours after ischemic PC [6–8], a time when heat-shock protein content is still increasing [44]. Heat-shock proteins may, however, be involved in a second window of protection that reappears 24 hours after ischemic preconditioning [45].

Kitakaze and colleagues have proposed that upregulation of 5′ nucleotidase may be responsible for PC's protection [46]. The evidence for this theory is that activity of both the endo and the exo forms of this enzyme is increased by about 50% in preconditioned myocardium and that a blocker of the exo form of 5′-nucleotidase prevents the antiinfarct effect of PC. This enzyme is responsible for adenosine production during ischemia, and therefore hearts in the PC state would be expected to make more adenosine during ischemia and thereby protect themselves. A major shortcoming of this theory is that it does not explain how adenosine actually protects. Measurements of interstitial myocardial adenosine with microdialysis reveal two other problems with the theory [47]. First, adenosine levels during ischemia quickly approach the micromolar level, which should cause 100% occupancy of adenosine receptors in even non-PC hearts. Secondly, PC hearts actually appear to release less rather than more adenosine during ischemia, suggesting that upregulation of 5′-nucleotidase is not the only change that has occurred in these hearts.

Intracellular Signaling Pathways for PC: Potein Kinase C

Most recently we have concentrated on the intracellular signaling pathways involved in PC. We found that treating rabbits with pertussis toxin abolished protection from ischemic PC [48], indicating that the adenosine receptors couple to their effector via a pertussis toxin–sensitive G protein. That would support a role for either G_i or G_o [49]. Recently it has been reported that A_1 receptors can activate phospholipase C in heart cells [50]. The product of phospholipase C is diacylglycerol (DAG), an activator of protein kinase C (PKC). We, therefore, chose to test whether the PKC pathway might be involved. We found that treatment with either staurosporine or polymyxin B could selectively inhibit protection from PC [51]. While staurosporine is not very selective among the protein kinases, polymyxin B has no known inhibitory action against any of the protein kinases except PKC. Figure 50-1 shows the staurosporine data. We (unpublished observtion) as well as others [52,53] have found the new and highly specific PKC inhibitor chelerythrine [54] to also be a potent blocker of PC's protective effect.

We next examined direct PKC activators. Using isolated hearts we found that a 5-minute exposure to the phorbol ester, 4β-phorbol 12-myristate 13-acetate (PMA), or the diacylglycerol analogue, 1-oleoyl-2-acetyl glycerol (OAG), was as effective as ischemia for putting the heart into a protected state (Fig. 50-1). *Based on the above observations we proposed that activation of PKC is a critical step in PC's protection.*

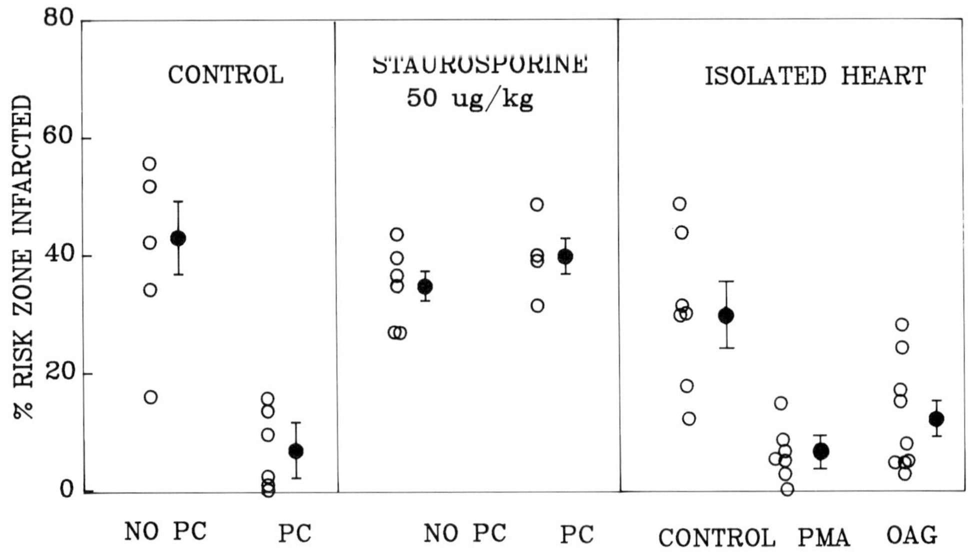

FIGURE 50-1. The effect of modulators of protein kinase C (PKC) on infarct size. Open symbols depict individual hearts, while solid symbols show group means and SEM. Thirty minutes of regional ischemia was used throughout. Note that the PKC blocker staurosporine selectively blocks the ability of ischemic preconditioning (PC) to protect the in situ heart. Intracoronary administration of 0.05 nmoles over 5 minutes of 4β-phorbol 12-myristate 13-acetate (PMA) or 50 nmoles of 1-oleoyl-2-acetyl glycerol (OAG) mimicked ischemic PC in isolated rabbit hearts. Ten minutes of washout was allowed for each of the PKC activators (From Ytrehus et al. [51], with permission.)

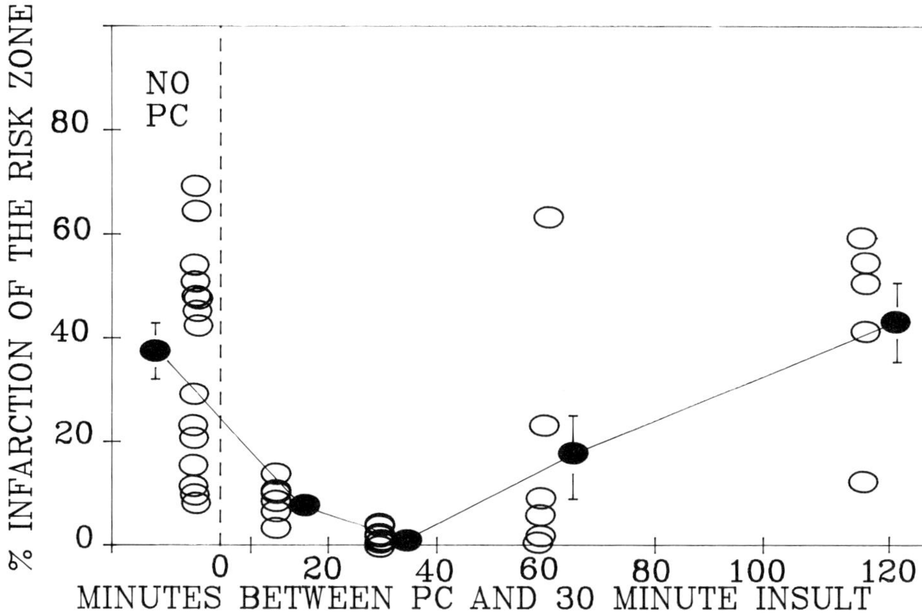

FIGURE 50-2. The infarct-sparing effect of preconditioning is seen to wane after about 1 hour in the open-chest rabbit. Thirty minutes of regional ischemia was used throughout. Protection is completely gone after 2 hours. The mechanism of this memory is a matter of speculation. (From Van Winkle et al. [6], with permission.)

What Is PC's Memory?

Any theory that explains the mechanism of PC must account for its memory. A transient exposure to ischemia or adenosine puts the heart into the PC state, which persists for about 1 hour thereafter (Fig. 50-2). How does the heart remember that it is in a PC state? Our first hypothesis was that some protein had been phosphorylated by PKC, which then is slowly dephosphorylated over the next hour. Timing studies with PKC blockers, however, argue against that theory. In Figure 50-1 staurosporine was administered 5 minutes into the reflow period *folllowing the 5-minute PC occlusion*. The fact that it blocked protection indicates that protein phosphorylation was required during the 30-minute ischemia in order to achieve an antiinfarct effect. The same was seen with polymyxin B. When staurosporine was administered 5 minutes prior to the ischemic PC period, protection was not blocked [55]. This indicates that kinase activity is not even required during the PC ischemia, which strongly argues against phosphorylation of substrate constituting the memory. Rather, *it would appear that PC acts to somehow upregulate the coupling between the adenosine receptors and PKC*. It is important to note that staurosporine blocks at the ATP binding site of PKC and does not alter the ability of the enzyme to physically translocate into the membranes, an important step in its activation sequence.

PKC Translocation Theory of PC

In light of the above studies, we were forced to consider other mechanisms by which PKC could be upregulated in the absence of protein synthesis. Most of the myocyte's PKC resides in the cytosol, where it is inactive. Upon generation of DAG by phospholipases, PKC that is in the membranes is activated. PKC in the cytosol is

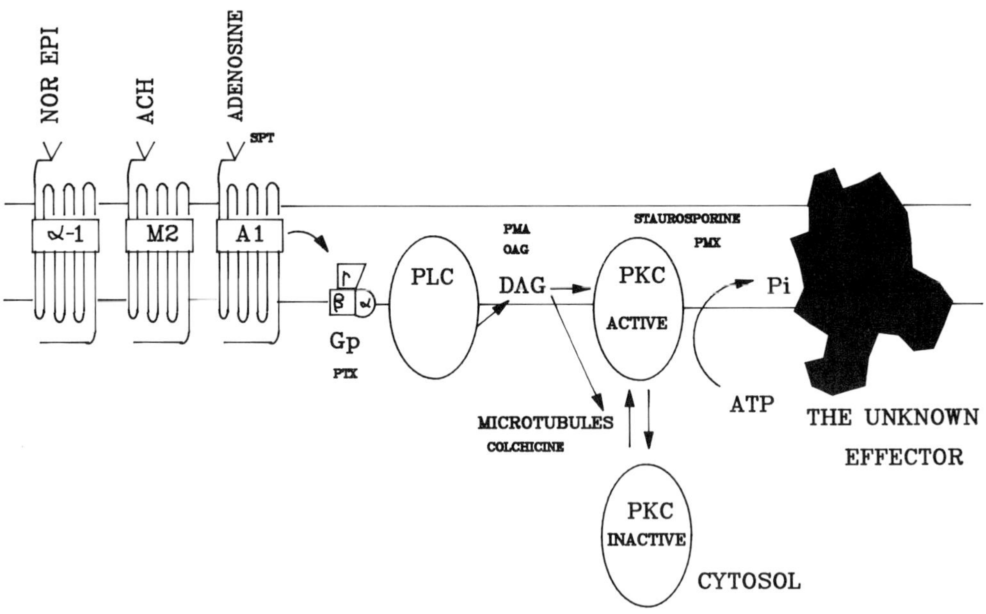

FIGURE 50-3. A diagram of the hypothesized signal transduction system in PC. Substances indicated in the smaller font represent interventions used to test each step. PMA = 4β-phorbol 12-myristate 13-acetate; OAG = 1-oleoyl-2-acetyl glycerol; PLC = phospholipase C; PMX = polymyxin B; PTX = pertussis toxin; G_p = G protein that couples to PLC.

stimulated to translocate into the membranes, where it can associate with the lipid cofactor required for activation. This translocation can be relatively slow and, depending on the mode of activation, may take up to 10 minutes [56] and has been shown to account for the delayed response often seen when PKC is stimulated. We propose that if substrate can be phosphorylated very early in a prolonged ischemia, then protection will occur. Because of the delay associated with translocation, phosphorylation would be delayed during ischemia in non-PC hearts. If, however, the heart already had PKC translocated into the membranes, then phosphorylation could occur shortly after adenosine receptors were reactivated. *We therefore suggest that the only difference between a PC and a non-PC heart is that the former has PKC translocated into the membranes.* There is supporting evidence for the translocation theory. Ischemia is known to induce this translocation [57]. We have also found that disruption of the microtubules required for translocation with colchicine completely prevents PC's protection [55].

The translocation theory also explains the dual role adenosine seems to play in PC. Adenosine receptor activation during the short ischemia clearly triggers PC in the rabbit heart. If, however, the receptors are blocked only during the long ischemia, protection is still aborted, indicating that adenosine receptors mediate protection as well [58]. This behavior would be predicted by the translocation theory. The first adenosine exposure serves to translocate PKC into the membranes and can be considered to be the upregulation phase of PC. Adenosine exposure during the subsequent prolonged ischemia acts to reactivate PKC so that substrate can be phosphorylated early in ischemia to mediate the actual protection. Figure 50-3 shows a diagram of the proposed signal transduction pathway involved in preconditioning.

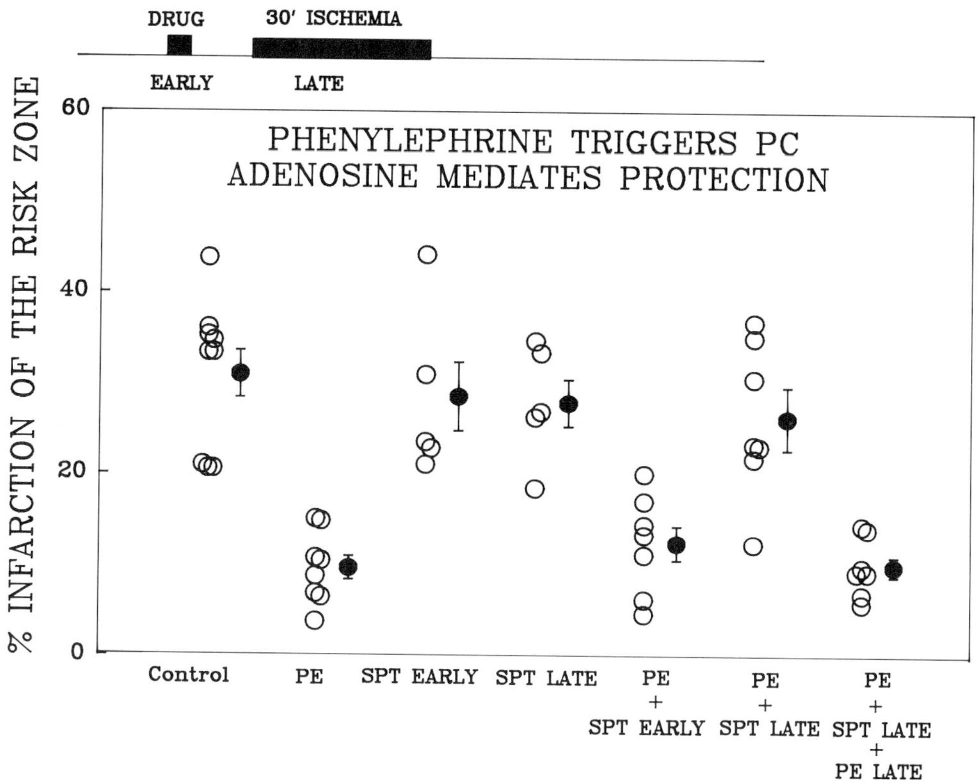

FIGURE 50-4. A plot of infarct size in isolated rabbit hearts treated with the alpha-adrenergic agonist phenylephrine (PE) and/or the adenosine antagonist [8-(p-sulfophenyl)]theophylline (SPT). The diagram at the top shows how the protocol was divided into an early period representing the 5-minute period of drug exposure, a 10-minute washout period, and a late period representing the 30 minutes of regional ischemia. SPT blocked PE's protection only when it was present during the late period. Administering PE along with the SPT during the late period overcame the block. See text for an interpretation. From Tsuchida et al. [62], with permission.

Other Receptors in the Heart that Activate PKC also Mimic PC

Several receptors in the heart, in addition to adenosine A_1, are known to activate PKC. These include the α_1-adrenergic [50], the muscarinic M_2 [50], the angiotensin AT_1 [59], and the endothelin [60] receptors. In theory any one of these should be capable of putting the heart into a PC state. We [48] had indeed reported that the M_2 agonist carbachol could mimic PC. We also had previously reported [61] that activation of the alpha$_1$ receptor with tyramine-induced catecholamine release mimicked PC, although at the time we attributed this effect to augmented adenosine production. In light of our recent PKC data, however, it seems more likely that transient alpha$_1$ activation initiated translocation of PKC, and then adenosine receptor activation during the subsequent ischemia turned kinase activity back on. That would explain why an adenosine receptor antagonist could block the tyramine-induced protection in that study.

We recently tested the above hypothesis using isolated rabbit hearts [62]. Figure 50-4 reveals that 5-minute exposure to the alpha$_1$ agonist phenylephrine followed by 10-minute washout before ischemia caused a potent antiinfarct effect ($p < 0.05$ vs. untreated). Protection persisted when the adenosine blocker 8-(p-sulfophenyl) theophylline (SPT) was present with the phenylephrine, indicating that the $alpha_1$ agonist's protection was independent of adenosine. However, SPT infused just prior to the 30-minute ischemic period did block phenylephrine's protection, revealing that endogenous adenosine is needed to reactivate PKC during the ischemic period, just as was seen in ischemic PC. The protection from the brief PE exposure could be restored if PE were included in the perfusate along with SPT during the 30-minute ischemic period. This indicates that despite the adenosine receptor blockade, PKC was being reactivated through the $alpha_1$ receptors. Thus, it does not seem to make any difference which receptor is activated at each of the two critical times.

It would appear that the PC pathway is highly redundant in that multiple potential activators of PKC are liberated during an ischemia/reperfusion cycle. These would include norepinephrine, oxygen-derived free radicals, adenosine, endothelin, and angiotensin II. A single 5-minute period of ischemia is very close to the threshold for triggering the PC response, and in the rabbit, pig, and dog adenosine appears to be the predominant activator of PKC. It is conceivable that more vigorous PC protocols could liberate additional substances such that adenosine blockade alone might not completely eliminate the protection. That was recently seen when we preconditioned rabbit hearts with a 10-minute perfusion with deoxygenated buffer [63]. In that protocol neither SPT nor the alpha blocker phenoxybenzamine could eliminate the protective effect. When the two blockers were combined, however, protection was completely abolished.

Adenosine may not be the only participant in the 5-minute PC protocol. While alpha-receptor blockade by itself does not seem to block PC's protection in the rabbit heart [62], the free radical scavenger N-2-mercaptopropinyl glycine reportedly does [64]. Thus the additive effect of adenosine receptor stimulation plus free radical generation, which can also stimulate phospholipases, may be required to reach a threshold sufficient to trigger PC. Blocking either pathway would result in a subthreshold response and loss of protection.

PC May Be the Result of a Single Isoform of PKC

At present seven isoforms of PKC have been described in the heart. It is unknown what their individual functions are, but it is assumed that each has preferred substrates and different activation kinetics. Three of these are calcium dependent (α, β, and γ), and the rest are not. Bogoyevitch et al. [65] recently reported that ε is the predominant isoform in rat heart. Also, Morgan's group reports that ε translocates and activates CAM kinase to initiate contraction in vascular smooth muscle [66]. We suspect that a single isoform of PKC may be responsible for PC. It is interesting to note that we find low doses of phorbol ester mimic PC [51], but protection is lost when the dose is increased by an order of magnitude (unpublished observation). A possible explanation for the loss of effect at high doses is that low doses of phorbol ester only recruit certain PKC isoforms.

Rats Are Different

Until now the mechanism of PC in the rat heart has remained an enigma. While the rat exhibits PC-induced protection against infarction identical to that seen in rabbits and other species, we [15] and others [16] have been unable to block protection in that species with an adenosine receptor blocker. Pertussis toxin does not block protection either [67]. Banerjee et al. [68] reported that blockade of $alpha_1$ receptors in rats with BE 2254 blocks PC's protection when postischemic function was used as the end point. That experiment could not be duplicated by Asimakis and Inners-McBride, however, when they used the alpha blocker prazosin [69]. Furthermore, BE 2254 does not block PC's effect against infarction in our rat model (unpublished observation). In light of the recent findings that PKC blockers abolish protection in the rabbit [51], a likely explanation is that in rat heart either a receptor other than the adenosine or the $alpha_1$ is responsible for activating PKC or that two or more parallel pathways to PKC are involved, making any single blocker insufficient to block protection. If one of those receptors coupled to the phospholipase via G_q, which is not pertussis toxin sensitive, then that would explain why pertussis toxin was incapable of blocking PC's protection in the rat heart. Another intriguing possibility is that hydrogen peroxide generated during reperfusion might activate PKC. Hydrogen peroxide has been reported to be a direct phospholipase activator [70] and the rat differs from the rabbit in that it has high levels of xanthine oxidase in its heart, which produces H_2O_2 upon reperfusion [71].

Mitchell et al. were the first to report that PKC blockers could block PC's protection against postischemic dysfunction in rat hearts, suggesting that activation of PKC is involved in this species [52]. Speechly-Dick et al. have reported that PKC inhibitors block the anti-infarct effect of PC in rat as well [53]. The most likely explanation is that something besides adenosine, or that adenosine in conjunction with something else, participates in activating PKC. While adenosine did not mimic PC in rat hearts when postischemic function constituted the end point [72,73], the A_1-selective adenosine agonist 2-chloro-N^6-cyclopentyladenosine (CCPA), however, did mimic PC in rats when infarct size was measured [15]. The effect of PC against postischemic dysfunction and against infarction may well involve different mechanisms.

Can the PC State Be Maintained Over an Extended Period of Time?

Adenosine must be given as a pretreatment in order to realize an antiinfarct effect. However, in the patient with acute myocardial infarction pretreatment is seldom possible. We, therefore, asked whether we could maintain a PC state indefinitely by long-term treatment with an adenosine A_1 agonist? If so then such a treatment could be used in high-risk patients. Accordingly we infused CCPA (0.043 mg/kg/hr, i.v.) into rabbits for up to 72 hours [74]. The CCPA lowered the heart rate by about 100 beats/min, a documented A_1 effect. At the end of 6 hours we found good protection when the hearts were challenged with 30 minutes of regional ischemia. After 72 hours of infusion, however, the bradycardia had disappeared and protection against infarction was no longer present, suggesting that the animals had become tolerant to the drug (Fig. 50-5). An interesting aspect of this study was that we were unable to protect these 72 hour-infusion hearts with ischemic PC either (column 6, Fig. 50-5), indicating that the hearts had become tolerant to the endogenous adenosine associated with PC as well. *Thus, tolerance is a serious obstacle to maintaining the heart in a PC state indefinitely using an adenosine agonist.*

We next wondered whether the heart might become adenosine tolerant from its own endogenous adenosine following repeated ischemic episodes. The first step in this study was to test whether PC could be renewed after the protective effects of a single episode had worn off. We exposed hearts to a 5-minute PC cycle and then reperfused for 2 hours to allow the protection to wane. Hearts exposed to a second PC protocol after the 2-hour reperfusion did show complete protection [75]. *Thus, in the rabbit PC can be reinstated at least once.*

We then tested whether repeated intermittent ischemia would also result in attenuation of protection [76]. Rabbits were fitted with a pneumatic occluder around a coronary branch and allowed to recover. A 30-minute coronary occlusion followed by 3 hours of reperfusion in the conscious state resulted in 37% infarction of the risk area. PC with a single cycle of 5-minute occlusion/10-minute reperfusion

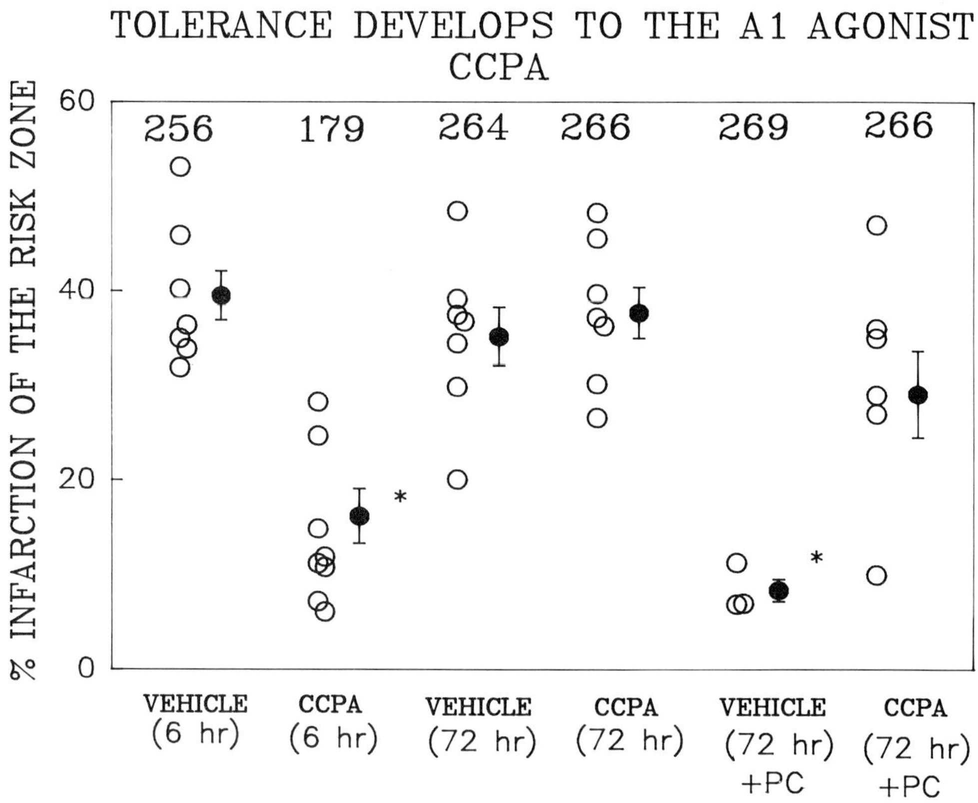

FIGURE 50-5. Infarct size in hearts subjected to prolonged infusions of the A_1-selective adenosine agonist 2-chloro-N^6-cyclopentyladenosine (CCPA). All hearts experienced 30-minute regional ischemia and received either CCPA or vehicle for either 6 or 72 hours prior to study. The mean heart rate at the time of study appears above each group. Note that bradycardia was present and hearts were protected after 6 hours of CCPA infusion, but neither bradycardia nor protection was present after 72 hours of CCPA infusion. The left-most groups indicate that ischemic PC could no longer protect hearts after a 72-hour infusion of CCPA, further indicating that a tolerance to adenosine had developed. (From Tsuchida et al. [74], with permission.)

decreased infarction in these conscious animals to only 5%. However, after exposure to 40–65 five-minute occlusions (one every 30 minutes for 8 hours per day for 3–4 days), infarction after ischemic PC significantly increased to 26%. *Because protection is greatly attenuated after multiple episodes of PC, we speculated that patients with recurrent angina may also become unresponsive to PC or adenosine-based interventions.*

At What Level Does Desensitization to Adenosine Occur?

We have recently examined the mechanism by which desensitization to A_1 agonists occurs. Previous reports suggest that the desensitization may occur at either the A_1 receptor itself or at the level of the G proteins. Many investigators have reported that A_1 receptor density is reduced after prolonged PIA exposure, as revealed by receptor binding assays in rat adipocyte [77–81], in chick embryo myocytes [82,83], and in hamster DDT_1 MF-2 cells [84]. Longabaugh et al. [78], Liang et al. [82], and Ramkumar et al. [84] all reported that the percentage of A_1 receptors in the high-affinity state was also reduced after desensitization with PIA. Because the high-affinity state is thought to occur when the receptor is attached to its G protein, desensitization could be the result of depletion of the G proteins or failure of the G proteins to couple properly to the receptor. Indeed, reduced levels of the inhibitory G protein, G_i, have been reported with desensitization. Green et al. [81, 85] and Longabaugh et al. [78] reported reductions of three of the alpha subunits of G_i protein ($G_{i\alpha 1}$, $G_{i\alpha 2}$, $G_{i\alpha 3}$) in adipocytes desensitized by PIA.

To answer the question as to where adenosine receptor downregulation occurs in rabbit hearts, we examined rabbits receiving a 72-hour infusion of PIA [86]. Again, esensitization resulted in the inability to be protected with ischemic PC (32% infarction of the risk zone in non-PC hearts vs. 29% with ischemic PC). The hearts were removed and perfused with buffer on a Langendorff apparatus, where they could be pharmacologically preconditioned with either carbachol or phenylephrine. Phenylephrine was able to protect with only 4% infarction. This indicated that PKC was intact in these hearts. $Alpha_1$ receptors couple to PKC via G_q [87], while adenosine and muscarinic receptors are thought to couple only to G_i [88]. We tested whether G_i was intact by attempting to precondition the PIA-treated hearts with carbachol. Carbachol was also highly protective, with only 9% infarction. Thus we conclude that the downregulation had occurred at the A_1 receptor itself, and hence patients who have become tolerant to adenosine may still respond to therapy with some other stimulant of PKC.

Concluding Remarks

Fortunately, the discovery of the preconditioning phenomenon occurred at the same time that great strides had been made in revealing the signal transduction pathways within the cell. By applying that knowledge to the problem of PC, we and others have been able to gain insight into the cellular mechanism of the preconditioning phenomenon. A particularly surprising finding in these studies was the revelation of how many pathways are present in the heart to ensure that protection will follow an ischemic episode. This redundancy implies that such an adaptation may be vital to the heart's survival.

A major problem yet to be solved is the identification of the actual end effector of this protection. Its elucidation should give us valuable insight into why ischemia kills cells, as well as suggest possible therapies. Unfortunately, to utilize PC's protection the heart must be treated prior to the onset of ischemia, which means that either a prophylactic treatment must be used for long-term protection or therapy must be limited to settings such as cardiac surgery, where ischemia is anticipated. While we now know how to institute PC through a variety of receptor systems, tolerance clearly presents an obstacle to long-term treatment. Finally, it should be recognized that patients with a large ischemic burden may become tolerant to their own endogenous adenosine. If that is the case then they may be refractory to an adenosine-based therapy as well. These then are the challenges that lie ahead as we try to harness the therapeutic power of ischemic preconditioning.

References

1. Kloner RA, Przyklenk K, Whittaker P (1989) Deleterious effects of oxygen radicals in ischemia/reperfusion: Resolved and unresolved issues. Circulation 80:1115–1127.
2. Reimer KA, Murry CE, Richard VJ (1989) The role of neutrophils and free radicals in the ischemic-reperfused heart: Why the confusion and controversy? J Mol Cell Cardiol 21:1225–1239.
3. Hearse DJ, Yellon DM, Downey JM (1986) Can beta blockers limit myocardial infarct size? Eur Heart J 7:925–930.
4. Murry CE, Jennings RB, Reimer KA (1986) Preconditioning with ischemia: A delay of lethal cell injury in ischemic myocardium. Circulation 74:1124–1136.

5. Li GC, Vasquez JA, Gallagher KP, Lucchesi BR (1990) Myocardial protection with preconditioning. Circulation 82:609–619.
6. Van Winkle DM, Thornton JD, Downey DM, Downey JM (1991) The natural history of preconditioning: Cardioprotection depends on duration of transient ischemia and time to subsequent ischemia. Coron Artery Dis 2:613–619.
7. Murry CE, Richard VJ, Jennings RB, Reimer KA (1991) Myocardial protection is lost before contractile function recovers from ischemic preconditioning. Am J Physiol 260:H796–H804.
8. Miura T, Adachi T, Ogawa T, Iwamoto T, Tsuchida A, Iimura O (1992) Myocardial infarct size-limiting effect of ischemic preconditioning: Its natural decay and the effect of repetitive preconditioning. Cardiovasc Pathol 1:147–154.
9. Ely SW, Mentzer RM Jr, Lasley RD, Lee BK, Berne RM (1985) Functional and metabolic evidence of enhanced myocardial tolerance to ischemia and reperfusion with adenosine. J Thorac Cardiovasc Surg 90:549–556.
10. Liu GS, Thornton J, Van Winkle DM, Stanley AWH, Olsson RA, Downey JM (1991) Protection against infarction afforded by preconditioning is mediated by A_1 adenosine receptors in rabbit heart. Circulation 84:350–356.
11. Van Winkle DM, Chien GL, Wolff RA, Soifer BE, Davis RF (1992) Intracoronary infusion of R-phenylisopropyl adenosine prior to ischemia/reperfusion reduces myocardial infarct size in swine (abstr). Circulation 86 (Suppl I):I-213.
12. Schwarz ER, Mohri M, Sack S, Arras M (1991) The role of adenosine and its A1-receptor in ischemic preconditioning (abstr). Circulation 84(Suppl II):II-191.
13. Auchampach JA, Gross GJ (1993) Adenosine A_1 receptors, K_{ATP} channels, and ischemic preconditioning in dogs. Am J Physiol 264: H1327–H1336.
14. Kitakaze M, Hori M, Kamada T (1993) Role of adenosine and its interaction with α adrenoceptor activity in ischaemic and reperfusion injury of the myocardium. Cardiovasc Res 27: 18–27.
15. Liu Y, Downey JM (1992) Ischemic preconditioning protects against infarction in rat heart. Am J Physiol 263:H1107–H1112.
16. Li Y, Kloner RA (1993) The cardioprotective effects of ischemic "preconditioning" are not mediated by adenosine receptors in rat hearts. Circulation 87:1642–1648.
17. Thornton JD, Liu GS, Olsson RA, Downey JM (1992) Intravenous pretreatment with A_1-selective adenosine analogues protects the heart against infarction. Circulation 85:659–665.
18. Liu GS, Richards SC, Olsson RA, Mullane K, Walsh RS, Downey JM (1994) Evidence that the adenosine A_3 receptor may mediate the protection afforded by preconditioning in the isolated rabbit heart. Cardiovasc Res 28:1057–1061.
19. Zhou Q-Y, Li C, Olah ME, Johnson RA, Stiles GL, Civelli O (1992) Molecular cloning and characterization of an adenosine receptor: The A_3 adenosine receptor. Proc Natl Acad Sci USA 89:7432–7436.
20. Sajjadi FG, Firestein GS (1993) cDNA cloning and sequence analysis of the human A3 adenosine receptor. Biochim Biophys Acta 1179: 105–107.
21. Murry CE, Richard VJ, Reimer KA, Jennings RB (1990) Ischemic preconditioning slows energy metabolism and delays ultrastructural damage during a sustained ischemic episode. Circ Res 66:913–931.
22. Vander Heide RS, Reimer KA, Jennings RB (1992) Adenosine mimics ischemic preconditioning and slows ischemic metabolism, possibly by inhibiting the effects of endogenous catecholamines (abstr). FASEB J 6:A940.
23. Wiltse AL, Toombs CF, Shebuski RJ (1993) Ischemic preconditioning fails to limit infarct size in reserpinized rabbit myocardium (abstr). FASEB J 7:A418.
24. Suzuki K, Miura T, Goto M, Endoh A, Urabe K, Sakamoto J, Iimura O (1993) Reserpine pretreatment attenuates the infarct size-limiting effect of preconditioning in the rabbit (abstr). Circulation 88(Suppl I):I-41.
25. Ardell JL, Yang X-M, Thornton JD, Swafford A, Cohen MV, Downey JM (1994) Depletion of norepinephrine by chronic surgical sympathectomy does not block protection from ischemic preconditioning (abstr). Circulation 90 (Suppl I):I-108.
26. Gross GJ, Auchampach JA (1992) Blockade of ATP-sensitive potassium channels prevents myocardial preconditioning in dogs. Circ Res 70:223–233.
27. Van Winkel DM, Chien GL, Wolff RA, Soifer BE, Kuzume K, Davis RF (1994) Cardioprotection provided by adenosine receptor activation is abolished by blockade of the K_{ATP} channel. Am

J Physiol 266:H829–H839.
28. Thornton JD, Thornton CS, Sterling DL, Downey JM (1993) Blockade of ATP-sensitive potassium channels increases infarct size but does not prevent preconditioning in rabbit hearts. Circ Res 72:44–49.
29. Toombs CF, Moore TL, Shebuski RJ (1993) Limitation of infarct size in the rabbit by ischaemic preconditioning is reversible with glibenclamide. Cardiovasc Res 27:617–622.
30. Downey JM (1993) An explanation for the reported observation that ATP dependent potassium channel openers mimic preconditioning. Cardiovasc Res 27:1565.
31. Snell KS, Tsuchida A, Van Wylen DGL, Downey JM, Cohen MV (1994) ATP-sensitive potassium channels fail to regulate cardiac interstitial fluid adenosine levels (abstr). FASEB J 8:A560.
32. Vegh A, Papp JG, Szekeres L, Parratt JR (1993) Are ATP sensitive potassium channels involved in the pronounced antiarrhythmic effects of preconditioning? Cardiovasc Res 27: 638–643.
33. Kirsch GE, Codina J, Birnbaumer L, Brown AM (1990) Coupling of ATP-sensitive K^+ channels to A_1 receptors by G proteins in rat ventricular myocytes. Am J Physiol 259: H820–H826.
34. Wang YG, Lipsius SL (1994) Acetylcholine pre-conditioning stimulates acetylcholine to activate a glibenclamide-sensitive K current dependent on SR Ca^{2+} release and PKC (abstr). Biophys J 66:A429.
35. de Weille JR, Schmid-Antomarchi H, Fosset M, Lazdunski M (1989) Regulation of ATP-sensitive K^+ channels in insulinoma cells: Activation by somatostatin and protein kinase C and the role of cAMP. Proc Natl Acad Sci USA 86:2971–2975.
36. Vander Heide RS, Hill ML, Steenbergen C, Reimer KA, Jennings RB (1991) Effect of reversible ischemia on mitochondrial ATPase activity in canine myocardium (abstr). Circulation 84 (Suppl II):II-192.
37. Turrens JF, Thornton J, Barnard ML, Snyder S, Liu G, Downey JM (1992) Protection from reperfusion injury by preconditioning hearts does not involve increased antioxidant defenses. Am J Physiol 262:H585–H589.
38. Omar BA, Hanson AK, Bose SK, McCord JM (1991) Reperfusion with pyruvate eliminates ischemic preconditioning in the isolated rabbit heart: An apparent role for enhanced glycolysis. Coron Artery Dis 2:799–804.
39. Liu GS, Stanley AWH, Downey JM (1992) Ischaemic preconditioning is not dependent on neutrophils or glycolytic substrate at reperfusion in rabbit heart. Cardiovasc Res 26:1195–1198.
40. Liu GS, Stanley AWH, Downey JM (1992) Cyclooxygenase products are not involved in the protection against myocardial infarction afforded by preconditioning in rabbit. Am J Cardiovasc Pathol 4:56–63.
41. Li Y, Kloner RA (1992) Cardioprotective effects of ischaemic preconditioning are not mediated by prostanoids. Cardiovasc Res 26:226–231.
42. Schott R, Nao B, Strieter R, Groh M, Kunkel S, McClanahan T, Schaper W, Gallagher K (1990) Heat shock does not "precondition" canine myocardium (abstr) Circulation 82(Suppl III):III-464.
43. Thornton J, Striplin S, Liu GS, Swafford A, Stanley AWH, Van Winkle DM, Downey JM (1990) Inhibition of protein synthesis does not block myocardial protection afforded by preconditioning. Am J Physiol 259:H1822–H1825.
44. Knowlton AA, Brecher P, Apstein CS (1991) Rapid expression of heat shock protein in the rabbit after brief cardiac ischemia. J Clin Invest 87:139–147.
45. Baxter GF, Marber MS, Patel VC, Yellon DM (1993) A "second window of protection" 24 hours after ischemic preconditioning may be dependent upon adenosine receptor activation (abstr). Circulation 88(Suppl I):I-101.
46. Kitakaze M, Hori M, Takashima S, Sato H, Inoue M, Kamada T (1993) Ischemic preconditioning increases adenosine release and 5′-nucleotidase activity during myocardial ischemia and reperfusion in dogs: Implications for myocardial salvage. Circulation 87:208–215.
47. Van Wylen DGL (1994) Effect of ischemic preconditioning on interstitial purine metabolite and lactate accumulation during myocardial ischemia. Circulation 89:2283–2289.
48. Thornton JD, Liu GS, Downey JM (1993) Pretreatment with pertussis toxin blocks the protective effects of preconditioning: Evidence for a G-protein mechanism. J Mol Cell Cardiol 25: 311–320.
49. Fleming JW, Wisler PL, Watanabe AM (1992) Signal transduction by G proteins in cardiac tissues. Circulation 85:420–433.
50. Kohl C, Linck B, Schmitz W, Scholz H, Scholz J, Tóth M (1990) Effects of carbachol and (−)-N^6-phenylisopropyladenosine on myocardial

inositol phosphate content and force of contraction. Br J Pharmacol 101:829–834.
51. Ytrehus K, Liu Y, Downey JM (1994) Preconditioning protects ischemic rabbit heart by protein kinase C activation. Am J Physiol 266: H1145–H1152.
52. Mitchell MB, Brew EC, Harken AH, Banerjee A (1993) Does protein kinase C mediate functional preconditioning (abstr) J Mol Cell Cardiol 25(Suppl III):S.64.
53. Speechly-Dick ME, Mocanu MM, Yellon DM (1994) Protein kinase C: its role in ischemic preconditioning in the rat. Circ Res 75:586–590.
54. Herbert JM, Augereau JM, Gleye J, Maffrand JP (1990) Chelerythrine is a potent and specific inhibitor of protein kinase C. Biochem Biophys Res Commun 172:993–999.
55. Liu Y, Ytrehus K, Downey JM (1994) Evidence that translocation of protein kinase C is a key event during ischemic preconditioning of rabbit myocardium. J Mol Cell Cardiol 26:661–668.
56. Yuan S, Sunahara FA, Sen AK (1987) Tumor-promoting phorbol esters inhibit cardiac functions and induce redistribution of protein kinase C in perfused beating rat heart. Circ Res 61: 372–378.
57. Strasser RH, Braun-Dullaeus R, Walendzik H, Marquetant R (1992) α_1 receptor-independent activation of protein kinase C in acute myocardial ischemia: Mechanisms for sensitization of the adenylyl cyclase system. Circ Res 70:1304–1312.
58. Thornton JD, Thornton CS, Downey JM (1993) Effect of adenosine receptor blockade: Preventing protective preconditioning depends on tiem of initiation. Am J Physiol 265:H504–H508.
59. Pfeilschifter J, Bauer C (1986) Pertussis toxin abolishes angiotensin II-induced phosphoinositide hydrolysis and prostaglandin synthesis in rat renal mesangial cells. Biochem J 236: 289–294.
60. Habuchi Y, Tanaka H, Furukawa T, Tsujimura Y, Takahasi H, Yoshimura M (1992) Endothelin enhances delayed potassium current via phospholipase C in guinea pig ventricular myocytes. Am J Physiol 262:H345–H354.
61. Thornton JD, Daly JF, Cohen MV, Yang X-M, Downey JM (1993) Catecholamines can induce adenosine receptor-mediated protection of the myocardium but do not particiate in ischemic preconditioning in the rabbit. Circ Res 73:649–655.
62. Tsuchida A, Liu Y, Lin GS, Cohen MV, Downey JM (1994) α_1-adrenergic agonists precondition rabbit ischemic myocardium independent of adenosine by direct activation of protein kinase C. Circ Res 75:576–585.
63. Cohen MV, Goto M, Walsh RS, Downey JM (1995) Hypoxia preconditions rabbit myocardium via adenosine and catecholamine release. J Mol Cell Cardiol, in press.
64. Tanaka M, Fujiwara H, Yamasaki K, Sasayama S (1994) Superoxide dismutase and N-2-merceptopropionylglycineattenuate infarct size limitation effect of ischemic preconditioning in the rabbit. Cardiovasc Res 28:980–986.
65. Bogoyevitch MA, Parker PJ, Sugden PH (1993) Characterization of protein kinase C isotype expression in adult rat heart: Protein kinase C-ε is a major isotype present and it is activated by phorbol esters, epinephrine and endothelin. Circ Res 72:757–767.
66. Khalil RA, Morgan KG (1993) PKC-mediated redistribution of mitogen-activated protein kinase during smooth muscle cell activation. Am J Physiol 265:C406–C411.
67. Liu Y, Downey JM (1993) Preconditioning against infarction in the rat heart does not involve a pertussis toxin sensitive G protein. Cardiovasc Res 27:608–611.
68. Banerjee A, Locke-Winter C, Rogers KB, Mitchell MB, Brew EC, Cairns CB, Bensard DD, Harken AH (1993) Preconditioning against myocardial dysfunction after ischemia and reperfusion by an adrenergic mechanism. Circ Res 73:656–670.
69. Asimakis GK, Inners-McBride K (1993) Attenuated postischemic contractile dysfunction associated with ischemic preconditioning is not mediated by alpha$_1$ adrenoreceptor stimulation (abstr). Circulation 88 (Suppl I):I-43.
70. Natarajan V, Taher MM, Roehm B, Parinandi NL, Schmid HHO, Kiss Z, Garcia JGN (1993) Activation of endothelial cell phospholipase D by hydrogen peroxide and fatty acid hydroperoxide. J Biol Chem 268:930–937.
71. Downey JM, Hearse DJ, Yellon DM (1988) The role of xanthine oxidase during myocardial ischemia in several species including man. J Mol Cell Cardiol 20(Suppl II):55–63.
72. Cave AC, Collis CS, Downey JM, Hearse DJ (1993) Improved functional recovery by ischaemic preconditioning is not mediated by adenosine in the globally ischaemic isolated rat heart. Cardiovasc Res 27:663–668.
73. Asimakis GK, Inners-McBride K, Medellin G,

Conti VR (1992) Ischemic preconditioning attenuates acidosis and postischemic dysfunction in isolated rat heart. Am J Physiol 263:H887–H894.

74. Tsuchida A, Thompson R, Olsson RA, Downey JM (1994) The anti-infarct effect of an adenosine A_1-selective agonist is diminished after prolonged infusion as is the cardioprotective effect of ischaemic preconditioning in rabbit heart. J Mol Cell Cardiol 26:303–311.
75. Yang X-M, Arnoult S, Tsuchida A, Cope D, Thornton JD, Daly JF, Cohen MV, Downey JM (1993) The protection of ischaemic preconditioning can be reinstated in the rabbit heart after the initial protection has waned. Cardiovas Res 27:556–558.
76. Cohen MV, Yang X-M, Downey JM (1994) Conscious rabbits become tolerant to multiple episodes of ischemic preconditioning. Circ Res 74:998–1004.
77. Parsons WJ, Stiles GL (1987) Heterologous desensitization of the inhibitory A_1 adenosine receptor-adenylate cyclase system in rat adipocytes. J Biol Chem 262:841–847.
78. Longabaugh JP, Didsbury J, Spiegel A, Stiles GL (1989) Modification of the rat adipocyte A_1 adenosine receptor-adenylate cyclase system during chronic exposure to an A_1 adenosine receptor agonist: Alterations in the quantity of $G_{s\alpha}$ and $G_{i\alpha}$ are not associated with changes in their mRNAs. Mol Pharmacol 36:681–688.
79. Hoffman BB, Chang H, Dall'Aglio E, Reaven GM (1986) Desensitization of adenosine receptor-mediated inhibition of lipolysis: The mechanism involves the development of enhanced cyclic adenosine monophosphate accumulation in tolerant adipocytes. J Clin Invest 78:185–190.
80. Green A (1987) Adenosine receptor down-regulation and insulin resistance following prolonged incubation of adipocytes with an A_1 adenosine receptor agonist. J Biol Chem 262: 15702–15707.
81. Green A, Milligan G, Dobias SB (1992) G_i down-regulation as a mechanism for heterologous desensitization in adipocytes. J Biol Chem 267:3223–3229.
82. Liang BT, Donovan LA (1990) Differential desensitization of A_1 adenosine receptor-mediated inhibition of cardiac myocyte contractility and adenylate cyclase activity: Relation to the regulation of receptor affinity and density. Circ Res 67:406–414.
83. Shryock J, Patel A, Belardinelli L, Linden J (1989) Downregulation and desensitization of A_1 adenosine receptors in embryonic chicken heart. Am J Physiol 256:H321–H327.
84. Ramkumar V, Olah ME, Jacobson KA, Stiles GL (1991) Distinct pathway of desensitization of A_1- adenosine and A_2-adenosine receptors in DDT_1 MF-2 cells. Mol Pharmacol 40:639–647.
85. Green A, Johnson JL, Milligan G (1990) Down-regulation of G_i sub-types by prolonged incubation of adipocytes with an A_1 adenosine receptor agonist. J Biol Chem 265:5206–5210.
86. Hashimi W, Thornton JD, Downey JM, Cohen MV (1994) Chronic adenosine A_1-agonist exposure results in adenosine receptor rather than G protein downregulation (abstr). Circulation 90 (Suppl I):I-107.
87. Karliner JS, Kagiya T, Simpson PC (1990) Effects of pertussis toxin on α_1-agonist-mediated phosphatidylinositide turnover and myocardial cell hypertrophy in neonatal rat ventricular myocytes. Experientia 46:81–84.
88. Belardinelli L, Linden J, Berne RM (1989) The cardiac effects of adenosine. Prog Cardiovasc Dis 32:73–97.

B. NERVOUS SYSTEM

51. ANTICONVULSANT ACTIONS OF NOVEL AND REFERENCE ADENOSINE AGONISTS

Lars J.S. Knutsen
Jesper Lau
Malcolm J. Sheardown
Karen Eskesen
Christian Thomsen
Jan U. Weis
Martin E. Judge
Henrik Klitgaard

A host of potential mechanisms exist by which manipulation of neurotransmission in the central nervous system (CNS) can lead to inhibition of seizures in mammals. This reflects the complex nature of the diverse disease states that result in the group of symptoms collectively known as epilepsy. The research group at Novo Nordisk has previously discovered agents with clinical anticonvulsant activity, such as Tiagabine (structure 1, Fig. 51-1), an inhibitor of the synaptosomal uptake of gamma-aminobutyric acid [1] and Abecarnil (2, Fig. 51-1), a β-carboline benzodiazepine partial agonist [2], both of which facilitate inhibitory neurotransmission. Antagonists at the excitatory amino acid (EAA) receptor subtypes, such as the AMPA receptor, the metabotropic glutamate receptor [3], and glycine receptors [4] (3, 4 and 5, Fig. 51-1), have also been shown to exhibit anticonvulsant properties in rodent seizure models.

Administration of purinergic agents, which result in activation of adenosine receptors, can be seen as an extension of the above approaches to epilepsy treatment. This is due at least in part to the ability of adenosine agonists to inhibit neurotransmitter release, especially the release of the excitatory neurotransmitter glutamate [5,6]. Adenosine has a wide variety of regulatory functions and physiologic effects. In recent years the CNS effects of this purine have received greater prominence. It is now clear that in the mammalian CNS, adenosine acts as a neuromodulator and has been implicated both as an endogenous anticonvulsant agent [7] and a neuroprotectant [8].

The considerable therapeutic potential of agonists at adenosine receptors has been reviewed

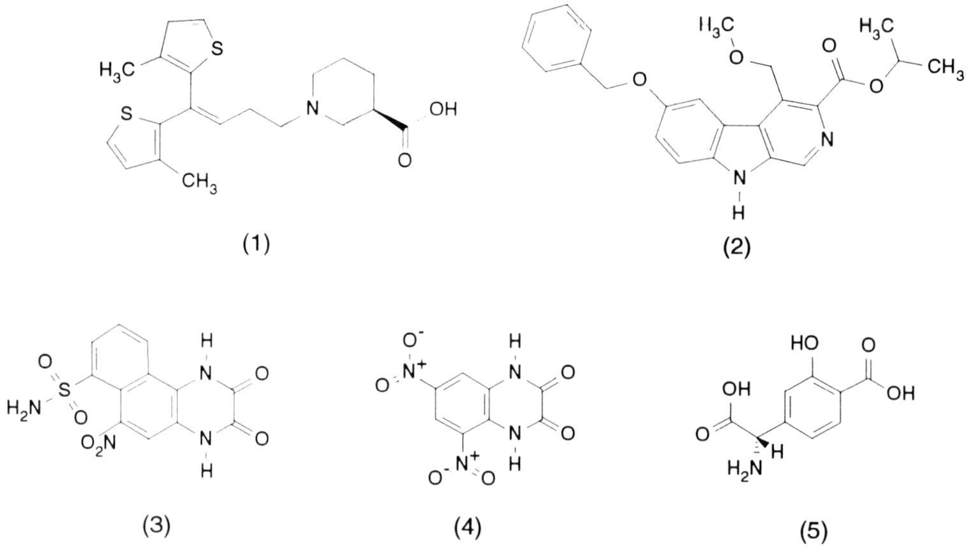

FIGURE 51-1. Other anticonvulsant agents.

[9,10], including their potential for treatment of neurologic diseases [11]. Some of the early investigations of the anticonvulsant potential of adenosine agonists [12,13] took place around the time when the neuromodulatory effect of adenosine [14] was first being established. It has also been illustrated more recently that purinergic agents hold considerable potential as anticonvulsants [15–19] in the face of evidence that status epilepticus is caused by loss of adenosine anticonvulsant mechanisms [7]. The curiosity of investigators has not been diminished by the notion of a purinergic component to the mechanism of action of marketed anticonvulsant drugs such as carbamazepine [20].

However, the supposition that adenosine receptor agonists exclusively exhibit anticonvulsant properties does not always hold. The nature of the chemoconvulsive agent used in the particular seizure model has a major influence. For example, where seizures are induced by administration of xanthine amine congener (XAC), an A_1 receptor antagonist, reference adenosine A_1 agonists such as (R)-phenylisopropyladenosine ((R)-PIA) and N^6-cyclopentyladenosine (CPA) are proconvulsant [21]. Likewise, adenosine A_2 agonists such as N-[2-(3,5-dimethoxyphenyl)-2-(2-methylphenyl) ethyl]adenosine (DPMA) and N-[(2-methylphenyl)-methyl]adenosine (metrifudil) are proconvulsant [21] when seizures are induced by 3,7-dimethyl-propargylxanthine (DMPX), an A_2 receptor antagonist. Similar observations have been made in the case of the mixed A_1/A_2 agonist 5′-N-ethylcarboxamidoadenosine (NECA), which is proconvulsant when seizures are induced by caffeine [22].

Materials and Methods

DRUGS

The novel 2-halo-N-substituted adenosine derivatives (compounds 1–17, Tables 51-2 and 51-3) were prepared by the published methods [23,24]. Metrifudil, N-[1S,*trans*]-2-hydroxycyclopentyladenosine (GR 79236) [25] and 2-chloro-N-cyclopentyladenosine (CCPA) were prepared by existing procedures [26,27]. CPA, R-PIA, S-PIA, 8-cyclopentyl-1,3-dimethylxanthine (CPT), and glutamate were obtained from Research Biochemicals (Natick, Ma). Methyl-6,7-dimethoxy-4-ethyl-β-carboline-3-carboxylate (DMCM) was a generous gift of Dr. L. Turski (Schering AG, Berlin, Germany). Pentylenetetrazole (PTZ) and pilocarpine were obtained from Sigma (St. Louis, MO). The novel adenosine receptor agonists featured in Tables 51-2 and 51-3 1/M CPA, R-PIA, S-PIA, CPT, and metrifudil 1/M were dissolved in 5% cremophore (BASF) prior to treatment. DMCM was dissolved in 0.02 N HCl; PTZ, pilocarpine, and glutamate were dissolved in a 0.9% saline solution.

ADENOSINE A_1 AND A_2 RECEPTOR BINDING

These determinations were carried out using the experimental procedures previously cited [21] and are only described briefly here.

[^{3}H]-(R)-PHENYLISOPROPYL-ADENOSINE BINDING TO CORTICAL MEMBRANES

[^{3}H]-(R)-PIA receptor binding assays were performed in crude membrane preparations from male Wistar rat whole forebrain. Washed tissue homogenates were incubated for 60 minutes in 50 mM TIS-HCl (pH 7.7) + 0.1 U/l adenosine deaminase (Sigma) at 25°C with 1 nM [^{3}H]-R-PIA (50 Ci/mmol, Amersham UK). Free and bound ligands were separated as previously described.

[^{3}H]-CGS 21680 BINDING TO STRIATAL MEMBRANES

Ligand binding with [^{3}H]CGS 21680 (2-p-(2-carboxyethyl)phenylethylamino-5′-N-ethylcarboxamidoadenosine) (48.1 Ci/mmol, Amersham) was measured as previously described [21], except that the concentration of radioligand was 2 nM in displacement studies. IC_{50} values were calculated by a nonlinear regression analysis using the GRAPHPAD program (I.S.I., Philadelphia, PA). Inhibitory constants (K_i) were calculated from the equation $K_i = IC_{50}/1 + [L]/K_d$, where [L] is the concentration of radioligand and K_d is the equilibrium dissociation constant for the ligand.

INHIBITION OF ISOPRENALINE-STIMULATED cAMP PRODUCTION IN DDT_1-MF2 CELLS

The smooth muscle cell line DDT_1-MF2 contains both A_1- and A_2-adenosine receptors, which appear to be functionally coupled to inhibition and stimulation of cAMP production, respectively. The assay protocol was a modification of a previously described method, to which we have earlier referred [21].

Cells were resuspended in DMEM + 20 mM HEPES, pH 7.4 and preincubated with test compounds 5 minutes prior to the addition of 10 μM isoprenaline in a final volume of 500 μl. Following 10-minute incubation at 37°C, the tubes were placed on ice and 1 ml ice-cold ethanol was added. The protein was removed by centrifugation and the ethanol was evaporated under a stream of nitrogen at 70°C. The cAMP content was determined using a radioimmunoassay kit (Amersham).

It was found that isoprenaline stimulated cAMP production to a multiple of 50- to 80-fold at the basal level. No stimulation of cAMP production was observed with the A_2 agonist CGS 21680 alone, and CGS 21680 was found to have a weak effect on stimulated cAMP release (IC_{50} = 1260 nM). The A_1 antagonists CPT and DPCPX did not reduce stimulated cAMP production at concentrations up to 10 μM.

ANIMALS

Male NMRI mice (Moellegaard's Breedings Labs, LI. Skensved, Denmark) weighing 18–27 g were used. The mice were kept on a 12-hour light/12-hour dark cycle (light from 06.00 to 18.00 hours) and were housed at a temperature of 20 ± 1°C and at a humidity of 55 ± 10%. The animals had free access to standard pellet food and water before testing. All testing was performed between 09.00 and 15.00 hours.

TABLE 1

Test/seizure model	ED_{50} value (mg/kg) after i.p. dose	
	NNC 90-1515	CPA
MES	**8**	**4.2**
PTZ s.c.	>**10**	**5.3**
DMCM i.p.	**2.0**	**0.17**
Pilocarpine	**0.2**	>**10**
Glutamate i.c.v.	**0.3**	**0.4**
Rotarod	**0.66**	**0.1**

ROTAROD PERFORMANCE

In order to determine if the adenosine analogues selected for this study produced ataxia (a measure of unwanted CNS effects in this context), following systemic administration a rotarod (Ugo Basile) test was performed. After a pretraining period, the mice were injected intraperitoneally (i.p.) with the test compound and after 30 minutes were placed on a rod rotating at a rate of 6 rpm. Performance was considered impaired if the animal fell from the rod more than twice during a 2-minute test period. From these results, doses below and above the ED_{50} value (see Tables 51-1 to 51-3) were selected for seizure testing.

SEIZURE TESTS

Seizures involving the GABA/benzodiazepine receptor complex were induced by DMCM (18 mg/kg i.p.) or PTZ (160 mg/kg, s.c.), and cholinergic receptors were activated by pilocarpine (400 mg/kg i.p.). The animals were placed individually in Perspex cages and observed for the presence of clonic convulsions during the following 15 minutes (DMCM) or 30 minutes (PTZ, pilocarpine). Animals were pretreated with the adenosine agonist via intraperitoneal administration 30 minutes before testing.

Seizures were also induced by activation of EAA receptors by glutamate, which was infused intracerebroventricularly for 1 minute at a concentration of 100 μg/5 μl and with a speed of infusion of 5 μl/min (Basic Microprocessor Syringe Pump, Harvard Apparatus), according to the method previously described [28]. Briefly, a cannula was inserted through the skull of the animal at a position approximately 1 mm posterior and 1 mm lateral to bregma. Placement and depth of the cannula was fixed by a plate positioned 4.3 mm from the end of the cannula. The animals were observed for 150 seconds after the start of infusion or until the appearance of clonic convulsions for more than 5 seconds, with 150 seconds being the end point. Maximal electroshock (MES) seizures were induced by corneal electrodes. A 60 Hz alternating current of 50 mA was applied for 0.2 seconds. Immediately after, the animals were observed for 10 seconds for the presence of tonic hindlimb extension.

STATISTICAL ANALYSIS OF IN VIVO EFFECTS

All ED_{50} values are determined from groups of six to eight animals. A dose impairing the performance of 50% of the animals in the rotarod test for ataxia (ED_{50}) and an anticonvulsive effective dose protecting 50% of the animals against clonic convulsions (ED_{50}) were calculated using a log-probit analysis. These values were calculated from the effect of three to four different doses of each compound.

Results

The in vitro binding profiles of the novel and reference agonists to A_1 and A_{2a} rat brain adenosine receptors are shown in Tables 51-2 and 51-3, expressed as K_i values (nM). Evidence is also given of their functional effect at A_1 receptors in terms of inhibition of isoprenaline-

TABLE 2

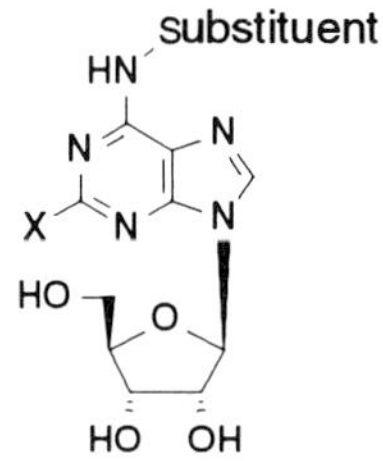

X	Adenosine N-substituent	Receptor binding K_i, nM A_1	Receptor binding K_i, nM A_{2a}	cAMP IC_{50} (nM)	DMCM-induced seizures, mice ED_{50}, mg/kg	Mouse rotarod ED_{50}, mg/kg	ED_{50} ratio rotarod/seizures	Cpd. #
Cl	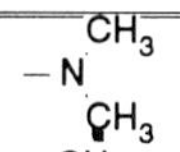	18	1500	3.2	0.09	0.067	0.8	1
Cl	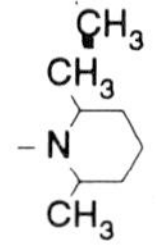	4.0	33	-	0.46	0.07	0.2	2
Cl	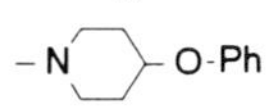	1.4	1100	0.03	0.9	0.62	0.7	3
Cl	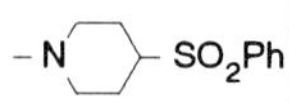	17	220	0.7	0.94	0.94	1.0	4
Cl	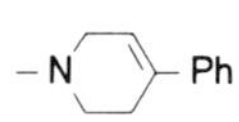	6.1	640	1.1	0.95	1.1	1.2	5
Cl	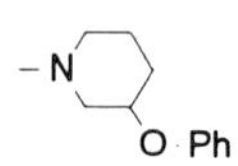	15	1700	4.8	1.0	-	-	6
Cl	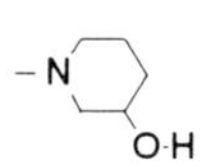	11	720	4.6	1.2	2.8	2.3	7
Cl	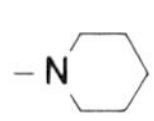	4.2	1500	2.1	2.0	0.66	0.3	8
Cl	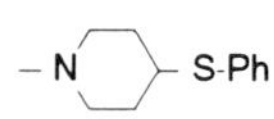	0.84	470	0.3	3.2	1.4	0.4	9
Cl	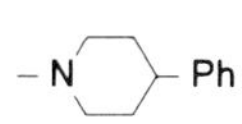	11	1200	2.5	6.7	6.3	0.9	10
Cl	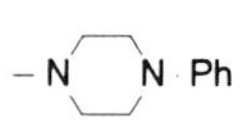	55	5600	3.8	46	39	0.8	11

stimulated cAMP production, expressed as IC_{50} values (nM).

A comparison of the in vivo effects of NNC 90-1515 (**8**, Table 51-2) and the reference A_1 agonist CPA in a range of seizure models and in the rotarod test are shown in Table 51-1, expressed as ED_{50} values (mg/kg). The in vivo effects of a larger range of compounds in inhibiting clonic DMCM-induced seizures and inhibition of rotarod performance are shown on

TABLE 3

X	Adenosine N-substituent	Receptor binding K_i, nM A_1	A_{2a}	cAMP IC_{50} (nM)	DMCM-induced seizures, mice ED_{50}, mg/kg	Mouse rotarod ED_{50}, mg/kg	ED_{50} ratio rotarod/seizures	cpd. #
F	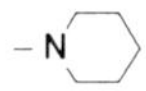	3.6	1100	3.8	0.31	0.73	2.4	12
Cl	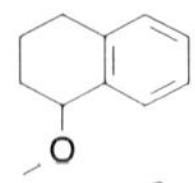	46	290	35	2.6	22	8.5	13
Cl	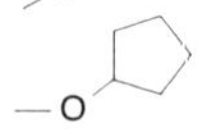	6.7	2400	4.2	2.7	39	14.4	14
Cl	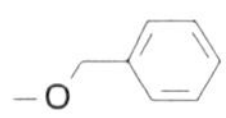	45	350	110	5.6	13	2.3	15
H	$-OCH_3$	110	2600	60	>100	>100	-	16
Cl	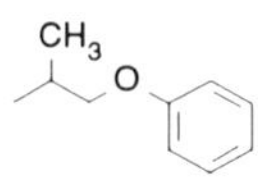	43	1200	4.5	3.4	17.2	5.1	17
H	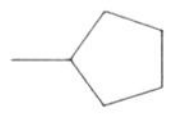	1.2	320	0.57	0.17	0.10	0.5	CPA
Cl	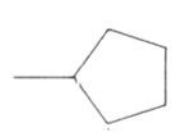	1.5	500	-	0.093	0.30	3.2	CCPA
H	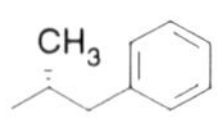	2.0	120	0.31	0.5	0.092	0.2	(R)-PIA
H	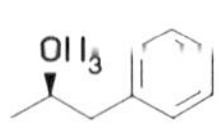	18	6700	27	4.7	12.0	2.6	(S)-PIA
H	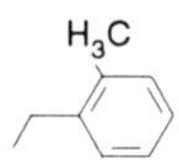	74	28	55	0.54	9.2	17.0	Metrifudil
H	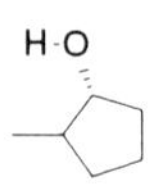	3.1	1320	2.6	0.30	0.20	0.7	GR79236

Tables 51-2 and 51-3, expressed as ED_{50} values (mg/kg). The 2-chloroadenosine derivatives in Table 51-2 are ranked by ascending potency versus DMCM seizures.

In order for a compound to be considered for further evaluation as a potential drug candidate, the ratio between rotarod ED_{50} and anticonvulsant ED_{50} should clearly be as high as possible, because this indicates a good separation of efficacious doses and those producing side effects. These ratios are included in Tables 51-2 and 51-3. This is the first publication of the in vivo effects of Novo Nordisk adenosine agonists.

Discussion

The main finding of this study of new adenosine agonists is that a range of substituted adenosines with heteroatoms on the purine 6-amine show very potent anticonvulsive effects, especially when seizures are induced by DMCM, an inverse benzodiazepine agonist. Furthermore, a compound with a mild cardiovascular profile when compared to NNC 90-1515 or CPA, 2-chloro-N-(1-phenoxy-2-propyl)adenosine (NNC 21-0041, 17; Table 51-3), retains a seizure inhibiting effect in this model.

The advantage of epilepsy models in which seizures are chemically induced is that there are mechanistic implications depending on the type of chemoconvulsant used. Given the prominence of mechanism-based strategies for drug discovery, these techniques can be very useful in the pharmaceutical industry. Adenosine agonists appear to have unique potency in seizure models where convulsions are induced by chemoconvulsants acting at the benzodiazepine receptor complex. We have previously discussed literature evidence for this modulation of the benzodiazepine receptor complex by purinergic ligands [21]. The novel agonists featured in the present study provide further evidence for a modulation of the benzodiazepine complex by adenosine agonists, given the particularly high potency of these compounds against convulsions induced by DMCM, an inverse agonist at benzodiazepine receptors. In contrast, we have previously shown that adenosine agonists are not especially potent against chemoconvulsants modulating the GABA/benzodiazepine receptor complex at other sites than the benzodiazepine receptor [21].

Our conclusion from the data in Table 51-1 is that the novel agonist NNC 90-1515 (**8**) and CPA have related anticonvulsant profiles. The table highlights the atypical profile of adenosine A_1 agonists in animal models of epilepsy. Both NNC 90-1515 and CPA are not so strongly active against seizures induced by MES or subcutaneous administration of PTZ. These two models are the classic tests for identification of agents with anticonvulsant activity [29]. By contrast, both agents show a more potent action against seizures induced by the chemoconvulsants DMCM (an inverse agonist at benzodiazepine receptors), pilocarpine (a cholinergic receptor agonist), and glutamate (the endogenous agonist at glutamate receptors). Furthermore, previous studies have shown an effect of the adenosine A_1 agonist CHA against both the development and expression of seizures in amygdala-kindled rats [18]. This suggests, given the atypical profile of adenosine A_1 agonists in epilepsy models involving benzodiazepine, cholinergic, and glutamate receptor-mediated seizures, that these purine ligands may be active in complex partial seizures [29]. We would therefore suggest that purinergic methodology, in particular utilizing adenosine A_1 agonists, holds promise for a novel approach to untreatable epilepsy.

The data in Tables 51-2 and 51-3 demonstrates how wide a range of anticonvulsant potencies is apparent in adenosine A_1 agonists, which otherwise appear to be structurally similar. Compounds **1**–**12** are substituted N-aminoadenosine derivatives, selected from a range of new compounds, the preparation and in vitro profiles of which have been described [23,24]. Compounds **13**–**16** are substituted N-alkoxyadenosine derivatives [24].

The number of previous observations of the anticonvulsive effects of adenosine A_1 agonists, as reviewed in the introduction, led us to expect that there would be a correlation between ED_{50} values in DMCM-induced seizures and A_1 receptor binding. This correlation is indeed observed, with a p value of less than 0.0001, for the series of 23 compounds featured on Tables 51-2 and 51-3. Although the correlation is highly significant, A_1 binding potency is by no means an ideal predictor of anticonvulsant efficacy in this model. For example, structure 1 (Table 51-2) has a DMCM ED_{50} value of 0.09 mg/kg and is the most potent agent in this series against this seizure type. It is not, however, the most potent compound in terms of binding to A_1 receptors, having a K_i value of 18 nM. In contrast, the most potent example in terms of A_1-receptor binding, 9 (Table 51-2) has a seizure ED_{50} of 3.2 mg/kg. However, there are many uncertainties lying between the peripheral dosing of a novel adenosine agonist and its interaction at central adenosine receptors. For example, ambiguous in vivo results can be caused by plasma protein binding, rapid metabolism, difficulty in penetration of the blood-brain barrier, and transport within the CNS. Some of the variations seen in Tables

51-1 and 51-2 may reflect these uncertainties. Paradoxically, we have previously observed that the A_1 antagonist 8-cyclopentyltheophylline (CPT) also displayed anticonvulsant properties in the DMCM model (ED_{50} = 23 mg/kg) [21,30]. This unusual effect may be due to a metabolite of CPT, and illustrates the fine and complex balance involved in the regulation of neuronal excitability by the adenosine A_1 receptor.

The compounds in this series with the best ratio between rotarod ED_{50} and seizure ED_{50} are compounds 14 (14.4) and the reference compound metrifudil [17]. The only agonists having ratios of between 5 and 10 are 13 and 17 (Table 51-3). There is therefore a tendency for the structurally novel agonists such as 13 and 14 with substituted hydroxylamines at the purine 6-position to have the best rotarod/seizure ratios.

A_1 receptor agonists generally possess cardiovascular effects in rodents, resulting in dose-dependent hypotension. The agonists 1–16 induce profound hypotension in anesthetized rats when assessed 5 minutes after bolus intravenous administration at 0.1 mg/kg. The average decrease in blood pressure observed for 1–16 at this dose was >50%.

These pronounced cardiovascular effects led us to synthesize a new series of agonists, an example of which is NNC 21-0041 (17) [27]. This agonist has a milder CV profile in anesthetized rats (no change in blood pressure at 0.1 mg/kg, 5 minutes after i.v. bolus dose) when compared to (R)-PIA, CPA, and compounds 1–16, while maintaining a strong A_1 functional effect and protection from DMCM-induced seizures in mice (ED_{50} = 3.4 mg/kg, rotarod ED_{50} = 17.2 mg/kg, i.p., −30 minutes).

Conclusions

We have demonstrated that a range of novel 2-substituted adenosine analogs with alkylated nitrogen and oxygen atoms on the 6-amino group have anticonvulsant effects, in some cases with high potency, in the DMCM-induced clonic seizure model in mice after i.p. administration. However, the potent cardiovascular effects of the above agonists led us to examine another range of adenosine agonists, represented by NNC 21-0041 (17), with milder CV effects. NNC 21-0041, however, maintained a potent effect in the mouse DMCM-induced clonic seizure model, as well as a separation between anticonvulsant and ataxic doses, and therefore represents a prototype adenosine agonist for future CNS drug development in this field.

Acknowledgments

The authors wish to thank Paw Bloch, Claus Jensen, and Henrik Stephensen for the skilled synthesis of novel and reference adenosine agonists as well Lisbeth Hansen for biochemical studies. In vivo testing was performed by Charlotte Halle, Lene Klein, Anne Lund, Kira Meyhoff, and Karina Nielsen.

References

1. Andersen KE, Braestrup C, Grønwald FC, Jørgensen AS, Nielsen EB, Sonnewald U, Sørensen PO, Suzdak PD, Knutsen LJS (1993) The synthesis of novel GABA uptake inhibitors. 1. Elucidation of the structure activity studies leading to the choice of R-1-[4,4-bis(3-methyl-2-thienyl)-3-butenyl]-3-piperidine carboxylic acid (Tiagabine) as an anticonvulsant drug candidate. J Med Chem 36:1716–1725.
2. Turski L, Stevens DN, Jensen LH, Petersen EN, Meldrum BS, Patel S, Bondo Hansen J, Löscher W, Schneider HH, Scmeichen R (1990) J Pharm Exp Ther 253:344–352.
3. Thomsen C, Klitgaard H, Sheardown M, Jackson H, Eskesen K, Jakobsen P, Treppendahl S, Suzdak P (1994) (S)-4-Carboxy-3-hydroxyphenylglycine, an antagonist of metabotropic glutamate receptor (mGluR) 1a and an agonist of mGluR2 protect against audiogenic seizures in DBA/2 mice. J Neurochem 62:2492–2495.
4. Sheardown MJ, Drejer J, Jensen LH, Stidsen CE, Honore T (1989) A potent antagonist of the strychnine insensitive glycine receptor has anticonvulsant properties. Eur J Pharmacol 174:197–204.

5. Dolphin AC, Prestwich SA (1985) Pertussis toxin reverses adenosine inhibition of neural glutamate release. Nature 316:148–150.
6. Simpson RE, O'Regan MH, Perkins LM, Phillis JW (1992) Excitatory transmitter release from the ischemic rat cerbral cortex: Effects of adenosine receptor agonists and antagonists. J Neurochem 58:1683–1690.
7. Young D, Dragunow M (1994) Status epilepticus may be caused by loss of adenosine anticonvulsant mechanisms. Neuroscience 58: 245–261.
8. Rudolphi KA, Schubert P, Parkinson FE, Fredholm BB (1992) Neuroprotective role of adenosine in cerebral ischaemia. Trends Pharmacol Sci 439.
9. Daval J-L, Nehlig A, Nicolas F (1991) Physiological and pharmacological properties of adenosine: Therapeutic implications. Life Sci 49:1435–1453.
10. Jacobson KA, van Galen PJM, Williams M (1992) Adenosine receptors: Pharmacology, structure activity relationships, and therapeutic potential. J Med Chem 35:407–422.
11. Marangos PJ (1991) Potential therapeutic roles for adenosine in neurologic disease. In Stone TW (ed.), Adenosine in the Nervous System. London: Academic Press, pp. 217–226.
12. Dunwiddie TV, Worth T (1982) Sedative and anticonvulsant effects of adenosine analogs in mouse and rat. J Pharm Exp Ther 220:70–76.
13. Murray TF, Sylvester D, Schultz CS, Szot P (1985) Purinergic modulation of the seizure threshold for pentylenetetrazol in the rat. Neuropharmacology 24:761–766.
14. Williams M (1984) Adenosine—a selective neuromodulator in the mammalian CNS? Trends Neurosci 7:164–168.
15. Berman RF, Jarvis M, Lupica C (1990) Adenosine involvement in kindled seizures. In Wada JA (ed.), Kindling 4. New York: Plenum Press, pp. 423–440.
16. Petersen EN (1991) Selective protection by adenosine receptor agonists against DMCM-induced seizures. Eur J Pharmacol 195:261–265.
17. Murray TF, Franklin PH, Zhang G, Tripp E (1992) A_1 adenosine receptors express seizure-suppressant activity in the rat preperiform cortex. In Avanzini G, Engel J Jr, Fariello R, Heinemann U (eds.), Neurotransmitters in Epilepsy. (Epilepsy Res Suppl 8). Amsterdam: Elsevier, pp. 255–261.
18. Herberg LJ, Rose IC, Mintz M (1993) Effect of an adenosine A_1 agonist injected into substantia nigra on kindling of epileptic seizures and convulsion duration. Pharmacol Biochem Behav 44:113–117.
19. Von Lubitz DKJE, Paul IA, Carter M, Jacobson KA (1993) Effects of N^6-cyclopentyladenosine and 8-cyclopently-1,3-dipropylxanthine on N-methyl-D-aspartate induced seizures in mice. Eur J Pharmacol 249:265–270.
20. Marangos PJ, Weiss SRB, Montgomery P, Patel J, Narang PK, Cappabianca AM, Post RM (1985) Chronic carbamazepine treatment increases brain adenosine receptors. Epilepsia 26:493–498.
21. Klitgaard H, Knutsen LJS, Thomsen C (1993) Contrasting effects of adenosine A_1 and A_2 receptor ligands in different chemoconvulsive models. Eur J Pharmacol 224:221–228.
22. Morgan PF, Durcan MJ (1990) Caffeine-induced seizures: Apparent proconvulsant action of *N*-ethyl carboxamidoadenosine (NECA). Life Sci 47:1.
23. Knutsen LJS, Lau J, Sheardown MJ, Thomsen C (1993) The synthesis and biochemical evaluation of new A_1 selective adenosine receptor agonists containing a 6-hydrazinopurine moiety. Biomed Chem Lett 3:2661–2666.
24. Knutsen LJS, Lau J (1993) Novel adenosine derivatives. WO 93/23417, November 25.
25. Gurden MF, Coates J, Ellis F, Evans B, Foster M, Hornby E, Kennedy I, Martin DP, Strong P, Vardey CJ, Wheeldon A (1993) Functional characterization of three adenosine receptor types. Br J Pharmacol 109:693–698.
26. Evans B (1991) 2,N-6-disubstituted adenosines and their antihypertensive methods of use. US Patent 5,032,583, July 16, 1991.
27. Knutsen LJS, Lau J (1993) Purine derivatives; WO 93/23418, November 25, 1993.
28. Steppuhn KG, Turski L (1993) Modulation of the seizure threshold for excitatory amino acids in mice by antiepileptic drugs and chemoconvulsants. J Pharmacol Exp Ther 265:1063–1070.
29. Löscher W, Schmidt D (1988) Which animal models should be used in the search for new antiepileptic drugs? A proposal based on experimental and clinical considerations. Epilepsy Res 2:145–181.
30. Klitgaard H, Knutsen LJS (1992) 8-Cyclopentyl-1,3-dimethylxanthine (CPT), an adenosine A_1 receptor antagonist, with anticonvulsant properties. Int J Purine Pyrimidine Res 3:45.

52. BEHAVIORAL EFFECTS OF ADENOSINE RECEPTOR STIMULATION

Dag K.J.E von Lubitz
Kenneth A. Jacobson

Although Western civilization has known and cherished the invigorating effects of a freshly brewed cup of tea or coffee for several centuries and has recognized the harmful or even lethal consequences of overindulgence in either for at least one century [1], the understanding of the involved mechanisms and their complexities is the result of vigorous research of the last few decades. Even more amazingly, behavioral consequences of these world's two most ubiquitous beverages are inseparably linked to one of the most ubiquitous molecules of a living organism, adenosine, and to its receptors.

Currently, three main adenosine receptor classes, A_1, A_2, and A_3, have been characterized (for the detailed classification of main types and subtypes, see Chapter 19). All are found throughout the mammalian body, and all are involved in regulation of physiologic processes as far apart as reproduction and protection against tissue injury and death. In the context of its involvement in the operation of the central nervous system, arguably the best characterized receptor is the A_1 receptor. It has been shown that its stimulation results in reduction of calcium influx and release of several neurotransmitters, depression of neuronal excitability and firing rate, and, consequently, decrease of neuronal metabolic rate [2]. Recent discoveries of a close functional relationship between adenosine and glutamate receptors [3–5] provide additional, fascinating evidence on the importance of adenosine in modulatory control of neuronal functions.

Studies of adenosine receptors have been substantially delayed by the lack of sufficiently selective ligands. Unsurprisingly, therefore, while there is an abundance of literature on behavioral effects of caffeine and related methylxanthines [6–8], relatively few studies on the effects of selective agonists and antagonists have been published. For this reason, behavioral actions of methylxanthines will be reviewed very briefly, and the interested reader is referred to the existing reviews and their comprehensive reference lists, while the present chapter will concentrate on the central effects of highly selective drugs.

Nonselective Adenosine Agents: Caffeine and Related Methylxanthines

Paradoxically, both the mechanisms of action and the effects caused by the most commonly ingested methylxanthine, caffeine, and its relatives are frequently controversial. Moreover, creation of a "unified theory of coffee" is made even more difficult by species-dependent responses and the fact that all examined animals metabolize caffeine in a manner different from that observed in humans [8]. Despite these complications, four theories on methylxanthine's mode of action have been proposed, that is, mobilization of calcium, phosphodiesterase inhibition, interaction with benzodiazepine receptors, and antagonism of adenosine receptors. Because the first three of the proposed mechanisms are active only at nonphysiologic concentrations of methylxanthines, the most likely is that involving adenosine receptors [8].

It is now clear that methylxanthines affect the formation and release of several neurotransmitters. Caffeine's effects on dopamine release in the mesolimbic/mesocortical and caudate neurons may provide an explanation for its effects on attention, vigilance, and locomotor activity [8,9]. Caffeine-mediated reduction of serotonin at postsynaptic receptors may be also related to the operation of sleep mechanisms, locomotor functions, and cerebral vasoregulation [8,10,11]. The latter effect is particularly curious because while both caffeine and theophylline cause peripheral vasodilation, in the brain vasoconstriction occurs [12]. Thus, while increasing local glucose metabolism [8], methylxanthines appear to uncouple cerebral blood flow from cerebral metabolism [8]. In view of the broad spectrum of methylxanthine actions on basic neuronal functions, it is only to be expected that they will be also reflected in the electric activity of the brain. However, although stimulation of cortical and hippocampal electroencephalographic (EEG) activity has been repetitively described in animals [13,14], in humans the effects of caffeine on resting EEG are inconsistent [8].

Similar inconsistencies are also typical of the reports on behavioral effects of methylxanthines. In animals, locomotor stimulation by methylxanthines is well known [15]. Interestingly, enhancement of locomotor activity appears to peak at approximately 30 mg/kg. At higher doses, a decrease has been observed [15]. Much higher doses may result in convulsions [16]. In humans, however, locomotor activity may increase or may remain unchanged [8]. In addition, the effect appears to be dependent on the amount of coffee habitually ingested by the subject [8].

Several authors agree that simple methylxanthines do not increase learning ability as such. On the other hand, their impact on heightened vigilance, attention, locomotion, and overall performance may be instrumental in the reported effect on memory acquisition [8]. Species differences are, however, quite pronounced. For example, while some species of rats and monkeys show an increased number of avoidance responses, in hamsters and at least one rat strain the number actually decreases [8,17]. In humans, the effect of caffeine on cognitive functions is subtle and, therefore, difficult to measure objectively. The overall impact appears to be affected by the mood at the time of the test, personality type, and sex of the subject [18,19].

Sleep appears to be the function most affected in humans [8]. Caffeine delays the onset, increases the duration of stage 2 (light), and decreases the duration of stage 3 and 4 (deep) sleep. No effect on rapid eye movement has been observed [8]. In rats, higher doses of caffeine either reduce the overall duration of sleep or change the length of its different phases [20,21]. Moreover, the duration of barbiturate-induced sleep is also shortened by caffeine in both rats and mice [22,23].

Selective Adenosine Agonists and Antagonists—Acute Administration

The advent of a broad range of ligands acting with a very high selectivity at either A_1 or A_2 receptors [24] opened the possibility to determine the role played by both receptors in mediation of physiologic and behavioral effects of adenosine. Regrettably, hopes for a solution to problems that emerged during studies of nonselective methylxanthines have not been entirely fulfilled. The "dissection" of adenosine-mediated effects, while providing many answers, led to just as many new questions.

LOCOMOTOR ACTIVITY

The locomotor depressant consequences of intraperitoneal administration of highly selective A_1 receptor agonists such as N^6-cyclohexyladenosine (CHA) or (R)-phenylisopropyladenosine (R-PIA) are very well known [6]. Both drugs exert powerful peripheral actions that may adversely affect spontaneous locomotor activity (e.g, depression of cardiac rate and depression of blood pressure, hypothermia). Yet, since blood-brain barrier–impermeable antagonists such as 8-PST fail to reverse behavioral effects of A_1 receptor agonists, while highly selective, blood-brain barrier–crossing antagonists such as 8-cyclopentyl-1,3,-dimethylxanthine (CPT) block these effects effectively [25], it appears that locomotor depression is centrally mediated [6,25].

The potencies of adenosine agonists causing reduction of spontaneous locomotor activity correlate much better with the potencies at A_2 than A_1 receptors, which may indicate a possibility of predominant A_2 receptor involvement [26]. This conclusion has been confirmed by a profound locomotor depression caused by a highly selective A_2 agonist, 2-[(2-aminoethyl-amino) carbonylethylphenylethylamino]-5′-*N*-ethylcarboxamidoadenosine (APEC) [25]. On the other hand, selective manipulation of A_1 or A_2 reception that are distinctly correlated with either receptor type [27]. Moreover, although synergistic effects of A_1 and A_2 receptors are absent in vitro, potentiation of A_1 motor depressant effect by coadministration of APEC has been recently reported in vivo [25].

These somewhat conflicting results indicate that specific innervation and/or arrangements of A_1 and A_2 receptors within brain structures involved in movement control may be critical in determining the overall behavioral effect. Studies of Barraco et al. [28] support this hypothesis by showing that although injections of 5′-N-ethylcarboxamidoadenosine (NECA) into nucleus accumbens of mice have no effect on stereotypies, a robust locomotor depression is elicited. Conversely, intraaccumbal N^6-cyclopentyladenosine (CPA) has no effect at all. However, when R-PIA is injected into caudoputamen, the effects are opposite, that is, the potency of locomotor reduction decreases significantly while stereotypies are enhanced [28,29].

The distinctive behavior patterns observed following intrastriatal injections of A_1 and A_2 agonists and antagonists may reflect the functional architecture of the striatum. Thus, the A_2-receptor–rich ventral striatum [30] appears to be critical in the coordination of the limbic and motor systems [31], while caudoputamen, in which A_1 receptors predominate (A_2 sites are also present), is involved in the mediation of dopamine-dependent stereotypic behavior [32]. Nonetheless, these considerations do not explain why selective A_1 agonists injected systemically induce a uniform decrease of spontaneous locomotor activity, even in the presence of blood-brain barrier–impermeable antagonists. It is possible that, in view of their potent inhibitory effects on neurotransmitter release [33], systematically applied A_1 agonists cause an overall reduction in the excitatory input to the locomotor centers of the brain. Combined with simultaneous perturbances of dopaminergic modulation [34] of the "common effector pathway," these influences may consequently express themselves as a depression of spontaneous locomotor activity.

Very recently locomotor depressant effects of the A_3 agonist N^6-(3-iodobenzyl)-5′-*N*-methylcarboxamidoadenosine (IB-MECA) have been described [35]. The biologic role of this newly discovered receptor is unknown. However, the presence of A_3 receptors in the cerebellum and striatum of mice [35] indicates the possibility of their direct involvement in either the control or modulation of processes affecting spontaneous locomotion.

SEIZURES

Anticonvulsant properties of adenosine are another example of its powerful depressant influences on the activity of the central nervous system. It is known that seizures elevate intracerebral levels of adenosine [36,37], and manipulation of endogenous adenosine in the prepiriform cortex modulates seizure susceptibility in rats [38]. Also, both acute and repetitive seizure episodes increase A_1 receptor density in the cortex, cerebellum, and hippocampus [33, 39]. Following the first description of adenosine-mediated blockade of audiogenic seizures in mice [40], several studies demonstrated that adenosine and its stable analogues either reduce the intensity of, or prevent seizures elicited by chemical, behavioral, and electric stimuli in a wide range of animal models [41].

Neurologically, anticonvulsant properties of adenosine and its anlogues are related to inhibition of seizure initiation and propagation [41]. The rank order of potency of agonists acting at the A_1 receptor indicates critical involvement of this receptor in the modulation of seizure activity [42]. Yet, as with many other actions of adenosine, antiepileptogenic effects of selective A_1-receptor agonists are not uniform. Thus, in methyl-6,7-dimethoxy-4-ethyl-β-carboline-3-carboxylate (DMCM)-induced seizures, R-PIA is effective when given at 30 μg/kg, although as much as 1 mg/kg CHA is required in order to elicit the same seizure-depressant effect [43]. Paradoxically, a selective A_1 receptor antagonist CPT attenuates DMCM– and pentylenetetrazol-elicited convulsions [44]. This somewhat surprising effect may, however, be a reflection of A_1/GABA receptor interactions [45] rather than a result of A_1 receptor stimulation per se. It has been shown that despite its potent anticonvulsant properties, another selective A_1 agonist 2-chloro-N-cyclopentyladenosine (CCPA) reduces the function of the GABA-coupled chloride channel [46]. Therefore, it is quite possible that CPT simply reverses this action.

Equally surprising is the potentiation of xanthine amine congener (XAC)-evoked seizures following administration of CPA or R-PIA at 1–10 mg/kg doses [47]. Quite likely, however, cardiac and cerebrovascular interactions caused by simultaneous systemic presence of both an A_1 agonist and antagonist may be greatly contributory to the apparent potentiation [47]. Finally, while R-PIA effectively antagonizes only quisqualate-induced seizures, CPA is just as effective following stimulation of NMDA [48], AMPA, and metabotropic glutamate receptors [44]. Thus, the effect of A_1 agonists on seizures elicited via glutamatergic mechanisms is complex and may involve mechanisms beyond simple neurotransmitter release inhibition [49].

Whether A_2 receptors participate in the control of epileptic discharges is less clear. Phillis [50] showed potent depressant effects of CGS 21680 on electric brain activity, and Klitgaard et al. [44] antagonized DMCM-induced seizures using the A_2-selective agonist N-[2-(3,5-dimethoxyphenyl)-2-(2-methylphenyl)ethyl] adenosine (DPMA). On the other hand, Zhang et al. [51] demonstrated low potency of CGS 21680 in antagonizing bicuculine seizures and suggested that the observed depressant effect is related to A_1 receptor activation by high concentrations of CGS 21680.

The role of acute A_3 receptor stimulation in seizure antagonism is unclear. Von Lubitz et al. (in press) showed protective effect of a selective A_3 agonist IB-MECA in NMDA-induced seizures. However, acutely administered IB-MECA may cause cerebral vasoconstriction [52]. Hence, it is likely that the observed protection resulted from subconvulsant amounts of NMDA delivered to their target sites in the brain.

The confusing results described above underline the difficulty in interpreting studies of complex behavioral phenomena elicited by the simultaneous presence of agents causing potent central and peripheral effects. There is no doubt that adenosine is indeed involved in modulation of seizure activity, but further studies are necessary in order to reveal the exact nature of the mechanisms involved.

SLEEP

The role of endogenous adenosine in sleep has not been determined [53]. However, intracerebral administration of adenosine in rats [54] and oral administration of mioflazine (nucleoside transport inhibitor) in dogs enhanced REM and slow wave sleep [55], indicating that such involvement is possible. Extensive studies of Radulovacki and his colleagues [53] point toward a very significant role played by A_1 receptors in the regulation of sleep activity. Very low nanomolar (0.3 μmol/kg) doses of R-PIA, CHA, or CPA injected into the preoptic area of rat brain result in an increase of deep slow-wave sleep, and it has been suggested that stimulation of A_1 receptors causes enhancement of total sleep but does not affect sleep latencies [56]. Conversely, injection of a selective A_1-receptor antagonist 8-cyclopentyl-1,3-dimethylxanthine (CPT) produces sleep suppression [21]. Interestingly, higher doses of R-PIA, CHA, and CPA (≥0.9 μmol/kg) appear to reduce the duration of REM sleep, while R-PIA increases waking as well [53]. Correlation of brain adenosine and A_1-receptor density with wakefulness and sleep has also been demonstrated in rats [53].

Quite consistently with other behavioral measures described earlier, the impact of A_2 receptor stimulation on sleep and hypnotic phenomena is ill defined. When injected into the preoptic area at comparatively high doses, NECA increases both total sleep and deep slow-wave sleep [53]. However, because the effect of administration of alloxazine (a nonspecific A_1/A_2 antagonist) produces sleep effects that do not differ from those observed with CPT, the A_1-mediated actions appear to be predominant.

PAIN—NOCICEPTIVE AND ANTINOCICEPTIVE EFFECTS

Nociceptive and antinociceptive properties of adenosine and its analogues are related to their effects at the level of sensory nerve endings, spinal cord, and the brain. Because this review concentrates on central actions of adenosine, the reader interested in the involvement of adenosine in mediation and control of pain is referred to recent reviews by Sawynok and her colleagues [57,58]. The followig outline serves merely to underline the variety of behavioral responses caused by the stimulation of adenosine receptors.

Both A_1 and A_2 receptors are present in the spinal cord [57], and systemic, intrathecal, and supraspinal administration of selective A_1 receptor agonists, for example, R-PIA, CHA, or CPA, result in measurable antinociceptive responses in a variety of species and tests. Somewhat paradoxically, adenosine is also involved in the stimulation of pain through interaction with the sensory nerve endings [57,58]. The dual role of adenosine in pain signaling and pain suppression constrains the usefulness of adenosine-based local analgesics. However, their development as general analgesic agents with systemic administration routes is more likely [57].

COGNITION

Although behavioral studies of methylxanthines indicated that their influence on learning is probably indirect [8], the depressant effects of selective adenosine agonists on development and consolidation of long-term potentiation [59] suggest, nonetheless, that adenosine and its receptors may be intimately involved in cognition.

Somewhat surprisingly, only a limited number of studies have been devoted to investigations of cognitive properties of selective agents acting at A_1 sites. Their results, like those of others mentioned in previous sections, are also inconsistent. Thus, Normile and Barraco [60] showed that micromolar doses of the agonist (CPA) impaired retention in passive avoidance tests. Yet, the working memory of rats remained unaffected by R-PIA [61]. Variation characterizes the effect of antagonists as well. For example, CPX (A_1 antagonist) was ineffectual in passive avoidance tests [60], whereas a selective A_1 antagonist KFM 19 produced an increase in memory acquisition in the Y maze [62]. Another selective A_1 antagonist, MDL 102234 reversed scopolamine-induced amnesia in rats exposed to water maze tests [63].

The nature of processes that link adenosine to the formation, retention, and expression of memory remains to be discovered, but discrepancies in response to A_1 receptor antagonists may indicate the complex nature of the studied phenomena. Establishment of different forms of memory and the subsequent behavioral responses to the environment involve different neural pathways connecting different parts of the brain. In turn, these are characterized by adenosine receptor populations of different composition. Long-term potentiation observed in a hippocampal slice is, unquestionably, a vastly simplified version of reality. It is, therefore, in this context that the previously mentioned studies of Barraco and his colleagues [28] are particularly seminal in demonstrating that the nature of a clearly identifiable behavioral response is intimately related to the activated brain area and to the composition of its adenosine receptor population.

SELECTIVE ADENOSINE AGONISTS AND ANTAGONISTS—CHRONIC ADMINISTRATION

Investigations of the effects of adenosine in the central nervous system, as rewarding as they may be, would lose some of their luster if not for the fact of the immense therapeutic possibilities opened by continually improving understanding of the breadth of physiologic functions that are affected. It is therefore hardly surprising that adenosine-based therapies have been proposed for such disorders as Alzheimer's, Huntington's, and Parkinson's diseases, seizures, stroke, and cerebral and spinal trauma; and a host of pathologies involving peripheral organs. The paucity of studies involving chronic administration of adenosine agonists and antagonists comes therefore as a surprise. This surprise is compounded by the fact that among all disorders in which adenosine therapies have been suggested, only stroke and cardiac arrest are amenable to the acute treatment.

Adaptive changes of A_1 receptors following prolonged exposure to their ligands are a well-known phenomenon [64]. It follows, therefore, that chronic treatment with such agents may result in behavioral responses that are different from those seen following their acute administration. Von Lubitz et al. [65] demonstrated such regimen dependence. Thus, contrary to the procognitive effect of selective A_1 antagonists, chronic administration of 8-cyclopentyl-1,3-dipropylxanthine (CPX) is ineffective, while chronic treatment with agonist CPA improves memory acquisition and retention in the water maze. Similar results have been also obtained with R-PIA [66]. Moreover, while acutely administered CPA is protective and acutely administered CPX enhances the lethality of NMDA-induced seizures, converse effects result following chronic administration of these drugs [48,49]. The effects of A_3 receptor stimulation are also the subject of a regimen dependent shift [52].

In at least two rodent species, neither upregulation nor downregulation of A_1 receptors (and probably A_3 as well) are involved in the process of regimen-dependent effect inversion [49,67], whose nature still remains unknown. It has been proposed, however, that desensitization of G proteins may be involved [49,68]. Yet, the contribution of other phenomena can not be discounted. For example, Abbracchio et al. [64] showed that prolonged exposure of rat brain slices to CPA results in functional "unmasking" of striatal A_2 sites. Moreover, chronic exposure to 2-chloroadenosine (CADO) resulted in a decrease, while α,β-methylene ATP (ATP/P_2 receptor agonist) resulted in a robust increase of astrocytes in rat striatum in vitro [69]. Thus, there is a distinct probability that the results obtained following chronic exposure to selective agents acting at adenosine receptors in vivo may be the consequence of very complex, dynamic interactions between receptors, G proteins, and, ultimately, second messenger systems.

Conclusions

A simple truth emerges from this brief description of the involvement of adenosine and its receptors in behavior: It is unquestionable and it is important. Possibly, it is critical. Yet, while the intimate role of adenosine in either the modulation or control of a very wide range of behavioral patterns has been demonstrated beyond any doubt, questions remain. As with other receptor systems, answering these questions will ultimately lead to better understanding of normal and pathologic processes and, perhaps, to the development of therapies for many of currently untreatable diseases. One aspect makes studies of adenosine much more fascinating than studies of any other system—its involvement in virtually every function of a living organism, be it simple or immensely complex. Thus, although much needs to be learned and many laborious experiments involving different approaches and different disciplines are yet to be performed, there is no doubt that the results may prove to be immensely rewarding.

PER ARDUA AD ASTRA.

References

1. Mendel F (1889) Die schädlichen Folgen des chronischen Kaffeemisbrauch. Berl Klin Wochenschr 26:877.
2. Daval J-L, Nehlig A, Nicolas F (1991) Physiological and pharmacological properties of adenosine. Life Sci 49:1435.
3. Hoehn K, White TD (1990) N-methyl-D-aspartate, kainate, and quisqualate release endogenous adenosine from rat cortical slices. Neuroscience 2:441.
4. Poli A, Lucchi R, Vibio M, Barnabei O (1991) Adenosine and glutamate modulate each other's release from rat hippocampal synaptosomes. J Neurochem 57:298.
5. Winder DG, Conn PJ (1993) Activation of metabotropic glutamate receptors increases cAMP accumulation in hippocampus by potentiating responses to endogenous adenosine. J Neurosci 1:38.
6. Barraco RA (1991) Behavioral actions of adenosine and related substances. In Phillis JW (ed.), Adenosine and Adenine Nucleotides as Regulators of Cellular Function. Boca Raton, FL: CRC Press, p. 339.
7. Daly JW (1993) Mechanism of action of caffeine. In Garattini C (ed.), Coffee and Health. New York: Raven Press, p. 97.
8. Nehlig A, Daval J-L, Debry G (1992) Caffeine and the central nervous system: Mechanisms of action, biochemical, metabolic and psychostimulant effects. Brain Res Rev 17:139.
9. Stoner GR, Skirboll LR, Werkman S, Hommer DW (1988) Preferential effects of caffeine on limbic and cortical dopamine systems. Biol Psychiatry 23:761.
10. Reith MEA, Sershen H, Lajtha A (1987) Effects of caffeine on monoaminergic systems in mouse brain. Acta Biochim Biophys Hung 22:149.
11. Warbritton JD, Stewart RM, Baldessarini RJ (1980) Increased sensitivity to intracerebroventricular infusion of serotonin and deaminated indoles after lesioning rat with dihydroxytryptamine. Brain Res 177:355.
12. Mathew RJ, Wilson WH (1985) Caffeine-induced changes in cerebral circulation. Stroke 16:814.
13. Ritchie JM (1975) Central nervous system stimulants II. The xanthines. In Goodman LS, Gilman A (eds.), The Pharmacological Basis of Therapeutics, New York: MacMillan, p. 358.
14. Popoli P, Sagratella S, Scotti de Carolis A (1987) An EEG and behavioural study of the excitatory properties of caffeine in rabbits. Arch Int Pharmacodyn Ther 290:5.
15. Nikodijević O, Jacobson KA, Daly JW (1993) Locomotor activity in mice during chronic treatment with caffeine and withdrawal. Pharmacol Biochem Behav 44:199.
16. Rall TW (1980) Central nervous system stimulants. The xanthines. In Goodman A, Gilman LS, Gilman A (eds.), The Pharmacological Basis of Therapeutics. New York: MacMillan, p. 592.
17. Satinder KP (1971) Genotype dependent effects of d-amphetamine sulphate and escape-avoidance behavior of rats. J Comp Physiol Psychol 76:359.
18. Smith AP, Rusted JM, Eaton-Williams P, Savory M, Leatherwood P (1990) Effects of caffeine given before and after lunch on sustained attention. Neuropsychobiology 23:160.
19. Smith BD, Wilson RJ, Jones BE (1983) Extraversion and multiple levels of caffeine-

induced arousal: Effects on overhabituation and dishabituation.
20. Radulovacki M, Mieltich RS, Green RD (1982) N^6 (L-phenylisopropyl)adenosine (L-PIA) increases slow wave sleep (S_2) and decreases wakefulness in rats. Brain Res 246:178.
21. Virus RM, Ticho S, Pilditch M, Radulowacki M (1990) A comparison of the effects of caffeine, 8-cyclopenthyltheophylline, and alloxazine on sleep in rats. Possible roles of central nervous system adenosine receptors. Neuropsychopharmacology 3:243.
22. Aeschbacher HU, Atkinson J, Domahidy B (1975) The effect of caffeine on barbiturate sleeping time and brain level. J Pharmacol Exp Ther 192:635.
23. Alleva E, Castellano C, Oliviero A (1978) Individual differences in barbiturate-induced sleeping time in the mouse. Prog Neuropsychopharmacol 2:451.
24. Jacobson KA, Van Galen PJN, Williams M (1992) Perspective, adenosine receptors: Pharmacology, structure activity relationships and therapeutic potential. J Med Chem 35:407.
25. Nikodijević O, Sarges R, Daly JW, Jacobson KA (1991) Behavioral effects of A_1- and A_2-selective adenosine agonists and antagonists: Evidence for synergism and antagonism. J Pharmacol Exp Ther 259:286.
26. Durcan MJ, Morgan PF (1989) Evidence for adenosine A_2 receptor involvement in the hypomotility effects of adenosine analogs in mice. Eur J Pharmacol 168:285.
27. Popoli P, Pèzzola A, Scotti de Carolis A (1994) Modulation of striatal adenosine A_1 and A_2 receptors induces rotational behaviour in response to dopaminergic stimulation in intact rats. Eur J Pharmacol 257:21.
28. Barraco RA, Martens KA, Parizon M, Normile HJ (1993) Adenosine A_{2A} receptors in the nucleus accumbens mediate locomotor depression. Brain Res Bull 31:397.
29. Barraco R, Bryant SD (1987) Depression of locomotor activity following bilateral injections of adenosine analogs into the striatum of mice. Med Sci Res 15:421.
30. Jarvis MF, Williams M (1989) Direct autoradiographic localization of adenosine A_2 receptors in the rat brain using the selective A_2 agonist, [^{3}H]CGS 21680. Eur J Pharmacol 168:243.
31. Mogenson GJ (1987) Limbic-motor integration. Prog Psychobiol Physiol Psychol 12:117.
32. Swerdlow NR, Johansson AM, Magnusson T, Carlsson A (1986) (+)-AJ76 and (+)-UH232: Central stimulants acting as preferential dopamine autoreceptor antagonists. Naunyn Schmiedbergs Arch Pharmacol 334:234.
33. Daval J-L, Sarfati A (1987) Effects of bicuculine-induced seizures on bensodiazepine and adenosine receptors in developing brain. Life Sci 41:1685.
34. Wood PL, Kim HS, Boyar WC, Hutchinson A (1989) Inhibition of nigrostriatal release of dopamine in the rat by adenosine receptor agonists: A_1 receptor mediation. Neuropharmacology 28:21.
35. Jacobson KA, Nikodijević O, Shi D, Gallo-Rodriguez C, Olah M, Stiles GL, Daly JW (1993) A role for central A_3-adenosine receptors. Mediation of behavioral depressant effects. FEBS Left 336:77.
36. Winn HR, Welsh JE, Rubio R, Berne RM (1980) Changes in brain adenosine during bicuculine-induced seizures in rats: Effects of hypoxia and altered blood pressure. Circ Res 47:568.
37. Schrader J, Wahl M, Kuschinsky W, Kreutzberg GW (1980) Increase of adenosine content in cerebral cortex of the cat during bicuculine-induced seizure. Pflügers Arch 387:245.
38. Zhang G, Franklin PH, Murray TF (1993) Manipulation of endogenous adenosine in the rat prepiriform cortex modulates seizure susceptibility. J Exp Pharmacol Ther 264:1415.
39. Psarropoulou C, Matsokis N, Angelatou F, Kostopoulos G (1994) Pentylenetetrazol-induced seizures decrease gamma-aminobutyric acid-mediated recurrent inhibition and enhance adenosine-mediated depression. Epilepsia 35:12.
40. Maitre M, Ciesielski L, Lehmann A, Kempf E, Mandel P (1974) Protective effect of adenosine and nicotinamide against audiogenic seizures. Biochem Pharmacol 23:2807.
41. Dragunow M (1991) Adenosine and epileptic seizures. In Phillis JW (ed.), Adenosine and Adenine Nucleotides as Regulators of Cellular Function. Boca Raton, FL: CRC Press, p. 367.
42. Murray TF, Sylvester D, Schultz CS, Szot P (1985) Purinergic modulation of the seizure threshold for pentylenetetrazol in the rat, Neuropharmacology 24:761.
43. Petersen EN (1991) Selective protection by adenosine receptor agonists against DMCM-induced seizures. Eur J Pharmacol 26:261.
44. Klitgaard H, Knutsen LJ, Thomsen C (1993) Contrasting effects of adenosine A_1 and A_2 receptor ligands in different chemoconvulsive rodent models. Eur J Pharmacol 242:221.

45. Wojcik WJ, Cavalla D, Neff NH (1985) Co-localized adenosine A_1 and gamma aminobutyric acid B ($GABA_B$ receptors of cerebellum may share a common adenylate cyclase catalytic unit. J Pharamacol Exp Ther 232:62.
46. Concas A, Santoro G, Mascia MP, Maciocco E, Dazzi L, Ongini E, Biggio G (1993) Anticonvulsant doses of 2-chloro-N^6-cyclopentyladenosine, an adenosine A_1 receptor agonist, reduce GABAergic transmission in different areas of the mouse brain. J Pharmacol Exp Ther 267:844.
47. Morgan P, Deckert J, Jacobson KA, Marangos PJ, Daly JW (1989) Potent convulsant actions of the adenosine receptor antagonist, xanthine amine congener (XAC). Life Sci 45:719.
48. von Lubitz DKJE, Paul IA, Carter M, Jacobson KA (1993) Effects of N^6-cyclopentyl adenosine and 8-cyclopentyl-1,3-dipropylxanthine on N-methyl-D-aspartate induced seizures in mice. Eur J Pharmacol 249:265.
49. von Lubitz DKJE, Paul IA, Ji X-D, Carter M, Jacobson KA (1994) Chronic adenosine A_1 receptor agonist and antagonist: Effect on receptor density and N-methyl-D-aspartate seizures in mice. Eur J Pharmacol 253:95.
50. Phillis JW (1990) The selective A_2 receptor agonist, CGS 21680, is a potent depressant of cerebral neuronal activity. Brain Res 509: 328.
51. Zhang G, Franklin PH, Murray TF (1994) Activation of adenosine A_1 receptors underlies anticonvulsant effect of CGS 21680. Eur J Pharmacol 255:239.
52. von Lubitz DKJE, Lin RC-S, Popik P, Carter MF, Jacobson KA (1994) Adenosine A_3 receptor stimulation and cerebral ischemia. Eur J Pharmacol 263:59.
53. Radulovacki M (1991) Adenosine and sleep. In Phillis JW (ed.), Adenosine and Adenine Nucleotides as Regulators of Cellular Function. Boca Raton, FL: CRC Press, p. 381.
54. Radulovacki M, Virus RD, Rapoza D, Crane RC (1985) A comparison of the dose response effects of pyrimidine ribonucleosides and adenosine on sleep in rats, Psychopharmacology (Berl) 87:136.
55. Wauquier A, Van Belle H, Van den Brock WAE, Janssen PAJ (1987) Sleep improvement in dogs after oral administration of mioflazine, a nucleoside transport inhibitor. Psychopharmacology (Berl) 91:434.
56. Ticho SR, Radulovacki M (1991) Role of adenosine in sleep and temperature regulation in the preotic area of rats. Pharmacol Biochem Behav 40:33.
57. Sawynok J (1991) Adenosine and pain. In Phillis JW (ed.), Purines and Purine Nucleosides as Regulators of Cellular Function. Boca Raton FL: CRC Press, p. 391.
58. Sawynok J, Sweeney MI (1989) The role of purines in nociception. Neuroscience 32:557.
59. Arai A, Kessler M, Lynch G (1990) The effects of adenosine on the development of long-term potentiation. Neurosci Lett 99:107.
60. Normile HJ, Barraco RA (1991) N^6-cyclopentyladenosine impairs passive avoidance retention by selective action at A_1 receptors. Brain Res Bull 27:101.
61. Pontecorvo MJ, Clissold DB, White MF, Ferkany JW (1991) N-methyl-D-aspartate antagonists and working memory performance: Comparison with the effects of scopolamine, propanolol, diazepam and phenylisopropyladenosine. Behav Neurosci 105:521.
62. Schingnitz G, Küfner-Mühl U, Ensinger H, Lehr E, Kuhn FJ (1991) Selective A_1-antagonists for treatment of cognitive deficits. Nucleosides Nucleotides 10:1067.
63. Dudley M, Hitchock J, Sorensen S, Chaney S, Zwolshen J, Lentz N, Borcherding D, Peet N (1994) Adenosine A_1 receptor antagonists as cognitive enhancers. Drug Dev Res 31:266 (1342).
64. Abbracchio MP, Brambilla R, Rovati GE, Cattabeni F (1993) Prolonged agonist exposure induces imbalance of A_1 and A_2 receptor-mediated functions in rat brain slices. Drug Dev Res 28:364.
65. von Lubitz DKJE, Paul IA, Bartus RT, Jacobson KA (1993) Effects of chronic administration of adenosine A_1 receptor agonist and antagonist on spatial learning and memory. Eur J Pharmacol 249:271.
66. Molinengo L, Scordo I, Pastorello B (1994) Action of caffeine, L-PIA, and their combination on memory retention in the rat. Life Sci 54: 1247.
67. von Lubitz DKJE, Lin RC-S, Melman N, Ji X-D, Carter MF, Jacobson KA (1994) Chronic administration of selective adenosine A_1 receptor agonist or antagonist in cerebral ischemia. Eur J Pharmacol 256:161.
68. Georgiev V, Johansson B, Fredholm BB (1993) Long-term caffeine treatment leads to a decreased susceptibility to NMDA-induced clonic seizures in mice without changes in adenosine A_1 receptor number. Brain Res 612:271.

69. Abbracchio MP, Saffrey MJ, Höpker V, Burnstock G (1994) Modulation of astroglial cell proliferation by analogues of adenosine and ATP in primary cultures of rat striatum. Neuroscience 59:67.

53. EVIDENCE FOR THE EXISTENCE OF ANTAGONISTIC INTRAMEMBRANE ADENOSINE A_{2a}/DOPAMINE D_2 RECEPTOR INTERACTIONS IN THE BASAL GANGLIA: ANALYSIS FROM THE NETWORK TO THE MOLECULAR LEVEL

Kjell Fuxe
Sergi Ferré
Sarmila Dasgupta
William T. O'Connor
Per Snaprud
Ernest Arenas
Håkan Persson
Urban Ungerstedt
Bertil Fredholm

During recent years it has become increasingly clear that adenosine receptors importantly interact with dopamine (DA) mechanisms in the regulation of the efferent GABAergic striatal pathways from the dorsal and ventral striatum [1]. One important anatomic basis for this adenosine/DA interaction is the demonstration of the existence of A_{2a} receptors and DA D_2 receptors in the GABA/enkephalin (ENK) striopallidal neurons as revealed by in situ hybridization histochemistry in combination with retrograde tracing [2,3]. In addition, A_2 receptors may exist on the DA nerve terminals of the striatum together with A_1 receptors, based on studies using in situ hybridization histochemistry on the regulation by adenosine A_2 receptors of ENK/substance P gene expression in the basal ganglia after caffeine treatment in the absence or presence of dopaminergic inputs into the striatum [3]. There exists no morphologic evidence for a colocalization of A_1 and D_1 receptors in striatal neurons.

In 1991 the first indications were obtained for the existence of an antagonistic intramembrane A_{2a}/D_2 receptor interaction in neostriatal membranes [4], leading to the introduction of the hypothesis that such an interaction in the striopallidal GABA/ENK neurons could be the basis for a major part of the adenosine/DA interactions observed in the brain as revealed in behavioral experiments [1]. In this chapter we summarize recent work giving further evidence for this hypothesis based on studies at the network level (microdialysis experiments), at the membrane level, and at the molecular level (studies on transiently and stably cotransfected COS-7 and fibroblast cell lines).

Analysis of the A_{2a} Receptor/Dopamine D_2 Receptor Interaction at the Network Level

BEHAVIORAL STUDIES

There exists a substantial amount of evidence that DA transmission in the brain can participate in mediating the psychomotor stimulant actions of caffeine as well as the psychomotor depressant actions of analogues of adenosine [1]. As an example it may be mentioned that the cataleptic action of central administration of the A_{2a} agonist CGS21680 is counteracted by a DA D_2 agonist [5]. Furthermore, the D_2 agonist-induced activation of locomotor activity in mice is reduced by A_{2a} receptor agonists and potentiated by the blockade of A_{2a} receptors [6,7]. It has also been observed that methylxanthines can substantially increase the actions of DA receptor agonists at supersensitive DA receptors in the striatum [8–10]. In addition, we have recently demonstrated that theophylline can produce a higher motor activation in rats pretreated with haloperidol, leading to upregulation of DA D_2 receptors [11]. All these results are compatible with an antagonistic A_{2a}/D_2 receptor interaction within the striatum, which appears to be stronger after the development of DA receptor supersensitivity.

MICRODIALYSIS EXPERIMENTS

The in vivo microdialysis technique was used in the awake freely moving rat to obtain further functional evidence for the existence of antagonistic A_{2a}/D_2 receptor interaction in the striopallidal pathway (Table. 53-1) [12]. One microdialysis probe was implanted in the striatum and one was implanted into the globus pallidus of the ipsilateral side, where the GABA/ENK striopallidal neurons terminate. The pallidal probe was used to determine GABA extracellular levels, while the striatal probe was used both to measure the extracellular levles of DA and GABA and also to infuse DA and adenosine agonists and antagonists (for details on the DA and GABA analysis, see Ferré et al. [12]).

When infused into the striatum, the DA D_2 agonist pergolide substantially and significantly reduced the pallidal GABA levels without changing the GABA levels within the striatum. All measurements were made on the ipsilateral side. In contrast, the A_{2a} agonist infused alone into the striatum failed to alter the pallidal GABA extracellular levels. It was of substantial interest, however, to see that the A_{2a} agonist CGS21680 (10^{-5} M) fully counteracted the actions of coinfused pergolide (10^{-5} M) with regard to the action on pallidal GABA extracellular levels.

The adenosine A_1/A_2 antagonist theophylline infused into the neostriatum produced a concentration-related increase of extracellular DA levels and a concentration-related reduction of the pallidal GABA extracellular levels. It was

TABLE 53-1. A_{2a}/D_2 receptor interactions at the network level

Microdialysis	GABA change	
	Neostriatum	Globus pallidus
DORSAL STRIOPALLIDAL NEURONS		
D_2 agonist pergolide (10^{-5} M)	—	↓
CGS21680 (10^{-5} M)	—	—
CGS21680 (10^{-5} M) + D_2 agonist (10^{-5} M)	NT	—
D_2 agonist pergolide (10^{-7} M)	—	—
Theophyllamine (10^{-4} M)	NT	—
Pergolide (10^{-7} M) + theophyllamine (10^{-4} M)	NT	↓
	Nuc. accumbens	Ventral pallidum
VENTRAL STRIOPALLIDAL SYSTEM		
CGS21680	—	↑
D_2/D_3 antagonist (raclopride)	—	↑
Threshold doses of CGS21680 + raclopride	—	↑

↑ increase; ↓ decrease; — no change; NT = not tested.

of substantial interest to note that theophylline (10^{-4} M), when infused together with a threshold concentration of pergolide (10^{-7} M), could significantly reduce the extracellular GABA pallidal levels, although the extracellular DA striatal levels were still significantly below the basal values (due to the effect of pergolide on DA autoreceptors). All these results appear to be best explained on the basis of a postsynaptic A_{2a}/D_2 receptor interaction at the cell body–dendritic level of the striopallidal GABA/ENK neurons, which thus may be a major target for the neuromodulator adenosine in the brain. The ability of theophylline to markedly increase extracellular DA levels only develops with the highest doses, when a full blockade of presynaptic A_1 receptors takes place. These are probably located on the DA nerve terminals. Based on the existence of the antagonistic A_{2a}/D_2 receptor interactions in the striopallidal neurons, A_2 antagonists can be useful therapeutic tools for the treatment of basal ganglia disorders such as Parkinson's disease.

STUDIES ON ADENOSINE/DOPAMINE INTERACTIONS IN THE VENTRAL STRIOPALLIDAL SYSTEM

In these microdialysis experiments using the dual-probe approach, different DA agonists and antagonists were infused into the nucleus accumbens and the effects were studied on DA extracellular levels in the nucleus accumbens and on GABA extracellular levels in the nucleus accumbens and in the ipsilateral ventral pallidal [13]. The A_{2a} agonist CGS21680 was also infused into the nucleus accumbens with or without coinfusion with the D_2 receptor antagonist raclopride using low concentrations of the compound, which when given alone were ineffective. The results demonstrated that local infusion of the D_1 and the D_2 receptor agonists SKF38393 and pergolide, respectively, reduced the extracellular DA levels in the nucleus accumbens, while infusion in the nucleus accumbens with the D_1 antagonist SCH23390 and the D_2 antagonist raclopride resulted in increases of extracellular DA levels in this nucleus. Furthermore, in agreement with previous work in the caudate putamen [14], the A_{2a} agonist CGS21680 failed to modulate

TABLE 53-1. Striatal A_{2a}/D_2 receptor interactions at the membrane level

Experimental conditions	^{3}H-Raclopride			^{3}H-NPA	
	K_H	K_L	R_H	K_D	B_{max}
Striatal membranes					
CGS21680	↑	↑	↑	↑	—
CTP analogue + CGS21680	↑	↑	↑	NT	NT
DA denervation + CGS21680	↑ ↑	↑ ↑	↑ ↑	NT	NT
Chronic haloperidol + CGS21680	↑ ↑	↑ ↑	↑ ↑	NT	NT
Transiently A_{2a}/D_2 receptor transfected COS-7 cell membranes					
CGS21680	—	—	—	—	—
Stably A_{2a}/D_2 receptor transfected fibroblast cell line membranes					
CGS21680	↑	↑	↑	NT	NT

↑ increase; — no change; NT = not tested.

the extracellular DA levels in the nucleus accumbens. In contrast to A_1 receptors [15], A_{2a} receptors do not appear to regulate DA release in the nucleus accumbens.

In the analysis of the actions of the extracellular GABA levels in the ipsilateral ventral pallidum, the D_1 and D_2 receptor antagonist SCH23390 and raclopride, respectively, but not the D_1 and D_2 receptor agonists, were found capable of modulating the extracellular GABA levels in the ventral pallidum. Thus, both SCH23390 and raclopride increased the GABA levels in the ventral pallidum. These results may be explained on the basis that DA may exert a more tonic action in the ventral striopallidal system than in the dorsal striopallidal system, where clear-cut reductions in GABA levels in the globus pallidus were found by the use of pergolide. It was of substantial interest that the infusion of the A_{2a} agonist CGS21680 mimicked the actions of the D_2 antagonist raclopride in significantly increasing the pallidal GABA extracellular levels without influencing the GABA extracellular levels in the nucleus accumbens. It must also be pointed out that the D_1 antagonist, in contrast with the D_2 antagonist, increased extracellular GABA levels both in the nucleus accumbens and in the ventral pallidum. An interesting result was that an ineffective concentration of CGS21680 coinfused with an ineffective concentration of raclopride led to the development of significant increases in pallidal GABA extracellular levels (on the order of 30%). These results can again be explained on the basis of the existence of an A_{2a}/D_2 antagonistic receptor interaction at the somadendritic level of the ventral striopallidal GABAergic neurons. In view of the current hypothesis that antipsychotics may at least in part act by blocking D_2 receptors linked to the ventral striopallidal system, the present findings do underline the view that the A_{2a} agonists can enhance the antipsychotic actions of low doses of D_2 receptor antagonists, so that unwanted Parkinsonianlike side effects may be avoided. In line with the present findings and with this hypothesis of a strong antagonistic A_{2a}/D_2 receptor interaction in the ventral striopallidal system are the observations by Barraco et al. [16] of a strong psychomotor depressant action of CGS21680 when infused into the nucleus accumbens. Taken together the antagonistic A_{2a}/D_2 receptor interaction should be considered as a basis for the development of new treatments for both Parkinson's disease and schizophrenia.

Studies on the A_{2a}/D_2 Receptor Interaction at the Membrane Level

STUDIES IN STRIATAL MEMBRANE PREPARATIONS

The intramembrane A_{2a}/D_2 receptor interactions have been repeatedly demonstrated in striatal membrane preparations [4,11,17,18]. Thus, the selective activation of adenosine A_{2a} receptors by means of the agonist CGS21680 reproducibly and substantially increases the high- and low-affinity dissociation constants of DA D_2 receptors and increases the proportion of high-affinity DA D_2 receptor sites. Also, opposing actions of the A_{2a} agonist and the GTP analogue were demonstrated in the regulation of DA D_2 receptors in rat neostriatal membranes [18]. Thus, while the GTP analogue reduces the proportion of D_2 receptors in the high-affinity state, the A_{2a} agonist increases the proportion in the high-affinity state, an action that appears to be independent of the effect of GTP. Thus, the GTP antagonist GDP-β-S failed to influence this action of the A_{2a} agonist. Based on these binding data, it also becomes clear that A_{2a} receptor activation also leads to a decreased G-protein activation, which is associated with the low-affinity state of the D_2 receptor. Further, an increased formation of β and γ subunits from G_s, which takes place upon A_{2a} receptor activation, may not be critically involved in these interactions, because, as pointed out, the action of the A_{2a} agonist was still found in the presence of a GTP antagonist. These results are well in line with the behavioral data and the neurochemical findings in the microdialysis experiments, which also demonstrated indications of antagonistic interactions.

Recently it was also possible to demonstrate that DA denervation results in an increase in the intramembrane interaction between adenosine A_{2a} and DA D_2 receptors in the neostriatal membranes [17]. In these experiments a unilateral DA denervation was produced by means of a intranigral injections of 6-OHDA. It was found that a low dose of CGS21680 (3 nmol), ineffective in membrane preparations from the neostriatum of naive rats, was active in membranes from the denervated neostriatum. This observation may reflect the existence of a compensatory development of a stronger interaction between the A_{2a} and D_2 receptors within the membrane due to the DA D_2 receptor supersensitivity development. It may be that the supersensitive D_2 receptors are more mobile and therefore can interact more easily with the A_{2a} receptors. These observations also explain the behavioral results demonstrating the contralateral rotational behavior induced by methylxanthines in unilaterally DA-denervated rats [8–10] and the stronger turning behavior induced by caffeine when injected into the denervated striatum [19]. Recently, we have also examined if DA-receptor supersensitivity development induced by chronic treatment with haloperidol can also be associated with strenghtening of intramembrane A_{2a}/D_2 receptor interaction in striatal membranes [11]. Again, it could be demonstrated that the low dose of CGS21680 (3 nmol), which by itself is ineffective in nontreated animals, can increase the equilibrium dissociation constants of high- and low-affinity DA D_2 binding sites and increase the proportion of high-affinity binding sites in striatal membranes from haloperidol-pretreated animals. Our findings obtained with the unselective adenosine antagonist theophylline (20 mg/kg) match these biochemical binding studies, because a high motor activation was obtained in the haloperidol-treated rats [11]. Thus, again a compensatory endogenous mechanism represented by the antagonistic A_{2a}/D_2 receptor interaction has become activated, probably in order to counteract the development of DA receptor supersensitivity, a phenomena that may reduce the development of tardive dyskinesias.

RECEPTOR AUTORADIOGRAPHIC STUDIES

Recently it was possible to demonstrate by quantitative receptor autoradiography using the D_2 receptor antagonist ^{125}I-iodosulpiride as a radioligand that the A_{2a} agonist CGS21680 could reduce the ability of DA to compete for these binding sites in the rat ventral striatum and especially in the nucleus accumbens [13]. Within the nucleus accumbens the IC_{50} values were increased by 73%, while in the olfactory tubercle only a 23% increase was observed and no changes were found within the caudate putamen. The failure to see any significant changes in the IC_{50} values within the caudate putamen is in line with previous results obtained on striatal membrane preparations. Thus, the increases in the equilibrium dissociation constants of high- and low-affinity D_2 binding sites with the concurrent increase in the proportion of high-affinity D_2 binding sites modify the competitive inhibition curves of ^{3}H-raclopride versus DA without changing IC_{50} values [4,17,18]. The results are compatible with a stronger A_{2a}/D_2 receptor interaction within the nucleus accumbens than in the caudate putamen [16], which at least in part may reflect the presence of D_3 receptors within the nucleus accumbens [13].

STUDIES ON MEMBRANE PREPARATIONS IN COS-7 CELLS TRANSIENTLY TRANSFECTED WITH THE DOPAMINE D_2 AND ADENOSINE A_{2a} RECEPTOR cDNA

A 2419-base cDNA sequence was used containing the full-length dog adenosine A_{2a} receptor cDNA [20,21]. The full-length cDNA coding for the rat short-type DA D_2 receptor was a 2511-base sequence [22]. The plasmid vector pXM containing the D_2 receptor gene and the A_{2a} receptor gene, respectively, was cotransfected by the DEAE dextran chloroquin method in the COS-7 cell [23]. High expression of D_2 and A_{2a} receptor proteins was demonstrated in membrane preparations from the COS-7 cells transiently cotransfected with DA D_2 and A_{2a} receptor cDNAs. ^{3}H-raclopride and ^{3}H-NPA were used as radioligands for the D_2 receptors, and the radioligand ^{3}H-CGS21680 was used for the A_{2a} receptors. At least 10 times higher densities of A_{2a} receptors versus D_2 receptors were demonstrated in membrane preparations. In contrast, in striatal membrane preparations similar A_{2a} and D_2 receptor densities are found. Nevertheless, the major finding of these experiments was that in these membrane preparations an optimal concentration of CGS21680 (30 nM) failed to modulate the binding characteristics of the D_2 receptors, which were shown to exhibit low- and high-affinity binding sites, in agreement with previous studies [4,17,18]. There may exist several explanations for the failure to demonstrate an intramembrane interaction between the high densities of A_{2a}/D_2 receptors of the membrane preparations of the cotransfected COS-7 cells. One explanation may be a low degree of coupling to G proteins, in view of the failure of the GTP analogue to reduce the RH values of the D_2 receptors. In line with the existence of an abnormal coupling in the membranes was the observation of only weak increases of cyclic AMP (cAMP) formation when CGS21680 was added to the COS-7 cells [24]. In view of the high amounts of receptor proteins present within the membrane preparations of the cotransfected cells, these may even be expected because only a small fraction of the receptors may become coupled to the G proteins [25,26]. Thus, this explanation would focus on the importance of G proteins for this type of receptor-receptor interaction to occur, although previous findings suggested that the modulation induced by A_{2a} receptor activation opposes the effects of the GTP mechanism in the presence of a GTP antagonist [18] as well.

However, there also exists an alternative or complementary explanation that the stoichiometry of the A_{2a} and D_2 receptors may be crucial for the interaction to take place. Thus, it may be speculated that with such high densities of A_{2a} receptors dimerization of these receptors can take place in the membranes [27], so that the D_2 receptors can no longer interact with the A_{2a} receptors. Thus, there exists evidence from coexpression studies with mutant muscarinic/adrenergic receptors for the existence of intermolecular crosstalk between G-protein–linked receptors [28], as revealed in analysis of chimeric α_2/M_3 and M_3/α_2 interactions as originally postulated by Agnati et al. [29] and Fuxe et al. [30]. Also some of the D_2 and A_{2a} receptors can exist in intracellular membranes. These results also illustrate the problem inherent with the use of transiently transfected cell lines as a model to analyze receptor-receptor interactions. Therefore the work has been continued by the use of stable fibroblast cell lines expressing A_{2a} and the D_2 receptors.

STUDIES IN MEMBRANE PREPARATIONS ON STABLY TRANSFECTED FIBROBLAST CELLS CONTAINING A_{2a} AND D_2 RECEPTORS

In these experiments [Dasgupta et al., unpublished observations] the same full-length dog adenosine A_{2a} receptor cDNA was used as in experiments on transiently transfected COS-7 cells. It was transfected into fibroblasts containing the long form of the DA D_2 receptors by means of the calcium phosphate precipitation method. The A_{2a}/D_2 receptor interaction was analyzed in membrane preparations from these cells. In binding experiments no interaction could be demonstrated in fibroblast cell lines containing only the D_2 receptors. However, in the cotransfected cells it became possible to demonstrate a marked modulation by the A_{2a} agonist CGS21680 of the binding characteristics of the D_2 receptors in the same manner as observed in the striatal membrane preparations, although only a low number of A_{2a} agonist binding sites were found. Thus, in competition experiments with ^{3}H-raclopride with DA as a competing agent, CGS21680 increased the K_H and K_L and the R_H of the DA D_2 receptors.

These preliminary findings suggest that the A_{2a}/D_2 receptor interaction can be demonstrated in membranes from fibroblast cell lines after stable cotransfection and that no modulation takes place when D_2 receptors alone are present. In this manner modulatory sites of the A_{2a} type on the D_2 receptors themselves can be excluded. In view of the low numbers of A_{2a} agonist binding sites present with a high density of D_2 receptors, it may not at first sight seem likely that a direct protein-protein interaction is involved in this modulation, but that a G-protein mechanism or another mobile membrane protein may mainly be involved. However, it must still be considered that in fibroblast cell membranes the G-protein receptors can be highly mobile and the interaction, at least in part, may occur via clustering of D_2 receptors around the A_{2a} receptors. Also, the choleratoxin that activates the G_s cannot mimick the action of CGS21680. Whatever the mechanism, these observations are important, because they do provide evidence for an antagonistic intramembrane interaction between the A_{2a} and D_2 receptors.

Acknowledgments

This work was supported by a grant (04X-715) from the Swedish Medical Research Council and by a grant from the Knut and Alice Wallenbergs Stiftelse. We are grateful to Mrs. Anne Edgren for excellent secretarial assistance.

References

1. Ferré S, Fuxe K, von Euler G, Johansson B, Fredholm B (1992) Adenosine-dopamine interactions in the brain. Neuroscience 51:501–512.
2. Schiffmann S, Jacobs O, Vanderhaeghen J-J (1991) Striatal restricted adenosine A2 receptor (RDC8) is expressed by enkephalin but not by substance P neurons: An in situ hybridization histochemistry study. J Neurochem 57:1062–1067.
3. Schiffmann S, Vanderhaeghen J-J (1993) Adenosine A2 receptors regulate the gene

expression of striatopallidal and striatonigral neurons. J Neurosci 13:1080–1087.
4. Ferré S, von Euler G, Johansson B, Fredholm B, Fuxe K (1991) Stimulation of high affinity adenosine A2 receptors decreases the affinity of dopamine D2 receptors in rat striatal membranes. Proc Natl Acad Sci USA 88:7238–7241.
5. Ferré S, Rubio A, Fuxe K (1991) Stimulation of adenosine A2 receptors induces catalepsy. Neurosci Lett 130:162–164.
6. Ferré S, Herrera-Marschitz M, Grabowska-Andén M, Ungerstedt U, Casas M, Andén N-E (1991) Postsynaptic dopamine/adenosine interaction: I. Adenosine analogues inhibit a D2 mediated behaviour in short-term reserpinized mice. Eur J Pharmacol 192:30–35.
7. Ferré S, Herrera-Marschitz M, Grabowska-Andén M, Ungerstedt U, Casas M, Andén N-E (1991) Postsynaptic dopamine/adenosine interaction: II. Postsynaptic dopamine agonism and adenosine antagonism of methylxanthines in short-term reserpinized mice. Eur J Pharmacol 192:36–42.
8. Fuxe K, Ungerstedt U (1974) Action of caffeine and theophyllamine on supersensitive dopamine receptors: Considerable enhancement of receptor response to treatment with dopa and dopamine receptor agonists. Med Biol 52:48–54.
9. Fredholm B, Fuxe K, Agnati L (1976) Effect of some phosphodiesterase inhibitors on central dopamine mechanisms. Eur J Pharmacol 38:31–38.
10. Fredholm B, Herrera-Marschitz M, Jonzon B, Lindström K, Ungerstedt U (1983) On the mechanisms by which methylxanthines enhance apomorphine-induced behaviour in the rat. Pharmacol Biochem Behav 19:535–541.
11. Ferré S, Schwarcz R, Li X, Snaprud P, Ögren S, Fuxe K (1994) Chronic haloperidol treatment leads to an increase in the intramembrane interaction between adenosine A2 and dopamine D2 receptor in the neostriatum. Psychopharmacology 116:279–284.
12. Ferré S, O'Connor W, Fuxe K, Ungerstedt U (1993) The striopallidal neuron: A main locus for adenosine-dopamine interactions in the brain. J Neurosci 13:5402–5406.
13. Ferré S, O'Connor W, Snaprud P, Ungerstedt U, Fuxe K (1994) Antagonistic interaction between adenosine A2a receptors and dopamine D2 receptors in the ventral striopallidal neuron. Neuroscience, in press.
14. Lupica C, Cass W, Zahniser N, Dunwiddie V (1990) Effects of the selective adenosine A2 receptor agonist CGS 21680 on in vitro electrophysiology, cAMP formation and dopamine release in rat hippocampus and striatum. J Pharmacol Exp Ther 252:1134–1141.
15. Cass W, Zahniser N (1991) Potassium channel blockers inhibit D2 dopamine, but not A1 adenosine, receptor-mediated inhibition of striatal dopamine release. J Neurochem 57:147–152.
16. Barraco RA, Martens KA, Parizon M, Normile HJ (1993) Adenosine A2a receptors in the nucleus accumbens mediate locomotor depression. Brain Res Bull 31:397–404.
17. Ferré S, Fuxe K (1992) Dopamine denervation induces a stronger intramembrane interaction between adenosine A2 and dopamine D2 receptors in the neostriatum. Brain Res 594:124–130.
18. Ferré S, Snaprud P, Fuxe K (1993) Opposing actions of an adenosine A2 agonist and a GTP analogue on the regulation of dopamine D2 receptors in rat neostriatal membranes. Eur J Pharmacol Mol Pharmacol Section 244:311–315.
19. Herrera-Marschitz M, Casas M, Ungerstedt U (1988) Caffeine produces contralateral rotation in rats with unilateral dopamine denervation: Comparisons with apomorphine-induced responses. Psychopharmacology 94:38–45.
20. Libert F, Parmentier M, Lefort A, Dinsart C, Van Sande J, Maenhaut C, Simons M-J, Dumont JE, Vassart G (1989) Selective amplication and cloning of four members of the G-protein-coupled receptor family. Science 244:569–572.
21. Maenhaut C, Van Sande J, Libert F, Abramowicz M, Parmentier M, Vanderhaeghen J-J, Dumont J, Vassart G, Shiffmann S (1990) RDC8 codes for an adenosine A2 receptor with physiological constitutive activity. Biochem Biophys Res Comm 173:1169–1178.
22. Bunzow J, Van Tol H, Grady D, Albert P, Salon J, Christie M, Machhida C, Neve K, Civelli O (1988) Cloning and expression of rat D2 dopamine receptor cDNA. Nature 336:783–787.
23. Luthman H, Magnusson G (1983) High efficiency polyoma DNA transfection of chloroquine treated cells. Nucleic Acids Res 11:1295–1305.
24. Snaprud P, Gerwins P, Caron M, Persson H, Fredholm B, Fuxe K (1994) A2a/D2 receptor interactions are not observed in COS-7 cells

transiently transfected with dopamine D2 and adenosine A2a receptor cDNA. Biochem Pharmacol, in press.
25. Senogles S, Spiegel A, Padrell E, Iyengar R, Caron M (1990) Specificity of receptor-G protein interactions. J Biol Chem 265: 4507–4514.
26. Monsma F, Shen Y, Ward R, Hamblin M, Sibley D (1992) Cloning and expression of a novel serotonin receptor with high affinity for tricyclic psychotrophic drugs. Mol Pharm 43:320–327.
27. Zoli M, Agnati L, Hedlund P, Li X, Ferré S, Fuxe K (1993) Receptor-receptor interactions as an integrative mechanism in nerve cells. Mol Neurobiol 7:1–37.
28. Maggio R, Vogel Z, Wess J (1993) Coexpression studies with mutant muscarinic/adrenergic receptors provide evidence for intermolecular "cross-talk" between G-protein-linked receptors. Proc Natl Acad Sci, USA 90: 3103–3107.
29. Agnati LF, Fuxe K, Zini I, Lenzi P, Hökfelt T (1980) Aspects on receptor regulation and isoreceptor identification. Med Biol 58:182–187.
30. Fuxe K, Agnati LF, Benfenati F, Cimino M, Algeri S, Hökfelt T, Mutt V (1981) Modulation by cholecystokinin of ^{3}H-spiroperidol binding in rat striatum: Evidence for increased affinity and reduction in the number of binding sites. Acta Physiol Scand 113:567–569.

C. OTHER APPLICATIONS

54. ROLE OF ADENOSINE AND ADENOSINE TRIPHOSPHATE AS ANESTHETIC ADJUVANTS

Atsuo F. Fukunaga
Tadaomi A. Miyamoto
Yoshinori Kikuta
Yuzuru Kaneko
Tatsuya Ichinohe

As medical and surgical techniques progress, more complex and aggressive surgeries are performed. Furthermore, older and sicker patients are scheduled for surgery. Hence, anesthesia providing merely sleep or unconsciousness, no matter how deep, seem insufficient to adequately protect the patient from multiple noxious stimuli during the perioperative period. In addition to the autonomic nervous system side effects produced by the anesthetic drugs themselves, very invasive surgical procedures inflicting severe stressful stimuli of all kinds and degrees may at times cause physiologic decompensation, which is usually reflected in exaggerated hemodynamic alterations. Thus, a principal concern of the anesthesiologist is to manage the acute hemodynamic changes that accompany major interventions such as induction of anesthesia, tracheal intubation, painful surgical incisions, clamping, bony resection, bowel tractions, and postoperative pain. Despite advances in anesthetic techniques and the multiple anesthetic drugs available, rendering the patient protection from the pain and stress of surgical intervention and preserving physiologic homeostasis intraoperatively still remains a great challenge to the anesthesiologist.

The ubiquitous nature and multiple roles of endogenous adenosine (ADO) and adenosine triphosphate (ATP) in the body acting as physiologic regulators and/or neuromodulators of both somatic and autonomic systems, which are so vital for controlling the body's adaptative homeostatic mechanisms during stressful conditions, has been attributed to these substances [1–3]. In particular, the retaliatory activity of ADO, such as antinociception, anticonvulsion, antiadrenal, antiarrhythmic, antiinflammatory, antithrombotic, and antiischemic, postulated by many researchers [1,4,5] seems relevant for the management of the intraoperative surgical insult. Surgical stimulation and stress responses, such as acute increases in sympathetic activity, circulating catecholamines, heart rate, blood pressure, and platelet aggregability, as well as decreases in fibrinolytic activity, can lower the threshold for tissue ischemia by decreasing blood supply while increasing cellular oxygen demand.

Adenosine's overall functional role of maintaining tissue oxygen supply and oxygen demand balance, which is protective to the vital organs, in particular the brain and heart, may be one of the ultimate goals for the use of ADO and ATP in the intraoperative period.

The broad depressant effects on the central nervous system (CNS) of exogenously administered ADO, its analogues, and ATP are well documented [6]. The potent physiologic effects on the cardiovascular system (CVS) and those related to antinociception have been extensively reviewed [7,8]. Intravenous ATP and ADO have been used clinically for the safe, deliberate induction of hypotension to decrease the risks of bleeding during orthognatic, orthopedic, and cerebral aneurysm surgeries [9,10], and for the reduction of the cardiovascular responses to tracheal intubation [11]. We have observed the potent sympatholytic and analgesic activity of ATP and ADO during deliberately induced hypotension, both experimentally and clinically. These observations led us to further evaluate the analgesic properties in a more specific and systematic manner in experimental animals [12–16]. The data demonstrated that continuous intravenous infusion of ADO or ATP, without causing hypotension, had more potent analgesic activity than morphine; and, contrary to morphine, these analgesic effects were not accompanied by respiratory depression [12,13]. Such antinociceptive effects enabled a significant reduction of the inhaled anesthetic (halothane, enflurane) requirements without compromising cardiopulmonary functions [14]. The analgesic properties of ADO and ATP were thus clearly established and were characterized as having primarily potent autonomic nervous system blocking or sympatholytic analgesic effects, with relatively weak hypnotic activity [15,16].

The use of ATP during clinical anesthesia is not new, and the safety of ADO and ATP in inducing and maintaining deliberate hypotension for considerably long periods has been clinically proved. Thus, we hypothesized that analgesic and sympatholytic effects similar to those observed in our experimental animals could be obtained in humans without causing significant hypotension. Therefore, we sought to determine 1) whether intravenous infusion of ATP can effectively decrease the anesthetic requirements of the inhaled agents (enflurane [ENF] and/or nitrous oxide [N_2O]) during surgical stimulation, and 2) whether ATP possesses the ability to block the autonomic responses to stressful, noxious stimuli and if ATP can achieve intraoperative hemodynamic stability in patients. This latter component of anesthesia, although vital, is often compromised by the presently used anesthetics because of the potent and life-threatening respiratory and/or myocardial depression that results when sufficient dosages are utilized.

Materials and Methods

Following institutional approval and informed consent, two pilot studies were carried out during the maintenance of anesthesia: Study A (Groups I and II) was conducted to determine if ATP could decrease the inhaled anesthetic requirements while maintaining cardiopulmonary functions, and Study B (Groups I and II) focused on hemodynamic changes. All patients were healthy, of ASA-1 status, and received no additional analgesic treatment intraoperatively. Patients with asthma or gout were not included in either study.

STUDY A. METHODS

Patients undergoing extensive orthognathic and maxillofacial surgery were randomly divided into two groups of seven patients each, after establishing surgical levels of anesthesia with the conventional gas mixture of ENF/N_2O/O_2. In Group I, ATP/ENF/O_2 (n = 7), the N_2O was substituted by ATP. In Group II, ATP/N_2O/O_2 (n = 7), the ENF was either totally replaced or almost completely substituted with ATP. After standard premedication (atropine 0.5 mg and 50–70 mg hydroxyzine) given intramuscularly 1 hour earlier, anesthesia was induced with intravenous thiopental (5 mg/kg). Anesthesia was initially maintained with 1.2–2.0% ENF in 60% N_2O in oxygen. The patients breathed spontaneously, with manually assisted ventilation when needed, to maintain an end-expiratory CO_2 concentration of 5–6%. No additional muscle relaxant was given after intubation. Monitoring and/or measurements included inspired and expired end-tidal concentrations of ENF, N_2O, O_2, and CO_2; continuous pulse oxymeter; ECG lead II; heart rate; radial artery blood pressure via an indwelling catheter; arterial blood gases; plasma catecholamines; cardiac output measurements; and timed urinary output. Rectal temperatures were maintained between 35.8 and 37.0°C with a warming blanket.

Assessment of Intraoperative Analgesic Effects of ATP. The standard MAC (minimum alveolar concentration of inhaled anesthetic to prevent movement in response to skin incision in 50% of patients [17]) scale was adapted to gauge the intraoperative inhaled anesthetic requirement to prevent movement in response to the multiple dynamic surgical stimuli during anesthesia. The calculated MAC values for the measured end-tidal concentration of ENF and N_2O before and after the administration of ATP were used to indirectly gauge the analgesic effects of ATP intraoperatively. The adequacy of anesthesia was judged based on the following end points: 1) unconsciousness (response to verbal stimulation), 2) analgesia (movement response), 3) inhibition of autonomic responses (hemodynamic variables), and 4) muscle tone.

Following endotracheal intubation, anesthesia was first maintained with the conventional inhaled anesthetic gas mixture of ENF/N_2O/O_2 (approximately 1.3–1.4 MAC). The control values were determined when steady-state equilibration of the alveolar (end-tidal) anesthetic concentration was achieved for at least 20 minutes. Thereafter, a 2% solution of ATP (Adephos®, Kowa, Japan) was infused into a peripheral vein via a continuous infusion pump in the range of 80–150 μg/kg/min and titrated to maintain the arterial pressure (BP) and heart rate (HR) within ±20% of the preanesthetic values. The end-tidal concentrations of ENF or N_2O were gradually decreased to slightly above their predicted concentrations for MAC-awake (alveolar anesthetic concentration midway between the value permitting and that preventing the response to verbal command [18]) values to ensure unconsciousness. Measurements and recordings of the concentrations of inhaled anesthetics during the ATP infusion for their respective MAC values (ATP/ENF MAC in Group I, and ATP/N_2O/ENF MAC in Group II) were taken after 20 minutes of steady-state conditions. These concentrations of both inhaled anesthetic agents and the intravenous ATP infusion remained constant for the rest of the operation.

This strategy allowed us to define the effects of ATP with regard to movement in response to surgical stimulation (analgesic end point) as well as the CVS responses (autonomic response end point). We assumed that sleep (unconsciousness) would be ensured by maintaining an alveolar concentration of the inhaled anesthetic agents at slightly above the predicted MAC-awake concentrations that are known to produce unconsciousness. This was accomplished by continuously monitoring the end-tidal anesthetic concentrations and by constant verbal stimulation with a lack of response. For evaluation of the autonomic activity, sympathetic responses such as sweating, grimacing, tearing, plasma catecholamine concentrations, as well as hemodynamic responses and ventilatory status, respiratory rate, and arterial blood gases, were carefully measured. If during the operation movement responses were elicited, the timing of the appearance was noted, and these responses were immediately controlled by intravenous thiopental (50–100 mg) or diazepam (10–15 mg) in Group I, or by increasing the concentration of ENF transiently in Group II. Gross bodily movement responses included bucking or any movement of the extremities or the head in response to surgical stimuli. Data are presented as the mean ± standard deviation (SD). Statistical significance was tested by paired or unpaired Student's t-tests, where applicable, and $p < 0.05$ was considered significant.

STUDY A. RESULTS

Reduction of Inhaled Anesthetic Requirements by ATP Infusion. Insofar as total replacement of N_2O and complete or almost total substitution of ENF from the conventional gas mixture were made possible by ATP, and in the absence of opioid supplementation, the analgesic effects of ATP at nonhypotensive levels are evident. These results are summarized in Figure 54-1. In Group I (ATP/ENF/O_2), infusion of ATP at 103 ± 23 µg/kg/min substituted completely for N_2O and allowed for a 36% reduction in ENF, resulting in a net reduction of anesthetic requirements of 61%, that is, from the control value of 1.4 ± 0.27 MAC to 0.57 ± 0.07 MAC, slightly below the predicted MAC-awake value of 0.58 MAC. Accordingly, movement responses were observed in four patients (4/7 = 57%) at 28, 28, 25, and 40 minutes, respectively, into the operative procedure after ATP administration. These patients received either thiopental or diazepam in adherance with the protocol. Because there was no further movement response, ATP and the low concentration of ENF were kept unchanged thereafter. In Group II (ATP/N_2O/ENF/O_2), infusion of ATP at 113 ± 19 µg/kg/min practically replaced ENF in five patients, and in the last two patients the ENF was totally removed, resulting in a total net reduction in the anesthetic requirement of 50%, that is, from 1.38 ± 0.13 MAC to 0.60 ± 0.09 MAC, slightly above the predicted MAC-awake levels. Movement response was observed in the first three patients (3/7 = 43%) at 42, 25, and 25 minutes, respectively. In these patients the ENF was transiently increased and then returned to the previous low concentrations, with the ATP infusion maintained at a constant level. Thus, the potent depressant effects of ATP enabled us to replace either one of the anesthetic gases or to markedly reduce the total inhaled anesthetic requirement while maintaining unconsciousness.

Cardiorespiratory and Adrenergic Responses. During ATP infusion all patients experienced significant inhibition of the autonomic responses to surgical stimulation in both groups (Table 54-1). Despite the intense noxious stimuli inflicted by the ongoing surgical manipulation, minimal changes in BP and HR were evidenced. In neither group of patients was myocardial depression observed with ATP. On the contrary, stroke volume was significantly increased (by 66% in Group I and 59% in Group II) without a significant change in HR and a minimal decrease in MAP in Group I (ATP/ENF/O_2). Although plasma levels of catecholamines (norepinephrine [NE] and epinephrine [EPI]) were moderately elevated during ATP infusion, only NE changes reached statistical significance (+97% in Group I, as compared with +45% in Group II). However, these increases were not sufficient to induce concomitant changes in BP or HR in either group. ATP appears to have stimulated ventilation because $PaCO_2$ values remained lower in both ATP groups. No patient developed respiratory depression during or after ATP infusion. Therefore, no mechanically assisted ventilation was required throughout the surgical period in either group.

It is noteworthy that in all patients respiratory and hemodynamic stability was easily achieved during ATP infusion, and significant inhibition of autonomic responses was attained even in those patients who showed transient movement responses. Furthermore, intraoperative hemodynamic stability was accomplished without the use of alpha- or beta-adrenergic receptor antagonists, or any other vasoactive agent(s). The degree of skeletal muscle relaxation was not monitored, but there was no

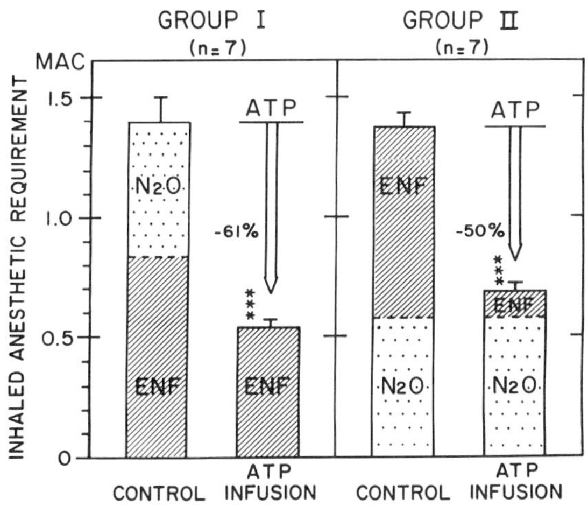

FIGURE 54-1. In both groups of surgical patients, continuous infusion of ATP substantially reduced the inhaled anesthetic requirements from 1.40 ± 0.10 MAC to 0.54 ± 0.03 MAC in Group I ($n = 7$, $p < 0.001$), and from 1.38 ± 0.05 MAC to 0.69 ± 0.03 MAC in Group II ($n = 7$, $p < 0.001$), respectively. Mean ± SEM ***$p < 0.001$. N_2O = nitrous oxide; ENF = enflurane.

evidence of muscle rigidity. Normal urine output was maintained in both groups throughout the studies (119 ± 49 ml/hr in Group I, and 113 ± 75 ml/hr in Group II). No patient had any recollection of intraoperative events, bad dreams, or of an unpleasant anesthetic experience. All 14 patients had a smooth emergence from anesthesia and were fully awake within 5–10 minutes after discontinuing the anesthetics, allowing endotracheal tube removal within that time frame. All patients had an uneventful postoperative course without experiencing nausea or vomiting.

STUDY B. METHODS AND RESULTS

The methods used in this study were similar to Study A. The group of 34 patients was divided into two groups: Group I ($n = 20$), patients not requiring muscle relaxant undergoing superficial surgery of the head, neck, and extremities; Group II ($n = 14$), patients who had abdominal surgery requiring muscle relaxants and mechanical ventilation. After standard premedication (atropine 0.5 mg, hydroxyzine 50–75 mg intramuscularly), anesthesia was induced with intravenous thiopental (5 mg/kg). Following intubation, anesthesia was initially maintained with the conventional inhaled anesthetic gas mixture of ENF (1.5–2.0%) and N_2O (67%) in oxygen, which was kept constant until surgery began. Shortly after the skin incision, a 2% solution of ATP (Adephos®, Kowa) was administered via a peripheral vein of the arm. The ATP infusion dose was titrated to maintain BP and HR within ±20% of the preanesthetic levels, while the ENF concentration was gradually decreased. The results were similar to those in Study A. Continuous infusion of ATP effectively inhibited the movement and autonomic (sympathetic: sweating, grimacing, and cardiovascular) responses to surgical stimulation in spite of the low dose of inhaled anesthetics. In particular, the hemodynamic and respiratory changes were carefully observed and recorded. These variables were well controlled during ATP infusion in both spontaneously breathing (Fig. 54-2) and mechanically ventilated patients (Fig. 54-3). In contrast, before ATP administration, there were large blood pressure fluctuations during induction of anesthesia and the beginning of surgery with the conventional anesthetics (thiopental/ENF/N_2O) (Figs. 54-2 and 54-3). In terms of the concentration of inhaled ENF anesthetic required, ATP enabled a decrease from 1.25 ± 0.24 to 0.48 ± 0.17 vol% in Group I and from 1.28 ± 0.19 to 0.24 ± 0.13 vol% in Group II (Figs. 54-2 and 54-3). ENF administration was completely discontinued approximately 30–60 minutes before surgery was concluded. After discontinuation of N_2O and ATP, all patients emerged from anesthesia smoothly, and almost immediately responded and followed verbal commands, such as opening their eyes, opening their mouth, and squeezing "my hand." No agitation was observed. Postanesthesia recovery was smooth, without nausea, vomiting, or shivering being

TABLE 54-1. Hemodynamic, respiratory, and plasma catecholamine data before (control) and during ATP infusion

	Group I		Goup II	
	Control ($N_2O/ENF/O_2$)	ATP Infusion ($ATP/ENF/O_2$)	Control ($N_2O/ENF/O_2$)	ATP Infusion ($ATP/N_2O/ENF/O_2$)
Hemodynamics				
HR (beat/min)	82 ± 11	97 ± 17	99 ± 16	95 ± 12
SBP (mmHg)	114 ± 14	100 ± 19	113 ± 21	129 ± 18
MBP (mmHg)	84 ± 15	61 ± 12[a]	79 ± 22	78 ± 16
DBP (mmHg)	69 ± 14	44 ± 8[a]	63 ± 18	54 ± 15
CO (l/min)	5.4 ± 2.5	10.5 ± 1.3[a]	4.3 ± 1.6	6.7 ± 2.4[a]
SV (ml)	65 ± 10	108 ± 23[a]	44 ± 21	70 ± 23[a]
Respiratory				
Rate (breath/min)	19 ± 3	19 ± 2	16 ± 4	19 ± 3[a]
pH	7.37 ± 0.02	7.35 ± 0.02	7.33 ± 0.04	7.34 ± 0.06
$PaCO_2$ (mmHg)	47 ± 7	44 ± 7[a]	45 ± 4	39 ± 3[a]
PaO_2 (mmHg)	215 ± 23	568 ± 50[a]	202 ± 28	191 ± 57
BE	−0.4 ± 0.9	−0.5 ± 0.9	−2.0 ± 2.0	−2.4 ± 1.2
Plasma catecholamine				
NE (pg/ml)	178 ± 102	351 ± 98[a]	296 ± 166	429 ± 270[a]
EPI (pg/ml)	86 ± 19	150 ± 130	92 ± 90	153 ± 76

Mean ± SD, [a] $p < 0.05$.
HR = heart rate; SBP = systemic blood pressure; MBP = mean blood pressure; DBP = diastolic blood pressure; CO = cardiac output; SV = stroke vloume; $PaCO_2$ = arterial carbon dioxide tension; PaO_2 = arterial oxygen tension; BE = base excess; NE = norepinephrine; EPI = epinephrine.
Group I, Control: 60% N_2O in O_2, ENF: 1.39 ± 0.43%; ATP infusion: 103 ± 23 μg/kg/min; ENF: 0.91 ± 0.12%.
Group II, Control: 60% N_2O in O_2; ENF: 1.34 ± 0.19%; ATP infusion: 113 ± 19 μg/kg/min; ENF: 0.18 ± 0.17%.

experienced. Within 10 minutes after extubation, the patients were well oriented, clear-headed, and remained calm without any sign of respiratory depression. Intraoperative urine output was 58.7 ± 43.3 ml/hr in Group I and 71.8 ± 44.6 ml/hr in Group II. All patients were visited postoperatively and asked about their anesthesia experience after 24 and 48 hours. None of the patients complained of intraoperative awareness or an unpleasant experience.

Discussion

ADO and ATP are endogenous metabolites that are physiologically ubiquitous, affecting practically every cell of the body, but playing particularly conspicuous roles in the CNS and the CVS. More importantly, these purines appear to participate in numerous local regulatory mechanisms, including those at synapses in the CNS and at the neuroeffector junctions in the periphery. Despite the wide spectrum of profound, interesting, and potentially practical effects that these purines exert in various tissues, only a few therapeutic usages have been indicated.

Surgery and some anesthetic procedures (laryngoscopy, endotracheal intubation) constitute major noxious stimuli, and as such, will affect numerous physiologic activities at once. The body responds to surgical trauma with a complex series of neuroendocrine and metabolic changes. Pharmacologic effects of the currently available analgesic or anesthetic drugs that inhibit pain or produce sleep do not seem to adequately counteract all of these sudden phy-

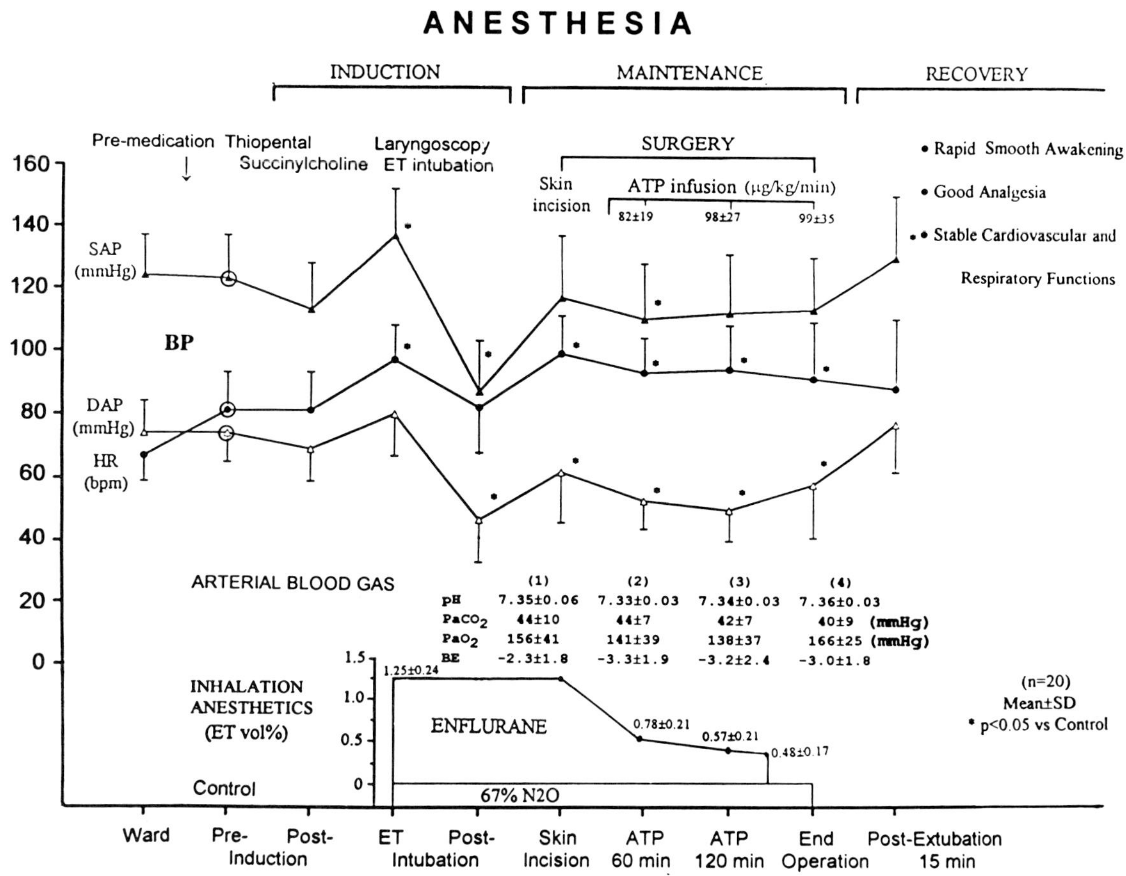

FIGURE 54-2. Intraoperative hemodynamic changes during superficial surgical procedures of the head, neck, and extremities in patients breathing spontaneously. Notice the large fluctuations (up and down) in blood pressure (BP) and heart rate (HR) after endotracheal (ET) intubation and skin incision during thiopental/ENF/N_2O anesthesia. ATP infusion achieved hemodynamic and respiratory stability despite the low dose of ENF. The decrease in ENF (vol%) is indicated at the bottom. The arterial blood gas data indicate normal ventilatory status throughout ATP infusion.

siologic disturbances. It is possible that endogenous purines (ATP and/or ADO) may have the intrinsic ability to attenuate or block such disturbances, but their physiologic concentrations may not be sufficient to successfully block all the excessive responses induced by the massive stressful stimuli. We hypothesize that when purines are given in the form and dosages that are pharmacologically adequate, that is, as a continuous infusion of ADO or ATP, sufficient tissue levels may be attained in order to achieve such effects by restoring the most effective and comprehensive natural protective and self-preservation mechanisms in a well-orchestrated and effective manner that helps to attain homeostatic adaptation.

Intravenously administered ATP is converted rapidly into ADO in the blood, and it is eliminated from the bloodstream without an intervening metabolizing or excreting organ, without causing apparent toxic effects or accumulation in the blood. Because of the extremely short half-life of less than 10 seconds for ATP and 1–2 seconds for ADO, it seems likely that these purines must be continuously infused at

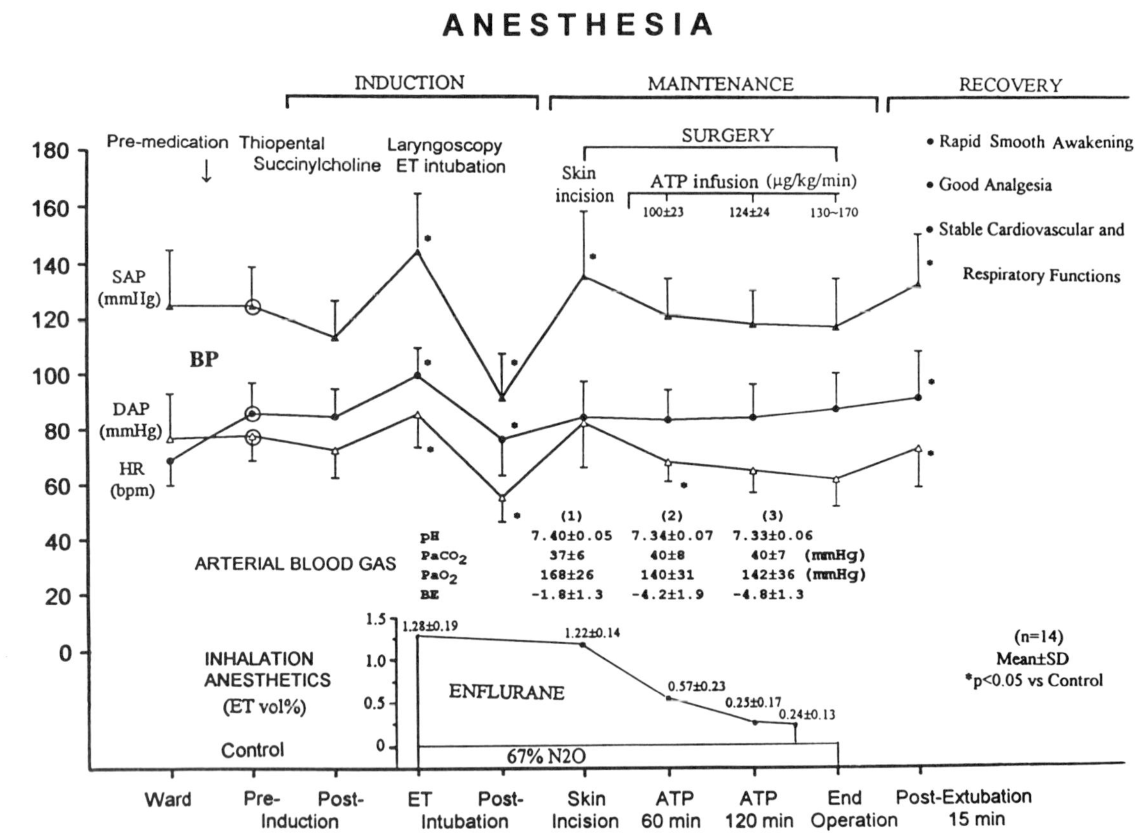

FIGURE 54-3. Intraoperative hemodynamic changes during abdominal surgery in patients with mechanical ventilation. Notice the large fluctuations (up and down) of blood pressure (BP) and heart rate (HR) after endotracheal (ET) intubation and skin incision during thiopental/ENF/N_2O anesthesia. ATP infusion achieved hemodynamic and respiratory stability despite the low ENF dose. The decrease in ENF (vol%) is indicated at the bottom. The arterial blood gas data indicate normal ventilatory status throughout ATP infusion.

effective doses if pharmacologic responses are to be achieved. Many standard pharmacokinetic descriptors, such as clearance, volume of distribution, and therapeutic plasma concentrations, may be difficult to determine or may not apply, posing problems for understanding the mechanisms of action. However, from a practical standpoint, a rapid, short-lasting effect and the absence of accumulation, tachyphylaxis, and toxic effects to the patient may constitute highly desirable characteristics for an intravenous drug for use in acute conditions such as clinical anesthesia.

In spite of technological advances, no presently available method has proven to be reliable in predicting or assessing the depth of anesthesia or the potency of anesthetic drugs during surgery. MAC has been considered the standard index of anesthetic potency [17], and it has been used to compare the anesthetic effects of various drugs, including intravenous agents. The measurement of MAC, however, represents only a somatic movement response to painful (skin incision) stimulation, and it does not take into account the autonomic responses (respiratory, circulatory, or hormonal). Further studies defining MAC derivatives such as MAC-awake (for

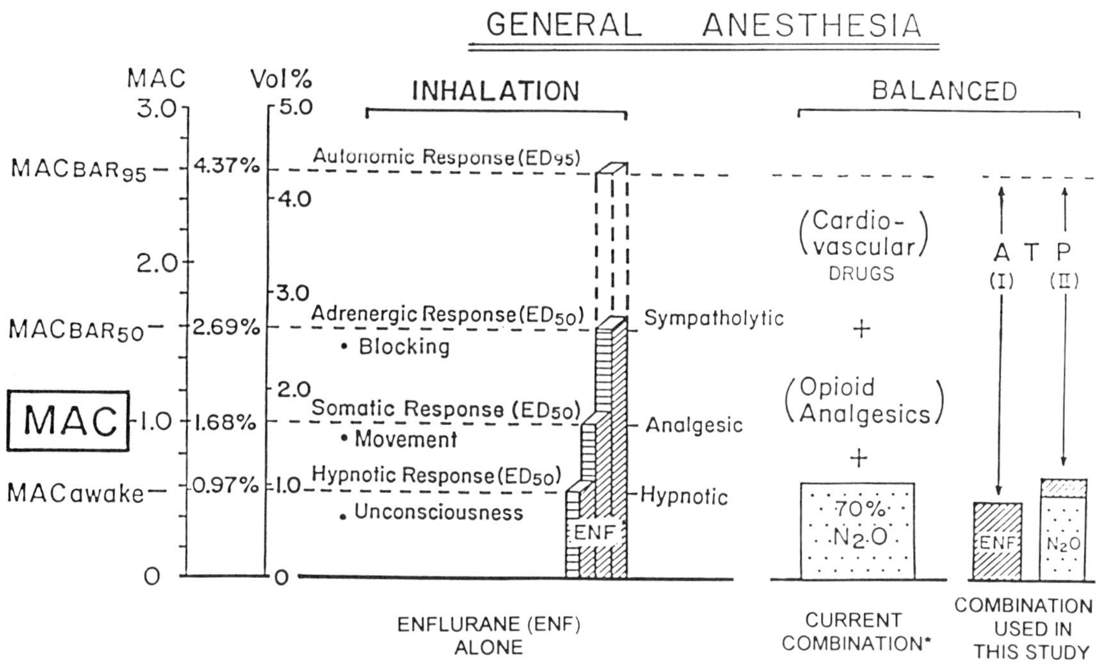

FIGURE 54-4. Schematic representation of the clinical end points in general anesthesia constructed based on the MAC concept and derivatives of MAC: MAC-awake (unconsciousness), MAC-movement (analgesia), MAC-BAR (blocking adrenergic response). The two parallel vertical scales on the left side indicate increasing doses of inhaled anesthetic in MAC and vol% of enflurane (ENF). In the middle, the anesthetic effects (hypnotic, analgesic, and sympatholytic) are represented by the ENF bars. Balanced anesthesia can be the combination of a variety of intravenous drugs and inhhaled drugs (for example, isoflurane, etc.). On the right side, the MAC concept was adapted to characterize the pharmacologic (sympatholytic analgesic) effects of ATP during surgical stimulation.

hypnotic index), MAC for intubation, and MAC-BAR (for blocking adrenergic responses) have been reported [19]. However, clinical studies have shown that anesthetics might blunt motor responses without blunting autonomic responses [20]. In clinical practice, the amount of anesthesia required depends on the intensity of stimulation that results from surgery, and, more importantly, lack of movement does not ensure lack of hemodynamic response to surgical stimulation. In addition, although skin incision has been considered the supramaximal stimulus, this stimulus may not cause the most autonomic responses when compared with other surgical stimuli, such as visceral stimulation. Thus, MAC is clinically useful but limited. We have observed that intraoperative surgical manipulation such as bony resection or visceral traction can inflict noxious stimulation, evoking intense somatic and/or hemodynamic responses. We used the MAC scale as a convenient guide to gauge the inhibitory analgesic effects of ATP by measuring the decrease in the amount of inhaled anesthetic requirements during intraoperative, continuously changing surgical stimulation. In addition, we adopted expanded multiples of MAC and used them as indices for different clinical end points of anesthesia, such as unconsciousness (MAC-awake), analgesia (MAC-movement), and blockade of autonomic responses (MAC-BAR) in order to characterize the pharmacologic effects of ATP (Fig. 54-4).

In the daily clinical setting, movement response cannot be used as a judgement for the adequacy of anesthesia, because muscle-paralyzing drugs are extensively utilized to facilitate surgical procedures. Thus, cardiovascular responses such as BP and/or HR have become mainstays in assessing the state of anesthesia. Although it has been said that BP and HR are used to judge "qualitative rather than quantitative" depths of anesthesia [19], the need to consider autonomic responses as an additional quantitative gauge to assess the depth of anesthesia gave rise to the concept of MAC-BAR [21]. The authors defined MAC-BAR_{50} as MAC to achieve blockade of autonomic responses in 50% of patients during skin incision; and MAC BAR_{95} as blockade in 95% of patients, as calculated by "logit" transformation. Accordingly, the ENF concentrations [ENF] (in volume %) to achieve MAC BAR_{50} = 1.6 MAC = [2.69% ENF], and for MAC BAR_{95} = 2.6 MAC = [4.4% ENF]. The latter concentrations are already toxic and produce severe, often fatal, cardiopulmonary depression and/or CNS seizures and convulsions, even when used in combination with 60% N_2O. In an attempt to avoid the potential risks of inhaled anesthetics at high concentrations, the so-called balanced anesthesia technique has been popularized. This technique uses N_2O in combination with opioid analgesics and muscle-relaxant drugs. In addition, it often requires several types of peripherally acting cardiovascular drugs, including alpha- and beta-adrenergic receptor agonists and antagonists, calcium channel blockers, and other nitrites, such as nitroglycerine or nitroprusside, to mimic the BAR (blocking autonomic responses) effect by peripheral manipulation. These considerations are graphically depicted and illustrated in Figure 54-4.

Although this preliminary study evaluated a small number of patients, the results demonstrated that in both groups intravenous ATP infusion substantially decreased the inhaled anesthetic requirements. Moreover, ATP inhibited the autonomic responses equivalent to MAC-BAR levels at low concentrations of inhalation agents. In the absence of any other analgesic used, it is likely that such levels of autonomic blockade would have required the administration of high dosages of inhaled anesthetic agents, or high doses of opioid analgesics plus several vasoactive drugs in the case of balanced anesthesia, as illustrated in Figure 54-4. It should be noted that hemodynamic stability was maintained throughout ATP administration (see Figs. 54-2 and 54-3). Moreover, unlike the opioid analgesics, ATP did not cause respiratory depression or muscular rigidity, as evidenced by the respiratory data in Table 54-1. This agreed with our previously observed experimental results [12–14]. Indeed, human studies have shown that ADO stimulates ventilation via carotid chemoreceptor activation [22]. In addition, it is also possible that the low concentration of inhaled anesthetics used enabled the safe maintenance of surgical anethesia with spontaneous breathing, obviating the need for mechanical ventilator assistance in patients not requiring muscle relaxants. The results from these pilot studies confirmed our previous experimental studies in which the inhibitory effects of purines were characterized as having mainly sympatholytic analgesic properties [15,16].

The accumulated evidence seems to implicate these purines (ADO and/or ATP) as mediators in some of the mechanisms modulating pain perception or analgesia, sedation, and sleep [23]. Hence, it should not be surprising that some of the actions of drugs commonly used in clinical anesthesia might be mediated in part by purinergic mechanisms. The potent cardiovascular effects of ADO and ATP have been extensively reviewed by Olsson et al. [7]. These effects may be exerted directly on the CVS via several purinoceptors: A_1, A_2 for adenosine and P_{2x}, P_{2y} for ATP. They are characterized as causing negative chronotropic, dromotropic, and inotropic effects in the atria and conduction tissues of the heart, mediated by A_1 receptors in atrial myocytes, with minimal effects on ventricular myocytes. Effects on regional blood flow to various organs are mediated by A_2 and P_{2y}, P_{2x} receptors at endothelial (P_{2y}) or smooth muscle (A_2) cells of various vascular beds, with vasodilation generally, or vasoconstriction (P_{2x}, A_1) in some afferent preglomerular vessels of the kidney. Most importantly from the anesthesia perspective is the indirect control of central or neural cardiovascular regulation, that is, modulation of the autonomic nervous system by the activation of presynaptic A_1 receptors that inhibit the release of a number of excitatory and perhaps inhibitory neurotransmitters, ultimately determining sympathetic as well as parasympathetic nerve activities. In addition, secretory granules contained at adrenergic or cholinergic terminals release ATP as a cotransmitter simultaneously with norepinephrine or acetylcholine release, respectively, thus modulating autonomic activity, even at the peripheral neuroeffector level, raising the possibility of the existence of a purinergic division of the autonomic nervous system [24]. Of particular interest is the evidence of ADO being involved centrally in blood pressure regulation at the level of the nucleus tractus solitarius, an area involved in cardiovascular baroreceptor integration [25]. Both effects together, that is, central and peripheral, seem to confer on these purines the rather marked autonomic nervous system blocking or sympatholytic activity observed in the present study.

With respect to preserving physiologic homeostasis during anesthesia and surgery, inhibition of central autonomic nervous system responses may have implications far more important than just the peripheral CVS effector manipulation. Central mediation will modulate effectors of not only the CVS but also those dependent on the hypothalamus-pituitary-endocrine effector axis. These clinical studies do not address the question of whether ATP achieved blockade of all these neurohumoral responses. However, judging by the stable hemodynamics and gross cardiovascular responses, ATP can achieve significant inhibition of the autonomic responses during surgical stimulation. The systemic BP in Study A, Group I, when ATP was combined with ENF, was lower than in Group II with N_2O (Table 54-1). The difference in BP responses might be due to the antiadrenergic effect of ATP and/or attributable to the vasodilating effects of ENF itself and the sympathomimetic vasoconstrictive property of N_2O. Although moderate increases of EPI and NE were observed during surgery, whether such a catecholamine response can be abolished completely by increasing the ATP dosage or the inhaled anesthetic concentration needs to be evaluated in future studies.

The quick onset and short-lasting action without accumulation of the purines provides the possibility of actually titrating "the levels of inhibition" on a real-time basis to constantly changing anesthetic requirement levels according to the intensity of surgical stimuli during the operative procedure. Moreover, the recent findings of sustained analgesia [26] produced by these purines may have potential value for patient care beyond the perioperative period into the recovery period and for the postoperative management of pain, where the use of opioid narcotics may be decreased, thereby preventing the incidence of undesirable respiratory complications (particularly in elderly patients). Although the specific sites of actions or ideal dosages are not clearly defined as yet, our preliminary data suggest that intravenous administration of ATP, when combined with low concentrations of inhaled anesthetics (ENF, N_2O), achieved effectively all the required end points for general anesthesia: unconsciousness, analgesia, control of autonomic, hemodynamic responses to surgical stimulation, and some muscle relaxation.

Clinically, dose reduction of other anesthetics has definite advantages of decreasing the risk of overdosage, toxicity, and side effects and speeding recovery from anesthesia. Intravenously administered ATP decreases the inhaled anesthetic requirement effectively and safely. Because of ADO's and ATP's rapid acting, predominantly sympatholytic analgesic properties, these substances may play important roles as anesthetic adjuvants in clinical anesthesia.

References

1. Williams M (1989) Adenosine: The prototypic neuromodulator. Neurochem Int 14:249–264.
2. Fredholm BB, Hedqvist P (1980) Modulation of neurotransmission by purine nucleotides and nucleosides. Biochem Pharmacol 29: 1635–1643.
3. McIlwain H, Poll JD (1986) Adenosine in cerebral homeostatic role: Appraisal through actions of homocysteine, colchicine, and dipyridamole. J Neurobiol 17:39–49.
4. Newby AC (1984) Adenosine and the concept of "retaliatory metabolites." Open question. Trends Biochem Sci 9:42–44.
5. Dunwiddie TV (1985) The physiological role of adenosine in the central nervous system. Int Rev Neurobiol 27:63–139.
6. Phillis JW, Wu PH (1983) Roles of adenosine and adenine nucleotides in the central nervous system. In Daly JW, Kuroda Y, Phillis JW, Shimizu H, Ui M (eds.), Physiology and Pharmacology of Adenosine Derivatives. New York: Raven Press, pp. 219–235.
7. Olsson RA, Pearson JD (1990) Cardiovascular purinoceptors. Physiol Rev 70:761–845.
8. Sawynok J, Sweeney MI (1989) The role of purines in nociception. Neuroscience 32: 557–569.
9. Maruyama M, Sato K, Shimoji K (1979) Hypotensive anesthesia using ATP. J Clin Anesth (Japan) 3:279–284.
10. Öwall A, Gordon E, Lagerkranser M, Lindquist C, Rudehill A, Sollevi A (1987) Clinical experience with adenosine for controlled hypotension during cerebral aneurysm surgery. Anesth Analg 66:229–234.
11. Mikawa K, Maekawa N, Kaetsu H, Goto R, Yaku H, Obara H (1991) Effects of adenosine triphosphate on the cardiovascular response to tracheal intubation. Br J Anaesth 67:410–415.
12. Ginsburg R, Fukunaga B, Sasaki S, Kikuta Y, Fukunaga AF (1990) Analgesic activity of intravenous adenosine: A comparison of potency with morphine sulphate. Anesthesiology 73(3A): A362.
13. Fukunaga A, Kikuta Y, Ginsburg R, Fukunaga B, Sasaki S (1991) Analgesic effects of intravenous ATP compared to those of morphine sulphate in the rabbit. Nihon Masui Gakkai, 38th: I-H-3.
14. Kikuta Y, Fukunaga AF, Ginsburg R, Fukunaga B, Okada K (1990) Effect of intravenous ATP on enflurane-N_2O MAC in spontaneously breathing rabbits: Assessment of cardiorespiratory effects. Anesthesiology 73(3A):A401.
15. Fukunaga AF, Fukunaga BM, Kikuta Y (1992) Assessment and characterization of the anesthetic effects of intravenous adenosine in the rabbit. Anesth Analg 74(2S):S103.
16. Fukunaga AF, Fukunaga B, Kikuta Y (1992) Anesthetic efficacy of adenosine and ATP compared to N_2O in the experimental animal model: Sympatholytic vs. sympathomimetic. Anesthesiology 77(3A):A349.
17. Quasha AL, Eger EI II, Tinker JH (1980) Determination and applications of MAC. Anesthesiology 53:315–334.
18. Stoelting RK, Longnecker DE, Eger EI II (1970) Minimum alveolar concentrations in man on awakening from methoxyflurane, halothane, ether and fluroxene anesthesia: MAC awake. Anesthesiology 33:5–9.
19. Stanski DR (1990) Monitoring depth of anesthesia. In Miller RD (ed.), Anesthesia, 3rd ed., Vol 1. New York: Churchill Livingstone, pp. 1001–1029.
20. Zbinden AM, Petersen-Felix S, Thomson DA (1994) Anesthetic depth defined using multiple noxious stimuli during isoflurane/oxygen anesthesia. II. Hemodynamic responses. Anesthesiology 80:261–267.
21. Roizen MF, Horrigan RW, Frazer BM (1981) Anesthetic doses blocking adrenergic (stress) and cardiovascular responses to incision-MAC BAR. Anesthesiology 54:390–398.
22. Biaggioni I, Olafsson B, Robertson RM, Hollister AS, Robertson D (1987) Cardiovascular and respiratory effects of adenosine in conscious man. Evidence for chemoreceptor activation. Circ Res 61:779–786.
23. Daval JL, Nehlig A, Nicolas F (1991) Physiological and pharmacological properties of adenosine: Therapeutic implications. Life Sci 49: 1435–1453.
24. Burnstock G (1982) The co-transmitter hypothesis, with special reference to the storage and release of ATP with noradrenaline and acetylcholine. In Cuello AC (ed.), Co-transmission. London: Macmillan Press, pp. 151–163.
25. Barraco RA (1988) Adenosinergic modulation of brainstem mechanisms involved in cardiovas-

cular control. In Paton DM (ed.), Adenosine and Adenine Nucleotides Physiology and Pharmacology. London: Taylor & Francis, pp. 233–240.

26. Fukunaga AF, Fukunaga B, Ichinohe T (1993) Intravenous adenosine produces profound and sustained analgesia in spontaneously breathing rabbits. Anesthesiology 79(3A):A423.

55. MECHANISMS AND THERAPEUTIC ACTIONS OF URIDINE TRIPHOSPHATE IN THE LUNG

R.C. Boucher
M.R. Knowles
K.N. Olivier
W. Bennett
S.J. Mason
M.J. Stutts

Lung Defense Mechanisms

The lung's primary function is gas exchange. However, during the process of respiration the lung is exposed to large quantities of potentially toxic and infectious agents from the ambient environment. To meet this challenge, the lung has evolved efficient ways of defending itself from such insults. In the alveolar region, the lung primarily defends itself from inhaled foreign material by the surveillance activity of a migratory cell, the alveolar macrophage. In the airways, the primary defense against inhaled foreign material is the mucociliary system, which traps foreign inhaled materials and clears them from the lung. The importance of this system to humans is demonstrated by the fact that genetic defects in components of this system, for example, cystic fibrosis and dysmotile cilial syndromes, lead to chronic, often fatal, pulmonary diseases.

Regulation of Airways Defense by Extracellular Nucleotides (ATP/UTP)

MUCOCILIARY TRANSPORT SYSTEM

A diagram depicting the components of the mucociliary clearance system is shown in Figure 55-1 [1]. There are three key elements to this system. First, there is a mucin layer, which may be in the form of a blanket or a tangled net, which serves to trap inhaled foreign material. The biophysical characterization of this mucin layer with respect to the variables that regulate its transportability is still incomplete. However, it has become clear that both the macromolecular and the salt and water content of mucus are important for transportability. Second, there is a periciliary liquid layer, which is regulated in part by movement of liquids axially from distal surfaces to more proximal surfaces, and locally by ion transport activities of the superficial epithelial cells. Because the composition and

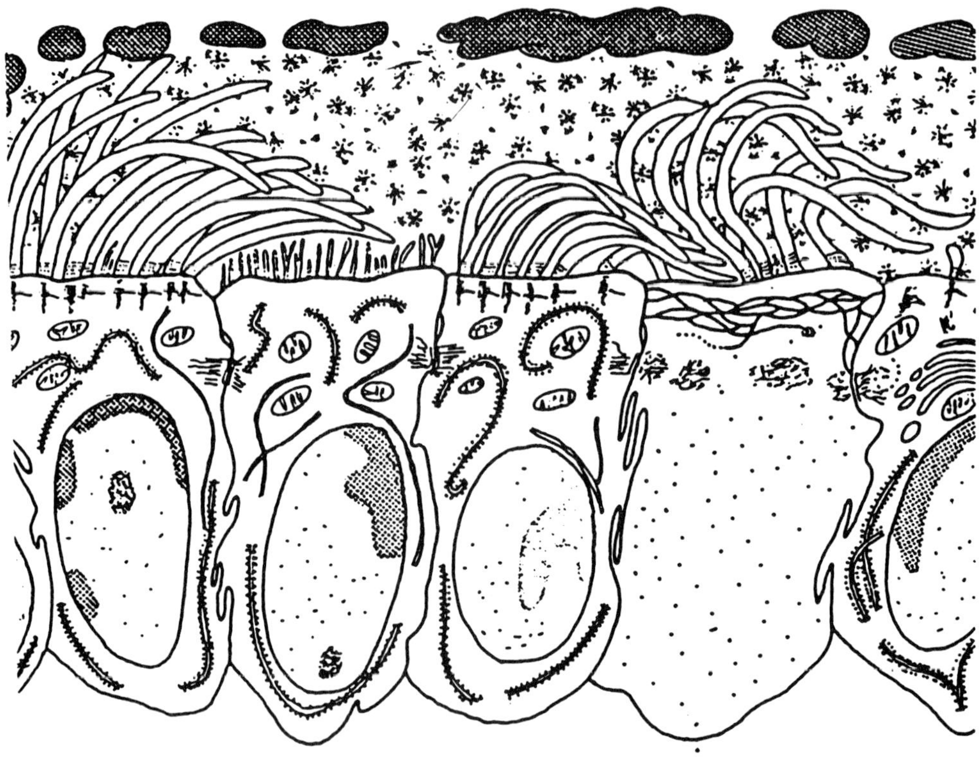

FIGURE 55-1. Cross-section of airway epithelium, depicting components of the mucociliary clearance system: the mucin layer, the periciliary liquid layer, and the cilia.

volume of the airway surface liquid (ASL) must be tightly regulated, airway epithelial cells exhibit both absorptive (active Na^+ transport) and secretory (Cl^- secretory) capacities. Third, the cilia (approximately 200/ciliated cell) provide propulsive energy for clearance of the mucin layer. The cilia beat with an effective stroke in one direction and beat typically at a relatively high basal rate (8–20 Hz). It appears that considerable regulation of cilial beat frequency can be effected by a variety of neurohumoral/inflammatory agents. All the components of this mucociliary transport system are regulated by extracellular triphosphate nucleotides, and these functions are described below.

UTP REGULATION OF MUCIN SECRETION

A series of studies have been performed in isolated canine and human tissues to examine nucleotide-regulated release of mucin from the superficial epithelial goblet (mucin-secreting) cell [2,3]. These studies have demonstrated that, in contrast to a variety of tachykinins and neurohumors, extracellular triphosphate nucleotides are almost uniquely capable of eliciting granule discharge from this cell type. The granule discharge appears to be most readily triggered by application of uridine triphosphate (UTP)/adenosine triphosphate (ATP) to the lumenal surface as compared to the basolateral surface of these cells. ATP/UTP-triggered exocytosis occurs in a millisecond time frame. Although not yet rigorously studied, it appears that the exocytosis regulatory pathway involves the activation in mucous cells of signals that raise intracellular calcium and proteinkinase C (PKC) activity (see below).

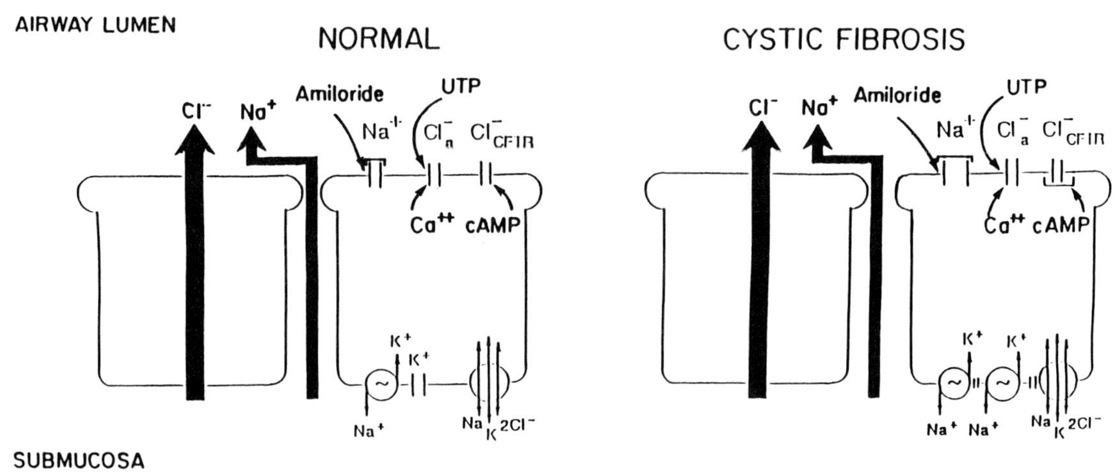

FIGURE 55-2. Electrolyte activities of proximal human epithelia: in cystic fibrosis, Na^+ absorption is increased, while cystic fibrosis transmembrane regulator (CFTR) mediated Cl^- secretion is absent.

REGULATION OF AIRWAY EPITHELIAL ELECTROLYTE AND WATER CONCENTRATION BY TRIPHOSPHATE NUCLEOTIDES

Figure 55-2 is a diagram of the electrolyte activities of proximal human epithelia. Under basal conditions, human airway epithelia absorb Na^+ through an active electrogenic mechanism that is proportional to the activity of amiloride-sensitive apical membrane Na^+ channels. Under conditions under which Na^+ entry into the cell is blocked, either pharmacologically by amiloride or potentially by in vivo regulators, the airway cell can convert to a Cl^- secretory tissue that normally secretes Cl^- via two molecularly distinct channels on the apical membrane. The best studied Cl^- channel is the cAMP-regulated cystic fibrosis transmembrane regulator (CFTR) Cl^- channel. The other Cl^- channel is a calcium-regulated "alternative" Cl^- channel ($Cl^-{}_a$).

A series of studies measuring the bioelectric correlates of ion transport have been reported that demonstrate that extracellular triphosphate nucleotides regulate both Na^+ and Cl^- transport pathways. First, when extracellular ATP/UTP is applied to the basolateral membrane in tissues that have not been pretreated with amiloride, there appears to be an acceleration of Na^+ transport [4]. This response appears to be a general mechanism exhibited by receptors linked to phospholipase C on the basolateral membrane and involves activation of calcium-sensitive basolateral K^+ channels, intracellular hyperpolarization, and an increase in the driving force for Na^+ entry. In contrast, when the apical Na^+ channel is blocked, ATP or UTP activate Cl^- secretion. From a series of studies, including studies in the cystic fibrosis mouse, it appears that UTP/ATP activates principally the non-CFTR Cl^- channel in normal tissues and only this channel in cystic fibrosis tissues [5,6]. Further, the mode of activation of $Cl^-{}_a$ may be complex. A component of the activation certainly involves an increase in intracellular Ca^{2+} levels [4]. However, it appears that direct signaling mechanisms may also be active [7]. Recent studies measuring net water transport across excised airway tissues have confirmed inferences, based on electric measurements, that UTP can stimulate net water secretion in the presence of amiloride [8].

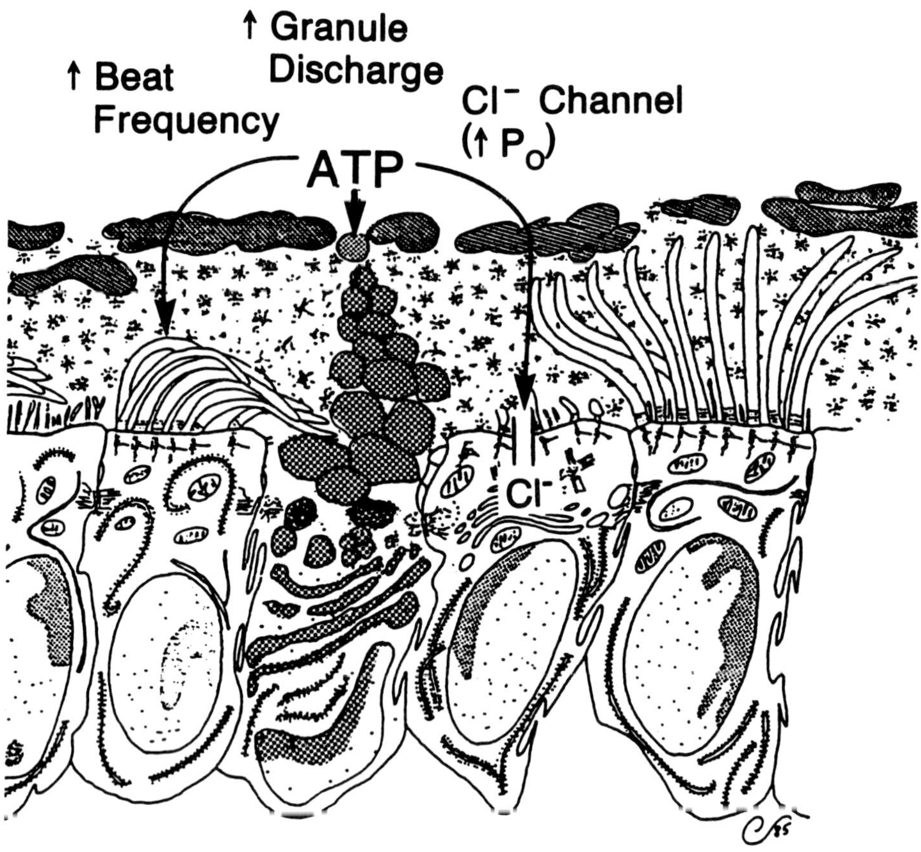

FIGURE 55-3. Cross-section of airway epithelium, depicting elements of airways defense regulated by endogenous ATP: ciliary beat frequency, mucin release from goblet cells, and Cl^- secretion.

CILIAL BEAT FREQUENCY REGULATION BY EXTRACELLULAR UTP/ATP

The effects of extracellular UTP/ATP on the regulation of beat frequency of cilia in human airway epithelial cells has been recently characterized [9; Geary et al., unpublished communication]. For these studies ciliated cells were detached from freshly excised nasal specimens, plated on coverslips, and cilial beat frequency measured by video imaging techniques. These studies demonstrated that extracellular ATP/UTP was highly effective in acutely increasing cilial beat frequency. UTP approximately doubled beat frequency, and the response was sustained for more than 20 minutes during continual application of UTP to the cell. Direct comparative studies with other agents that regulate beat frequency established that UTP was approximately twice as effective as agents that raised cellular cAMP or cGMP in this process. Although the precise signaling mechanisms for UTP with respect to regulating ciliary beat frequency are not yet known, they likely involve the activation of cellular processes that raise intracellular Ca^{2+} and protein kinase C activity.

INTEGRATING THE EFFECTS OF ATP/UTP IN THE MUCOCILIARY CLEARANCE SYSTEM

Extracellular triphosphate nucleotides regulate all three major components of the mucociliary transport system. These observations have led to the hypothesis that ATP endogenously released into the airway lumenal compartment may serve as a local autocrine/paracrine factor that coordinates the activities of the airway mucosa to respond to inhaled toxic foreign materials (Fig. 55-3). In this hypothesis, inhaled foreign material deposited on the airway surface activates, possibly through stress or pressure, the release of a small amount of ATP from airway surface epithelial cells. The released ATP triggers the secretion of mucin from goblet cells, which serves to trap inhaled foreign material on the airway surface. In concert, ATP triggers the secretion of salt and water by airway epithelial cells to hydrate the mucin and render it biophysically optimal for clearance from the lung. Finally, extracellular ATP increases the frequency of the cilial beat. The net effect of this ATP-regulated activity is to trap the foreign material and speed its removal from the lung. Ultimately, it may require a series of models, including models in which ATP release mechanisms and potentially ATP/UTP receptors are molecularly deleted from "knockout mice," to test this hypothesis fully.

Human UTP Receptor

BIOCHEMICAL CHARACTERIZATION OF THE HUMAN AIRWAY P_{2U} RECEPTOR

A series of biochemical studies have been performed to characterize the rank-order potency of nucleotides at the human airway epithelial receptor and to characterize the signal transduction mechanisms triggered by an activated receptor. Studies from cultured human airway epithelial cells and transformed human airway epithelial cell lines revealed that extracellular triphosphate nucleotides increase intracellular Ca^{2+} and inositol phosphate levels in G-protein–mediated interactions with phospholipase C (PLC) [10]. It appears that the G protein that links the receptor to PLC is minimally pertussis toxin sensitive and is therefore not G_i or G_o but, based on analogy with other systems, is probably G_q or G_{21}. The PLC isoform activated by the P_{2U} receptor also is not known. Recent studies have shown that as a consequence of the increase in intracellular Ca^{2+} levels following receptor activation, another intracellular enzyme, phospholipase A_2, is activated [11]. Finally, as mentioned above, it is likely that extracellular ATP/UTP can directly activate ion channels in the plasma membrane of airway epithelial cells [7].

Using both intracellular Ca^{2+} measurements and measurements of inositol triphosphate (IP_3) formation, a pharmacologic characterization of the human extracellular triphosphate nucleotide receptor has been performed. These studies revealed that the rank-order potency of agonists for this receptor is UTP $\geqslant$ ATP $>$ $ATP_{\gamma}S$ $\gg$ ADP $>$ AMP $>$ adenosine [4,10]. These studies have established that the receptor homologously desensitizes with respect to ATP and UTP, although it should be noted that desensitization has not been studied in detail. In a series of other studies it was found that human airway epithelial cells also express adenosine receptors of the A_{2b} class. In summary, these studies identified a receptor that responded to UTP with equal potency and effectiveness as to ATP, and the receptor has been termed the P_{2U} receptor or, alternatively, the 5′-nucleotide receptor.

MOLECULAR CHARACTERIZATION OF THE HUMAN P_{2U} RECEPTOR

The work of Lustig et al. identified in mouse/rat hybridoma cells a cDNA that translated into a protein that had the pharmacologic characterization of a P_{2U} receptor [12]. Using sequence information from this clone, a cDNA was isolated from a human CF airway epithelial (CF/T43) library and characterized [13]. The cDNA was approximately 90% homologous to the murine cDNA for this receptor. When the cDNA was stably expressed in a null cell line (^{121}N astrocytoma cells), these cells expressed intracellular calcium and inositol phosphate responses to extracellular UTP/ATP that followed the pharmacologic profile of the receptor in CF/T43 cells. Further, this receptor appeared to be highly expressed at the message level in lung epithelial cells, congruent with physiologic studies, but not in the brain. An interesting phenomenon that was observed during the expression of the heterologous receptor in ^{121}N astrocytoma cells was that the cells appeared to be basally activated with respect to the inositol phosphate system. Follow-up studies tested the hypothesis that this phenomenon reflected continual release of ATP from cells into the extracellular compartment that activated the transduced receptor. In brief, these studies demonstrated that administration of enzymes that metabolized ATP (apyrase, hexokinase) to the cell culture medium abolished the tonic activation of the inositol phosphate system.

Therapeutic Potential of Extracellular UTP

GENERAL

Because extracellular triphosphate nucleotides regulate normal mucociliary defense mechanisms, it has been postulated that these agents may have therapeutic potential for the treatment of human pulmonary diseases that reflect abnormalities of airway mucus clearance. These diseases would include well-known genetic diseases, for example, cystic fibrosis and dysmotile cilia syndromes, and also include diseases in which inhaled foreign materials, for example cigarette smoke, have damaged airway mucosa to generate chronic bronchitis. There is good evidence that all of these diseases are characterized by slowing of the mucociliary clearance mechanism, and therefore measurements of this activity may serve as a surrogate endpoint to estimate the actions of these compounds on disease activity. The premise that the University of North Carolina group has followed is that these agents must be delivered via aerosol to dose the lumenal surface of airway epithelial cells. A second premise has been that extracellular UTP will be a potentially better agonist for therapeutic purposes than ATP because the nucleoside breakdown product of ATP, adenosine, is a potent bronchoconstrictor of human airways, particularly in patients with airways diseases. Recent studies have shown that the nucleoside breakdown product of UTP, uridine, exhibits none of these activities [14]. A series of animal and human toxicity and efficacy studies have been performed and reported in preliminary fashion. These are summarized in the following discussion.

ANIMAL TOXICITY

A series of acute studies have been performed in dogs to test whether inhaled UTP has acute effects on the mechanical properties of the lung and the hemodynamic properties of the pulmonary and systemic circulations. In these studies dogs were exposed to inhaled UTP by itself or in combination with amiloride, and resistance and dynamic compliance of the lung and circulatory pressures were measured. In brief, no acute toxic effects were seen with respect to those lung functions, nor were systemic or pulmonary circulatory changes noted. To test whether aerosolized UTP may have actions on the circulation if absorbed rapidly in high quantities from diseased lungs, animals were exposed to high concentrations of injected UTP or its metabolite, uridine. In contrast to ATP or adenosine, UTP and uridine appear to exhibit no hemodynamic effects on the pulmonary or systemic circulations. Thus, it appears that these compounds would likely be quite safe in humans.

Prior to administration in humans, a small study was performed to test the semichronic effects of inhaled UTP on the mammalian lung. In a series of studies with hamsters, the effects of inhaled UTP at the projected therapeutic concentration (10^{-4} M) and six times higher than this value were tested over a 1-week interval. In brief, these studies identified no evidence of pulmonary toxicity, as indexed by pulmonary inflammatory responses.

ACUTE STUDIES IN HUMANS OF SAFETY AND EFFICACY

The effects of inhaled UTP in humans were first tested with respect to safety. In a series of studies focusing on pulmonary function and systemic hemodynamic responses, normal subjects were exposed to aerosolized UTP in concentrations designed to deliver from 10^{-7} to 10^{-4} M to airway surfaces, with or without amiloride. In a series of studies over a 3-day interval, UTP was delivered via aerosol to patients in log-increasing concentrating on day 1, midrange concentrations on day 2, and maximal concentrations on day 3. In summary, no adverse effects were noted with regard to spirometric, blood gas, or pulmonary systemic measures. This safety characterization was then extended to cystic fibrosis patients and similarly, no untoward effects were noted.

Inhaled triphosphate nucleotides (UTP) were then tested with respect to efficacy in the human lung. For this assessment, a technique was utilized to measure the clearance of an inhaled radiolabeled particle from the lung by radionuclide scanning techniques. In this technique, ^{99M}Tc-labeled sulphur-sulphur colloid particles are inhaled. These particles approximate the size of inhaled bacteria and/or other commonly inhaled foreign materials. For these studies, a four-cell randomized protocol was used to test the effects of UTP alone, amiloride alone, UTP plus amiloride, and vehicle. In normal human subjects, it appeared that UTP alone or with amiloride was highly effective in increasing the rate of mucociliary clearance from the lung. The increase in clearance rates was approximately threefold, and the effects on mucociliary clearance rates appeared to last for approximately 30–40 minutes after a 10-minute inhalation of UTP.

In cystic fibrosis patients, the major defects that were detected in mucociliary clearance were in the peripheral airways of the lung, likely the site of early pathogenesis of this disease. Initial studies showed that the rate of clearance of mucus from this compartment was approximately 50% reduced compared with normal human subjects. Further, preliminary studies in 14 patients revealed that the combination of amiloride plus UTP restored mucociliary transport rates in this peripheral region to near-normal levels. Thus, it appears that UTP is highly effective in regulating mucociliary clearance rates in normal and cystic fibrosis subjects and sets the stage for the broad application of this type of therapy for the treatment of cystic fibrosis and other human lung diseases that affect this fundamental defense system of the airways.

Future Prospects

Recent studies demonstrate that the triphosphate nucleotides have profound effects on the normal regulation of airway defense by mucociliary clearance mechanisms and that these compounds may be exploited for therapy of diseases that affect this region. It is clear that a number of fundamental studies must be performed to test this thesis. First, it will be highly important to test the hypothesis that ATP (or other nucleotides) may be locally released to coordinate this system as described above. Second, it will be important to test the consequences of the molecular deletion of key elements of this system by the generation of appropriate mouse models. Third, it will be important to more fully explore the potential therapeutic benefits of inhaled UTP with respect to extending the observation from surrogate end points to clinically relevant end points. Given the initial data describing this system in the human airways, the prospects for these studies appear exciting, and the potential for developing novel therapies in the lung based on the UTP/P_{2U} system are broad. What is important to generate in the future are potentially more specific ligands for cell-selective regulation of various elements of the mucociliary system, development of longer acting agonists to perturb the system clinically for longer periods of time, and potentially the development of potent antagonists for certain indications. Finally, it appears likely that if the P_{2U} receptor system is important in the regulation of airway mucosal systems, it will be important in the regulation of mucosal defense in other systems of the body, for example, the ocular and genitourinary systems, and it will be important to test these concepts in these systems in a timely fashion.

References

1. Boucher RC (1994) Human airway ion transport (Part 1) (state-of-the-art review). Am J Respir Crit Care Med 150:271–281.
2. Davis CW, Dowell ML, Lethem MI, Van Scott M (1992) Goblet cell degranulation in isolated canine tracheal epithelium: Response to exogenous ATP, ADP, and adenosine. Am J Physiol 262:C1313–C1323.
3. Lethem MI, Dowell ML, Van Scott M, Yankaskas JR, Egan T, Boucher RC, Davis CW (1993) Nucleotide regulation of goblet cells in human airway epithelium, in vitro. Am J Respir Cell Mol Biol 9:315–322.
4. Mason SJ, Paradiso AM, Boucher RC (1991) Regulation of transepithelial ion transport and intracellular calcium by extracellular adenosine triphosphate in human normal and cystic fibrosis airway epithelium. Br J Pharmacol 103: 1649–1656.
5. Clarke LL, Boucher RC (1992) Chloride secretory response to extracellular ATP in normal and cystic fibrosis nasal epithelia. Am J Physiol 263:C348–C356.
6. Clarke LL, Grubb BR, Yankaskas JR, Cotton CU, McKenzie A, Boucher RC (1994) Relationship of a non-CFTR mediated chloride conductance to organ-level disease in *cftr*(−/−) mice. Proc Natl Acad Sci USA 91:479–483.
7. Stutts MJ, Chinet TC, Mason SJ, Fullton JM, Clarke LL, Boucher RC (1992) Regulation of chloride channels in normal and cystic fibrosis airway epithelial cells by extracellular ATP. Proc Natl Acad Sci USA 89:1621–1625.
8. Jiang C, Finkbeiner WE, Widdicombe JH, McCray PB Jr, Miller SS (1993) Altered fluid transport across airway epithelium in cystic fibrosis. Science 262:424–427.
9. Weiss T, Gheber L, Shoshan-Barmatz V, Priel Z (1992) Possible mechanism of ciliary stimulation by extracellular ATP: Involvement of calcium-dependent potassium channels and exogenous Ca^{2+}. J Membr Biol 127:185–193.
10. Brown HA, Lazarowcki ER, Boucher RC, Harden TK (1991) Evidence that UTP and ATP regulated phospholipase C through a common extracellular 5′-nucleotide receptor in human airway epithelial cells. Mol Pharmacol 40: 648–655.
11. Lazarowski ER, Boucher RC, Harden TK (1994) Calcium-dependent release of arachidonic acid in response to purinergic receptor activation in airway epithelium. Am J Physiol 266: C406–C415.
12. Lustig KD, Shiau AK, Brake AJ, Julius D (1993) Expression cloning of an ATP receptor from mouse neuroblastoma cells. Proc Natl Acad Sci USA 90:5113–5117.
13. Parr CE, Sullivan DM, Paradiso AM, Lazarowski ER, Burch LH, Olsen JC, Erb L, Weisman GA, Boucher RC, Turner JT (1994) Cloning and expression of a human P_{2U} nucleotide receptor, a target for cystic fibrosis pharmacotherapy. Proc Natl Acad Sci USA 91:3275–3279.
14. Lazarowski ER, Mason SJ, Clarke LL, Harden TK, Boucher RC (1992) Adenosine receptors on human airway epithelia and their relationship to chloride secretion. Br J Pharmacol 106: 774–782.

INDEX